9/1/95
MM
W9-ATL-157
THE NEW YORK PUBLIC LIBRARY ONLY.

Fundamentals of Rock Mechanics

J. C. Jaeger
Emeritus Professor of Geophysics in the Australian National University

N. G. W. Cook
Research and Development Consultant Chamber of Mines of South Africa and Adjunct Professor in the Department of Civil Engineering, University of Minnesota

SECOND EDITION

LONDON
CHAPMAN AND HALL

A Halsted Press Book
John Wiley & Sons, Inc., New York

First published 1969
by Methuen & Co Ltd
Reprint 1971
published by Chapman and Hall Ltd
11 New Fetter Lane, London EC4P 4EE
First published as a Science Paperback 1971
Second edition 1976

Printed in Great Britain by
Fletcher & Son Ltd, Norwich

ISBN 0 412 21410 5

Distributed in the U.S.A. by Halsted Press,
a Division of John Wiley & Sons, Inc., New York

Library of Congress Cataloging in Publication Data

Jaeger, John Conrad, 1907–
Fundamentals of rock mechanics.

"A Halsted Press book."
Includes bibliographical references and index.
1. Rock mechanics. I. Cook, H. G. W., joint author. II. Title.
TA706. J32 1976 – 624′. 1513 76–5444
ISBN 0–470–15063–7

Foreword

by PROFESSOR CHARLES FAIRHURST
School of Mineral and Metallurgical Engineering, University of Minnesota

The problems of rock deformation which face the designer of excavations and structures in and on rock are somewhat different from those confronting the scientist seeking to explain the origin of geological structures. Although still intimately concerned with the same natural material, the designer is primarily interested in those changes in mechanical behaviour of the rock brought about by engineering activities. Inasmuch as these changes can be directly observed and their causes stated with greater confidence, it should be possible to provide a more quantitative specification of them than can be hoped for in most problems of structural geology.

Encouraged by this hope and spurred by the demand for more rational design procedures, scientists and engineers have, over the past three decades, become increasingly interested in understanding the short term mechanical behaviour of rock. Sufficient experience has now been gained to realize that problems of rock behaviour involve considerably more than selecting appropriate elastic constants and strength parameters and inserting them into solutions from elementary theories of continuum mechanics; a view that was at one time commonly held.

Although rock, of course, obeys the same laws of mechanics as other materials there are sufficient differences in behaviour and emphases in methods of approach to problems that the distinctive term 'rock mechanics' is warranted.

It is now realized, for example, that initiation of tensile failure of 'brittle' rock within a compressive stress field is not synonymous with structural collapse. This is in contrast to such failure in a tensile stress field where the distinction is generally unimportant; and often tacitly ignored in studies and texts concerned with situations where tensile stress fields predominate. Inelastic deformations are, in fact, often of major significance in determining the stability and safety of rock structures.

Discontinuities and inhomogeneities exert a dominant influence on rock deformation and failure both in the laboratory and the field. Knowledge of their properties is essential to a correct analysis of observed behaviour.

Inability to preserve the scale of discontinuities and other structural features of a rock mass in a laboratory specimen introduces considerable uncertainty into the extrapolation of laboratory measured 'strengths' to field values, and large variations may be expected. The 'constants' demanded by simple theory are, in fact, statistical functions of the properties of the discontinuities, the gradient of the applied stress field, and possibly other variables.

The probability of unpredictable variations in geological conditions is another factor that is always to be considered.

Such complications indicate a need for care in the application of established theories of mechanics and an awareness of the validity of the assumptions involved with respect to rock. Mere increase in the sophistication of mathematical formulation of a problem is of little value if the correspondingly required physical details are not available. Conversely, acquisition of experimental data without guidance from theoretical hypothesis is at best expensive and wasteful, and often misleading.

Realistic solutions to rock mechanics problems must involve a combination of general principles of mechanics with a mature physical insight developed through intelligent observation and experiment both in the laboratory and the field.

This approach is very well demonstrated in *Fundamentals of Rock Mechanics*.

Emphasis is placed on the mechanical behaviour of rock and those principles of mechanics necessary for the rational interpretation of this behaviour. The treatment of friction as a fundamentally important topic in rock mechanics is a good illustration of this emphasis.

When faced with lack of definitive evidence the authors do not hesitate to hypothesize and make assumptions in order to allow them to complete an analysis and reveal its principal implications. This is consequently a vital and thought-provoking book. It will undoubtedly inspire additional research. A book such as this has been much needed and will be enthusiastically received.

It is appropriate that the authors are Professor Jaeger and Dr Cook. Although they modestly claim to have assembled existing information, it should be recognized that they have both contributed much to this body of knowledge. The reader will also perceive that the book contains many new thoughts which appreciably advance our understanding.

The appearance of *Fundamentals of Rock Mechanics* represents a significant step forward for the science of rock mechanics.

CHARLES FAIRHURST

University of Minnesota
August, 1968

Contents

Preface to the First Edition

Rock must have been used as a structural material by man for a far longer period than any other single material. The experience progressively accumulated during this longstanding use has made man so familiar with its properties that he has not until recently felt the need to investigate it specifically. In fact, the structural adequacy of rock is taken so much for granted that few consciously realize that most structures are founded upon or built within the rocks of the earth's crust.

Only in recent decades, with the advent of larger structures situated in less-favoured circumstances than had been the case in the past, has the need for more than an intuitive grasp of the properties of rock been felt. This trend can be expected to continue as increasing demands for services and materials call for increased mining activity and for the construction of more buildings, dams, underground excavations, and large open cuts.

The need to investigate the mechanical properties of rock has stimulated research in a field which has become known as 'rock mechanics'. The results of this research, and other theory and experiment relevant to rock mechanics, are described in a great number of papers scattered over many scientific and engineering journals and texts, and the proceedings of several conferences and symposia on the subject of rock mechanics. We felt that sufficient information was now available to provide a usefully complete description of the mechanical properties of rock. This book sets out the mathematical and experimental foundations of the mechanical behaviour of rock. It does not attempt to describe practical solutions to engineering problems of rock mechanics, mainly because much of the knowledge in rock mechanics is so recent that there is as yet little experience of its application to the solution of such problems. We believe that a prerequisite for the useful application of rock mechanics to engineering problems is that knowledge of the behaviour of rock be made readily available to engineers concerned with these problems.

The science of rock mechanics is a whole composed of parts taken from a number of different subjects. Much of the theory of elasticity is continually needed, and this book covers the relevant parts of this theory and its applications in sufficient detail, it is hoped, to enable the reader to do similar calculations. In the same way, the theories of perfectly plastic and other rheological materials, Coulomb aggregate, and porous elastic materials are set out. Laboratory studies of the mechanical properties of rock and the mechanism of, and criteria for, its failure are described in considerable detail. An essential link between the behaviour of rock specimens in the laboratory and rock masses in the field is provided by the study of friction. Rock masses are broken up by faults, joints, and other planes of weakness, and relative movement and friction on these is of great importance.

On the more practical side, the principles of the methods for measuring stress and rock properties in the field are described. The estimation of stresses and displacements around excavations is discussed, together with the principles involved in a number of engineering and mining problems.

Rock mechanics borders on two allied subjects: structural geology and soil mechanics. The description of faults, joints, and rock failure is an essential part of structural geology and an important part of rock mechanics. Deeply weathered or fragmented rocks behave much as do soils, and this side of rock mechanics thus grades into soil mechanics, a subject with which it has many affinities but also very great differences.

Much of the material given here is basic for soil mechanics and structural geology and is set out more fully than is usual in texts on these subjects. It is therefore hoped that the book may be useful to students and teachers of these subjects as well as to those of rock mechanics.

One innovation has been made. In soil mechanics, structural geology, and rock mechanics, compressive stresses are most common, while tensile stresses are exceptional (although they are of great importance when they do occur). In mathematical and engineering practice tensile stresses are reckoned positive, and the use of this convention in rock and soil mechanics and structural geology entails a great deal of circumlocution, particularly when friction is involved. In these subjects compression is frequently reckoned positive by a rather half-hearted reversal of signs. The present treatment is based throughout on the choice of a positive sign for compressive stresses. A further innovation, which it is hoped will be useful, is a dual notation to reconcile the discrepancy between the 'engineering' and tensor definitions of shear strain.

In conclusion we have to thank the Chamber of Mines of South Africa and the Snowy Mountains Hydro-Electric Authority for much assistance. We are also grateful to Miss Rose Port and Mrs Claire Richardson for their work on the typing and bibliography.

Finally, we are much indebted to Professor Charles Fairhurst for contributing a Foreword in which he makes so many valuable comments on this subject.

1969

J.C.J.
N.G.W.C.

Preface to the Second Edition

Since the first editions of this text appeared, recognition of the importance of rock mechanics to understanding phenomena observed in geology, geophysics and seismology has grown enormously. Accordingly, substantive additions have been made to the chapters on Friction, Elasticity and Strength of Rock and The State of Stress Underground. Also, Chapter 17 now includes both Geological and Geophysical Applications with the addition of three new sections.

The treatment of Granular Materials has been extended by a major revision of this chapter. The application of rock mechanics to the solution of problems of underground mining has been advanced by more quantitative data than were available in the past and, especially, by the development of digital methods of computation. New sections dealing with these questions and that of rockbursts have been added. Rock mechanics research has led to an improved understanding of blasting, and mechanical and other non-explosive methods of rock breaking are subjects of much greater interest than in the past; sections dealing with these matters have been added to the chapter on Mining and Other Engineering Applications.

Throughout the text most of the units of measurement used are the same as those in the original sources from which the data were drawn. As a result, there is to be found a mixture of current American and old British Imperial units, those of the cgs metric system and, in the new sections, units of the Système International. In science and engineering there is a growing tendency, throughout the world, to adopt SI (Système International) units. For this reason an Appendix on units has been added, in which the basic SI units and derivatives of them used frequently in rock mechanics are defined. Tables of conversion factors between the SI units and those of the other systems used in the text are given for convenience.

Attention must be drawn to a potential source of confusion arising out of the mixed use of the comma (,). In the SI, the comma is used to separate units from decimals in place of the point and spaces are left between groups of three digits to separate thousands, millions and so on. The older usage of the comma for this latter purpose and of the decimal point together with the new SI usage, could cause confusion, but it is felt that, knowing of this possibility, the context of the usage will make the matter clear to readers.

Finally, the authors wish to thank Dr M. D. G. Salamon for his assistance with the section on digital computation and Miss P. M. E. Antink for preparing the revised typescript, references and index.

August 1975

J.C.J.
N.G.W.C.

Notation

Most symbols are defined each time they occur, but the following are commonly used.

C_0	Uniaxial compressive strength
E	Young's modulus
e_x, e_y, e_z	Strain deviations
e_1, e_2, e_3	Principal strain deviations
e	Mean normal strain
G	Modulus of rigidity
g	Acceleration of gravity
$I_1, I_2, \ldots$	Invariants of stress
$J_1, J_2, \ldots$	Invariants of stress deviation
K	Bulk modulus
s_x, s_y, s_z	Stress deviations
s_1, s_2, s_3	Principal stress deviations
s	Mean normal stress
S_0	Intrinsic shear strength
T_0	Uniaxial tensile strength
u, v, w	Displacements
X, Y, Z	Body forces
$\gamma_{xy}, \gamma_{yz}, \gamma_{zx}$	Components of shear strain
$\Gamma_{xy} = \frac{1}{2}\gamma_{xy}, \Gamma_{yz} = \frac{1}{2}\gamma_{yz}, \Gamma_{zx} = \frac{1}{2}\gamma_{zx}$	
Δ	Volumetric strain
$\varepsilon_x, \varepsilon_y, \varepsilon_z$	Components of strain
$\varepsilon_1, \varepsilon_2, \varepsilon_3$	Principal strains
η	Viscosity
λ, G	Lame's parameters
λ	Quadratic elongation
μ	Coefficient of friction or internal friction
ν	Poisson's ratio
ρ	Density
$\sigma_x, \sigma_y, \sigma_z$	Components of normal stress

$\sigma_1, \sigma_2, \sigma_3$	Principal stresses
σ_0	Yield stress
$\sigma_x', \sigma_y', \sigma_z'$	Effective stresses
$\tau_{xy}, \tau_{yz}, \tau_{zx}$	Components of shear stress
$\omega_x, \omega_y, \omega_z$	Components of rotation

A stress is described as 'hydrostatic' if $\sigma_1 = \sigma_2 = \sigma_3$. Compressive stresses and strains are reckoned positive.

Chapter One

Rock as a Material

1.1 Introduction

Rock mechanics was defined by the Committee on Rock Mechanics of the Geological Society of America in the following terms: 'Rock mechanics is the theoretical and applied science of the mechanical behaviour of rock; it is that branch of mechanics concerned with the response of rock to the force fields of its physical environment' (Judd, 1964). For practical purposes it is mostly concerned with rock masses on the scale which appears in engineering and mining work and so might be regarded as the study of the behaviour and properties of accessible rock masses under stress or change of conditions. Since these rocks may be weathered or fragmented, it grades at one extreme into soil mechanics. On the other hand, at depths at which the rocks are no longer accessible to mining or drilling, it grades into the mechanical aspects of structural geology.

Historically, it has been very much influenced by these two subjects. At conferences, for a long time it was directly associated with soil mechanics, and there is a great similarity between much of the theory and many of the problems. On the other hand, the demand from structural geologists for a knowledge of the behaviour of rocks under the conditions which occur deep in the earth's crust has stimulated a great deal of research at high pressures and temperatures and a great deal of study of the experimental deformation of both rocks and single crystals.

A most important feature of accessible rock masses is that they are broken up by joints and faults and that fluid under pressure is frequently present both in open joints and in the pores of the rock itself. It also happens that, because of the conditions controlling mining and the siting of structures in civil engineering, several lithological types may occur in any one investigation.

Thus from the outset two distinct problems are always involved: (i) a study of the directions and properties of the joints, and (ii) a study of the properties and fabric of the rocks between them.

In any practical investigation in rock mechanics the first stage is a geological and geophysical investigation to establish the lithology and boundaof tries he rock types involved; the second, by means of drilling or investi-

excavations, to establish the detailed pattern of the jointing, and to ne the mechanical and petrological properties of the rocks con- rom samples; the third, in some cases, is to measure the 'virgin rock present in the unexcavated rock. With this information it should be possible to predict the response of the rock mass to excavation or loading.

1.2 Joints and faults

Joints are by far the most common type of geological structure. They are defined as cracks or fractures in rock along which there has been little or no displacement, Price (1966). They usually occur in sets which are more or less parallel and regularly spaced; also there are usually several sets in very different directions so that the rock mass is broken up into a blocky structure.

This is the reason for the importance of joints in rock mechanics: the material is not a mathematical continuum, but is divided into a number of parts by joint surfaces along which sliding may take place.

Joints occur on all scales. Those of the most important set, referred to as major joints, can usually be traced for tens or hundreds of feet and are usually more or less plane and parallel. The sets of joints which intersect major joints, known as cross-joints, are usually of less importance and more likely to be curved and irregularly spaced; however, in some cases two sets are of equal importance. The spacing between joints may vary from several feet to the microscopic scale; however, here, very closely spaced joints will be regarded as a property of the rock fabric, and it is spacings of the order of a foot which are mainly in question.

Joints may be 'filled' with various minerals, such as calcite, dolomite, quartz, or clay minerals, or they may be 'open', in which case they may contain water under pressure.

Joint systems are affected by the lithological nature of the rock, so that they may change with a change of rock type.

Jointing, as described above, is a phenomenon common to all rocks, sedimentary and igneous. Despite its importance and widespread occurrence, little is known about the mechanism by which it is produced; a full discussion of the various possibilities is given by Price (1966).

Another and quite distinct type of jointing is the columnar jointing, which is best developed in basalts and dolerites, but occasionally occurs in granites and some metamorphic rocks. This is of some importance in rock mechanics, since igneous dykes and sheets are frequently encountered in mining and engineering practice. The effects are very well defined, Tomkeieff (1940), Spry (1961). The material is divided into columns whose side length is of the order of a foot and which are most commonly hexagonal. The columns are intersected by cross-joints which are less regular towards the interior of a large body. The primary cause of columnar jointing appears to be the tensile stresses set up by thermal contraction during cooling. At an external surface the columns run normal to the surface, and Jaeger (1961a) and others have suggested that in the interior of an irregularly shaped body they run normal to the isotherms during cooling. The detailed mechanism of columnar jointing has been discussed by Lachenbruch (1961, 1962); it has affinities to the formation of cracks in soil and mud on drying out, and to some effects in permafrost.

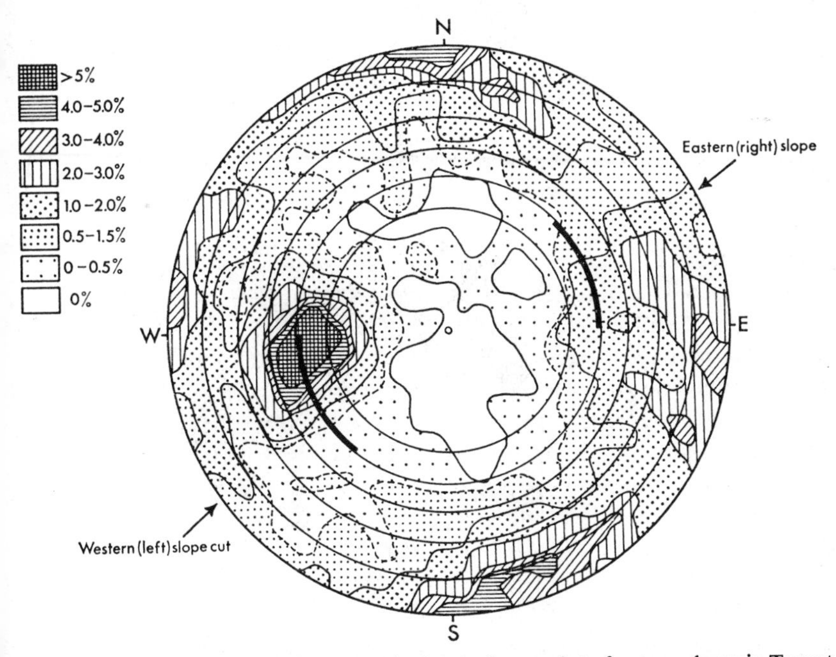

Fig. 1.2.1 Stereographic plot (lower hemisphere) of normals to fracture planes in Tumut 3 Headrace Channel. Contours enclose areas of equal density of poles.

Faults are surfaces of fracture on which there has been relative displacement. They are usually unique structures, but a large number may be merged into a fault zone. They are usually approximately plane, and so provide important planes on which sliding can take place. Joints and faults may have a common origin, de Sitter (1956), and it is often observed underground that joints become more frequent as a fault is approached. Faults are regarded as the equivalent, on a geological scale, of the laboratory shear fractures studied in Chapter 4. The criteria for fracture developed there are applied to faulting in § 17.2.

From the point of view of rock mechanics, the importance of joints and faults is that they cause the existence of fairly regularly spaced, approximately plane surfaces, separating blocks of 'solid' rock which may slide on one another. The essential procedure in practice is to measure the orientation of all joint planes and similar features, either in an exploratory tunnel or in a set of boreholes, and to plot the directions of their normals on a stereographic projection. Some typical examples are shown in the following

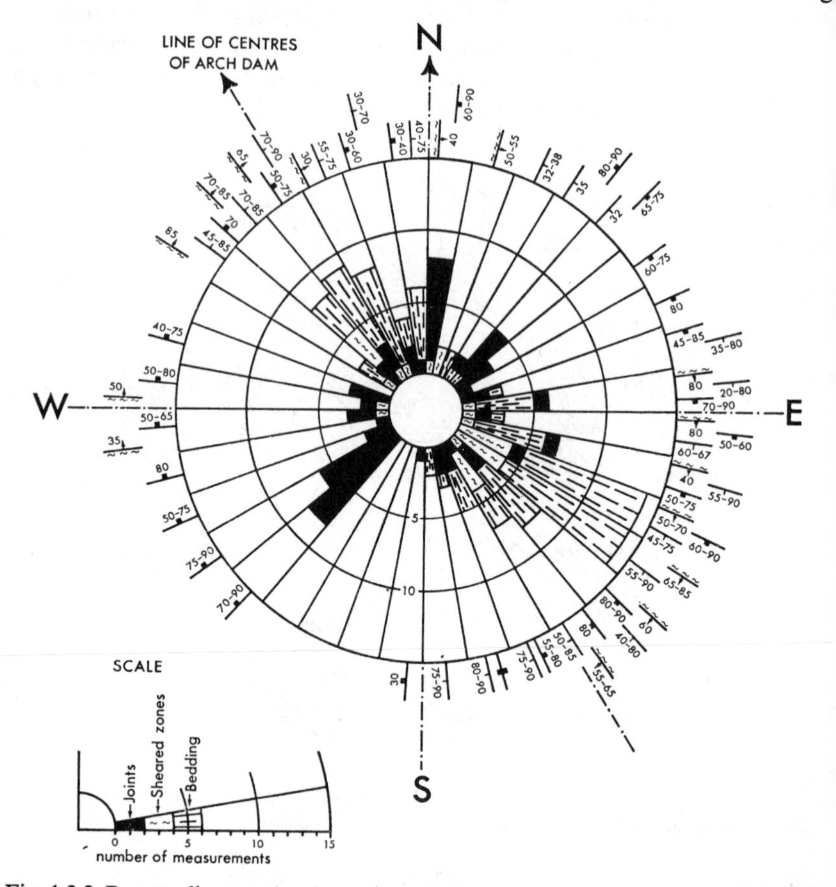

Fig. 1.2.2 Rosette diagram showing strikes of joints, sheared zones, and bedding planes at Murray 2 damsite. The predominant dip for each strike is also shown.

figures taken from investigations of the Snowy Mountains Hydro-electric Authority.

Fig. 1.2.1 is a plot on the stereographic projection of the normals to fracture planes in the headrace channel for the Tumut 3 Project. The thick lines show the position of proposed slope cuts. In this case 700 normals were measured.

Fig. 1.2.2 shows the important geological features at Murray 2 dam site on a different representation. Here the directions of strike of various features are plotted as a rosette, the angles of dip of the dominant features at each strike being given numerically. The features recorded are joints, sheared zones, and bedding planes, any or all of which may be of importance.

Finally, Fig. 1.2.3 gives a simplified representation of the situation at the intersection of three important tunnels. There are three sets of joints whose dips and strikes are shown on the diagram.

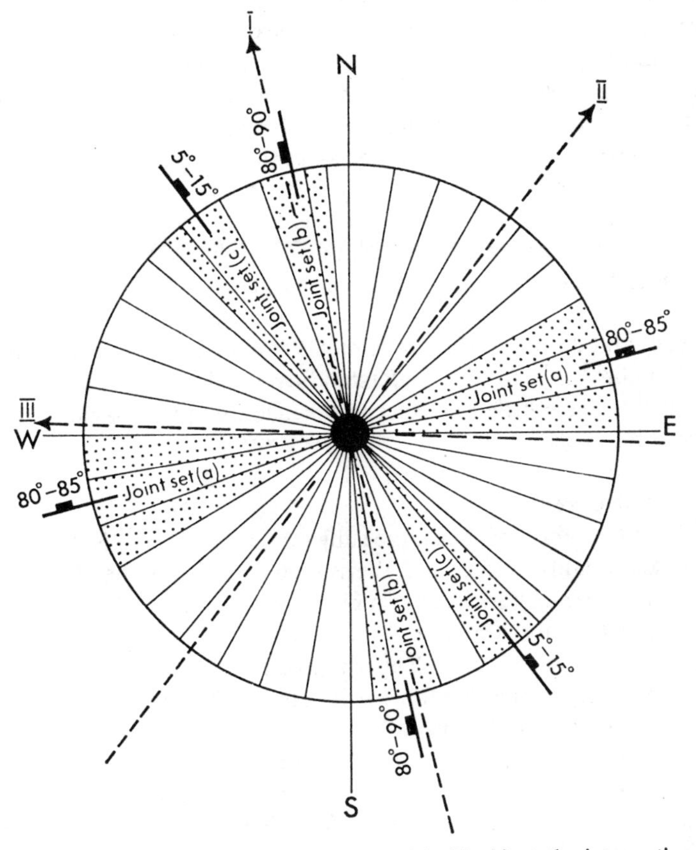

Fig. 1.2.3 Dips and strikes of three joint sets, (*a*), (*b*), (*c*) at the intersection of three tunnels: I, Island Bend intake; II, Eucumbene–Snowy tunnel; III, Snowy–Geehi tunnel.

1.3 Rock-forming minerals

Igneous rocks consist of a completely crystalline assemblage of minerals such as quartz, plagioclase, pyroxene, mica, etc. Sedimentary rocks consist of an assemblage of detrital particles and possibly pebbles from other rocks in a matrix of materials such as clay minerals, calcite, quartz, etc. From their nature, sedimentary rocks contain voids or empty spaces, some of which may form an interconnected system of pores. Metamorphic rocks are produced by the action of heat, stress, or heated fluids on other rocks, sedimentary or igneous.

All these minerals are anisotropic, and the elastic constants of the common ones, defined as in § 5.12, are known numerically. If in a polycrystalline rock there is any preferred orientation of crystals this will lead to anisotropy of the rock itself. If the orientations of the crystals are random the rock itself will be isotropic and its elastic constants may be estimated by the methods referred to in § 5.12.

There are a number of general statistical correlations between the elasticity and strength of rocks and their petrography, and it is desirable to include a full petrographic description with all measurements, as is done by the U.S. Bureau of Reclamation (1953). Grain size also has an effect on mechanical properties. In sedimentary rocks there is, as would be expected, a correlation between mechanical properties and porosity, § 12.3.

A great deal of systematic research has been done on the mechanical properties of single crystals both with regard to their elasticity and plastic deformation. Single crystals show preferred planes for slip and twinning, and these have been studied in great detail, for example, calcite by Turner *et al.* (1954) and dolomite Handin *et al.* (1955). Such measurements are an essential preliminary to the understanding of the fabric of deformed rocks, but they have little relevance to the macroscopic behaviour of large polycrystalline specimens.

1.4 The fabric of rocks

The study of the fabric of rocks, the subject of petrofabrics, is described in many books, notably Turner and Weiss (1963). All rocks have a fabric of some sort. Sedimentary rocks have a primary depositional fabric, of which the bedding is the most obvious element, but other elements may be produced by currents in the water. Superimposed on this primary fabric, and possibly obscuring it, may be fabrics determined by subsequent deformation, metamorphism, and recrystallization.

The study of petrofabrics comprises the study of all fabric elements, both microscopic and macroscopic, on all scales. From the present point of view, the study of the larger elements, faults and relatively widely spaced joints, is an essential part of rock mechanics. Microscopic elements and very closely spaced features such as cleats in coal are regarded as determining the fabric of the rock elements between joints. These produce an anisotropy in the elastic properties and strength of the rock elements. In principle, this anisotropy can be measured completely by mechanical experiments on rock samples, but petrofabric measurements can provide considerable useful information, in particular about preferred directions. Also they are much more rapid to make, and so are much more susceptible to statistical analysis. Studies of rock fabric are therefore better made by a combination of mechanical and petrofabric measurements, but the latter cannot be used as a substitute for the former. Combination of the two methods has led to the use of what may be regarded as standard anisotropic rocks. For example, Yule marble, in which the calcite is known, Turner (1949), to have a strong preferred orientation, has been used in a great many studies of rock deformation, Turner *et al.* (1956), Handin *et al.* (1960).

A second application of petrofabric measurements in rock mechanics arises from the fact that some easily measured fabric elements, such as twin lamellae in calcite and dolomite, quartz deformation lamellae, kink bands, and translation or twin gliding in some crystals, may be used to infer the directions of the principal stresses under which they were generated. These directions, of course, may not be the same as those at present existing, so that they form an interesting complement to underground stress measurements. Again, such measurements are relatively easy to make and to study statistically. The complete fabric study of joints and fractures on all scales is frequently used both to indicate the directions of the principal stresses and the large-scale fabric of the rock mass as a whole, Gresseth (1964).

A great deal of experimental work at the present time is being concentrated on the study of the fabrics produced in rocks in the laboratory under conditions of high temperature and pressure. In some cases rocks of known fabric are subjected to prescribed laboratory conditions and the changes in fabric studied, e.g. Turner *et al.* (1956) on Yule marble and Friedman (1963) on sandstone.

Alternatively, specific attempts to produce important types of fabric are made, e.g. Carter *et al.* (1964) on the deformation of quartz, Paterson and Weiss (1966) on kink bands, and Means and Paterson (1966) on the production of minerals with a preferred orientation.

Useful reviews of the application of petrofabrics to rock mechanics and engineering geology are given by Friedman (1964) and Knopf (1957).

1.5 The mechanical nature of rock

The mechanical structure of rock presents several different appearances, depending upon the scale and detail with which it is studied.

Most rocks comprise an aggregate of crystals and amorphous particles joined by varying amounts of cementitious materials. The chemical composition of the crystals may be relatively homogeneous, as in some limestones, or very heterogeneous, as in a granite. Likewise, the size of the crystals may be uniform or variable, but they generally have dimensions of the order of inches and small fractions thereof. These crystals generally represent the smallest scale on which the mechanical properties are studied. On the one hand, the boundaries between crystals represent weaknesses in the structure of the rock which can otherwise be regarded as continuous. On the other hand, the deformation of the crystals themselves provides interesting evidence concerning the deformation to which the rock has been subjected.

On a scale with dimensions ranging from feet to hundreds of feet the structure of some rocks is continuous, but it is more often interrupted by cracks, joints, and bedding planes separating different strata. It is this scale and these discontinuities which are of most concern in engineering, where structures founded upon, or built within, rock have similar dimensions.

The overall mechanical properties of rock depend upon every one of its structural features. However, individual features have varying degrees of importance in different circumstances.

At some stage it becomes necessary to attach numerical values to the mechanical properties of rock. These values are most readily obtained from laboratory experiments on specimens of rock. These specimens usually have dimensions of inches and contain a sufficient number of structural particles for them to be regarded as grossly homogeneous. Thus, although the properties of the individual particles in such a specimen may differ widely from one particle to another and although the individual crystals themselves are often anisotropic, they and the grain boundaries between them interact in a sufficiently random manner to imbue the specimen with average homogeneous properties. These average properties are not necessarily isotropic, because the processes of rock formation or alteration often align the structural particles so that their interaction is random with respect to size, composition, and distribution, but not with respect to their anisotropy. Nevertheless, specimens of such rock have gross anisotropic properties which can be regarded as homogeneous.

On a larger scale, the presence of cracks, joints, bedding, and minor faulting raises an important question concerning the continuity of a rock mass. These disturbances may interrupt the continuity of displacements in a rock mass if they are subjected to tension, fluid pressure, or shear in excess of their frictional resistance to sliding. Where such disturbances are small in relation to the dimensions of a structure in a rock, their effect is to alter the mechanical properties of the rock mass, but this may still be treated as a continuum. Where they have significant dimensions they must be treated as part of the structure or as a boundary.

The loads applied to a rock mass are generally due to gravity, and compressive stresses are encountered more often than not. Under these circumstances a most important factor in connection with the properties and continuity of a rock mass is the friction between the surfaces of cracks and joints of all sizes in the rock. If conditions are such that sliding is not possible on any surfaces the system may be treated to a good approximation as a continuum of rock with the properties of the average test specimen. If sliding is possible on any surfaces the system has to be treated as a system of discrete elements separated by these surfaces and with frictional boundary conditions over them.

Chapter Two

Analysis of Stress and Infinitesimal Strain

2.1 Introduction

The analysis of stress and strain is fundamental for all work on rock mechanics. It is dealt with in detail in many works on the theory of elasticity, notably, Love (1927), Timoshenko and Goodier (1951), Nadai (1950), Sokolnikoff (1956), Durelli, Phillips and Tsao (1958), Southwell (1941). These works, however, are more concerned with the solution of problems in elasticity than with the study of stresses themselves. In rock mechanics it frequently happens that the stresses are measured, or may be derived from a known solution, and that it is the detailed study of the stress field which is of importance. Similarly, strains and displacements are frequently measured, and it is desired to derive information about the stress field from them.

The analysis of stress is a matter of pure statics, quite independent of the properties assumed for the material, which may be elastic, plastic, viscous, or of any other type.

The analysis of strain is fundamental for the study of the movement of any material. If the movements can be large this is a much more sophisticated matter than the analysis of stress. However, if the strains are infinitesimal the theory runs exactly parallel to that for stress, *mutatis mutandis*, and this theory, which is all that is needed for the classical theory of linear elasticity and many other purposes, will be given in this chapter. Finite strain, which is fundamental for structural geology, will be discussed briefly in Chapter 17.

2.2 Definition of stress

The forces acting at a point O in the interior of a body may be described in the following way. For each direction OP through O we suppose the body to be cut across a small area δA in a plane through O normal to OP, Fig. 2.2 (*a*). The surface of this cut on the side of P will be called the positive side, while that on the side opposite to P will be called the negative side. The effect of the internal forces which exist in the body across the surface δA is equivalent to a force $\delta \mathbf{F}$ exerted by the material on the positive side of the surface on that on the negative side (and an equal, oppositely directed, force exerted by the material on the negative side on that on the positive side). There will also be a couple, but this may be neglected, since the area δA is supposed to be very small.

The limiting value of the ratio $\delta\mathbf{F}/\delta A$ as δA tends to zero is called the *stress vector* at the point O across the plane whose normal is in the direction OP. It is a vector $\mathbf{p}_{OP}$ defined as

$$\mathbf{p}_{OP} = \lim_{\delta A \to 0} \frac{\delta\mathbf{F}}{\delta A}. \tag{1}$$

It is now necessary to introduce a convention of sign, and the one which will be used here is that forces are reckoned positive when compressive, that is, in the direction shown by $\delta\mathbf{F}$ in Fig. 2.2 (*a*). This is opposite to the convention adopted in works on the theory of elasticity and continuum

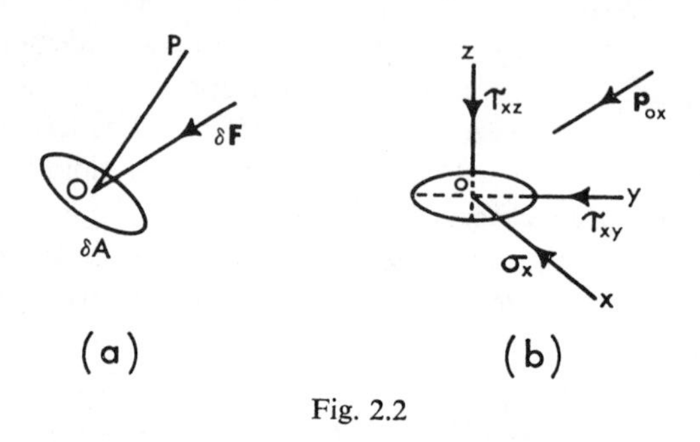

Fig. 2.2

mechanics in which stresses are usually reckoned positive when tensile. In rock mechanics, however, it is more convenient to have compressive stresses positive for the following reasons: (i) the environmental stresses, such as stress due to depth of burial, confining pressure in apparatus, and fluid pressure in pores, are always compressive; (ii) this convention is universal in the closely related subject of soil mechanics, cf. Scott (1963), and has been much used in structural geology; (iii) many problems in rock mechanics involve friction over surfaces, and in this case the normal stress across the surfaces is necessarily compressive. This change of convention leaves all formulae unaltered, but when using results from works on the theory of elasticity (which use the convention that stresses are positive when tensile) it has to be remembered that all signs have to be changed.

If we take a right-handed system of rectangular axes Ox, Oy, Oz at O, Fig. 2.2 (*b*), and take OP in the direction of Ox the vector $\mathbf{p}_{Ox}$ will have components in the x-, y-, and z-directions which will be written

$$\sigma_x, \ \tau_{xy}, \ \tau_{xz}. \tag{2}$$

Here σ_x is called a *normal stress*, since the element of area δA is perpendicular to Ox, while τ_{xy} and τ_{xz} are in the plane of the area δA and are called *shear stresses*, since they represent forces tending to slide or shear the material in the plane of δA. Similarly, if OP is taken in the direction Oy the components of $\mathbf{p}_{Oy}$ will be τ_{yx}, σ_y, τ_{yz}, and if OP is in the direction Oz they will be τ_{zx}, τ_{zy}, σ_z. These nine quantities

$$\begin{bmatrix} \sigma_x & \tau_{xy} & \tau_{xz} \\ \tau_{yx} & \sigma_y & \tau_{yz} \\ \tau_{zx} & \tau_{zy} & \sigma_z \end{bmatrix} \tag{3}$$

are called the *stress-components* at the point O. It will be shown later that $\tau_{xy} = \tau_{yx}$, $\tau_{yz} = \tau_{zy}$, $\tau_{zx} = \tau_{xz}$, so that they reduce to six, and also that the stress vector $\mathbf{p}_{OP}$ for any direction OP may be expressed in terms of the six stress-components.

This notation, in which σ is used for normal stress and τ for shear stress, is convenient and common in engineering work: it is used by Timoshenko and Goodier (1951) and Nadai (1950). Many other notations are current, for example, Sokolnikoff (1956) preserves symmetry by writing τ_{xx} in place of σ_x, so that (2) is replaced by τ_{xx}, τ_{xy}, τ_{xz}, and so on. Love (1927) and other writers frequently use the notation $\widehat{xx}$, $\widehat{xy}$, $\widehat{xz}$ and also X_x, Y_x, Z_x for the quantities in (2).

Finally, the tensor notation in which the axes, instead of being denoted by x, y, z, are represented by the suffixes 1, 2, 3 must be mentioned. In this case the quantities (2) are replaced by σ_{11}, σ_{12}, σ_{13} and the whole set of stress components (3) is given by σ_{rs}, where r and s take values from 1 to 3. It can be shown that the quantities (3) are the components of a mathematical entity known as a tensor (the stress tensor), and the theory can be studied in terms of tensor analysis, cf. Green and Zerna (1954). Tensor analysis will not be used here, since its additional sophistication is unnecessary for the applications in view, but occasional reference will be made to the tensor notation so that formulae written in it may be understood.

2.3 Stress in two dimensions

The analysis of stress in two dimensions is relatively easy and well worth considering in its own right, also many of the problems which will arise later are effectively two-dimensional. In this section the theory of stresses in two dimensions will be developed *ab initio*: the plane considered will be the xy-plane, and all quantities will be independent of z, which therefore need not be considered at all.

Taking the xy-plane as the plane of the paper, the stress-components σ_x, τ_{xy}, σ_y, τ_{yx}, will be as shown in Fig. 2.3.1 (*a*) and (*b*).

The relationship between the stress-components, as well as the variation of stress with direction, can now be studied by considering the equilibrium of a very small region of the material. If the material is in equilibrium and at rest the forces exerted by the stresses over the surface of this region must be in equilibrium. Consider a square $OABC$ of very small side length a, then, neglecting the variation of σ_x, σ_y, etc., with position, the forces per unit area exerted on the sides of the square by the material beyond it are as shown in Fig. 2.3.1 (c). It appears that the force on the square in the x- and y-directions is zero, but that there is a couple

$$a^2(\tau_{xy} - \tau_{yx}),$$

per unit length perpendicular to the plane of the paper, exerted on the square in the plane of the paper. Thus for the forces exerted on the square to be in equilibrium it is necessary that

$$\tau_{xy} = \tau_{yx}. \tag{1}$$

When the variation of σ_x, σ_y, etc., with position is taken into account, a calculation of the same type but retaining higher powers of a leads to the stress-equations of equilibrium – this is carried out in § 5.5.

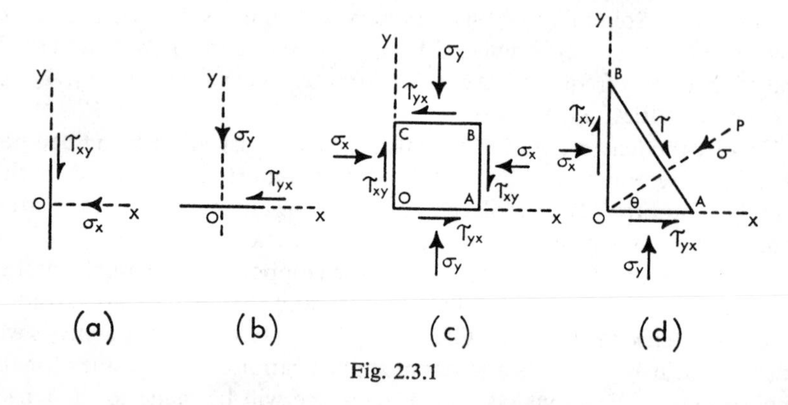

Fig. 2.3.1

The effect of equation (1) is to reduce the four stress-components in two dimensions to three. However, it is often convenient to continue to use both τ_{xy} and τ_{yx} in order to preserve symmetry in formulae.

Next, we determine the components p_x, p_y of the stress-vector $\mathbf{p}_{OP}$ corresponding to the direction OP inclined at θ to Ox. To do this we consider the equilibrium of the small triangular region (of unit thickness perpendicular to the plane of the paper) OAB whose sides are $AB = a$, $OB = a \cos \theta$, $OA = a \sin \theta$, Fig. 2.3.1 (d). Resolving parallel to Ox, we have

$$AB \,.\, p_x = OB \,.\, \sigma_x + OA \,.\, \tau_{yx},$$

or, cancelling a factor a,

$$p_x = \sigma_x \cos \theta + \tau_{yx} \sin \theta. \tag{2}$$

Similarly, resolving parallel to Oy,

$$p_y = \sigma_y \sin \theta + \tau_{xy} \cos \theta. \tag{3}$$

These may most conveniently be expressed in terms of the normal and shear stress σ and τ across the plane AB. Here the positive directions of σ and τ are chosen as in Fig. 2.3.1 (*d*) to be consistent with those for $\theta = 0$ in Fig. 2.3.1 (*a*). It follows from (1), (2), and (3) that

$$\begin{aligned}\sigma &= p_x \cos\theta + p_y \sin\theta,\\ &= \sigma_x \cos^2\theta + 2\tau_{xy}\sin\theta\cos\theta + \sigma_y \sin^2\theta. \qquad (4)\end{aligned}$$

$$\begin{aligned}\tau &= p_y \cos\theta - p_x \sin\theta\\ &= (\sigma_y - \sigma_x)\sin\theta\cos\theta + \tau_{xy}(\cos^2\theta - \sin^2\theta), \qquad (5)\\ &= \tfrac{1}{2}(\sigma_y - \sigma_x)\sin 2\theta + \tau_{xy}\cos 2\theta. \qquad (6)\end{aligned}$$

σ and τ as found above may be regarded as the stresses $\sigma_{x'}$ and $\tau_{x'y'}$ for a system of axes $Ox'y'$ rotated through θ from Oxy as in Fig. 2.3.2 (*a*). The

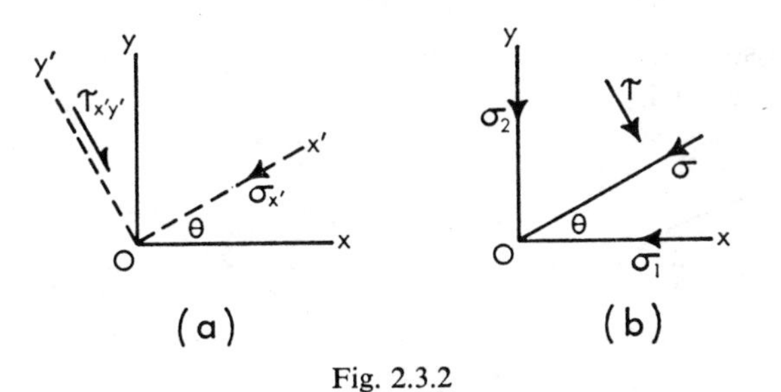

Fig. 2.3.2

stress $\sigma_{y'}$ for these axes is found by replacing θ by $\theta + \frac{1}{2}\pi$ in (4). Thus the formulae for change of axes are

$$\sigma_{x'} = \sigma_x \cos^2\theta + 2\tau_{xy}\sin\theta\cos\theta + \sigma_y \sin^2\theta, \qquad (7)$$

$$\sigma_{y'} = \sigma_x \sin^2\theta - 2\tau_{xy}\sin\theta\cos\theta + \sigma_y \cos^2\theta, \qquad (8)$$

$$\tau_{x'y'} = \tfrac{1}{2}(\sigma_y - \sigma_x)\sin 2\theta + \tau_{xy}\cos 2\theta. \qquad (9)$$

Adding (7) and (8) gives

$$\sigma_{x'} + \sigma_{y'} = \sigma_x + \sigma_y, \qquad (10)$$

so that this sum is *invariant* or unchanged by rotation of axes.

It follows from (9) that $\tau_{x'y'} = 0$ if

$$\tan 2\theta = \frac{2\tau_{xy}}{\sigma_x - \sigma_y}. \qquad (11)$$

Thus there is one value of θ for which the axes $Ox'y'$ are such that the shear stress vanishes. These axes are called *principal axes of stress*, and the stress-components relative to them are called *principal stresses* and denoted by σ_1 and σ_2.

If θ is given by (11) it follows that

$$\left.\begin{aligned}\sin 2\theta &= \pm[1+\cot^2 2\theta]^{-\frac{1}{2}} = \pm\tau_{xy}[\tau_{xy}^2+\tfrac{1}{4}(\sigma_x-\sigma_y)^2]^{-\frac{1}{2}},\\ \cos 2\theta &= \pm[1+\tan^2 2\theta]^{-\frac{1}{2}} = \pm\tfrac{1}{2}(\sigma_x-\sigma_y)[\tau_{xy}^2+\tfrac{1}{4}(\sigma_x-\sigma_y)^2]^{-\frac{1}{2}}.\end{aligned}\right\} \quad (12)$$

Using these values in (7), which may be written

$$\sigma_{x'} = \tfrac{1}{2}(\sigma_x+\sigma_y)+\tfrac{1}{2}(\sigma_x-\sigma_y)\cos 2\theta+\tau_{xy}\sin 2\theta,$$

this becomes

$$\sigma_{x'} = \tfrac{1}{2}(\sigma_x+\sigma_y) \pm [\tau_{xy}^2+\tfrac{1}{4}(\sigma_x-\sigma_y)^2]^{\frac{1}{2}}, \quad (13)$$

and (8) gives the same pair of values for $\sigma_{y'}$. The two values (13) are thus the two principal stresses, and we have

$$\left.\begin{aligned}\sigma_1 &= \tfrac{1}{2}(\sigma_x+\sigma_y)+[\tau_{xy}^2+\tfrac{1}{4}(\sigma_x-\sigma_y)^2]^{\frac{1}{2}},\\ \sigma_2 &= \tfrac{1}{2}(\sigma_x+\sigma_y)-[\tau_{xy}^2+\tfrac{1}{4}(\sigma_x-\sigma_y)^2]^{\frac{1}{2}}.\end{aligned}\right\} \quad (14)$$

Here the signs have been chosen so that $\sigma_1 \geqslant \sigma_2$, and this convention will always be used.

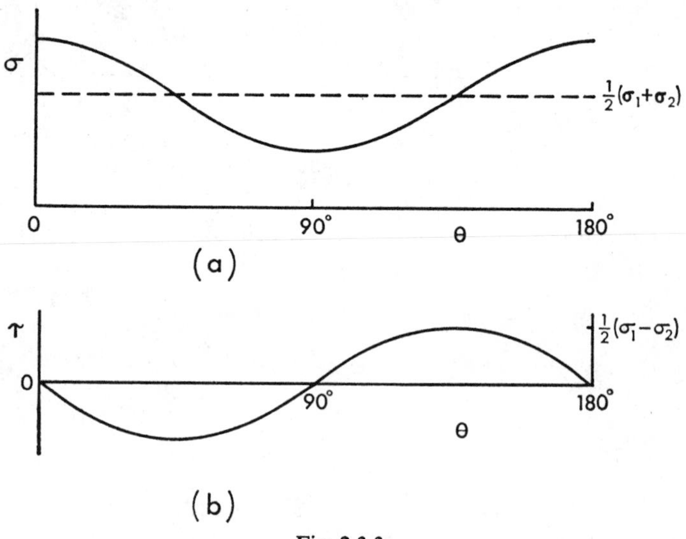

Fig. 2.3.3

It is frequently convenient to choose the *x*- and *y*-axes in the directions of σ_1 and σ_2 as in Fig. 2.3.2 (*b*), and if this is done the normal and shear stresses, σ, τ, across a plane whose normal is inclined at θ to σ_1 are by (4) and (6)

$$\sigma = \sigma_1\cos^2\theta+\sigma_2\sin^2\theta, \quad (15)$$

$$= \tfrac{1}{2}(\sigma_1+\sigma_2)+\tfrac{1}{2}(\sigma_1-\sigma_2)\cos 2\theta, \quad (16)$$

$$\tau = -\tfrac{1}{2}(\sigma_1-\sigma_2)\sin 2\theta. \quad (17)$$

The variation of σ and τ with θ is shown graphically in Fig. 2.3.3. It appears that the magnitude of the shear stress has its greatest value of $\frac{1}{2}(\sigma_1 - \sigma_2)$ when $\theta = \pi/4$ or $3\pi/4$.

Referred to principal axes, the formulae (2) and (3) for the components of the stress vector for a plane whose normal is inclined at θ to σ_1 become

$$p_x = \sigma_1 \cos\theta, \quad p_y = \sigma_2 \sin\theta, \tag{18}$$

so that

$$\frac{p_x^2}{\sigma_1^2} + \frac{p_y^2}{\sigma_2^2} = 1, \tag{19}$$

that is, p_x and p_y lie on an ellipse, sometimes called the ellipse of stress, whose semi-axes are σ_1 and σ_2.

There are many graphical methods of representing the variation of stress

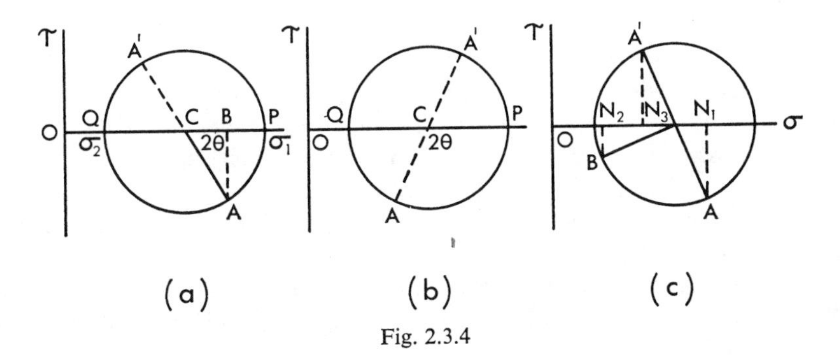

Fig. 2.3.4

in two dimensions, of which *Mohr's circle* diagram is by far the most important. This follows immediately from (16) and (17). Suppose that in the σ, τ plane, Fig. 2.3.4 (*a*), we mark off lengths $OP = \sigma_1$, $OQ = \sigma_2$, where $\sigma_1 > \sigma_2$, on the axis $O\sigma$ and draw a circle on PQ as diameter with its centre C at $\sigma = \frac{1}{2}(\sigma_1 + \sigma_2)$. This circle is called Mohr's circle. The co-ordinates of a point A on the circle for which the angle PCA, measured clockwise from OP, is 2θ are

$$\sigma = OB = OC + CB = \tfrac{1}{2}(\sigma_1 + \sigma_2) + \tfrac{1}{2}(\sigma_1 - \sigma_2)\cos 2\theta,$$
$$\tau = -AB = -\tfrac{1}{2}(\sigma_1 - \sigma_2)\sin 2\theta,$$

which are just the values of σ and τ given by (15) and (17). This construction holds for all values, positive or negative, of σ_1 and σ_2, provided $\sigma_1 > \sigma_2$, and for all values of θ. In particular, the point A' on the diameter ACA', Fig. 2.3.4 (*a*) gives σ and τ for a plane perpendicular to that corresponding to A.

The use of the Mohr circle permits simple graphical solutions of many important problems. First, suppose σ_x, σ_y, τ_{xy} are known, and it is required to find the principal stresses and the directions of the principal axes. In the σ, τ diagram of Fig. 2.3.4 (*b*), σ_x, τ_{xy} for the plane whose normal is Ox will give a point such as A. The corresponding quantities for the perpendicular plane whose normal is Oy will be σ_y, $-\tau_{xy}$, which will give a point A': the reason for the negative sign before τ_{xy} is that τ_{yx}, as defined in Fig. 2.3.1 (*b*), is in the opposite direction to that of the convention of Fig. 2.3.1 (*d*) which is being used here. By the preceding argument, A and A' must be at opposite ends of a diameter of the Mohr circle, and if this circle, Fig. 2.3.4 (*b*), is drawn with centre C and intersecting $O\sigma$ in P and Q, the principal stresses are given by $OP = \sigma_1$, $OQ = \sigma_2$, and the angle ACP is twice the angle between Ox and $O\sigma_1$. Since, from the geometry of the figure,

$$AA' = \sigma_1 - \sigma_2 = [(\sigma_x - \sigma_y)^2 + 4\tau_{xy}^2]^{\frac{1}{2}}, \tag{20}$$

the formulae (14) for the principal stresses follow from this construction.

As a second important example suppose that $\sigma' = ON_1$, Fig. 2.3.4 (*c*), is the normal stress across some plane, $\sigma'' = ON_2$ is that across a plane inclined at 45° to it, and $\sigma''' = ON_3$ is that across a plane inclined at 90° to it. The stresses corresponding to these three planes must be represented by points A, B, A' on a circle whose centre C is mid-way between N_1 and N_3. Further, from the geometry of this figure, the triangles $A'N_3C$ and CN_2B must be equal in all respects, so that $BN_2 = CN_3 = \frac{1}{2}(\sigma' - \sigma''')$. This gives the position of the point B, and a circle of centre C and radius CB will be the Mohr circle. Analytically, the radius CB which is $\frac{1}{2}(\sigma_1 - \sigma_2)$ may be found from $CB^2 = BN_2^2 + CN_2^2$, so that

$$(\sigma_1 - \sigma_2)^2 = (\sigma' - \sigma''')^2 + (\sigma' + \sigma''' - 2\sigma'')^2. \tag{21}$$

This provides an alternative graphical or analytical treatment of the problems of this type discussed in greater detail in § 2.13.

Other representations of stress in two dimensions are described by Durelli, Phillips and Tsao (1958). Of these it is necessary to mention one which is of little practical interest but provides an alternative method of developing the theory by a procedure which is common in applied mathematics. Referring to (4), suppose we plot a point P in the xy-plane at distance r in the direction θ where r is chosen to be proportional to $\sigma^{-\frac{1}{2}}$, so that $\sigma = \lambda/r^2$ where λ is a constant. Then (4) gives

$$\sigma_x r^2 \cos^2\theta + 2\tau_{xy} r^2 \sin\theta\cos\theta + \sigma_y r^2 \sin^2\theta = \lambda,$$

and since $x = r\cos\theta$, $y = r\sin\theta$ are the coordinates of P, this point lies on the conic

$$\sigma_x x^2 + 2\tau_{xy} xy + \sigma_y y^2 = \lambda. \tag{22}$$

This conic is called the *stress conic*, and its axes, which may be found by the methods of coordinate geometry, are the principal axes. Referred to them, its equation will have the form

$$\sigma_1 x^2 + \sigma_2 y^2 = \lambda, \tag{23}$$

which is an ellipse if σ_1 and σ_2 have the same sign, and an hyperbola if their signs are opposite.

To specify the complete condition of stress in a two-dimensional body it is necessary to know the values of σ_x, σ_y, τ_{xy} at every point, or, alternatively, the principal stresses σ_1, σ_2 and the directions of the principal axes. While it is difficult to display all this information, there is a number of simple representations which may be used to give a very useful picture of a stress-field. These are:

(i) *Isobars*, which are curves of constant principal stress, there being one set for σ_1 and another for σ_2.

(ii) *Isochromatics*, which are curves of constant maximum shear stress $(\sigma_1 - \sigma_2)/2$. They are directly obtained by photoelastic methods, § 10.19.

(iii) *Isopachs*, which are lines of constant mean stress, $(\sigma_1 + \sigma_2)/2$. Since this quantity satisfies Laplace's equation if the material is elastic (§ 5.5 (16)), it is frequently determined by analogue methods, notably the use of conducting paper. A combination of this with the photoelastic determination of $\sigma_1 - \sigma_2$ is the most usual experimental method of studying stresses in irregularly shaped regions.

(iv) *Stress-trajectories*, or isostatics, are an orthogonal system of curves whose directions at any point are the directions of the principal axes. They, therefore, intersect a free boundary at right angles. Their separation gives an immediate indication of the intensity of stress.

(v) *Slip-lines*, or lines of maximum shear stress, are an orthogonal system of curves whose directions at any point bisect the angles between the principal axes at that point.

(vi) *Isoclinics* are curves on which the principal axes make a constant angle with a reference direction. They are obtained directly from photoelastic measurements.

2.4 Stress in three dimensions

Let σ_x, τ_{xy}, τ_{xz}, etc., be the nine stress-components defined as in § 2.2 for rectangular axes Ox, Oy, Oz.

Considering the equilibrium of a small cube, Fig. 2.4 (a), it follows precisely as in § 2.3 that the condition that there be no couple on the cube about any of the three axes is

$$\tau_{xy} = \tau_{yx}, \quad \tau_{yz} = \tau_{zy}, \quad \tau_{zx} = \tau_{xz}, \tag{1}$$

so that the nine stress-components reduce to six. To keep expressions symmetrical, both τ_{xy} and τ_{yx}, etc., will continue to be used.

Next, we wish to calculate the stress vector corresponding to a plane whose normal OP is inclined to the axes Ox, Oy, Oz, Fig. 2.4 (*b*), at angles α, β, γ, respectively. These angles are called *direction angles*, and their cosines $l = \cos\alpha$, $m = \cos\beta$, $n = \cos\gamma$, are known as the *direction cosines* of OP. They are connected by the relation (Bell, 1920)

$$l^2 + m^2 + n^2 = 1. \tag{2}$$

Numbers a, b, c known to be proportional to the direction cosines of a line, so that $a = kl$, $b = km$, $c = kn$, will be called *direction ratios*, and it follows from (2) that the constant k must be $(a^2 + b^2 + c^2)^{\frac{1}{2}}$, so that

$$l = a[a^2 + b^2 + c^2]^{-\frac{1}{2}}, \quad m = b[a^2 + b^2 + c^2]^{-\frac{1}{2}}, \quad n = c[a^2 + b^2 + c^2]^{-\frac{1}{2}}. \tag{3}$$

A further result (Bell, 1920) which is frequently needed is that the angle ψ

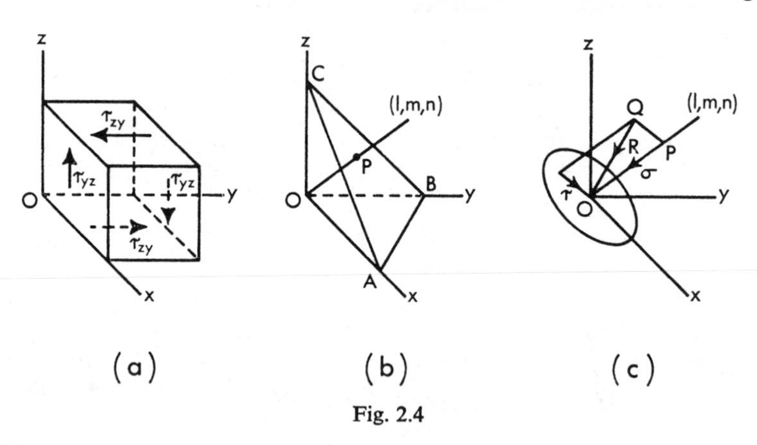

Fig. 2.4

between two lines whose direction cosines are l, m, n, and l', m', n' is given by

$$\cos\psi = ll' + mm' + nn'. \tag{4}$$

We now consider the equilibrium of the small tetrahedron $OABC$, Fig. 2.4 (*b*), whose face ABC has area ω and is normal to the line OP of direction cosines l, m, n. Then the areas of the faces OAB, OBC, OCA, respectively, are ωn, ωl, and ωm.

Suppose that p_x, p_y, p_z are the components of the stress across the face ABC, then, considering the equilibrium of the tetrahedron $OABC$, and resolving in the x-direction we get

$$\left.\begin{aligned} \omega p_x &= \omega l\sigma_x + \omega m\tau_{yx} + \omega n\tau_{zx}, \\ \text{or} \quad p_x &= l\sigma_x + m\tau_{yx} + n\tau_{zx}. \end{aligned}\right\} \tag{5}$$

Similarly, resolving in the directions Oy and Oz, we get

$$p_y = l\tau_{xy} + m\sigma_y + n\tau_{zy}, \tag{6}$$

$$p_z = l\tau_{xz} + m\tau_{yz} + n\sigma_z. \tag{7}$$

The component of the stress normal to the plane is

$$\sigma = lp_x + mp_y + np_z, \tag{8}$$
$$= l(l\sigma_x + m\tau_{yx} + n\tau_{zx}) + m(l\tau_{xy} + m\sigma_y + n\tau_{zy}) + n(l\tau_{xz} + m\tau_{yz} + n\sigma_z),$$
$$= l^2\sigma_x + m^2\sigma_y + n^2\sigma_z + 2mn\tau_{yz} + 2nl\tau_{zx} + 2lm\tau_{xy}. \tag{9}$$

The shear stress may be found in the same way and will be discussed later.

First we consider the variation of the normal stress with l, m, n, and seek its maxima and minima. The analysis is a little complicated, since l, m, n are not independent but are connected by (2). Choosing l, m as two independent variables, the conditions for σ to be stationary are

$$\frac{\partial\sigma}{\partial l} = 0, \quad \frac{\partial\sigma}{\partial m} = 0, \tag{10}$$

together with, by (2),

$$l + n\frac{\partial n}{\partial l} = 0, \quad m + n\frac{\partial n}{\partial m} = 0. \tag{11}$$

Differentiating (8) partially with respect to l and m and using these results in (10) gives

$$\frac{\partial\sigma}{\partial l} = p_x + p_z\frac{\partial n}{\partial l} = 0, \quad \frac{\partial\sigma}{\partial m} = p_y + p_z\frac{\partial n}{\partial m} = 0. \tag{12}$$

Using (11) in (12) gives

$$\frac{p_x}{l} = \frac{p_y}{m} = \frac{p_z}{n}. \tag{13}$$

This states that, when the normal stress across a plane is a maximum or a minimum the total stress p_x, p_y, p_z across the plane is actually in the direction of the normal, so that the shear stress across the plane is zero. If Σ is the (unknown) value of the normal stress across this plane (13) gives

$$p_x = l\Sigma, \quad p_y = m\Sigma, \quad p_z = n\Sigma, \tag{14}$$

or, using the values (5) to (7) of p_x, p_y, p_z,

$$l(\sigma_x - \Sigma) + m\tau_{yx} + n\tau_{zx} = 0, \tag{15}$$

$$l\tau_{xy} + m(\sigma_y - \Sigma) + n\tau_{zy} = 0, \tag{16}$$

$$l\tau_{xz} + m\tau_{yz} + n(\sigma_z - \Sigma) = 0. \tag{17}$$

Equations (15)–(17) are a set of homogeneous linear equations for l, m, n, and it is known (Bell, 1920) that they have a non-zero solution only if Σ is a root of the equation

$$\begin{vmatrix} \sigma_x - \Sigma & \tau_{yx} & \tau_{zx} \\ \tau_{xy} & \sigma_y - \Sigma & \tau_{zy} \\ \tau_{xz} & \tau_{yz} & \sigma_z - \Sigma \end{vmatrix} = 0. \tag{18}$$

This is a cubic in Σ, and it is known that its roots are all real (Bell, 1920): let them be $\sigma_1, \sigma_2, \sigma_3$. Corresponding to σ_1, solving any two of (15)–(17), say (15) and (16), gives a set of direction cosines l_1, m_1, n_1. Similarly, a direction l_2, m_2, n_2 is associated with σ_2, and l_3, m_3, n_3 with σ_3. That these three directions are mutually perpendicular may be shown in the following way. Multiplying the set of equations (15)–(17) corresponding to σ_1, l_1, m_1, n_1 by l_2, m_2, n_2, respectively, gives

$$\left.\begin{aligned} l_2l_1(\sigma_x - \sigma_1) + l_2m_1\tau_{yx} + l_2n_1\tau_{zx} &= 0, \\ m_2l_1\tau_{xy} + m_2m_1(\sigma_y - \sigma_1) + m_2n_1\tau_{zy} &= 0, \\ n_2l_1\tau_{xz} + n_2m_1\tau_{yz} + n_2n_1(\sigma_z - \sigma_1) &= 0. \end{aligned}\right\} \tag{19}$$

Similarly, multiplying the set corresponding to σ_2, l_2, m_2, n_2, by l_1, m_1, n_1, respectively, gives

$$\left.\begin{aligned} l_1l_2(\sigma_x - \sigma_2) + l_1m_2\tau_{yx} + l_1n_2\tau_{zx} &= 0, \\ m_1l_2\tau_{xy} + m_1m_2(\sigma_y - \sigma_2) + m_1n_2\tau_{zy} &= 0, \\ n_1l_2\tau_{xz} + n_1m_2\tau_{yz} + n_1n_2(\sigma_z - \sigma_2) &= 0. \end{aligned}\right\} \tag{20}$$

Subtracting the sum of equations (20) from the sum of equations (19) gives

$$(\sigma_1 - \sigma_2)(l_1l_2 + m_1m_2 + n_1n_2) = 0, \tag{21}$$

so that, if $\sigma_1 \neq \sigma_2$, (l_1, m_1, n_1) and (l_2, m_2, n_2) are perpendicular by (4). The same argument applies to the other pairs of directions.

We have thus found a set of orthogonal axes, the *principal axes of stress*, such that the stresses in these directions, the *principal stresses* $\sigma_1, \sigma_2, \sigma_3$, are purely normal. We shall always take $\sigma_1 \geqslant \sigma_2 \geqslant \sigma_3$. It is possible for two, or even three, principal stresses to be equal, and in such cases there is an obvious indeterminacy in the directions of the principal axes.

When the principal axes are known it is convenient to adopt them as axes of reference, since a very great simplification results. For the present, then, suppose that the axes $Oxyz$ of Fig. 2.4 (*b*) are principal axes corresponding to principal stresses $\sigma_1, \sigma_2, \sigma_3$. Then equations (5)–(7) for the components of the stress across a plane whose normal has direction cosines (l, m, n) become

$$p_x = l\sigma_1, \quad p_y = m\sigma_2, \quad p_z = n\sigma_3. \tag{22}$$

It follows from (2) that

$$\frac{p_x^2}{\sigma_1^2}+\frac{p_y^2}{\sigma_2^2}+\frac{p_z^2}{\sigma_3^2}=1, \tag{23}$$

so that p_x, p_y, p_z lie on an ellipsoid of semi-axes $\sigma_1, \sigma_2, \sigma_3$. This is sometimes called the ellipsoid of stress.

The magnitude R of the resultant stress $\mathbf{R}$ across the plane is

$$R=(p_x^2+p_y^2+p_z^2)^{\frac{1}{2}}=(l^2\sigma_1^2+m^2\sigma_2^2+n^2\sigma_3^2)^{\frac{1}{2}}. \tag{24}$$

The normal stress σ across the plane is by (22) or (9)

$$\sigma=lp_x+mp_y+np_z=l^2\sigma_1+m^2\sigma_2+n^2\sigma_3. \tag{25}$$

The shear stress across the plane lies in the plane of $\mathbf{R}$ and the normal to the surface, Fig. 2.4 (*c*), and its magnitude τ is given by

$$\begin{aligned}\tau^2&=R^2-\sigma^2\\&=l^2\sigma_1^2+m^2\sigma_2^2+n^2\sigma_3^2-(l^2\sigma_1+m^2\sigma_2+n^2\sigma_3)^2\\&=(\sigma_1-\sigma_2)^2l^2m^2+(\sigma_2-\sigma_3)^2m^2n^2+(\sigma_3-\sigma_1)^2n^2l^2.\end{aligned} \tag{26}$$

The variation of τ with l, m, n may now be discussed; in particular, it is desirable to find its maximum and minimum values. To do this, we put $n^2=1-l^2-m^2$ in (26), which becomes

$$\begin{aligned}\tau^2&=(\sigma_1-\sigma_2)^2l^2m^2+[(\sigma_2-\sigma_3)^2m^2+(\sigma_3-\sigma_1)^2l^2](1-l^2-m^2)\\&=l^2(\sigma_1^2-\sigma_3^2)+m^2(\sigma_2^2-\sigma_3^2)+\sigma_3^2-\\&\qquad[l^2(\sigma_1-\sigma_3)+m^2(\sigma_2-\sigma_3)+\sigma_3]^2.\end{aligned} \tag{27}$$

This is now a function of two independent variables l and m, and so the condition for τ to be stationary is

$$\frac{\partial\tau}{\partial l}=0,\quad \frac{\partial\tau}{\partial m}=0. \tag{28}$$

Differentiating (27) gives

$$\tau\frac{\partial\tau}{\partial l}=l(\sigma_1^2-\sigma_3^2)-2l(\sigma_1-\sigma_3)[l^2(\sigma_1-\sigma_3)+m^2(\sigma_2-\sigma_3)+\sigma_3], \tag{29}$$

$$\tau\frac{\partial\tau}{\partial m}=m(\sigma_2^2-\sigma_3^2)-2m(\sigma_2-\sigma_3)\,[l^2(\sigma_1-\sigma_3)+m^2(\sigma_2-\sigma_3)+\sigma_3]. \tag{30}$$

To make both (29) and (30) vanish as required by (28) we could take $l=m=0$, but this gives a principal axis on which $\tau=0$, and so is not the required solution.

Next we could take $l=0$, so that $\partial\tau/\partial l=0$, and putting $l=0$ in (30) gives

$$\tau\frac{\partial\tau}{\partial m}=m(\sigma_2-\sigma_3)^2(1-2m^2)$$

which vanishes if $m = 2^{-\frac{1}{2}}$. Thus if

$$l = 0, \quad m = 2^{-\frac{1}{2}}, \quad n = 2^{-\frac{1}{2}}, \tag{31}$$

τ is stationary and its value is $\frac{1}{2}(\sigma_1 - \sigma_2)$.

Similarly, taking $m = 0$ to give $\partial\tau/\partial m = 0$, substituting in (29) requires $1 - 2l^2 = 0$ so that τ is stationary if

$$l = 2^{-\frac{1}{2}}, \quad m = 0, \quad n = 2^{-\frac{1}{2}}, \tag{32}$$

and its value is $\frac{1}{2}(\sigma_1 - \sigma_3)$.

Finally, by an independent calculation (eliminating l or m in place of n) a third set of values

$$l = 2^{-\frac{1}{2}}, \quad m = 2^{-\frac{1}{2}}, \quad n = 0, \quad \tau = \tfrac{1}{2}(\sigma_1 - \sigma_2), \tag{33}$$

is found.

These directions of maximum shear stress bisect the angles between the principal stresses. Since $\sigma_1 > \sigma_2 > \sigma_3$, the greatest value of the shear stress is $\frac{1}{2}(\sigma_1 - \sigma_3)$ for the direction (32) which bisects the angle between the greatest and least principal stresses σ_1 and σ_3.

The stationary values of the shear stress are sometimes called the principal shear stresses and denoted by τ_1, τ_2, τ_3, so that

$$\tau_1 = \tfrac{1}{2}(\sigma_2 - \sigma_3), \quad \tau_2 = \tfrac{1}{2}(\sigma_3 - \sigma_1), \quad \tau_3 = \tfrac{1}{2}(\sigma_1 - \sigma_2). \tag{34}$$

The values of the normal stress corresponding to these are, respectively,

$$\tfrac{1}{2}(\sigma_2 + \sigma_3), \quad \tfrac{1}{2}(\sigma_3 + \sigma_1), \quad \tfrac{1}{2}(\sigma_1 + \sigma_2). \tag{35}$$

So far, only the magnitude of the shear stress has been discussed. Its direction may be found as follows: from Fig. 2.4 (*c*) the direction OT of the shear stress is perpendicular to OP, (l, m, n) and to the normal to the plane of OP and OQ. Suppose (λ, μ, ν) are the direction cosines of this normal, then, since it is perpendicular to (l, m, n) and to OQ, whose direction ratios are $l\sigma_1$, $m\sigma_2$, $n\sigma_3$ by (22), it follows from (4) that

$$\left.\begin{aligned} \lambda l + \mu m + \nu n &= 0, \\ \lambda\sigma_1 l + \mu\sigma_2 m + \nu\sigma_3 n &= 0. \end{aligned}\right\} \tag{36}$$

It follows that

$$\frac{\lambda}{mn(\sigma_3 - \sigma_2)} = \frac{\mu}{nl(\sigma_1 - \sigma_3)} = \frac{\nu}{lm(\sigma_2 - \sigma_1)}. \tag{37}$$

Similarly, since OT is perpendicular to (l, m, n) and (λ, μ, ν), it is found in the same way that its direction ratios are

$$l\{m^2(\sigma_2 - \sigma_1) - n^2(\sigma_1 - \sigma_3)\}, \quad m\{n^2(\sigma_3 - \sigma_2) - l^2(\sigma_2 - \sigma_1)\}, \quad n\{l^2(\sigma_1 - \sigma_3) - m^2(\sigma_3 - \sigma_2)\}. \tag{38}$$

Thus, for example, the direction cosines of the direction of the shear stress when it has its absolute maximum value (32) are $-2^{-\frac{1}{2}}$, 0, $2^{-\frac{1}{2}}$.

The determination of principal axes for any general system of stresses depends on the solution of the cubic (18), which, when the determinant is expanded, may be written

$$\Sigma^3 - I_1\Sigma^2 - I_2\Sigma - I_3 = 0, \tag{39}$$

where

$$I_1 = \sigma_x + \sigma_y + \sigma_z, \tag{40}$$

$$I_2 = -(\sigma_y\sigma_z + \sigma_z\sigma_x + \sigma_x\sigma_y) + \tau_{yz}^2 + \tau_{zx}^2 + \tau_{xy}^2, \tag{41}$$

$$I_3 = \sigma_x\sigma_y\sigma_z + 2\tau_{yz}\tau_{zx}\tau_{xy} - \sigma_x\tau_{yz}^2 - \sigma_y\tau_{zx}^2 - \sigma_z\tau_{xy}^2. \tag{42}$$

Since the roots of (39) give the principal stresses which are independent of the original choice of axes, the coefficients I_1, I_2, I_3 must also be independent of the choice of axes and they are therefore *invariant* with respect to change of axes and have the same value for all systems of rectangular axes. Thus, equating the general values (40) to (42) with those for principal axes, it follows that

$$I_1 = \sigma_x + \sigma_y + \sigma_z = \sigma_1 + \sigma_2 + \sigma_3, \tag{43}$$

$$I_2 = -(\sigma_y\sigma_z + \sigma_z\sigma_x + \sigma_x\sigma_y) + \tau_{yz}^2 + \tau_{zx}^2 + \tau_{xy}^2 = -(\sigma_2\sigma_3 + \sigma_3\sigma_1 + \sigma_1\sigma_2), \tag{44}$$

$$I_3 = \sigma_1\sigma_2\sigma_3. \tag{45}$$

The useful result

$$\sigma_x^2 + \sigma_y^2 + \sigma_z^2 + 2\tau_{yz}^2 + 2\tau_{zx}^2 + 2\tau_{xy}^2 = \sigma_1^2 + \sigma_2^2 + \sigma_3^2 \tag{46}$$

follows from (43) and (44).

The quantities I_1, I_2, I_3 are called the *invariants of stress* and are of great importance, since, for example, it is obviously desirable to express criteria for failure in terms of them.

A related set of quantities which have been much used are the normal and shear stress on the plane whose normal

$$l = m = n = 3^{-\frac{1}{2}} \tag{47}$$

is equally inclined to the principal axes. This plane is frequently called the *octahedral plane*, since it is parallel to a face of an octahedron with vertices on the principal axes.

By (25) the *octahedral normal stress*, σ_{oct}, is

$$\sigma_{oct} = \frac{1}{3}(\sigma_1 + \sigma_2 + \sigma_3) = \frac{1}{3}I_1. \tag{48}$$

By (26) the *octahedral shear stress*, τ_{oct}, is

$$\tau_{oct} = \frac{1}{3}\{(\sigma_1 - \sigma_2)^2 + (\sigma_2 - \sigma_3)^2 + (\sigma_3 - \sigma_1)^2\}^{\frac{1}{2}}, \tag{49}$$

$$= \frac{\sqrt{2}}{3}\{\sigma_1^2 + \sigma_2^2 + \sigma_3^2 - \sigma_1\sigma_2 - \sigma_2\sigma_3 - \sigma_3\sigma_1\}^{\frac{1}{2}}, \tag{50}$$

$$= \frac{\sqrt{2}}{3}\{I_1^2 + 3I_2\}^{\frac{1}{2}}, \tag{51}$$

where I_1 and I_2 are the invariants defined in (43) and (44).

The octahedral shear stress has the interesting physical interpretation that it is $(5/3)^{\frac{1}{2}}$ times the root mean square shear stress, all directions being weighted equally.

It follows from (38) that the direction ratios of the octahedral shear stress are

$$\sigma_2 + \sigma_3 - 2\sigma_1, \quad \sigma_3 + \sigma_1 - 2\sigma_2, \quad \sigma_1 + \sigma_2 - 2\sigma_3. \tag{52}$$

Finally, an alternative representation of stresses in three dimensions similar to that leading to § 2.3 (22) should be mentioned. If a length $r = (\lambda/\sigma)^{\frac{1}{2}}$ is plotted in the direction l, m, n, where λ is a constant and σ the normal stress across a plane with direction cosines l, m, n, it follows from (9) that

$$r^2(l^2\sigma_x + m^2\sigma_y + n^2\sigma_z + 2mn\tau_{yz} + 2nl\tau_{zx} + 2lm\tau_{xy}) = \lambda,$$

so that the point $x = lr$, $y = mr$, $z = nr$ lies on the quadric

$$x^2\sigma_x + y^2\sigma_y + z^2\sigma_z + 2yz\tau_{yz} + 2zx\tau_{zx} + 2xy\tau_{xy} = \lambda. \tag{53}$$

This is a quadric, known as the *stress quadric*, and the theory of this section could have been developed in terms of this quadric and its reduction to principal axes.

2.5 Stress calculations in three dimensions

In § 2.4 most of the discussion was referred to principal axes, and we return now to the general case with particular reference to change of axes.

Suppose that Ox, Oy, Oz are a system of rectangular axes relative to which the stress-components are σ_x, τ_{xy}, etc., and that Ox', Oy', Oz' are another set of rectangular axes whose direction cosines relative to $Oxyz$ are, respectively, (l_1, m_1, n_1), (l_2, m_2, n_2), (l_3, m_3, n_3), Fig. 2.5 (*a*).

Considering the plane whose normal is Ox', the normal stress across this will be $\sigma_{x'}$ and so by § 2.4 (9)

$$\sigma_{x'} = l_1^2\sigma_x + m_1^2\sigma_y + n_1^2\sigma_z + 2m_1n_1\tau_{yz} + 2n_1l_1\tau_{zx} + 2l_1m_1\tau_{xy}. \tag{1}$$

To find the shear stresses, we note that by § 2.4 (5)–(7) the components of the resultant stress across the plane are

$$l_1\sigma_x + m_1\tau_{yx} + n_1\tau_{zx}, \quad l_1\tau_{xy} + m_1\sigma_y + n_1\tau_{zy}, \quad l_1\tau_{xz} + m_1\tau_{yz} + n_1\sigma_z. \tag{2}$$

Resolving this in the direction Oy' gives the component of shear stress $\tau_{x'y'}$ in the direction Oy': this is

$$\begin{aligned}\tau_{x'y'} &= l_2(l_1\sigma_x + m_1\tau_{yx} + n_1\tau_{zx}) + m_2(l_1\tau_{xy} + m_1\sigma_y + n_1\tau_{zy}) + \\ &\qquad n_2(l_1\tau_{xz} + m_1\tau_{yz} + n_1\sigma_z) \\ &= l_1l_2\sigma_x + m_1m_2\sigma_y + n_1n_2\sigma_z + (m_1n_2 + m_2n_1)\tau_{yz} + \\ &\qquad (n_1l_2 + n_2l_1)\tau_{zx} + (l_1m_2 + l_2m_1)\tau_{xy}. \end{aligned}\tag{3}$$

Similarly, resolving in the direction Oz' gives

$$\begin{aligned}\tau_{x'z'} &= l_1l_3\sigma_x + m_1m_3\sigma_y + n_1n_3\sigma_z + (m_1n_3 + m_3n_1)\tau_{yz} + \\ &\qquad (n_1l_3 + n_3l_1)\tau_{zx} + (l_1m_3 + l_3m_1)\tau_{xy}. \end{aligned}\tag{4}$$

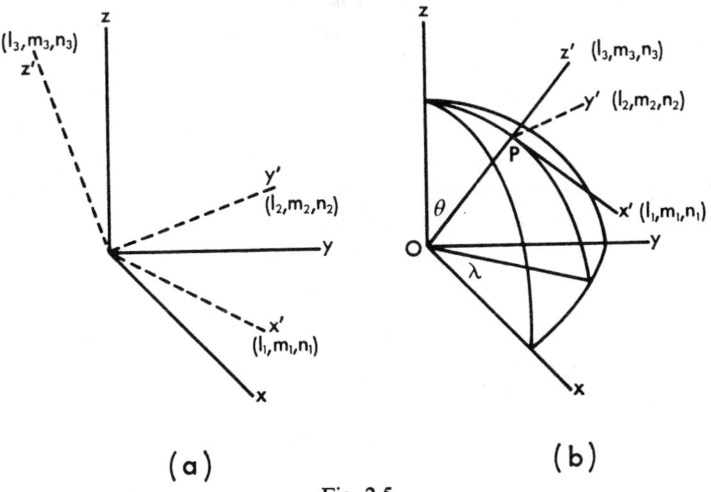

Fig. 2.5

There will, of course, be similar relations for the other stress-components, so that

$$\sigma_{y'} = l_2^2\sigma_x + m_2^2\sigma_y + n_2^2\sigma_z + 2m_2n_2\tau_{yz} + 2n_2l_2\tau_{zx} + 2l_2m_2\tau_{xy}, \tag{5}$$

$$\sigma_{z'} = l_3^2\sigma_x + m_3^2\sigma_y + n_3^2\sigma_z + 2m_3n_3\tau_{yz} + 2n_3l_3\tau_{zx} + 2l_3m_3\tau_{xy}, \tag{6}$$

$$\begin{aligned}\tau_{y'z'} &= l_2l_3\sigma_x + m_2m_3\sigma_y + n_2n_3\sigma_z + (m_2n_3 + m_3n_2)\tau_{yz} + \\ &\qquad (n_2l_3 + n_3l_2)\tau_{zx} + (l_2m_3 + l_3m_2)\tau_{xy}. \end{aligned}\tag{7}$$

It follows from these relations that the stress-components in any plane transform according to the two-dimensional laws of § 2.3. For example, putting $l_1 = \cos\theta$, $m_1 = \sin\theta$, $n_1 = 0$, $l_2 = -\sin\theta$, $m_2 = \cos\theta$, $n_2 = 0$ in (1), (5), and (3) gives § 2.3 (7)–(9). The maxima and minima of stress in a plane found in this way are called *subsidiary principal stresses.*

For the simple case in which $Oxyz$ are principal axes, the relations (1)–(7) become

$$\sigma_{x'} = l_1^2\sigma_1 + m_1^2\sigma_2 + n_1^2\sigma_3, \tag{8}$$

$$\sigma_{y'} = l_2^2\sigma_1 + m_2^2\sigma_2 + n_2^2\sigma_3, \tag{9}$$

$$\sigma_{z'} = l_3^2\sigma_1 + m_3^2\sigma_2 + n_3^2\sigma_3, \tag{10}$$

$$\tau_{x'y'} = l_1l_2\sigma_1 + m_1m_2\sigma_2 + n_1n_2\sigma_3, \tag{11}$$

$$\tau_{y'z'} = l_2l_3\sigma_1 + m_2m_3\sigma_2 + n_2n_3\sigma_3, \tag{12}$$

$$\tau_{z'x'} = l_3l_1\sigma_1 + m_3m_1\sigma_2 + n_3n_1\sigma_3. \tag{13}$$

This determination of shear stress is, of course, an alternative to that of § 2.4 (38) and to other methods which will be given later.

This discussion in terms of direction cosines, while very convenient mathematically, will frequently be found to be rather unsuitable in many practical cases in which other parameters are more natural. For example, a direction is frequently specified by its longitude λ and zenith angle θ, Fig. 2.5 (*b*). The axes associated with these coordinates are Pz' radial, Px' in the plane OPz and associated with θ, and Py' chosen to make a right-handed system and in the direction of λ increasing. Let (l_3, m_3, n_3), (l_1, m_1, n_1), (l_2, m_2, n_2), respectively, be the direction cosines of these. To express them in terms of θ and λ we note that Oz' intersects a unit sphere with centre O in a point of coordinates $x = \sin\theta\cos\lambda$, $y = \sin\theta\sin\lambda$, $z = \cos\theta$, so that these are the direction cosines of Pz' that is,

$$l_3 = \sin\theta\cos\lambda, \quad m_3 = \sin\theta\sin\lambda, \quad n_3 = \cos\theta. \tag{14}$$

For a line through O parallel to Px' the result will be the same as (14) with θ replaced by $\frac{1}{2}\pi + \theta$ so that

$$l_1 = \cos\theta\cos\lambda, \quad m_1 = \cos\theta\sin\lambda, \quad n_1 = -\sin\theta. \tag{15}$$

Finally, a line through O parallel to Py' is perpendicular to Oz and makes angles $\frac{1}{2}\pi + \lambda$ and λ, respectively, with Ox and Oy. Therefore the direction cosines of Py' are

$$l_2 = -\sin\lambda, \quad m_2 = \cos\lambda, \quad n_2 = 0. \tag{16}$$

If the axes $Oxyz$ are principal axes, the normal and shear stresses across the plane $Px'y'$ may then be written down by (10), (12), and (13). They are

$$\sigma_{z'} = [\sigma_1\cos^2\lambda + \sigma_2\sin^2\lambda]\sin^2\theta + \sigma_3\cos^2\theta, \tag{17}$$

$$\tau_{y'z'} = -\tfrac{1}{2}(\sigma_1 - \sigma_2)\sin\theta\sin 2\lambda, \tag{18}$$

$$\tau_{x'z'} = \tfrac{1}{2}[\sigma_1\cos^2\lambda + \sigma_2\sin^2\lambda - \sigma_3]\sin 2\theta. \tag{19}$$

These relations are used by Bott (1959) for defining tectonic regimes.

2.6 Mohr's representation of stress in three dimensions

This representation, because of its simplicity, has been much used in theoretical discussion, in particular by Nadai (1950). The original work is given in Mohr (1900, 1914).

We begin with the equations § 2.4 (25) and (26) for the normal stress σ and the magnitude τ of the shear stress across a plane whose normal has direction cosines l, m, n, relative to principal axes $Oxyz$. These are

$$l^2\sigma_1 + m^2\sigma_2 + n^2\sigma_3 = \sigma, \tag{1}$$

$$l^2\sigma_1^2 + m^2\sigma_2^2 + n^2\sigma_3^2 = \sigma^2 + \tau^2, \tag{2}$$

where

$$l^2 + m^2 + n^2 = 1. \tag{3}$$

Solving (1)–(3) for l^2, m^2, n^2 gives

$$l^2 = \frac{(\sigma_2 - \sigma)(\sigma_3 - \sigma) + \tau^2}{(\sigma_2 - \sigma_1)(\sigma_3 - \sigma_1)}, \tag{4}$$

$$m^2 = \frac{(\sigma_3 - \sigma)(\sigma_1 - \sigma) + \tau^2}{(\sigma_3 - \sigma_2)(\sigma_1 - \sigma_2)}, \tag{5}$$

$$n^2 = \frac{(\sigma_1 - \sigma)(\sigma_2 - \sigma) + \tau^2}{(\sigma_1 - \sigma_3)(\sigma_2 - \sigma_3)}. \tag{6}$$

These equations, if they lead to real values of l, m, n, give the direction cosines of the normal to a plane across which the normal and shear stress would have specified values.

Now suppose that one direction cosine, say n, is fixed, so that the normal to the plane in question makes a fixed angle $\theta = \cos^{-1} n$ with Oz, Fig. 2.6.1 (*a*), and that its intersection with a unit sphere lies on a small circle $F'E'D'$. By (6), σ and τ for such a plane are related by

$$\tau^2 + (\sigma_1 - \sigma)(\sigma_2 - \sigma) = n^2(\sigma_1 - \sigma_3)(\sigma_2 - \sigma_3),$$

or

$$\tau^2 + [\sigma - \tfrac{1}{2}(\sigma_1 + \sigma_2)]^2 = \tfrac{1}{4}(\sigma_1 - \sigma_2)^2 + n^2(\sigma_1 - \sigma_3)(\sigma_2 - \sigma_3). \tag{7}$$

That is, plotted on the σ, τ plane, Fig. 2.6.1 (*b*), σ and τ lie on a circle whose centre is at the point A, $(\tfrac{1}{2}(\sigma_1 + \sigma_2), 0)$, and whose radius is

$$\{\tfrac{1}{4}(\sigma_1 - \sigma_2)^2 + n^2(\sigma_1 - \sigma_3)(\sigma_2 - \sigma_3)\}^{\frac{1}{2}}. \tag{8}$$

This radius varies from $AQ = \tfrac{1}{2}(\sigma_1 - \sigma_2)$ for $n = 0$ to

$$AR = \tfrac{1}{2}(\sigma_1 + \sigma_2) - \sigma_3$$

for $n = 1$, a typical circle of the family being DEF.

In the same way, taking l constant in (4) gives the family of circles

$$\tau^2 + [\sigma - \tfrac{1}{2}(\sigma_2 + \sigma_3)]^2 = \tfrac{1}{4}(\sigma_2 - \sigma_3)^2 + l^2(\sigma_2 - \sigma_1)(\sigma_3 - \sigma_1) \tag{9}$$

with centres at B, $(\frac{1}{2}(\sigma_2 + \sigma_3), 0)$ and radii varying from $BQ = \frac{1}{2}(\sigma_2 - \sigma_3)$ for $l = 0$ to BP for $l = 1$, a typical circle being GEH. In Fig. 2.6.1 (*a*), $l =$ constant corresponds to a cone of axis Ox and angle $\phi = \cos^{-1} l$, meeting the unit sphere in $G'E'H'$.

Finally, taking m constant in (5) gives the circles

$$\tau^2 + [\sigma - \tfrac{1}{2}(\sigma_1 + \sigma_3)]^2 = \tfrac{1}{4}(\sigma_1 - \sigma_3)^2 + m^2(\sigma_3 - \sigma_2)(\sigma_1 - \sigma_2) \quad (10)$$

with centres at the point C and radii decreasing from CR for $m = 0$ to CQ for $m = 1$. These circles are not needed and are not shown on the diagram to avoid confusion.

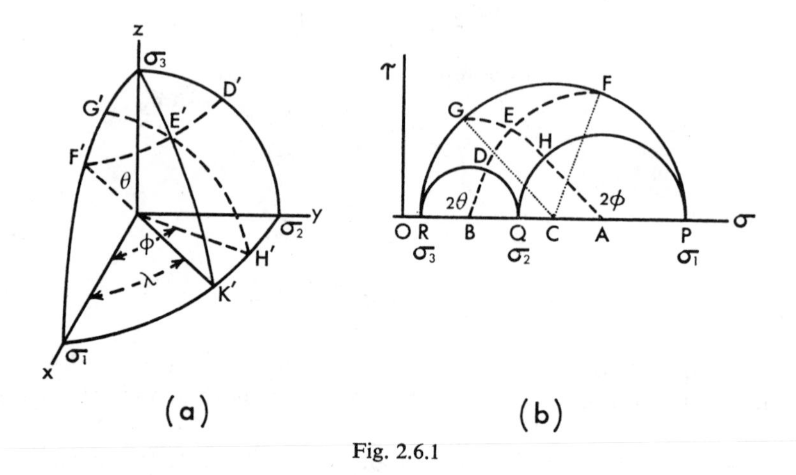

Fig. 2.6.1

In Fig. 2.6.1 (*b*) the point E, corresponding to the intersection of the circles (7) and (9), gives σ and τ for the direction OE' of Fig. 2.6.1 (*a*) of direction cosines $l = \cos \phi$, $n = \cos \theta$.

The point D of the curve DEF corresponds to the point D' of Fig. 2.6.1 (*a*) for which $l = 0$, $m = \sin \theta$, $n = \cos \theta$ so that by (1) and (2)

$$\begin{aligned} \sigma &= \sigma_2 \sin^2 \theta + \sigma_3 \cos^2 \theta \\ &= \tfrac{1}{2}(\sigma_2 + \sigma_3) - \tfrac{1}{2}(\sigma_2 - \sigma_3) \cos 2\theta, \end{aligned} \quad (11)$$

$$\begin{aligned} \tau^2 &= \sigma_2^2 \sin^2 \theta + \sigma_3^2 \cos^2 \theta - \sigma^2 \\ &= \tfrac{1}{4}(\sigma_2 - \sigma_3)^2 \sin^2 2\theta. \end{aligned} \quad (12)$$

It follows that the angle RBD is 2θ and, by a similar argument, that the angle HAP is 2ϕ. These results, of course, follow from the fact that the circles of centres B and A are the Mohr circles for the planes of σ_2, σ_3 and σ_1, σ_2, respectively.

In this way, a diagram, Fig. 2.6.2, can be constructed from which σ and τ can be read off for any direction specified by angles θ and ϕ.

Mohr's figure, which essentially consists of three related families of circles with centres A, B, C, has many interesting geometrical properties which are sometimes useful. For example, in Fig. 2.6.1 (*b*), BD and CF are parallel. This follows immediately, since $AC = \frac{1}{2}(\sigma_2 - \sigma_3) = BD$, $AB = CF$, and $AD = AF$, so that the triangles ABD and FCA are equal in all respects. Similarly, AH and CG are parallel. Further relations appear if all three sets of circles are drawn.

In the above, only the magnitude τ of the shear stress has been considered, and this is all that is usually necessary. However, the components

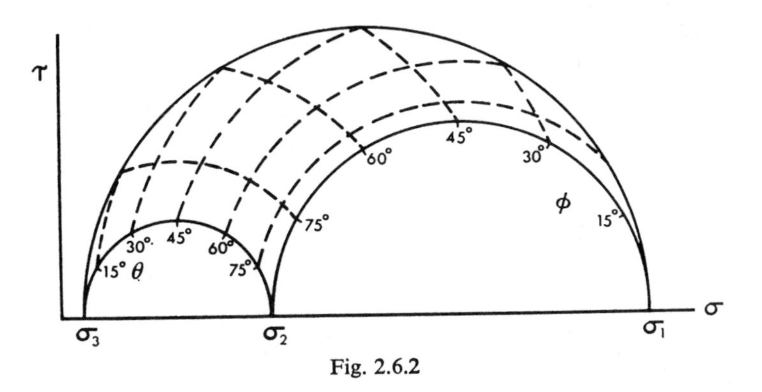

Fig. 2.6.2

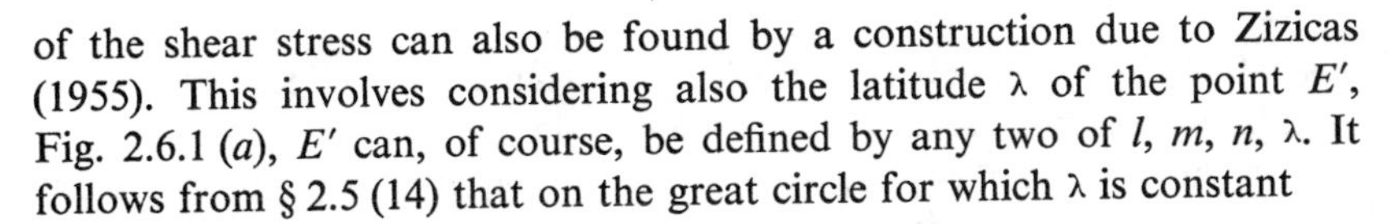
of the shear stress can also be found by a construction due to Zizicas (1955). This involves considering also the latitude λ of the point E', Fig. 2.6.1 (*a*), E' can, of course, be defined by any two of l, m, n, λ. It follows from § 2.5 (14) that on the great circle for which λ is constant

$$l^2 = (1 - n^2) \cos^2 \lambda. \tag{13}$$

Using (4) and (6) in this, it corresponds to the locus in the σ, τ plane

$$(\sigma_2 - \sigma_3)[(\sigma_2 - \sigma)(\sigma_3 - \sigma) + \tau^2] + (\sigma_2 - \sigma_1)[(\sigma_1 - \sigma_3)(\sigma_2 - \sigma_3) - (\sigma_1 - \sigma)(\sigma_2 - \sigma) - \tau^2] \cos^2 \lambda = 0. \tag{14}$$

This is a circle with centre on the σ-axis which passes through the point R, $\sigma = \sigma_3$, $\tau = 0$, Fig. 2.6.1 (*b*). Also when $n = 0$ corresponding to the point K', $l = \cos \lambda$, $m = \sin \lambda$, and it follows from (1) and (2) as in the argument leading to (11) and (12) that

$$\sigma = \tfrac{1}{2}(\sigma_1 + \sigma_2) + \tfrac{1}{2}(\sigma_1 - \sigma_2) \cos 2\lambda, \tag{15}$$

$$\tau^2 = \tfrac{1}{4}(\sigma_1 - \sigma_2)^2 \sin^2 2\lambda. \tag{16}$$

That is, the point K corresponding to K' lies on the circle PQ of Fig. 2.6.3, and is such that the angle KAP is 2λ. The fact that this circle passes through the points *R* and *K* and has its centre on the σ-axis is sufficient to define it. It must pass through the point *E*, and its intersection with either of the circles *GEH* or *DEF* of Fig. 2.6.1 (*b*) is sufficient to define the point *E*. The lines of the original construction of Fig. 2.6.1 (*b*) are shown dotted in Fig. 2.6.3.

Consider, now, stresses in the plane *OK'Z*, Fig. 2.6.1 (*a*). The normal stresses in the directions *Oz* and *OK'* are $OR = \sigma_3$ and *OL* in Fig. 2.6.3. Therefore a circle on *RL* as diameter is the Mohr circle for this plane. When the normal stress is *OM*, corresponding to the point *E'*, the shear stress is *NM*. We have thus found the component of shear stress in the plane *zOK'*, and the total shear stress given by *EM* must make an angle

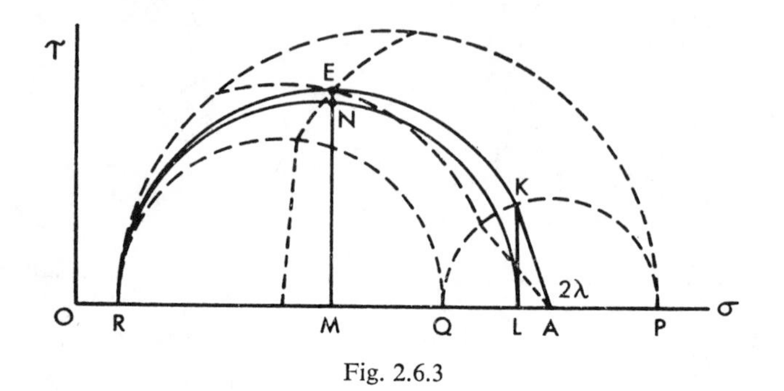

Fig. 2.6.3

$\cos^{-1}[NM/EM]$ with this. There is still an ambiguity of sign, since the calculation has only dealt with the magnitude of τ.

The above calculation and Figs. 2.6.1 (*a*) and 2.6.3 have been set out using latitude and longitude coordinates with the σ_3-axis as polar direction. Clearly, a similar construction could be set out for either of the other principal directions.

There are many other applications of Mohr theory. In particular, Werfel (1965) shows how to use it to determine principal axes and stresses.

2.7 The use of the stereographic projection

In practical underground problems, the directions of the principal stresses, if known, would be represented by angles measured from rectangular axes which might be vertical, North and East. The most important problem will be to determine the direction and magnitude of the normal and shear stresses across some plane, which again will be specified by its dip and strike or by the angular coordinates of its normal. Goodman (1964) has shown that this problem may be very effectively treated by the use of the stereographic projection, combined with a small amount of calculation.

Essentially the stereographic projection consists of projecting a grid of great and small circles (such as circles of longitude and latitude) on a hemisphere on to the flat surface of the hemisphere from a pole of the sphere on the axis of the hemisphere but on the opposite side to it. The actual situation used in structural geology, Fig. 2.7.1 (*a*), has a unit sphere with its axes *NS* and *EW* in the horizontal plane. The lower portion of the sphere is projected on to the plane *NSEW* from a pole at Z'. The curves on the sphere taken to form a net, Fig. 2.7.1 (*a*), are its intersections with planes *NPDS* dipping at δ and small circles *APB* formed by the intersection with cones of axis *NS* and semi-vertical angle θ. The point *P* defined by these two curves is specified further in Fig. 2.7.1 (*b*).

When a grid at regular intervals, say 10°, in θ and δ is projected, the result is the Wulff net or Meridional stereonet Fig. 2.7.1 (*c*).

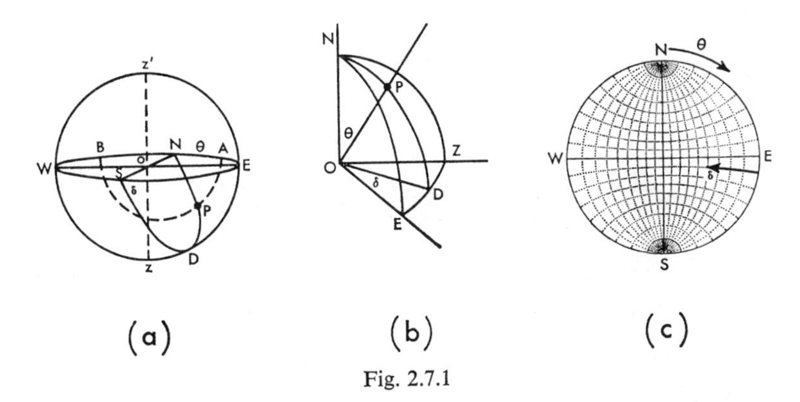

Fig. 2.7.1

The essential properties of the projection are that: (i) great circles on the sphere project into circles, and also that (ii) small circles on the sphere project into circles. The simplest and most important technique is that of superimposing a transparent Wulff net on another and rotating the transparent net so that angles can be measured along great circles. Full details of the method and its applications are given by Phillips (1954).

Suppose, now that the values of δ and θ for the directions of the principal stresses are known. These can immediately be set out as the points X, Y, Z, on the stereogram, Fig. 2.7.2. If the direction of the normal to the plane across which the stresses are required is known the corresponding point P can also be set out. The angles α, β, γ between the direction of P and X, Y, Z can then be measured by rotating the stereonet. The direction cosines of the normal to the plane relative to the principal axes are then found from

$$l = \cos\alpha, \quad m = \cos\beta, \quad n = \cos\gamma.$$

The magnitude R of the resultant stress across the plane is then calculated from § 2.4 (24), namely

$$R = (l^2\sigma_1{}^2 + m^2\sigma_2{}^2 + n^2\sigma_3{}^2)^{\frac{1}{2}}.$$

The direction cosines of the direction of the resultant stress are by § 2.4 (22)

$$l\sigma_1/R, \quad m\sigma_2/R, \quad n\sigma_3/R.$$

From these the direction angles α_1, β_1, γ_1 of the direction of the resultant stress are found from tables. The projection of the direction of the resultant stress is then found by drawing circles which are the projection of small circles of angles α_1 about X, β_1 about Y, γ_1 about Z; these are shown dotted in Fig. 2.7.2. These should intersect in a point R which gives the direction of the resultant stress. The great circle through R and P corresponds to the

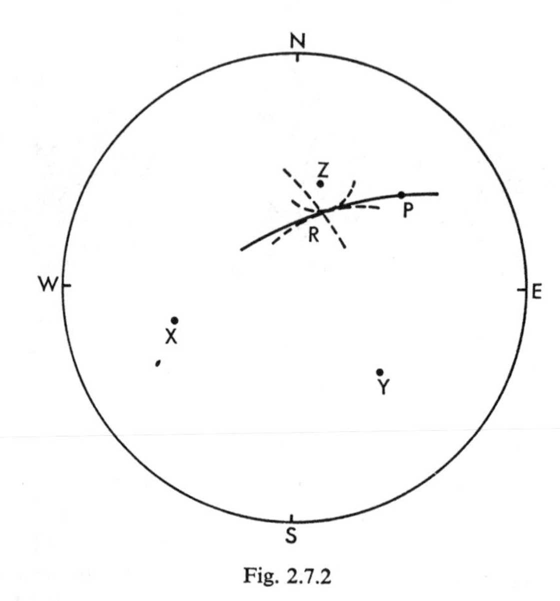

Fig. 2.7.2

plane containing the normal to the plane and the resultant stress across it. The angle χ between R and P can be read off, and the normal and shear stresses are $R \cos \chi$ and $R \sin \chi$.

2.8 Stress deviation

If σ_x, σ_y, σ_z, τ_{yz}, τ_{zx}, τ_{xy} are the stress-components referred to axes $Oxyz$, the mean normal stress s is defined as

$$s = \tfrac{1}{3}(\sigma_x + \sigma_y + \sigma_z) = \tfrac{1}{3}(\sigma_1 + \sigma_2 + \sigma_3) = \tfrac{1}{3}I_1. \qquad (1)$$

This quantity is the invariant $\frac{1}{3}I_1$ by § 2.4 (43). It is also the octahedral normal stress σ_{oct} defined in § 2.4 (48).

For many purposes it is convenient to subtract s from the stress-components and to describe the resulting quantities as the *components of stress deviation*, so that

$$\left.\begin{aligned} s_x = \sigma_x - s, \quad s_y = \sigma_y - s, \quad s_z = \sigma_z - s, \\ s_{yz} = \tau_{yz}, \quad s_{zx} = \tau_{zx}, \quad s_{xy} = \tau_{xy}. \end{aligned}\right\} \tag{2}$$

The principal axes of stress deviation will be the same as those of stress, and the principal stress deviations will be

$$s_1 = \sigma_1 - s = (2\sigma_1 - \sigma_2 - \sigma_3)/3, \quad s_2 = \sigma_2 - s = (2\sigma_2 - \sigma_1 - \sigma_3)/3,$$
$$s_3 = \sigma_3 - s = (2\sigma_3 - \sigma_1 - \sigma_2)/3. \tag{3}$$

The reason for this separation is that essentially s determines uniform compression or dilatation, while the stress deviation determines distortion. Since many criteria of failure are concerned primarily with distortion, and since they must be invariant with respect to rotation of axes, it appears that the invariants of stress deviation will be involved. These will be denoted by J_1, J_2, J_3 and are found as in the case of stress, cf. § 2.4 (40)–(42), to be

$$J_1 = s_x + s_y + s_z = 0, \tag{4}$$

$$J_2 = -(s_y s_z + s_z s_x + s_x s_y) + s_{yz}^2 + s_{zx}^2 + s_{xy}^2, \tag{5}$$

$$J_3 = s_x s_y s_z + 2 s_{yz} s_{zx} s_{xy} - s_x s_{yz}^2 - s_y s_{zx}^2 - s_z s_{xy}^2. \tag{6}$$

Using (4) in (5) gives various alternative forms of J_2,

$$J_2 = \tfrac{1}{2}(s_x^2 + s_y^2 + s_z^2) + s_{yz}^2 + s_{zx}^2 + s_{xy}^2, \tag{7}$$

$$= \tfrac{1}{6}[(\sigma_y - \sigma_z)^2 + (\sigma_z - \sigma_x)^2 + (\sigma_x - \sigma_y)^2] + \tau_{yz}^2 + \tau_{zx}^2 + \tau_{xy}^2, \tag{8}$$

$$= \tfrac{1}{6}[(\sigma_2 - \sigma_3)^2 + (\sigma_3 - \sigma_1)^2 + (\sigma_1 - \sigma_2)^2], \tag{9}$$

$$= \tfrac{1}{2}(s_1^2 + s_2^2 + s_3^2). \tag{10}$$

It may be noted from § 2.4 (49) that the octahedral shear stress

$$\tau_{\text{oct}} = (2J_2/3)^{\frac{1}{2}}. \tag{11}$$

2.9 Displacement and strain

The fundamental concept of continuum mechanics is that of the displacement of all particles of the material. The initial position x, y, z of every one of these is supposed to be known, and the forces applied to the system cause it to be displaced to a final position. The convention of sign in use has to be introduced here. In § 2.2 stresses were defined as positive when compressive, that is, in the negative directions of the axes. To have displacements follow the same pattern, that is for positive displacements to correspond to positive stresses, displacements must be reckoned positive when in the negative directions of the axes. If u, v, w are the displacements of the particle initially at x, y, z, its final position will be

$$x^* = x - u, \quad y^* = y - v, \quad z^* = z - w. \tag{1}$$

The final object of the theory is to determine u, v, w for every particle from the stresses and boundary conditions. To do this, intermediate quantities called strains are introduced, and much of the theory of elasticity concerns itself with strains rather than displacements. In practical rock mechanics, however, displacements assume a very considerable importance, since they are frequently the quantities actually observed, either in experiments or in practical situations such as closure in mining.

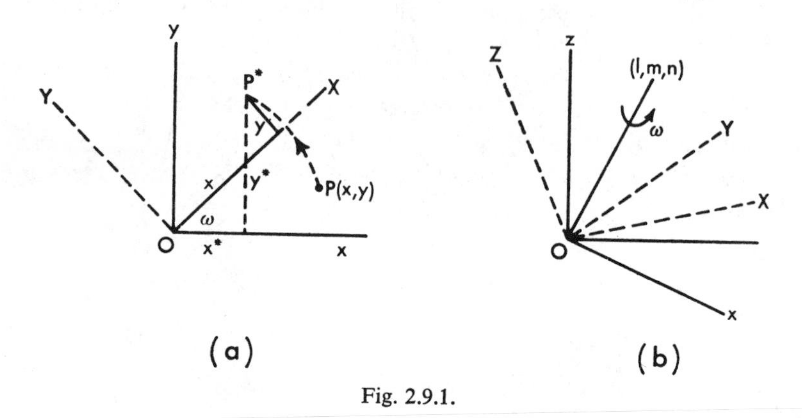

Fig. 2.9.1.

One simple form of displacement in which the relative positions of the particles of the body are not altered is translation as a rigid body in which

$$x^* = x - a, \quad y^* = y - b, \quad z^* = z - c, \tag{2}$$

where a, b, c are constants.

Another is rotation as a rigid body about a fixed axis. In two dimensions, suppose that Oxy are the axes of reference and that the whole body, carrying the point P with it, is rotated through an angle ω so that P moves to P^*. The particles of the body which lay along the axes Ox, Oy before rotation will lie in the directions OX, OY. The coordinates x^*, y^* of P^* relative to the original axes Ox, Oy may be written down from Fig. 2.9.1 (*a*) and are

$$x^* = x \cos \omega - y \sin \omega; \quad y^* = x \sin \omega + y \cos \omega. \tag{3}$$

If ω is small these become

$$x^* = x - \omega y; \quad y^* = y + \omega x. \tag{4}$$

The expression for x, y in terms of x^*, y^* is

$$x = x^* \cos \omega + y^* \sin \omega; \quad y = y^* \cos \omega - x^* \sin \omega. \tag{5}$$

In three dimensions suppose that a rigid body is rotated through an angle ω about an axis whose direction cosines are l, m, n, relative to axes $Oxyz$, Fig. 2.9.1 (*b*). The lines in the body which before rotation lay in the directions Ox, Oy, Oz will be moved by the rotation into directions OX, OY, OZ, whose direction cosines with respect to $Oxyz$ are l_1, m_1, n_1; l_2, m_2, n_2; l_3, m_3, n_3; respectively. It can be shown, Thomson and Tait (1962), that these are given by

$$l_1 = l^2(1 - \cos\omega) + \cos\omega; \quad m_1 = lm(1 - \cos\omega) + n\sin\omega; \quad n_1 = ln(1 - \cos\omega) - m\sin\omega, \tag{6}$$

$$l_2 = lm(1 - \cos\omega) - n\sin\omega; \quad m_2 = m^2(1 - \cos\omega) + \cos\omega; \quad n_2 = mn(1 - \cos\omega) + l\sin\omega, \tag{7}$$

$$l_3 = ln(1 - \cos\omega) + m\sin\omega; \quad m_3 = mn(1 - \cos\omega) - l\sin\omega; \quad n_3 = n^2(1 - \cos\omega) + \cos\omega. \tag{8}$$

The final position of any point can then be found from the transformation formulae given in § 2.14. Thomson and Tait (1962) show that the most general displacement of a rigid body may be described by translation of some point O to O^* followed by rotation about an axis through O^*. They also give a very general discussion of strain and its specification.

For a small rotation ω about the axis l, m, n, the final position x^*, y^*, z^* of the point x, y, z is

$$x^* = x + \omega(mz - ny); \quad y^* = y + \omega(nx - lz); \quad z^* = z + \omega(lx - my). \tag{9}$$

This follows from the fact that if $\mathbf{r}$ is the vector of components (x, y, z) and $\boldsymbol{\omega}$ that of components $(\omega l, \omega m, \omega n)$ the displacement of the point $\mathbf{r}$ in the rotation is given in magnitude and direction by the vector product

$$\boldsymbol{\omega} \wedge \mathbf{r}. \tag{10}$$

If the relative positions of the particles of a body are changed so that their initial and final positions cannot be made to correspond by movement as a rigid body the body is said to be strained, and quantitative methods of measuring the state of strain have to be devised. The most convenient ones are changes of length, and of the angles between lines.

Suppose that l is the distance between two points O, P in the unstrained state and that l^* is the corresponding distance between them in the strained state, Fig. 2.9.2 (*a*). Then

$$\varepsilon = (l - l^*)/l \tag{11}$$

is defined as the *elongation* corresponding to the point O and the direction OP. It is positive when a contraction in accordance with the convention of sign stated earlier in this section. This is the quantity used in the discussion of infinitesimal strain, but it is by no means the only measure used, and in fact the *quadratic elongation* λ defined as

$$\lambda = (l^*/l)^2 = (1 - \varepsilon)^2, \tag{12}$$

and the natural strain, defined in terms of $(l - l^*)/l^*$, will appear later.

The other quantity used as a measure of the strain at a point O is the change in angle between two perpendicular directions OP, OQ. If in the strained state the angle $P^*O^*Q^*$, Fig. 2.9.2 (*b*), is $\frac{1}{2}\pi + \psi$ the quantity

$$\gamma = \tan \psi \tag{13}$$

is called the *shear strain* for the point O and the directions OP, OQ. The reason for the name may be seen from the simple case of Fig. 2.9.2 (*b*), in which straight lines transform into straight lines. Here, the lines parallel

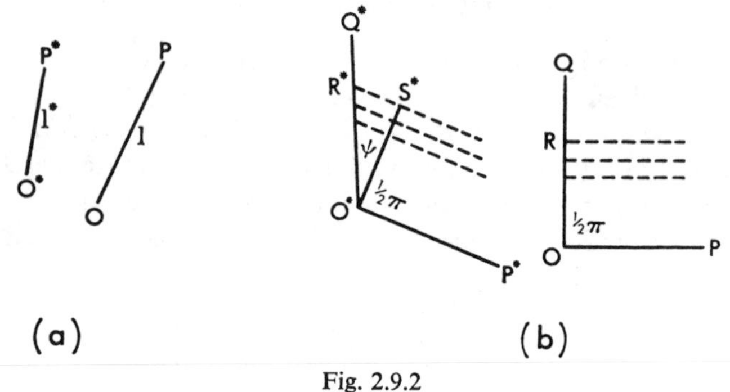

Fig. 2.9.2

to OP, shown dotted, may be regarded as having slid or been sheared across one another to attain their final positions, and the amount of this movement, R^*S^* in Fig. 2.9.2 (*b*), is just $OR \tan \psi = \gamma \,.\, OR$. In the general case, straight lines through O will transform into curves through O^*, but the same considerations will apply for points very near to O and O^*.

This definition of γ has been in use in the literature for a very long time. However, it will be seen in § 2.10 that the quantity $\frac{1}{2}\gamma$ occurs much more naturally in the mathematics. It is also $\frac{1}{2}\gamma$ which is always used when the subject is developed by tensor analysis. For this reason it seems worth while using separate symbols for both quantities, and in future Γ will frequently be used for $\frac{1}{2}\gamma$, so that

$$\Gamma = \tfrac{1}{2}\gamma = \tfrac{1}{2} \tan \psi. \tag{14}$$

In the above discussion no mention of the size or nature of the strains has been made. The general case, mentioned briefly in § 17.6, is extremely complicated, but two special cases are of great importance and easily developed. The first of these is the theory of *infinitesimal strain*, in which

ε and γ are supposed to be so small that their squares and product can be neglected: this is the assumption of the classical theory of elasticity and will be discussed in this chapter. The second case is that of *homogeneous strain*, in which the nature of the strain is supposed to be the same at all points and so parallel straight lines remain parallel after straining. The case of finite homogeneous strain is discussed in §§ 17.7, 17.8. Infinitesimal strain, as defined above, is homogeneous in a small region around any point.

The displacements u, v, w relative to rectangular axes are the components of a vector and transform as such. In two dimensions, if u, v are components of displacement relative to rectangular axes Oxy, and u', v' are those relative to axes $Ox'y'$ rotated through θ, it follows as in (3) and (5) that

$$u' = u \cos\theta + v \sin\theta; \quad v' = v \cos\theta - u \sin\theta, \tag{15}$$

$$u = u' \cos\theta - v' \sin\theta; \quad v = v' \cos\theta + u' \sin\theta. \tag{16}$$

The corresponding formulae in three dimensions are given in § 2.14 (11), (12).

2.10 Infinitesimal strain in two dimensions

Consider two neighbouring points $P(x, y)$ and $Q(x + x', y + y')$, Fig. 2.10 (*a*), which are so close that the squares and products of x', y' may be

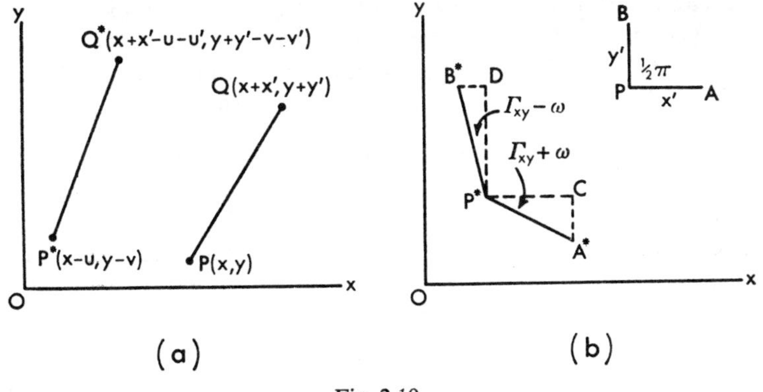

Fig. 2.10

neglected. Let the displacement of P be u, v, which it is assumed varies continuously with x and y. Then by Taylor's theorem the displacement of Q will be $u + u'$, $v + v'$, where

$$u' = x'\frac{\partial u}{\partial x} + y'\frac{\partial u}{\partial y}, \tag{1}$$

$$v' = x'\frac{\partial v}{\partial x} + y'\frac{\partial v}{\partial y}, \tag{2}$$

neglecting terms in x'^2, $x'y'$, etc.

The fundamental notation is now defined, namely,

$$\varepsilon_x = \frac{\partial u}{\partial x}, \quad \varepsilon_y = \frac{\partial v}{\partial y}, \tag{3}$$

$$\Gamma_{xy} = \Gamma_{yx} = \tfrac{1}{2}\gamma_{xy} = \tfrac{1}{2}\gamma_{yx} = \tfrac{1}{2}\left(\frac{\partial v}{\partial x} + \frac{\partial u}{\partial y}\right), \tag{4}$$

$$\omega = \tfrac{1}{2}\left(\frac{\partial v}{\partial x} - \frac{\partial u}{\partial y}\right). \tag{5}$$

In terms of these, (1) and (2) become

$$u' = x'\varepsilon_x + y'\Gamma_{xy} - y'\omega, \tag{6}$$

$$v' = x'\Gamma_{xy} + y'\varepsilon_y + x'\omega. \tag{7}$$

The quantities ε_x, ε_y, Γ_{xy} are called the *components of strain.* ω is called the *component of rotation*, since the terms $-y'\omega$, $x'\omega$ are exactly those found in § 2.9 (4) to be caused by a small rotation ω.

The assumption will now be made that the components of strain and rotation, ε_x, ε_y, Γ_{xy}, are so small that their squares and products can be neglected. This is called the assumption of infinitesimal strain.

The physical significance of ε_x, ε_y, Γ_{xy}, and ω can be seen as follows. Suppose that P, A, B are the points (x, y), $(x + x', y)$, $(x, y + y')$, Fig. 2.10 (*b*), and that their final positions are P^*, A^*, B^*. Suppose that P^*C is the projection of P^*A^* on a line parallel to Ox and that P^*D is the projection of P^*B^* on a line parallel to Oy. Then it follows from (7) and (8) that the configuration is that shown in Fig. 2.10 (*b*) with

$$P^*C = x'(1 - \varepsilon_x), \quad A^*C = x'\Gamma_{xy} + x'\omega, \tag{8}$$

$$P^*D = y'(1 - \varepsilon_y), \quad B^*D = y'\Gamma_{xy} - y'\omega. \tag{9}$$

The elongation ε at P corresponding to the x-direction PA as defined in § 2.9 (11) is

$$\varepsilon = (PA - P^*A^*)/PA \tag{10}$$

$$= \{x' - x'[(1 - \varepsilon_x)^2 + (\Gamma_{xy} + \omega)^2]^{\frac{1}{2}}\}/x' = \varepsilon_x, \tag{11}$$

neglecting squares and products of the small quantities ε_x, Γ_{xy}, ω. Similarly, the elongation in the y-direction is ε_y.

The angle A^*P^*C is by (8)

$$\tan^{-1}\left(\frac{\Gamma_{xy} + \omega}{1 - \varepsilon_x}\right) = \Gamma_{xy} + \omega, \tag{12}$$

neglecting squares of small quantities, and the angle B^*P^*D is

$$\tan^{-1}\left(\frac{\Gamma_{xy} - \omega}{1 - \varepsilon_y}\right) = \Gamma_{xy} - \omega. \tag{13}$$

Thus the angle $A^*P^*B^*$ is $\frac{1}{2}\pi + 2\Gamma_{xy}$, and from the definition § 2.9 (14) the shear strain associated with the directions PA, PB will be $2\Gamma_{xy}$. These two directions, originally at right angles, make an angle $\frac{1}{2}\pi + 2\Gamma_{xy}$ with one another in the final state and are rotated through an angle ω in a clockwise direction.

The elongation in a direction inclined at θ to Ox can now be expressed in terms of ε_x, ε_y, and Γ_{xy}. Suppose that P^*, $(x - u, y - v)$ and Q^*, $(x + x' - u - u', y + y' - v - v')$ are the final position of the points $P(x, y)$ and $Q(x + x', y + y')$, Fig. 2.10 (a), where u' and v' are given by (7) and (8) and

$$x' = r\cos\theta, \quad y' = r\sin\theta. \tag{14}$$

Then

$$\begin{aligned} P^*Q^{*2} &= (x' - u')^2 + (y' - v')^2 \\ &= [x'(1 - \varepsilon_x) - y'(\Gamma_{xy} - \omega)]^2 + [y'(1 - \varepsilon_y) - x'(\Gamma_{xy} + \omega)]^2 \qquad (15) \\ &= r^2\{[(1 - \varepsilon_x)\cos\theta - (\Gamma_{xy} - \omega)\sin\theta]^2 + \\ &\qquad [(1 - \varepsilon_y)\sin\theta - (\Gamma_{xy} + \omega)\cos\theta]^2\} \\ &= r^2\{1 - 2\varepsilon_x\cos^2\theta - 2\varepsilon_y\sin^2\theta - 4\Gamma_{xy}\sin\theta\cos\theta\}, \qquad (16) \end{aligned}$$

neglecting squares and products of ε_x, ε_y, Γ_{xy}, ω. Neglecting them again, it follows from (16) that

$$P^*Q^* = r\{1 - \varepsilon_x\cos^2\theta - \varepsilon_y\sin^2\theta - 2\Gamma_{xy}\sin\theta\cos\theta\}. \tag{17}$$

Therefore the elongation ε in the direction θ defined as in § 2.9 (11) is

$$\varepsilon = (r - P^*Q^*)/r = \varepsilon_x\cos^2\theta + \varepsilon_y\sin^2\theta + 2\Gamma_{xy}\sin\theta\cos\theta. \tag{18}$$

Next, we wish to calculate the shear strain associated with PQ and a perpendicular direction PR. The final position P^*Q^* of PQ will be inclined to Ox at an angle θ_1 given by

$$\begin{aligned} \tan\theta_1 &= \frac{y' - v'}{x' - u'} = \frac{y'(1 - \varepsilon_y) - x'(\Gamma_{xy} + \omega)}{x'(1 - \varepsilon_x) - y'(\Gamma_{xy} - \omega)} \\ &= \frac{1 - \varepsilon_y - (\Gamma_{xy} + \omega)\cot\theta}{1 - \varepsilon_x - (\Gamma_{xy} - \omega)\tan\theta}\cdot\tan\theta \\ &= \tan\theta + \{\varepsilon_x - \varepsilon_y + (\Gamma_{xy} - \omega)\tan\theta - \\ &\qquad (\Gamma_{xy} + \omega)\cot\theta\}\tan\theta, \qquad (19) \end{aligned}$$

neglecting squares and products of small quantities. Since θ_1 will be nearly equal to θ, we may put $\theta_1 = \theta + \alpha_1$, where α_1 is small and then

$$\begin{aligned} \tan\theta_1 = \tan(\theta + \alpha_1) &= \frac{\alpha_1 + \tan\theta}{1 - \alpha_1\tan\theta} \\ &= \tan\theta + \alpha_1\sec^2\theta, \qquad (20) \end{aligned}$$

neglecting α_1^2, etc. Comparing (19) and (20) gives

$$\alpha_1 = \tfrac{1}{2}\{\varepsilon_x - \varepsilon_y + (\Gamma_{xy} - \omega)\tan\theta - (\Gamma_{xy} + \omega)\cot\theta\}\sin 2\theta. \quad (21)$$

Similarly, the final position P^*R^* of PR will be inclined to Ox at an angle $\tfrac{1}{2}\pi + \theta + \alpha_2$, where α_2 may be found by replacing θ by $\tfrac{1}{2}\pi + \theta$ in (21) and so is

$$\alpha_2 = -\tfrac{1}{2}\{\varepsilon_x - \varepsilon_y - (\Gamma_{xy} - \omega)\cot\theta + (\Gamma_{xy} + \omega)\tan\theta\}\sin 2\theta. \quad (22)$$

The shear strain $\gamma = 2\Gamma$ corresponding to the directions PQ, PR is just

$$2\Gamma = \alpha_2 - \alpha_1 = (\varepsilon_y - \varepsilon_x)\sin 2\theta + 2\Gamma_{xy}\cos 2\theta. \quad (23)$$

(18) and (23) give the elongation and shear strain associated with a direction inclined at θ to Ox.

It will be seen that they take exactly the same form as § 2.3 (4) and (6) for the normal stress σ and shear stress τ with σ_x, σ_y, τ_{xy} replaced by ε_x, ε_y, Γ_{xy}, so that the theory may be developed precisely as in § 2.3 and the whole of the results of that section taken over. If the quantity $\gamma_{xy} = \tfrac{1}{2}\Gamma_{xy}$ had been used this would not have been so simple.

As in § 2.3 (11), the shear strain Γ vanishes when

$$\tan 2\theta = \frac{2\Gamma_{xy}}{\varepsilon_x - \varepsilon_y}, \quad (24)$$

and for these directions the elongation is a maximum or minimum. These directions are the *principal axes of strain*, and the elongations ε_1 and ε_2 in these directions are the *principal strains*. By § 2.3 (14) they are

$$\tfrac{1}{2}(\varepsilon_x + \varepsilon_y) \pm [\Gamma_{xy}^2 + \tfrac{1}{4}(\varepsilon_x - \varepsilon_y)^2]^{\frac{1}{2}}. \quad (25)$$

Referred to principal axes, the elongation ε and shear strain Γ for a direction inclined at θ to the ε_1– direction are

$$\varepsilon = \varepsilon_1\cos^2\theta + \varepsilon_2\sin^2\theta, \quad (26)$$

$$\Gamma = -\tfrac{1}{2}(\varepsilon_1 - \varepsilon_2)\sin 2\theta. \quad (27)$$

The Mohr representation in the ε, Γ plane and the constructions based on it follow precisely as in Fig. 2.3.4.

It follows from (26) that a circle of unit radius is deformed into an ellipse of axes $(1 - \varepsilon_1)$ and $(1 - \varepsilon_2)$ and area $\pi(1 - \varepsilon_1 - \varepsilon_2)$. The ratio of the change in area to the original area is called the *volumetric strain* Δ, so that Δ is the invariant

$$\Delta = \varepsilon_1 + \varepsilon_2 = \varepsilon_x + \varepsilon_y. \quad (28)$$

With the present notation, Δ is positive when the area is decreased. Δ is more usually referred to as the dilatation.

One new point of importance arises. The *three* components of strain ε_x, ε_y, Γ_{xy} were derived in (3) and (4) from *two* components of displacement, so that they are not independent. In fact, it follows by differentiating (3) and (4) that

$$2\frac{\partial^2 \Gamma_{xy}}{\partial x \partial y} = \frac{\partial^2 \varepsilon_x}{\partial y^2} + \frac{\partial^2 \varepsilon_y}{\partial x^2}. \tag{29}$$

This relation is called the compatibility condition for strain in two dimensions. Its significance will appear in § 5.7.

2.11 Infinitesimal strain in three dimensions

We consider a set of rectangular axes Ox, Oy, Oz and suppose that the initial position of a point P is x, y, z and that the displacement at x, y, z has components u, v, w, measured positively in the negative x-, y-, z-directions, so that the final position P^* of P is

$$x^* = x - u, \quad y^* = y - v, \quad z^* = z - w. \tag{1}$$

Now suppose that Q is a point $x + x'$, $y + y'$, $z + z'$ near to P and such that x', y', z' are small and their squares and products are negligible. The displacement at Q will be $u + u'$, $v + v'$, $w + w'$ where, by Taylor's theorem,

$$u' = \frac{\partial u}{\partial x}x' + \frac{\partial u}{\partial y}y' + \frac{\partial u}{\partial z}z', \tag{2}$$

$$v' = \frac{\partial v}{\partial x}x' + \frac{\partial v}{\partial y}y' + \frac{\partial v}{\partial z}z', \tag{3}$$

$$w' = \frac{\partial w}{\partial x}x' + \frac{\partial w}{\partial y}y' + \frac{\partial w}{\partial z}z', \tag{4}$$

neglecting x'^2, $x'y'$, etc. The final position Q^* of Q will be

$$x + x' - u - u', \quad y + y' - v - v', \quad z + z' - w - w'. \tag{4a}$$

As in § 2.10, we define the *components of strain* as

$$\varepsilon_x = \frac{\partial u}{\partial x}, \quad \varepsilon_y = \frac{\partial v}{\partial y}, \quad \varepsilon_z = \frac{\partial w}{\partial z}, \tag{5}$$

$$2\Gamma_{yz} = 2\Gamma_{zy} = \gamma_{yz} = \gamma_{zy} = \frac{\partial w}{\partial y} + \frac{\partial v}{\partial z}, \tag{6}$$

$$2\Gamma_{zx} = 2\Gamma_{xz} = \gamma_{zx} = \gamma_{xz} = \frac{\partial u}{\partial z} + \frac{\partial w}{\partial x}, \tag{7}$$

$$2\Gamma_{xy} = 2\Gamma_{yx} = \gamma_{xy} = \gamma_{yx} = \frac{\partial v}{\partial x} + \frac{\partial u}{\partial y}, \tag{8}$$

and the components of rotation

$$2\omega_x = \frac{\partial w}{\partial y} - \frac{\partial v}{\partial z}, \quad 2\omega_y = \frac{\partial u}{\partial z} - \frac{\partial w}{\partial x}, \quad 2\omega_z = \frac{\partial v}{\partial x} - \frac{\partial u}{\partial y}. \tag{9}$$

If $\boldsymbol{\omega}$ is the vector of components $\omega_x, \omega_y, \omega_z$ specifying the rotation, and $\mathbf{u}$ is the vector of components (u, v, w), (9) may be written

$$2\boldsymbol{\omega} = \text{curl } \mathbf{u}. \tag{10}$$

Using the notation (5)–(9), (2)–(4) become

$$u' = x'\varepsilon_x + y'\Gamma_{xy} + z'\Gamma_{xz} + z'\omega_y - y'\omega_z, \tag{11}$$

$$v' = x'\Gamma_{yx} + y'\varepsilon_y + z'\Gamma_{yz} + x'\omega_z - z'\omega_x, \tag{12}$$

$$w' = x'\Gamma_{zx} + y'\Gamma_{zy} + z'\varepsilon_z + y'\omega_x - x'\omega_y, \tag{13}$$

The last terms, namely

$$z'\omega_y - y'\omega_z, \quad x'\omega_z - z'\omega_x, \quad y'\omega_x - x'\omega_y, \tag{14}$$

are by § 2.9 (10) just the components of the displacement of (x', y', z') due to a small rotation of components $\omega_x, \omega_y, \omega_z$. As before, the assumption of infinitesimal strain is that all components of strain and rotation are so small that their squares and products may be neglected.

Next we calculate the change in length of the line PQ. From (1) to (13), neglecting squares and products of small quantities,

$$\begin{aligned} P^*Q^{*2} &= (x' - u')^2 + (y' - v')^2 + (z' - w')^2 \\ &= [x'(1 - \varepsilon_x) - y'(\Gamma_{xy} - \omega_z) - z'(\Gamma_{xz} + \omega_y)]^2 \\ &\quad + [y'(1 - \varepsilon_y) - x'(\Gamma_{yx} + \omega_z) - z'(\Gamma_{yz} - \omega_x)]^2 \\ &\quad + [z'(1 - \varepsilon_z) - y'(\Gamma_{zy} + \omega_x) - x'(\Gamma_{zx} - \omega_y)]^2 \\ &= x'^2 + y'^2 + z'^2 - 2x'^2\varepsilon_x - 2y'^2\varepsilon_y - 2z'^2\varepsilon_z \\ &\quad - 4x'y'\Gamma_{xy} - 4y'z'\Gamma_{yz} - 4z'x'\Gamma_{xz} \end{aligned} \tag{15}$$

neglecting terms in ε_x^2, etc.

Now suppose PQ has length r and direction cosines l, m, n so that $x' = lr$, $y' = mr$, $z' = nr$, then (15) gives

$$\begin{aligned} P^*Q^{*2} = r^2\{1 - 2l^2\varepsilon_x - 2m^2\varepsilon_y - 2n^2\varepsilon_z - \\ 4mn\Gamma_{yz} - 4nl\Gamma_{zx} - 4lm\Gamma_{xy}\}, \end{aligned} \tag{16}$$

and, again neglecting squares and products of small quantities,

$$P^*Q^* = r\{1 - l^2\varepsilon_x - m^2\varepsilon_y - n^2\varepsilon_z - 2mn\Gamma_{yz} - 2nl\Gamma_{zx} - 2lm\Gamma_{xy}\}. \tag{17}$$

Thus the elongation ε defined as $(PQ - P^*Q^*)/r$ is

$$\varepsilon = \varepsilon_x l^2 + \varepsilon_y m^2 + \varepsilon_z n^2 + 2mn\Gamma_{yz} + 2nl\Gamma_{zx} - 2lm\Gamma_{xy}. \tag{18}$$

This has precisely the same form as § 2.4 (9), so that all its consequences follow. There are three perpendicular directions in which ε is stationary as a function of l, m, n: these are the *principal axes of strain*, and the strains in these directions, the *principal strains* $\varepsilon_1, \varepsilon_2, \varepsilon_3$, are the roots of

$$\varepsilon^3 - I_1\varepsilon^2 - I_2\varepsilon - I_3 = 0, \tag{19}$$

where I_1, I_2, I_3 are the invariants

$$I_1 = \varepsilon_x + \varepsilon_y + \varepsilon_z, \tag{20}$$

$$I_2 = -(\varepsilon_y\varepsilon_z + \varepsilon_z\varepsilon_x + \varepsilon_x\varepsilon_y) + \Gamma_{yz}^2 + \Gamma_{zx}^2 + \Gamma_{xy}^2, \tag{21}$$

$$I_3 = \varepsilon_x\varepsilon_y\varepsilon_z + 2\Gamma_{yz}\Gamma_{zx}\Gamma_{xy} - \varepsilon_x\Gamma_{yz}^2 - \varepsilon_y\Gamma_{zx}^2 - \varepsilon_z\Gamma_{xy}^2. \tag{22}$$

These correspond to § 2.4 (39)–(42).

By (18), a sphere of radius 1 is deformed into an ellipsoid (the strain ellipsoid, cf. § 17.8) of axes $(1 - \varepsilon_1,\ 1 - \varepsilon_2,\ 1 - \varepsilon_3)$ and volume $4\pi(1 - \varepsilon_1 - \varepsilon_2 - \varepsilon_3)/3$, so that the volumetric strain Δ which is the ratio of the change in volume to the original volume is

$$\Delta = \varepsilon_1 + \varepsilon_2 + \varepsilon_3 = \varepsilon_x + \varepsilon_y + \varepsilon_z. \tag{23}$$

To calculate the shear strain corresponding to PQ of length r and direction cosines l_1, m_1, n_1 and a perpendicular line PR of length r and direction cosines l_2, m_2, n_2 we proceed as follows. The direction ratios of P^*Q^* are

$$x' - u', \quad y' - v', \quad z' - w',$$

or, from (11) to (13) with $x' = l_1r$, $y' = m_1r$, $z' = n_1r$,

$$\begin{aligned} &r\{l_1(1 - \varepsilon_x) - m_1(\Gamma_{xy} - \omega_z) - n_1(\Gamma_{xz} + \omega_y)\},\\ &r\{-l_1(\Gamma_{yx} + \omega_z) + m_1(1 - \varepsilon_y) - n_1(\Gamma_{yz} - \omega_x)\},\\ &r\{-l_1(\Gamma_{zx} - \omega_y) - m_1(\Gamma_{zy} + \omega_x) + n_1(1 - \varepsilon_z)\}, \end{aligned} \tag{24}$$

and the direction cosines of P^*Q^* are found by dividing the quantities in (24) by the square root of the sum of their squares (cf. § 2.4 (3)), and this is just the quantity P^*Q^* which has been evaluated in (15)–(17). The direction cosines of P^*R^* are given by the same formulae with l_1, m_1, n_1 replaced by l_2, m_2, n_2.

The angle ϕ between P^*Q^* and P^*R^* is then by § 2.4 (4) given by

$$\begin{aligned} &r^{-2}\,.\,P^*Q^*\,.\,P^*R^*\cos\phi\\ &= \{l_1(1 - \varepsilon_x) - m_1(\Gamma_{xy} - \omega_z) - n_1(\Gamma_{xz} + \omega_y)\}\{l_2(1 - \varepsilon_x) -\\ &\quad m_2(\Gamma_{xy} - \omega_z) - n_2(\Gamma_{xz} + \omega_y)\} + \{-l_1(\Gamma_{yx} + \omega_z) +\\ &\quad m_1(1 - \varepsilon_y) - n_1(\Gamma_{yz} - \omega_x)\}\{-l_2(\Gamma_{yx} + \omega_z) + m_2(1 - \varepsilon_y) -\\ &\quad n_2(\Gamma_{yz} - \omega_x)\} + \{-l_1(\Gamma_{zx} - \omega_y) - m_1(\Gamma_{zy} + \omega_x) +\\ &\quad n_1(1 - \varepsilon_z)\}\{-l_2(\Gamma_{zx} - \omega_y) - m_2(\Gamma_{zy} + \omega_x) + n_2(1 - \varepsilon_z)\}\\ &= l_1l_2(1 - 2\varepsilon_x) - (l_1m_2 + m_1l_2)(\Gamma_{xy} - \omega_z) - (l_1n_2 + n_1l_2)(\Gamma_{xz} + \omega_y) +\\ &\quad m_1m_2(1 - 2\varepsilon_y) - (l_1m_2 + m_1l_2)(\Gamma_{yx} + \omega_z) -\\ &\quad (m_1n_2 + n_1m_2)(\Gamma_{yz} - \omega_x) + n_1n_2(1 - 2\varepsilon_z) -\\ &\quad (l_1n_2 + n_1l_2)(\Gamma_{zx} - \omega_y) - (m_1n_2 + n_1m_2)(\Gamma_{zy} + \omega_x), \end{aligned} \tag{25}$$

neglecting squares and products of ε_x, Γ_{xy}, etc.

Now since PQ and PR are perpendicular

$$l_1 l_2 + m_1 m_2 + n_1 n_2 = 0. \tag{26}$$

Also $\phi = \frac{1}{2}\pi + 2\Gamma$, where 2Γ is the shear strain for the pair of directions PQ, PR, so that

$$\cos\phi = -2\Gamma. \tag{27}$$

Using (17), (26), and (27) in (25), and again neglecting squares and products of small quantities, gives finally

$$\Gamma = l_1 l_2 \varepsilon_x + m_1 m_2 \varepsilon_y + n_1 n_2 \varepsilon_z + (l_1 m_2 + m_1 l_2)\Gamma_{xy} + (m_1 n_2 + m_2 n_1)\Gamma_{yz} + (n_1 l_2 + l_1 n_2)\Gamma_{zx}. \tag{28}$$

Since (18) and (28) have precisely the same form as the equations § 2.5 (1), (3) for transformation of stress, with stress-components replaced by strain-components, it follows that all the formulae deduced for stresses may be taken over with this change of notation.

Finally the *compatibility conditions* for strains have to be stated. Nine quantities, six components of strain and three components of rotation, were defined in terms of the differential coefficients of the components of displacement so that they cannot be independent. The relations between them may be stated as six relations between components of stress, namely three of type § 2.12 (29), namely

$$2\frac{\partial^2\Gamma_{xy}}{\partial x\,\partial y} = \frac{\partial^2\varepsilon_x}{\partial y^2} + \frac{\partial^2\varepsilon_y}{\partial x^2}; \quad 2\frac{\partial^2\Gamma_{yz}}{\partial y\,\partial z} = \frac{\partial^2\varepsilon_y}{\partial z^2} + \frac{\partial^2\varepsilon_z}{\partial y^2};$$

$$2\frac{\partial^2\Gamma_{zx}}{\partial z\,\partial x} = \frac{\partial^2\varepsilon_z}{\partial x^2} + \frac{\partial^2\varepsilon_x}{\partial z^2}, \tag{29}$$

and three others, namely

$$\frac{\partial^2\varepsilon_x}{\partial y\,\partial z} = \frac{\partial}{\partial x}\left(-\frac{\partial\Gamma_{yz}}{\partial x} + \frac{\partial\Gamma_{zx}}{\partial y} + \frac{\partial\Gamma_{xy}}{\partial z}\right), \tag{30}$$

$$\frac{\partial^2\varepsilon_y}{\partial z\,\partial x} = \frac{\partial}{\partial y}\left(\frac{\partial\Gamma_{yz}}{\partial x} - \frac{\partial\Gamma_{zx}}{\partial y} + \frac{\partial\Gamma_{xy}}{\partial z}\right), \tag{31}$$

$$\frac{\partial^2\varepsilon_z}{\partial x\,\partial y} = \frac{\partial}{\partial z}\left(\frac{\partial\Gamma_{yz}}{\partial x} + \frac{\partial\Gamma_{zx}}{\partial y} - \frac{\partial\Gamma_{xy}}{\partial z}\right). \tag{32}$$

These may be verified easily by differentiation. For a more fundamental treatment, and the connection with the components of rotation, see Love (1927).

2.12 Strain deviation

Just as in the case of stress, the components of strain can be decomposed into the mean normal strain e defined as

$$e = \tfrac{1}{3}(\varepsilon_x + \varepsilon_y + \varepsilon_z) = \tfrac{1}{3}(\varepsilon_1 + \varepsilon_2 + \varepsilon_3) = \tfrac{1}{3}\Delta, \tag{1}$$

using § 2.11 (23).

The components e_x, e_y, e_z, e_{yz}, e_{zx}, e_{xy} of strain deviation are then defined as

$$e_x = \varepsilon_x - e, \quad e_y = \varepsilon_y - e, \quad e_z = \varepsilon_z - e, \tag{2}$$

$$e_{yz} = \Gamma_{yz}, \quad e_{zx} = \Gamma_{zx}, \quad e_{xy} = \Gamma_{xy}. \tag{3}$$

The principal axes of strain deviation coincide with the principal axes of strain, and the principal strain deviations are

$$e_1 = \varepsilon_1 - e, \quad e_2 = \varepsilon_2 - e, \quad e_3 = \varepsilon_3 - e, \tag{4}$$

so that

$$e_1 + e_2 + e_3 = 0. \tag{5}$$

2.13 Determination of principal stresses or strains from measurements

In practice, in underground measurements, the normal stress in any direction may be measured by a flat-jack perpendicular to that direction, or the elongation in any direction may be measured by strain gauges or displacement meters in that direction.

Since there are three components of stress or strain in two dimensions, or six in three dimensions, it may be expected that three measurements would be necessary in two dimensions, and six in three dimensions, to determine the complete state of stress or strain. In two dimensions the theory is the well-known theory of strain rosettes (Hetényi, 1950; Durelli, Phillips and Tsao, 1958). Since the equations in terms of stress and strain are the same, the problem will be set out in terms of stress.

In two dimensions the normal stress σ across a plane whose normal is inclined at θ to Ox is given by § 2.3 (4), namely

$$\sigma = \sigma_x \cos^2 \theta + 2\tau_{xy} \sin \theta \cos \theta + \sigma_y \sin^2 \theta, \tag{1}$$

so that if $\theta = 45°$ the normal stress σ_{45} is

$$2\sigma_{45} = \sigma_x + \sigma_y + 2\tau_{xy}. \tag{2}$$

It follows that if σ_x, σ_y, and σ_{45} are measured, τ_{xy} is given by (2). It follows from § 2.3 (11) that the principal axes are inclined at

$$\tfrac{1}{2} \tan^{-1}\left\{\frac{2\sigma_{45} - \sigma_x - \sigma_y}{\sigma_x - \sigma_y}\right\} \tag{3}$$

to Ox, and from § 2.3 (14) that the principal stresses are

$$\tfrac{1}{2}(\sigma_x + \sigma_y) \pm \tfrac{1}{2}\{(2\sigma_{45} - \sigma_x - \sigma_y)^2 + (\sigma_x - \sigma_y)^2\}^{\frac{1}{2}}. \tag{4}$$

This is not the most usual treatment, which is as follows. Suppose $O\sigma_1$, $O\sigma_2$, Fig. 2.13 (*a*), are the directions of the principal stresses σ_1 and σ_2 and that normal stresses σ_P, σ_Q, σ_R are measured in the directions OP at an unknown angle θ to $O\sigma_1$, and OQ and OR at known angles α and $\alpha + \beta$ to OP. Then by § 2.3 (16)

$$\sigma_P = \tfrac{1}{2}(\sigma_1 + \sigma_2) + \tfrac{1}{2}(\sigma_1 - \sigma_2)\cos 2\theta, \tag{5}$$

$$\sigma_Q = \tfrac{1}{2}(\sigma_1 + \sigma_2) + \tfrac{1}{2}(\sigma_1 - \sigma_2)\cos 2(\theta + \alpha), \tag{6}$$

$$\sigma_R = \tfrac{1}{2}(\sigma_1 + \sigma_2) + \tfrac{1}{2}(\sigma_1 - \sigma_2)\cos 2(\theta + \alpha + \beta). \tag{7}$$

These equations can be solved for σ_1, σ_2, and θ. For the important case $\alpha = \beta = \pi/4$ the solutions are

$$\sigma_1 + \sigma_2 = \sigma_P + \sigma_R, \tag{8}$$

$$\sigma_1 - \sigma_2 = \{(\sigma_P - 2\sigma_Q + \sigma_R)^2 + (\sigma_P - \sigma_R)^2\}^{\frac{1}{2}}, \tag{9}$$

$$\tan 2\theta = (\sigma_P - 2\sigma_Q + \sigma_R)/(\sigma_P - \sigma_R), \tag{10}$$

which are the same as (3) and (4).

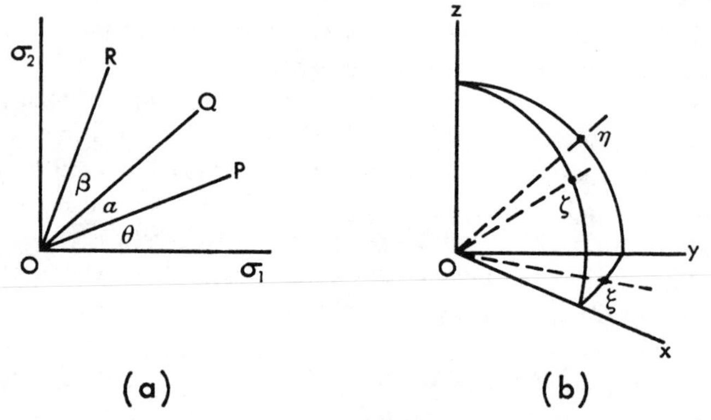

Fig. 2.13

For the other common case in which $\alpha = \beta = 60°$ the solutions are

$$\sigma_1 + \sigma_2 = \tfrac{2}{3}(\sigma_P + \sigma_Q + \sigma_R), \tag{11}$$

$$(\sigma_1 - \sigma_2)^2 = \tfrac{4}{3}(\sigma_Q - \sigma_R)^2 + \tfrac{4}{9}(2\sigma_P - \sigma_R - \sigma_Q)^2, \tag{12}$$

$$\tan 2\theta = 3^{\frac{1}{2}}(\sigma_Q - \sigma_R)/(\sigma_Q + \sigma_R - 2\sigma_P). \tag{13}$$

These and solutions for other angles, and graphical and analogue methods for finding them, are given by Hetényi (1950). The use of the Mohr circle in this connection has been described in § 2.3.

In three dimensions the normal stress across a plane whose direction cosines are l, m, n is by § 2.4 (9)

$$\sigma = l^2\sigma_x + m^2\sigma_y + n^2\sigma_z + 2mn\tau_{yz} + 2nl\tau_{zx} + 2lm\tau_{xy}. \tag{14}$$

Here, σ_x, σ_y, σ_z may be taken to have been measured, and we may choose three other axes, e.g. ξ of direction cosines $(2^{-\frac{1}{2}}, 2^{-\frac{1}{2}}, 0)$: η of direction cosines $(0, 2^{-\frac{1}{2}}, 2^{-\frac{1}{2}})$; and ζ of direction cosines $(2^{-\frac{1}{2}}, 0, 2^{-\frac{1}{2}})$, Fig. 2.13 (*b*). The normal stresses σ_ξ, σ_η, σ_ζ for these are, by (14),

$$2\sigma_\xi = \sigma_x + \sigma_y + 2\tau_{xy}, \tag{15}$$

$$2\sigma_\eta = \sigma_y + \sigma_z + 2\tau_{yz}, \tag{16}$$

$$2\sigma_\zeta = \sigma_x + \sigma_z + 2\tau_{zx}. \tag{17}$$

Measurement of σ_ξ, σ_η, σ_ζ gives τ_{xy}, τ_{yz}, τ_{zx}, and the principal axes and stresses are found by the methods described in § 2.4. The corresponding procedure in two dimensions has been used in (1)–(4) above.

As a second example suppose that axes $Oxyz$ are such that σ_y and σ_z are known and that $\tau_{yz} = 0$. Also suppose that for another set of rectangular axes $Ox'y'z'$ with direction cosines (l_1, m_1, n_1), (l_2, m_2, n_2), (l_3, m_3, n_3), $\sigma_{y'}$ and $\sigma_{z'}$ are known and $\tau_{y'z'} = 0$. Then by § 2.5 (5), (6), (7)

$$\sigma_{y'} = l_2{}^2\sigma_x + m_2{}^2\sigma_y + n_2{}^2\sigma_z + 2n_2l_2\tau_{zx} + 2l_2m_2\tau_{xy}, \tag{18}$$

$$\sigma_{z'} = l_3{}^2\sigma_x + m_3{}^2\sigma_y + n_3{}^2\sigma_z + 2n_3l_3\tau_{zx} + 2l_3m_3\tau_{xy}, \tag{19}$$

$$0 = l_2l_3\sigma_x + m_2m_3\sigma_y + n_2n_3\sigma_z + (n_2l_3 + n_3l_2)\tau_{zx} + (l_2m_3 + l_3m_2)\tau_{xy}, \tag{20}$$

which are three equations for the unknowns σ_x, τ_{zx}, τ_{xy}. However, the determinant of these equations vanishes, so that they are not independent. It follows that the complete state of stress cannot be found from measurements of the subsidiary principal stresses in two inclined boreholes. A third borehole must be used, which will give another set of equations similar to (18)–(20), and the result will be a set of four equations for three unknowns which will determine them, and, in addition, give an estimate of the accuracy of the measurement. If, as is possible with some instrumentation, σ_x can be measured as well as σ_y and σ_z, any two of (18)–(20) may be used to give τ_{zx} and τ_{xy}.

In the most general case in which the normal stresses in six directions are known, there will be six equations of type § 2.4 (9) to determine the stress-components.

2.14 Matrices and change of axes

The theory above has been set out using the most elementary methods possible. If a sufficient background of the theory of vectors, tensors, or matrices is available it may be developed more shortly and elegantly. Many of the results can be stated in matrix notation, and it is useful to be able to use this, partly because it facilitates repeated transformations, and partly because the matrix language is well suited to computers.

The set of numbers a_{rs}, $r = 1, 2, \ldots m$; $s = 1, 2, \ldots n$, which may be set out in the form

$$\begin{bmatrix} a_{11} & a_{12} \ldots a_{1n} \\ a_{21} & a_{22} \ldots a_{2n} \\ \cdots & \cdots \\ a_{m1} & a_{m2} \ldots a_{mn} \end{bmatrix} \tag{1}$$

is described as a matrix of m rows and n columns and may be written shortly $[a_{rs}]$ or $[a]$. If $m \neq n$ the matrix is called *rectangular*, and if $m = n$ it is *square*. If $m = n$ and $a_{rs} = a_{sr}$ the matrix is *symmetrical*. If $a_{rs} = a_{sr} = 0$ unless $r = s$ the matrix is *diagonal*, and in this case if $a_{rr} = 1$ for all r it is the *unit matrix*

$$[1] = \begin{bmatrix} 1 & 0 & 0 & 0 \\ 0 & 1 & 0 & 0 \\ - & - & - & - \\ 0 & 0 & - & 1 \end{bmatrix}. \tag{2}$$

Two matrices of the same number of rows and columns may be added and their sum is given by

$$[a_{rs}] + [b_{rs}] = [a_{rs} + b_{rs}]. \tag{3}$$

The importance of matrices in the present context lies in the law of multiplication, which states that the product of a matrix $[a_{rs}]$ of m rows and n columns with a matrix $[b_{rs}]$ of n rows and p columns is a matrix $[c_{rs}]$ of m rows and p columns whose elements are given by

$$c_{rs} = \sum_{k=1}^{n} a_{rk} b_{ks}, \tag{4}$$

which is obtained by multiplying the elements of the rth row of $[a]$ by those of the sth column of $[b]$ and adding. It is important to note that multiplication in general is not commutative, that is $[a] \times [b] \neq [b] \times [a]$.

The inverse $[a]^{-1}$ of a matrix $[a]$ is such that

$$[a] \times [a]^{-1} = [a]^{-1} \times [a] = [1]. \tag{5}$$

Solution of a system of linear equations is equivalent to finding the inverse of a matrix.

A set of n linear equations

$$\left.\begin{matrix} a_{11}x_1 + a_{12}x_2 + \ldots + a_{1n}x_n = b_1 \\ \cdots\cdots\cdots\cdots \\ a_{n1}x_1 + a_{n2}x_2 + \ldots + a_{nn}x_n = b_n \end{matrix}\right\} \tag{6}$$

may be written in matrix form

$$[a] \times [x] = [b], \tag{7}$$

and their solution

$$[x] = [a]^{-1} \times [b] \tag{8}$$

is equivalent to finding the inverse of the matrix $[a]$.

The matrix notation is particularly suitable for studying transformation of coordinates. For example § 2.9 (3) may be written

$$\begin{bmatrix} x^* \\ y^* \end{bmatrix} = \begin{bmatrix} \cos\omega & -\sin\omega \\ \sin\omega & \cos\omega \end{bmatrix} \times \begin{bmatrix} x \\ y \end{bmatrix}, \tag{9}$$

and similarly § 2.9 (5) are

$$\begin{bmatrix} x \\ y \end{bmatrix} = \begin{bmatrix} \cos\omega & \sin\omega \\ -\sin\omega & \cos\omega \end{bmatrix} \times \begin{bmatrix} x^* \\ y^* \end{bmatrix}. \tag{10}$$

The square matrices in (9) and (10) are, of course, inverse to one another.

Formulae for change of axes in three dimensions are frequently needed. If x, y, z are coordinates relative to rectangular axes $Oxyz$ and x', y', z' are the coordinates of the same point relative to rectangular axes $Ox'y'z$ whose direction cosines relative to $Oxyz$ are (l_1, m_1, n_1), (l_2, m_2, n_2), (l_3, m_3, n_3), respectively, it is shown by Bell (1920) that

$$\begin{bmatrix} x \\ y \\ z \end{bmatrix} = \begin{bmatrix} l_1 & l_2 & l_3 \\ m_1 & m_2 & m_3 \\ n_1 & n_2 & n_3 \end{bmatrix} \times \begin{bmatrix} x' \\ y' \\ z' \end{bmatrix}, \tag{11}$$

$$\begin{bmatrix} x' \\ y' \\ z' \end{bmatrix} = \begin{bmatrix} l_1 & m_1 & n_1 \\ l_2 & m_2 & n_2 \\ l_3 & m_3 & n_3 \end{bmatrix} \times \begin{bmatrix} x \\ y \\ z \end{bmatrix}. \tag{12}$$

Here, again, the two square matrices are inverse to one another, as may be verified from the properties of direction cosines.

Displacements, being the components of a vector, will transform according to (11) and (12).

The formulae for transformation of stresses in two dimensions § 2.3 (7)–(9) become

$$\begin{bmatrix} \sigma_{x'} \\ \sigma_{y'} \\ \tau_{x'y'} \end{bmatrix} = \begin{bmatrix} \cos^2\theta & \sin^2\theta & \sin 2\theta \\ \sin^2\theta & \cos^2\theta & -\sin 2\theta \\ -\frac{1}{2}\sin 2\theta & \frac{1}{2}\sin 2\theta & \cos 2\theta \end{bmatrix} \times \begin{bmatrix} \sigma_x \\ \sigma_y \\ \tau_{xy} \end{bmatrix}. \tag{13}$$

In three dimensions § 2.5 (1)–(7) may be written

$$\begin{bmatrix} \sigma_{x'} \\ \sigma_{y'} \\ \sigma_{z'} \\ \tau_{y'z'} \\ \tau_{z'x'} \\ \tau_{x'y'} \end{bmatrix} = \begin{bmatrix} l_1^2 & m_1^2 & n_1^2 & 2m_1n_1 & 2n_1l_1 & 2l_1m_1 \\ l_2^2 & m_2^2 & n_2^2 & 2m_2n_2 & 2n_2l_2 & 2l_2m_2 \\ l_3^2 & m_3^2 & n_3^2 & 2m_3n_3 & 2n_3l_3 & 2l_3m_3 \\ l_2l_3 & m_2m_3 & n_2n_3 & (m_2n_3 + m_3n_2) & (n_2l_3 + n_3l_2) & (l_2m_3 + l_3m_2) \\ l_3l_1 & m_3m_1 & n_3n_1 & (m_1n_3 + m_3n_1) & (n_1l_3 + n_3l_1) & (l_1m_3 + l_3m_1) \\ l_1l_2 & m_1m_2 & n_1n_2 & (m_1n_2 + m_2n_1) & (n_1l_2 + n_2l_1) & (l_1m_2 + l_2m_1) \end{bmatrix} \times \begin{bmatrix} \sigma_x \\ \sigma_y \\ \sigma_z \\ \tau_{yz} \\ \tau_{zx} \\ \tau_{xy} \end{bmatrix}. \tag{14}$$

Strains obey similar transformation laws. All these formulae simply rewrite the old ones in matrix notation. The advantage appears when further transformations are needed which can be handled readily by matrix algebra but which would otherwise be very complicated.

2.15 Stress and strain in polar and cylindrical coordinates

Polar coordinates are extensively used in two-dimensional problems. In them, the position of a point P is specified by its distance r from an origin O and the angle θ the direction OP makes with a reference direction OX, Fig. 2.15 (*a*). Corresponding to these coordinates there are two fundamental directions, radial, PR, and transverse, PT. The components of stress relative to this system will be designated σ_r, σ_θ, and $\tau_{r\theta}$. They can be

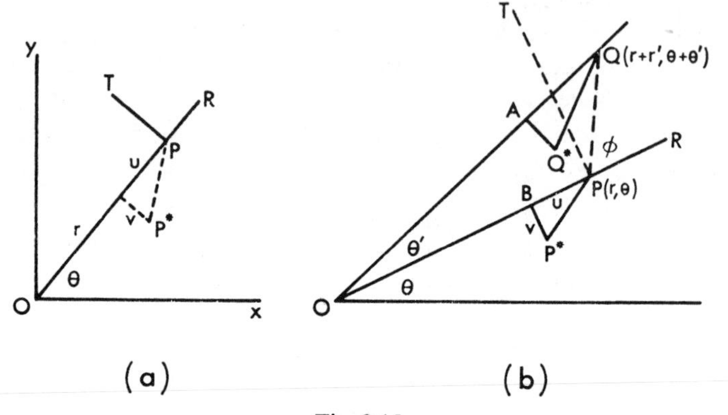

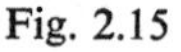
Fig. 2.15

expressed in terms of the stresses for the system OXY by the transformation formulae of § 2.3.

Displacements, again, will be measured as u in the radial direction OP and v in the perpendicular (transverse) direction so that P^*, Fig. 2.15 (*a*), is the final position of P.

To study strains, suppose that P, (r, θ) and Q, $(r + r', \theta + \theta')$, Fig. 2.15 (*b*), are two neighbouring points, where r' and $r\theta'$ are so small that their squares can be neglected. Then if (u, v) are the components of displacement of P, those of Q will be $u + u'$, $v + v'$ where, by Taylor's theorem, neglecting squares and products of r' and θ',

$$u' = \frac{\partial u}{\partial r}r' + \frac{\partial u}{\partial \theta}\theta', \quad v' = \frac{\partial v}{\partial r}r' + \frac{\partial v}{\partial \theta}\theta'. \tag{1}$$

Suppose now that the distance PQ is ρ and that PQ makes an angle ϕ with OP. Writing, for shortness, $l = \cos\phi$, $m = \sin\phi$, and neglecting terms in ρ^2, it follows that

$$r' = \rho\cos\phi = l\rho, \quad r\theta' = \rho\sin\phi = m\rho. \tag{2}$$

Next we determine the elongation ε corresponding to the direction PQ which is, by § 2.9 (11),

$$\varepsilon = (PQ - P^*Q^*)/PQ. \tag{3}$$

To find the length of P^*Q^* we calculate the projections of PQ^* on the directions PR and PT. These are, respectively,

$$PQ \cos\phi - QA \cos\theta' + AQ^* \sin\theta' = l\rho - (u + u') + (v + v')\theta', \tag{4}$$

and

$$PQ \sin\phi - QA \sin\theta' - AQ^* \cos\theta' = m\rho - (u + u')\theta' - (v + v'), \tag{5}$$

neglecting squares of the small quantity θ', and in (4) and (5) the quantities $v'\theta'$ and $u'\theta'$ can be neglected by (1). Since the coordinates of P^* relative to the same axes are $(-u, -v)$

$$\begin{aligned} P^*Q^{*2} &= (l\rho - u' + v\theta')^2 + (m\rho - u\theta' - v')^2 \\ &= \rho^2\left\{l\left(1 - \frac{\partial u}{\partial r}\right) - \frac{m}{r}\left(\frac{\partial u}{\partial \theta} - v\right)\right\}^2 + \rho^2\left\{m\left(1 - \frac{u}{r} - \frac{1}{r}\frac{\partial v}{\partial \theta}\right) - l\frac{\partial v}{\partial r}\right\}^2, \end{aligned}$$

using (1) and (2). Assuming that $\partial u/\partial r$, etc., are small and also that u/r, v/r are small, this becomes

$$P^*Q^{*2} = \rho^2\{1 - 2l^2\varepsilon_r - 4lm\Gamma_{r\theta} - 2m^2\varepsilon_\theta\}, \tag{6}$$

where

$$\varepsilon_r = \frac{\partial u}{\partial r}, \quad \varepsilon_\theta = \frac{1}{r}\left(u + \frac{\partial v}{\partial \theta}\right), \quad 2\Gamma_{r\theta} = \gamma_{r\theta} = \frac{1}{r}\left(\frac{\partial u}{\partial \theta} - v\right) + \frac{\partial v}{\partial r}. \tag{7}$$

Taking the square root of (6), it follows from (3) that the elongation ε in the direction PQ inclined at ϕ to PR is

$$\varepsilon = \varepsilon_r \cos^2\phi + 2\Gamma_{r\theta} \sin\phi \cos\phi + \varepsilon_\theta \sin^2\phi. \tag{8}$$

This is the old formula § 2.10 (18), but the components of strain ε_r, ε_θ, $\Gamma_{r\theta}$ are now given by (7). The reason for the appearance of the terms such as u/r in (7) is that displacements themselves, as well as their derivatives, contribute to the strains. For example, a radial displacement u of an arc of a circle of radius r gives a tangential strain u/r of the arc.

The volumetric strain Δ is

$$\Delta = \varepsilon_r + \varepsilon_\theta = \frac{\partial u}{\partial r} + \frac{u}{r} + \frac{1}{r}\frac{\partial v}{\partial \theta}. \tag{9}$$

The compatibility condition for strains is

$$2\frac{\partial^2(r\Gamma_{r\theta})}{\partial r\,\partial\theta} = r\frac{\partial^2(r\varepsilon_\theta)}{\partial r^2} - r\frac{\partial \varepsilon_r}{\partial r} + \frac{\partial^2 \varepsilon_r}{\partial \theta^2}, \tag{10}$$

which may be easily verified by differentiation from (7).

In many problems involving cylinders or axial symmetry in three dimensions the system of *cylindrical coordinates* r, θ, z, is used. These involve polar coordinates r, θ in the OXY plane and a z-coordinate perpendicular to it. The stress components will be

$$\sigma_r, \quad \sigma_\theta, \quad \sigma_z, \quad \tau_{\theta z}, \quad \tau_{zr}, \quad \tau_{r\theta}, \tag{11}$$

and the components of strain are

$$\varepsilon_r = \frac{\partial u}{\partial r}, \quad \varepsilon_\theta = \frac{1}{r}\left(u + \frac{\partial v}{\partial \theta}\right), \quad \varepsilon_z = \frac{\partial w}{\partial z}, \tag{12}$$

$$2\Gamma_{\theta z} = \frac{1}{r}\frac{\partial w}{\partial \theta} + \frac{\partial v}{\partial z}, \quad 2\Gamma_{rz} = \frac{\partial w}{\partial r} + \frac{\partial u}{\partial z}, \quad 2\Gamma_{r\theta} = \frac{1}{r}\left(\frac{\partial u}{\partial \theta} - v\right) + \frac{\partial v}{\partial r}, \tag{13}$$

where w is the displacement in the z-direction.

The volumetric strain Δ is

$$\Delta = \frac{\partial u}{\partial r} + \frac{1}{r}\left(u + \frac{\partial v}{\partial \theta}\right) + \frac{\partial w}{\partial z}. \tag{14}$$

Chapter Three

Friction

3.1 Introduction

The study of friction is of the greatest importance in rock mechanics. Its effects arise on all scales: (i) the microscopic scale in which friction is postulated between opposing surfaces of minute Griffith cracks; (ii) a larger scale in which it occurs between individual grains or pieces of aggregate; (iii) in friction on joint or fault surfaces in which the areas in question may vary from a few to very many square feet.

Our knowledge of frictional phenomena derives largely from the work of Bowden and his co-workers on metals, Bowden and Tabor (1950, 1964), Bowden (1954). Metals are especially suitable for fundamental work on friction because their electrical conductivity enables measurement of the actual area of contact to be made, and, in the case of dissimilar metals, thermoelectric effects enable the actual temperature at the contact to be measured. Much less work has been done on friction of minerals and rocks, and it is by no means clear that the same fundamental processes are operating. Nevertheless, the observed phenomena are much the same, so that it is useful to discuss metallic friction as a starting-point.

3.2 Amonton's law

Suppose that two bodies with an approximately plane surface of contact of apparent area A are pressed together by force W normal to the plane of contact, and the shear force F parallel to the surface of contact necessary to initiate sliding on it is measured. The relationship between F and W may be written

$$F = \mu W, \tag{1}$$

where μ is called the coefficient of friction. μ depends on the nature of the materials and the finish and state of the surfaces in contact. μ might also be expected to depend on A and W, but experiment has shown that, to a reasonable approximation, it is independent of both these quantities – this is known as Amonton's law (Bowden and Tabor, 1950). Dividing (1) by A, it becomes

$$\tau = \mu\sigma, \tag{2}$$

where σ is the normal stress across the surfaces in contact and τ is the shear stress across them necessary to initiate sliding.

A great deal of study has been devoted to the frictional properties of metals, and for them a simple and satisfactory explanation of the approximate truth of Amonton's law has emerged. It is postulated that even the best-prepared surfaces are not accurately flat and in contact over the whole area A, but are in fact in contact only at a small number of protuberances or 'asperities', Fig. 3.2 (*a*). The normal stress at these will be very great and will exceed the yield stress Y of the softer material, so that the actual area of contact A_c will be given by

$$W = YA_c. \tag{3}$$

At these junctions the contact is so intimate that the metals are regarded as being welded together, and the initiating of sliding requires these welds

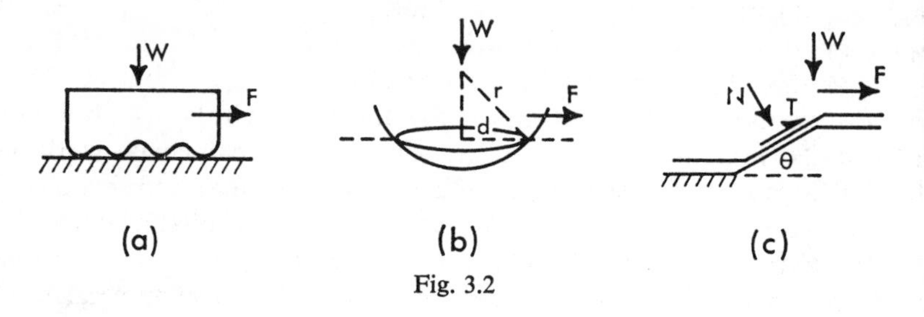

Fig. 3.2

to be sheared through, so that if S is the shear strength of the welded junction

$$F = SA_c. \tag{4}$$

(3) and (4) lead to (1) with $\mu = S/Y$, and Amonton's law is thus qualitatively explained. Because of work-hardening, the material at the actual junctions will probably be stronger than the surrounding material, so that failure will probably occur away from the junction and a fragment of material may be detached. This transfer of material during sliding is observed.

It is also found that if the shearing force is increased steadily some irreversible displacement takes place at forces less than those necessary to cause slip. This displacement is found to be associated with an increase in the area of the contact (which is measured very simply from the electrical resistance of the contact) and arises from the fact that the criterion for yield is more complicated than (3) and involves both the normal and shear stresses, Tabor (1959).

A further effect which has been extensively studied with metals is the separation of the frictional force into so-called shearing and ploughing terms. This is of particular importance, because many experiments are conducted with comparatively small sliding elements. Suppose that such an element is spherical, of radius of curvature r, and that it is sliding on softer material of yield stress Y, Fig. 3.2 (*b*). It will indent the material until the radius of contact is d, where

$$W = \pi d^2 Y. \tag{5}$$

During sliding the sphere will plough a groove through the softer material, whose cross-sectional area will be

$$r^2 \sin^{-1}(d/r) - d(r^2 - d^2)^{\frac{1}{2}} = 2d^3/3r, \tag{6}$$

approximately. The force necessary to plough this groove will be of the order of its area times the yield strength of the material, or $2Yd^3/3r$; this is the 'ploughing' term. The 'shearing' term will be the force necessary to shear the junctions at the surface of the sphere, and thus will be of order $\pi d^2 S$. Adding these and using (3),

$$\begin{aligned} F &= \pi d^2 S + 2Yd^3/3r \\ &= SW/Y + 2W^{3/2}/3\pi^{3/2} Y^{\frac{1}{2}} r, \end{aligned} \tag{7}$$

so that F is no longer proportional to W but increases more rapidly. If the load is equally distributed over n contacts the ploughing term contains an additional factor $n^{-\frac{1}{2}}$ and so becomes less important.

With irregular surfaces it may happen that sliding takes place up a plane inclined at a small angle θ to the direction of the force F and the macroscopic plane of the surfaces, Fig. 3.2 (*c*). In this case the normal and tangential forces N and T across the sliding surface are

$$N = W\cos\theta + F\sin\theta, \quad T = F\cos\theta - W\sin\theta, \tag{8}$$

and if $\quad T = \mu N,$

$$F = (\mu + \tan\theta)W/(1 - \mu\tan\theta) \tag{9}$$

$$= [\mu + \theta(1 + \mu^2)]W, \tag{10}$$

if θ is small, so that there is an increase in the apparent coefficient of friction.

While all the simplified events referred to here probably occur, the situation is much more complicated. There will probably be a large number of contacts at which the loads will be very different, so that, while the deformation at some may be plastic, at others it may be elastic. Archard (1958) has discussed the details of the process more fully and concludes that the law of friction may best be expressed by the power law

$$F = \mu_0 W^m, \tag{11}$$

where $2/3 < m < 1$. The case $m = 2/3$ corresponds to purely elastic contact in which the area of contact is proportional to $W^{2/3}$, Love (1927). The contact of rough surfaces is discussed by Greenwood and Tripp (1967).

While the above discussion is satisfactory for metals and much of it has been checked experimentally, it is of doubtful validity for rocks and minerals. First, most minerals tend to fail by brittle fracture rather than by plastic flow, and, secondly, the 'welding' postulated for metals may not hold for minerals. However, the concept of contact at a limited number of asperities remains valid, but no adequate theory of their behaviour has yet been developed. Byerlee (1967a) has used a truncated cone as a model of an asperity, which he assumes to fail by tensile fracture, and after making a number of assumptions deduces that Amonton's law should hold for a single asperity, with a small value of μ, of the order of 0·1. The fact that higher values of μ are commonly measured he attributes to the interlocking of asperities. Apart from these small-scale effects, the extended surfaces which occur in practical rock mechanics are probably very irregular, and effects of ploughing and irregular sliding similar to those described in (7) and (10) may be expected to occur.

All experiments on friction make a number of measurements of τ in (2) for a given pair of surfaces at various values of σ. These may be reduced individually to give $\mu = \tau/\sigma$, which may be plotted as a function of σ. The coefficient of friction is then found not to be accurately a constant as in Amonton's law but to vary with σ, usually being greater for small values of σ. This approach is frequently used by Bowden and Tabor (1950) and by Maurer (1965).

Alternatively, it is reasonable to see whether other simple laws may give a better fit to the experimental results. For example, (11) suggests the law

$$\tau = \mu_0 \sigma^m \tag{12}$$

where μ_0 and m are constants, and this has been found by Murrell (1965) to fit his observations over a wide range of σ.

Jaeger (1959) used the linear law

$$\tau = S_0 + \mu\sigma, \tag{13}$$

where S_0 and μ are constants, and found good agreement with experiment, particularly at low stresses, cf. § 3.3. This law will be used in most of the future discussion: it has the great mathematical advantage of being linear and has the same form as the expression used in soil mechanics. S_0 may be regarded as an inherent shear strength of the contact surface; it corresponds to the cohesion c of soil mechanics, and this term may also be used for it.

The discussion above has referred to the initiation of sliding. However, it is also an experimental fact that when sliding is in progress at constant speed the force F is proportional to the load W and independent of the area of the surfaces. In this case the constant of proportionality is denoted by μ' and called the *dynamic* coefficient of friction by contrast with the *static* coefficient of friction μ. There is no reason for these to be equal and, indeed, μ' would be expected to be less than μ and to vary with the velocity of sliding.

3.3 Friction of rocks

Frictional effects are of importance in rock mechanics mainly in two connections. First, on a very small scale, between the surfaces of minute Griffith cracks postulated in § 12.1. Secondly, on a large scale, between the surfaces of joints or fracture planes: here, two cases have to be distinguished, the surfaces in question may be new surfaces (such as a tension joint, on which no sliding has occurred) or they may be old surfaces, on which considerable sliding has already taken place. The scale and conditions of laboratory measurements lie between these.

Bowden and his co-workers made some experiments on minerals using the classical techniques involving small sliders and low values of the normal force W. Bowden (1954) reports results on rock salt in which it appears that there is considerable fragmentation of the surface, both on a crystalline and a micro-crystalline scale. For diamond on diamond, he finds that the frictional force is proportional to $W^{2/3}$, suggesting that only elastic deformation is involved. Bowden and Tabor (1950, 1964) also report very large effects due to surface contamination: for example, for freshly cleaved mica μ is of the order 1; for a surface which has become contaminated by exposure, μ falls to 0·3; while for a surface which has been outgassed *in vacuo* it may rise as high as 35. These results suggest that μ may have relatively high values in newly opened Griffith cracks. Horn and Deere (1962) and Penman (1953) also used small contacts of single minerals and demonstrated an important effect of wetting the surfaces, namely that in some cases, e.g. quartz, μ is raised, while in others, e.g. biotite, it is lowered. Some of these results are given in Table 3.3.1. Byerlee (1967a) has used rather larger specimens (approximately 2 cm^2 in area) of various minerals sliding on sapphire and finds an increase of μ from 0·1 to 0·4 with increasing roughness of the sapphire surface.

All the measurements described above use the simple system, Fig. 3.3.1 (*a*), in which the surfaces are pressed together by a normal force W. This has also been used on a much larger scale (with surfaces of the order of one foot square) to test natural joint surfaces. In this case elaborate precautions are necessary to ensure that the load W is applied uniformly over the surface. Other methods which have been used are shown in Fig. 3.3.1. Fig. 3.3.1 (*b*) shows a symmetrical system in which one block is pushed between two others; this is suitable for testing relatively large surfaces and for studying the effects of sliding for some distance over them. The load W can conveniently be applied by flat pressure cells and F by a testing machine, Hoskins *et al.* (1968). The rotating system of Fig. 3.3.1 (*c*) has the advantage that the same surfaces are in contact throughout the experiment, and so is suitable for studying the effects of very large amounts

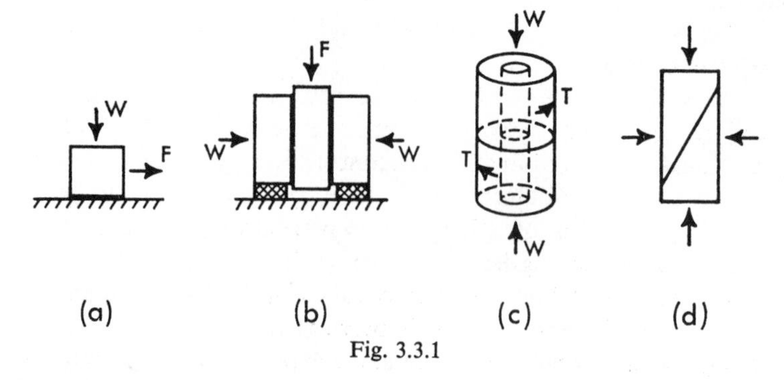

Fig. 3.3.1

of sliding on the surfaces. Finally, Fig. 3.3.1 (*d*) shows the system which has been most used up to the present time, namely that of sliding across inclined surfaces in a cylindrical specimen: this method, which will be described in detail in § 3.7, has the disadvantage that the geometry changes during sliding, so that long-continued experiments are not possible.

The results of a number of measurements by various workers and with

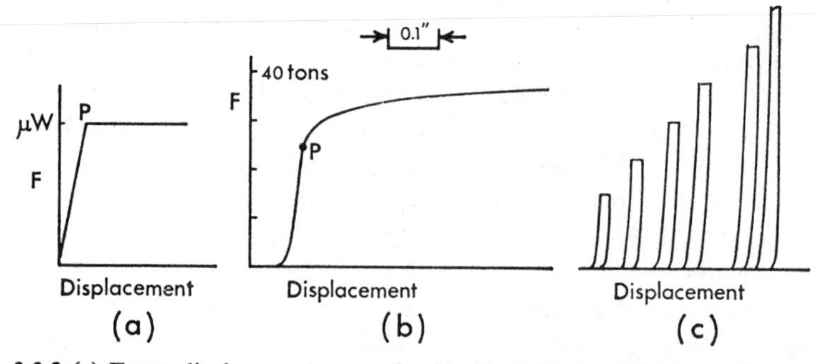

Fig. 3.3.2 (*a*) Force–displacement curve for idealized friction. (*b*) Force–displacement curve for moderately rough trachyte in the apparatus of Fig. 3.3.1 (*b*). (*c*) Force–displacement curve for the same surfaces but at normal loads 0·25, 0·5, 0·75, 1, 1·25, 1·5 times that in (*b*).

different types of apparatus are given in Table 3.3.1. In view of the number of variables involved, it is gratifying that the spread of values is not large, but it is clear that many more experiments under carefully defined conditions are necessary.

To illustrate the various points which have been made, Fig. 3.3.2

shows force–displacement curves for a block of trachyte of area 125 sq. in. pushed between two similar blocks with load $W = 28$ tons in the apparatus of Fig. 3.3.1 (*b*). The surfaces were commerically produced, approximately flat and parallel to about 0·5 mm. The behaviour shown is typical of that for similar surfaces of other rocks. The ideal behaviour at any load W would be that shown in Fig. 3.3.2 (*a*) with elastic displacement up to a point P, followed by sliding at constant force F. In the

Table 3.3.1

Coefficients of friction of rocks and minerals

Notation: Authors. B, Bowden and Tabor (1950, 1964); By, Byerlee (1967b); H, Hoskins *et al.* (1968); HS, Handin and Stearns (1964); HD, Horn and Deere (1962); J, Jaeger (1959); P, Penman (1953); R, Rae (1963); W, Wiebols *et al.* (1968). l, large surface; s, small surface; t, triaxial test; r, rough surface; g, coarsely ground surface; p, finely ground surface; n, natural shear surface; w, wet surface; c, clean surface.

Minerals	μ	Minerals	μ	μ (wet)
Na Cl, [B; s]	0·7	Quartz, [HD]	0·11	0·42
Pb S, [B; s]	0·6	Quartz, [P]	0·19	0·65
S, [B; s]	0·5	Feldspar, [HD]	0·11	0·46
Al_2O_3, [B; s]	0·4	Calcite, [HD]	0·14	0·68
Ice, [B; s]	0·5	Muscovite, [H, D]	0·43	0·23
Glass, [B; s]	0·7	Biotite, [HD]	0·31	0·13
Diamond, [B; s]	0·1	Serpentine, [HD]	0·62	0·29
Diamond, [B; s; c]	0·3	Talc, [HD]	0·36	0·16
Rocks		**Rocks**		
Sandstone, [R]	0·68	Dolomite, [HS; t; g]	0·4	
Sandstone, [J; t; n]	0·52	Trachyte, [H; l; p]	0·63	
Sandstone, [H; l; r]	0·51	Trachyte, [H; l; g]	0·68	
Sandstone, [H; l; r; w]	0·61	Trachyte, [H; l; g; w]	0·56	
Granite, [By; t; n; g]	0·60	Marble, [H; l; p]	0·75	
Granite, [By; t; n; g; w]	0·60	Marble, [J; t; n]	0·62	
Granite, [H; l; g]	0·64	Porphyry, [J; t; n]	0·86	
Quartzite, [W; t; g]	0·48	Gneiss, [J; t; n]	0·71	
Quartzite, [W; t; n]	0·67	Gneiss, [J; t; n; w]	0·61	
Dolerite, [W; t; g]	0·64	Gabbro, [H; l; p]	0·18	
Dolerite, [W; t; n]	0·95	Gabbro, [H; l; g]	0·66	

actual case, Fig. 3.3.2 (*b*), sliding begins at the point P, and the value of F/W at this point would be taken as μ by workers who only consider the initial sliding. However, if sliding over the surface is continued, F continues to rise, and in this case approaches a nearly constant value after a displacement of the order of an inch. This change is undoubtedly due to the development of a new surface of sliding frequently showing slickensides and detrital material. If sliding is recommenced at a changed normal load a constant value of F is attained much more rapidly, sometimes after a slight initial peak, Fig. 3.3.2 (*c*). This behaviour seems to be characteristic of moderately rough surfaces with fairly uniform contact. For surfaces with a great many larger interlocked asperities, an initial peak, caused by shearing of the asperities, might be expected. For inclined asperity surfaces, Patton (1966) has demonstrated that sliding on the asperity surfaces as in Fig. 3.2 (*c*) may occur at low normal loads, but that at higher normal loads the asperities may be sheared.

Next, reverting to the method of reducing and reporting results, Fig.

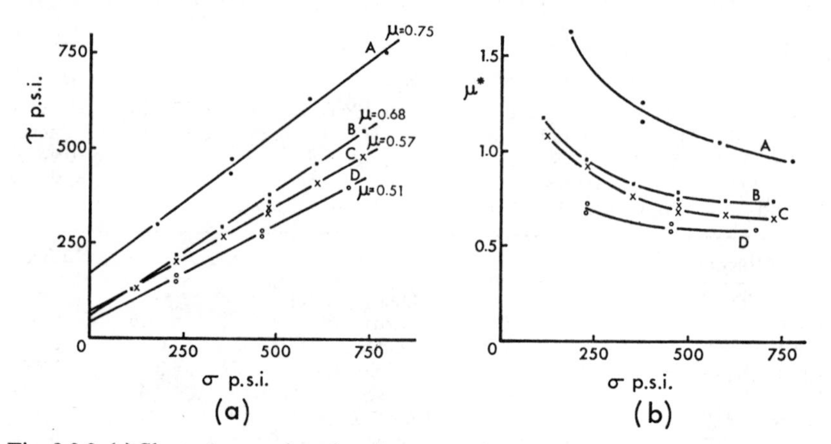

Fig. 3.3.3 (*a*) Shear stress τ plotted against normal stress σ for various sliding surfaces. (*b*) $\mu^* = \tau/\sigma$ plotted against σ for the same experiments. A: Marble. B: Trachyte. C: Trachyte with smoother surface. D: Sandstone.

3.3.3 (*a*) shows the values of τ plotted against σ for the various runs shown in Fig. 3.3.2 (*c*), as well as similar results for other materials. It appears that the law § 3.2 (13)

$$\tau = S_0 + \mu\sigma \tag{1}$$

is very well obeyed in all cases. Table 3.3.2 shows some values of S_0 and μ obtained for moderately rough surfaces in this way.

Table 3.3.2

Values of μ and S_0 for some materials

Rock	μ	S_0(psi)
Granite	0·64	45
Gabbro	0·66	55
Trachyte	0·68	60
Sandstone	0·51	40
Marble	0·75	160

In this method of reduction, μ derived from (1) would be described as the coefficient of friction for the surfaces. In the alternative method of reduction discussed in § 3.2 the value $\mu^* = \tau/\sigma$ would be calculated from the results and plotted against σ. The results of Fig. 3.3.2 (*a*), replotted in this way, are shown in Fig. 3.3.2 (*b*). Clearly, this representation is more complicated.

3.4 Phenomena with smooth surfaces

For very finely ground or polished surfaces, a new effect, usually described as 'stick–slip' appears. Fig. 3.4.1 (*a*) shows the variation of frictional force F with displacement for two new polished surfaces of granite at constant normal load W of 28 tons. F steadily increases up to a point P at which a sudden slip occurs between the surfaces, which subsequently

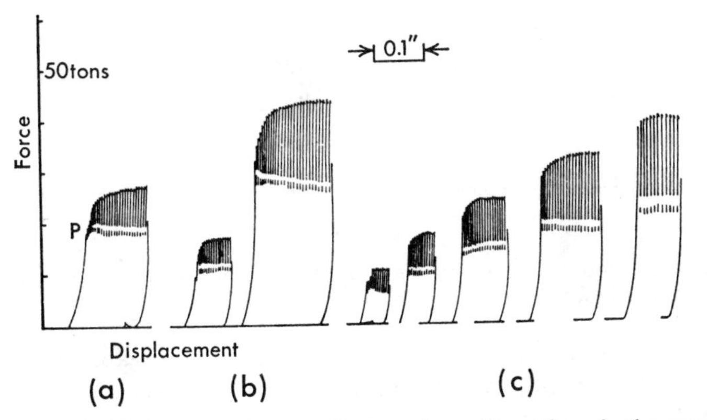

Fig. 3.4.1 (*a*) Force–displacement curve for smooth granite surfaces in the apparatus of Fig. 3.3.1 (*b*); stick–slip oscillations begin at the point *P*. (*b*), (*c*) Force–displacement curves at other normal loads.

lock together again. If loading is continued at a constant rate a regular set of oscillations sets in. As may be seen from Fig. 3.4.1 (*a*), the amplitude of these increases with displacement along the surface and the maximum force F attained in each oscillation increases also, both tending asymtotically to fixed values as the displacement increases. The same applies if the experiment is repeated at different normal forces W, Fig. 3.4.1 (*b*), (*c*). If the asymptotic values τ and τ_1 of the maximum and minimum tangential stress are plotted against the normal stress σ as in § 3.3 straight lines of the form

$$\tau = S_0 + \mu\sigma, \tag{1}$$

$$\tau_1 = S_{01} + \mu_1\sigma, \tag{2}$$

are again obtained, Fig. 3.4.2. Here μ will be the coefficient of static friction, and it will appear from the discussion of § 3.5 that the coefficient of dynamic friction μ' is

$$\mu' = \tfrac{1}{2}(\mu + \mu_1). \tag{3}$$

A typical curve of displacement against time is given in Fig. 3.4.3 and shows how sudden the slipping is.

Regular periodic oscillations of this type were reported by Bowden and Leben (1939) for the sliding of dissimilar metals. They were described as *stick–slip* oscillations and were attributed to the welding together of the

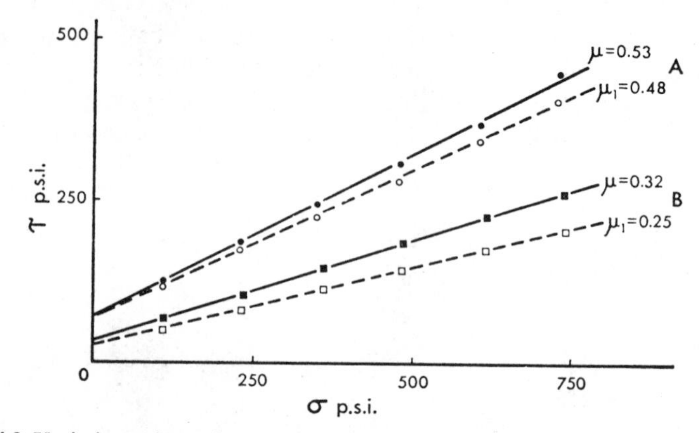

Fig. 3.4.2 Variation of maximum and minimum shear stresses with normal load. A: Granite. B: Gabbro.

surfaces at their points of contact. Blok (1940), Morgan, Muskat, and Reed (1941), and others showed that relaxation oscillations could be produced by the sliding of surfaces for which the coefficient of dynamic friction is less than the coefficient of static friction, and that there was no

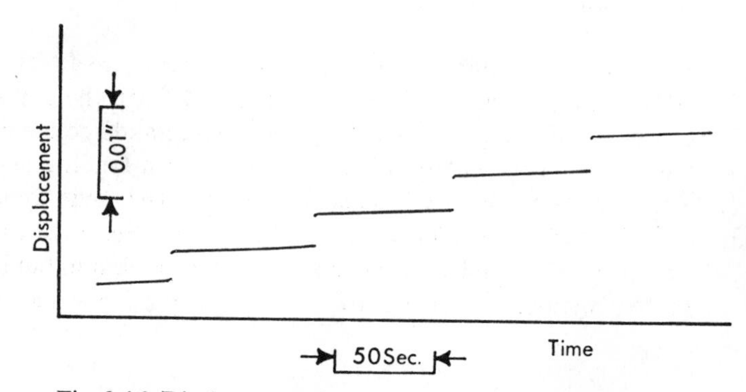

Fig. 3.4.3 Displacement–time curve in a stick–slip oscillation.

need to invoke concepts such as the welding or shearing of asperities to explain them. The oscillations are determined by the gross frictional properties of the surfaces and the mechanical stiffness of the system. A simplified theory is given in § 3.5.

Bowden and Leben (1939) also showed that regular periodic stick–slip oscillations in the above sense did not occur for sliding of similar metals but that there were irregular fluctuations in the frictional force.

At the present time the situation with regard to the sliding of rocks is not clear. Hoskins, Jaeger, and Rosengren (1968) observed periodic oscillations only with smooth surfaces and found that they could be inhibited by roughening the surfaces slightly. Jaeger (1959) observed them with ground surfaces, but did not observe them with natural surfaces of shear fracture. On the other hand, Brace and Byerlee (1966) describe stick–slip as occurring in both cases, and Byerlee (1967a) states that it is more pronounced for rough than smooth surfaces. The position may well be that the term 'stick–slip' is being used to describe two different types of phenomena, first, relaxation oscillations in the sense of § 3.5, and secondly, irregular fluctuations in frictional force which might be caused by shearing of irregularities during continued sliding.

3.5 Stick–slip oscillations

An elementary theory of the stick–slip relaxation oscillations described in § 3.4 may be given as follows.

Suppose that a mass M is pressed against a fixed surface by force W,

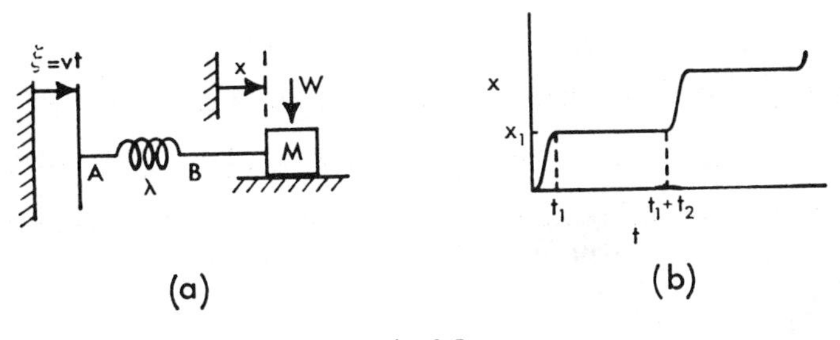

Fig. 3.5

the static and dynamic coefficients of friction at the surface being μ and μ', respectively. Suppose that the mass M is moved by force applied through a spring AB of stiffness λ, whose free end A is forced to move with constant velocity V, Fig. 3.5 (*a*). Suppose that initially the mass is at rest, so that the force of static friction is μW; the relative displacement between the ends of the spring necessary to overcome this force and initiate sliding would be $\lambda\xi_0$ given by

$$\lambda\xi_0 = \mu W. \qquad (1)$$

We shall take the instant of initiation of motion as the origin of time, $t = 0$, and measure the displacement $\xi = Vt$ of the end A of the spring from its position at this time, and the displacement x of the mass M from its position at this time. Then the force applied to the mass M by the spring is $\lambda(\xi + \xi_0 - x)$ and the equation of motion is

$$M\ddot{x} = \lambda(\xi + \xi_0 - x) - \mu' W, \tag{2}$$

where dots are used to note differentiation with respect to the time.

Writing $\lambda = Mn^2$ and using (1) this becomes

$$\ddot{x} + n^2 x = n^2 Vt + (\mu - \mu')W/M. \tag{3}$$

This has to be solved with $x = \dot{x} = 0$ when $t = 0$, and the solution is

$$x = Vt + (\mu - \mu')W/Mn^2 - (V/n)\sin nt - [(\mu - \mu')W/Mn^2]\cos nt, \tag{4}$$

$$\dot{x} = V - V\cos nt + [(\mu - \mu')W/Mn]\sin nt. \tag{5}$$

By (5), the mass comes to rest when $t = t_1$ given by

$$\tan \tfrac{1}{2}nt_1 = -\frac{(\mu - \mu')W}{MnV}, \tag{6}$$

or

$$t_1 = \frac{2\pi}{n} - \frac{2}{n}\tan^{-1}\left\{\frac{(\mu - \mu')W}{MnV}\right\}. \tag{7}$$

Using (6) and (7) in (4) gives for the displacement x_1 at this time

$$x_1 = Vt_1 + 2(\mu - \mu')W/Mn^2, \tag{8}$$

and the force exerted by the spring on the mass is then

$$\lambda(Vt_1 + \xi_0 - x_1) = \lambda[\xi_0 - 2(\mu - \mu')W/Mn^2] = (2\mu' - \mu)W. \tag{9}$$

The mass is now at rest, and the spring will continue to compress, and the force exerted by it will reach $\lambda\xi_0$ after an additional time t_2 given by

$$t_2 = 2(\mu - \mu')W/\lambda V \tag{10}$$

and slipping will commence and the cycle will be repeated. The whole period is $t_1 + t_2$. The curve of displacement against time is shown in Fig. 3.5 (*b*), which may be compared with the experimental curve of Fig. 3.4.3. It follows from (7) and (10) that

$$\tfrac{1}{2}nt_1 = \pi - \tan^{-1}[\tfrac{1}{2}nt_2]. \tag{11}$$

In the usual case the time of slipping t_1 is much smaller than the time of resting t_2, so $t_1 < t_2$ and by (8) the displacement X at each step is nearly

$$X = 2(\mu - \mu')W/\lambda = (F - F_1)/\lambda, \tag{12}$$

where F and F_1 are the maximum and minimum frictional forces in the oscillation as defined in § 3.4. The relation (12) is very well obeyed in practice, and values of λ obtained from it agree very well with stiffnesses of the mechanical system measured by other methods.

Brace and Byerlee (1966) have suggested that stick–slip may provide a mechanism for earthquakes. If so, there must be a relation similar to (12) connecting the displacement with the friction at the surface and the stiffness of the whole system. Further, if there is some form of continuous driving force this will determine the frequency of shocks.

The simple model of Fig. 3.5 (*a*) may be generalized in an interesting manner by replacing the single mass M by a chain of masses M, M_1, M_2, . . . connected in series by springs of stiffnesses λ_1, λ_2, . . . and with normal loads W_1, W_2, ... In this case some masses may move much more frequently than others and the pattern of energy release, corresponding to many small slips and occasional large ones, bears a striking resemblance to that in earthquake sequences, Burridge and Knopoff (1967).

3.6 Sliding on a plane of weakness: two-dimensional theory

In two dimensions, suppose that the material has a plane of weakness whose normal makes an angle β with the greatest principal stress, σ_1, Fig. 3.6.1 (*a*). The nature of the plane of weakness will be discussed later;

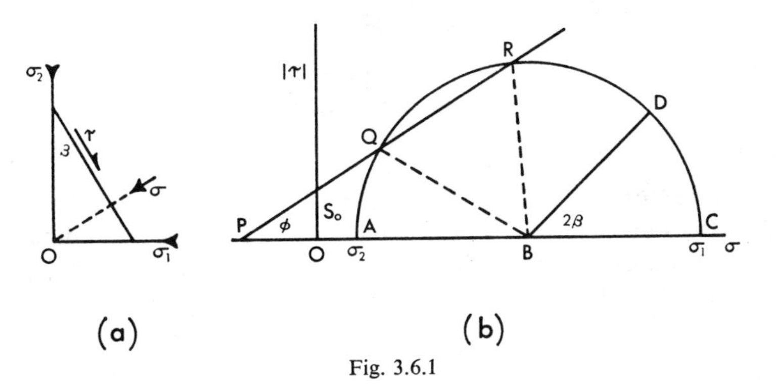

Fig. 3.6.1

for the present it will merely be assumed that, as in § 3.2 (13), the criterion for slip in the plane is

$$|\tau| = S_0 + \mu\sigma, \tag{1}$$

where σ and τ are the normal and shear stresses across the plane.

By § 2.3 (15), (17), σ and τ are given by

$$\sigma = \tfrac{1}{2}(\sigma_1 + \sigma_2) + \tfrac{1}{2}(\sigma_1 - \sigma_2)\cos 2\beta, \tag{2}$$

$$\tau = -\tfrac{1}{2}(\sigma_1 - \sigma_2)\sin 2\beta. \tag{3}$$

These may be put in the alternative form

$$\sigma = \sigma_m + \tau_m \cos 2\beta, \tag{4}$$

$$\tau = -\tau_m \sin 2\beta, \tag{5}$$

where σ_m is the mean stress and τ_m the maximum shear stress, so that

$$\sigma_m = \tfrac{1}{2}(\sigma_1 + \sigma_2), \quad \tau_m = \tfrac{1}{2}(\sigma_1 - \sigma_2). \tag{6}$$

Writing

$$\mu = \tan \phi \tag{7}$$

where ϕ is the *angle of friction*, and using (4) and (5) in (1) it becomes

$$\tau_m\{\sin 2\beta - \tan \phi \cos 2\beta\} = S_0 + \sigma_m \tan \phi, \tag{8}$$

or

$$\tau_m = (\sigma_m + S_0 \cot \phi) \tan \delta, \tag{9}$$

where

$$\tan \delta = \sin \phi \operatorname{cosec} (2\beta - \phi). \tag{10}$$

Alternatively, using the values (6) of σ_m and τ_m, (8) may be written in the form

$$\sigma_1[\sin (2\beta - \phi) - \sin \phi] - \sigma_2[\sin (2\beta - \phi) + \sin \phi] = 2S_0 \cos \phi. \tag{11}$$

Finally, (8) may be rewritten in the forms

$$\sigma_1 - \sigma_2 = \frac{2S_0 + 2\mu\sigma_2}{(1 - \mu \cot \beta) \sin 2\beta}, \tag{12}$$

and

$$\sigma_1 = \frac{2S_0 \cot \phi}{(1 - k) \sin (2\beta - \phi) \operatorname{cosec} \phi - (1 + k)}, \tag{13}$$

where

$$k = \sigma_2/\sigma_1. \qquad (14$$

All of (1), (9), (11), (12), (13) are useful expressions of the criterion of failure (1). For example, (12) shows the way in which the stress difference $\sigma_1 - \sigma_2$ necessary to cause failure varies with β for fixed σ_2 and μ. As $\beta \to \frac{1}{2}\pi$, that is, as the plane moves towards the direction of σ_1, $\sigma_1 - \sigma_2 \to \infty$.

Also as

$$\beta \to \tan^{-1} \mu = \phi, \tag{15}$$

$\sigma_1 - \sigma_2 \to \infty$. Between these values, $\phi < \beta < \frac{1}{2}\pi$, failure is possible and occurs at the value of $\sigma_1 - \sigma_2$ given by (12). Differentiating (12) shows that this has a minimum when

$$\tan 2\beta = -1/\mu, \tag{16}$$

and this minimum value of $\sigma_1 - \sigma_2$ is

$$2(S_0 + \mu\sigma_2)[(\mu^2 + 1)^{\frac{1}{2}} + \mu]. \tag{17}$$

The variation of σ_1 with β for the case $\mu = 0{\cdot}5$ is shown in Fig. 3.6.2 for various values of σ_2.

The situation can be seen very clearly from the Mohr diagram, Fig. 3.6.1 (*b*). In this, the criterion (1) for failure is represented by the line *PQR* inclined at ϕ to $O\sigma$ and making an intercept $OP = -S_0 \cot \phi$ on this axis. If σ_1 and σ_2 are the principal stresses, the normal and shear stresses across the plane whose normal is inclined at β to the σ_1-direction are represented by the point D on the Mohr circle on AC as diameter, cf. § 2.3. If D lies in either of the arcs AQ or RC these stresses will not be sufficient to cause slip, but if it lies in the arc QR they will. The limiting condition, corresponding to the points R and Q, may be found from either

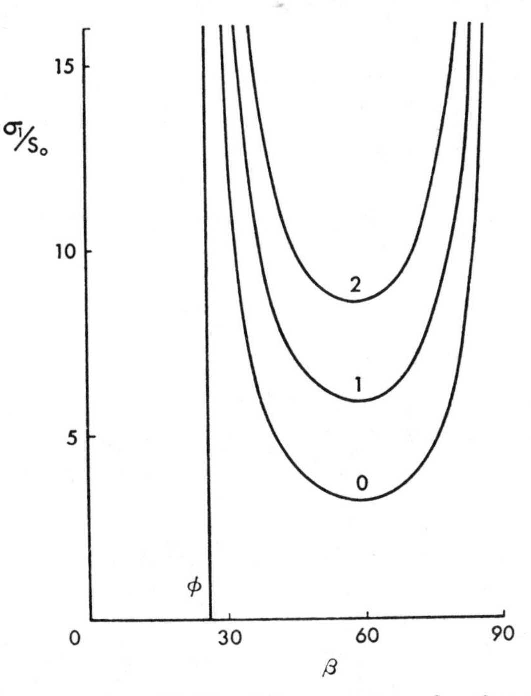

Fig. 3.6.2 The variation of σ_1 with β for sliding on a plane of weakness with $\mu = 0{\cdot}5$. Numbers on the curves are values of σ_2/S_0.

of the triangles PQB or PRB. Taking the latter triangle, the angle RBC will be 2β, so the angle PRB will be $2\beta - \phi$. Then

$$\frac{BR}{\sin\phi} = \frac{PB}{\sin(2\beta - \phi)},$$

or

$$\tau_m \sin(2\beta - \phi) = (\sigma_m + S_0 \cot\phi)\sin\phi, \tag{18}$$

which is (9), so that we have rederived this relation using the Mohr circle.

The values β_1 and β_2 of β corresponding to the points Q and R in Fig. 3.6.1 (*b*) can be found from (18). They are

$$2\beta_1 = \pi + \phi - \sin^{-1}\{[(\sigma_m + S_0 \cot\phi)/\tau_m]\sin\phi\}, \tag{19}$$

$$2\beta_2 = \phi + \sin^{-1}\{[(\sigma_m + S_0 \cot\phi)/\tau_m]\sin\phi\}. \tag{20}$$

One further result of importance follows from Fig. 3.6.1 (*b*). Since $OP = S_0 \cot \phi = S_0/\mu$, it appears that the geometrical situation is unaltered if the origin is moved to the left by this amount. That is, results for given principal stresses σ_1, σ_2 with a finite value of S_0 are the same as those for principal stresses $\sigma_1 + S_0/\mu$, $\sigma_2 + S_0/\mu$ for the case in which S_0 is absent.

This basic theory has three fundamental applications of great importance:

(i) The study of sliding across open joints or cut rock surfaces in which the criterion for slip is that found experimentally in § 3.3.

(ii) If the joints are filled with a weaker material the same theory may be assumed to apply. In this case S_0 would be the shear strength of the joint-filling materials, and μ will be a coefficient of internal friction in the sense of the Coulomb theory developed in § 4.6.

(iii) An anisotropic material with parallel planes of weakness will behave in the same way as the material in (ii) with a single plane of weakness.

3.7 The study of friction in triaxial apparatus

The simplest and most commonly used method for studying friction between rock surfaces is to use a cylindrical test piece with a cut at an

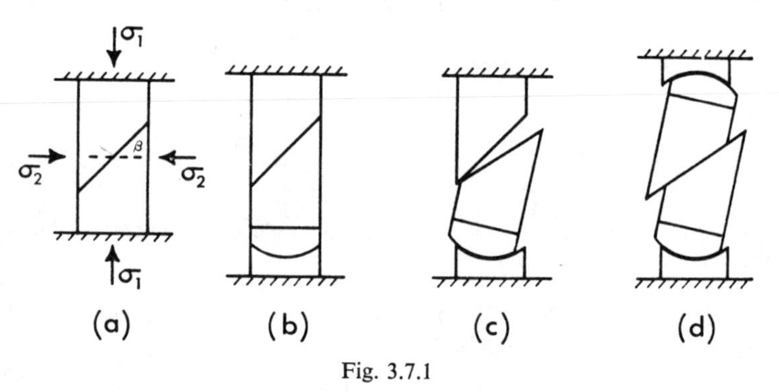

Fig. 3.7.1

angle $\frac{1}{2}\pi - \beta$ to its axis. The specimen is jacketed in rubber or copper, lateral pressure σ_2 is applied to its curved surface by oil pressure, and axial stress σ_1 from a testing machine. The general procedure is that for the triaxial test on cylinders of solid rock, § 6.4.

This method allows high normal and tangential stresses to be applied to the surface, but it has the disadvantage that a change of geometry, or stress, or both, occurs after any slip. Three systems may be used: (i) Rigid platens with no spherical seats, Fig. 3.7.1 (*a*) – in this system unknown lateral stresses are produced by slip; (ii) A single spherical seat, Fig. 3.7.1

(*b*) – in this case slip rotates half the specimen into the situation of Fig. 3.7.1 (*c*); (iii) Two spherical seats, Fig. 3.7.1 (*d*) – in this case the angle of contact is maintained, but the area and inclination of the contact change and lateral forces are produced. Thus none of these systems is strictly usable after the first slip occurs, but in fact it is possible to obtain consistent results with all of them for moderate amounts of displacement.

Two types of behaviour are again observed, corresponding to those of §§ 3.3, 3.4. For rough surfaces, for any fixed value of the oil pressure $\sigma_2 = \sigma_3$, the stress σ_1 tends asymptotically to a constant value. If σ_2 is then increased, a second constant value is then approached, and so on, leading

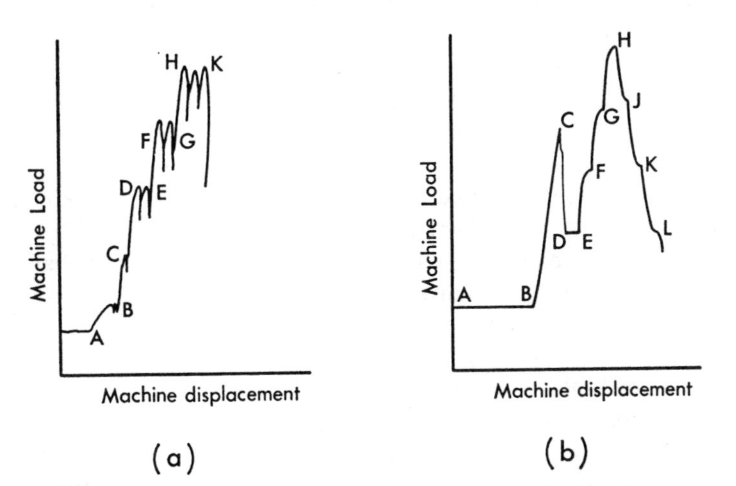

Fig. 3.7.2 (*a*) Stick–slip oscillations during sliding across ground surfaces of a cut at an angle of 25° to the axis of a cylinder at various confining pressures. (*b*) Shear fracture of a solid cylinder at *C* followed by sliding across the plane of fracture at various confining pressures.

to an experimental curve such as the portion *EFGH* of Fig. 3.7.2 (*b*). From such a curve, values of the tangential stress across the surface for various values of the normal stress can be calculated and curves similar to those of Fig. 3.3.3 (*a*) obtained, leading to values of μ and S_0. For finely ground or polished surfaces, stick–slip oscillations again appear; an example is shown in Fig. 3.7.2 (*a*).

These methods have been used by Jaeger (1959), Handin and Stearns (1964), Byerlee (1967b), and Lane and Heck (1964) using the system of Fig. 3.7.1 (*b*). Experiments can be made at angles β varying from about 30° to 65° and, in principle, an experiment at any angle gives its own value of μ. There is little variation between the values of μ so obtained, indicating that the method is valid despite the changes in geometry suggested in Fig. 3.7.1 (*c*). However, if the displacement is continued too far subsidiary fractures occur in the specimens in the regions of contact shown in Fig. 3.7.1 (*c*).

This method may also be used to study friction on natural surfaces of shear fracture. In this case a shear fracture is produced in a solid cylinder in the triaxial test of § 6.4, and sliding on this surface is studied by the present methods. It is necessary, subsequently, to examine the surface and see that it is reasonably flat and does not intersect either platen.

An example of this is shown in Fig. 3.7.2 (*b*), which shows the load–displacement curve for a cylinder of Rand quartzite treated in this way. The portion *AB* corresponds to the load on the piston at the initial confining pressure of 400 bars before contact is made with the specimen. *BC* corresponds to loading the specimen, which fails at *C*, producing (what is subsequently verified to be) a plane shear fracture. *DE* corresponds to sliding along this fracture plane at a confining pressure of 400 bars. At *E* the confining pressure is raised to 600 bars, producing the curve *EF*; *FG* corresponds to 800 bars, and *GH* to 1,000 bars. Lowering the confining pressure to 800, 600, 400 bars successively gives the regions *HJ*, *JK*, *KL*. It is seen that the loads at *E* and *L*, both corresponding to 400 bars, agree well, indicating that the change in position of the specimen during this amount of sliding has not had a marked effect.

3.8 Sliding on a plane of weakness: three-dimensional theory

Sliding on a plane of weakness is conveniently studied in three dimensions with the aid of Mohr's representation of stress, § 2.6. An appreciation of the situation may be gained very simply by considering the two extreme cases in which either the minor principal stresses are equal ($\sigma_1 > \sigma_2 = \sigma_3$) or the major principal stresses are equal ($\sigma_1 = \sigma_2 > \sigma_3$). In both these cases the stresses everywhere in the rock are represented by a single Mohr circle with centre $\sigma_m = \frac{1}{2}(\sigma_1 + \sigma_3)$ and radius $\tau_m = \frac{1}{2}(\sigma_1 - \sigma_3)$. As remarked in § 3.6, results for the case in which the plane of weakness has inherent shear strength S_w can be deduced from those for the case in which the inherent shear strength is zero, so we shall first consider the latter case, for which the criterion for sliding will be

$$\tau = \mu_w \sigma_n, \tag{1}$$

where σ_n is the normal stress across the plane of weakness and μ_w is its coefficient of sliding friction. This relationship is represented by a line on the Mohr diagram passing through the origin and of slope μ_w. Sliding can occur if the plane of weakness is so oriented that the shear stress on it is in excess of that allowed by (1), that is, if the normal to the plane makes an angle between β_1 and β_2 with the σ_1-direction where the angles $2\beta_1$ and $2\beta_2$ are as shown in Fig. 3.8.1 (*a*). Their values are easily calculated from the geometry of the figure or from § 3.6 (19), (20), which may be written

$$\beta_1 = \tfrac{1}{2}\pi - \tfrac{1}{2}\cos^{-1}[\mu_w/(\mu_w^2+1)^{\frac{1}{2}}] + \tfrac{1}{2}\cos^{-1}[\mu_w\sigma_m/\tau_m(\mu_w^2+1)^{\frac{1}{2}}], \quad (2)$$

$$\beta_2 = \tfrac{1}{2}\pi - \tfrac{1}{2}\cos^{-1}[\mu_w/(\mu_w^2+1)^{\frac{1}{2}}] - \tfrac{1}{2}\cos^{-1}[\mu_w\sigma_m/\tau_m(\mu_w^2+1)^{\frac{1}{2}}]. \quad (3)$$

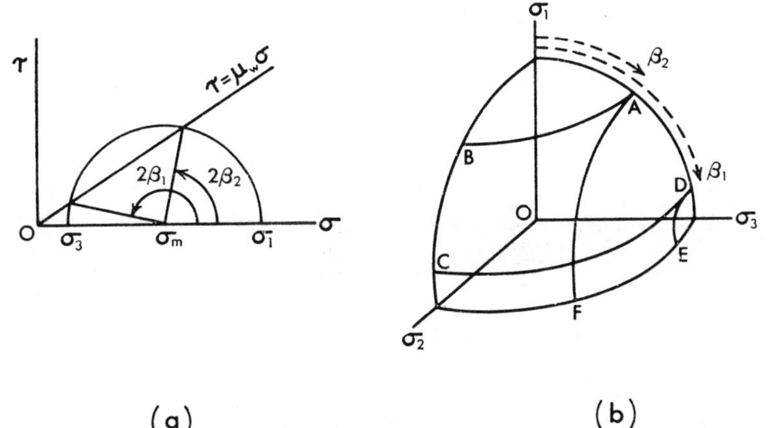

Fig. 3.8.1 (*a*) Mohr diagram for sliding across a plane of weakness. (*b*) Regions in which the normal to the plane must fall for sliding to be possible: *ABCD* for the case $\sigma_2 = \sigma_3$; *ADEF* for the case $\sigma_1 = \sigma_2$.

Values of β_1 and β_2 calculated from (2) and (3) for values 1/3, 2/3, and 1 of μ_w and various ratios of σ_3/σ_1 are given in Table 3.8. Small values of

Table 3.8

σ_3/σ_1	σ_m/τ_m	$\mu_w = \frac{1}{3}$		$\mu_w = \frac{2}{3}$		$\mu_w = 1$	
		β_1	β_2	β_1	β_2	β_1	β_2
0·01	1·02	89·7°	18·7°	89·7°	34·0°	89·5°	45·5°
0·02	1·04	89·5°	19·0°	89·0°	34·5°	89·0°	46·0°
0·05	1·11	89·0°	20·0°	88·0°	36·0°	86·5°	48·5°
0·10	1·22	88·0°	21·0°	85·5°	38·0°	82·5°	52·5°
0·15	1·35	86·5°	22·0°	82·5°	41·0°	76·0°	59·0°
0·20	1·50	85·0°	23·5°	78·5°	45·0°	—	—
0·25	1·67	83·0°	25·0°	73·0°	51·0°	—	—
0·30	1·86	81·0°	27·0°	—	—	—	—
0·35	2·08	78·5°	30·0°	—	—	—	—
0·40	2·33	75·5°	33·0°	—	—	—	—
0·45	2·64	71·0°	38·0°	—	—	—	—
0·50	3·00	63·4°	45·0°	—	—	—	—

σ_3/σ_1 are of importance in conditions near the surface of an excavation where one principal stress will be small.

The significance of the angles β_1, β_2 for stresses in three dimensions is shown in Fig. 3.8.1 (*b*), which shows the directions of the principal axes and one octant of a unit sphere. When the two minor principal stresses are equal, $\sigma_1 > \sigma_2 = \sigma_3$, slip can occur on any plane whose normal makes an angle of between β_1 and β_2 with $O\sigma_1$. These normals lie in a zone *ABCD* symmetrical about $O\sigma_1$. When the two major principal stresses are equal, $\sigma_1 = \sigma_2 > \sigma_3$, slip can occur on any plane whose normal makes angles between $\frac{1}{2}\pi - \beta_1$ and $\frac{1}{2}\pi - \beta_2$ with $O\sigma_3$. These normals lie in a zone *ADEF* symmetrical about $O\sigma_3$. This shows how the solid angle within which the normal to a plane of weakness must lie for it to slide

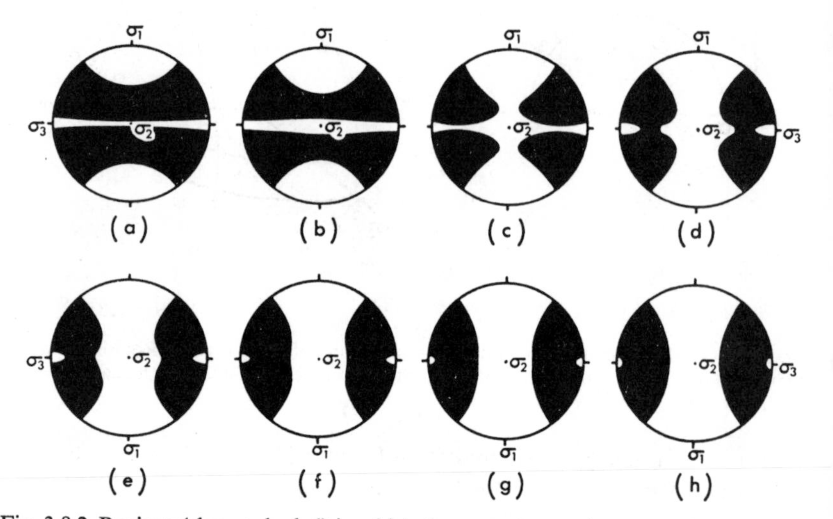

Fig. 3.8.2 Regions (shown shaded) in which the normal to a plane must fall for sliding on it to be possible. Figures are for $\mu = 0{\cdot}67$, $\sigma_3 = 0{\cdot}1\sigma_1$ and $\sigma_2/\sigma_1 = 0{\cdot}1, 0{\cdot}2, 0{\cdot}3, 0{\cdot}4, 0{\cdot}5, 0{\cdot}6, 0{\cdot}8, 1{\cdot}0$ in (*a*) . . . (*h*) respectively. The projection is stereographic.

decreases as the intermediate principal stress changes from $\sigma_2 = \sigma_3$ to $\sigma_2 = \sigma_1$.

Any three-dimensional case must have a value of the intermediate principal stress lying between the two extreme values considered above. The effect of this can most readily be seen by plotting on the stereographic projection the directions of the normals to planes on which slip can occur. Diagrams of this type are shown in Fig. 3.8.2 for the case $\mu = 0{\cdot}67$ and $\sigma_3 = 0{\cdot}1\sigma_1$, in which the shaded regions are those in which slip is possible. Fig. 3.8.2 (*a*) is for $\sigma_2 = \sigma_3$, corresponding to *ABCD* of Fig. 3.8.1 (*b*) and Fig. 3.8.2 (*h*) is for $\sigma_2 = \sigma_1$, corresponding to *ADEF* in Fig. 3.8.1 (*b*). Figs. 3.8.2 (*b*) . . . (*g*) are for $\sigma_2/\sigma_1 = 0{\cdot}2, 0{\cdot}3, 0{\cdot}4, 0{\cdot}5, 0{\cdot}6, 0{\cdot}8$, respectively, and show how the pattern changes with increase of σ_2, rapidly at first and then slowly.

Allowance must be made for the effects of water pressure and of the intrinsic shear strength S_w of the plane of weakness. The effect of S_w is to replace (1) by

$$\tau = S_w + \mu_w \sigma_n, \tag{4}$$

and as shown in § 3.6 the geometry becomes the same as that for (1) if S_w/μ_w is added to each of the principal stresses.

Also, it is shown in §§ 8.8, 8.9 that the effect of water pressure p in the plane of weakness may be allowed for by subtracting p from each of the principal stresses. The preceding analysis can be used for weaknesses with shear strength and water pressure by making the appropriate additions to or subtractions from the principal stress and using these corrected values in Table 3.8.1 or Fig. 3.8.2.

The purpose of this discussion has been to give a general description of sliding on a plane of weakness in three dimensions and to indicate the effect of different values of the intermediate principal stress. This discussion should be adequate for many practical purposes, since usually neither the directions nor the magnitudes of the principal stresses are known with any degree of precision.

Any specific case can be handled by determining the normal and shear stress across the plane by the Mohr construction of Fig. 2.6.2. If these correspond to a point which lies above the line (4), sliding across the plane is possible. If there is water pressure p in the plane the principal stresses are to be reduced by p before making the Mohr construction.

Finally, it should be stated that the present discussion has been concerned solely with questions of sliding across a plane, and it has appeared that in certain orientations this is impossible. In this case it is in fact possible for shear fracture of the rock to take place in planes cutting across the plane of weakness. This is discussed in § 4.10.

3.9 The effects of time and relative displacement on rock friction

The effects of the period for which surfaces are in stationary contact under high normal stresses have been studied by Dieterich (1972), using a fine- to medium-grained sandstone containing about 90 per cent quartz, a dense, fine-grained graywacke composed mainly of quartz and plagioclase in a silty micaceous matrix and coarse-grained granite and quartzite. Friction characteristics were found to be determined primarily by surface finish and only weakly related to the type of rock.

Severe stick–slip motion characterized the sliding between smooth surfaces at normal stresses from 20 bar to 850 bar. The coefficient of static friction was high and variable, 0,95 to 1,5. Sliding on rough surfaces was initially stable; the coefficient of friction increased with displacement. Sliding of the rough surfaces generated finely comminuted debris or gouge. After the coefficient of dynamic friction reached a maximum value as a result of continued displacement, the coefficient of static friction became very sensitive to the period of time for which the surfaces were held in stationary contact. This time-dependency gave rise to stick–slip motion because the coefficient of static friction becomes greater than that of dynamic friction for finite periods of stick. Stick–slip was observed for all rock samples at all normal stresses, 20 bar to 850 bar, only if there was an accumulation of gouge. Fig. 3.9.1 shows the variation in the coefficient of

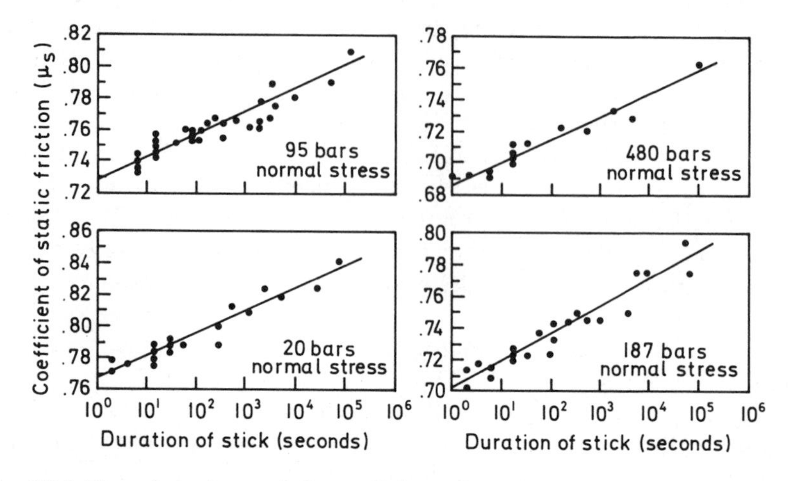

Fig. 3.9.1 Time dependence of the coefficient of static friction between sandstone separated by gouge (after Dieterich, 1972).

static friction as a function of the period of time for which surfaces of sandstone with gouge were in stationary contact. Similar results were obtained for the graywacke, the granite and the quartzite. The variation of the coefficient of static friction with normal stress for all the experiments has been summarized in Fig. 3.9.2. For periods of less than 1 second, the sandstone and graywacke moved by stable sliding for all values of normal stress; this may be due to the presence of mica in these rocks. The granite and quartzite showed stable sliding for periods of less than 1 second up to normal stresses of 200 bar only.

In stick–slip sliding experiments between finely ground surfaces of granite at high confining pressures Byerlee and Brace (1968) found the magnitude of the stress drop to be independent of the stiffness of the test arrangement. This suggests that the coefficient of dynamic friction is independent of the displacement between the two surfaces. It is difficult to explain how the coefficient of friction drops abruptly from its static to its dynamic value immediately there is relative displacement between the surfaces and then remains constant. Very careful experiments by Jaeger and Cook (1971), in which two surfaces were caused to stick by decreasing the

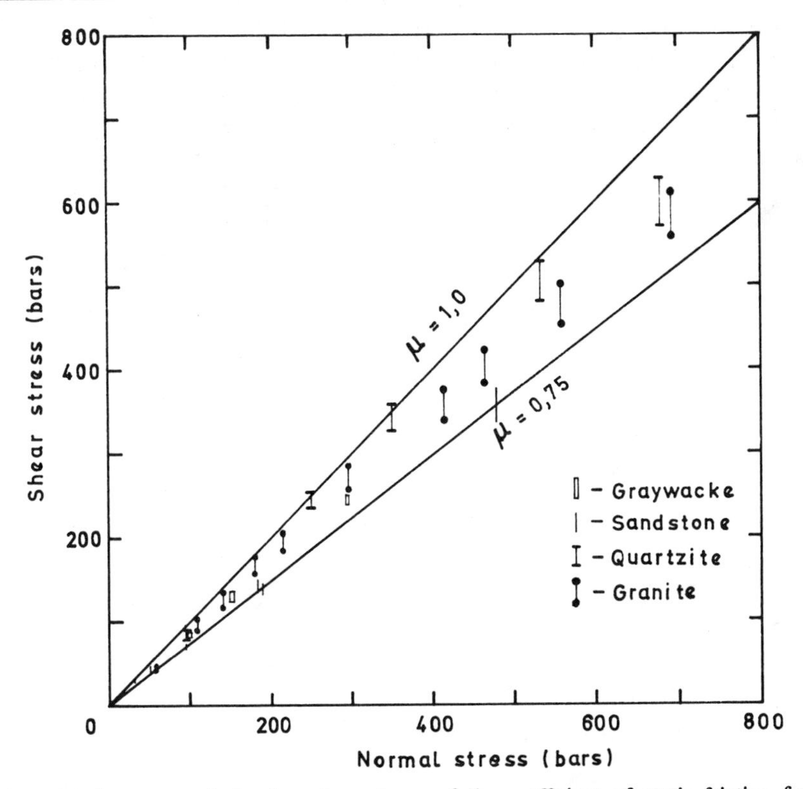

Fig. 3.9.2 Summary of the time dependence of the coefficient of static friction for graywacke, sandstone, quartzite and granite. The bottom of each bar shows the value after 15 seconds of static contact and the top after 10^5 seconds of contact (after Dieterich, 1972).

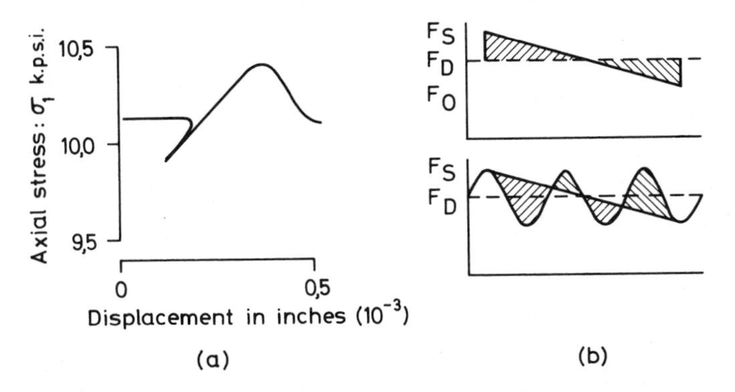

Fig. 3.9.3 (*a*) An enlargement of 'stable stick–slip' produced by a small stress reversal on a saw-cut through a specimen of quartzite (after Jaeger and Cook, 1971). (*b*) Diagrams showing the equivalence between a dynamic coefficient of sliding friction and one which varies with displacement.

shear stress slightly and then increasing it until sliding commenced again, suggest that this change is not abrupt, Fig. 3.9.3 (*a*). Byerlee (1967) proposed that when two surfaces are in contact asperities become locked together and at high normal stresses sliding can occur only when these locked asperities fracture in a brittle manner. This could account for the apparently abrupt change in the coefficient of sliding friction from its static value to a constant average dynamic value, Fig. 3.9.3 (*b*).

Chapter Four

Elasticity and Strength of Rock

4.1 Introduction

Metals were the type specimens of the classical theory of strength of materials, and most developments of it were based on their behaviour. While the same tests are made on both metals and rocks, their behaviour, both macroscopically and microscopically, is very different, so that concepts which may be of great importance to the one have much less relevance for the other. For this reason it is desirable in rock mechanics to base the discussion on the actual behaviour of rocks rather than to adapt ideas derived from the study of metals.

In §§ 4.2–4.4 the nature of the stress–strain curve for rocks will be described, and, based on it, definitions of the terms 'brittle', 'ductile', 'failure', 'fracture', and 'strength' will be given. It should be said that, partly for the historical reasons mentioned above, there is a great variety of usage in the literature, and the terms 'failure', 'fracture', and 'rupture' are frequently used in different senses. It seems preferable to follow the common practice of using the term 'fracture' in the sense of brittle fracture to imply a complete loss of cohesion across a surface. The distinction between brittle fracture and failure will be discussed in § 4.2. The term 'plastic' is frequently used to describe processes involving yield and ductility: this is undesirable, since its use should be restricted to true plastic phenomena in crystals; however, in Chapter 9 it will be used in the sense defined in the mathematical theory of plasticity.

In metals yield is more important than fracture, and its theory is usually discussed first. In rocks failure and fracture are of the greatest importance, and criteria for these will be introduced first in their own right in §§ 4.6–4.9, discussion of yield being relegated to a minor position in Chapter 9.

Most experimenting has been done using the so-called triaxial test, in which two principal stresses are equal, usually $\sigma_1 > \sigma_2 = \sigma_3 > 0$. Since the term 'triaxial' is used so frequently for this situation, the term *polyaxial* will be used in future for the general case in which $\sigma_1 > \sigma_2 > \sigma_3$. It will appear later that the intermediate principal stress does not appear in many of the criteria for failure, notably those of §§ 4.6–4.8. For this reason, although much of the following discussion is essentially two-dimensional using formulae developed in § 2.3, it will be written in terms of major and minor principal stresses, σ_1, σ_3, to facilitate comparison with the general case.

Both British and metric units are commonly in use, and both will be used here. For stress, the bar and kilobar are most usual in geophysically orientated work, while in engineering work lb/in² (psi) or kips (kpsi or 10^3 psi) are used. Conversion factors are: 1 bar $= 10^6$ dyne/cm² $= 1{\cdot}02$ kg/cm² $= 14{\cdot}5$ psi. Strain is conveniently expressed in millistrains (units of 10^{-3}).

4.2 The stress–strain curve

The most common method of studying the mechanical properties of rocks is by axial compression of a circular cylinder whose length is two to three times its diameter. For any stress applied to the cylinder, the axial and lateral strains may be measured either by strain gauges attached to the

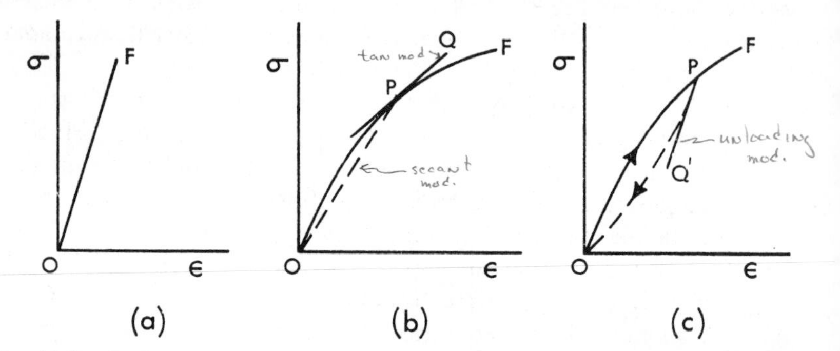

Fig. 4.2.1 (*a*) Linearly elastic material. (*b*) Perfectly elastic material showing tangent modulus *PQ* and secant modulus *OP*. (*c*) Elastic material with hysteresis, showing loading and unloading cycle.

cylinder or by measurement of displacements (for further details see § 6.2). If stress is plotted against strain a *stress–strain curve* is obtained.

The general nature of the stress–strain curve for rocks and the terminology to be used later will now be discussed by studying this case in detail. Behaviour under more complicated systems of stress is similar.

For most rocks the stress–strain curve takes approximately the linear form of Fig. 4.2.1 (*a*), ending abruptly in failure at *F*. This may be represented by

$$\sigma = E\varepsilon, \tag{1}$$

where the constant *E* is called *Young's modulus*. A material is described as *linearly elastic* if the relation (1) holds accurately, and the theory of linear elasticity, Chapter 5, is based on this assumption.

A material is said to be *perfectly elastic* if there is a unique relation

$$\sigma = f(\varepsilon) \tag{2}$$

between stress and strain which need not be linear, Fig. 4.2.1 (*b*). The term 'loading' will be used for the process of applying gradually increasing stress to the specimen as is done in testing, and 'unloading' for decreasing this stress. Perfect elasticity implies that if the material is loaded and subsequently unloaded the same path given by equation (2) is traversed; also all the energy stored in the specimen while loading is released during unloading. There is now no unique modulus, but for any value of σ, corresponding to a point P, the slope PQ of the tangent to the curve

$$d\sigma/d\varepsilon \tag{3}$$

is called the *tangent modulus*. The slope of the secant OP, which is σ/ε, is called the *secant modulus*.

A material is called *elastic* if, after loading and subsequent unloading to zero stress, the strain returns to zero, but possibly by a different path,

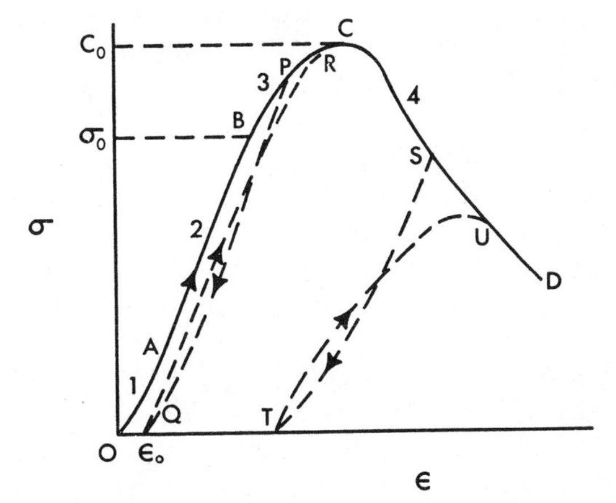

Fig. 4.2.2 The complete stress–strain curve for rock.

Fig. 4.2.1 (*c*). This effect is called *hysteresis*. If it occurs, more work is done on the body during loading than is done by it during unloading, so that energy is dissipated in the body during a cycle of loading and unloading. The slope PQ' of the tangent to the unloading curve at any point P is the *unloading modulus* corresponding to that stress.

The idealized materials of Fig. 4.2.1 all behave elastically in their various fashions until sudden failure takes place at some point F. The actual behaviour of rock may be described more nearly as follows. The stress–strain curve, Fig. 4.2.2, divides into four regions: (1) OA, in which it is slightly convex upwards; (2) AB, a very nearly linear portion; (3) BC, in which it is concave downwards, reaching a maximum at *C*; (4) a falling region *CD*.

In the first two regions, *OA*, *AB*, the behaviour is very nearly elastic; slight hysteresis may be observed, but loading and unloading in this region does not produce irreversible changes in the structure or properties of the rock. In the third region, *BC*, which usually begins at a stress of the order of two-thirds of the maximum value at *C*, the slope of the stress–strain curve decreases progressively to zero with increasing stress. In this region irreversible changes are induced in the rock, and successive cycles of loading and unloading trace out different curves. An unloading cycle, *PQ*, Fig. 4.2.2, leads to *permanent set* ε_0 at zero stress; if the material is reloaded a curve *QR* is traced which lies below the curve *OABC* but ultimately joins it.

The fourth region, *CD*, begins at the maximum, *C*, of the stress–strain curve and is characterized by a negative slope of the stress–strain curve. An unloading cycle, *ST*, often leads to large permanent set, and subsequent reloading *TU* approaches the curve *CD* at a stress lower than that corresponding to *S*. This region *CD* is characteristic of brittle behaviour, but it is usually totally obscured by the instability of the machine-specimen system which results in violent failure very near to the point *C*. The condition for this, which will be discussed in § 6.13, is that the downward slope of the stress–strain curve exceeds the resilience of the testing machine.

After this discussion a number of further fundamental definitions can be made.

A material is said to be in a ductile state or *ductile* under conditions in which it can sustain permanent deformation without losing its ability to resist load.

A material is said to be in a brittle state or *brittle* under conditions in which its ability to resist load decreases with increasing deformation. Thus, in Fig. 4.2.2 the material is in a ductile state in the region *BC* and in a brittle state in the region *CD*. The *brittleness* of a material will be defined as the magnitude of the greatest slope of the falling portion *CD*. If the load is cycled in the ductile region, as in *PQR*, higher stresses can be attained, whereas if it is cycled in the brittle region as in *STU* only lower stresses can be reached.

The maximum ordinate of the curve at *C* which marks the transition from ductile to brittle behaviour is known as the *uniaxial compressive strength* and denoted by C_0.

The process of *failure* is regarded as a continuous one which occurs progressively throughout the brittle region *CD*, in which the rock steadily deteriorates. Failure thus begins at the maximum ordinate *C* of the curve, and the *criteria for failure* which will be discussed in later sections attempt to predict the beginning of failure under more general conditions. In actual testing it frequently happens that sudden failure occurs at some point of the curve CD with complete loss of cohesion across a plane, and this is known as *brittle fracture*. In conventional testing machines, because of instability of the machine-specimen system, brittle fracture occurs spontaneously at a point very near to C, so that in such cases failure and fracture become synonymous. However, in stiff testing machines, and in underground rock systems in which stress is applied through a rock mass which behaves as a stiff system, the ability of partially failed rock in the region CD to withstand load is of the greatest importance.

The point B at which the transition from elastic to ductile behaviour takes place is known as the *yield point* and the corresponding stress σ_0 as

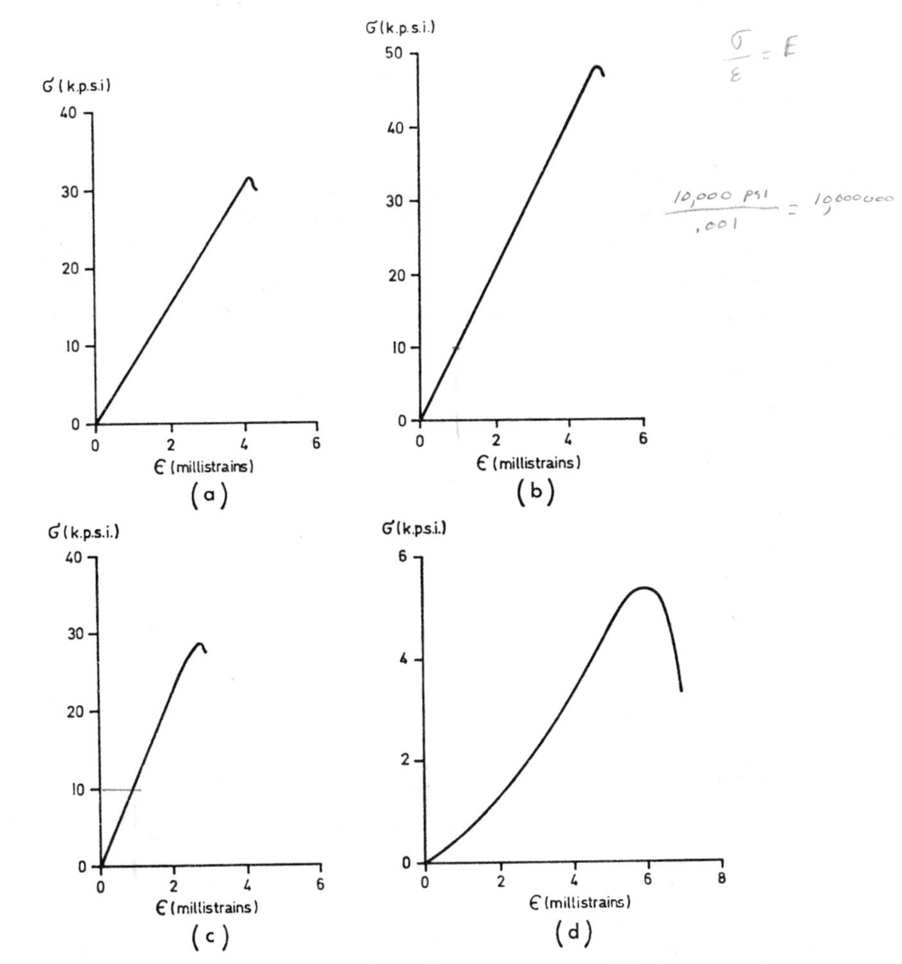

Fig. 4.2.3 Stress–strain curves for uniaxial compression in a stiff testing machine. (*a*) Solenhofen limestone. (*b*) Karroo dolerite. (*c*) Rand quartzite. (*d*) Gosford sandstone

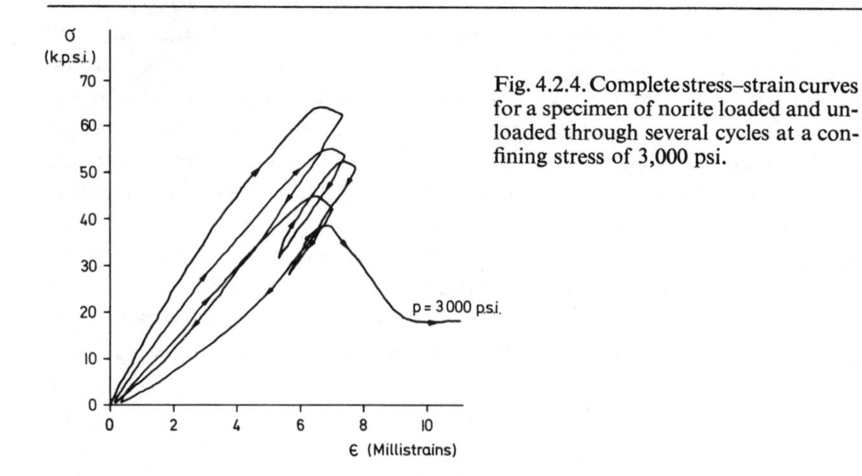

Fig. 4.2.4. Complete stress–strain curves for a specimen of norite loaded and unloaded through several cycles at a confining stress of 3,000 psi.

the *yield stress*. From its nature it is, of course, difficult to measure with any accuracy.

Fig. 4.2.3 shows some stress–strain curves of typical rocks for comparison with the idealized curves discussed above. They have been measured in a stiff testing machine and so a portion of the falling region *CD* is recorded. They show how unimportant the regions *OA* and *BC* are in many practical cases and therefore that the assumption of linear elasticity

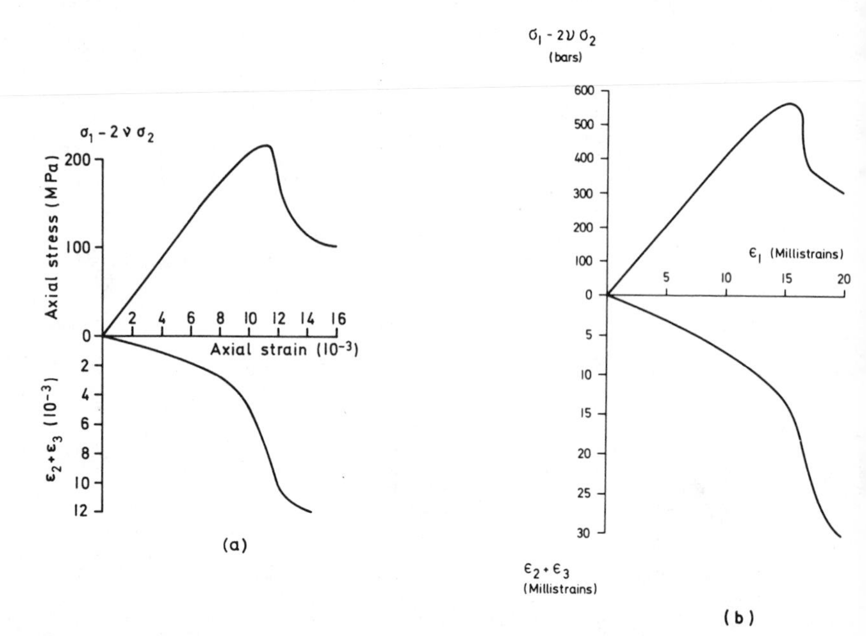

Fig. 4.2.5 Complete axial stress–strain lateral strains–axial strain curves for a triaxial test in a stiff testing machine for (*a*) sandstone and (*b*) coal.

up to failure really is a good one in many cases. As remarked earlier, rock under other systems of stress also shows the typical behaviour of Fig. 4.2.2. As an example, Fig. 4.2.4 shows a complete stress–strain curve for norite under triaxial conditions and illustrates the effect of repeated loading and unloading.

In (1) Young's modulus, *E*, is defined for uniaxial stress. However, the theory of elasticity, Chapter 5, is a two-constant theory and for the complete specification of a material another constant is needed. A compressive stress σ_1 causes a contraction giving rise to a strain ε_1 in the direction of this stress and expansions giving rise to strains $-\varepsilon_2$ and $-\varepsilon_3$ in directions normal to it. The other elastic constant usually used with Young's modulus is Poisson's ratio $\nu = -\varepsilon_2/\varepsilon_1$, § 5.2. The lateral expansion of a specimen in a triaxial test is readily measured by the volume of oil displaced by it, from which the sum $\varepsilon_2 + \varepsilon_3$ can be found. Examples of complete axial stress–strain and lateral strains–axial strain curves are shown in Fig. 4.2.5.

A suite of complete stress–strain curves, together with the corresponding radial strain–axial strain curves obtained from triaxial compression tests at different confining pressures on specimens of rock taken from a single piece of argillaceous Witwatersrand quartzite is shown in Fig. 4.2.6. It is

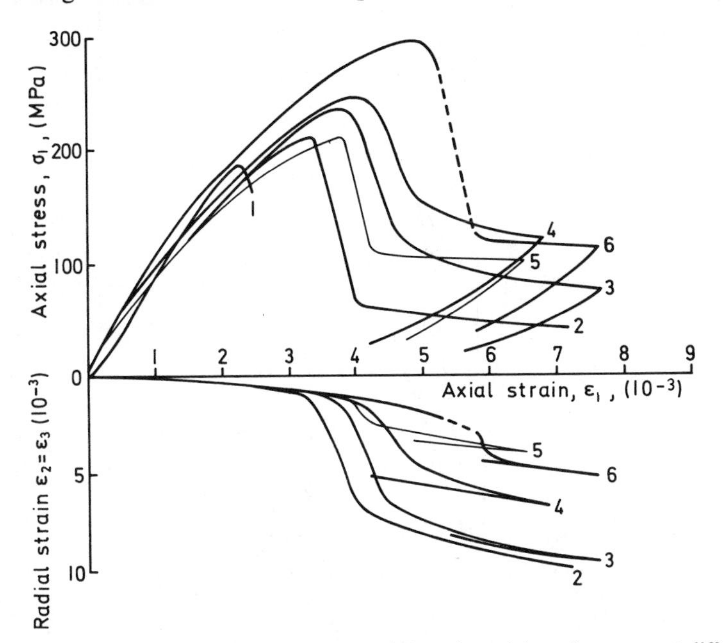

Fig. 4.2.6 Complete axial stress–strain and radial strain–axial strain curves at different confining pressures for a suite of specimens prepared from a single piece of argillaceous quartzite. Uniaxial compression: curve 1. Triaxial compression: curves; (2) $\sigma_2 = \sigma_3 =$ 3,45 MPa; (3) $\sigma_2 = \sigma_3 =$ 6,9 MPa; (4) $\sigma_2 = \sigma_3 =$ 13,8 MPa; (5) $\sigma_2 = \sigma_3 =$ 27,6 MPa, cracked specimen; (6) $\sigma_2 = \sigma_3 =$ 27,6 MPa (after Hojem *et al.*, 1975).

important to note how the radial strain–axial strain curves accentuate the departures from linearity evident in the axial stress–strain curves, particularly at low confining pressures. The method of measuring the radial strain–axial strain curve from the displacement of the confining fluid is not applicable to the uniaxial compression test, but it seems reasonable to assume that this curve would have exhibited the same phenomenon.

The sum of the three principal strains $\varepsilon_1 + \varepsilon_2 + \varepsilon_3$ equals the volumetric strain, $\Delta V/V_0$, where ΔV is the change in volume of the specimen during compression and V_0 is its original volume in the absence of any applied stress. For a linear, elastic material with constant values of Young's modulus and Poisson's ratio, the volumetric strain during compression is a straight line with a positive slope, that is, the volume decreases with increasing compression because $\varepsilon_1 > |\varepsilon_2 + \varepsilon_3|$. The behaviour of all three principal strains, $\varepsilon_1 + \varepsilon_2 + \varepsilon_3 = \Delta V/V_0$, has been studied by Brace *et al.* (1966) in the regions *OA*, *AB* and *BC* of the stress–strain curve, Fig. 4.2.7. From this figure it can be seen that the volumetric

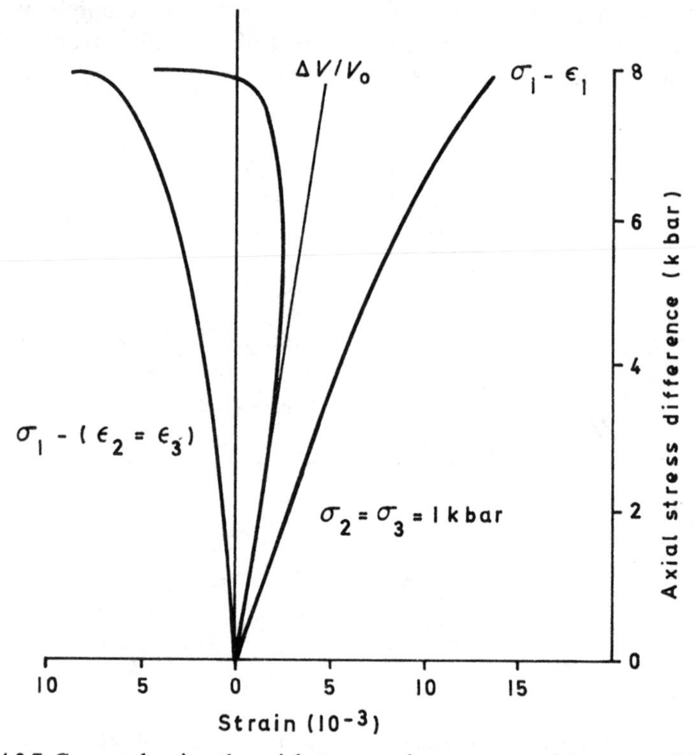

Fig. 4.2.7 Curves showing the axial stress–strain, $\sigma_1 - \varepsilon_1$, axial stress–radial strain, $\sigma_1 - (\varepsilon_2 = \varepsilon_3)$, and axial stress–volumetric strain, $\Delta V/V_0$, behaviour of Westerly granite in a triaxial test at a confining pressure of 1 kbar. The deviation of the $\Delta V/V_0$ curve from the dashed straight line represents dilatancy (after Brace *et al.*, 1966).

strain begins to deviate from the straight line of an elastic material when the stress reaches a value of about half the strength, and that the deviation from this straight line becomes so great near fracture that the volume of the rock at this stage of compression exceeds its original volume. This deviation represents an *increase* in volume with compression relative to the behaviour of a linear, elastic material, that is, a relative negative volumetric strain with compression, and is known as *dilatancy*. To account for this dilatancy, the sum of the lateral strains of the specimen near fracture must exceed its axial strain, that is, $|\varepsilon_2 + \varepsilon_3| > \varepsilon_1$. It follows also that Poisson's ratio for rock specimens in such compression tests is far from constant. By testing specimens of rock in the form of thick-walled tubes Cook (1970) showed that dilatancy is a pervasive volumetric phenomenon, because both the outside diameter of the specimen and that of the hole through it *increased* in the same proportions during compression; had the phenomenon of dilatancy been superficial the diameter of the hole would have *decreased*. To illustrate dilatancy throughout the four regions of the complete stress–strain curve, curve 3 of Fig. 4.2.6 has been re-plotted in Fig. 4.2.8.

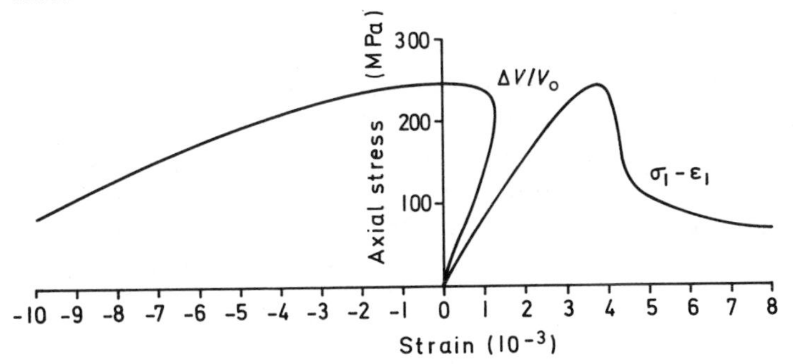

Fig. 4.2.8 Axial stress–strain and axial stress–volumetric strain curves prepared from curve 3 of Fig. 4.2.6 showing dilatancy for a complete stress–strain curve at low confining pressure.

Dilatancy can be ascribed to the formation and extensions of open micro-cracks within the rock specimens which have their long axes orientated parallel to the direction of the maximum principal stress, § 4.5.

The behaviour above is described as if it were independent of time. In fact, all rocks show time-dependent effects, even in the linearly elastic range, known as time-dependent elasticity or anelasticity. For example, if stress is applied suddenly the displacements do not take up their elastic values instantaneously but tend to them approximately exponentially, cf. § 11.3. This implies also that the rate of application of stress will also have an effect, and to simplify matters a standard rate of loading (for example, 100 psi/sec, cf. Obert *et al.* (1946)) is employed in ordinary laboratory testing. At very rapid rates of change, as in sonic or explosion work, and at very slow strain rates, results are rather different.

The other simple measurement involving a single stress is that of uniaxial tension of a circular cylinder. This has been much less studied because of the relative difficulty of performing the test accurately (cf. § 6.3). In this case, on the conventions of Chapter 2, both σ and ε are negative, and if $|\sigma|$ is plotted against $|\varepsilon|$ a stress–strain curve is obtained. Using the terminology of Fig. 4.2.2, the slope of the stress–strain curve in tension decreases continually between O and C and the modulus is usually less than that in compression, for reasons which are discussed in § 12.3. A material with different values of the modulus in tension and compression is described as *bilinear*, and this concept is often valuable in studying the behaviour of rocks, § 6.10. Usually failure in tension occurs as brittle fracture at C, and the *uniaxial tensile strength* is defined as the value of $|\sigma|$ at this point. The brittle region, CD, has not until recently been observed because soft testing machines have generally been used in tensile testing. However, theoretical complete stress–strain curves in tension and compression are similar, § 12.4 and Cook (1965), and Wawersik (1968) has made measurements of them.

4.3 The effects of confining pressure and temperature

It has been known qualitatively since the last century that if lateral displacement of a compression test specimen is resisted by applying pressure to its sides it becomes stronger and there is a tendency to greater ductility. Geologists were very clear about the possibility and importance of such effects (cf. Becker (1893)), and early workers developed a technique of

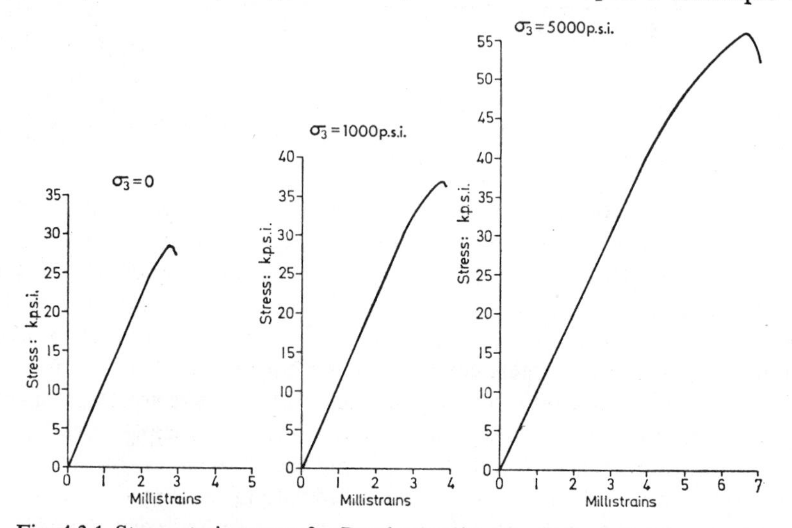

Fig. 4.3.1 Stress–strain curves for Rand quartzite at various confining pressures.

compressing cylindrical specimens inside closely fitting steel cylinders to obtain lateral restraint, Adams (1912).

The classical experiments were made by von Karman (1911) and Böker (1915), who used oil to apply *confining pressure* $\sigma_2 = \sigma_3$ to the sides of the specimen and increased the axial stress σ_1 steadily. This is the triaxial test discussed in detail in § 6.4. A revival of interest in the subject, particularly with reference to geophysical applications, dates from Griggs (1936).

The effect of increasing the confining pressure $\sigma_2 = \sigma_3$ can be seen in Fig. 4.3.1 for quartzite in a stiff testing machine. The curve for zero confining pressure (strictly 1 bar) has been discussed in § 4.2. Curves for confining pressures of 1 and 5 kpsi are also shown. These are of precisely the same form as before, showing a linear portion with the same Young's modulus as in the uniaxial case, a progressively better-defined yield point, and a maximum value followed by a short descending portion. For any value of the confining pressure σ_3 the maximum ordinate of the stress–strain curve will be defined as the *strength* of the rock at that confining pressure. In a conventional testing machine failure will occur at, or very near to, this stress. The conclusion from Fig. 4.3.1 is that as the confining pressure is increased the strength increases, and also that the amount of permanent set before fracture increases, though it remains small. This behaviour is characteristic of many rocks, even to confining pressures of the order of 50 kpsi, that is, throughout the range of importance in practical rock mechanics. Handin (1966) gives extensive tables of strength and it should be noted that, like many other writers, he defines strength in terms of the stress-difference $\sigma_1 - \sigma_3$ at failure. This definition is not used here since it ignores the effect of the intermediate principal stress σ_2.

A different type of behaviour is shown by some rocks, notably carbonates and some sediments, which was first studied in detail in the work of von Karman (1911) on Carrara marble. Fig. 4.3.2 shows his results. For confining pressures of up to about 500 bars, brittle fracture occurs as before with an increase of strength and a small increase in permanent set. The curve for $\sigma_3 = 685$ bars is completely different, since the material can now undergo strains of over 7 per cent with no loss of strength. This is generally known as ductile behaviour, but, since ductility has already been used in § 4.2 in connection with the relatively small amounts of permanent set which occur between the yield point and the maximum of the stress–strain curve, the term *fully ductile* will be used when necessary to denote conditions in which large amounts of strain can occur without loss of strength. In Fig. 4.3.2 the curve for a confining pressure of 235 bars shows transitional behaviour. The conclusion from Fig. 4.3.2 is that there is a rather ill-defined value of the confining pressure at which there is a transition from typical brittle behaviour to fully ductile behaviour. This is called the *brittle–ductile* transition. At still higher confining pressures,

1,650 and 3,260 bars in Fig. 4.3.2, σ_1 increases steadily with increasing strain after the yield point has been passed. This phenomenon is called work-hardening. Similar and more detailed results have been obtained by Heard (1960) for Solenhofen limestone.

The effect of increasing the temperature is to lower the brittle–ductile transition pressure. Fig. 4.3.3 shows curves for granite at a confining pressure of 5,000 bars and different temperatures, Griggs, Turner and Heard (1960). At room temperature the behaviour is brittle, but at higher temperatures substantial amounts of permanent deformation may be introduced without loss of load. At 800° the material is almost fully ductile. The brittle–ductile transition may be studied as a function of both pressure and temperature. Rate of strain also has an influence on it, though of less importance; at very slow strain rates the transition pressure is lowered.

The brittle–ductile transition is of great interest in geology and geophysics in connection with the behaviour of materials in the lower crust. As a result of this, a great deal of attention has been paid to materials in which it is readily observed. It is not of great interest in rock mechanics, since the necessary pressures and temperatures rarely occur. It may be of importance in carbonates, and sandstone shows the transition at around 1,400 bars at room temperature, Murrell (1965). Handin and Hager (1957) studied a number of sedimentary rocks at room temperature and (1958) at elevated temperatures. Griggs, Turner and Heard (1960) made extensive experiments on a variety of rocks at elevated temperatures.

4.4 Failure under polyaxial stresses

The remainder of this chapter will be concerned with brittle fracture, and the descending portion of the stress–strain curve will be neglected, so that failure is assumed to take place at the maximum ordinate of the stress–strain curve, which is the strength of the material as previously defined. This conception may be generalized into the statement that if any particular component of stress is increased under specified conditions until failure takes place, the magnitude of that stress at failure is known as the strength of the material under those conditions. Thus we may speak of the uniaxial tensile or compressive strength, the triaxial compressive strength at a certain confining pressure, and so on.

When polyaxial stresses are under consideration a still more general approach is necessary, and it will be assumed that failure takes place when a definite relation characteristic of the material

$$\sigma_1 = f(\sigma_2, \sigma_3), \tag{1}$$

is satisfied.

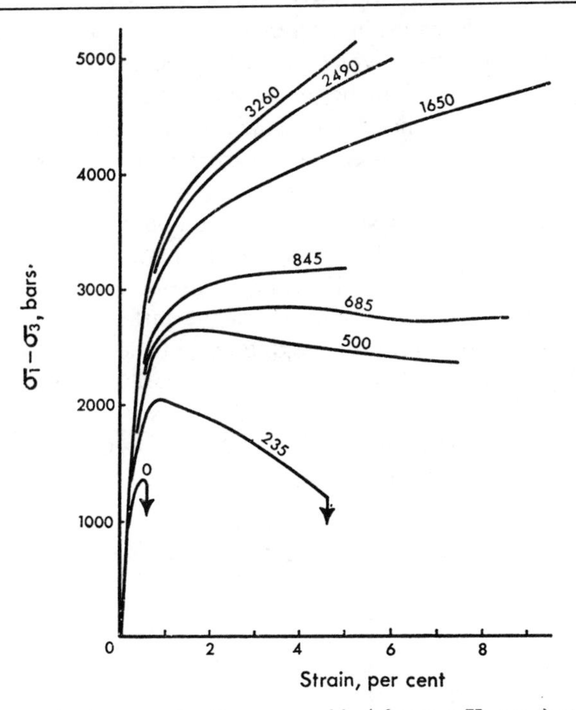

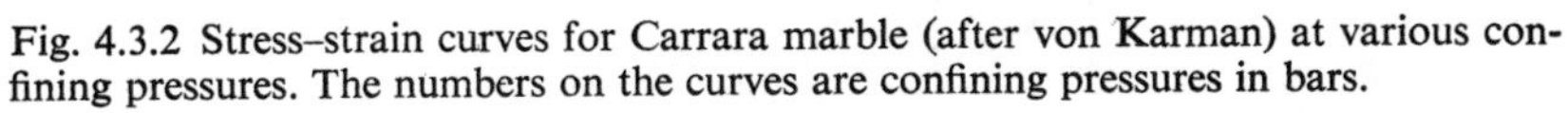

Fig. 4.3.2 Stress–strain curves for Carrara marble (after von Karman) at various confining pressures. The numbers on the curves are confining pressures in bars.

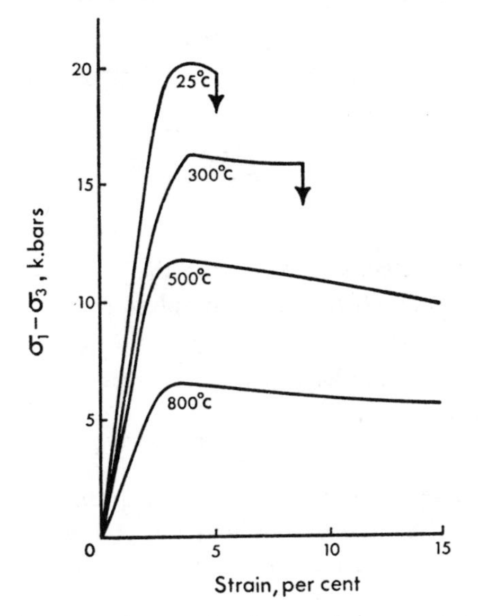

Fig. 4.3.3 Stress–strain curves for granite at a confining pressure of 5 kilobars and various temperatures (after Griggs, Turner, and Heard).

Such a relation is called a *criterion of failure*, and its geometrical representation is a surface which will be called the *failure surface*. The term 'fracture surface' might be preferable, but liable to confusion with the physical surface on which fracture takes place. Nadai (1950) calls it the limiting surface of rupture. Essentially, experimental measurements under different conditions should provide the form of this surface, but it has to be borne in mind that its existence is an assumption; for example, failure might depend not only on the stresses in the material but also on stress gradients or on the way in which the final state of stress is approached.

For the present it will be assumed that the surface (1) exists, and discussion will be confined to the region $\sigma_1 \geqslant 0$, since polyaxial tensile stresses rarely occur. Fig. 4.4 shows the information at present available about this surface, namely, the uniaxial tensile strength $\sigma_3 = -T_0$, $\sigma_1 = \sigma_2 = 0$; the uniaxial compressive strength $\sigma_1 = C_0$, $\sigma_2 = \sigma_3 = 0$; and values obtained in the triaxial test $\sigma_1 > \sigma_2 = \sigma_3$ which lie on a curve C_0T.

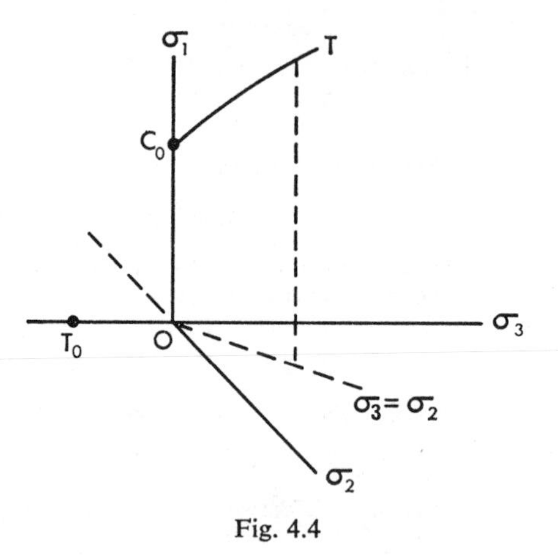

Fig. 4.4

Various empirical assumptions about the form of the criterion for failure have been made. Of these, the simplest are the following:

(i) *Maximum tensile stress.* The material is assumed to fail by brittle fracture in tension if the least principal stress σ_3 is equal to minus the uniaxial tensile strength,

$$\sigma_3 = -T_0. \tag{2}$$

This is, in fact, adequate under certain restricted conditions.

(ii) *Maximum shear stress.* The material is assumed to fail when the maximum shear stress is equal to a value S_0 characteristic of the material. That is

$$\sigma_1 - \sigma_3 = 2S_0. \tag{3}$$

This criterion, known as Tresca's, is definitely not strictly true, but is frequently useful as a special case of the Coulomb criterion of § 4.6.

(iii) *Maximum octahedral shear stress.* The material is assumed to fail when the octahedral shear stress τ_{oct} defined in § 2.4 (49) reaches a value k characteristic of the material, that is,

$$(\sigma_1 - \sigma_2)^2 + (\sigma_2 - \sigma_3)^2 + (\sigma_3 - \sigma_1)^2 = 9k^2. \tag{4}$$

This important criterion has frequently been applied to the study of experimental results on fracture, Hobbs (1962), Ely (1965). It has several interesting properties. First, it is unusual among simple criteria in that it involves all the principal stresses. Secondly, by § 2.8 (11) it may be written in the form

$$J_2 = 3k^2/2, \tag{5}$$

where J_2 is an invariant of the stress deviation. Since all criteria for failure must be invariant with respect to change of axes, it is natural to express them in terms of invariants and to study the simplest invariants first.

A linear relation between octahedral normal and shear stresses

$$\tau_{oct} = C_1\sigma_{oct} + C_2 \tag{6}$$

was used by Freudenthal (1951) and Bresler and Pister (1957). The failure surface corresponding to (6) is a circular cone with the line $\sigma_1 = \sigma_2 = \sigma_3$ as axis. Bresler and Pister (1957) found that (6) represented many results satisfactorily, but was inadequate for some on combined stresses. They suggested that a third invariant must be involved and proposed

$$\tau_{oct}/C_0 = f_1(I_1/C_0) + f_2(I_3/C_0^3), \tag{7}$$

where C_0 is the uniaxial compressive strength and I_1 and I_3 are invariants of stress. Scaling all quantities in terms of C_0, as in (7), is a useful method of making experimental results comparable.

The criteria for failure which have proved most useful have not been obtained, as above, from simple mathematical assumptions, but rather as the expressions of simple physical hypotheses. These are the Coulomb, Mohr, and Griffith criteria, which will be discussed in the following sections.

4.5 Types of fracture

In all discussions of brittle fracture the nature and description of the fractured surface is of the greatest importance. The types which occur may be illustrated from the behaviour of Solenhofen limestone at various confining pressures as described in § 4.3. In unconfined compression, Fig. 4.5 (*a*), irregular longitudinal splitting is observed: the explanation of this common phenomenon is still not altogether clear, and is discussed in §§ 6.2, 12.7. With quite a moderate amount of confining pressure the irregular behaviour of Fig. 4.5 (*a*) is replaced by a single plane of fracture, Fig. 4.5 (*b*), inclined at an angle of less than 45° to the direction of σ_1: this is the typical fracture under compressive stresses and will be described as a *shear fracture*. Its characteristic feature is a shearing displacement along the surface of fracture. Griggs and Handin (1960), who introduce a classification of fractures, call it a fault because of its correspondence with geological faulting, and they have been followed by many writers; however, it seems preferable to confine the term fault to the geological context and to retain the term shear fracture in the experimental context. If the confining pressure is increased so that the material becomes fully ductile, Fig. 4.5 (*c*), a network of shear fractures, accompanied by plastic deformation of the individual crystals, appears.

The second basic type of fracture, *an extension fracture*, appears typically in uniaxial tension. Its characteristic is a clean separation with no offset between the surfaces, Fig. 4.5 (*d*).

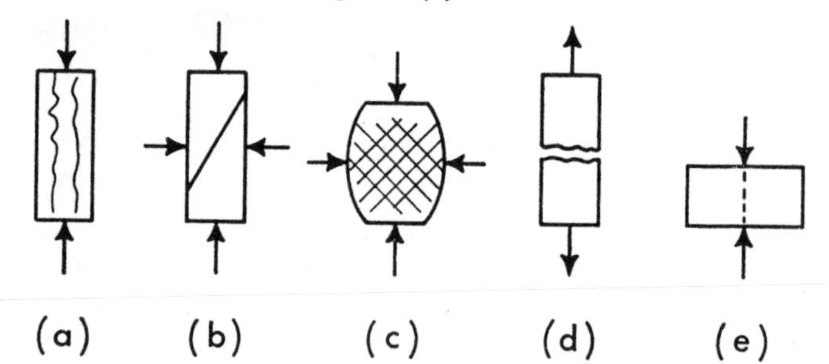

Fig. 4.5.1 (*a*) Longitudinal splitting in uniaxial compression. (*b*) Shear fracture. (*c*) Multiple shear fractures. (*d*) Extension fracture. (*e*) Extension fracture produced by line loads.

Under more complicated stress-systems fractures appear which may be regarded as belonging to one or other of these types. If a slab is compressed between line loads, Fig. 4.5 (*e*), an extension fracture appears between the loads. If these loads are caused by the squeezing of a jacket around the material into cracks into the material the fracture has been described as an *intrusion* fracture, Brace (1964b). When the fracture surfaces in the case of Fig. 4.5 (*a*) are examined parts of them will have the appearance of shear fractures, and other parts will apparently be extension fractures.

In § 4.2 attention has been directed to the important phenomenon of dilatancy which occurs during the unaxial and triaxial compression of rock specimens. Usually in triaxial tests the confining pressure is applied by a fluid through an impermeable membrane of negligible stiffness, § 6.4. It follows that in such tests the radial expansion and dilatation of the specimen is not resisted locally nor on average by increasing confinement, as would be the case if it were surrounded by more rock, as occurs in practical situations. The resistance provided by a surround of rock whether also failing or not, would be expected to have the effect of increasing the value of the minimum principal stress, so tending to inhibit failure and focus fracture into a limited volume. To approach the practical situation more closely than is done in normal triaxial testing, Hallbauer *et al.* (1973) jacketed specimens of a fine-grained, argillaceous quartzite in a copper tube of 1 mm wall thickness and tested them in a sealed triaxial cell with a small fluid volume. The copper tube provided resistance to local lateral expansion of the circumference of the specimens and the stiffness of the sealed cell caused the confining pressure to increase as 110,7 k bar $\times$ ε_2; this tended to inhibit radial expansion and dilatation. The tests were done in a stiff testing machine, § 6.13, using a suite of specimens made from one piece of quartzite and the compression of different specimens was stopped at predetermined points along the complete stress–strain curve. Careful macroscopic and microscopic studies of longitudinal sections through the axes of these specimens enabled the development of micro-cracks and the growth of fractures to be studied in relation to the complete stress–strain curve.

Some of the results are illustrated in Fig. 4.5.2. In the region *AB* of the complete stress–strain curve, § 4.2, the first visible structural damage appears as elongated microcracks distributed at random but concentrated in the central portion of the specimen with their long axes tending parallel to the direction of the maximum principal stress. Towards the end of the region *BC*, there is a pronounced increase in microcracking which tends to coalesce along a plane in the central portion of the specimen. At the point of maximum stress, *C*, a macroscopic fracture plane develops in the *central* portion of the specimen; this fracture grows towards the ends of the specimen by the step-wise joining of microcracks. Initially there is little or no movement across this fracture plane. Finally, in the region *CD*, the fracture plane extends towards the ends of the specimen where its direction changes so as to allow relative movement to take place across its surface between the two halves of the specimen, and a rapid drop in the resistance of the specimen to the applied load occurs.

Measurements of the microcracks made after the specimens had been unloaded and sectioned showed them to be about 300 μm long by 3 μm wide. When under stress their width is presumably substantially greater than 3 μm. The ratio between the volume of these microcracks as measured in the stress-free condition and the corresponding inelastic volumetric dilatancy at those positions on the complete stress–strain curve to which the specimens were loaded proved to be remarkably constant at about 16 per cent to 19 per cent.

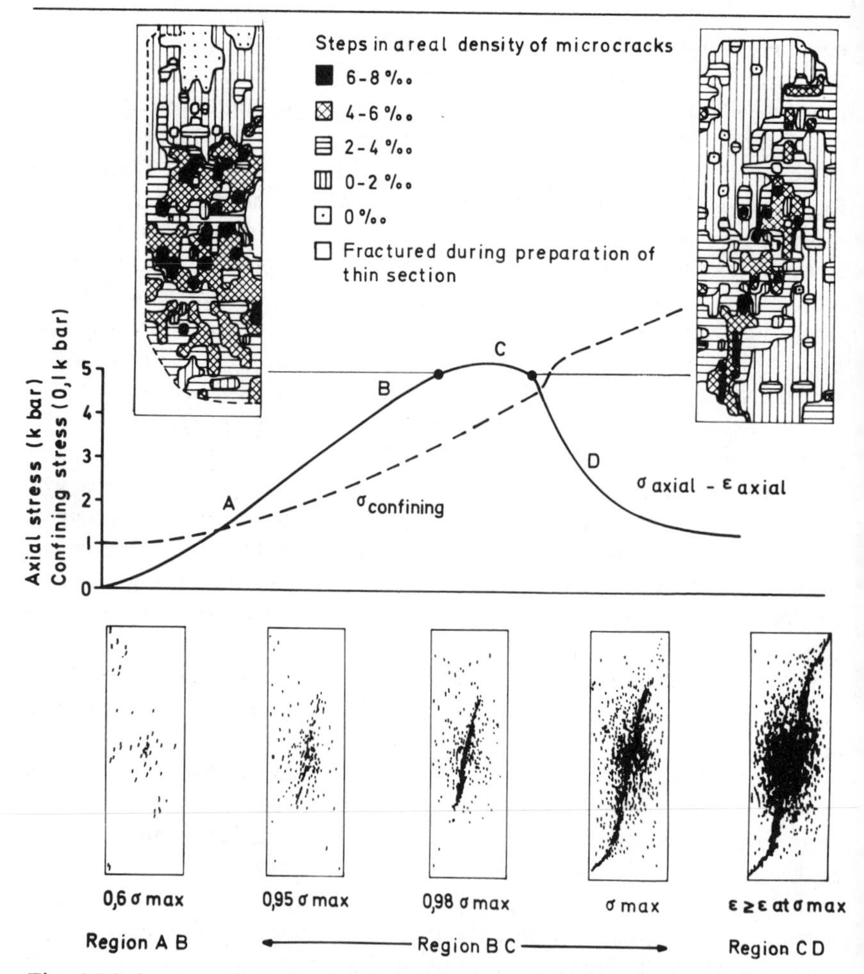

Fig. 4.5.2 A composite representation of the complete stress–strain curve and the incremental radial stress–axial strain curve for a suite of triaxial compression tests done in a stiff-testing machine and in a stiff, sealed triaxial cell, using specimens of argillaceous quartzite prepared from a single piece of rock. The axial sections through specimens stopped at various stages of compression show the structural changes associated with the complete stress–strain curve and associated dilatancy (after Hallbauer *et al.*, 1973).

The appearance of fracture surfaces has been little studied despite the importance of correlating laboratory effects with those observed in the field. Slickensides are commonly produced in shear fracture, and Paterson (1958) showed that for marble the direction of motion was that for which the steps would approach one another. When fracture occurs under relatively complicated stress systems a wide variety of surface markings can arise, Seldenrath and Gramberg (1958). These may be compared with those seen in the field, Roberts (1961), Hodgson (1961).

4.6 The Coulomb criterion

What remains the simplest and most important criterion was introduced by Coulomb (1773). Coulomb had made extensive researches into friction, and he suggested in connection with shear failure of rocks that the shear stress tending to cause failure across a plane is resisted by the cohesion (*adhérence*) of the material and by a constant times the normal stress across the plane. That is, that the criterion for shear failure in a plane is

$$|\tau| = S_0 + \mu\sigma, \tag{1}$$

or

$$|\tau| - \mu\sigma = S_0, \tag{2}$$

where σ and τ are the normal and shear stresses across the plane, S_0 is a constant which may be regarded as the inherent shear strength of the material, and μ is a constant which, by analogy with ordinary sliding, will be called the *coefficient of internal friction* of the material. Since the sign of the shear stress τ only affects the direction of sliding, only $|\tau|$ appears in the criterion (1). It appears that the relation (1) is precisely that which was found experimentally for sliding friction in §§ 3.2, 3.3, and it is also the criterion used in soil mechanics, S_0 being called the cohesion and denoted by c. This term and notation will occasionally be used.

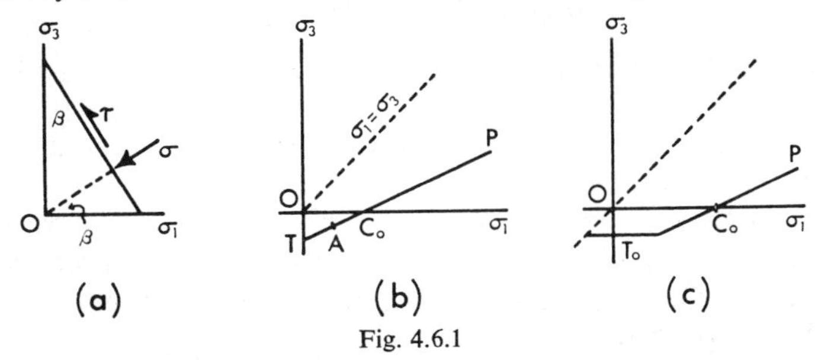

Fig. 4.6.1

Various names have been used for this criterion. Anderson (1951) calls it the Coulomb–Navier criterion and was followed by Jaeger (1962a). Navier did, indeed, use the criterion for yield, as did Guest (1940), and in this case S_0 is described as plasticity and μ is regarded as a sort of coefficient of dynamic friction. Other writers describe it as the Coulomb–Mohr criterion, but it is better to keep the Coulomb and Mohr theories separate because of their very different physical background.

The two-dimensional case will be considered first, but, as stated in § 4.1, the principal stresses will be written σ_1, σ_3 to facilitate comparison with the general case. The analysis is exactly that of Coulomb (1773). By § 2.3 (16), (17) the normal and shear stresses across a plane whose normal is inclined at β to σ_1, Fig. 4.6.1 (*a*), are

$$\sigma = \tfrac{1}{2}(\sigma_1 + \sigma_3) + \tfrac{1}{2}(\sigma_1 - \sigma_3)\cos 2\beta, \tag{3}$$

$$\tau = -\tfrac{1}{2}(\sigma_1 - \sigma_3)\sin 2\beta. \tag{4}$$

The quantity $|\tau| - \mu\sigma$ which occurs in (1) is then given by

$$|\tau| - \mu\sigma = \tfrac{1}{2}(\sigma_1 - \sigma_3)\ [\sin 2\beta - \mu\cos 2\beta] - \tfrac{1}{2}\mu(\sigma_1 + \sigma_3), \tag{5}$$

and its maximum value as a function of β is attained when

$$\tan 2\beta = -1/\mu, \tag{6}$$

so that 2β lies between 90° and 180° and

$$\sin 2\beta = (\mu^2 + 1)^{-\frac{1}{2}}, \quad \cos 2\beta = -\mu(\mu^2 + 1)^{-\frac{1}{2}}. \tag{7}$$

Using (7) in (5) gives for the maximum value of $|\tau| - \mu\sigma$,

$$\tfrac{1}{2}(\sigma_1 - \sigma_3)(\mu^2 + 1)^{\frac{1}{2}} - \tfrac{1}{2}\mu(\sigma_1 + \sigma_3),$$

and by (1) failure will not occur if this is less than S_0, and will occur if it is equal to S_0. That is, the criterion for failure is

$$\sigma_1[(\mu^2 + 1)^{\frac{1}{2}} - \mu] - \sigma_3[(\mu^2 + 1)^{\frac{1}{2}} + \mu] = 2S_0. \tag{8}$$

This is a straight line TC_0P in the σ_1, σ_3 plane which has intercept

$$C_0 = 2S_0[(\mu^2 + 1)^{\frac{1}{2}} + \mu] \tag{9}$$

on the σ_1-axis, and intercept $-2S_0[(\mu^2 + 1)^{\frac{1}{2}} - \mu]$ on the σ_3-axis, Fig. 4.6.1 (*b*).

The intercept C_0 on the σ_1-axis is the uniaxial compressive strength as defined in § 4.2, but the intercept on the σ_3-axis is not the uniaxial tensile strength (as is sometimes stated), since physical conditions restrict the criterion to only a portion of the line TC_0P. Essentially, a physical assumption implicit in (1) is that σ should be positive. Using the value (7) of β in (3), this requires that

$$\sigma_1[(\mu^2 + 1)^{\frac{1}{2}} - \mu] + \sigma_3[(\mu^2 + 1)^{\frac{1}{2}} + \mu] > 0, \tag{10}$$

and this, combined with (8), requires

$$\sigma_1 > S_0[(\mu^2 + 1)^{\frac{1}{2}} + \mu] = \tfrac{1}{2}C_0. \tag{11}$$

It follows that only the portion AC_0P, Fig. 4.6.1 (*b*), of the line represents a valid criterion.

For some negative (tensile) values of σ_3 it is known experimentally that extension fractures in planes perpendicular to σ_3 occur, in particular, in uniaxial tension at the tensile strength T_0. This behaviour is entirely different from the shear fracture which occurs with compressive stresses, and a simple and useful criterion, Paul (1961), is obtained by assuming

$$\sigma_1[(\mu^2 + 1)^{\frac{1}{2}} - \mu] - \sigma_3[(\mu^2 + 1)^{\frac{1}{2}} + \mu] = 2S_0,$$
$$\sigma_1 > C_0[1 - C_0T_0/4S_0^2], \tag{12}$$

$$\sigma_3 = -T_0, \quad \sigma_1 < C_0[1 - C_0T_0/4S_0^2]. \tag{13}$$

This is shown in Fig. 4.6.1 (*c*).

Considering, now, the angle β at which fracture takes place, (6) gives

$$\beta = \tfrac{1}{2}\pi - \tfrac{1}{2}\tan^{-1}(1/\mu). \tag{14}$$

If $\mu \to 0$, $\beta \to \pi/4$; if $\mu = 1$, $\beta = 3\pi/8$; if $\mu \to \infty$, $\beta \to \pi/2$. Thus the direction of shear fracture is always inclined at an acute angle to the direction of maximum stress. Since the previous discussion is independent of the sign of β, there are in fact two possible directions of fracture inclined at equal angles $\frac{1}{2}\tan^{-1}(1/\mu)$ to the direction of maximum principal stress and on either side of it. These are called conjugate directions.

The relation (8) may be put into several other useful forms. Introducing the *angle of internal friction* ϕ defined by

$$\mu = \tan\phi, \tag{15}$$

it follows that

$$(\mu^2 + 1)^{\frac{1}{2}} + \mu = \sec\phi + \tan\phi = \tan\alpha, \tag{16}$$

where

$$\alpha = (\pi/4) + \tfrac{1}{2}\phi. \tag{17}$$

It follows from (14) that the angle α defined in (17) is equal to the angle β which the normal to the plane of fracture makes with the direction of σ_1.

Using (9) and (16), the criterion (8) may be written in the useful forms

$$\sigma_1 = 2S_0\tan\alpha + \sigma_3\tan^2\alpha, \tag{18}$$

$$= C_0 + \sigma_3\tan^2\alpha, \tag{19}$$

$$= C_0 + q\sigma_3, \tag{20}$$

where $$q = \tan^2\alpha = [(\mu^2 + 1)^{\frac{1}{2}} + \mu]^2, \tag{21}$$

and C_0 is the uniaxial compressive strength. These give σ_1 at failure in terms of σ_3, Fig. 4.6.2 (*a*), and μ can be determined from the slope of this line using (17) or (21).

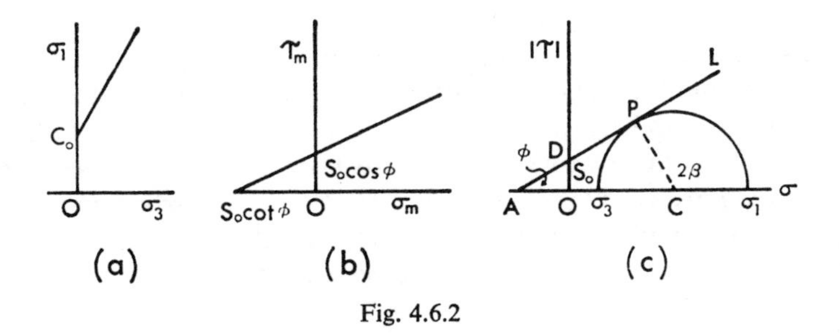

Fig. 4.6.2

Experiments on failure under triaxial stress conditions, § 6.4, show general agreement with the above results. The curve of σ_1 against σ_3 is approximately linear, and the plane of shear fracture is inclined at an angle of less than 45° to the direction of σ_1.

Another representation of the criterion (8) is obtained by introducing the mean stress σ_m and the maximum shear stress τ_m defined by

$$\sigma_m = \tfrac{1}{2}(\sigma_1 + \sigma_3), \quad \tau_m = \tfrac{1}{2}(\sigma_1 - \sigma_3); \tag{22}$$

using them, (8) becomes

$$(\mu^2 + 1)^{\frac{1}{2}}\tau_m - \mu\sigma_m = S_0,$$

or

$$\tau_m = \sigma_m \sin\phi + S_0 \cos\phi, \tag{23}$$

which is a line in the τ_m, τ_m plane of inclination $\tan^{-1}(\sin\phi)$ and intercept $S_0 \cos\phi$ on the τ_m — axis, Fig. 4.6.2 (*b*).

Finally, the way in which these results follow from the Mohr circle representation, § 2.3, may be given. In the plane of σ and $|\tau|$ the criterion (1) is represented by the straight line *ADL*, Fig. 4.6.2 (*c*) of slope $\mu = \tan\phi$ and intercept $OC = S_0$ on the $|\tau|$ axis. The normal and shear stresses across a plane are given by the Mohr circle on σ_1 and σ_3 as diameter. Failure occurs if this circle touches the line *ADL*. It follows from the geometry of the triangle *CAP* that

$$2\beta = \tfrac{1}{2}\pi + \phi$$

and

$$CP = (AO + OC)\sin\phi,$$

or

$$\tfrac{1}{2}(\sigma_1 - \sigma_3) = [S_0 \cot\phi + \tfrac{1}{2}(\sigma_1 + \sigma_3)]\sin\phi,$$

which is (23).

The similarity of this discussion to that of § 3.6 will be apparent. In that section there was only one plane of weakness, but if all planes were equally weak and the material was allowed to select its own plane of fracture the present theory would result. It may thus be regarded as applicable to soils and to rocks with random fracturing in all directions.

The case of a general three-dimensional system of stresses may be treated in the same way. If σ and τ are written down in terms of the direction cosines *l*, *m*, *n* of the normal to the plane relative to axes $O\sigma_1\sigma_2\sigma_3$ by § 2.4 (25), (26) and the maximum value of $|\tau| - \mu\sigma$ is sought as in § 2.4, it is found that it occurs when $m = 0$, $l = \cos\beta$, $n = \sin\beta$, where β is given by (6). As before, the sign of β can be changed without affecting the result. Thus there are *two possible conjugate planes of fracture passing through the direction* $O\sigma_2$ *of the intermediate principal stress* and making angles $\tfrac{1}{4}\pi - \tfrac{1}{2}\phi$ with the direction of maximum stress.

The criterion of failure does not involve σ_2 and takes any of the previous forms (8), (18), (20), (23).

The same result follows from the argument of Fig. 4.6.2 (*c*) using the Mohr circle, since by § 2.6 the extreme Mohr circle for which failure will first occur is that corresponding to the points, σ_1, σ_3.

4.7 Mohr's hypothesis

Mohr (1900) proposed that when shear failure takes place across a plane, the normal stress σ and the shear stress τ across this plane are related by a functional relation characteristic of the material,

$$|\tau| = f(\sigma). \tag{1}$$

Since the sign of τ only affects the direction of sliding, only the magnitude of τ is in question. The relation (1) will be represented by a curve such as *AB* in the σ, τ plane, Fig. 4.7.1 (*a*). If we have three unequal principal stresses, σ_1, σ_2, σ_3, the values of σ and τ can be found by the Mohr construction of § 2.6, and failure will not take place if the values of σ and τ so found lie below the curve *AB*. Failure will take place if the circle of centre *C* on σ_1 and σ_3 as diameter just touches *AB*.

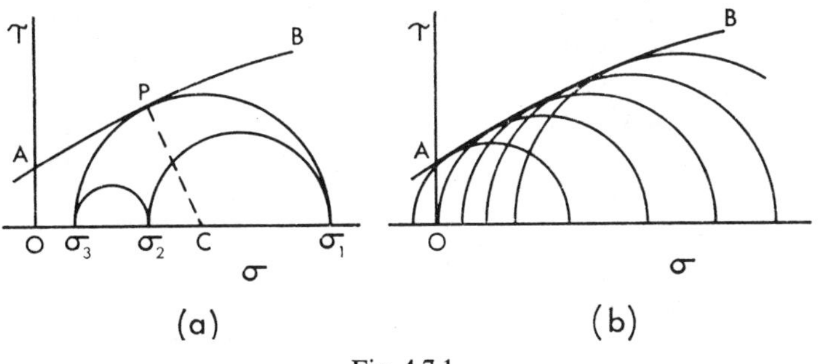

Fig. 4.7.1

This carries several important conclusions:

(i) the value of the intermediate principal stress σ_2 does not affect failure;

(ii) the plane of shear fracture passes through the direction of the intermediate principal stress, and its normal makes an angle β with the direction of maximum principal stress, where 2β is the angle $PC\sigma$ which the normal to the curve *APB* makes with the σ-axis.

The curve (1) is not in general thought of as defined by an explicit formula, but is assumed to be obtained experimentally as the envelope of the Mohr circles corresponding to failure under a variety of conditions, Fig. 4.7.1 (*b*). For this reason it is called the *Mohr envelope*, and the results of a set of experiments are usually represented by drawing the Mohr envelope as well as by plotting σ_1 against σ_3. It will be noted that the direction of fracture is determined by the normal to the Mohr envelope, and the observed angles should be (and in general are) consistent with this. The Mohr envelope is usually concave downwards as in Fig. 4.7.1, so that as the mean stress $\frac{1}{2}(\sigma_1 + \sigma_3)$ is increased the plane of fracture becomes inclined at an increasing angle to the σ_1-direction. This is observed in practice.

For a brittle material, $\sigma_1 - \sigma_3$ increases with σ_3, so that the curve *APB* increases steadily and will be open to the right. Its behaviour for negative σ will be discussed later.

A very important case is that of the Mohr envelope for the Coulomb criterion of § 4.6. This criterion was represented by the relationship § 4.6 (23)

$$\tau_m = \sigma_m \sin \phi + S_0 \cos \phi \tag{2}$$

between

$$\tau_m = \tfrac{1}{2}(\sigma_1 - \sigma_3) \quad \text{and} \quad \sigma_m = \tfrac{1}{2}(\sigma_1 + \sigma_3), \tag{3}$$

which are the maximum shear stress and the mean stress in the plane σ_1, σ_3. σ_m is to be distinguished from the three-dimensional mean stress, which is $(\sigma_1 + \sigma_2 + \sigma_3)/3$. The relation (2) corresponds to the straight line *ADE*, Fig. 4.7.2 (*a*), for which $OA = -S_0 \cot \phi$, $OD = S_0 \cos \phi$, and the angle *DAO* is $\tan^{-1} (\sin \phi)$.

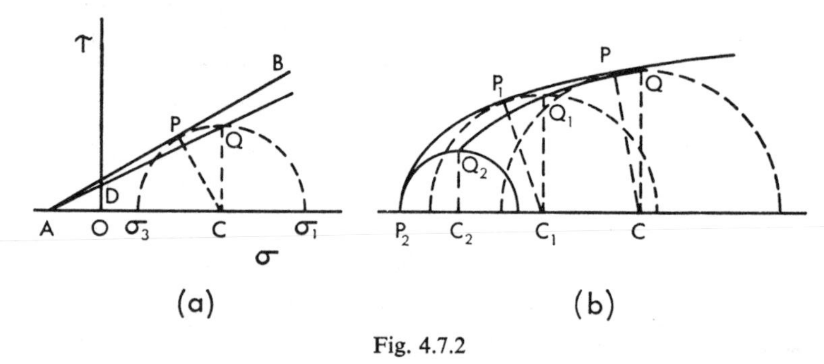

Fig. 4.7.2

The Mohr envelope for this criterion will be the envelope of circles such as *CQ* with centre at *C* and radius τ_m. This envelope will be the line *APB* which touches the circle. The inclination of this line is given by

$$\sin PAC = CP/AC = \tau_m/(\sigma_m + S_0 \cot \phi) = \sin \phi, \tag{4}$$

using (2), so that the angle *BAC* is ϕ, the angle of friction, and the equation of the Mohr envelope *AB* is

$$\tau = S_0 + \sigma \tan \phi = S_0 + \mu\sigma. \tag{5}$$

The Coulomb criterion is thus equivalent to the assumption of a linear Mohr envelope, which, accepting the Mohr hypothesis, would be the form to be tried first. The next most simple case, that of the parabolic Mohr envelope, has also been much used, cf. Leon (1934). It has the advantage that it allows the transition from compressive to tensile conditions to be discussed on a simple model and is consistent with some field observations, Muehlberger (1961). As remarked earlier, experimental Mohr envelopes are not strictly linear as in Fig. 4.7.2 (*a*) but are usually slightly concave downwards, though not necessarily parabolic. Griffith theory, § 4.8, leads to a parabolic envelope.

There are geometrical limitations to the simple concept of the Mohr envelope as stated above, which, while they are not important in practice, need to be understood. Essentially the position is that if σ_m, τ_m at failure are plotted they will lie on a curve Q_1Q, Fig. 4.7.2 (*b*), and the envelope of circles with centre C and radius CQ, centre C_1 and radius C_1Q_1, etc., is the Mohr envelope, so that either of these curves determines the other. Suppose, now, that the Mohr envelope has a finite radius of curvature C_2P_2 at the point where it cuts the σ-axis. In this case the curve QQ_1Q_2 cannot continue into the interior of the circle of centre C_2, since the Mohr circle corresponding to such a point would lie inside the Mohr envelope and so could not cause failure. The interesting case of the parabolic Mohr envelope is discussed in § 4.8.

4.8 The plane Griffith criterion

This is based on the hypothesis enunciated by Griffith in 1921 that fracture is caused by stress-concentrations at the tips of minute *Griffith cracks* which are supposed to pervade the material, and that fracture is initiated when the maximum stress near the tip of the most favourably orientated crack reaches a value characteristic of the material. The theory is discussed in detail in § 10.13, and the results are stated here as an example of an important criterion for failure. They are scaled in terms of the uniaxial tensile strength T_0 and state that, for principal stresses σ_1, σ_3 in two dimensions, failure takes place if

$$(\sigma_1 - \sigma_3)^2 = 8T_0(\sigma_1 + \sigma_3), \quad \text{if} \quad \sigma_1 + 3\sigma_3 > 0, \tag{1}$$

$$\sigma_3 = -T_0, \quad \text{if} \quad \sigma_1 + 3\sigma_3 < 0. \tag{2}$$

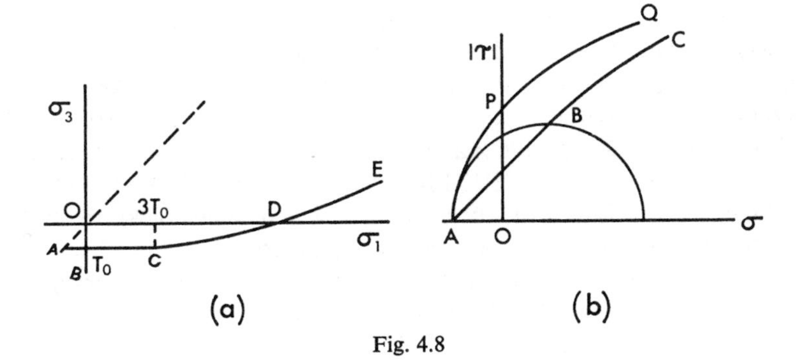

Fig. 4.8

Thus the criterion is represented in the (σ_1, σ_3) plane by portion of the straight line ABC, $\sigma_3 = -T_0$, for $-T_0 < \sigma_1 < 3T_0$, and portion CDE of the parabola (1) which touches the line ABC at C, $(3T_0, -T_0)$. The complete parabola (1) passes through the origin, Fig. 4.8.

When $\sigma_3 = 0$, uniaxial compression, $\sigma_1 = 8T_0$, so that the uniaxial compressive strength C_0 is

$$C_0 = 8T_0. \tag{3}$$

This result, which is unequivocally predicted by the theory, is reasonable in order of magnitude but not obeyed in detail.

Next, we determine the Mohr envelope corresponding to (1) and (2). Writing $\sigma_m = \frac{1}{2}(\sigma_1 + \sigma_3)$, $\tau_m = \frac{1}{2}(\sigma_1 - \sigma_3)$, as in § 4.7 (3), (1), and (2) become

$$\tau_m{}^2 = 4T_0\sigma_m, \quad \text{if} \quad 2\sigma_m > \tau_m, \tag{4}$$

$$\tau_m = \sigma_m + T_0, \quad \text{if} \quad 2\sigma_m < \tau_m. \tag{5}$$

This consists of the line AB of slope 45° for $-T_0 < \sigma_m < T_0$, and portion BC of a parabola for $\sigma_m > T_0$, Fig. 4.8 (*b*). The Mohr circles for all points of the line AB touch at the point A. To find the Mohr envelope for points on the parabola BC we have to find the envelope of the circles of centre $(\sigma_m, 0)$ and radius τ_m given by (4), whose equations are

$$f(\sigma_m) \equiv (\sigma - \sigma_m)^2 + \tau^2 - \tau_m{}^2 = (\sigma - \sigma_m)^2 + \tau^2 - 4T_0\sigma_m = 0, \tag{6}$$

where (4) has been used. To find the envelope, the usual procedure is to eliminate σ_m between (6) and $\partial f/\partial\sigma_m = 0$ which is

$$\sigma - \sigma_m + 2T_0 = 0. \tag{7}$$

It follows from (6) and (7) that the equation of the envelope is

$$\tau^2 = 4T_0(\sigma + T_0), \tag{8}$$

which is a parabola APQ passing through A.

The theory leading to the Griffith criterion (1), (2) neglects the fact that cracks may be expected to close under sufficiently high compressive stresses. If they do close, it may be expected that frictional forces will operate between the closed surfaces. A modification to take this into account, known as the *Modified Griffith Theory*, has been proposed by McClintock and Walsh (1962) and Brace (1960b). Its effect is to leave the early part of the curve $ABCDE$, Fig. 4.8 (*a*), for which all cracks are open, unaltered; a transition zone then occurs in which some cracks are closed; and, finally, when all cracks are closed, the theory becomes identical with the Coulomb theory of § 4.6 and gives a linear stress–strain curve and a linear Mohr envelope. The theory is discussed in § 10.14.

4.9 Murrell's extension of the Griffith criterion

The theory described in § 4.8 was two-dimensional only, derived from the study of flat elliptical cracks. A generalization would require a complete study of flat ellipsoidal cracks, and this has not yet been given. Murrell (1963) has given a logical extension to three dimensions of the two-dimensional theory of § 4.8, which is very valuable as indicating the sort of properties which a fracture criterion in three dimensions might be expected to have.

The Griffith criterion in two dimensions, § 4.8, has two geometrical properties:

(i) Part of the failure curve consists of a portion of the straight line $\sigma_3 = -T_0$.

(ii) The remainder of the failure curve consists of a portion of a parabola which has the line $\sigma_1 = \sigma_3$ for axis, passes through the origin, and touches the line $\sigma_3 = -T_0$.

If these properties only are assumed, the equations § 4.8 (1), (2) of the failure curve may be deduced. By symmetry, the whole parabola may be used, which then touches the lines $\sigma_1 = -T_0$ and $\sigma_3 = -T_0$.

In three dimensions the obvious generalization of (ii) is that the failure surface shall be a portion of a paraboloid of revolution whose axis is the line $\sigma_1 = \sigma_2 = \sigma_3$. The generalization of (i) may be stated in the form that the paraboloid shall be terminated by the pyramid of mutually perpendicular planes

$$\sigma_1 = -T_0, \quad \sigma_2 = -T_0, \quad \sigma_3 = -T_0. \tag{1}$$

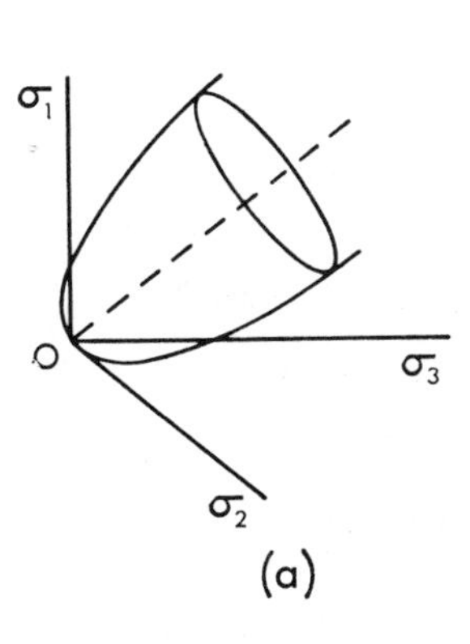

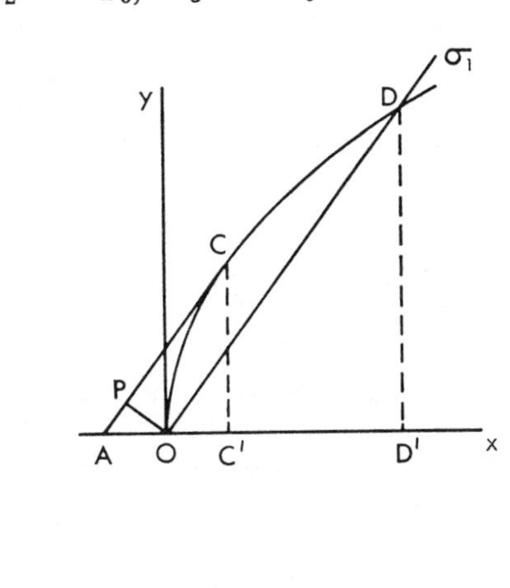

Fig. 4.9

Since it is essential that all generators of this pyramid shall intersect the paraboloid, the simplest assumption is that the edges of the pyramid, in particular

$$\sigma_2 = \sigma_3 = -T_0, \tag{2}$$

shall touch the paraboloid.

With these assumptions, the equation of the paraboloid is fully determined. Suppose that the axis of the paraboloid is Ox, Fig. 4.9 (*a*), which is the line $\sigma_1 = \sigma_2 = \sigma_3$ of direction cosines $l = m = n = 3^{-\frac{1}{2}}$ which is inclined at an angle $\alpha = \cos^{-1}(3^{-\frac{1}{2}}) = \tan^{-1}(2^{\frac{1}{2}})$ to the σ_1, σ_2, σ_3 axes. Suppose that the paraboloid is formed by the rotation of the parabola

$$y^2 = 4bx \tag{3}$$

about Ox. This parabola and its geometry is shown in Fig. 4.9 (*b*), which is a section of Fig. 4.9 (*a*) in the plane $xO\sigma_1$. In Fig. 4.9 (*b*), AC, which is the tangent inclined at α to Ox, will be the line (2), which is just to touch the parabola and whose distance from $O\sigma_1$ is $2^{\frac{1}{2}}T_0$, so that

$$OP = OA \sin \alpha = 2^{\frac{1}{2}}T_0. \tag{4}$$

Differentiating (3) gives for the slope of the tangent at C

$$\frac{dy}{dx} = \frac{2b}{y} = \tan \alpha = 2^{\frac{1}{2}}.$$

Therefore

$$CC' = 2^{\frac{1}{2}}b, \quad OC' = \tfrac{1}{2}b, \quad AC' = b, \quad OA = \tfrac{1}{2}b, \quad AC = 3^{\frac{1}{2}}b, \tag{5}$$

and for the point D for which $y = x \tan \alpha = 2^{\frac{1}{2}}x$

$$OD' = 2b, \quad OD = 2b\sqrt{3}. \tag{6}$$

Using the value (4) for OA in (5) gives

$$b = 2T_0\sqrt{3}. \tag{7}$$

Now x and y are connected with σ_1, σ_2, σ_3 by

$$x = (\sigma_1 + \sigma_2 + \sigma_3)3^{-\frac{1}{2}}, \tag{8}$$

$$y^2 = \tfrac{1}{3}\{(\sigma_2 - \sigma_3)^2 + (\sigma_3 - \sigma_1)^2 + (\sigma_1 - \sigma_2)^2\}, \tag{9}$$

and using these in (3) with the value (7) of b gives finally

$$(\sigma_2 - \sigma_3)^2 + (\sigma_3 - \sigma_1)^2 + (\sigma_1 - \sigma_2)^2 = 24T_0(\sigma_1 + \sigma_2 + \sigma_3), \tag{10}$$

or, in terms of the octahedral stresses § 2.4 (48), (49),

$$\tau_{oct}^2 = 8T_0\sigma_{oct}. \tag{11}$$

This theory is in reasonable agreement with experiment, Murrell (1963).

It has the merit of the simple expression (10) in terms of invariants. It follows from (10) that the uniaxial compressive strength C_0, OD in Fig. 4.9 (*b*), is

$$C_0 = 12T_0. \tag{12}$$

This is an improvement, since the value $8T_0$ derived from the plane theory is usually too low. However, the main advantage of the theory is that it provides a simple criterion for studying the effects of polyaxial stresses. It is of interest to study simple stress systems.

For the *triaxial test*, $\sigma_2 = \sigma_3$, (10) becomes

$$(\sigma_1 - \sigma_3)^2 = 12T_0(\sigma_1 + 2\sigma_3), \tag{13}$$

which is a parabola whose slope when $\sigma_3 = 0$ is $(d\sigma_1/d\sigma_3) = 4$.

For extension in the triaxial test, $\sigma_1 = \sigma_2$, (10) gives

$$(\sigma_1 - \sigma_3)^2 = 12T_0(2\sigma_1 + \sigma_3), \tag{14}$$

which is a parabola which intersects $\sigma_3 = 0$ when $\sigma_1 = 24T_0$ and at this point $(d\sigma_1/d\sigma_3) = 2{\cdot}5$.

For *biaxial stress*, $\sigma_3 = 0$, (10) gives

$$\sigma_1^2 - \sigma_1\sigma_2 + \sigma_2^2 = 12T_0(\sigma_1 + \sigma_2), \tag{15}$$

which is an ellipse whose slope when $\sigma_2 = 0$ is $(d\sigma_1/d\sigma_2) = 2$. It intersects the line $\sigma_1 = \sigma_2$ when $\sigma_2 = 24T_0$.

If $\sigma_2 = \frac{1}{2}(\sigma_1 + \sigma_3)$, (10) becomes

$$(\sigma_1 - \sigma_3)^2 = 24T_0(\sigma_1 + \sigma_3), \tag{16}$$

which has the same form as the relation in two dimensions.

The intersection of (10) with the plane $\sigma_3 = -T_0$ is the ellipse

$$\sigma_1^2 - \sigma_1\sigma_2 + \sigma_2^2 - 11T_0(\sigma_1 + \sigma_2) + 13T_0^2 = 0. \tag{17}$$

This touches the line $\sigma_2 = -T_0$ when $\sigma_1 = 5T_0$, corresponding to C, Fig. 4.9 (*b*).

Relations of type (10) have been used in connection with yield, Nadai (1950, p. 227).

Jaeger (1967b) has suggested that experimental results under different conditions of polyaxial stress may be compared by using as failure surface the surface formed by rotating the experimentally determined curve for the triaxial test about the line $\sigma_1 = \sigma_2 = \sigma_3$.

Finally, having set out all these criteria, the question arises of whether any one of them should be preferred. This question cannot be answered until more experimental results of the types discussed in Chapter 6 become available, but a brief indication of the situation may be given here. There are two major differences between the prediction of the theories.

First, the Coulomb and modified Griffith criteria predict a linear variation of σ_1 with σ_3 in the triaxial test (at least for large σ_3 in the latter case), while the von Mises and Griffith criteria give a parabolic variation. Neither of these fits the experimental results particularly well, and the empirical relation (Murrell, 1965),

$$\sigma_1 = C_0 + b\sigma_3{}^m,$$

where C_0, b, and m are constants, probably is better.

Secondly, the Coulomb, Mohr, and Griffith theories all predict that the intermediate principal stress has no effect on the strength, and only the von Mises and three-dimensional Griffith criteria show one. It is, of course, a matter of very great practical importance that this question should be settled. Experiments using inhomogeneous stress (Murrell (1963); Jaeger and Hoskins (1966b)) suggest that the intermediate principal stress does have an important effect, but such experiments may involve the effects of stress gradient. From comparison of triaxial extension and compression tests, Brace (1964b) concluded that the intermediate principal stress has no effect; however, Wiebols and Cook (1968) have found one with unequal homogeneous principal stresses. Workers on materials such as ceramics which can be obtained in the form of thin-walled hollow cylinders have been able to compare the various criteria for failure by studying failure under a wide range of (approximately) homogeneous stresses, Ely (1965); Broutman and Cornish (1965); Cornet and Grassi (1961), and recently Handin, Heard, and Magouirk (1967) have been able to use this method for rocks. It seems likely that more sophisticated criteria for failure, based on the actual mechanism of fracture, have yet to be developed: one such is developed in §§ 6.15, 12.7 and does show an effect of the intermediate principal stress.

4.10 The effect of anisotropy on strength

Since most sedimentary and metamorphic rocks are anisotropic, the effect of anistropy on strength is of great importance. Results for the simplest case of planar anisotropy in which the rock has a set of parallel planes of weakness can be written down from the theory of § 3.6. If S_w is the inherent shear strength in the planes of weakness and μ_w is their coefficient of internal friction in the sense of § 4.6, any one of § 3.6 (8), (11), (12), (13) may be used as the criterion for failure in a plane of weakness. Of these, § 3.6 (12) is probably the most useful and, in the present notation, is

$$\sigma_1 - \sigma_3 = \frac{2(S_w + \mu_w \sigma_3)}{(1 - \mu_w \cot \beta) \sin 2\beta}, \tag{1}$$

where β is the angle between σ_1 and the normal to the planes of weakness.

As shown in § 3.6 (15)–(17), $\sigma_1 - \sigma_3$ given by (1) tends to infinity as $\beta \to \frac{1}{2}\pi$, and as $\beta \to \tan^{-1}\mu_w = \phi_w$. Between these values, failure is possible with a value of σ_1 depending on β, and this has a minimum value

$$\sigma_{\min} = \sigma_3 + 2(S_w + \mu_w\sigma_3)[(\mu_w^2 + 1)^{\frac{1}{2}} + \mu_w] \tag{2}$$

which occurs when β has the value β_w given by

$$\tan 2\beta_w = -1/\mu_w. \tag{3}$$

This is the inclination most favourable for failure. It appears from (1) that the variation of $\sigma_1 - \sigma_3$ with β has the same form for all values of σ_3, but with a different multiplying factor. The nature of this variation for the case $\mu_w = 0{\cdot}5$ has been given in Fig. 3.6.2.

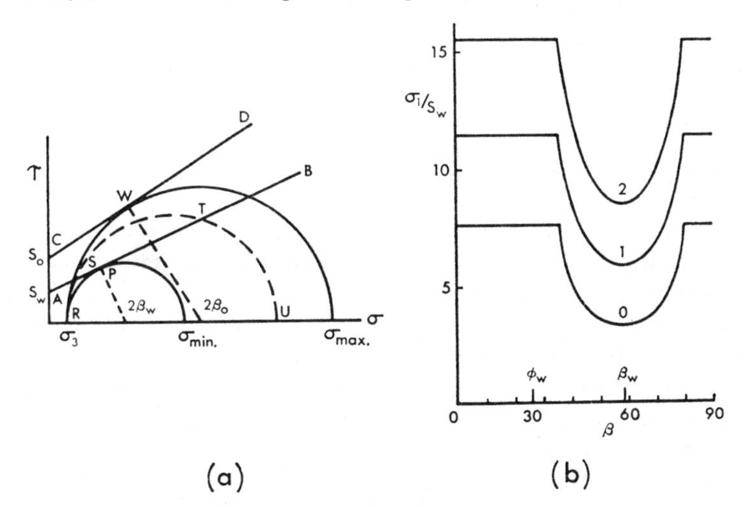

Fig. 4.10 (*a*) Mohr diagram for discussion of fracture in and across parallel planes of weakness in a material. (*b*) Variation of σ_1 with β for the case $\mu_w = 0{\cdot}5$, $\mu_0 = 0{\cdot}7$, $S_0 = 2S_w$. Numbers on the curves are values of σ_3/S_w.

So far, only the possibility of the material failing in a plane of weakness has been considered. However, the possibility remains that for unfavourable values of β failure may take place through the material in a plane which intersects the planes of weakness. The criterion for this will be taken to be the Coulomb criterion with inherent shear strength S_0 and coefficient of internal friction μ_0. It may be assumed that $S_0 > S_w$ and $\mu_0 > \mu_w$.

The situation can most easily be seen from the Mohr diagram, Fig. 4.10 (*a*). In this, the line AB, $\tau = S_w + \sigma\mu_w$ corresponds to failure in the planes of weakness, and CD, $\tau = S_0 + \sigma\mu_0$, corresponds to failure through the body of the material. Suppose, now, that σ_3 is held fixed and σ_1 is increased. At some value, $\sigma_{\min}$, of σ_1, the Mohr circle on σ_3 and $\sigma_{\min}$ will touch AB at P, and failure first becomes possible in the planes of weakness, but only if their inclination is β_w given by (3). If σ_1 is increased to the point U, failure in the planes of weakness becomes possible for values of 2β in the arc ST, but failure cannot occur if 2β lies in the arcs RS or TU. If σ_1 is increased further, it will reach a value σ_{max} for which the Mohr circle touches CD, and failure through the body of the material becomes possible. By § 4.6 (8) the condition for this is

$$\sigma_{max} = \sigma_3 + 2(S_0 + \mu_0\sigma_3)[(\mu_0^2 + 1)^{\frac{1}{2}} + \mu_0]. \tag{4}$$

(2) and (4) give the greatest and least stresses which the materials can withstand; for $\sigma_1 < \sigma_{max}$ failure is only possible for a restricted range of inclinations of the planes of weakness. Fig. 4.10 (*b*) shows the variation of σ_1 with β calculated from (1) and (4) for the case $\mu_w = 0{\cdot}5$, $\mu_0 = 0{\cdot}7$, $S_0 = 2S_w$, for various values of σ_3/S_w. This may be compared with Fig. 3.6.2, which only takes into account failure in the planes of weakness.

Experimental results, Donath (1961, 1964), Hoek (1964b), Jaeger (1965), all show general behaviour similar to that of Fig. 4.10 (*b*).

Alternative discussions of the effect of anisotropy have been given by Walsh and Brace (1964) and Hoek (1964b) based on Griffith crack theory. They assume two sets of cracks, the most important set lying in the planes of weakness and the other set being randomly orientated. The final results are essentially the same as those set out above.

Jaeger (1960b) generalized the Coulomb theory of § 4.6 to the case in which the inherent shear strength of the rock was not constant but varied with β according to the formula

$$S_1 - S_2 \cos 2(\beta - \beta_0)$$

so that it has a minimum of $S_1 - S_2$ when $\beta = \beta_0$ and a maximum of $S_1 + S_2$ in the perpendicular direction. The value of μ was assumed to be independent of direction. With this model it was found that there was a single plane of fracture which lay between the plane of minimum shear strength and the nearer to it of the two conjugate directions possible on the simple Coulomb theory. Casagrande and Carrillo (1953) have studied graphically the case in which both S_0 and μ vary with direction.

Chapter Five

Linear Elasticity

5.1 Introduction

In this chapter the one-dimensional definition of linear elasticity in which uniaxial strain is proportional to uniaxial stress will be generalized to three dimensions by the assumption that each component of strain is a linear function of the components of stress. This is done in § 5.2 for isotropic bodies and in § 5.12 for some cases of anisotropic solids. Some important special cases are studied in § 5.3.

The next step is to set up the equations of equilibrium in terms of stresses. These are partial differential equations which the stresses must satisfy as a matter of pure statics. In three dimensions they are three equations connecting the six components of stress, and they hold for all media, whether elastic or not. They are derived in §§ 5.5, 5.6.

The final equations necessary for solving problems on the behaviour of a medium with specified mechanical properties are obtained by combining the stress-equations of equilibrium with equations specifying the mechanical behaviour of the solid. In the case of the theory of linear elasticity these latter are the stress–strain relations and the compatibility conditions for strain. The solution of most problems in elasticity is deferred until Chapter 10, but relatively simple problems which arise in connection with laboratory testing of materials are discussed in §§ 5.9–5.11.

While the theory strictly applies only to the case of a material with a stress–strain curve which is linear and completely reversible under all conditions, it is essentially set up simply on the basis of linearity, and many materials which are not linearly elastic in the large do behave linearly for moderate departures from some condition which may be regarded as standard (for example, the unloading modulus from a given stress). The results are therefore probably fairly widely applicable, with the appropriate constants and reservations.

5.2 The stress–strain relations for an isotropic linearly elastic solid

In an isotropic solid the principal axes of stress and the principal axes of strain, which were defined quite independently in §§ 2.4, 2.11, must coincide. This may be seen as follows: along a principal axis of stress the stresses are purely normal and the stresses vary symmetrically with departure from this direction. These symmetrical stresses must produce a symmetrical system of strains: now symmetrical strains exist only about the principal axes of strain, so these must coincide with the principal axes of stress. An alternative proof of this result may be given by the methods used for anisotropic solids in § 5.12.

The general assumption of the theory of linear elasticity is that the components of stress are linear functions of the components of strain. For principal stresses and strains this may be expressed by the relation

$$\sigma_1 = (\lambda + 2G)\varepsilon_1 + \lambda\varepsilon_2 + \lambda\varepsilon_3, \tag{1}$$

$$\sigma_2 = \lambda\varepsilon_1 + (\lambda + 2G)\varepsilon_2 + \lambda\varepsilon_3, \tag{2}$$

$$\sigma_3 = \lambda\varepsilon_1 + \lambda\varepsilon_2 + (\lambda + 2G)\varepsilon_3, \tag{3}$$

where λ and G are constants known as Lamé's parameters. This form is chosen to express the fact that one constant, $\lambda + 2G$, connects stress and strain in the same direction, while a different one, λ, relates stress and strain in the two perpendicular directions which must be on the same footing.

Introducing the volumetric strain, § 2.11 (23),

$$\Delta = \varepsilon_1 + \varepsilon_2 + \varepsilon_3, \tag{4}$$

(1) to (3) take the form

$$\sigma_1 = \lambda\Delta + 2G\varepsilon_1, \quad \sigma_2 = \lambda\Delta + 2G\varepsilon_2, \quad \sigma_3 = \lambda\Delta + 2G\varepsilon_3. \tag{5}$$

While λ and G are frequently used in theoretical work, other important constants are also used.

Young's modulus E is defined as the ratio of stress to strain for uniaxial stress $\sigma_2 = \sigma_3 = 0$. In this case (1)–(3) become

$$\sigma_1 = (\lambda + 2G)\varepsilon_1 + \lambda\varepsilon_2 + \lambda\varepsilon_3, \tag{6}$$

$$0 = \lambda\varepsilon_1 + (\lambda + 2G)\varepsilon_2 + \lambda\varepsilon_3, \tag{7}$$

$$0 = \lambda\varepsilon_1 + \lambda\varepsilon_2 + (\lambda + 2G)\varepsilon_3. \tag{8}$$

It follows that

$$\varepsilon_2 = \varepsilon_3 = -\frac{\lambda}{2(\lambda + G)}\varepsilon_1, \tag{9}$$

and

$$E = \frac{\sigma_1}{\varepsilon_1} = \frac{G(3\lambda + 2G)}{\lambda + G}. \tag{10}$$

In this case if σ_1 is positive (compression), ε_1 is positive (contraction), and ε_2 and ε_3 are negative (expansion). The ratio $(-\varepsilon_2/\varepsilon_1)$ of lateral expansion to longitudinal contraction is Poisson's ratio ν, and is by (9)

$$\nu = \frac{\lambda}{2(\lambda + G)}. \tag{11}$$

The bulk modulus or incompressibility K is defined as the ratio of hydrostatic pressure p to the volumetric strain Δ which it produces. If $\sigma_1 = \sigma_2 = \sigma_3 = p$, adding the three equations (5) gives

$$3p = (3\lambda + 2G)\Delta,$$

so

$$K = \frac{p}{\Delta} = \lambda + \tfrac{2}{3}G. \tag{12}$$

The reciprocal of K, $\beta = 1/K$, is called the compressibility.

There are many relations between the quantities λ, G, E, ν, K, of which the following are the most important:

$$\lambda = \frac{E\nu}{(1+\nu)(1-2\nu)}, \quad G = \frac{E}{2(1+\nu)}, \tag{13}$$

$$K = \frac{2(1+\nu)G}{3(1-2\nu)} = \frac{E}{3(1-2\nu)}, \tag{14}$$

$$E = \frac{9KG}{3K+G}, \quad \nu = \frac{(3K-2G)}{2(3K+G)}, \quad \frac{\lambda}{G} = \frac{2\nu}{1-2\nu}. \tag{15}$$

The assumed stress–strain relations refer to principal axes, and general relations may be found as follows. Suppose that mutually perpendicular axes Ox, Oy, Oz have direction cosines l_1, m_1, n_1; l_2, m_2, n_2; l_3, m_3, n_3, respectively, relative to principal axes $0\sigma_1$, $0\sigma_2$, $0\sigma_3$.

Then by § 2.5 (8)–(13),

$$\sigma_x = l_1^2\sigma_1 + m_1^2\sigma_2 + n_1^2\sigma_3, \quad \tau_{xy} = l_1l_2\sigma_1 + m_1m_2\sigma_2 + n_1n_2\sigma_3, \tag{16}$$

with four similar equations for σ_y, etc. Also, as in § 2.11, there will be corresponding relations for the components of strain, namely

$$\varepsilon_x = l_1^2\varepsilon_1 + m_1^2\varepsilon_2 + n_1^2\varepsilon_3, \quad \Gamma_{xy} = l_1l_2\varepsilon_1 + m_1m_2\varepsilon_2 + n_1n_2\varepsilon_3. \tag{17}$$

Substituting (5) in (16) and using (17) gives

$$\begin{aligned}\sigma_x &= (l_1^2 + m_1^2 + n_1^2)\lambda\Delta + 2G(l_1^2\varepsilon_1 + m_1^2\varepsilon_2 + n_1^2\varepsilon_3)\\ &= \lambda\Delta + 2G\varepsilon_x,\end{aligned} \tag{18}$$

$$\begin{aligned}\tau_{xy} &= (l_1l_2 + m_1m_2 + n_1n_2)\lambda\Delta + 2G(l_1l_2\varepsilon_1 + m_1m_2\varepsilon_2 + n_1n_2\varepsilon_3)\\ &= 2G\Gamma_{xy} = G\gamma_{xy}.\end{aligned} \tag{19}$$

Similar equations hold for the other components, and so the general stress–strain relations for any axes are

$$\sigma_x = \lambda\Delta + 2G\varepsilon_x, \quad \sigma_y = \lambda\Delta + 2G\varepsilon_y, \quad \sigma_z = \lambda\Delta + 2G\varepsilon_z, \tag{20}$$

$$\begin{gathered}\tau_{yz} = 2G\Gamma_{yz} = G\gamma_{yz}, \quad \tau_{zx} = 2G\Gamma_{zx} = G\gamma_{zx},\\ \tau_{xy} = 2G\Gamma_{xy} = G\gamma_{xy}.\end{gathered} \tag{21}$$

Adding the three equations (20) gives the important relation

$$\sigma_x + \sigma_y + \sigma_z = (3\lambda + 2G)\Delta. \tag{22}$$

For a simple shear in which τ_{xy} is the only non-vanishing component of stress, (21) gives

$$\tau_{xy} = G\gamma_{xy} = G\psi, \tag{23}$$

where ψ by § 2.9 (14) is the small angle of shear. For this reason, G is called the *modulus of rigidity*.

The equations (20) may be solved for strains in terms of stresses, and the result is found to be, using (10) and (11)

$$E\varepsilon_x = \sigma_x - \nu(\sigma_y + \sigma_z), \quad E\varepsilon_y = \sigma_y - \nu(\sigma_x + \sigma_z),$$
$$E\varepsilon_z = \sigma_z - \nu(\sigma_x + \sigma_y), \tag{24}$$

$$E\gamma_{yz} = 2E\Gamma_{yz} = 2(1 + \nu)\tau_{yz}, \quad E\gamma_{zx} = 2E\Gamma_{zx} = 2(1 + \nu)\tau_{zx},$$
$$E\gamma_{xy} = 2E\Gamma_{xy} = 2(1 + \nu)\tau_{xy}. \tag{25}$$

In cylindrical coordinates, § 2.16 (11)–(13), the stress–strain relations will be

$$\sigma_r = \lambda\Delta + 2G\varepsilon_r, \quad \sigma_\theta = \lambda\Delta + 2G\varepsilon_\theta, \quad \sigma_z = \lambda\Delta + 2G\varepsilon_z, \tag{26}$$

$$\tau_{\theta z} = 2G\Gamma_{\theta z}, \quad \tau_{zr} = 2G\Gamma_{zr}, \quad \tau_{r\theta} = 2G\Gamma_{r\theta}. \tag{27}$$

In the further development of the subject all the quantities λ, G, E, ν, K will appear. Some formulae, such as (1)–(3), are expressed most naturally in terms of λ and G; others, such as the inverse formulae (24), involve E and ν. Further, different experimental situations lead to measurement of different quantities: uniaxial compression or tension gives E and ν, while torsion gives G. It follows that the connecting formulae, (13)–(15), are of great importance and will continually be referred to.

There are certain restrictions imposed on ν by these formulae. It follows from (15) that $\nu < \frac{1}{2}$, and from (13) that $\nu > -1$.

The theory is a two-constant theory, and for the complete specification of a material two of λ, G, E, ν, K are needed. However, certain simplifying assumptions are frequently made which reduce the number of parameters to 1. The most important of these is *Poisson's relation*, $\lambda = G$, and if this is assumed it follows, from (13)–(15), that

$$\lambda = G, \quad K = 5G/3, \quad E = 5G/2, \quad \nu = \tfrac{1}{4}. \tag{28}$$

This assumption gives considerable simplification to theoretical results and has some experimental status, since Poisson's ratio is of the order of $\frac{1}{4}$ for many substances.

Another simplifying assumption sometimes made is that the material is incompressible, in this case

$$\lambda = K = \infty, \quad \nu = \tfrac{1}{2}, \quad G = E/3. \tag{29}$$

Finally, the limiting case of a compressible fluid is that in which the rigidity $G \rightarrow 0$, and so

$$G \rightarrow 0, \quad E \rightarrow 0, \quad \nu \rightarrow \tfrac{1}{2}, \quad \lambda \rightarrow K \rightarrow G/(1 - 2\nu). \tag{30}$$

The dimensions of all the elastic moduli, λ, G, E, K are those of stress, since strain, being a ratio, is dimensionless. The metric units most commonly used are dyne cm^{-2}, kg cm^{-2}, and the bar equal to 10^6 dyne cm^{-2}, which is approximately equal to 1 kg cm^{-2}. The usual English units are lb in^{-2} (psi) and kpsi equal to 1,000 psi. This latter unit is of convenient magnitude and will usually be employed here. To convert from psi to dyne cm^{-2} multiply by $6{\cdot}897 \times 10^4$ and to convert from psi to kg cm^{-2} multiply by 0·0703.

The determination of elastic moduli will be discussed subsequently, and some typical values are given in § 6.2 for reference. It will be apparent from this that there are large variations in ν and that two quantities must be measured to specify the behaviour of a rock adequately.

5.3 Special cases

There are a number of special cases of great importance in practice:

(i) *Uniaxial stress,* $\sigma_1 \neq 0$, $\sigma_2 = \sigma_3 = 0$

This is the case of a specimen uniformly loaded in one direction and free in the two others which has already been considered in § 5.2 (6)–(10). There is a contraction $\varepsilon_1 = \sigma_1/E$ in the direction of σ_1 and an expansion $\varepsilon_2 = \varepsilon_3 = -\nu\varepsilon_1$ in the perpendicular directions. The fractional change in volume is

$$\Delta = (1 - 2\nu)\sigma_1/E, \tag{1}$$

and so, since $\nu < \frac{1}{2}$, there is a decrease in volume if $\sigma_1 > 0$ and an increase if $\sigma_1 < 0$.

(ii) *Uniaxial strain,* $\varepsilon_1 \neq 0$, $\varepsilon_2 = \varepsilon_3 = 0$

Here from § 5.2 (1)–(3)

$$\sigma_1 = (\lambda + 2G)\varepsilon_1, \quad \sigma_2 = \sigma_3 = \lambda\varepsilon_1 = [\nu/(1 - \nu)]\sigma_1, \tag{2}$$

using § 5.2 (15).

The assumption is that there is no displacement perpendicular to the σ_1 – axis; stresses σ_2, σ_3 perpendicular to this axis are called into play to prevent displacement. This case arises in the simplest attempt to calculate stress below the earth's surface on the assumption that there are no lateral displacements.

(iii) *The case,* $\sigma_1 \neq 0$, $\varepsilon_2 = 0$, $\sigma_3 = 0$

that is, zero stress and zero strain in two directions perpendicular to the stress σ_1. By § 5.2 (24)

$$E\varepsilon_1 = (1 - \nu^2)\sigma_1, \quad \sigma_2 = \nu\sigma_1, \quad \varepsilon_3 = -[\nu/(1 - \nu)]\varepsilon_1. \tag{3}$$

(iv) *Biaxial stress or plane stress,* $\sigma_1 \neq 0$, $\sigma_2 \neq 0$, $\sigma_3 = 0$
By § 5.2 (24)

$$E\varepsilon_1 = \sigma_1 - \nu\sigma_2, \quad E\varepsilon_2 = \sigma_2 - \nu\sigma_1, \quad E\varepsilon_3 = -\nu(\sigma_1 + \sigma_2). \tag{4}$$

Also, it follows from § 5.2 (1) and (2) on eliminating ε_3 by using § 5.2 (3)

$$(\lambda + 2G)\sigma_1 = 4G(\lambda + G)\varepsilon_1 + 2\lambda G\varepsilon_2, \tag{5}$$

$$(\lambda + 2G)\sigma_2 = 2\lambda G\varepsilon_1 + 4G(\lambda + G)\varepsilon_2. \tag{6}$$

From (4), there is expansion in the σ_3 — direction if $\sigma_1 + \sigma_2 > 0$, and contraction if $\sigma_1 + \sigma_2 < 0$. For pure shear, $\sigma_1 + \sigma_2 = 0$, $\varepsilon_3 = 0$. The fractional change in volume is

$$\Delta = (\sigma_1 + \sigma_2)(1 - 2\nu)/E. \tag{7}$$

This case occurs when a thin plate is stressed in its own plane. It also occurs in the analysis of stress at any free surface, since, if the x- and y-axes are taken in the surface, the condition that there be no stress on the surface is

$$\sigma_z = 0, \quad \tau_{xz} = \tau_{yz} = 0. \tag{8}$$

(v) *Biaxial strain or plane strain,* $\varepsilon_1 \neq 0$, $\varepsilon_2 \neq 0$, $\varepsilon_3 = 0$
From § 5.2 (1)–(3)

$$\sigma_1 = (\lambda + 2G)\varepsilon_1 + \lambda\varepsilon_2, \quad \sigma_2 = (\lambda + 2G)\varepsilon_2 + \lambda\varepsilon_1, \tag{9}$$

$$\sigma_3 = \lambda(\varepsilon_1 + \varepsilon_2) = [\lambda/2(\lambda + G)](\sigma_1 + \sigma_2) = \nu(\sigma_1 + \sigma_2). \tag{10}$$

Solving (9), or using § 5.2 (24) and (10), gives

$$E\varepsilon_1 = (1 - \nu^2)\sigma_1 - \nu(1 + \nu)\sigma_2, \quad E\varepsilon_2 = (1 - \nu^2)\sigma_2 - \nu(1 + \nu)\sigma_1. \tag{11}$$

In this case a stress (10) must be provided perpendicular to the plane of σ_1 and σ_2 to give $\varepsilon_3 = 0$.

If the x- and y-axes are not principal axes, (10) and § 5.2 (24), (25) give

$$E\varepsilon_x = (1 - \nu^2)\sigma_x - \nu(1 + \nu)\sigma_y, \quad E\varepsilon_y = (1 - \nu^2)\sigma_y - \nu(1 + \nu)\sigma_x, \tag{12}$$

$$E\gamma_{xy} = 2E\Gamma_{xy} = 2(1 + \nu)\tau_{xy}. \tag{13}$$

The assumption of plane strain is frequently made in studying stresses around drill holes and two-dimensional openings.

(vi) *Combined formulae for plane stress and plane strain*
It may be verified that if in (9) for plane strain λ is replaced by $2\lambda G/(\lambda + 2G)$, these equations take the form (5) and (6) for plane stress. This suggests that the stress–strain relations might be set out in a form suitable for either case.

The equations may be combined in the form

$$8G\varepsilon_1 = (\varkappa + 1)\sigma_1 + (\varkappa - 3)\sigma_2, \tag{14}$$

$$8G\varepsilon_2 = (\varkappa - 3)\sigma_1 + (\varkappa + 1)\sigma_2, \tag{15}$$

where

$$\varkappa = 3 - 4\nu \text{ for plane strain,} \tag{16}$$

$$\varkappa = (3 - \nu)/(1 + \nu) \text{ for plane stress.} \tag{17}$$

This is easily verified by substituting the appropriate value of $\varkappa$ and using $E = 2G(1 + \nu)$.

If the axes in the σ_1, σ_2 plane are not principal axes, the appropriate equations are

$$8G\varepsilon_x = (\varkappa + 1)\sigma_x + (\varkappa - 3)\sigma_y, \tag{18}$$

$$8G\varepsilon_y = (\varkappa - 3)\sigma_x + (\varkappa + 1)\sigma_y, \tag{19}$$

$$\tau_{xy} = G\gamma_{xy} = 2G\Gamma_{xy}. \tag{20}$$

It is important to be able to convert solutions for plane stress to plane strain and vice-versa, since many solutions in the literature are written out for one or the other. This is readily done if the solutions are expressed in terms of G and ν. Then solutions for plane strain may be converted to the case of plane stress by replacing $3 - 4\nu$ by $(3 - \nu)/(1 + \nu)$, that is, by replacing ν by $\nu/(1 + \nu)$. Similarly, solutions for plane stress may be converted to plane strain by replacing $(3 - \nu)/(1 + \nu)$ by $3 - 4\nu$, i.e. by replacing ν by $\nu/(1 - \nu)$.

(vii) *The case of constant strain* ε *along the z-axis*

It is assumed that

$$\frac{\partial w}{\partial z} = \varepsilon, \text{ constant,}$$

that w is independent of x and y, and that u and v are independent of z. In these circumstances $\Gamma_{yz} = \Gamma_{xz} = 0$.

Equations § 5.2 (24) become

$$E\varepsilon_x = \sigma_x - \nu(\sigma_y + \sigma_z), \quad E\varepsilon_y = \sigma_y - \nu(\sigma_x + \sigma_z), \tag{21}$$

$$E\varepsilon = \sigma_z - \nu(\sigma_x + \sigma_y). \tag{22}$$

Substituting from (22) in (21) gives

$$E\varepsilon_x = (1 - \nu^2)\sigma_x - \nu(1 + \nu)\sigma_y - E\nu\varepsilon, \tag{23}$$

$$E\varepsilon_y = -\nu(1 + \nu)\sigma_x + (1 - \nu^2)\sigma_y - E\nu\varepsilon, \tag{24}$$

and § 5.2 (25) gives

$$E\gamma_{xy} = 2E\Gamma_{xy} = 2(1 + \nu)\tau_{xy}. \tag{25}$$

These provide a simple generalization of (12) and (13) for plane strain. They are frequently used in underground situations in which there is often considerable uncertainty about the nature of the stresses. For example, in considering the stresses and displacements around a horizontal tunnel, it is equivalent to assuming that there exists a constant stress in the direction of the tunnel and studying variations in the plane perpendicular to them. The more general case in which Γ_{yz} and Γ_{xz} do not vanish is considered in § 10.4.

5.4 Stress–strain relations in terms of stress and strain deviations

The stress–strain relations take an especially simple form when expressed in terms of the stress and strain deviations defined in §§ 2.8, 2.12.

Adding the three equations § 5.2 (20) and using § 2.8 (1) and § 2.12 (1) gives

$$3s = 3\lambda\Delta + 6Ge = (9\lambda + 6G)e,$$

so, using § 5.2 (12)

$$s = (3\lambda + 2G)e = 3Ke. \tag{1}$$

Subtracting (1) from § 5.2 (20) in turn gives

$$s_x = 2Ge_x, \quad s_y = 2Ge_y, \quad s_z = 2Ge_z \tag{2}$$

where s_x and e_x are defined in § 2.8 (2) and § 2.12 (2). Also § 5.2 (21) in the notation § 2.8 (3) and § 2.12 (2) are

$$s_{yz} = 2Ge_{yz}, \quad s_{zx} = 2Ge_{zx}, \quad s_{xy} = 2Ge_{xy}. \tag{3}$$

In the tensor notation in which the x-, y-, and z-directions are denoted by suffixes 1, 2, 3, s_x is s_{11}, s_{yz} is s_{23}, and so on, and the equations (2) and (3) take the compact form

$$s_{rs} = 2Ge_{rs}. \tag{4}$$

s_{rs} and e_{rs} as defined here are in fact components of second order tensors. If e_{yz} is defined as γ_{yz} in place of $\Gamma_{yz} = \frac{1}{2}\gamma_{yz}$, this is not the case.

5.5 The equations of equilibrium

The general problem of elasticity is that of determining the stresses and displacements in a body of known shape subject to prescribed conditions at its surface. Usually, either the displacements at the surface or the stresses applied to it are prescribed. In addition to these there may be *body forces*, such as gravity, which act throughout the interior of the body. We shall write X, Y, Z for the components of the body forces per unit mass at the point x, y, z. The equations of equilibrium are partial differential equations which describe the way in which stresses or displacements vary in the interior of the body, and they have to be solved subject to the specified conditions at the surface.

To study the variation of the stresses with position, we consider a small rectangular parallelepiped with centre at x, y, z, faces parallel to the axes $Oxyz$ and small side lengths $AA' = \delta x$, $AB = \delta y$, $AD = \delta z$, Fig. 5.5. The body forces will be assumed to have components X, Y, Z per unit mass, the positive directions of these being, in accordance with the present convention, in the negative x-, y-, and z-directions.

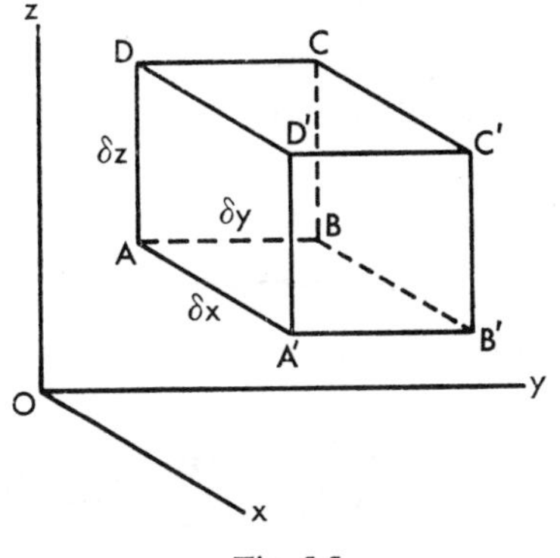

Fig. 5.5

Considering forces on the parallelepiped in the negative x-direction, those on the faces $ABCD$ and $A'B'C'D'$, respectively, will be

$$-\left\{\sigma_x - \tfrac{1}{2}\frac{\partial \sigma_x}{\partial x}\delta x\right\}\delta y\,\delta x \quad \text{and} \quad \left\{\sigma_x + \tfrac{1}{2}\frac{\partial \sigma_x}{\partial x}\delta x\right\}\delta y\,\delta z. \tag{1}$$

Those on the faces $AA'D'D$ and $BB'C'C$ will be

$$-\left\{\tau_{yx} - \tfrac{1}{2}\frac{\partial \tau_{yx}}{\partial y}\delta y\right\}\delta x\,\delta z \quad \text{and} \quad \left\{\tau_{yx} + \tfrac{1}{2}\frac{\partial \tau_{yx}}{\partial y}\delta y\right\}\delta x\,\delta z. \tag{2}$$

Those on the faces $AA'B'B$ and $DD'C'C$ will be

$$-\left\{\tau_{zx} - \tfrac{1}{2}\frac{\partial \tau_{zx}}{\partial z}\delta z\right\}\delta x\,\delta y \quad \text{and} \quad \left\{\tau_{zx} + \tfrac{1}{2}\frac{\partial \tau_{zx}}{\partial z}\delta z\right\}\delta x\,\delta y. \tag{3}$$

The component of body force in this direction is

$$\rho X\,\delta x\,\delta y\,\delta z, \tag{4}$$

where ρ is the density of the body.

Adding (1) to (4), and cancelling the factor $\delta x \delta y \delta z$, gives the condition for equilibrium

$$\frac{\partial \sigma_x}{\partial x} + \frac{\partial \tau_{yx}}{\partial y} + \frac{\partial \tau_{zx}}{\partial z} + \rho X = 0, \tag{5}$$

and similarly for the y- and z-directions,

$$\frac{\partial \tau_{xy}}{\partial x} + \frac{\partial \sigma_y}{\partial y} + \frac{\partial \tau_{zy}}{\partial z} + \rho Y = 0, \tag{6}$$

$$\frac{\partial \tau_{xz}}{\partial x} + \frac{\partial \tau_{yz}}{\partial y} + \frac{\partial \sigma_z}{\partial z} + \rho Z = 0. \tag{7}$$

(5)–(7) are the *equations of equilibrium in terms of stresses.* They may be expressed in terms of strains, using the stress–strain relations, and then in terms of displacements using the definitions of the components of strain. Thus using § 5.2 (18), (19), (5) becomes

$$\frac{\partial}{\partial x}(\lambda\Delta + 2G\varepsilon_x) + \frac{\partial}{\partial y}(2G\Gamma_{yx}) + \frac{\partial}{\partial z}(2G\Gamma_{zx}) + \rho X = 0, \tag{8}$$

where
$$\Delta = \varepsilon_x + \varepsilon_y + \varepsilon_z. \tag{9}$$

Using § 2.11 (5) to (8) and assuming that λ and G are constants, (8) becomes

$$\lambda\frac{\partial \Delta}{\partial x} + G\left\{2\frac{\partial^2 u}{\partial x^2} + \frac{\partial}{\partial y}\left(\frac{\partial v}{\partial x} + \frac{\partial u}{\partial y}\right) + \frac{\partial}{\partial z}\left(\frac{\partial u}{\partial z} + \frac{\partial w}{\partial x}\right)\right\} + \rho X = 0,$$

or
$$\lambda\frac{\partial \Delta}{\partial x} + G\left\{\frac{\partial^2 u}{\partial x^2} + \frac{\partial^2 v}{\partial x\,\partial y} + \frac{\partial^2 w}{\partial x\,\partial z} + \frac{\partial^2 u}{\partial x^2} + \frac{\partial^2 u}{\partial y^2} + \frac{\partial^2 u}{\partial z^2}\right\} + \rho X = 0,$$

or
$$(\lambda + G)\frac{\partial \Delta}{\partial x} + G\nabla^2 u + \rho X = 0, \tag{10}$$

where
$$\nabla^2 u = \frac{\partial^2 u}{\partial x^2} + \frac{\partial^2 u}{\partial y^2} + \frac{\partial^2 u}{\partial z^2}. \tag{11}$$

In the same way from (6) and (7),

$$(\lambda + G)\frac{\partial \Delta}{\partial y} + G\nabla^2 v + \rho Y = 0, \tag{12}$$

$$(\lambda + G)\frac{\partial \Delta}{\partial z} + G\nabla^2 w + \rho Z = 0. \tag{13}$$

(10)–(13) are the equations of equilibrium in terms of displacements. The relative status and usefulness of the two sets will be discussed in § 5.7.

Some general results follow immediately from (10) to (13). Considering, for simplicity, only the case of no body forces $X = Y = Z = 0$, differentiating (10), (12), and (13) partially with respect to x, y, z, respectively, and adding gives

$$(\lambda + G)\nabla^2\Delta + G\nabla^2\left\{\frac{\partial u}{\partial x} + \frac{\partial v}{\partial y} + \frac{\partial w}{\partial z}\right\} = 0,$$

so that, using (9),

$$\nabla^2\Delta = 0, \tag{14}$$

or the volumetric strain Δ satisfies Laplace's equation.

Using § 5.2 (22), namely

$$(3\lambda + 2G)\Delta = \sigma_x + \sigma_y + \sigma_z = 3s, \tag{15}$$

where s is the mean normal stress defined in § 2.8 (1), it follows that

$$\nabla^2(\sigma_x + \sigma_y + \sigma_z) = 3\nabla^2 s = 0, \tag{16}$$

so that the mean normal stress also satisfies Laplace's equation. Differentiating (10), again with $X = 0$, with respect to x gives

$$(\lambda + G)\frac{\partial^2 \Delta}{\partial x^2} + G\nabla^2 \epsilon_x = 0,$$

and using the result $\sigma_x = \lambda\Delta + 2G\epsilon_x$ and (14) and (15) in this gives

$$\nabla^2 \sigma_x = -2(\lambda + G)\frac{\partial^2 \Delta}{\partial x^2} = -\frac{6(\lambda + G)}{(3\lambda + 2G)}\frac{\partial^2 s}{\partial x^2}, \tag{17}$$

with two similar equations for y and z. In the same way from (12) and (13)

$$\nabla^2 \tau_{yz} + \frac{6(\lambda + G)}{(3\lambda + 2G)}\frac{\partial^2 s}{\partial y \partial z} = 0, \tag{18}$$

with two similar equations for τ_{zx} and τ_{zy}.

For the two-dimensional case in which u and v are independent of z and there is at most a constant strain in the z-direction, so that $\Gamma_{yz} = \Gamma_{xz} = 0$, (5) and (6) become

$$\frac{\partial \sigma_x}{\partial x} + \frac{\partial \tau_{yx}}{\partial y} + \rho X = 0, \tag{19}$$

$$\frac{\partial \tau_{xy}}{\partial x} + \frac{\partial \sigma_y}{\partial y} + \rho Y = 0. \tag{20}$$

The equations of motion are derived in precisely the same way by equating the resultant force on the parallelepiped in any direction to its mass times the component of acceleration in that direction. Provided the velocities are not large, the components of acceleration are

$$\frac{\partial^2 u}{\partial t^2}, \quad \frac{\partial^2 v}{\partial t^2}, \quad \frac{\partial^2 w}{\partial t^2},$$

so that (5)–(7) give for the equations of motion

$$\rho\frac{\partial^2 u}{\partial t^2} = \frac{\partial \sigma_x}{\partial x} + \frac{\partial \tau_{yx}}{\partial y} + \frac{\partial \tau_{zx}}{\partial z} + \rho X, \tag{21}$$

$$\rho\frac{\partial^2 v}{\partial t^2} = \frac{\partial \tau_{xy}}{\partial x} + \frac{\partial \sigma_y}{\partial y} + \frac{\partial \tau_{zy}}{\partial z} + \rho Y, \tag{22}$$

$$\rho\frac{\partial^2 w}{\partial t^2} = \frac{\partial \tau_{xz}}{\partial x} + \frac{\partial \tau_{yz}}{\partial y} + \frac{\partial \sigma_z}{\partial z} + \rho Z. \tag{23}$$

Similarly, as in (10)–(12), the equations of motion in terms of displacements are

$$\rho \frac{\partial^2 u}{\partial t^2} = (\lambda + G)\frac{\partial \Delta}{\partial x} + G\nabla^2 u + \rho X, \tag{24}$$

$$\rho \frac{\partial^2 v}{\partial t^2} = (\lambda + G)\frac{\partial \Delta}{\partial y} + G\nabla^2 v + \rho Y, \tag{25}$$

$$\rho \frac{\partial^2 w}{\partial t^2} = (\lambda + G)\frac{\partial \Delta}{\partial z} + G\nabla^2 w + \rho Z. \tag{26}$$

5.6 The equations of equilibrium in cylindrical coordinates

The equations of equilibrium in cylindrical coordinates may be written down as in § 5.5. Fig. 5.5 is replaced by Fig. 5.6, consisting of portion of a wedge bounded by radial planes $AA'D'D$ and $BB'C'C$ making angles

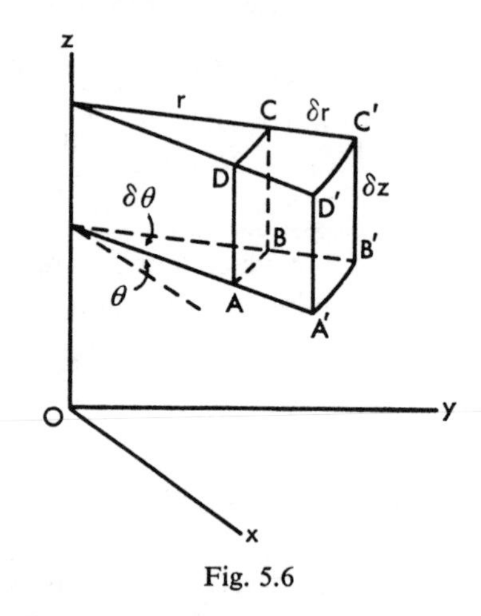

Fig. 5.6

$\theta \mp \frac{1}{2}\delta\theta$ with Ox, by planes $AA'B'B$ and $DD'C'C$ at $z \mp \frac{1}{2}\delta z$, and by portions of cylinders $ABCD$ and $A'B'C'D'$ of radii $r \mp \frac{1}{2}\delta r$. Suppose that the components of body free per unit mass in cylindrical coordinates are R, Θ, Z.

Considering the radial direction, the forces applied to the element of the solid in the direction radially inwards over the faces $ABCD$ and $A'B'C'D'$ are, respectively,

$$-\left\{\sigma_r - \tfrac{1}{2}\frac{\partial \sigma_r}{\partial r}\delta r\right\}(r - \tfrac{1}{2}\delta r)\,\delta\theta\,\delta z \quad \text{and} \quad \left\{\sigma_r + \tfrac{1}{2}\frac{\partial \sigma_r}{\partial r}\delta r\right\}(r + \tfrac{1}{2}\delta r)\,\delta\theta\,\delta z. \tag{1}$$

The forces over the faces $AA'B'B$ and $DD'C'C$ are

$$-\left\{\tau_{zr} - \tfrac{1}{2}\frac{\partial \tau_{zr}}{\partial z}\delta z\right\}r\,\delta\theta\,\delta r \quad \text{and} \quad \left\{\tau_{zr} + \tfrac{1}{2}\frac{\partial \tau_{zr}}{\partial z}\delta z\right\}r\,\delta\theta\,\delta r. \tag{2}$$

The situation on the planes $AA'D'D$ and $BB'C'C$ is more complicated, since both the shear stresses $\tau_{\theta r}$ and the normal stresses σ_θ give components in the r-direction. Neglecting squares of $\delta\theta$, the shear stresses give

$$-\left\{\tau_{\theta r} - \tfrac{1}{2}\frac{\partial \tau_{\theta r}}{\partial \theta}\,\delta\theta\right\}\delta r\,\delta z \quad \text{and} \quad \left(\tau_{\theta r} + \tfrac{1}{2}\frac{\partial \tau_{\theta r}}{\partial \theta}\,\delta\theta\right)\delta r\,\delta z, \tag{3}$$

while the normal stresses give

$$-\left\{\sigma_\theta - \tfrac{1}{2}\frac{\partial \sigma_\theta}{\partial \theta}\,\delta\theta\right\}\delta r\,\delta z\,[\tfrac{1}{2}\,\delta\theta] \quad \text{and} \quad -\left\{\sigma_\theta + \tfrac{1}{2}\frac{\partial \sigma_\theta}{\partial \theta}\,\delta\theta\right\}\delta r\,\delta z[\tfrac{1}{2}\,\delta\theta]. \tag{4}$$

The component of body force is

$$\rho R\,r\,\delta r\,\delta\theta\,\delta z. \tag{5}$$

Adding (1) to (5) and dividing by $r\,\delta r\,\delta\theta\,\delta z$ gives

$$\frac{\partial \sigma_r}{\partial r} + \frac{1}{r}\frac{\partial \tau_{\theta r}}{\partial \theta} + \frac{\partial \tau_{zr}}{\partial z} + \frac{\sigma_r - \sigma_\theta}{r} + \rho R = 0, \tag{6}$$

and similarly

$$\frac{\partial \tau_{r\theta}}{\partial r} + \frac{1}{r}\frac{\partial \sigma_\theta}{\partial \theta} + \frac{\partial \tau_{z\theta}}{\partial z} + \frac{2\tau_{r\theta}}{r} + \rho\Theta = 0, \tag{7}$$

$$\frac{\partial \tau_{rz}}{\partial r} + \frac{1}{r}\frac{\partial \tau_{\theta z}}{\partial \theta} + \frac{\partial \sigma_z}{\partial z} + \frac{\tau_{rz}}{r} + \rho Z = 0. \tag{8}$$

In two dimensions, if $\Gamma_{rz} = \Gamma_{\theta z} = 0$, (6) and (7) give

$$\frac{\partial \sigma_r}{\partial r} + \frac{1}{r}\frac{\partial \tau_{\theta r}}{\partial \theta} + \frac{\sigma_r - \sigma_\theta}{r} + \rho R = 0, \tag{9}$$

$$\frac{\partial \tau_{r\theta}}{\partial r} + \frac{1}{r}\frac{\partial \sigma_\theta}{\partial \theta} + \frac{2\tau_{r\theta}}{r} + \rho\Theta = 0. \tag{10}$$

Introducing the stress–strain relations § 5.2 (26), (27) and the strain components and volumetric strain Δ defined in § 2.15 (12)–(14), the equations of equilibrium (6)–(8) become when expressed in terms of displacements.

$$(\lambda + G)\frac{\partial \Delta}{\partial r} + G\left\{\frac{\partial^2 u}{\partial r^2} + \frac{1}{r}\frac{\partial u}{\partial r} - \frac{u}{r^2} + \frac{1}{r^2}\frac{\partial^2 u}{\partial \theta^2} - \frac{2}{r^2}\frac{\partial v}{\partial \theta} + \frac{\partial^2 u}{\partial z^2}\right\} + \rho R = 0, \tag{11}$$

$$(\lambda + G)\frac{\partial \Delta}{r\,\partial \theta} + G\left\{\frac{\partial^2 v}{\partial r^2} + \frac{1}{r}\frac{\partial v}{\partial r} - \frac{v}{r^2} + \frac{1}{r^2}\frac{\partial^2 v}{\partial \theta^2} + \frac{2}{r^2}\frac{\partial u}{\partial \theta} + \frac{\partial^2 v}{\partial z^2}\right\} + \rho\Theta = 0, \tag{12}$$

$$(\lambda + G)\frac{\partial \Delta}{\partial z} + G\left\{\frac{\partial^2 w}{\partial r^2} + \frac{1}{r}\frac{\partial w}{\partial r} + \frac{1}{r^2}\frac{\partial^2 w}{\partial \theta^2} + \frac{\partial^2 w}{\partial z^2}\right\} + \rho Z = 0. \tag{13}$$

5.7 Compatibility conditions for stress and the Airy stress function

The stress-equations of equilibrium and boundary conditions are not sufficient to determine a solution, and two alternative methods of solution are possible, or various combinations of them.

First, the equations of equilibrium in terms of displacements § 5.5 (10)–(13) may be regarded as fundamental and solved; strains then being found by differentiation of u, v, w and stresses from the stress–strain relations. This is the most direct, but not always the most convenient, method.

Alternatively, stresses and strains may be regarded as fundamental. In this case, as remarked in §§ 2.10, 2.11, the components of strain are not independent, and, since they are connected with the components of stress by the stress–strain relations, the components of stress cannot also all be independent.

In two dimensions the three components of strain are derived by differentiation from two components of strain and are connected by the compatibility condition § 2.10 (29), namely,

$$2\frac{\partial^2 \Gamma_{xy}}{\partial x\,\partial y} = \frac{\partial^2 \varepsilon_x}{\partial y^2} + \frac{\partial^2 \varepsilon_y}{\partial x^2}. \tag{1}$$

Substituting the two-dimensional stress–strain relations 5.3 (18)–(20) in (1) gives the compatibility condition for stress

$$(\varkappa + 1)\left[\frac{\partial^2 \sigma_x}{\partial y^2} + \frac{\partial^2 \sigma_y}{\partial x^2}\right] + (\varkappa - 3)\left[\frac{\partial^2 \sigma_x}{\partial x^2} + \frac{\partial^2 \sigma_y}{\partial y^2}\right] = 8\frac{\partial^2 \tau_{xy}}{\partial x\,\partial y}, \tag{2}$$

where $\varkappa = 3 - 4\nu$ for plane strain and $\varkappa = (3 - \nu)/(1 + \nu)$ for plane stress. Here the stress-components must satisfy the equations of equilibrium § 5.5 (19), (20), and differentiating these gives

$$\frac{\partial^2 \tau_{xy}}{\partial x\,\partial y} = -\frac{\partial^2 \sigma_x}{\partial x^2} - \rho\frac{\partial X}{\partial x} = -\frac{\partial^2 \sigma_y}{\partial y^2} - \rho\frac{\partial Y}{\partial y}. \tag{3}$$

Substituting (3) in (2) gives

$$(\varkappa + 1)\left(\frac{\partial^2}{\partial x^2} + \frac{\partial^2}{\partial y^2}\right)(\sigma_x + \sigma_y) + 4\rho\left(\frac{\partial X}{\partial x} + \frac{\partial Y}{\partial y}\right) = 0. \tag{4}$$

If the forces X, Y are derived from a potential V, so that $\nabla^2 V = 0$ and

$$X = -\frac{\partial V}{\partial x}, \quad Y = -\frac{\partial V}{\partial y}, \tag{5}$$

(4) becomes

$$\nabla^2(\sigma_x + \sigma_y) = 0. \tag{6}$$

Thus stresses in two dimensions have not merely to satisfy the stress-equations of equilibrium but also (2) or (6).

Now suppose that there is a function U such that

$$\sigma_x = \frac{\partial^2 U}{\partial y^2} + \rho V, \quad \sigma_y = \frac{\partial^2 U}{\partial x^2} + \rho V, \quad \tau_{xy} = -\frac{\partial^2 U}{\partial x \partial y}, \tag{7}$$

so that it follows on substitution that the stress-equations of equilibrium § 5.5 (19), (20) are satisfied automatically. Then substituting in (6) requires

$$\nabla^2(\nabla^2 U) = 0, \tag{8}$$

or

$$\frac{\partial^4 U}{\partial x^4} + 2\frac{\partial^4 U}{\partial x^2 \partial y^2} + \frac{\partial^4 U}{\partial y^4} = 0. \tag{9}$$

(8) is the so-called *biharmonic equation*, and we have shown that any solution of it will automatically satisfy the stress-equations of equilibrium and the compatibility conditions. Its solution and applications to special problems will be discussed in Chapter 10. The function U defined in (7) is called *Airy's stress function*.

A similar discussion can be given in plane polar coordinates. Here, taking R and Θ zero for simplicity,

$$\sigma_r = \frac{1}{r}\frac{\partial U}{\partial r} + \frac{1}{r^2}\frac{\partial^2 U}{\partial \theta^2}, \quad \sigma_\theta = \frac{\partial^2 U}{\partial r^2}, \quad \tau_{r\theta} = -\frac{\partial}{\partial r}\left(\frac{1}{r}\frac{\partial U}{\partial \theta}\right) \tag{10}$$

automatically satisfy the stress equations § 5.6 (9), (10). To satisfy the compatibility condition for strains, U must satisfy

$$\left(\frac{\partial^2}{\partial r^2} + \frac{1}{r}\frac{\partial}{\partial r} + \frac{1}{r^2}\frac{\partial^2}{\partial \theta^2}\right)\left(\frac{\partial^2 U}{\partial r^2} + \frac{1}{r}\frac{\partial U}{\partial r} + \frac{1}{r^2}\frac{\partial^2 U}{\partial \theta^2}\right) = 0, \tag{11}$$

which is the biharmonic equation in polar coordinates.

The situation is similar in three dimensions. There are six compatibility conditions for strain, § 2.11 (29)–(32), and compatibility conditions for stress may be obtained as above by substituting the stress–strain relations in them. The results so obtained have been derived by another method in § 5.5 (17), (18).

5.8 Strain energy

The strain energy or the potential energy stored in a strained body is of fundamental importance, particularly in rock mechanics.

To determine it, consider a small cube of material of side a with faces perpendicular to the principal stresses σ_1, σ_2, σ_3. The principal strains associated with these will be ε_1, ε_2, ε_3, and we may suppose this state of stress to be built up by a gradual increase from zero, so that at any time the principal stresses are $k\sigma_1$, $k\sigma_2$, $k\sigma_3$, where k increases from 0 to 1. As k increases from k to $k + \delta k$ the displacement of the face perpendicular to σ_1 is $a\varepsilon_1\delta k$, and since the force on this face is $a^2\sigma_1 k$, the work done by the force is $\sigma_1\varepsilon_1 a^3 k\delta k$, so the total work done by the forces in the σ_1-direction in establishing the final state of strain is

$$\sigma_1\varepsilon_1 a^3 \int_0^1 k\,dk = \tfrac{1}{2}\sigma_1\varepsilon_1 a^3. \tag{1}$$

There will be similar contributions from the other two directions, so that, dividing by the volume a^3, the *potential energy per unit volume* W is given by

$$W = \tfrac{1}{2}(\sigma_1\varepsilon_1 + \sigma_2\varepsilon_2 + \sigma_3\varepsilon_3). \tag{2}$$

This may be put in many useful forms. Using § 5.2 (5) it becomes

$$W = \tfrac{1}{2}\lambda\Delta^2 + G(\varepsilon_1{}^2 + \varepsilon_2{}^2 + \varepsilon_3{}^2) \tag{3}$$

$$= \tfrac{1}{2}\{(\lambda + 2G)(\varepsilon_1{}^2 + \varepsilon_2{}^2 + \varepsilon_3{}^2) + 2\lambda(\varepsilon_1\varepsilon_2 + \varepsilon_2\varepsilon_3 + \varepsilon_3\varepsilon_1)\}. \tag{4}$$

Using § 5.2 (24) for principal axes, (2) becomes

$$W = \frac{1}{2E}\{\sigma_1{}^2 + \sigma_2{}^2 + \sigma_3{}^2 - 2\nu(\sigma_2\sigma_3 + \sigma_3\sigma_1 + \sigma_1\sigma_2)\}, \tag{5}$$

and using the formulae § 2.4 (44), (46) derived from the invariants of stress this becomes

$$W = \frac{1}{2E}\{\sigma_x{}^2 + \sigma_y{}^2 + \sigma_z{}^2 - 2\nu(\sigma_y\sigma_z + \sigma_z\sigma_x + \sigma_x\sigma_y) + 2(1 + \nu)(\tau_{yz}{}^2 + \tau_{zx}{}^2 + \tau_{xy}{}^2)\} \tag{6}$$

$$= \tfrac{1}{2}(\sigma_x\varepsilon_x + \sigma_y\varepsilon_y + \sigma_z\varepsilon_z + 2\tau_{yz}\Gamma_{yz} + 2\tau_{zx}\Gamma_{zx} + 2\tau_{xy}\Gamma_{xy}), \tag{7}$$

on using § 5.2 (24), (25).

Using § 5.2 (20), (21), equation (7) becomes

$$W = \tfrac{1}{2}\{\lambda\Delta^2 + 2G(\varepsilon_x{}^2 + \varepsilon_y{}^2 + \varepsilon_z{}^2) + 4G(\Gamma_{yz}{}^2 + \Gamma_{zx}{}^2 + \Gamma_{xy}{}^2)\}. \tag{8}$$

If W is regarded as a function of $\sigma_x, \sigma_y, \sigma_z, \tau_{yz}, \tau_{zx}, \tau_{xy}$, it follows from (6) and § 5.2 (24), (25) that

$$\frac{\partial W}{\partial \sigma_x} = \frac{1}{E}[\sigma_x - \nu(\sigma_y + \sigma_z)] = \varepsilon_x, \tag{9}$$

$$\frac{\partial W}{\partial \tau_{yz}} = \frac{2(1 + \nu)}{E}\tau_{yz} = 2\Gamma_{yz}. \tag{10}$$

Similarly from (8) and § 5.2 (18), (19),

$$\frac{\partial W}{\partial \varepsilon_x} = \sigma_x, \quad \frac{\partial W}{\partial \Gamma_{yz}} = 2\tau_{yz}. \tag{11}$$

A number of relations of type

$$\frac{\partial \sigma_x}{\partial \varepsilon_y} = \frac{\partial^2 W}{\partial \varepsilon_x\,\partial \varepsilon_y} = \frac{\partial \sigma_y}{\partial \varepsilon_x} \tag{12}$$

follow formally from (9)–(11).

The results (9)–(12), which have been deduced here by formal manipulation, are in fact very general results of conservation of energy and express the fact that the strain energy is independent of the way in which the forces are applied, cf. Southwell (1941, p. 15). They are made fundamental in more general treatments, Love (1927, §§ 61, 62).

The result (2) for the strain energy may be rewritten in terms of the stress and strain deviations defined in § 2.8 (2) and § 2.12 (2). It becomes

$$\begin{aligned}
2W &= (s_1 + s)(e_1 + e) + (s_2 + s)(e_2 + e) + (s_3 + s)(e_3 + e), \\
&= s_1e_1 + s_2e_2 + s_3e_3 + 3se + s(e_1 + e_2 + e_3) + e(s_1 + s_2 + s_3), \\
&= s_1e_1 + s_2e_2 + s_3e_3 + 3se, \\
&= \frac{1}{2G}(s_1^2 + s_2^2 + s_3^3) + \frac{s^2}{K}, && (13) \\
&= 2G(e_1^2 + e_2^2 + e_3^3) + 9Ke^2, && (14)
\end{aligned}$$

where § 5.4 (1) and (2) have been used.

Thus the strain energy per unit volume may be split into a part

$$s^2/2K \tag{15}$$

associated with the mean normal stress, and a part

$$W_d = \frac{1}{4G}(s_1^2 + s_2^2 + s_3^2) = G(e_1^2 + e_2^2 + e_3^2) \tag{16}$$

associated with stress deviation. This is called the *strain energy of distortion*. It follows from § 2.8 (10), (11) that

$$W_d = J_2/2G = (3/4G)\tau_{oct}^2. \tag{17}$$

The total strain energy or potential energy stored in a body under stress is obtained by integrating over the whole volume of the body and is

$$\iiint W\,dx\,dy\,dz. \tag{18}$$

The evaluation of this integral is frequently difficult, particularly in solids extending to infinity, and two important results (Love, 1927, §§ 120, 121) are frequently used to avoid it and to make further deductions.

The first of these is that *the total strain energy in a body which is in equilibrium under the action of given forces (including body forces) is equal to half the work done by these forces acting through their displacements from the unstrained state to the position of equilibrium.*

As a simple example, consider a cube of side a with stress σ perpendicular to one pair of faces, the resulting strain being ε. The total strain energy in the cube is $\frac{1}{2}a^3\sigma\varepsilon$ by (2). The displacement of the strained faces is $a\varepsilon$, so that the work done by constant forces $a^2\sigma$ acting through this displacement is $a^3\sigma\varepsilon$.

A general proof is as follows. Suppose that (p_x, p_y, p_z) are the components of the stress applied to the surface of the body at any point and that (X, Y, Z) are the components of body force per unit mass in its interior. Then the work Q done by these forces moving through displacements (u, v, w) is

$$Q = \iiint \rho(uX + vY + wZ)dV + \iint (up_x + vp_y + wp_z)dS, \tag{19}$$

where the first integral is taken over the volume of the body and the second over its surface. Now the surface forces p_x, p_y, p_z may be related to the stresses at the surface by § 2.4 (5)–(7), so that the second integral in (19) becomes

$$\iint \{l(u\sigma_x + v\tau_{yx} + w\tau_{zx}) + m(u\tau_{xy} + v\sigma_y + w\tau_{zy}) + n(u\tau_{xz} + v\tau_{yz} + w\sigma_z)\}dS. \tag{20}$$

This may be transformed by Green's theorem, which states that (in the present notation and under fairly general conditions on ξ, η, ζ)

$$\iint (l\xi + m\eta + n\zeta)dS = \iiint \left(\frac{\partial \xi}{\partial x} + \frac{\partial \eta}{\partial y} + \frac{\partial \zeta}{\partial z}\right)dV. \tag{21}$$

Using (21) in (20) and carrying out the differentiations, (19) becomes

$$Q = \iiint \left\{u\left(\rho X + \frac{\partial \sigma_x}{\partial x} + \frac{\partial \tau_{yx}}{\partial y} + \frac{\partial \tau_{zx}}{\partial z}\right) + \ldots\right\} dV + \iiint (\sigma_x\varepsilon_x + \sigma_y\varepsilon_y + \sigma_z\varepsilon_z + 2\tau_{yz}\Gamma_{yz} + 2\tau_{zx}\Gamma_{zx} + 2\tau_{xy}\Gamma_{xy})\, dV, \tag{22}$$

where the dots in the first integral stand for two similar terms. The first term in (22) vanishes by the equations of equilibrium, § 5.5 (5)–(7), and by (7) the integrand of the second term is just $2W$. Thus, finally, (19) becomes

$$Q = 2\iiint W\, dV. \tag{23}$$

A second important result, the *reciprocal theorem*, is proved in much the same way. Suppose that we have two different possible states of equilibrium of the body under different surface and body forces. Let $\sigma_x, \ldots, \varepsilon_x, \ldots, u, v, w, X, Y, Z, p_x, p_y, p_z$ be the stresses, strains, displacements, body forces, and surface stresses in the first set and let $\sigma_x', \ldots, \varepsilon_x', \ldots, u', \ldots, X', \ldots, p_x', \ldots$ be the corresponding quantities in the second set. The reciprocal theorem states that *the work Q_1 done by the forces of the first set acting through the displacements of the second set is equal to the work Q_2 done by the forces of the second set acting through the displacements* *of the first set.* As in (19), the work done by the forces of the first set acting over the displacements of the second set is

$$Q_1 = \iiint \rho(u'X + v'Y + w'Z)\,dV + \iint (u'p_x + v'p_y + w'p_z)\,dS.$$

By precisely the same steps as those leading from (19) to (22) this transforms into

$$Q_1 = \iiint \{\sigma_x\varepsilon_x' + \sigma_y\varepsilon_y' + \sigma_z\varepsilon_z' + 2\tau_{yz}\Gamma_{yz}' + 2\tau_{zx}\Gamma_{zx}' + 2\tau_{xy}\Gamma_{xy}'\}\,dV.$$

Using the stress–strain relations § 5.2 (20), (21) this becomes

$$\begin{aligned} Q_1 &= \iiint \{\lambda\Delta\Delta' + 2G(\varepsilon_x\varepsilon_x' + \varepsilon_y\varepsilon_y' + \varepsilon_z\varepsilon_z') + \\ &\qquad 4G(\Gamma_{yz}\Gamma_{yz}' + \Gamma_{zx}\Gamma_{zx}' + \Gamma_{xy}\Gamma_{xy}')\}\,dV, \\ &= Q_2, \end{aligned} \tag{24}$$

since (24) is symmetrical with respect to the two sets, and the theorem is proved.

5.9 Torsion of circular cylinders

Suppose that a circular cylinder of radius a and length l is twisted so that the displacements u, v, w in cylindrical coordinates are

$$u = 0, \quad v = \Omega rz/l, \quad w = 0, \tag{1}$$

so that when $z = l$, $v = \Omega r$, corresponding to a relative twist between the ends by an angle Ω.

The components of strain for the displacements (1) are, by § 2.15 (12), (13),

$$\varepsilon_r = \varepsilon_\theta = \varepsilon_z = 0, \quad \Gamma_{\theta z} = \Omega r/2l, \quad \Gamma_{r\theta} = \Gamma_{rz} = 0. \tag{2}$$

The stress–strain relations § 5.2 (26), (27) give

$$\tau_{\theta z} = G\Omega r/l, \tag{3}$$

and all other stress components vanish. It is easy to verify that (2) and (3) satisfy the equations of equilibrium and the compatibility conditions. The boundary conditions $\sigma_r = \tau_{r\theta} = 0$ at $r = a$ are also satisfied. The torque M on the cylinder is

$$M = 2\pi\int_0^a r^2\tau_{\theta z}\,dr = \frac{\pi G\Omega a^4}{2l}. \tag{4}$$

Measuring Ω for a given applied couple provides the most direct method of measuring G.

Using (4) in (3) gives

$$\tau_{\theta z} = \frac{2Mr}{\pi a^4}, \tag{5}$$

so that the shear stress increases linearly from zero at $r = 0$ to a maximum of

$$\tau_m = 2M/\pi a^3 \tag{6}$$

when $r = a$.

Referred to axes Px, radial, Py tangential, and Pz, Fig. 5.9, at the surface, the principal stresses at the surface are, by § 2.3 (14),

$$\pm\tau_m \tag{7}$$

in directions Py_1, Pz_1, in the plane Pyz and inclined at 45° to Py.

In many experiments an additional confining pressure p is applied to

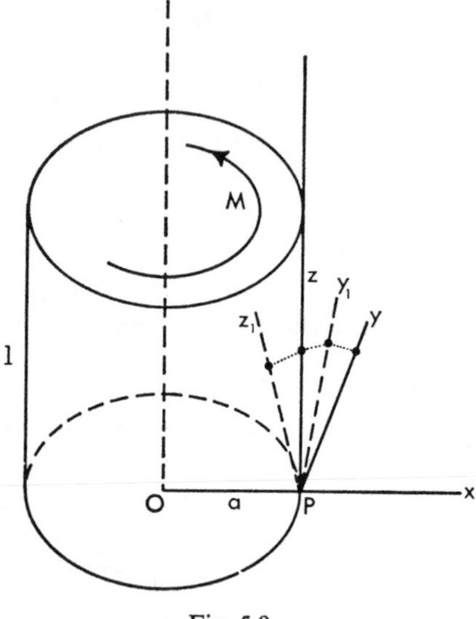

Fig. 5.9

the circumference of the cylinder and an additional axial pressure $p + p_a$, so that the stresses at P are

$$\sigma_x = p, \quad \sigma_y = p, \quad \sigma_z = p + p_a, \quad \tau_{yz} = \tau_m, \quad \tau_{xy} = \tau_{xz} = 0. \tag{8}$$

In this case the principal axes Py_1, Pz_1 at P are inclined at

$$-\tfrac{1}{2}\tan^{-1}\left(\frac{2\tau_m}{p_a}\right) \tag{9}$$

to Py by § 2.3 (11), and the principal stresses are, by § 2.3 (14),

$$\tfrac{1}{2}(2p + p_a) + [\tau_m^2 + \tfrac{1}{4}p_a^2]^{\frac{1}{2}} \quad \text{along } Py_1, \tag{10}$$

and

$$\tfrac{1}{2}(2p + p_a) - [\tau_m^2 + \tfrac{1}{4}p_a^2]^{\frac{1}{2}} \quad \text{along } Pz_1, \tag{11}$$

and p along Px.

Many solutions for torsion of cylinders of non-circular sections are available, cf. Love (1927), but they have been little used for studying rocks.

5.10 Bending and buckling

Bending is used in the testing of rocks, both for measurement of E and tensile strength. It is also a very sensitive method of studying creep and transient behaviour of rocks. The simple engineering theory is very well known, but will be given here briefly for reference, cf. Timoshenko and Goodier (1951), Morley (1923).

It will be assumed for simplicity that the cross-section of the beam is

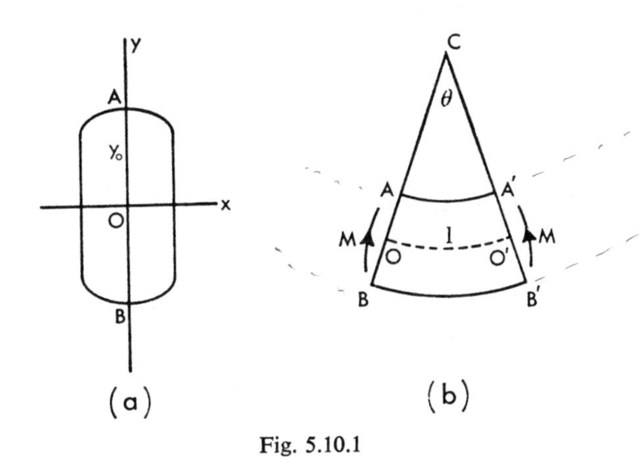

Fig. 5.10.1

symmetrical about the axes Ox, Oy, Fig. 5.10.1 (*a*), and that the greatest depth of the beam is $2y_0$. The simple Euler–Bernouilli theory assumes that plane sections of the beam remain plane after bending, and that each longitudinal fibre of the beam behaves as if in uniaxial compression or tension independent of the others.

Suppose that length l of the beam in the z-direction perpendicular to the plane Oxy of Fig. 5.10.1 (*a*) is bent by couples M in the plane Oyz so that plane sections BOA, $B'O'A'$ make an angle θ with one another, Fig. 5.10.1 (*b*). The region $y > 0$ will be compressed and the region $y < 0$ extended, and, since the cross-section is symmetrical and there is no longitudinal stress on the beam, the length OO' in the plane $y = 0$ must be l. If R is the radius of curvature of OO', the shortening of a fibre at y must be $R\theta - (R - y)\theta = y\theta$, and the longitudinal strain in it must be $y\theta/R\theta = y/R$, so that the stress in it must be Ey/R. The couple M necessary to maintain this state is

$$M = \frac{E}{R}\iint y^2\, dx\, dy, \tag{1}$$

where the double integral is taken over the cross-section of the beam and is called the *moment of inertia I* of the cross-section. Thus (1) becomes

$$M = \frac{EI}{R} = EI\frac{d^2y}{dz^2} \tag{2}$$

using the approximate value d^2y/dz^2 for the radius of curvature.

The stress f in the extreme fibre $y = y_0$ is given by

$$f = \frac{Ey_0}{R} = \frac{My_0}{I}. \tag{3}$$

When the beam is loaded in a known fashion M can be calculated and the extreme fibre stress found from (3), and the deflection by solving (2).

Two systems of loading are in use for testing.

(i) *Three-point loading*

Fig. 5.10.2 (*a*), in which the beam is supported on two knife edges distant $2l$ apart and loaded by W at its mid-point.

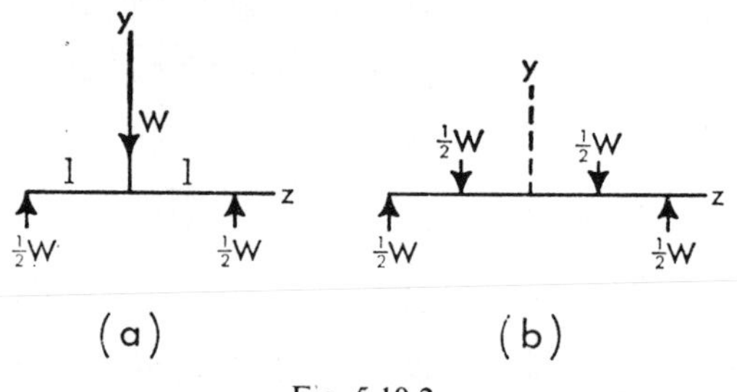

Fig. 5.10.2

In this case the bending moment M at the point z is

$$M = \tfrac{1}{2}W(l - z). \tag{4}$$

The extreme fibre stress at the centre of the beam, $z = 0$, is by (3)

$$f = Wly_0/2I, \tag{5}$$

and is used for measuring tensile strength.

The differential equation for the deflection (2) becomes

$$EI\frac{d^2y}{dz^2} = \tfrac{1}{2}W(l - z),$$

which, if the origin O is at the centre of the beam, has to be solved with $y = dy/dz = 0$ at $z = 0$. So, integrating twice,

$$EIy = \tfrac{1}{2}W(\tfrac{1}{2}lz^2 - \tfrac{1}{6}z^3). \tag{6}$$

The value of y at $z = l$ is equal to the deflection at the centre and is

$$y = Wl^3/6EI. \tag{7}$$

Measurement of this gives E.

(ii) *Four-point loading*

The system of Fig. 5.10.2 (*a*) is open to criticism, since there is a stress concentration at the point at which the load is applied, and the maximum tensile stress occurs directly below this. This is obviated by the system of Fig. 5.10.2 (*b*). In this,

$$\left.\begin{aligned} M &= \tfrac{1}{2}W(l - z), \quad a < z < l \\ M &= \tfrac{1}{2}W(l - a), \quad 0 < z < a \end{aligned}\right\} \tag{8}$$

so that M is constant in the region $| z | < a$, and the extreme fibre stress throughout this region is the same as that at the mid-point of the beam and is

$$f = Wy_0(l - a)/2I. \tag{9}$$

The deflection at the centre in this case is

$$\frac{W}{12EI}(2l^3 - 3a^2l + a^3). \tag{10}$$

(iii) *Distributed loading*

If the load on the beam is distributed, being $W(z)$ per unit length, bending moments can be calculated as before from statical principles. The part of the bending moment at z due to the distributed load (i.e. excluding concentrated loads and reactions) is

$$M = \int_z W(z')(z' - z)\, dz',$$

$W(z)$ being measured positively in the direction of the y-axis. It follows that

$$\frac{dM}{dz} = -\int_z W(z')\, dz',$$

$$\frac{d^2M}{dz^2} = W(z). \tag{11}$$

This result holds in regions away from concentrated loads. Using it in (2) gives, if E and I are independent of z,

$$EI\frac{d^4y}{dz^4} - W(z) = 0. \tag{12}$$

This result will be needed later.

(iv) *A column with elastic restraint at its mid-point*
Suppose that a beam, AOB, of length $2l$ is freely hinged at its ends A, B and subjected to axial compression, P. The undeflected solution $y = 0$ is always possible, but it will appear that for sufficiently high values of P it becomes unstable. It will be supposed, in addition, that lateral movement of the centre point O is restricted by a spring of stiffness k, Fig. 5.10.3.

Consider the situation in which the centre point O has been displaced by an amount y_0. The restraining force exerted by the spring will be ky_0, so that reactions $-\frac{1}{2}ky_0$ must be supplied at the hinges.

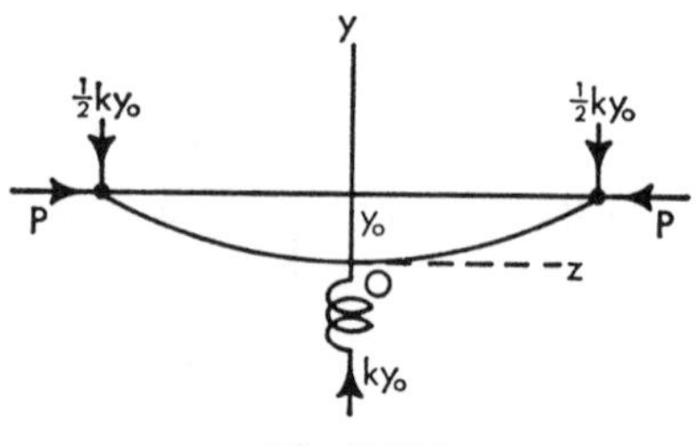

Fig. 5.10.3

Then, taking the origin in the displaced position as in Fig. 5.10.3, (2) becomes

$$EI\frac{d^2y}{dz^2} = P(y_0 - y) - \tfrac{1}{2}ky_0(l - z), \tag{13}$$

or

$$\frac{d^2y}{dz^2} + \omega^2 y = \omega^2 y_0 - (k/2P)\omega^2 y_0(l - z), \tag{14}$$

where

$$\omega^2 = P/EI. \tag{15}$$

The general solution of (14) is

$$y = A \sin \omega z + B \cos \omega z + y_0 - (k/2P)y_0(l - z), \tag{16}$$

and the boundary conditions to be satisfied are

$$y = dy/dz = 0, \quad z = 0,$$

$$y = y_0, \quad z = l.$$

Substituting (16) in these gives

$$0 = B + y_0 - (kly_0/2P), \quad 0 = \omega A + (ky_0/2P), \tag{17}$$

$$y_0 = A \sin \omega l + B \cos \omega l + y_0. \tag{18}$$

Substituting from (17) in (18) gives

$$y_0\{\sin \omega l - \omega l(1 - 2EI\omega^2/kl) \cos \omega l\} = 0.$$

It follows that the only solution is $y_0 = 0$, that is the beam is undeflected, unless ωl is a root of the equation

$$\tan \omega l = \omega l(1 - 2EI\omega^2/kl). \tag{19}$$

If $k = 0$, (17) and (18) reduce to $\cos \omega l = 0$ or $\omega l = \pi/2$, so that from (15)

$$P = \pi^2 EI/4l^2. \tag{20}$$

This is called the Euler load for the freely hinged column, and the position is that if P is less than this value the only solution is the undeflected one, $y = 0$. If P has the value (20) the solution $y = A \cos (\pi z/2l)$ becomes possible and the solution $y = 0$ is unstable.

If k is not zero, that is if a stabilizing spring is included, as in Fig. 5.10.3, the load which causes instability is

$$P = \alpha^2 EI/l^2, \tag{21}$$

where α is the least root of (19) or

$$\tan \alpha = \alpha(1 - 2EI\alpha^2/kl^3). \tag{22}$$

The case in which deflection of the column is resisted by an elastic medium which supplies restraining force proportional to displacement at all points may be treated in the same way, Goldstein (1926).

(v) *Buckling with viscous restraint*

The previous problem could have been discussed in a different and more general manner. Consider an infinitely long column along the z-axis with axial load P. Suppose the column to be deflected so that its displacement at z is y and that this displacement is resisted by a force $q(y)$. Then by (2)

$$EI\frac{d^2y}{dz^2} = -Py + M_1, \tag{23}$$

where M_1 is the bending moment due to the restraining forces q. Using (11) this becomes

$$EI\frac{d^4y}{dz^4} + P\frac{d^2y}{dz^2} = -q. \tag{24}$$

This result is true for any type of restraining force, and here q will be taken to be that of viscous damping through a dash-pot (so that the surrounding medium may be regarded as a Maxwell substance § 11.3) for which

$$q = \eta\frac{dy}{dt}, \tag{25}$$

where t is the time and η is a constant. Then (24) becomes

$$EI\frac{d^4y}{dz^4} + P\frac{d^2y}{dz^2} = -\eta\frac{dy}{dt}. \tag{26}$$

We now seek a solution of (26) with period $2\pi/\omega$ in z, so that we assume $y = Y \exp(i\omega z)$ and (26) gives for Y

$$\eta \frac{dY}{dt} = \omega^2(P - EI\omega^2)Y. \tag{27}$$

The solution of (27) is

$$Y = A \exp\{\omega^2 t(P - EI\omega^2)/\eta\}. \tag{28}$$

If $P < EI\omega^2$, Y dies away exponentially and the system is stable. The wavelength λ of the disturbance is $\lambda^2 = 4\pi^2/\omega^2 = 4\pi^2 EI/P$; this corresponds to the result (20), since the length l in (20) is one-quarter of a wavelength. This value of λ is called the Euler wavelength.

If $P > EI\omega^2$, any displacement increases exponentially. Also the exponent $\omega^2(P - EI\omega^2)$ in (28) has a maximum value when $\omega^2 = P/2EI$. Thus for this value of ω or a wavelength λ_d given by

$$\lambda_d^2 = 8\pi^2 EI/P \tag{29}$$

the rate of increase will be greatest. λ_d will be called the dominant wavelength, and it is $\sqrt{2}$ times the Euler wavelength. If the displacement of the column is studied in detail, starting from any prescribed initial position, the form of the column will tend to become progressively nearer to a sinusoid of the dominant wavelength.

Biot (1961) gave this discussion as a preliminary to the discussion of the folding of a geological system. He considers next the case of an elastic plate of Young's modulus E, Poisson's ratio ν, and thickness h in infinite medium of viscosity η. In this case the dominant wavelength is $\pi h\{E/(1 - \nu^2)P\}^{\frac{1}{2}}$. Finally, he considers a plane sheet of material of thickness h and viscosity η immersed in an infinite medium of viscosity η_1, and in this case finds the dominant wavelength to be $2\pi h(\eta/6\eta_1)^{1/3}$. This theory can be extended to the case of several layers and to include the effects of gravity. It has been developed by Biot (1957, 1961, 1965) in a series of papers and verified experimentally by Biot, Odé, and Roever (1961). Other discussions have been given by Ramberg (1964) and Currie, Patnode, and Trump (1962).

5.11 Hollow cylinders with internal or external pressure

This system is of very great importance in many applications. Suppose that the internal radius is R_1 and the external radius R_2, and that hydrostatic pressure p_1 is applied at the interior R_1, and p_2 at the exterior R_2.

Since all quantities are independent of r and z, the stress-equation § 5.6 (9) gives

$$\frac{\partial \sigma_r}{\partial r} + \frac{\sigma_r - \sigma_\theta}{r} = 0. \tag{1}$$

Also the stress–strain relations § 5.2 (26) give for the case of plane strain

$$\sigma_r = (\lambda + 2G)\varepsilon_r + \lambda\varepsilon_\theta = (\lambda + 2G)\frac{du}{dr} + \frac{\lambda u}{r}, \tag{2}$$

$$\sigma_\theta = (\lambda + 2G)\varepsilon_\theta + \lambda\varepsilon_r = \lambda\frac{du}{dr} + \frac{(\lambda + 2G)u}{r}, \tag{3}$$

using § 2.15 (12) with $v = 0$.

Using (2) and (3) in (1) gives

$$\frac{d^2u}{dr^2} + \frac{1}{r}\frac{du}{dr} - \frac{u}{r^2} = 0,$$

which may be written

$$\frac{d}{dr}\left(\frac{du}{dr} + \frac{u}{r}\right) = 0. \tag{4}$$

The general solution of (4) is

$$u = Ar + \frac{B}{r}, \tag{5}$$

where A and B are constants which have to be found from the boundary conditions

$$\sigma_r = p_1, \text{ when } r = R_1; \quad \sigma_r = p_2, \text{ when } r = R_2. \tag{6}$$

Substituting (5) in (2) gives, using (6),

$$2A(\lambda + G) - 2GB/R_1^2 = p_1; \quad 2A(\lambda + G) - 2GB/R_2^2 = p_2. \tag{7}$$

Solving gives

$$A = \frac{p_2R_2^2 - p_1R_1^2}{2(\lambda + G)(R_2^2 - R_1^2)}, \quad B = \frac{(p_2 - p_1)R_1^2R_2^2}{2G(R_2^2 - R_1^2)},$$

so that finally from (5), (2), and (3),

$$u = \frac{(p_2R_2^2 - p_1R_1^2)r}{2(\lambda + G)(R_2^2 - R_1^2)} + \frac{(p_2 - p_1)R_1^2R_2^2}{2G(R_2^2 - R_1^2)r}, \tag{8}$$

$$\sigma_r = \frac{p_2R_2^2 - p_1R_1^2}{(R_2^2 - R_1^2)} - \frac{(p_2 - p_1)R_1^2R_2^2}{r^2(R_2^2 - R_1^2)}, \tag{9}$$

$$\sigma_\theta = \frac{p_2R_2^2 - p_1R_1^2}{R_2^2 - R_1^2} + \frac{(p_2 - p_1)R_1^2R_2^2}{r^2(R_2^2 - R_1^2)}. \tag{10}$$

Many important results follow from these relations. If $R_1 = 0$, $u = p_2r/2(\lambda + G) = (1 + \nu)(1 - 2\nu)rp_2/E$, and $\sigma_r = \sigma_\theta = p_2$. This is the case of a solid cylinder.

If $R_2 \to \infty$,

$$u = \frac{p_2 r}{2(\lambda + G)} + \frac{(p_2 - p_1)R_1^2}{2Gr}, \tag{11}$$

$$\sigma_r = p_2\left(1 - \frac{R_1^2}{r^2}\right) + \frac{p_1 R_1^2}{r^2}; \quad \sigma_\theta = p_2\left(1 + \frac{R_1^2}{r^2}\right) - \frac{p_1 R_1^2}{r^2}. \tag{12}$$

If $p_2 = 0$, the displacement u_{R_1} at $r = R_1$ caused by internal pressure p_1 is

$$u_{R_1} = -p_1 R_1/2G = -p_1 R_1(1 + \nu)/E. \tag{13}$$

This provides a commonly used method of measuring G from boreholes, cf. Talobre (1957). Similarly if $p_1 = 0$ and $p_2 = -P_2$ the displacement at $r = R_1$ is

$$-(1 - \nu)P_2 R_1/G; \tag{14}$$

this is the displacement which would be measured at $r = R_1$ if stress P_2 at infinity is released.

Equations (12) show the way in which the stresses vary from $\sigma_r = p_1$, $\sigma_\theta = 2p_2 - p_1$ at $r = R_1$ to $\sigma_r = \sigma_\theta = p_2$ as $r \to \infty$.

In the general case of a finite cylinder, if $p_1 = 0$ the displacement u_{R_1} at $r = R_1$ is

$$\frac{(\lambda + 2G)p_2 R_1 R_2^2}{2G(\lambda + G)(R_2^2 - R_1^2)} = \frac{(1 - \nu)p_2 R_1 R_2^2}{G(R_2^2 - R_1^2)}. \tag{15}$$

As in § 5.3 (v), these results for plane strain may be converted to plane stress by replacing ν by $\nu/(1 + \nu)$, so that (15) becomes for plane stress

$$\frac{p_2 R_1 R_2^2}{G(1 + \nu)(R_2^2 - R_1^2)} = \frac{2p_2 R_1 R_2^2}{E(R_2^2 - R_1^2)}. \tag{16}$$

(16) has been used for measurement of E on hollow cores by Fitzpatrick (1962). It will be noticed that it measures it in a direction transverse to the axis of the cylinder.

It follows from (9) and (10) that

$$\sigma_r + \sigma_\theta = 2(p_2 R_2^2 - p_1 R_1^2)/(R_2^2 - R_1^2), \tag{17}$$

$$\sigma_\theta - \sigma_r = 2(p_2 - p_1)R_1^2 R_2^2/(R_2^2 - R_1^2)r^2. \tag{18}$$

Thus the magnitude of the stress difference decreases from a maximum of $2|p_2 - p_1|R_2^2/(R_2^2 - R_1^2)$ when $r = R_1$ to a minimum of $2|p_2 - p_1|R_1^2/(R_2^2 - R_1^2)$ when $r = R_2$.

If $p_1 = 0$, external pressure only, (9) and (10) become

$$\sigma_r = \frac{p_2 R_2^2}{R_2^2 - R_1^2}\left(1 - \frac{R_1^2}{r^2}\right), \quad \sigma_\theta = \frac{p_2 R_2^2}{(R_2^2 - R_1^2)}\left(1 + \frac{R_1^2}{r^2}\right). \tag{19}$$

If $p_2 = 0$, internal pressure only, they give

$$\sigma_r = -\frac{p_1 R_1^2}{R_2^2 - R_1^2}\left(1 - \frac{R_2^2}{r^2}\right), \quad \sigma_\theta = -\frac{p_1 R_1^2}{R_2^2 - R_1^2}\left(1 + \frac{R_2^2}{r^2}\right). \quad (20)$$

The variation of σ_r and σ_θ for these cases is shown in Fig. 5.11 (*a*), (*b*) for the case $R_2 = 2R_1$.

The strain energy per unit length of the cylinder may be calculated for the case of plane strain either by using § 5.8 (2) or (4) with the stress–strain relations § 5.3 (9), (10) and integrating over the volume of the cylinder, or, more simply, by using the general theorem § 5.8 (23).

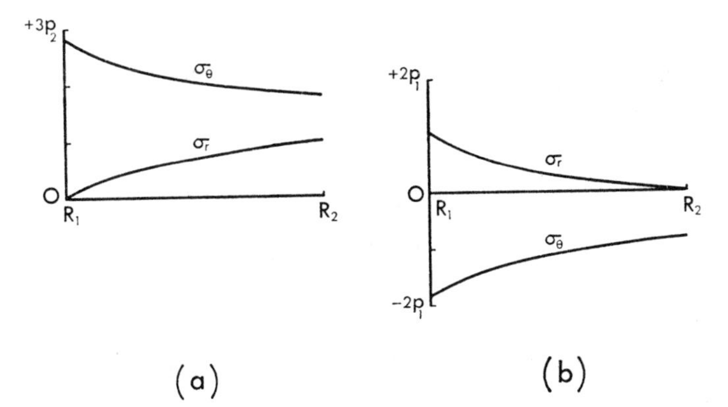

Fig. 5.11 Stress distribution in a hollow cylinder of radii R_1 and $R_2 = 2R_1$. (*a*) For external pressure p_2 only. (*b*) For internal pressure p_1 only.

Using the displacements for the forces p_1 and p_2 given by (8), the theorem states that the total strain energy per unit length of the cylinder is

$$\tfrac{1}{2}p_2 \,.\, 2\pi R_2\left\{\frac{(p_2R_2^2 - p_1R_1^2)R_2}{2(\lambda + G)(R_2^2 - R_1^2)} + \frac{(p_2 - p_1)R_1^2R_2}{2G(R_2^2 - R_1^2)}\right\} -$$

$$\tfrac{1}{2}p_1 \,.\, 2\pi R_1\left\{\frac{(p_2R_2^2 - p_1R_1^2)R_1}{2(\lambda + G)(R_2^2 - R_1^2)} + \frac{(p_2 - p_1)R_1R_2^2}{2G(R_2^2 - R_1^2)}\right\}$$

$$= \frac{\pi(1 + \nu)}{E(R_2^2 - R_1^2)}\{(1 - 2\nu)(p_2R_2^2 - p_1R_1^2)^2 + (p_2 - p_1)^2R_1^2R_2^2\}. \quad (21)$$

The case of a circular cylinder with an eccentric bore is discussed by Jeffery (1921).

5.12 Stress–strain relations for anisotropic materials

Most rocks are more or less anisotropic, the most common cases being sediments which have different properties in and perpendicular to the bedding planes, and metamorphic rocks, such as slates, with a well-defined plane of cleavage. The most general case, that of crystals, is very fully discussed by Love (1927), Nye (1957), or Voigt (1910).

In the most general case we take axes $Oxyz$ related to the material concerned rather than to the principal axes of stress or strain so that all six components of both will appear. The most general linear assumption then is that each component of stress is a linear combination of the six components of strain, that is,

$$\sigma_x = c_{11}\varepsilon_x + c_{12}\varepsilon_y + c_{13}\varepsilon_z + c_{14}\gamma_{yz} + c_{15}\gamma_{zx} + c_{16}\gamma_{xy}, \tag{1}$$

$$\sigma_y = c_{21}\varepsilon_x + c_{22}\varepsilon_y + c_{23}\varepsilon_z + c_{24}\gamma_{yz} + c_{25}\gamma_{zx} + c_{26}\gamma_{xy}, \tag{2}$$

$$\sigma_z = c_{31}\varepsilon_x + c_{32}\varepsilon_y + c_{33}\varepsilon_z + c_{34}\gamma_{yz} + c_{35}\gamma_{zx} + c_{36}\gamma_{xy}, \tag{3}$$

$$\tau_{yz} = c_{41}\varepsilon_x + c_{42}\varepsilon_y + c_{43}\varepsilon_z + c_{44}\gamma_{yz} + c_{45}\gamma_{zx} + c_{46}\gamma_{xy}, \tag{4}$$

$$\tau_{zx} = c_{51}\varepsilon_x + c_{52}\varepsilon_y + c_{53}\varepsilon_z + c_{54}\gamma_{yz} + c_{55}\gamma_{zx} + c_{56}\gamma_{xy}, \tag{5}$$

$$\tau_{xy} = c_{61}\varepsilon_x + c_{62}\varepsilon_y + c_{63}\varepsilon_z + c_{64}\gamma_{yz} + c_{65}\gamma_{zx} + c_{66}\gamma_{xy}. \tag{6}$$

The 36 constants c_{rs} are called *elastic constants* or *stiffnesses*, and in the matrix notation of § 2.14 the equations (1)–(6) may be written as the matrix product

$$[\sigma] = [c] \times [\varepsilon], \tag{7}$$

where $[\sigma]$ and $[\varepsilon]$ are matrices of six rows and one column and $[c]$ is the matrix $[c_{rs}]$.

Equations (1)–(6) may be solved for the components of strain in terms of those of stress, the result being

$$[\varepsilon] = [s] \times [\sigma], \tag{8}$$

where $[s] = [s_{rs}]$ is the matrix inverse to $[c]$ and the equations run

$$\varepsilon_x = s_{11}\sigma_x + s_{12}\sigma_y + s_{13}\sigma_z + s_{14}\tau_{yz} + s_{15}\tau_{zx} + s_{16}\tau_{xy}, \tag{9}$$

etc. The quantities s_{rs} are called *elastic moduli* or *compliances*, and it is they, in general, rather than c_{rs} which are measured experimentally. The nomenclature is a source of great confusion, since the quantities E, K, G described historically as moduli are, on the present definition, not moduli but their reciprocals. If any ambiguity arises it is better to use the terms stiffness and compliance, which are usual in mechanics.

Next we show that

$$c_{rs} = c_{sr}, \tag{10}$$

so that the matrix $[c_{rs}]$ is in fact symmetrical, and so there are at most 21 independent constants. It follows, of course, that $s_{rs} = s_{sr}$. These results follow from general considerations of strain energy. In particular, as in § 5.8 (12), it is true in general that

$$\frac{\partial \sigma_x}{\partial \varepsilon_y} = \frac{\partial^2 W}{\partial \varepsilon_x \partial \varepsilon_y} = \frac{\partial \sigma_y}{\partial \varepsilon_x}, \tag{11}$$

and using (1) and (2) in this gives $c_{12} = c_{21}$. Similarly, all other pairs are equal.

The general triclinic material has in fact the full 21 elastic constants. For all other crystal classes further relations between the constants follow from the symmetry properties. The general cases are dealt with in the works referred to earlier. For rock materials, the most important is that of orthorhombic symmetry, in which there are three mutually perpendicular planes of symmetry which will be taken to be perpendicular to the directions *Oxyz*.

It follows from the nature of the material that the stress–strain relations must be the same if, instead of looking at it in the *x*-direction, Fig. 5.12 (*a*),

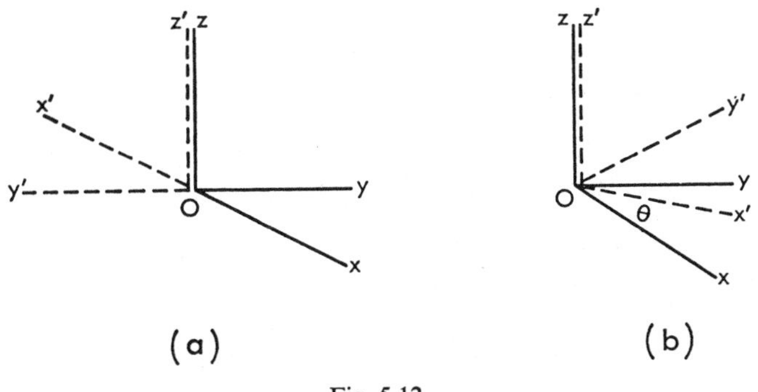

Fig. 5.12

the opposite direction Ox' is chosen, the axes then being $Ox'y'z'$. Relative to these axes, (1) will give

$$\sigma_{x'} = c_{11}\varepsilon_{x'} + c_{12}\varepsilon_{y'} + c_{13}\varepsilon_{z'} + c_{14}\gamma_{y'z'} + c_{15}\gamma_{z'x'} + c_{16}\gamma_{x'y'} \tag{12}$$

The direction cosines of Ox', Oy', Oz' relative to $Oxyz$ are, respectively, $(-1, 0, 0)$, $(0, -1, 0)$, $(0, 0, 1)$, so it follows from § 2.5 (1)–(7) for stress, and the similar relations for strain, that

$$\sigma_{x'} = \sigma_x, \quad \sigma_{y'} = \sigma_y, \quad \sigma_{z'} = \sigma_z, \quad \tau_{y'z'} = -\tau_{yz}, \quad \tau_{z'x'} = \tau_{zx}, \quad \tau_{x'y'} = \tau_{xy}, \tag{13}$$

$$\varepsilon_{x'} = \varepsilon_x, \quad \varepsilon_{y'} = \varepsilon_y, \quad \varepsilon_{z'} = \varepsilon_z, \quad \gamma_{y'z'} = -\gamma_{yz}, \quad \gamma_{z'x'} = -\gamma_{zx}, \quad \gamma_{x'y'} = \gamma_{xy}. \tag{14}$$

Using (13) and (14) in (12), it becomes

$$\sigma_x = c_{11}\varepsilon_x + c_{12}\varepsilon_y + c_{13}\varepsilon_z - c_{14}\gamma_{yz} - c_{15}\gamma_{zx} + c_{16}\gamma_{xy}. \tag{15}$$

Comparing this with (1) gives

$$c_{14} = c_{15} = 0. \tag{16}$$

Similarly, the equations for the other stress-components require

$$c_{24} = c_{25} = c_{34} = c_{35} = c_{46} = c_{56} = 0. \tag{17}$$

Applying the same argument to the axis Oz in place of Ox gives in addition

$$c_{16} = c_{26} = c_{36} = c_{45} = 0. \tag{18}$$

Thus, finally, the relations for orthorhombic symmetry become

$$\sigma_x = c_{11}\varepsilon_x + c_{12}\varepsilon_y + c_{13}\varepsilon_z, \tag{19}$$

$$\sigma_y = c_{12}\varepsilon_x + c_{22}\varepsilon_y + c_{23}\varepsilon_z, \tag{20}$$

$$\sigma_z = c_{13}\varepsilon_x + c_{23}\varepsilon_y + c_{33}\varepsilon_z, \tag{21}$$

$$\tau_{yz} = c_{44}\gamma_{yz}, \quad \tau_{zx} = c_{55}\gamma_{zx}, \quad \tau_{xy} = c_{66}\gamma_{xy}. \tag{22}$$

It appears that there are nine independent elastic constants. Solving (19)–(21) gives for the compliances

$$s_{11} = (c_{22}c_{33} - c_{23}^2)/D, \quad s_{12} = (c_{13}c_{23} - c_{12}c_{33})/D, \\ s_{13} = (c_{12}c_{23} - c_{22}c_{13})/D, \tag{23}$$

$$s_{22} = (c_{11}c_{33} - c_{13}^2)/D, \quad s_{23} = (c_{21}c_{13} - c_{11}c_{23})/D, \\ s_{33} = (c_{11}c_{22} - c_{12}^2)/D, \tag{24}$$

$$s_{44} = 1/c_{44}, \quad s_{55} = 1/c_{55}, \quad s_{66} = 1/c_{66}, \tag{25}$$

where

$$D = \begin{vmatrix} c_{11} & c_{12} & c_{13} \\ c_{21} & c_{22} & c_{23} \\ c_{31} & c_{32} & c_{33} \end{vmatrix} \tag{26}$$

Young's modulus for uniaxial stress in the x-direction will be

$$\sigma_x/\varepsilon_x = 1/s_{11} = D/(c_{22}c_{33} - c_{23}^2), \tag{27}$$

and there will be different Poisson's ratios for the y- and z-directions, since

$$\frac{\varepsilon_y}{\varepsilon_x} = \frac{c_{23}c_{13} - c_{12}c_{33}}{c_{22}c_{33} - c_{23}^2}, \quad \frac{\varepsilon_z}{\varepsilon_x} = \frac{c_{12}c_{23} - c_{22}c_{13}}{c_{22}c_{33} - c_{23}^2}. \tag{28}$$

Each direction of stress will have its own Young's modulus and its own variation of Poisson's ratio about it which can be calculated from the above formulae.

A further simplification occurs if there is an axis of symmetry, say Oz, so that the properties of the material in all directions at right angles to it are the same. In this case the stress–strain relations must be independent of the angle θ, Fig. 5.12 (*b*). Here proceeding in the same way using the results of § 2.5, it is found that

$$\sigma_x = c_{11}\varepsilon_x + (c_{11} - 2c_{66})\varepsilon_y + c_{13}\varepsilon_z, \tag{29}$$

$$\sigma_y = (c_{11} - 2c_{66})\varepsilon_x + c_{11}\varepsilon_y + c_{13}\varepsilon_z, \tag{30}$$

$$\sigma_z = c_{13}\varepsilon_x + c_{13}\varepsilon_y + c_{33}\varepsilon_z, \tag{31}$$

$$\tau_{yz} = c_{44}\gamma_{yz}, \quad \tau_{zx} = c_{44}\gamma_{zx}, \quad \tau_{xy} = c_{66}\gamma_{xy}, \tag{32}$$

involving five independent coefficients.

This is the case of a sedimentary rock with z-axis perpendicular to the bedding, and the increase of the number of elastic constants from two for the isotropic case to five is formidable. There is no great difficulty in handling many mathematical problems involving such materials, cf. Hearmon (1961), Savin (1961); the difficulty for practical purposes is in obtaining and using realistic values of the elastic constants.

For many purposes in rock mechanics two-dimensional solutions are adequate, and the simplest case is that of the *two-dimensional orthotropic solid*, which has perpendicular axes of symmetry Ox, Oy. The stress–strain relations for this case may be derived from (19)–(21) or (29)–(31) by specializing them to the case of plane stress or strain, but it is simpler to write down two-dimensional relations *ab initio*. These will be

$$\varepsilon_x = s_{11}\sigma_x + s_{12}\sigma_y, \quad \varepsilon_y = s_{12}\sigma_x + s_{22}\sigma_y, \quad \gamma_{xy} = s_{66}\tau_{xy}, \tag{33}$$

where it may be verified as before that all other coefficients vanish by symmetry.

If $$\sigma_y = 0, \quad \varepsilon_x = s_{11}\sigma_x, \quad \varepsilon_y/\varepsilon_x = s_{12}/s_{11}; \tag{34}$$

if $$\sigma_x = 0, \quad \varepsilon_y = s_{22}\sigma_y, \quad \varepsilon_x/\varepsilon_y = s_{12}/s_{22}. \tag{35}$$

We may write $s_{11} = 1/E_1$, Young's modulus in the x-direction; $s_{22} = 1/E_2$, Young's modulus in the y-direction; $\nu_{12} = -E_1 s_{12}$ is Poisson's ratio for stress in the x-direction; and $\nu_{21} = -E_2 s_{12}$ is Poisson's ratio for stress in the y-direction.

It follows from (34) and (35) that $E_2\nu_{12} = E_1\nu_{21}$. $1/s_{66}$ may also be written G, the modulus of rigidity in the plane.

For stress σ in a direction inclined at θ to Ox it follows from § 2.10 (18) that

$$\varepsilon = \varepsilon_x \cos^2\theta + \gamma_{xy}\sin\theta\cos\theta + \varepsilon_y\sin^2\theta. \tag{36}$$

Then, using

$$\sigma_x = \sigma\cos^2\theta, \quad \sigma_y = \sigma\sin^2\theta, \quad \tau_{xy} = \sigma\sin\theta\cos\theta, \text{ in (33) and (36),}$$

$$\varepsilon/\sigma = (s_{11}\cos^2\theta + s_{12}\sin^2\theta)\cos^2\theta + s_{66}\sin^2\theta\cos^2\theta + (s_{12}\cos^2\theta + s_{22}\sin^2\theta)\sin^2\theta,$$

so that if σ/ε is E_θ, Young's modulus for the direction θ,

$$\frac{1}{E_\theta} = \frac{1}{E_1}\cos^4\theta + \frac{1}{E_2}\sin^4\theta + \left\{\frac{1}{G} - \frac{2\nu_{12}}{E_1}\right\}\sin^2\theta\cos^2\theta. \tag{37}$$

(33) may be solved for σ_x and σ_y to give

$$\sigma_x = c_{11}\varepsilon_x + c_{12}\varepsilon_y, \quad \sigma_y = c_{12}\varepsilon_x + c_{22}\varepsilon_y, \quad \tau_{xy} = c_{66}\gamma_{xy}, \tag{38}$$

where

$$c_{11} = s_{22}/\Delta' = E_1/\mu, \quad c_{22} = s_{11}/\Delta' = E_2/\mu, \quad c_{12} = -s_{12}/\Delta' = \nu_{21}E_1/\mu = \nu_{12}E_2/\mu, \tag{39}$$

and $$\Delta' = s_{11}s_{22} - s_{12}^2, \quad \mu = 1 - \nu_{21}\nu_{12}, \quad c_{66} = 1/s_{66} = G. \tag{40}$$

Finally, the case of polycrystalline aggregates should be referred to. If the crystals in a polycrystalline aggregate are randomly orientated the resulting material will be isotropic, and it should be possible to estimate its elastic properties from those of its constituent crystals. Two methods are possible, Hearmon (1961), the Voigt method, which takes the stiffness of the aggregate to be the space-average of the stiffnesses of the constituent crystals; and the Reuss method, which takes the compliance of the aggregate to be the space-average of the compliances of the individual crystals. Both models are approximate only, and it can be shown that the true values should lie between the two.

Chapter Six

Laboratory Testing

6.1 Introduction

The structure of rock and its mechanical nature are mentioned in Chapter 1. Its mechanical properties depend both upon the interaction between the crystals, particles, and cementitious material of which it is composed and such cracks, joints, bedding, and minor faults as exist. On the one hand, the task of specifying the mechanical properties of rock, especially its strength, in terms of the properties of its constituent particles is formidable. On the other hand, the distribution of cracks, joints, bedding, and faults is so variable that the mechanical properties of any particular large volume of rock influenced by such separations have little general relevance for any other large volume of rock. Therefore, the most basic mechanical properties of rock are those of a specimen of a size sufficient to contain a large number of constituent particles but small enough to exclude major structural discontinuities, so that it possesses gross homogeneous properties. Specimens of rock with dimensions of inches are usually adequate for this purpose and can conveniently be tested in a laboratory.

Laboratory tests usually consist of simple experiments appropriate to the nature of rock in which important quantities, often stress and strain, are determined. The general relation between these quantities emerges as a result of hypothesis and experiment. It is then expressed in an idealized mathematical form, on the basis of which it can be extended to cover more complicated situations than those of the experiments. A point which deserves more attention than it has generally received is the extent to which the measured behaviour of rock specimens depends upon the experimental system. The extrapolation of behaviour induced by the system to different circumstances can be most misleading, and every effort must be made to distinguish between the properties of the rock and those of the experimental system.

Geologically, millions of years may be required for strains in the rock to change significantly, or they may change in seconds following an earthquake. In mining, and other engineering operations, the strains may change in milliseconds during blasting or over years as mining proceeds or a structure is built and loaded. In laboratory testing it is convenient to use strain rates in the range from 0·1 to 0·001 millistrains per second, corresponding to loading rates in the range 1,000–10 psi per second. In special cases it is possible to use strain rates as high as 10^4 millistrains per second and as low as 10^{-7} millistrains per second. Between these extremes differences in the behaviour of rock are observed; most rock being weaker and more ductile at low strain rates. However, the differences are only of the order of a hundred per cent for ten orders of magnitude change in the rate of strain, and the properties of most rocks can be regarded as being virtually independent of strain rate in the range covered by the usual laboratory tests.

6.2 Uniaxial compression

The uniaxial compression test, in which right circular cylinders or prisms of rock are compressed parallel to their longitudinal axis, is the oldest and simplest test, and continues to be one of the most convenient and useful

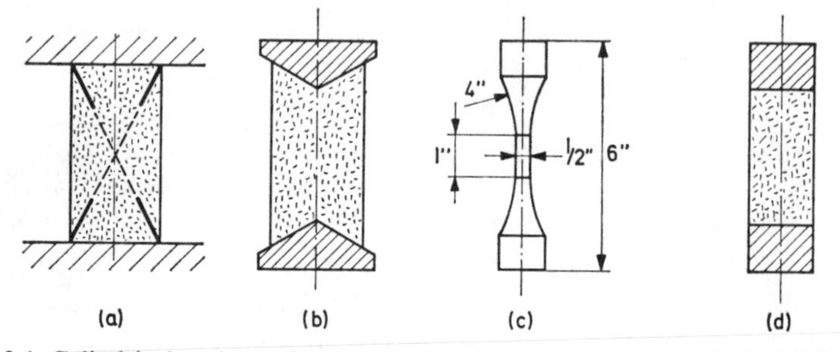

Fig. 6.2.1 Cylindrical rock specimens in uniaxial compression. (*a*) Failure initiation at the circumferential contact with the platen. (*b*) Conical end-pieces to eliminate frictional restraint. (*c*) Shaped specimen with central gauge length. (*d*) Matched end-pieces.

ways for determining the properties of rock. In the past tests were usually made on cubes, as is still the case with tests on concrete (B.S. 1881, 1952), or with cylinders having a length equal to their diameter, which is still the practice with coal, Evans and Pomeroy (1958).

The behaviour of rock specimens which are short in relation to their diameter is affected strongly by contact with the platens between which they are compressed. Even when the surfaces of the specimen and the platens are flat and parallel, the rigidity of the platens restricts the lateral expansion of the ends of the specimen. The recent tendency is to use relatively longer cylinders with a ratio between length, L, and diameter, D, which ranges from 2 to 3. Obert *et al.* (1946) found that the uniaxial strength of rock cylinders varied as

$$C_0 = C_1[0{\cdot}778 + 0{\cdot}222\, D/L], \qquad (1)$$

where C_1 is the value of the strength when $D/L = 1$. This indicates that the strength decreases with increasing length in relation to the diameter,

As a result, it might be thought that the true strength of rock can be determined from tests on cylinders with a ratio between length and diameter in excess of 2, where the stresses in the central portion of the specimen are affected only slightly by contact with the platens. However, this is not so, as experiment shows that failure usually starts where the circumference of the cylinder is in contact with the platen, Fig. 6.2.1 (*a*). Filon (1902) studied the stress distribution in cylinders with a ratio between length and diameter of unity and showed that the stress concentration at the circumferential contact is 1·69. Thus, the effect of lateral restraint is often to initiate failure at the contact between the specimen and the platen at a value of the load less than that corresponding to the true uniaxial strength of the specimen. This gives rise to the conical fragments based on each platen which are so often observed in uniaxial compression tests.

Various expedients have been adopted to avoid the effects of the discontinuity in section and properties at the contact between the specimen and a platen. As the platens are usually more rigid than the specimen, these effects are due more to frictional restraint than to bending of the surface of the platen. Some appreciation of the magnitude of this problem is gained from the solution of Siebel and Pomp (1927), who proposed that specimens be made with hollow conical ends and be compressed between conical end-pieces, the surfaces of which are inclined to the diameter of the specimen at the angle of friction, Fig. 6.2.1 (*b*).

Attempts to remove frictional restraint by lubricating the contact with graphite, molybdenum disulphide, and other solid lubricants have been made, while paper, lead, and other ductile materials have been inserted between the specimen and the platen to allow of lateral expansion. All these techniques must be treated with circumspection, as intrusion of the lubricant into the ends of the specimen or extrusion of the insert are likely to set up radial tensile stresses which promote longitudinal splitting of the specimen.

At present the most common method of avoiding these problems is to use shaped specimens, Barnard (1964), Brace (1964b), Hoek (1965), Murrell (1965), Fig. 6.2.1 (*c*). The stresses in such a specimen have their greatest values in the necked gauge length of the specimen. The shape of the specimen is determined from photoelastic studies so as to ensure a uniform distribution of the axial stress across the section of the specimen in this length. These specimens are troublesome to prepare, and the dimensions of the gauge length tend to be small.

Instead of compressing a specimen between wide platens, it may be compressed between metal end-pieces with the same section as that of the specimen, Fig. 6.2.1 (*d*). In this way, any effects due to bending of the interface are eliminated, and those due to lateral restraint are diminished because the end-pieces also expand radially. If the metal of the end-pieces is chosen so that their lateral expansion is the same as that of the rock specimen, that is, the ratio ν/E is the same for the metal of the end-piece and for the rock, then end-effects should disappear during the elastic portion of compression. Most rocks can be matched by suitable metals, and it might even be possible to extend this concept to the ductile region at the stress–strain curve. The method has been used by Cook (1962) and Hoek (1965).

The true mode of fracture in uniaxial compression is obscured by several factors. Shear fractures and conical or wedge-shaped fragments are often noticed, but these are associated with end-effects. Where steps have been taken to eliminate end-effects, longitudinal splitting has been observed, but this may often be the result of radial tension induced by the material inserted between the specimen and platen to diminish friction. Jaeger (1960a) studied the failure of rock specimens down to very low confining pressures and noted that the values for the strength decreased continuously and the shear mode of fracture persisted down to confining pressures as low as 50 psi, from which he deduced that failure is also by shear fracture in uniaxial compression.

The main difficulty is that failure in uniaxial compression is generally

Table 6.2.1

The uniaxial compressive strength, C_0, loading Young's modulus, E, and Poisson's ratio, ν, for a number of rocks.

Rock	C_0 (kpsi)	E (10^6 psi)	ν	Reference
Granite, Westerly	33·2	8·1	0·11	Brace (1964b) Walsh (1965c)
Quartzite, Cheshire	66·7	11·4		
Diabase, Frederick	70·6	14·4		
Marble, Tennessee	22·1	6·9		Cook (1965)
Granite, Charcoal	25·1	6·4		
Shale, Witwatersrand	24·9	9·8	0·23	Cook *et al.* (1966)
Granite Aplite (Chert)	85·2	12·0	0·20	
Quartzite, Witwatersrand	29·0	11·3		Wiebols *et al.* (1968)
Dolerite, Karroo	48·0	12·2		
Marble, Wombeyan	11·2	9·4		
Sandstone, Gosford	5·36	1·4		
Limestone, Solenhofen	32·5	7·7		

violent as a result of the rapid release of energy stored in the resilience of most testing machines, and its primary nature is obscured by subsequent phenomena. In a stiff testing machine with matched end-pieces Fairhurst and Cook (1966) have observed both longitudinal splitting and shear fractures in the uniaxial compression of specimens of quartzite.

Values of typical uniaxial compressive strengths, C_0, and Young's modulus, E, for the linear portion of the loading curve are given in Table 6.2.1 for a variety of rock types.

6.3 Uniaxial tension

Despite the importance of the tensile strength of rock in practice and in connection with theories of failure, direct measurements of tensile strength are difficult and are not commonly made. Early measurements were made by cementing cylindrical specimens of rock into grips which gave rise to stress concentrations at the grip in a manner analogous to those at the contact between specimen and platen in the compression test. Fairhurst (1961) overcame much of this difficulty by cementing the ends of the rock specimen with epoxy resin direct to a steel end-piece having the same cross-section as the specimen. These end-pieces were attached to flexible cables, through which the tensile load was applied, so as to minimize the bending stress transmitted to the specimen. Other investigators have used universal- or ball-joints for this purpose. Carefully shaped specimens have also been used to ensure that failure occurs in a region where the axial tensile stress is uniform across the section of the specimen.

It is known that minor scratches on the surfaces of tensile specimens may have a considerable effect on their strength, and the surfaces are usually finely ground or polished to avoid this effect. However, it is probably of less significance in rock than in the case of glass or metals, because of the large number of mechanical defects inherent in rock and its heterogeneous structure.

The tensile strength of rock is more variable and more influenced by specimen size than any other mechanical property of rock. This factor is discussed in § 7.4 and makes it necessary to conduct large numbers of tensile tests and interpret the results carefully in the light of the manner in which specimens are selected.

Indirect tests for determining the tensile strength of rock, §§ 6.10, 6.11, 6.12, are more common than the uniaxial tensile test, but suffer from the defect that they involve inhomogeneous stresses. Extension of shaped specimens in triaxial apparatus avoids many of these difficulties, Brace (1964b), Murrell (1965), but is not a true uniaxial tensile test.

6.4 Triaxial compression

The triaxial compression test has proved to be the most useful test in the study of the mechanical properties of rock over a wide range of values for stress and at different temperatures. Tests have been conducted with temperatures ranging from room temperature to 800° C and with confining pressures from less than 100 psi to over 1,000 kpsi. All these tests conform to the same basic arrangement, Fig. 6.4.1 (*a*), in which the major principal stress is applied along the axis of a cylindrical rock specimen by the testing machine and equal minor principal stresses are applied to the curved surfaces of the specimen through an impervious metal or rubber jacket by fluid-confining pressure.

Specimens usually have a ratio between length and diameter from 2 : 1 to 3 : 1, with diameters in the range from 6 in. to less than an inch. In general, the smallest specimens are used for tests at high pressures and temperatures where confining pressure is applied by argon or carbon dioxide instead of the hydraulic oil usually used at low temperatures and pressures. Sometimes, shaped specimens are used, Brace (1964b), and

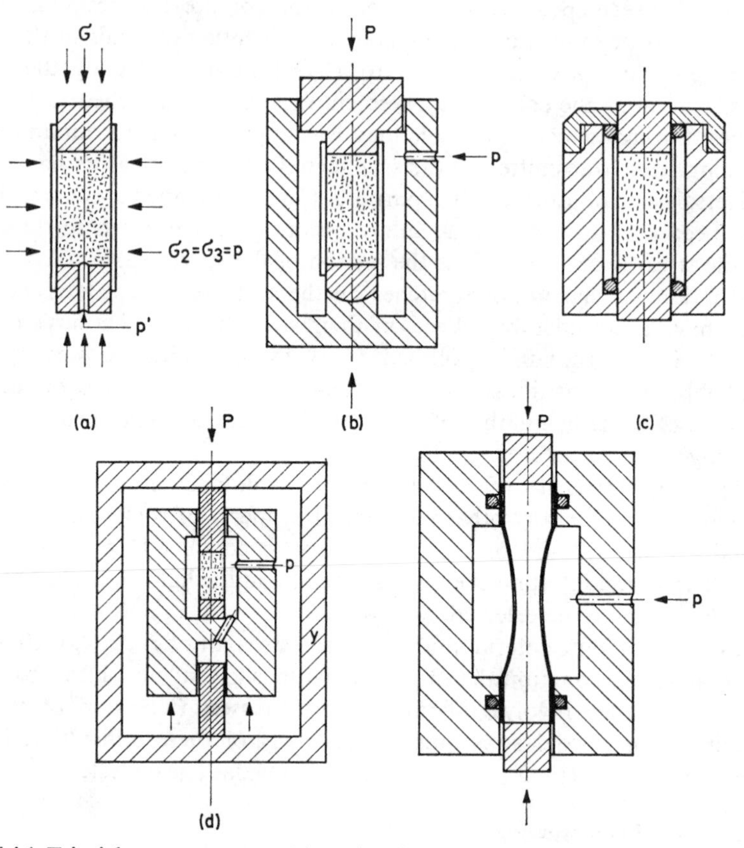

Fig. 6.4.1 Triaxial test apparatus. (*a*) Jacketed cylindrical rock specimen and end-pieces with provision for pore pressure. (*b*) and (*c*) Simple triaxial cells. (*d*) Constant-volume triaxial cell. (*e*) Cell for shaped specimens.

where the effects of pore pressure are being studied, provision is made through the end-pieces to admit pore fluid and control its pressure.

A typical triaxial cell for use in a conventional testing press is that of the U.S. Bureau of Reclamation, Fig. 6.4.1 (*b*). This cell suffers from two disadvantages. First, the confining pressure generates an axial component of force which reacts against the load applied by the press. Second, as the specimen compresses, the volume of the cell decreases due to the relatively large piston area, which makes control of the confining pressure difficult. This effect is greatly diminished in the simple triaxial cell, Fig. 6.4.1 (*c*), which has combined pistons and end-pieces with the same cross-section as that of the specimen. Donath (1966) has designed a very compact cell of this type which is coupled coaxially to its own jack by a steel collar and has provision for studies including pore pressure, p'. Griggs *et al.* (1960) and Paterson (1964) avoid interaction between axial displacement of the end-pieces and the confining pressure completely by using two pistons coupled together by a yoke, *Y*, Fig. 6.4.1 (*d*), so that the volume of confining fluid remains constant. A triaxial cell for use with shaped specimens, which also allows triaxial extension tests to be made, is illustrated in Fig. 6.4.1 (*e*).

Most triaxial cells have end-pieces with a cross-section equal to that of the specimen, so as to facilitate fitting a jacket over the specimen. Since the platens of the testing machine may not be parallel, it is usual to incorporate a spherical seat on one end-piece. This introduces an undesirable degree of asymmetry into the system which may influence the nature of the ultimate fracture and is certainly undesirable where lateral displacements occur, as in the study of friction, § 3.7. This difficulty can be overcome by the use of a spherical seat at each end, which complicates the system and increases the chances of buckling phenomena arising. Careful construction and alignment of the cell and press, together with careful grinding of the ends of the specimen to ensure that these are parallel, probably provide the least uncertain method.

Because the stresses in a triaxial test are homogeneous and the geometry is symmetrical the results are easier to interpret than those of other experiments. However, one important factor is frequently overlooked. The strength of most rock specimens varies from one part of the specimen to another. The major principal stress is applied to the ends of the specimen through steel end-pieces which are more rigid than the rock. In consequence, when a weak part near the end of a specimen begins to fail there is a tendency for the stress on this part to diminish and that on the stronger parts surrounding it to increase. This effect also tends to occur within rock, but to a lesser extent, because the rigidity of the rock is less than that of the end-pieces. The minor principal stresses are applied to the circumferential surface by a fluid with negligible rigidity. When a weak part near this surface of the specimen begins to fail, the confining stress applied to it does not decrease as it would on a weak part within the rock. The results of triaxial tests must, therefore, be susceptible to surface effects as well as end-effects. Certain reservations must be applied to results of tests on small specimens where the surface-to-volume ratio is high. Strain measurements with gauges cemented to the surface of small specimens are especially suspect.

Hardy (1966a) and others have done experiments with several strain gauges cemented around the circumference of a specimen and have observed that the strains measured by different gauges are very variable, even to the extent that some gauges measured tensile strains in a compression test. This could be due to eccentric loading or bending. The effect of eccentric loading is to superpose a couple on the axial compression. If the axial load, P, is applied at a distance, e, from the axis of a specimen the resultant axial stress in any position is

$$\sigma = P/A + Pey/I, \tag{1}$$

where A = cross-sectional area of the specimen;
I = moment of inertia of the cross-section;
y = distance from the axis of the specimen in the direction of the eccentricity.

The minimum axial stress for a cylindrical specimen is

$$\sigma_{\min} = P(1 - 8e/D)/A, \tag{2}$$

so that $8e > D$ for the surface strain to be tensile. Surface observations of tensile strain may equally well be due to surface effects, which allow parts of the surface to deform differently from the average deformation in the specimen as a whole, or inelastic effects which are not uniformly distributed through the specimen, such as incipient fracture. It would therefore seem desirable to determine the strain in a specimen indirectly by measuring the compression and expansion of the specimen as a whole, as does Donath (1966), especially where inelastic behaviour is being studied. This can introduce errors due to the elasticity of the end-pieces and contact effects, which must be corrected by calibration using standard metal cylinders and rock specimens of different lengths.

6.5 Extension in the triaxial test

In the usual triaxial tests $\sigma_1 > \sigma_2 = \sigma_3$, but it is possible to make the confining pressure greater than the axial stress, so that $\sigma_1 = \sigma_2 > \sigma_3$. Böker (1915) and Heard (1960) have made experiments under conditions where the axial stress $\sigma_3 > 0$, while Brace (1964b), using shaped specimens, was able to make the axial stress tensile. If A_2 is the maximum cross-sectional area of a shaped specimen and A_1 its area over the gauge length, the confining pressure, p, generates an axial tensile stress, $-kp$, where

$$k = (A_2 - A_1)/A_1. \tag{1}$$

If an axial load, P, is applied, the resultant axial stress in the gauge length is

$$\sigma_3 = [P - (A_2 - A_1)p]/A_1. \tag{2}$$

A state of uniform compression, $\sigma_1 = \sigma_2 = \sigma_3 = p$, exists in the gauge length when

$$P = pA_1(1 + k). \tag{3}$$

In practice, P and p are increased together in accordance with (3) until some pre-determined values are reached. The axial load P is then decreased, so that σ_3 follows the path DC in Fig. 6.5 until failure occurs at the point C. In this way the failure surface for the condition $\sigma_1 = \sigma_2$ can be measured as a function of σ_3. Extension fractures are observed in this type of experiment even when σ_3 is compressive. Failure occurs at the point B in biaxial compression, $\sigma_1 = \sigma_2 > 0$, $\sigma_3 = 0$. Depending upon

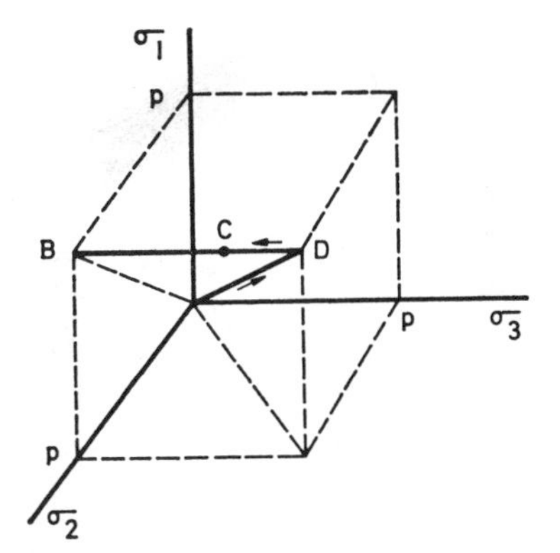

Fig. 6.5 The stress path in a triaxial extension test.

whether the value of the stress at failure is dependent upon the intermediate stress or not, the value of p will be different from or equal to the uniaxial compressive strength of the rock C_0. However, the different types of fracture in these two tests and the differences in geometry and the mode of application of the major principal stress by rigid end-pieces and fluid pressure must be noted.

6.6 Results of triaxial tests

The classical triaxial experiments were made by von Kármán (1911) on specimens of Carrara marble in copper jackets at room temperature. The results of these experiments typify most of the salient features of rock behaviour in triaxial tests, Fig. 6.6.1. At zero and low confining stress the marble failed in a brittle way along a single inclined shear fracture. The amount of ductile deformation and the strength increased progressively with increasing confining pressure until fully ductile deformation occurred with apparent work-hardening, giving rise to barrell-shaped specimens. This deformation is largely the result of sliding across a multiplicity of intersecting shear planes and is not true plastic deformation. Von Kármán's experiments and many of the subsequent ones by geophysicists such as Griggs (1936), Handin and Hager (1957), Paterson (1958), Heard (1960), and others have been directed towards studying the transition from brittle failure to ductile deformation. At high temperatures, above 500° C, many rocks become extremely ductile, and true plastic deformation of their constituent crystals occurs.

Work of the greatest value for rock mechanics was done by the U.S.

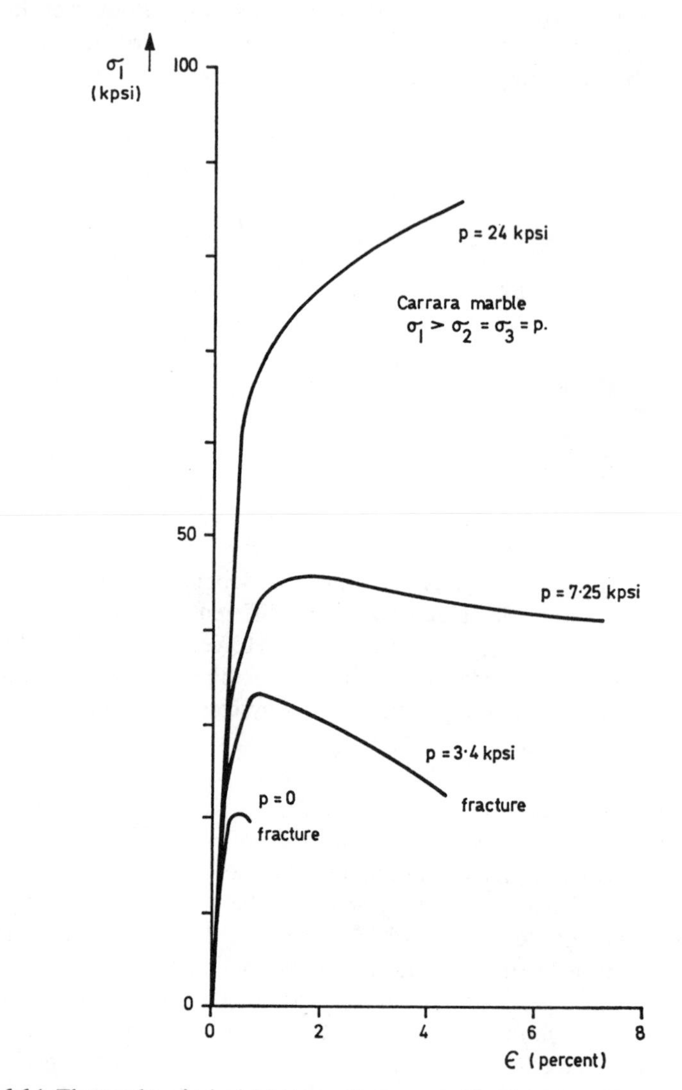

Fig. 6.6.1 The results of triaxial tests on Carrara marble (from von Kármán, 1911).

Bureau of Reclamation (1953) in studies of a wide range of foundation rocks. Since then a great number of investigations have been made. Hoek (1965) has surveyed the published data on the strength of rock in triaxial tests, Table 6.6.1. By normalizing the values of both the major and minor

Table 6.6.1
Summary of triaxial test results on rock used in Fig. 6.6.2

Rock	C_0* psi	Letter	Reference
Marble	20,000	a	Von Kármán (1911)
Carthage Marble	10,000	b	Bredthauer (1957)
Carthage Marble	7,500	c	Bredthauer (1957)
Wombeyan Marble	10,000	d	Jaeger (1960)
Granite Gneiss	25,500	e	Jaeger (1960)
Barre Granite	24,200	f	Robertson (1955)
Westerly Granite	33,800	g	Brace (1964)
Iwaki Sandstone	1,780	h	Horibe and Kobayashi (1960)
Rush Springs Sandstone	26,000	i	Bredthauer (1957)
Pennant Sandstone	22,500	j	Price (1960)
Darley Dale Sandstone	5,780	k	Price (1960)
Sandstone	9,000	l	Jaeger (1960)
Oil Creek Sandstone	†	m	Handin
Dolomite	24,000	n	Bredthauer (1957)
White Dolomite	12,000	o	Bredthauer (1957)
Clear Fork Dolomite	†	p	Handin
Blair Dolomite	†	q	Handin
Webtuck Dolomite	22,000	r	Brace (1964)
Chico Limestone	10,000	s	Bredthauer (1957)
Virginia Limestone	48,000	t	Bredthauer (1957)
Limestone	20,000	u	Jaeger (1960)
Anhydrite	6,000	v	Bredthauer (1957)
Knippa Basalt	38,000	w	Bredthauer (1957)
Sandy shale	8,000	x	Bredthauer (1957)
Shale	15,000	y	Bredthauer (1957)
Porphyry	40,000	z	Jaeger (1960)
Sioux Quartzite	†	A	Handin
Frederick Diabase	71,000	B	Brace (1964)
Cheshire Quartzite	68,000	C	Brace (1964)
Chert dyke material	83,000	D	Hoek (1963)
Quartzitic shale (dry)	30,900	E	Colback and Wiid (1965)
Quartzitic sandstone (dry)	9,070	F	Colback and Wiid (1965)

* Uniaxial compressive strength.
† Presented in dimensionless form by McClintock and Walsh (1962).

principal stresses at failure with respect to the value of the uniaxial compressive strength he was able to plot all the data together, Fig. 6.6.2, and compare it with the predictions of the various criteria for failure, §§ 4.8, 4.9, 10.14, 12.7, using several values for the coefficient of friction in the modified Griffith criterion, which is essentially similar to the Coulomb criterion § 4.6. The results of this comparison lend credence to these

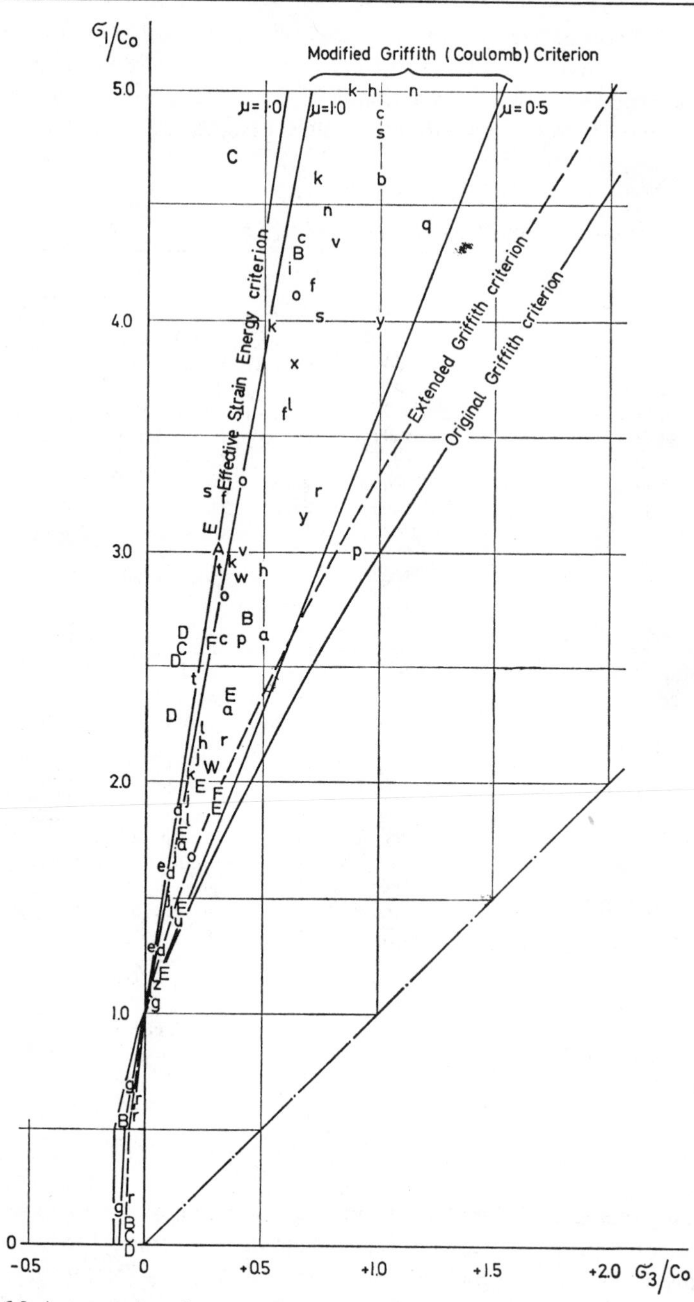

Fig. 6.6.2 A comparison between the measured strengths of rock and various failure criteria.

criteria as predictions of the triaxial strength of rock, and indicate that the coefficients of internal friction generally lie between 0·5 and 1·0, though some values appear to be in excess of unity.

In general, failure in triaxial tests results in a single, roughly plane, shear fracture inclined at a narrow angle to the direction of the major principal stress, which is usually parallel to the axis of the specimen. There is a tendency for this angle to increase as the values of the principal stresses at failure increase. This phenomenon and the values of these angles are roughly consistent with the predictions of § 4.6. However, the ratio between the length and diameter of triaxial specimens is usually 2 : 1, and the fracture often runs into one platen or diagonally across the specimen from one end to the other. Paul and Gangal (1967) has suggested that the angle of fracture is determined by the fact that it runs diagonally across the specimen. However, similar but smaller angles of fracture have been observed in specimens with a length-to-diameter ratio of 3 : 1. Mogi (1966) and Wiebols *et al.* (1968), using matched end-pieces, found some fractures which intersected the circumference of the specimens away from the ends. Nevertheless, it should be noted that in the absence of buckling one or more inclined fractures are necessary to allow the platens of the testing machine to converge as the specimen fails, so that this mode of fracture may be a result of the geometry of the triaxial test. According to §§ 4.6, 4.7, two planes of shear fracture parallel to the intermediate principal stress are possible. On account of the axial symmetry of the triaxial test, any number of planes should be possible. Usually only one fracture plane is observed, but multiple fracture planes develop under conditions when the rock deforms in a fully ductile manner, as in von Kármán's experiments at high confining pressure.

The preceding discussion is concerned with dry triaxial tests. Triaxial tests are also made on saturated specimens with pore-pressure (undrained specimens) and with zero pore-pressure (drained specimens), Heard (1960), Handin *et al.* (1963), Donath (1966), and Neff (1966). In general, the results of triaxial tests with pore-pressure on all but exceptionally impermeable rocks may be interpreted in accordance with the concept of effective stress, § 8.8. According to this, the principal stresses, σ_1, σ_2, σ_3, in the failure criterion are replaced by the effective principal stresses, $\sigma_1' = \sigma_1 - p'$, $\sigma_2' = \sigma_2 - p'$, $\sigma_3' = \sigma_3 - p'$, where p' is the pore-pressure.

During the last decade the phenomenon of dilatancy, §§ 4.2 and 4.5, has emerged as an important result of triaxial tests. Dilatancy appears to be caused by the formation and extension of open microcracks within a rock specimen undergoing compression, especially as the value of the maximum principal applied stress approaches the strength of the specimen. These cracks seem to have a high aspect ratio and to have their long axes orientated parallel to the direction of the maximum principal stress.

Dilatancy affects not only the quasi-static deformability of the rock, but also its electrical resistivity (Brace and Orange, 1966a, b) and the velocities of propagation of seismic waves (Hadley, 1975), particularly in directions normal to that of the maximum principal stress.

A summary of the effects of stress during a triaxial test up to the peak of the stress–strain curve only of hard, brittle rock is given in Fig. 6.6.3. Among other things, these properties are of interest in connection with the mechanism of earthquakes, and as precursors for their potential prediction (Brace, 1975).

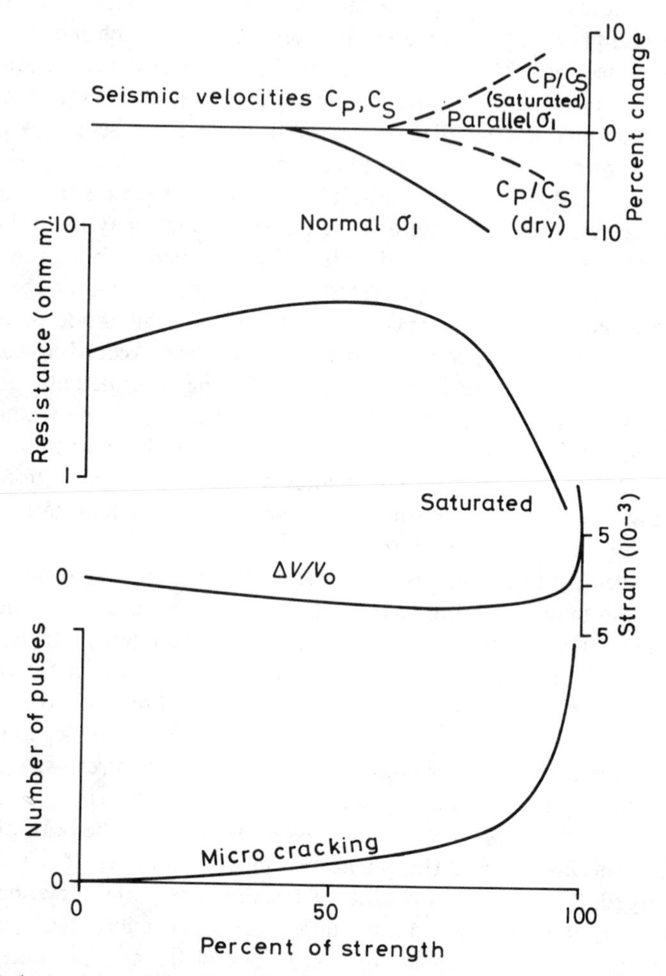

Fig. 6.6.3 A summary of the effect of increasing maximum principal compressive stress on the microseismicity, dilatancy, electrical resistivity and velocities of propagation of seismic waves on a hard, brittle rock such as Westerly granite (after Brace, 1971 and Hadley, 1975).

In most triaxial tests dilatancy occurs under artificial conditions where the confining stress is not changed by the radial expansion of the specimen. If the specimen were part of a rock mass, its radial expansion would generate a reaction in the surrounding rock in the form of increased confining stress. Accordingly, triaxial tests have been made in a sealed, stiff cell so that the pressure of the confining fluid increases and resists both Poisson expansion and inelastic dilatancy, § 4.5. The results of such tests show that dilatancy still occurs giving rise to an increase in confining pressure. This increase in confining pressure is similar in shape to the dilatant radial expansion in conventional triaxial tests. However, the increase in confining pressure with compression tends to make the axial stress–strain curve less linear; deformation up to maximum stress is more ductile and fracture beyond it is less brittle in a stiff confining cell than it would be in a conventional cell.

The nature of the change in resistance of a material to deformation with increasing deformation is of critical importance where problems involving instability are concerned (Nur, 1974), such as in earthquakes, §§ 17.10 and 17.11 and rockbursts, § 18.12. 'Work-hardening' materials where the resistance to deformation always increases with increasing deformation, are essentially stable, and any local increase in deformation gives rise to a local increase in resistance to deformation which, therefore, tends to spread uniformly throughout such materials. However, in 'work-softening' materials an increase in deformation results in a decrease in resistance to further deformation, so that any local concentration of deformation softens the material in its vicinity, focusing further deformation into the same area. Since this results in the concentration of deformation at the edges of the softened area, this tends to spread outwards to form a fracture surface. Such 'work-softening' behaviour is potentially unstable. The complete stress–strain curve for rock, § 4.2, exhibits elastic, 'work-hardening' behaviour up to its peak stress, or the strength, and brittle 'work-softening' behaviour at strains greater than that corresponding to the strength.

Triaxial experiments are now being extended to studies of the behaviour and properties of rock specimens at strains greater than that corresponding to the peak of the complete stress–strain curve. Instabilities involving the fracture of originally intact specimens and 'stick–slip' sliding, §§ 3.4 and 3.5, on these new fractures and on pre-existing faults or cuts have been made (Brace, 1972; Byerlee and Brace, 1968; Byerlee, 1967 and Byerlee, 1970). Most of these laboratory studies of 'work-softening' behaviour involving instabilities have been complicated by the relatively low stiffness of the testing machines used. However, the advent of very stiff and servo-controlled testing machines, § 6.13, is likely to result in a much better understanding of these phenomena than has been possible in the past.

6.7 Tests with homogeneous polyaxial stress

The term triaxial has become so closely identified with the type of test in which two equal principal stresses are applied to a specimen by confining pressure, §§ 6.4, 6.5, 6.6, that the term polyaxial has been chosen to describe tests in which the values of the three principal stresses differ from one another. In principal, polyaxial homogeneous stresses, $\sigma_1 > \sigma_2 > \sigma_3$, can be obtained by loading a rectangular parallepiped across its three pairs of mutually perpendicular surfaces. If this is done with end-pieces, or platens, little reliance can be placed on the results, quantitatively, because of the unknown effects of friction between the surfaces of the specimen and the end-pieces. Difficulties arising from this were shown to be severe in the case of uniaxial compression, § 6.2. In the case of polyaxial stresses, where at least two sets of surfaces on the specimen must be loaded, the effects of frictional restraint pervade the whole specimen and prove to be formidable.

Föppl (1900) loaded rock specimens biaxially, $\sigma_1 > \sigma_2$, $\sigma_3 = 0$, and

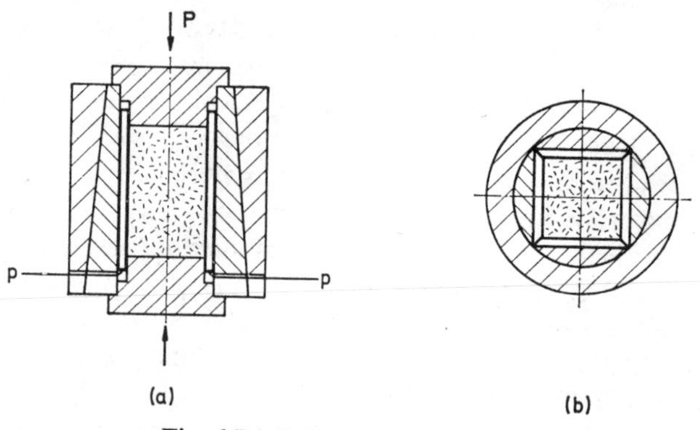

Fig. 6.7.1 Polyaxial test apparatus.

observed splitting perpendicular to σ_3, while Hobbs (1962) found shear fractures parallel to the direction of the intermediate principal stress with specimens of coal.

In the triaxial test the effect of friction between the specimen and the end-pieces is kept within acceptable limits mainly by choosing the geometry of the specimen so that only a small proportion of its surface is loaded through these rigid components, the remainder being loaded by fluid pressure to avoid frictional restraint. The impermeable jackets used to exclude the confining fluid from the specimen introduce a small degree of lateral and axial restraint. Even where metal jackets are used, these effects are small, provided that the jacket is thin and adequate corrections for them can be made by calculation. Following this line of reasoning, Hojem and Cook (1968) have constructed a polyaxial cell for testing rectangular specimens of rock 1 in. square by 3 in. long. As in the triaxial test, an axial load is applied through steel end-pieces having the same cross-section as that of the rock specimen, while opposite sides of the specimen are confined between pairs of thin, copper flat-jacks. The flat-jacks react against segmental pieces of brass restrained by a steel cylinder, Fig. 6.7.1 (*a*), the bore of which is tapered to facilitate assembling and dismantling of the apparatus. The mitred joints between adjacent flat-jacks, Fig. 6.7.1 (*b*), ensure that virtually the whole surface of the specimen is subjected to flat-jack pressure.

A comparison between the strengths of cylindrical specimens 1 in. in diameter by 3 in. long of Karroo Dolerite measured in a conventional triaxial cell using a latex jacket and the strengths of 1 in. square by 3 in.

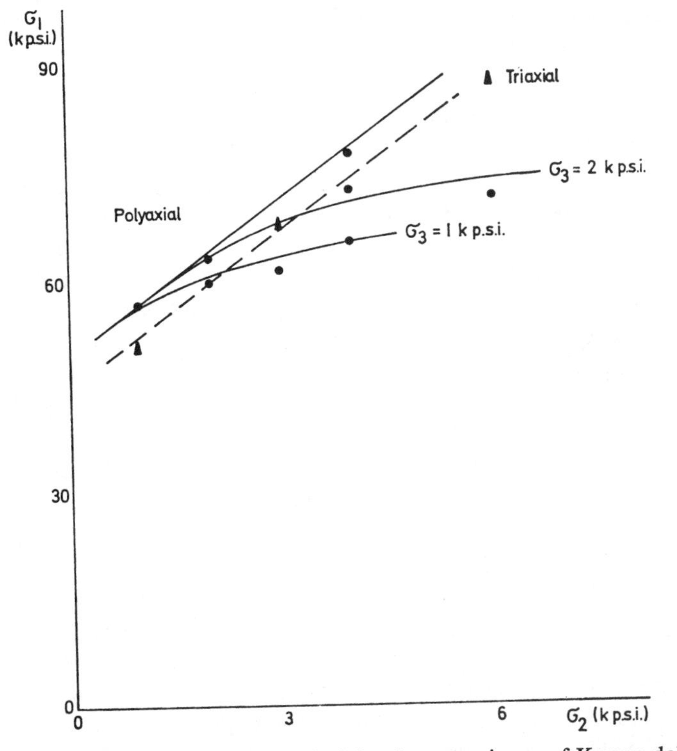

Fig. 6.7.2 Results of polyaxial and triaxial tests on specimens of Karroo dolerite.

long specimens of the same rock measured in the polyaxial cell with equal pressures in all four flat-jacks showed that the former strengths were consistently lower than the latter by a small amount, Fig. 6.7.2. As the difference in strength was independent of confining pressure and persisted down to zero confining pressure, it must be ascribed to a geometrical effect or a slight difference in the strength of the rock from which the two suites of specimens were prepared. With equal pressures in the four flat-jacks the fracture surface is curved about the axis of the specimen, and its intersection with the sides of the specimen is axially symmetric. When different pressures are used in the two pairs of opposite jacks a single or two intersecting shear fractures in the plane of the intermediate stress result. For Karroo Dolerite, Fig. 6.7.2, and Sölenhofen limestone the effect of increasing the value of the intermediate principal stress is to increase the strength of the rock.

6.8 Hollow cylinders

Hollow cylinders with axial load and internal or external fluid pressure provide the most ready method of studying the strength and fracture of rock under a wide variety of unequal principal stresses. In thin-walled cylinders, or tubes, the stresses are almost homogeneous, and this is one of the standard methods for studying the strength of metals under combined, usually biaxial but sometimes polyaxial, stresses, Lode (1926), Nadai (1950), and Drucker (1962). More recently, it has been used for studying plastics, ceramics, and graphite. Sedlacek and Holden (1962) have used the hoop stress in rings to study uniaxial tension, and Ely (1965) has used graphite tubes with expanded ends which can be held in grips to provide axial tension or compression with internal fluid pressure. Broutman and Cornish (1965) have studied ceramic tubes with internal and external fluid pressure and tension generated by the action of the internal pressure on plugs cemented into the ends of the tubes. Cornet and Grassi (1961) studied the brittle fracture of cast iron using tubes and give a bibliography of work on this material.

The preparation of thin-walled cylindrical specimens of rock is obviously difficult. The wall thickness, and hence the specimen size, are determined by the grain size of the rock. Handin, Heard, and Magouirk (1967) have succeeded in preparing and testing thin-walled cylinders of fine-grained dolerite and limestone, to which they applied torsion in addition to surface loads. With rock, most workers have used thick-walled cylinders and, consequently, inhomogeneous stresses. Using hollow cylinders with axial loads and an external steel jacket carefully fitted to provide confining pressure, Adams (1912) observed failure by spalling at the inner surface of the cylinder. His results and the systems of fracture described below were discussed by King (1912). Robertson (1955) used cylinders of rock with a range of ratios between their internal and external diameters stressed by fluid pressure applied to their ends and outer surfaces. He discussed his results, in which failure started on the interior surface, in terms of elastic–plastic theory. The system of hollow cylinders with axial load and external fluid pressure has been used by Hobbs (1962) for coal, and Obert and

Stephenson (1965), and Jaeger and Hoskins (1966b) for a variety of rock types. The theory of the stresses in hollow cylinders subjected to internal and external pressure is developed in § 5.11.

Consider first the most usual case of a jacketed hollow cylinder with inner and outer radii of R_1 and R_2, respectively, subjected to an axial load P and an external fluid pressure, p_2. The radial and tangential stresses in the cylinder are given by § 5.11 (9), (10). The principal stresses, axial, σ_z, tangential, σ_θ, and radial, σ_r, at the inner and outer surfaces of the cylinder, respectively, are:

$$\sigma_{z1} = P/\pi(R_2^2 - R_1^2), \quad \sigma_{\theta 1} = 2p_2R_2^2/(R_2^2 - R_1^2), \quad \sigma_{r1} = 0, \qquad (1)$$

and

$$\sigma_{z2} = P/\pi(R_2^2 - R_1^2), \quad \sigma_{\theta 2} = p_2(R_2^2 + R_1^2)/(R_2^2 - R_1^2), \quad \sigma_{r2} = p \qquad (2)$$

Equations (1) and (2) show that the maximum and minimum values of the principal stresses are found at the inner radius and that the value of the least stress is zero. On the basis of the criteria developed in §§ 4.6, 4.7, 4.8,

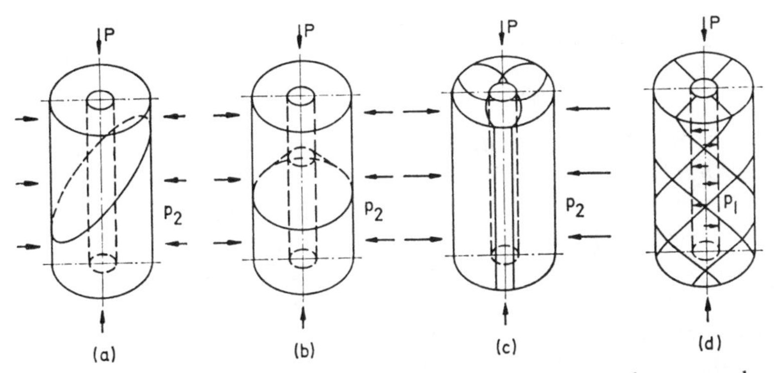

Fig. 6.8.1 Different systems of fracture in hollow cylinders subjected to external pressure and axial load, (*a*), (*b*), and (*c*), and internal pressure and axial load (*d*).

failure would be expected to start at the inner surface at a value of the maximum principal stress which differs from the uniaxial strength of the rock by an amount determined by the influence, if any, of the intermediate principal stress.

If $P/p_2 > 2\pi R_2^2$ the principal stresses at the inner surface are:

$$\sigma_1 = P/\pi(R_2^2 - R_1^2), \quad \sigma_2 = 2p_2R_2^2/(R_2^2 - R_1^2), \quad \sigma_3 = 0. \qquad (3)$$

For relatively small values of p_2 and R_1 it is found that failure takes place much as it does in a solid cylinder in triaxial compression, Fig. 6.8.1 (*a*), forming a single shear fracture across the whole cylinder at some narrow angle to its axis. The results of § 6.7 suggest that the effect of a small intermediate stress is to enhance the strength. Thus, the axial stress necessary to produce failure is increased through the body of the cylinder by the confining pressure, p_2, and at the inner surface by the tangential stress, σ_2. Even if failure does first start at the inner surface, it is unlikely to have an overriding influence on the resulting fracture unless R_1 is a large fraction of R_2. For relatively larger values of p_2 and R_2 the characteristic type of fracture forms a portion of a conical surface with a small semi-apical angle coaxial with the specimen, Fig. 6.8.1 (*b*). The theory developed in § 12.7 suggests that once the value of the intermediate principal stress exceeds some optimum value, further increases reduce the strength. Thus, failure can be expected to start at the inner surface and control the resultant fracture if R_1 is a sufficiently large fraction of R_2. It should be noted that the conical fracture surface is tangential to the direction of the intermediate principal stress.

The transition from a single shear fracture to a conical fracture can be made at similar values of the intermediate principal stress by varying the ratio R_1/R_2. This would seem to provide an interesting method, which has as yet not been exploited, of exploring the effects of size and stress gradient.

If $P/p_2 < 2\pi R_2^2$ the principal stresses at the inner surface become:

$$\sigma_1 = 2p_2R_2^2/(R_2^2 - R_1^2), \quad \sigma_2 = P/\pi(R_2^2 - R_1^2), \quad \sigma_3 = 0 \qquad (4)$$

because the value of the tangential compressive stress exceeds that of the axial compressive stress. In this case failure produces fractures along spiral surfaces parallel to the axis of the cylinder and to the direction of the intermediate principal stress. A special case of this system which happens to be experimentally convenient is that where the axial load is generated by the pressure p_2 acting on caps covering the ends of the cylinder, so that $P = p_2\pi R_2^2$ and $P/p_2 = \pi R_2^2$, Robertson (1955), Fig. 6.8.1 (*c*).

Second, consider the case of a hollow cylinder with an internal jacket which is subjected to an internal fluid pressure p_1 and an axial load P. If $P/p_1 < \pi(R_2^2 - R_1^2)$ the principal stresses at the inner surface are:

$$\sigma_1 = p_1, \quad \sigma_2 = P/\pi(R_2^2 - R_1^2), \quad \sigma_3 = -p_1(R_2^2 + R_1^2)/R_2^2 - R_1^2), \qquad (5)$$

and failure usually occurs as a plane diametral extension fracture. Values of the tensile strength calculated from (5) are usually greater than those found from direct measurements in uniaxial tension. This can probably be attributed to the effects of stress gradient, § 7.6.

If $P/p_1 > \pi(R_2^2 - R_1^2)$ the principal stresses at the inner surface become:

$$\sigma_1 = P/\pi(R_2^2 - R_1^2), \quad \sigma_2 = p_1, \quad \sigma_3 = -p_1(R_2^2 + R_1^2)/(R_2^2 - R_1^2). \qquad (6)$$

In this case the intermediate principal stress is radial and helicoidal fractures, Fig. 6.8.1 (*d*), are observed.

Assuming, as is suggested in § 4.4, that there is a unique surface

$$\sigma_1 = f(\sigma_2, \sigma_3) \tag{7}$$

defining the stresses at which failure occurs, much of this surface can be mapped using various combinations of p_1, p_2, and P, provided that the effects due to inhomogeneous stresses can be kept relatively small. Certain sections of the surface defined by (7) are well known, Fig. 6.8.2. The triaxial curve, C_0, C, extends from C_0 on the σ_1-axis with a projection OC', $\sigma_2 = \sigma_3$, in the σ_2–σ_3 plane.

The biaxial curve, C_0B, starts from the same point and extends in the σ_1–σ_2 plane. This curve is traced in the systems defined by (3) and (4). It should be noted that the major principal stress in the former system is the

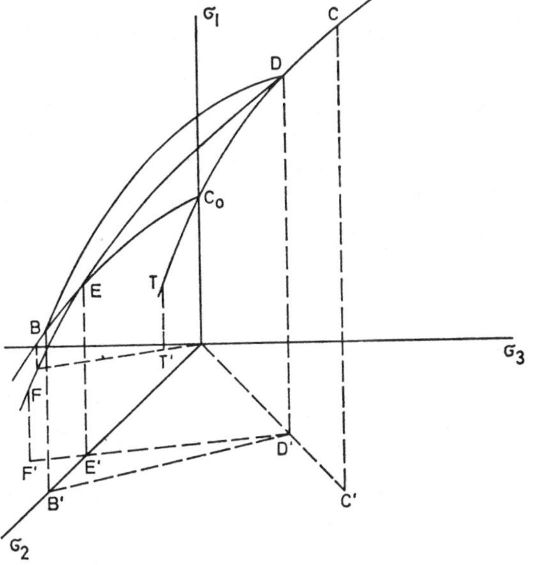

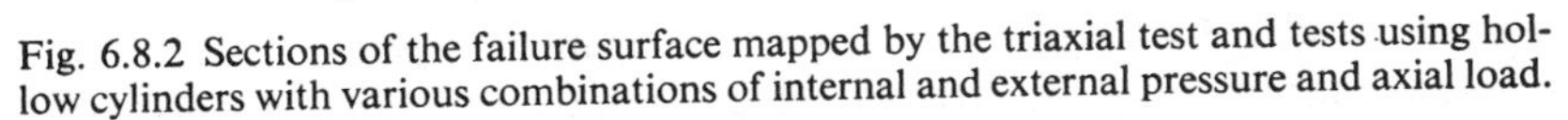
Fig. 6.8.2 Sections of the failure surface mapped by the triaxial test and tests using hollow cylinders with various combinations of internal and external pressure and axial load.

only homogeneous stress, while in the latter system it is only the intermediate principal stress which is homogeneous. If stress-gradient has an effect, the two curves, C_0B, traced by these systems should differ from one another. The systems defined by (6) trace curves such as C_0T with projections in the σ_2–σ_3 plane defined by

$$(R_2^2 + R_1^2)\sigma_2 + (R_2^2 - R_1^2)\sigma_3 = 0. \tag{8}$$

In each of the systems (5) and (6) only one of the principal stresses is homogeneous: the intermediate principal stress in the first system and the major principal stress in the second. If stress-gradient has an effect the curves traced by these systems will not intersect one another.

By using an axial load concurrently with both internal and external pressure further regions of the failure surface can be studied.

If $P/\pi > [2p_2R_2^2 - p_1(R_2^2 - R_1^2)]$ and $p_2 > p_1$ all the principal stresses are compressive and those at the inner surface are given by:

$$\sigma_1 = P/\pi(R_2^2 - R_1^2),$$
$$\sigma_2 = [2p_2R_2^2 - p_1(R_1^2 + R_2^2)]/(R_2^2 - R_1^2), \quad \sigma_3 = p_1. \tag{9}$$

This system traces curves such as BD between the triaxial curve and the plane $\sigma_3 = 0$ with projections in the σ_2–σ_3 plane defined by

$$(R_2^2 - R_1^2)\sigma_2 + (R_2^2 + R_1^2)\sigma_3 = 2p_2R_2^2. \tag{10}$$

When $p_2 < p_1$ the principal stresses at the inner surface are

$$\sigma_1 = P/\pi(R_2^2 - R_1^2),$$
$$\sigma_2 = p_1, \quad \sigma_3 = [2p_2R_2^2 - p_1(R_1^2 + R_2^2)]/(R_2^2 - R_1^2), \tag{11}$$

and σ_3 may become tensile. This system traces curves such as DEF on the failure surface with projections in the σ_2–σ_3 plane defined by

$$(R_2^2 - R_1^2)\sigma_3 + (R_2^2 + R_1^2)\sigma_2 = 2p_2R_2^2, \tag{12}$$

which are parallel to (8).

These systems allow most of the failure surface to be studied. Experiments indicate that σ_1 varies continuously over much of the failure surface, certainly for positive values of σ_3. The results of experiments on hollow cylinders of trachyte, Hoskins (1967), 4·5 in. in length with inside and outside diameters of 1 in. and 2 in., respectively, are shown in Fig. 6.8.3.

Anomalously high values of tensile strength have been found when failure occurs as an extension fracture, and Robertson (1955) observed different modes of failure for ratios of R_2/R_1 less than 3 and greater than 3, respectively. All the results of these experiments are affected in some degree by inhomogeneous stresses, which requires that they be interpreted with some care. The theory of elasticity holds reasonably well only up to the initiation of failure, and the preceding equations in this section should be used only to calculate the stresses at failure from those values of P, p_1, and p_2 at which failure first starts at the inside surface. These values do not necessarily coincide with the maximum values at which the whole specimen collapses, which are ordinarily used in testing to determine the stresses at failure. The start of failure at the inner surface should be readily detectable as a change in the slope of the curve of internal volume strain of the cylinder as a function of the applied loads. If the effect of inhomogeneous stresses is negligible, then the start of failure and collapse of the specimen may be almost coincident. An alternative approach is to treat the problem as one combining elastic and inelastic behaviour of the specimen. However, this requires a knowledge of the stress conditions at the transition from elastic to inelastic behaviour, which is precisely the information sought in the experiment. Robertson (1955) treated the problem on the basis of ideal elastic–plastic behaviour and calculated the position of the elastic–plastic boundary and the stresses at this boundary when the

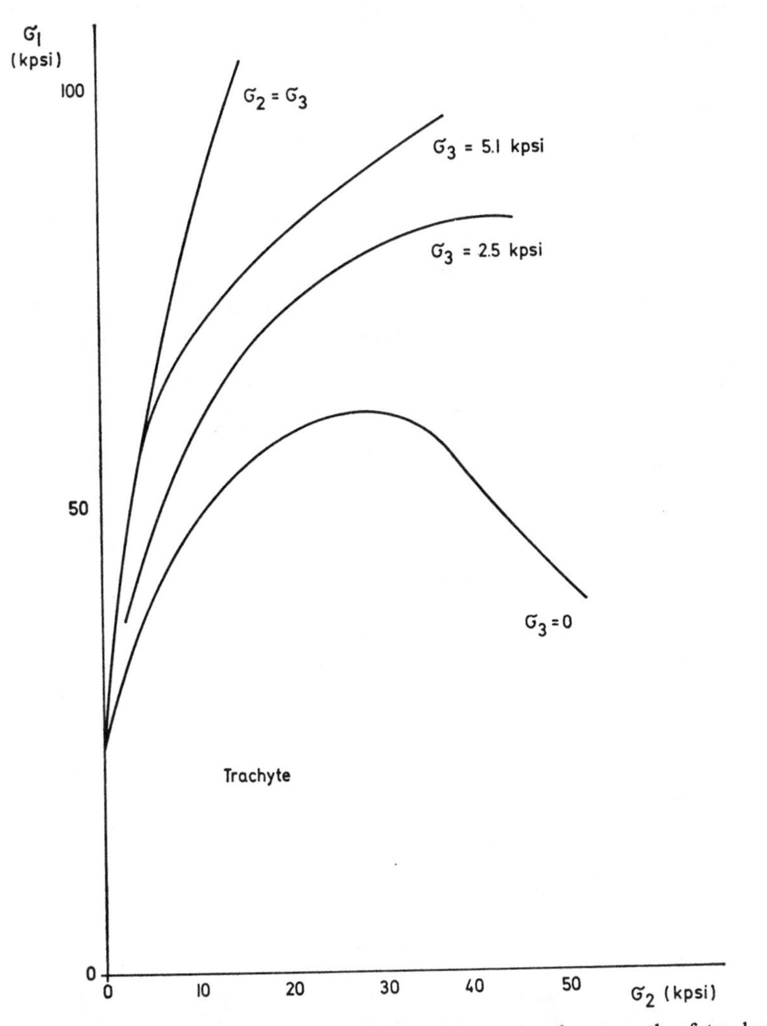

Fig. 6.8.3 The effect of the intermediate principal stress on the strength of trachyte (from results of tests on hollow cylinders by Hoskins, 1967).

specimen collapsed. However, the ability of an ideally elastic–plastic cylinder to resist external pressure increases continuously until the whole cylinder has become plastic, unless buckling occurs. The results of experiments on hollow cylinders where collapse occurs before the whole thickness of the cylinder has become plastic are not consistent with elastic–plastic behaviour. Inelastic behaviour which allows instability to occur when the elastic–inelastic boundary lies within the thickness of the cylinder is discussed in § 18.2. In view of these complications, it seems desirable to calculate the stress at failure on the basis of the theory of elasticity, using those values of the applied loads which correspond to the start of failure at the inner surface.

6.9 Torsion

Torsion of hollow or solid circular cylinders combined with axial load and fluid-confining pressure provides another method of studying failure under conditions of unequal principal stresses, Böker (1915), Handin, Higgs, and O'Brien (1960), Handin, Heard, and Magouirk (1967).

If the axial load is P, the fluid pressure, p, the couple, M, and the inner and outer radii of the cylinder are R_1 and R_2 respectively, then by § 5.9 (4) the maximum value of the shear stress due to torsion occurs at the outer surface and is

$$\tau_2 = 2M/\pi R_2^3(1 - R_1^4/R_2^4), \tag{1}$$

while the axial stress due to P is

$$p_a = P/\pi(R_2^2 - R_1^2), \tag{2}$$

so that the principal stresses become

$$\begin{aligned} \sigma_1 &= \tfrac{1}{2}(p_a + p) + \tfrac{1}{2}[(p_a - p)^2 + 4\tau_2^2]^{\frac{1}{2}}, \\ \sigma_2 &= p, \\ \sigma_3 &= \tfrac{1}{2}(p_a + p) - \tfrac{1}{2}[(p_a - p)^2 + 4\tau_2^2]^{\frac{1}{2}}. \end{aligned} \tag{3}$$

The maximum principal stress is inclined to the axis of the cylinder at an angle θ defined by

$$\tan 2\theta = 2\tau_2/(p_a - p), \tag{4}$$

and the principal stresses are connected by

$$\sigma_1 - \sigma_2 + \sigma_3 = p_a. \tag{5}$$

This system can be used to trace a curve on the failure surface given by the intersection of the plane (5) with this surface.

In pure torsion, $p_a = p = 0$, failure is by helicoidal extension fractures at 45° to the axis of the cylinder as it is also at small values of p_a and p. At intermediate values of p_a and p, which result in high stress differences, failure occurs by shear fracture at small angles to the direction of the major principal stress. Handin *et al.* (1967) conclude that both the stress at which shear fracture occurs and the angle of these fractures are affected by the value of the intermediate principal stress.

6.10 Bending

The values of the tensile strength determined from bending tests on rock are significantly greater than the uniaxial tensile strength. While this effect was not understood, it was well recognized, and the term *modulus of rupture* is used for the extreme tensile stress determined from a bending test. Obert, Windes, and Duvall (1946) include three-point bending of 6-in. lengths of drill core in their standardized tests for studying mine rock, and Pomeroy and Morgans (1956), Berenbaum and Brodie (1959a), and Evans (1961) have used bending to study the tensile strength of coal, which is, of course, extremely difficult to test in direct uniaxial tension. Fairhurst (1961) discusses the effects of different values for Young's modulus in tension and compression on bending tests and describes apparatus for four-point loading. Bending is the simplest method of studying time-dependent behaviour of rock, such as creep, and has been used for this purpose by Phillips (1931, 1932) and Price (1964).

The simple theory of bending is developed in § 5.10. When a load, W, is applied at the centre of a simply supported beam of length $2l$, producing a central deflection y perpendicular to the axis of the beam the average Young's modulus in bending is, by § 5.10 (7),

$$E_{av} = Wl^3/6Iy, \tag{1}$$

and the modulus of rupture is, by § 5.10 (5),

$$F = Wly_0/2I, \tag{2}$$

where $I =$ the moment of inertia of the cross-section of the beam. (For a circular cross-section of radius R, $I = \pi R^4/4$ and for a rectangular cross-section, $h \times b$, $I = bh^3/12$.)

$y_0 =$ the distance from the neutral axis of the beam to the extreme point of the cross-section on the tensile side of the neutral axis.

In three-point loading the bending moment and the tensile stress reach maximum value immediately beneath the point at which the load, W, is applied. When two loads, each of $W/2$, separated by a distance $2a$, are applied symmetrically to a simply supported beam of length $2l$, the average Young's modulus and modulus of rupture are, respectively,

$$E_{av} = W(2l^3 - 3a^2l + a^3)/12Iy, \tag{3}$$

and

$$F = Wy_0(l - a)/2I. \tag{4}$$

Four-point loading generates a uniform bending moment and tensile stress over the length $2a$.

Loads are usually applied to a beam, and simple support is provided by round steel bars lying across the axis of the beam. It is important that these be free to rotate relative to one another about the axis of the beam to avoid torsional loading of the beam. This is achieved automatically with beams of circular cross-section, but is important with beams of rectangular cross-section.

For rocks, the value of Young's modulus in tension, E_T, is generally less than that in compression, E_C. This has important implications for bending tests, which can conveniently be studied in terms of a transformed beam section (Timoshenko, 1958). Consider a beam of rectangular cross-section, Fig. 6.10 (*a*). The effect of a decreased Young's modulus in tension is to move the position of the neutral axis towards the concave side of the beam. On the concave side of the neutral axis the modulus is E_C, and on the convex side it is E_T. This beam can be represented by a transformed beam with a uniform Young's modulus, E_T, but an increased width b

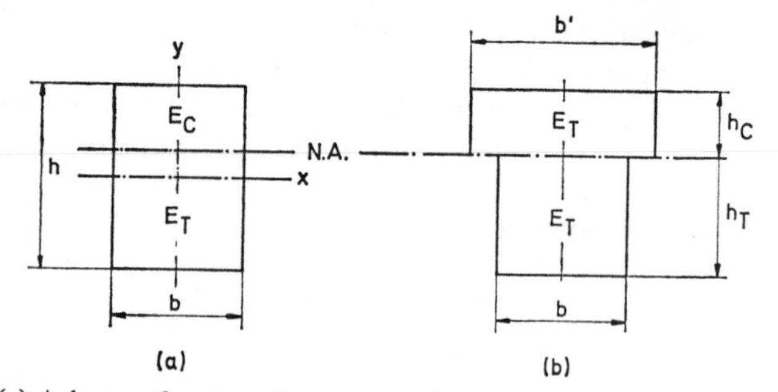

Fig. 6.10 (*a*) A beam of rectangular cross-section with Young's modulus in compression, E_c, greater than in tension, E_T. (*b*) The transformed section of the beam.

above the neutral axis, Fig. 6.10 (*b*). Assuming that cross-sections of the transformed beam also remain plane after bending, the relations between the moduli and sections are

$$b'/b = E_C/E_T = (h_T/h_C)^2. \tag{5}$$

The moment of inertia of the transformed section is

$$I = bh_T{}^3[1 + (E_T/E_C)^{\frac{1}{2}}]/3, \tag{6}$$

and a bending moment of M produces an extreme tensile stress of

$$F = 3M/bh_T{}^2[1 + (E_T/E_C)^{\frac{1}{2}}], \tag{7}$$

but $$h = h_T[1 + (E_T/E_C)^{\frac{1}{2}}]$$

so that $$F = 3M[1 + (E_T/E_C)^{\frac{1}{2}}]/bh^2. \tag{8}$$

The importance of (8) is that, for any given bending moment, a decrease in the value of Young's modulus in tension results in a decrease in the extreme tensile stress. In part, this explains why values for the modulus of rupture determined from (2) or (4) are greater than the uniaxial tensile strength. Griffith cracks which open in tension are probably the cause of the decreased value of Young's modulus in tension, § 12.3. When bending tests are used to study creep, time-dependent crack growth and nucleation are likely to occur, decreasing the value of E_T with time. The value of E_{av} is given by

$$E_{av} = 4E_T/[1 + (E_T/E_C)^{\frac{1}{2}}]^2, \tag{9}$$

so that the central deflection is

$$y = Wl^3[1 + (E_T/E_C)^{\frac{1}{2}}]^2/24E_TI \tag{10}$$

for three-point loading and

$$y = W(2l^3 - 3a^2l + a^3)[1 + (E_T/E_C)^{\frac{1}{2}}]^2/48E_TI \tag{11}$$

for four-point loading.

Measurements of increasing deflection with time in bending tests might therefore be due to time-dependent crack growth rather than viscous deformation.

6.11 Diametral compression of cylinders

The difficulties associated with performing a direct uniaxial tensile test on rock have led to a number of indirect methods for assessing the tensile strength. The most popular of these indirect tensile tests consists of applying diametral compression between the platens of a compression testing machine to a rock cylinder which, for convenience, is usually shorter than its diameter. This test is frequently referred to as the 'Brazilian' test and has been used to determine the elastic properties of concrete by Hondros (1959), the tensile strength of coal by Berenbaum and Brodie (1959a) and Evans (1961), and that of rocks by Berenbaum and Brodie (1959b) and Hobbs (1964c). It provides an easy method of studying anisotropy and has been used for this purpose by Evans (1961) on coal and by Hobbs (1964c) on rocks. In addition to its convenience, this test is of great interest in connection with the failure of rock under conditions of quite complicated stresses.

If a circular cylinder of radius R is compressed across a diameter between flat surfaces which apply concentrated loads of W per unit axial length of the cylinder, Fig. 6.11.1 (a), the stresses on this diameter, by

§ 10.7 (13), (14) are

$$\sigma_x = -W/\pi R, \quad \sigma_y = W(3R^2 + y^2)/\pi R(R^2 - y^2). \tag{1}$$

On the diameter perpendicular to this the stresses are, from § 10.7 (11), (12), found to be

$$\sigma_x = -W[(R^2 - x^2)/(R^2 + x^2)]^2/\pi R. \tag{2}$$

By symmetry these are principal stresses, and it can be seen that the greatest and smallest principal stresses are along and across the y-diameter respectively. The former stress varies slowly near the centre of the cylinder, but tends to infinity at the circumferential contacts, while the latter stress is uniform along the y-diameter and varies slowly across the x-diameter.

When a cylinder of rock is compressed along its diameter failure occurs

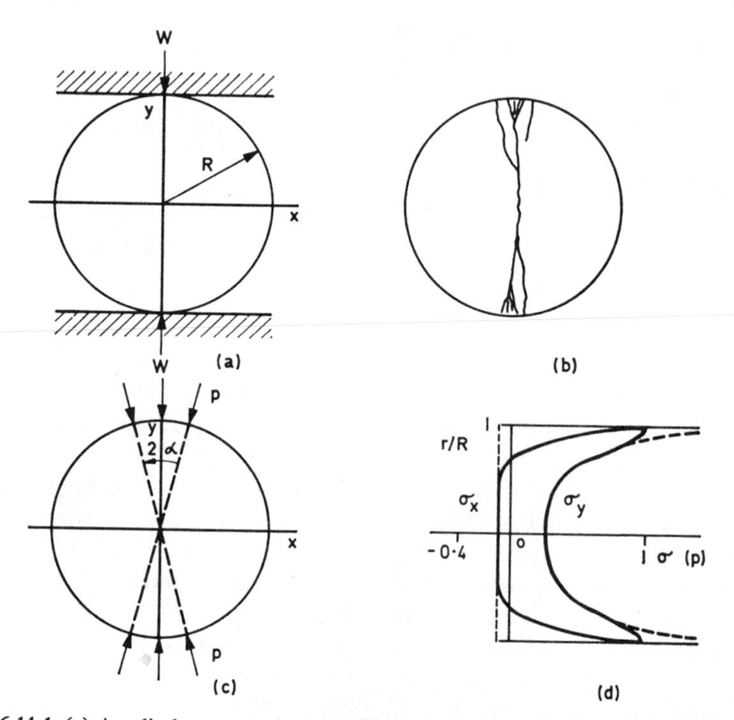

Fig. 6.11.1 (*a*) A cylinder compressed between parallel surfaces generating a line load, W, per unit length. (*b*) Typical fractures resulting from (*a*). (*c*) A cylinder compressed by a diametral load due to uniform pressure p over the arc 2α. (*d*) Stresses along the loaded diameter of a cylinder loaded as in (*c*) solid lines and as in (*a*) dotted lines.

by an extension fracture in, or close to, the loaded diametral plane at some value of the applied load W. It is generally assumed that failure is the result of the uniform tensile stress normal to this plane given by (1). Tensile strengths measured in this way are very reproducible and are in reasonable agreement with values obtained in uniaxial tension. While exceptional cases are found with a single diametral fracture which may not even extend to both platens, there are generally several fractures branching from the diametral plane, some of which appear to be wedges near the contacts, Fig. 6.11.1 (*b*). These subsidiary effects have given rise to doubts about the mechanism of failure in this test, and it has been suggested that failure starts by shear fracture in the region of high compressive stresses near the contacts.

If the load is applied to the circumference of the cylinder as a pressure, p, distributed over a small arc 2α, Fig. 6.11.1 (*c*), either by using shaped platens or inserting thick paper between the platens and the specimen, so that $W = 2p\alpha R$, then equal biaxial compression exists near the contacts with a value of p. This decreases the likelihood of failure by shear fracture at the contacts but has virtually no effect on the stresses in the body of the specimen, Fig. 6.11.1 (*d*). Experiments with distributed loads over a narrow arc, usually about 15°, yield values for the tensile strength little different from those obtained with line loads and give rise to similar diametral extension fractures.

In hard rocks fracture takes place with considerable violence, and the specimens may shatter into several pieces. It is likely that these multiple fractures are caused by the release of energy stored in the testing machine after the specimen has been divided into two hemi-cylinders by the primary extension fracture, Colback (1967). The sudden formation of an extension fracture is equivalent to the sudden superposition of a uniform compressive stress along the fracture surface. This tends to propel the two hemi-cylinders apart, against the frictional restraint at their contacts with the platens. By using a stiff testing machine and slipping a yoke over the specimen just prior to failure to restrain the hemi-cylinders from flying apart, Wiebols *et al.* (1968) were able to produce a single diametral extension fracture in cylinders of dolerite with no hint of shear or other subsidiary fractures, even when using line loads.

The tensile strength determined from (1) is not the true uniaxial tensile strength, but for plane stress is determined for the principal stresses

$$\sigma_1 = 3W/\pi R, \quad \sigma_2 = 0, \quad \sigma_3 = -W/\pi R. \tag{3}$$

From § 4.6 (11) the Coulomb criterion applies only for $\sigma_1 > C_0/2$, so that these conditions fall outside the realm of this criterion unless the ratio between the uniaxial compressive and tensile strengths is less than 6 : 1, which is seldom the case for rocks. From (3), $\sigma_1 + 3\sigma_3 = 0$, so that in terms of the Griffith criterion, failure occurs under conditions corresponding to the transition from tensile failure at a value of the minor principal stress equal to the uniaxial tensile strength to failure determined by § 4.8 (1). The Mohr circle corresponding to (3) has the same radius as the Mohr envelope of the plane Griffith criterion, § 4.8 (8), where both intersect the σ-axis at T_0, Fig. 6.11.2. The stresses along the loaded diameter in the plane normal to the axis of the cylinder are shown by Mohr circles, which lie inside the Griffith envelope. Therefore, even when the ratio between the uniaxial compressive and tensile strengths is only 8 : 1, as in the plane Griffith criterion, failure occurs at a value of the tensile stress equal to the true uniaxial tensile strength.

Hondros (1959) has described a method for determining Young's modulus and Poisson's ratio from strain measurements at the centre of the cylinder. This method assumes that these properties are the same in tension and compression, which is not true for most rocks, but may be adequate for determining these values at low stresses. For a short cylinder these properties are given by

$$E = 8W/\pi R(3\varepsilon_y + \varepsilon_x),$$
$$\nu = -(3\varepsilon_x + \varepsilon_y)/(3\varepsilon_y + \varepsilon_x), \tag{4}$$

where ε_y and ε_x are the strains along and normal to the loaded diameter at the centre of the cylinder. Since the stresses vary through the cylinder,

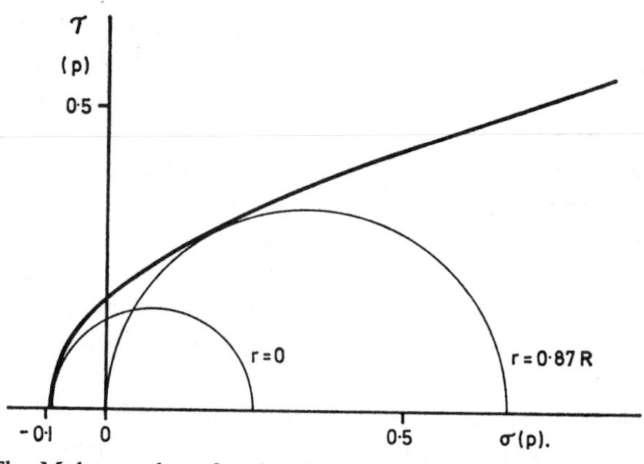

Fig. 6.11.2 The Mohr envelope for the plane Griffith criterion and the largest Mohr stress circles on the loaded diameter of a cylinder for uniform pressure p over arcs $2\alpha = 15°$.

errors arise from the use of gauges with a finite length, and the gauge length should be less than a tenth of the radius if a few per cent accuracy is required.

In view of the uncertainty associated with the origin of the primary fracture, several investigators have sought to promote diametral tensile failure by drilling a small hole along the axis of the cylinder. Using § 10.4

(10), it is seen that in the stress field (3) the tangential stress around a small hole ranges from $10W/\pi R$ perpendicular to the x-diameter to $-6W/\pi R$ perpendicular to the y-diameter, that is, a tensile stress concentration normal to the plane of the extension fracture of 6. In fact, the load W at failure is not greatly affected by the presence of a small axial hole, and tensile strengths determined by this method are apparently much greater than those determined with a solid cylinder, Hobbs (1965). Jaeger and Hoskins (1966a) have studied the general problem of failure in rings of rock loaded in diametral compression or tension, the theory of which is given in § 10.8, and have found that the calculated tensile stress at failure decreases as the ratio between the diameter of the hole and that of the cylinder increases, approaching a constant value of about twice the uniaxial tensile strength when this ratio is a half.

The occurrence of extension fractures along a diameter subjected to compression and tension is of special interest, as these stresses probably approximate those obtaining when extension fractures occur in the field more closely than does uniaxial tension. Jaeger (1965) suggested that if jacketed cylinders were subjected to fluid-confining pressure as well as diametral compression, failure could be studied as the minor principal stress varied from tension to compression. If the fluid pressure and the load at failure are p and W, respectively, the principal stresses become:

$$\sigma_1 = p + 3W/\pi R, \quad \sigma_2 = p, \quad \sigma_3 = p - W/\pi R \tag{5}$$

so that this test provides a means of studying failure when the value of σ_3 is small in relation to σ_2 and σ_1, as is often the case near the surface of underground excavations. The principal stresses are connected by the relation

$$\sigma_1 - 4\sigma_2 + 3\sigma_3 = 0, \tag{6}$$

so that this test determines a curve on the failure surface given by its intersection with (6). It is found that failure occurs by extension fracture along the loaded diameters, even when the value of p is such that all the principal stresses are compressive. Jaeger and Hoskins (1966b) found that the values of σ_1 and σ_3 from (5) fitted the results of triaxial tests on three different rocks well, though the values for σ_1 tended to be consistently higher than those in the triaxial test. This they ascribe to the strengthening influence of the intermediate principal stress.

6.12 Other tests

Bridgman's (1912) celebrated 'pinching-off' experiment, in which fluid pressure, p, is applied to a length of the surface of a long cylindrical rod, between the seals, S, of the pressure cell, Fig. 6.12 (a), is of considerable interest. The stresses in the specimen between the seals are $\sigma_1 = \sigma_2 = p$,

$\sigma_3 = 0$. If the rod is ductile, it 'pinches-off' and the ends extrude for values of p greater than the yield stress. In unjacketed specimens of brittle materials such as rock, Jaeger and Cook (1963), Jaeger (1963), and glass and plastic, Gurney and Rowe (1945a, b), failure is by diametral extension fracture, which occurs at any point between the seals at a value of p of the order of but greater than the uniaxial tensile strength of the specimen. This effect is presumably due to fluid pressure in the cracks or pores of the specimen. If a cylinder of rock is jacketed to prevent penetration of the pressure fluid, failure appears to occur first as diametral extension fractures at the seals and to be associated with the axial tensile stresses which occur near them, Tranter and Craggs (1945). The fluid pressure necessary to cause failure is of the order of the uniaxial compressive strength of the rock. Other diametral fractures, which may be subsidiary effects, are observed between the seals.

If a specimen of rock with a blind central hole containing a small projection, Fig. 6.12 (*b*), is subjected to biaxial compression normal to the axis of the hole the projection will break off at a value of the stress of the order of but less than the uniaxial strength. The fracture occurs at the bottom of the hole, is convex towards the solid, and has an appearance

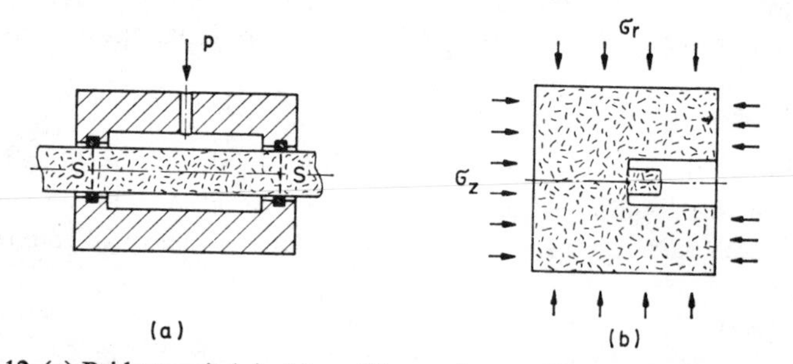

Fig. 6.12 (*a*) Bridgeman's 'pinching-off' experiment. (*b*) Core discing experiment.

intermediate between that of an extension and a shear fracture. This effect is possibly related to the 'pinching-off' in the Bridgman experiment and is relevant to the phenomenon of discing during core drilling with a diamond crown. When drilling into highly stressed hard rock it is common for the core to emerge as regular discs, perhaps as thin as a fifth of the core diameter and quite unrelated to the structure of the rock. In fact, the phenomenon is best observed when drilling parallel to the bedding in hard homogeneous Witwatersrand quartzite, Cook (1962). The thickness of the discs diminishes with increasing stress, Jaeger and Cook (1963), the first discs having a thickness of about their diameter. Obert and Stephenson (1965) report that discing occurs when

$$\sigma_r > 3{,}400 + 2S_0 + 0{\cdot}7\sigma_z, \qquad (1)$$

where σ_r = the radial field stress;
σ_z = the axial field stress;
S_0 = the inherent shear strength, § 4.6.

The diametral compression of spheres and cubes has been studied, Berenbaum and Brodie (1959b), Hiramatsu and Oka (1966), Jaeger (1967a), in relation to the measurement of tensile strength. Formulae for the stresses in a sphere have been given by Sternberg and Rosenthal (1952) and Abramian *et al.* (1964), but they do not derive numerical values for these quantities. If W is the diametral load at failure for a sphere of radius R the tensile strength is given by

$$T_0 = kW/R^2. \tag{2}$$

Jaeger (1967a) found the constant, k, to range from 0·23 to 0·45, while Hiramatsu and Oka (1966) calculated it to be about 0·225. D'Andrea *et al.* (1944) used a similar formula for the diametral compression of a long cylinder between point loads with $k = 0{\cdot}24$.

The case of a square prism of side $2R$ compressed by a diametral load W per unit axial length distributed over a width $2b$ on opposite sides has been used by Berenbaum and Brodie (1959b). They find the maximum tensile stress for $b/R = 0{\cdot}125$ to be

$$\sigma_r = 0{\cdot}366W/R, \tag{3}$$

which is very similar to the tensile stress in a Brazilian test § 6.11 (3). Goodier (1932) has shown that the tensile stress when a long rectangular prism is loaded across its short diameter is given by

$$\sigma_t = 0{\cdot}85W/\pi R. \tag{4}$$

These results are relevant to the failure of irregular pieces and of aggregates of such pieces. In connection with the latter, Jaeger (1967a) has studied the failure of cylinders compressed by four- and three-line loads. The relevant theory is given in § 10.6. This system provides a method of studying relatively stable crack growth, which is invariably in almost straight lines closely following the maximum tensile stress between the loading points.

A system which allows the growth of an extension fracture to be controlled and studied is provided by a rectangular prism of rock partially divided in two by longitudinal slots cut into two opposing sides, Gilman (1961), Perkins and Bartlett (1963). If the two halves are forced apart by a jack attached to one end of the beam it cleaves by an extension fracture in the plane of the slots. The growth of this fracture is relatively stable, and the energy absorbed by it can be determined from measurements of the jack load and extension.

Jaeger (1967a) has studied the failure of circular discs of rock simply loaded over a concentric ring of radius b and simply supported by another concentric ring of radius a. If the thickness of the plate is d and the total load is W the extreme tensile stress is constant in the annulus $0 \leqslant r \leqslant b$ and is given by

$$\sigma_r = \sigma_\theta = 3(1 + \nu)W\{2 \ln (a/b) + (1 - \nu)(1 - b^2/a^2)/(1 + \nu)\}/4\pi d^2, \quad (5)$$

Morley (1923), Sokolnikoff (1956). The values of these biaxial tensile stresses at failure are found to be similar to those in other systems with non-uniform uniaxial tensile stress, §§ 6.8, 6.10, 6.15, that is, about twice the uniaxial tensile strength. This suggests that with non-uniform stresses the biaxial tensile strength is not very different from the uniaxial tensile strength. Cracks occur inside the ring $r = b$, tending to form angles of 120° between them. Not more than three cracks meet at any point, although the plates often separate into more than three pieces.

Punching tests to measure shear strength have been made by supporting one side of a disc of rock on a hollow cylinder and applying a load to the opposite side by means of a cylindrical punch with a diameter slightly less than the bore of the cylinder, Robertson (1955), Jaeger (1962b), Maurer (1965). These experiments can be done with or without confining pressure.

McWilliams (1966) has compressed discs of rock axially between two indenters to produce extension fractures in studies of the effects of anisotropy and microstructure on fracture. Indentation tests are common in studying the hardness of metals, Mott (1956), Tabor (1951), but have not often been used to study rock properties, though Brace (1964a) has used them for this purpose. The indentation of rock surfaces by chisel bits under a wider range of conditions has been much studied in relation to rock failure associated with drilling, Cheatham and Gnirk (1967). In general, the force increases with penetration in a series of steps. Confining pressure tends to smooth these steps, and penetration proceeds by ductile deformation at high confining pressures. Pore pressure has little effect on force–penetration characteristics for static tests, indicating that the effective stress principle is applicable to this situation. However, in dynamic penetration tests pore pressure is observed to have an effect similar to confining pressure, suggesting that the fluid does not have time to flow into a fast-growing crack. Analysis of the interaction between the chisel and rock on the basis of plasticity, using a Mohr or Coulomb yield envelope, gives results in close agreement with experimental observations for ductile conditions.

Early workers, Adams (1912), provided confining stress in triaxial tests by jacketing specimens in thick metal cylinders. These give rise to frictional effects which are difficult to determine. Cook (1962) was able to provide moderate confining stresses in triaxial tests by wrapping specimens with steel wire under tension. If the wrapping pitch is slightly greater than the diameter of the wire this method avoids frictional effects altogether.

It also allows the effect of stiffness of the confining load, §§ 6.4, 6.15, to be studied.

6.13 Stiff and servo-controlled testing machines

In the past, most testing of materials to measure their mechanical properties has been done in 'soft-' or 'dead-weight' type testing machines. The extent to which the soft characteristic of such machines can obscure certain behaviour or properties of a material has come to be recognized in only relatively recent times. These effects are so pronounced that most compression tests on specimens of brittle materials such as rock and concrete terminate at, or just past, the peak of the stress–strain curve by violent, almost explosive, disintegration of the specimen. However, such behaviour is seldom observed in the failure of concrete structures or of rock around underground excavations such as mines.

In 1935, Speath postulated that the precise form of the stress–strain curve for iron during plastic deformation was influenced by the stiffness of the testing machines; descending portions of the stress–strain curve would be masked by the unloading characteristic of the machine because the stiffness of most machines was about 200 MN/m while the stiffness of a 'standard size' iron specimen was about 500 MN/m.

The first definitive explanation of the effect of the stiffness of the testing machine on the failure of test specimens was given by Whitney (1943). In discussing some stress–strain curves for concrete in a paper by Jensen (1943), he stated that 'the vertical descent of the stress–strain curves . . . is without basis. . . . High strength cylinders fail suddenly but as a matter of fact during the failure the elastic movement of the testing machine head imposes a large additional strain on the concrete. This strain may be considerably greater than the total strain up to the time that failure starts. At that time a large amount of elastic energy is stored in the cylinder and in the machine (about three times as much in the latter as in the former), and the release of this energy causes the breakdown of the cylinder.'

Whitney measured the stiffness of four testing machines and then compared these characteristics with the slopes of the stress–strain curves at failure. He concluded that 'shortly after the maximum load, the slope of the concrete curve becomes equal to that of the machine curve. The elastic recovery of the machine at that stage is rapid enough to maintain the load required to continue the straining of the cylinder without operating the machine. At this point the strain starts to increase automatically and rapid failure follows.'

Brock (1962) pointed out that the testing machine evolved as an apparatus for applying load rather than deformation and that it is traditional to regard the testing machine load as the independent variable. The natural outcome of a steadily increasing load is violent failure at the compressive strength of the material. Brock suggested that specimens should be tested with displacement control and loaded them through a steel beam (to increase the stiffness of the loading system). In this way, he was able to obtain complete stress–strain curves for concrete.

A testing machine consists of an assembly of cross-heads, tie bars, platens, screws and hydraulic cylinders and pipes, each with its own stiffness, which contribute to an overall stiffness for the testing machine. Hinde (1964) quotes figures for the stiffness of concrete testing machines ranging from 2×10^6 lb/in. to 7×10^6 lb/in., and Cook and Hojem (1966) quote $0{\cdot}54 \times 10^6$ lb/in. for a common 100 000 lb testing machine, noting that most of the resilience comes from the hydraulic oil and the tie bolts.

A stiff testing machine was developed by Turner and Barnard (1962) using a stiff frame and an oil column of large area and small length. They were also able to obtain complete stress–strain curves for concrete and this work was continued by Barnard (1964).

Studies of controlling the failure of rock specimens began at this time. Cook (1965) loaded a steel tube in parallel with a rock specimen in a conventional testing machine and reduced significantly the violence of the failure normally associated with specimens of Tennessee Marble. He showed that the failure process became unstable when the descending slope of the stress–strain curve of the specimen exceeded the slope of the machine stiffness. Paulding (1966) loaded specimens of granite through a steel beam to stiffen the testing machine and this enabled him to observe crack growth near the peak of the compressive stress–strain curve. Finally, in 1966, Cook and Hojem designed and built a stiff testing machine in which an hydraulic jack was used to pre-stress the specimen and the final deformation was induced by thermal contraction of the columns of the machine frame. They obtained the first complete stress–strain curve for rock, a specimen of Wombeyan Marble. Wawersick (Wawersick and Fairhurst, 1970) continued this work on a similar machine, measuring complete stress–strain curves for various rocks. They were able to do this, in part, by controlling, manually, the displacements of the machine with hydraulic jacks in parallel with the specimens. Though Wawersick did not appear to appreciate it at the time, the fact that he could effect manual servo-control proved that the processes of rock failure and fracture are much slower than had been thought. Fracture in brittle materials is normally thought to be a rapid process with terminal fracture velocities close to the Raleigh wave velocity, that is, about 3 km/s to 4 km/s in rock. The processes which Wawersick controlled manually obviously took place in seconds and not in microseconds, as would have been expected from their dimensions and brittle fracture theory. The relative slowness of the processes of failure and fracture in rock have been confirmed subsequently in numerous experiments; in fact, many investigators now use electrohydraulic, servo-controlled machines for measuring complete stress–strain curves and allied properties (Hudson *et al.*, 1971).

The stiffness, k, of an elastic member is defined as the force per unit displacement, that is

$$k = P/x, \tag{1}$$

where x is the displacement produced by P along the direction in which it acts. The amount of energy stored in an elastic member and recoverable from it, as P is decreased to zero, is

$$S = P^2/2k. \tag{2}$$

A rock specimen of area A, length L, and Young's modulus E has a stiffness

$$k_R = AE/L, \tag{3}$$

so that typical cylindrical specimens measuring 1 in. and 2 in. in diameter with $L/D = 2{,}5$ have a stiffness of about 3×10^6 lb/in. and 6×10^6 lb/in., respectively. The load, P, at failure of these specimens ranges from 10 000 lb for a small specimen of soft rock in unaxial compression to 200 000 lb for a large specimen of hard rock in triaxial compression.

If the load on a specimen of rock is decreased to zero it relaxes in an almost elastic manner and most of the energy stored by virtue of its stiffness is recovered. This applies even after ductile deformation of the specimen has taken place, but in this case the recovered energy is less than the total work done during loading the specimen. The elastic energy stored in the system comprising a testing machine and a specimen is

$$S_s = P^2(1/k_R + 1/k_M)/2, \tag{4}$$

as each part is subjected to the same load P. Taking typical values of $k_R = 4 \times 10^6$ lb/in. and $k_M = 1 \times 10^6$ lb/in., it is seen that the testing machine stores four times as much energy as the specimen. When failure occurs the energy given by (4) is released in the process of fracture, which must be influenced by the large quantity of energy released by the testing machine.

This effect can be understood by studying the relationship between machine stiffness and the complete stress–strain curve for rock, § 4.2, using a load-displacement diagram, Fig. 6.13.1 (*a*). The stiffness of a soft machine is represented by a flat line, k_1, in this diagram, while that of a stiff machine is represented by a steep line, k_2. The region around the point where the complete load-displacement curve of the specimen is tangent to k_1 is shown on an enlarged scale in Fig. 6.13.1 (*b*).

Consider the effect of a small additional compression, Δx, near the point of tangency. This compression decreases the ability of the rock

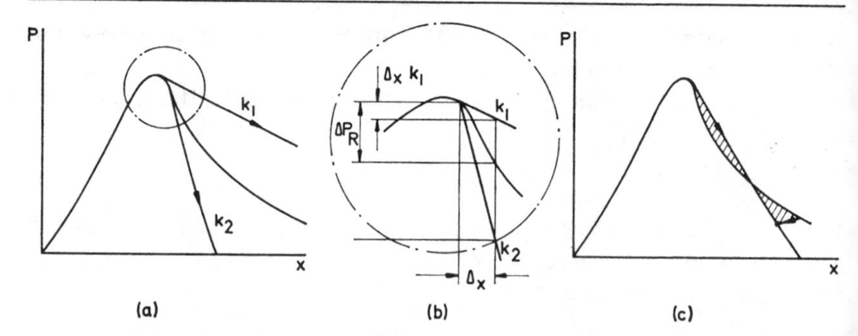

Fig. 6.13.1 (*a*) A complete load–displacement curve for a rock specimen with lines showing the behaviour of soft, k_1, and stiff, k_2, testing machines. (*b*) An enlargement of the peak of the load–displacement diagram showing the loss of specimen strength, ΔP_R, and machine load, $k\Delta x$, for an incremental compression, Δx. (*c*) Transient instability when the energy stored in the testing machine is less than that necessary for complete fracture.

specimen to resist the applied load by an amount $\Delta P_R = dP/dx\Delta x$. It also results in a decrease in the load applied by the machine of $k\Delta x$, provided energy is neither added to nor withdrawn from the system in this interval. If $|dp/dx| > |k|$, as is the case with k_1, the ability of the specimen to resist load at $x + \Delta x$ is less than the load applied to it by the testing machine in this part. This situation is unstable and produces a violent failure. Most testing machines are soft in relation to the slope of the load-displacement curve of rock specimens, and tests are terminated violently at a load and compression corresponding to the point of tangency between the stiffness of the machine and the load-displacement curve of the specimen. This point is usually very close to the maximum ordinate of the stress–strain curve, or the strength of the specimen, and corresponds to the transition from ductile to brittle behaviour, § 4.2.

If $|dP/dx| < |k|$, as is the case with a stiff machine, k_2, this unstable situation does not arise. The energy stored in the testing machine at any load, represented by the area beneath the line k_2, is always less than that required for further compression of the specimen, represented by the area beneath the load-displacement curve of the specimen, Fig. 6.13.1 (*a*). Compression of the specimen always requires energy to be added to the system, and the complete load-displacement curve of a specimen can be traced with such a machine. The non-linear nature of the load-displacement curve makes it possible for a transient instability to arise at a point where $|dP/dx| > |k|$, but the total amount of energy stored in the machine is less than that required for complete compression of the specimen, Fig. 6.13.1 (*c*).

The practical importance of the continued ability of rock to resist an applied load, in the region of brittle behaviour, is discussed in § 18.2. In conventional testing with soft machines the strength of a rock specimen is determined and the test is disrupted at the transition from ductile to brittle behaviour. The large amount of energy stored in the testing machine makes the test so sensitive to the change in the nature of the behaviour of the specimen from ductile, or increasing ability to resist load with increasing deformation, that the important capacity of a specimen to resist load in the brittle state becomes completely obscured.

Ideally a testing machine should be completely rigid so that the specimen can be subjected to any strain independent of its behaviour. The ultimate stiffness of any testing machine is probably represented by the indentation stiffness of its platens. Timoshenko and Goodier (1951) give the displacement of a rigid circular punch with a diameter D indenting a flat surface as

$$w = P(1 - \nu^2)/DE, \tag{6}$$

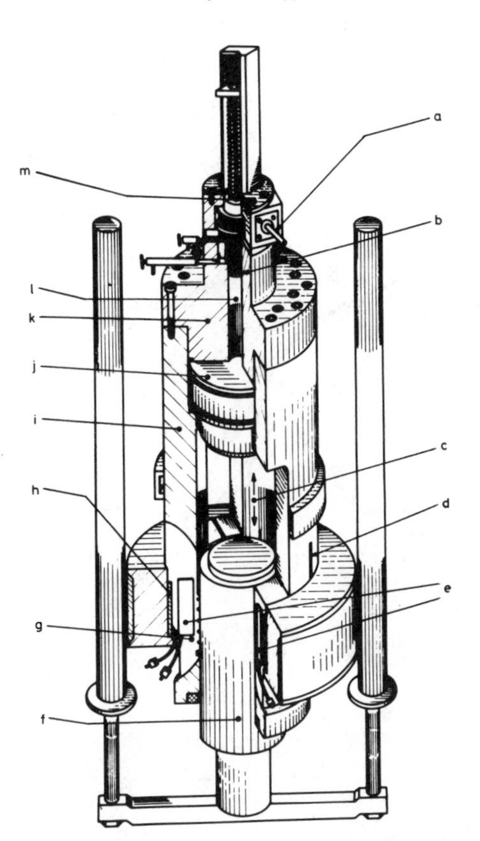

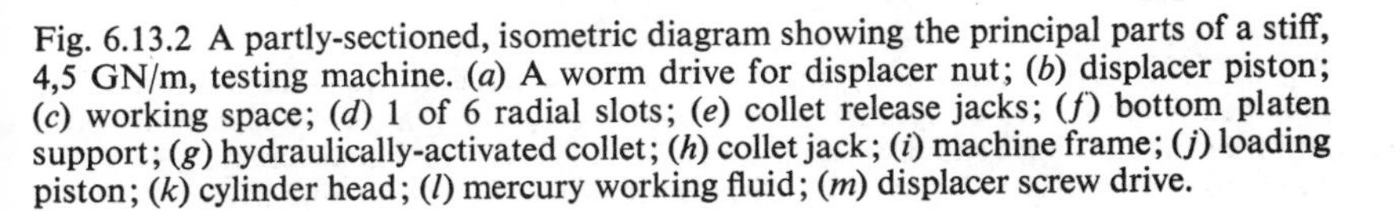
Fig. 6.13.2 A partly-sectioned, isometric diagram showing the principal parts of a stiff, 4,5 GN/m, testing machine. (*a*) A worm drive for displacer nut; (*b*) displacer piston; (*c*) working space; (*d*) 1 of 6 radial slots; (*e*) collet release jacks; (*f*) bottom platen support; (*g*) hydraulically-activated collet; (*h*) collet jack; (*i*) machine frame; (*j*) loading piston; (*k*) cylinder head; (*l*) mercury working fluid; (*m*) displacer screw drive.

from which it can be seen that the stiffness of two rigidly connected platens is of the order of $\frac{1}{2}DE$ or 15×10^6 lb/in. for steel platens with a specimen 1 in. in diameter.

For the same ratio of L/D the elastic stiffness of a specimen increases as its diameter, so that the maximum ratio between the machine stiffness and that of the specimen appears to be about 5:1 for all sizes of specimen.

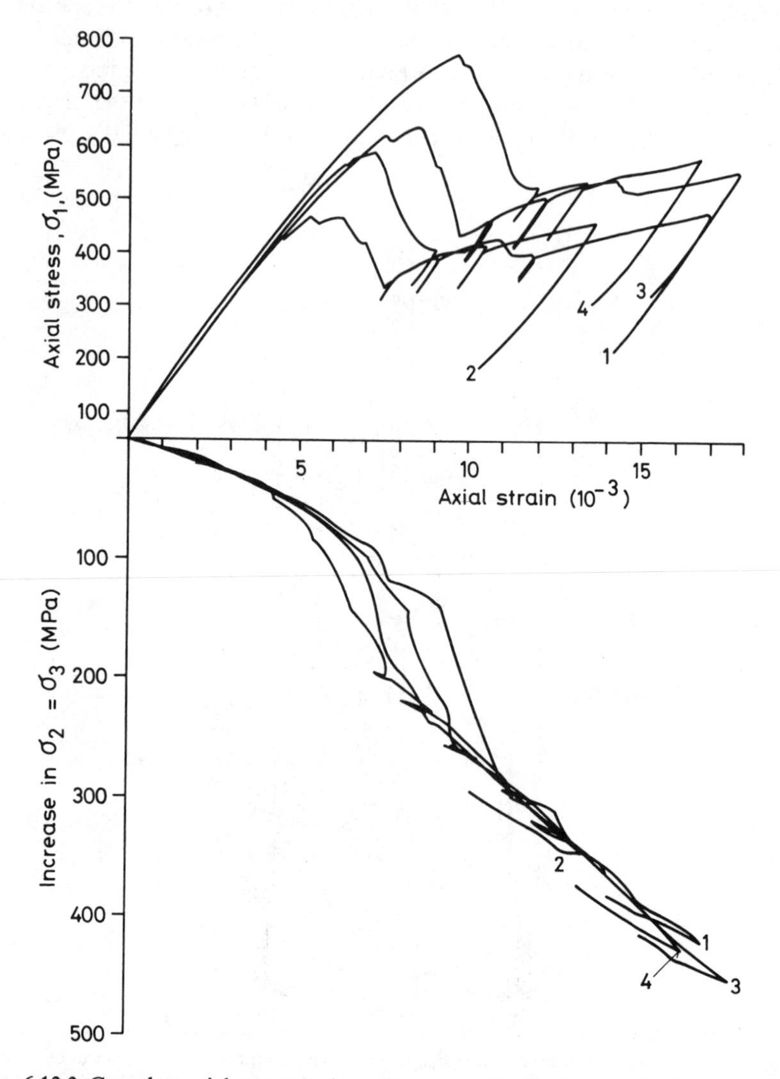

Fig. 6.13.3 Complete axial stress–strain and incremental radial stress–axial strain curves obtained from a suite of specimens of a hard, fine-grained andesitic lava tested in a stiff triaxial cell using a stiff testing machine. Curve (1) is for an initial confining pressure, $\sigma_2 = \sigma_3$, of 10,4 MPa; (2) for 20,7 MPa; (3) for 27,6 MPa; (4) for 34,5 MPa.

A very stiff testing machine incorporating several novel features, such as an hydraulic transformer filled with mercury to increase the stiffness of the mechanized drive system and a zero-compliance hydraulic collet to hold the adjustable bottom platen support, has been built by Hojem *et al.* (1975). This machine has an inherent design stiffness of about 4,5 GN/m and a load capacity of 2 MN. A diagram showing its main design and structural features is given in Fig. 6.13.2.

Ordinarily, this machine is used to test specimens 1 in. in diameter by 3 in. in length or $1\frac{1}{2}$ in. in diameter by $4\frac{1}{2}$ in. in length. To reduce end effects and to facilitate jacketing the specimens for triaxial tests they are compressed between end-pieces, § 6.2, with the same section as that of the specimen. These end-pieces are made of sintered tungsten carbide as are also the platens for the machine, so as to minimize the amount of compliance they introduce; Young's modulus for sintered tungsten carbide is about 600 GPa compared with 200 GPa for steel. The stiffness of the machine with the 1 in. diameter sintered tungsten carbide end-pieces and platens is 1,8 GN/m.

This machine has been used with a stiff triaxial cell, § 4.5, to measure complete stress–strain and axial strain–incremental radial strain curves for several hard rocks. An example from such a suite of tests is shown in Fig. 6.13.3.

6.14 Dynamic methods

Three types of experiment are made to determine the behaviour of rock under dynamic loading: resonance, pulse propagation, and impact. The U.S. Bureau of Mines, Obert *et al.* (1946), and the U.S. Bureau of Reclamation (1953) developed methods for determining the fundamental resonant frequencies of rock cylinders. Hughes and Jones (1951), Wyllie *et al.* (1958), Birch (1960), and Fairhurst (1961) have measured the velocities of propagation of pulses in rock specimens. Pulse and resonance methods are generally limited to small changes of stress. Attewell (1962) and Hakalehto (1967) have used Hopkinson bar apparatus (Kolsky, 1963) to study the behaviour of rocks when subjected to dynamic stresses in excess of their strength.

In long bars, L/D of about 10, waves of dilatation and distortion propagate with velocities of

$$C_{BP} = (E/\rho)^{\frac{1}{2}} \tag{1}$$

$$C_{BS} = (G/\rho)^{\frac{1}{2}} \tag{2}$$

respectively, § 13.2. Specimens of rock can be excited by suitable transducers, usually electromagnetic or piezoelectric, and made to resonate in longitudinal or torsional vibration. For long cylindrical specimens with free ends the longitudinal and torsional resonant frequencies, respectively, are

$$\omega_P = \pi C_{BP}/L, \tag{3}$$

$$\omega_S = \pi C_{BS}/L. \tag{4}$$

Corrections to (3) and (4) must be made if $L/D < 5$ or if the mass of the transducers added to the ends of the specimen is significant, Obert and Duvall (1967). If one end of a specimen is fixed rigidly to a large mass this has the effect of doubling the length of the specimen.

When the resonant frequencies ω_P and ω_S have been determined in an appropriate experiment the dynamic values of Young's modulus, E_D, and the modulus of rigidity, G_D, can be found from

$$E_D = \omega_P{}^2 L^2 \rho/\pi^2 \tag{5}$$

$$G_D = \omega_S{}^2 L^2 \rho/\pi^2 \tag{6}$$

provided that the density, ρ, is known. The dynamic value of Poisson's ratio, from § 5.2 is

$$\nu_D = (\omega_P{}^2/2\omega_S{}^2) - 1. \tag{7}$$

Unfortunately, (7) is the difference between two nearly equal quantities, and the dynamic determination of Poisson's ratio is therefore intrinsically inaccurate.

Resonant measurements of dynamic properties are generally limited to small changes of stresses about zero ambient stress, although Birch and Bancroft (1938a, 1940) used torsional resonance in a pressure cell. Measurements of the velocities of pulse propagation can be made to determine the dynamic properties for small changes of stress in specimens subjected to large values of uniaxial, triaxial, or hydrostatic stress. Pulses of dilatation or distortion can conveniently be generated and detected in triaxial specimens by dilatational or shear piezoelectric transducers built into the end pieces between which the specimen is loaded, Fairhurst (1961) and Simmons (1965).

The velocity of propagation of waves of dilatation in long bars, given by (1), is correct only when the ratio between the diameter of the bar and the wavelength is very small, Kolsky (1963). In practice, it is reasonably accurate provided that this ratio is less than a third, as is the case with the resonant method. The phase velocity of waves of dilatation in long bars decreases to a value close to the velocity of propagation of Rayleigh surface waves in a semi-infinite medium for wavelengths greater than the diameter. The velocity of waves of dilatation in an extended medium, § 13.2, is

$$C_P = [E(1-\nu)/\rho(1+\nu)(1-2\nu)]^{\frac{1}{2}}, \tag{8}$$

which is greater than the bar velocity (1). Triaxial specimens do not generally have the properties of a long bar, and a pulse contains a broad Fourier spectrum from short to long wavelengths, so that it is very difficult to decide with which velocity pulses of waves of dilatation propagate in these specimens. If small sources and sensors situated near the centre of the surfaces of a specimen are used so that the shortest path between them passes centrally through the specimen and great care is taken to detect the arrival of the first part of any pulse, which almost certainly does not have the largest amplitude, then it is likely that the velocity of propagation of the first part of a dilatational pulse is given by (8). Fortunately, the velocity of propagation of waves of distortion (2) applies exactly to bars of all proportions and extended media.

The dynamic values of Young's modulus and Poisson's ratio can be calculated from measurements of C_P and C_S, using

$$E_D = C_S^2\rho[3(C_P/C_S)^2 - 4]/[(C_P/C_S)^2 - 1], \tag{9}$$

$$\nu_D = \tfrac{1}{2}[(C_P/C_S)^2 - 2]/[(C_P/C_S)^2 - 1], \tag{10}$$

which are derived from (2) and (8) above and § 5.2 (13). Substituting (9) and (10) in § 5.2 (14) gives the dynamic bulk modulus as

$$K_D = \rho(C_P^2 - 4C_S^2/3). \tag{11}$$

In general, the dynamic values of Young's modulus are found to be significantly greater than the static values, Rinehart *et al.* (1961), even when the static modulus is determined from the tangent at zero stress.

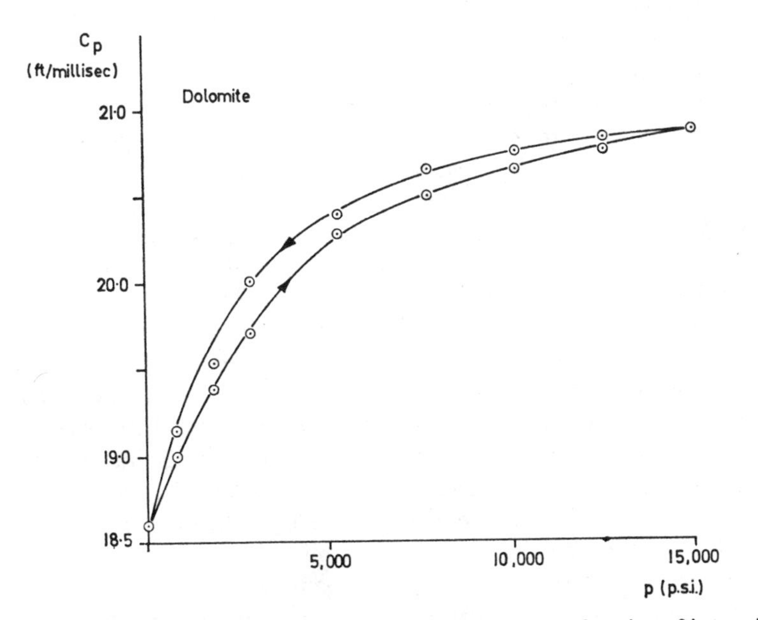

Fig. 6.14 The velocity of propagation of waves of dilatation as a function of increasing and decreasing confining pressures, p (from Hughes and Jones, 1951).

This discrepancy is most marked in rocks with small values of Young's modulus, which are, presumably, more cracked and porous, § 12.3, than those with large values. Measurements of the effect of triaxial stress on the velocities of propagation of waves of dilatation in rock specimens, Tocher (1957), Wyllie *et al.* (1958), and Birch (1960), show that these velocities increase by 10–30 per cent of their values at zero stress as the applied stress increases to 10,000 psi. Thereafter, velocities increase slowly at the rate of about 1 per cent for an increase of 10,000 psi in stress. This effect is slightly more marked on the velocities of propagation in the direction of the major principal stress than it is in a direction perpendicular to it. The increase in modulus with stress implied by the increase in velocity can be attributed to the closing under stress of cracks within the rock, § 12.3. This explanation is also consistent with the observation that the effect is most pronounced at low stresses and is more marked in the direction of the major principal stress. Sliding of the surfaces of closed cracks past one another has the effect of decreasing the value of Young's modulus. The small changes of stress associated with pulses are insufficient to produce sliding, and the dynamic modulus can therefore be expected to be greater on this account. Cook and Hodgson (1965) proposed that compressive pulses superposed on an increasing compressive stress could produce sliding and would therefore propagate at a lower velocity than compressive pulses superposed on a decreasing compressive stress. Hughes and Jones (1951) found that the velocities of propagation of waves of dilatation were greater while the stress was decreasing than they were while it was increasing, Fig. 6.14.

Velocities of wave propagation appear to reflect the degree of microcracking and porosity of rocks. The velocities increase with stress at low values of stress, but this effect becomes negligible at high values of stress comparable with the strength of rock. It is therefore unlikely to provide a means of measuring stresses as failure approaches, that is, when such measurements would be of most interest. Jones (1952) has observed that the velocity of propagation in a direction perpendicular to the major principal stress decreases in concrete specimens at values of this stress exceeding about two-thirds the strength. This phenomenon is presumably due to the onset of failure by crack growth, and independent confirmation of this is provided by observations, Cook (1965), that the tensile strength of rock specimens in a direction perpendicular to the major principal stress is decreased by subjecting them to stresses approaching their strength. Several investigators, Cook (1965), Chugh *et al.* (1967), and Scholz (1967), have monitored microseismic vibrations due to local cracking in parts of a rock specimen as it approaches failure. Microseismic activity of a characteristic pattern precedes, accompanies, and follows the failure of most rocks. In heterogeneous rocks it starts at a value of the stress of about half the strength and increases progressively to failure. In homogeneous rocks it starts only as the value of the stress approaches the strength. Microseismic activity appears to be associated with irreversible processes, presumably crack growth, in the third region of the stress–strain curve, § 4.2. Similar activity can be expected to an even greater extent in the fourth region.

The original Hopkinson (1914) apparatus for studying the propagation of stress pulses consisted of a steel bar about 1 in. in diameter and several feet in length. Measurements were made by catching a short 'time piece' of the same diameter and material as the bar, held weakly on to one end of the bar, in a ballistic pendulum. Davies (1948) determined the stress–time curve of pulses in a Hopkinson bar by measuring the displacement of its end, or the Poisson expansion along the bar, with electrostatic transducers. In the split Hopkinson bar test, Kolsky (1949), short specimens with the same cross-section as the bar are placed between two long bars and a stress pulse is propagated through the system from the end of one of the bars. The pulse may be generated by the impact of a hammer or by an explosive.

Let u be the axial displacement in the direction of pulse propagation and designate quantities in the first bar by subscript 1 and those in the second bar by subscript 2. Elastic strains in the Hopkinson bars are readily measured with electric resistance strain gauges and, from § 13.2, the displacement as a function of time is given by

$$u = C_B \int_0^t \varepsilon \, dt. \tag{12}$$

If the incident and reflected strain pulses in the first bar are ε_I and ε_R, respectively, and the transmitted pulse in the second bar is ε_T the displacements of the ends of these bars, between which the specimen is held, are

$$u_1 = C_B \int_0^t (\varepsilon_I - \varepsilon_R) \, dt \tag{13}$$

$$u_2 = C_B \int_0^t \varepsilon_T \, dt. \tag{14}$$

The average strain in the specimen of length L at any time, t, is

$$\varepsilon_S = (u_1 - u_2)/L = (C_B/L) \int_0^t (\varepsilon_I - \varepsilon_R - \varepsilon_T) \, dt. \tag{15}$$

Similarly, the average stress in the specimen at any time is given by

$$\sigma_S = E(\varepsilon_I - \varepsilon_R + \varepsilon_T)/2, \tag{16}$$

where $E =$ the Young's modulus of the bar.

Equations (12) and (13) ignore the effects of wave propagation in the

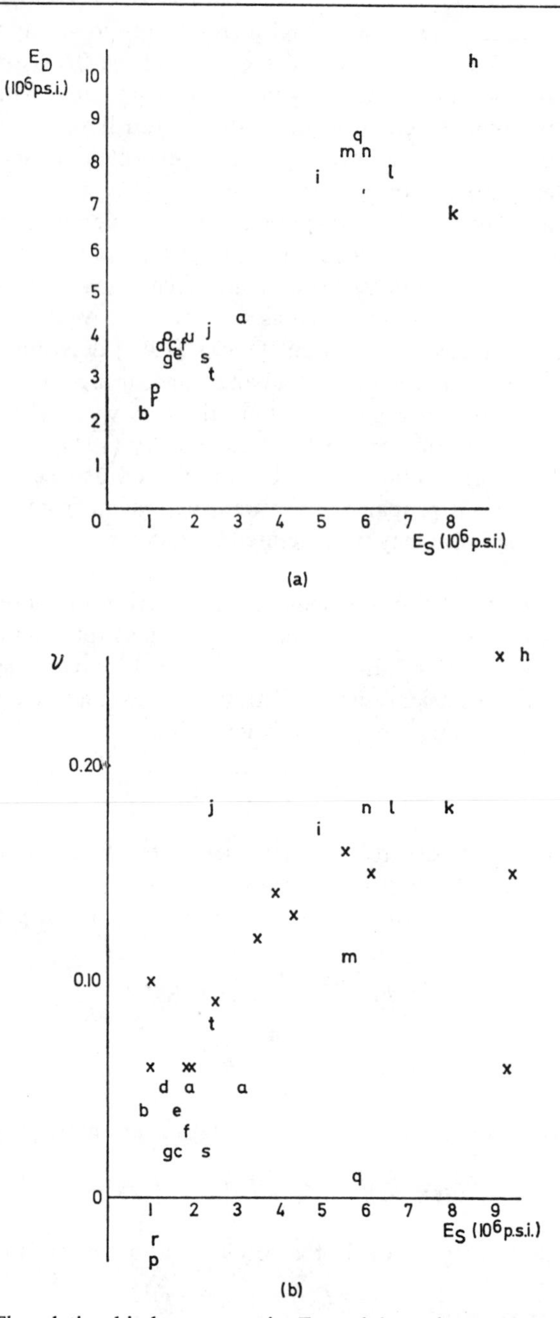

Fig. 6.15.1 (*a*) The relationship between static, E_S, and dynamic, E_D, Young's modulus for foundation rocks. (*b*) The relationship between E_S and Poisson's ratio, ν. (Data from tests by U.S. Bureau of Reclamation, 1953.)

specimen and are really applicable only to very short specimens. The behaviour of short specimens is affected strongly by end-effects, § 6.2, arising from contact with the bars. The use of longer specimens necessitates consideration of wave propagation in the specimen. As the purpose of these experiments is usually to study the behaviour of specimens at large values of the stress producing inelastic deformation, the problem becomes complex.

Complete stress–strain curves for Carrara marble have been obtained by Attewell (1962b), using a split Hopkinson bar and very short specimens. Hakalehto (1967) used specimens of Tennessee marble with L/D ratios from 0·67 to 2·67 and found that the amount of energy in the transmitted wave corresponded to a peak-pulse stress in the specimen equal to the static uniaxial compressive strength of specimens with the same proportions.

6.15 Comparison of results

Any one type of rock may have a wide range of mechanical properties, so that a geological description does not provide satisfactory information concerning these properties. Many workers have attempted to compare different properties of rocks in the hope of finding a method for predicting the strength and other properties from a few simple measurements. The U.S. Bureau of Mines, Obert *et al.* (1946), Windes (1949, 1950), and Blair (1955, 1956), have measured the properties, including the uniaxial compressive strength, the modulus of rupture, and the static and dynamic Young's moduli, of many rocks. Similar measurements have been made on a range of typical foundation rocks by the U.S. Bureau of Reclamation (1953). D'Andrea *et al.* (1964) and Judd (1964) have used regression analysis to compare measurements of compressive and tensile strengths, dynamic and static moduli, and velocities of wave propagation in a wide variety of rocks. Hobbs (1967) has studied the relationship between the Brazilian tensile strength, the uniaxial compressive strength, the compressive strength of irregular pieces, and the impact strength. In general,

Rock	*Key to letter on Fig. 6.15.1*	*Rock*	*Key to letter on Fig. 6.15.1*
Diorite (quartz)	a	Limestone (oolitic)	l
Granite (altered)	b	Limestone (stylolitic)	m
Graywacke (coarse grained)	c	Monzonite Porphyry	n
Graywacke (coarse grained)	d	Phyllite (graphitic)	o
Graywacke (fine grained)	e	Phyllite (quartzose)	p
Graywacke (medium grained)	f	Schist (biotite)	q
Graywacke (medium grained)	g	Schist (sericite)	r
Limestone (fine grained)	h	Shale (calcareous)	s
Limestone (medium grained)	i	Shale (quartzose)	t
Limestone (porous)	j	Siltstone	u
Limestone (chalcedonic)	k		

many properties, such as the dynamic and static Young's moduli, Fig. 6.15.1 (*a*), are related in a broad way. However, the aim of predicting most of the mechanical properties with some accuracy from a few simple measurements has not been realized. This is not altogether surprising, as various features affect different properties in varying degree. A feature important to the strength of rock may have little effect on its modulus. Therefore, it is important to study the relationships between various properties in the light of the theoretical knowledge concerning the features affecting them.

The Coulomb, the plane, the modified, and the extended Griffith criteria, §§ 4.6, 4.8, 4.9, are used to predict the triaxial strength of rock. The predictions of these and the effective strain-energy criterion are compared with measured data in Fig. 6.6.2. The Coulomb and modified Griffith criteria are essentially the same, but the latter incorporates the tensile strength and is therefore used in this discussion. The various Griffith criteria predict specific relations between the uniaxial compressive and tensile strengths of rock, which are important because of their mechanistic significance. The plane and extended Griffith criteria predict unequivocally that the value of the uniaxial compressive strength, C_0, is eight or twelve times that of the uniaxial tensile strength, T_0, respectively. According to the modified Griffith criterion, the ratio between these strengths is

$$C_0/T_0 = 4/[(\mu^2 + 1)^{\frac{1}{2}} - \mu], \tag{1}$$

where the coefficient of friction, μ, corresponds to the coefficient of internal friction of the Coulomb criterion. The uniaxial compressive and tensile strengths, each measured by the same investigators, Brace (1964b) and Jaeger and Hoskins (1966a, b), of six different rocks are shown in Table 6.15.1, together with the ratio between the measured values of these strengths and the ratio calculated on the basis of (1). It is seen that the

Table 6.15.1

Comparison of uniaxial compressive, C_0, and tensile, T_0, strengths

Rock	C_0 kpsi	T_0 psi	C_0/T_0	μ	$C_0/T_0 =$ $4[(\mu^2 + 1)^{\frac{1}{2}} + \mu)]$	Reference
Quartzite, Cheshire	66·8	4,060	16·5	0·9	8·5	Brace (1964)
Granite, Westerly	33·2	3,040	10·9	1·4	12·5	
Diabase, Frederick	70·5	5,800	12·1	1·7	15·0	
Sandstone, Gosford	7·25	520	13·5	0·5	5·5	Jaeger and Hoskins (1966a, b)
Marble, Carrara	13·0	1,000	13·0	0·7	7·5	
Trachyte, Bowral	21·75	1,990	10·9	1·0	9·5	
Quartzite, Witwatersrand	28·0	3,000	9·3	1·0	9·5	Hoek (1965)

ratio of the measured values is closer to 12, predicted by the extended Griffith criterion, than it is to 8, predicted by the plane Griffith criterion. The ratios calculated on the basis of the modified Griffith criterion bear little relationship to the measured ratios, although under conditions of compressive stress the straight line of this criterion appears to fit the measured value of strength better than does the parabola of the extended criterion, Fig. 6.6.1.

Direct measurements of the tensile strength of rocks are difficult and tedious to make. Indirect methods of measuring the tensile strength are described in §§ 6.10, 6.11, 6.12. Jaeger and Hoskins (1966a) compare the values of the direct uniaxial tensile strength of three different rocks with values obtained by different methods, Table 6.15.2. These appear to fall quite clearly into two sets with values differing by a factor of about two. The values obtained from the two tests with near uniform tension, namely, the direct uniaxial tensile test and the Brazilian test, are similar, while those obtained from tests involving non-uniform stresses are almost twice as great.

It would appear that a very useful description of the triaxial strength of rock is obtained from the extended Griffith criterion. Theoretically, measurement of the strength at one stress condition defines this criterion for a particular rock. The uniaxial compressive strength is obviously the most suitable measurement to make. From the preceding discussion, it would appear that the Brazilian test provides a simple means of estimating the uniaxial tensile strength which can conveniently be used as a check.

Table 6.15.2
Comparison of tensile strengths (psi)*

Method	Rock		
	Marble, Carrara	Sandstone, Gosford	Trachyte, Bowral
Uniaxial tension	1,000	520	1,990
Brazilian (15° arc)	1,265	540	1,740
Hollow Brazilian ($R_2/R_1 = 2$)	2,500	1,200	3,500
Bending (3 point)	1,710	1,140	3,659

* From Jaeger and Hoskins (1966).

The extended Griffith criterion has the further advantage that it describes the whole of the failure surface. The results of tests on hollow cylinders to determine the influence of the intermediate principal stress on the strength, Fig. 6.8.2, are qualitatively in agreement with the predictions of the extended criterion. These tests are complicated by the effects of non-uniform stress, but do not differ qualitatively from those using homogeneous polyaxial stress, Fig. 6.7.2.

The extended Griffith criterion predicts that the value of the intermediate principal stress has an effect on the strength, whereas the Coulomb and modified Griffith criteria suggest that it does not. Brace (1964a) remarks that the influence of the intermediate principal stress on the strength is probably negligible, because in some cases the same values of stress were found at failure for conditions of $\sigma_1 > 0$, $\sigma_2 = \sigma_3 = 0$ as were found when $\sigma_1 = \sigma_2 > 0$, $\sigma_3 = 0$. However, the nature of the fracture in pinching or extension tests is different from that in the uniaxial or triaxial compression test. There is evidence to suggest that failure in compression starts as cleavage parallel to the direction of the major principal stresses, §§ 6.4, 12.4. In triaxial compression and extension tests the major principal stress is applied with different geometries and properties, § 6.6. It is quite conceivable that primary cleavage leads to intrusion types of fracture and termination of the extension test at a value of the major stress less than that necessary to cause shear fracture and terminate a compression test. Jaeger and Hoskins (1966) have observed diametral extension fractures in confined Brazilian tests under conditions where all three principal stresses at failure are compressive. These failures occurred at values of the major principal stress about 10 per cent greater than those for failure of the same rock in a triaxial test, suggesting that the intermediate principal stress affects the strength. Presumably ultimate failure under conditions of compressive stress in these experiments occurs by extension fracture because shear fracture is not geometrically advantageous to the collapse of the test system, as it is in the triaxial compression test.

The effective strain-energy criterion, § 12.7, not only describes the whole failure surface for compressive stresses and predicts that the intermediate principal stress has an effect, Fig. 12.7.1, similar to that observed in experiments, Figs. 6.7.2 and 6.8.2, but also agrees better with data on triaxial strength, Fig. 6.6.2, than does the extended Griffith criterion. This close agreement between the predictions of the effective strain-energy criterion and experimental data and the fact that this criterion allows irreversible changes in the structure of the rock to take place before failure suggests that it is both the most useful and realistic criterion for the failure of rock in compression.

Failure in a stiff testing machine results in less damage to a specimen than occurs in a soft machine. This is expected on energetic grounds, but may mechanistically be related to buckling in compression tests. The proportions of most compression-test specimens are such as to preclude elastic buckling. However, the deformation in the third region of the stress–strain curve, § 4.2, is inelastic. The problem of inelastic buckling has been treated by von Kármán, and a complete account is given by Timoshenko and Gere (1961). When a column, compressed beyond the elastic limit, bends, the stress on the convex side decreases while that on the concave side continues to increase. In this case buckling is controlled by the *reduced modulus* of *elasticity*, E_r, which can be derived in a manner directly analogous to that used to derive the average Young's modulus in bending, § 6.10, using the tangent, E_t, and intrinsic Young's moduli, E, in place of the compressive, E_C, and tensile moduli, E_T, used in that case. The value of the reduced modulus is less than that of the ordinary elastic modulus, and the critical Euler buckling load decreases accordingly. In the limit, when the increase in axial load compensates for any decrease in load on the convex side due to buckling, the reduced modulus becomes equal to the tangent modulus. The critical buckling load, § 5.10, is given by

$$P_C = \pi^2 E_t I / 4L^2, \tag{2}$$

for a specimen with hinged ends and four times this for a specimen fixed at both ends. At the peak of the stress–strain curve $E_t = 0$, and it would appear that inelastic buckling is probable. Unlike elastic buckling, the load decreases with deflection when inelastic buckling occurs. If this decrease is greater than the decrease of the applied load with deflection due to the resilience of the testing machine the situation is unstable, § 6.13. This may occur with soft testing machines, but is unlikely to occur in a stiff machine. It may be concluded that a stiff testing machine and fixed ends are desirable in compression testing to avoid the effects of inelastic buckling.

Inelastic behaviour is also of concern in connection with matched end-pieces, § 6.2. During elastic deformation of a specimen a discontinuity between the specimen and end-piece results in stress concentrations which

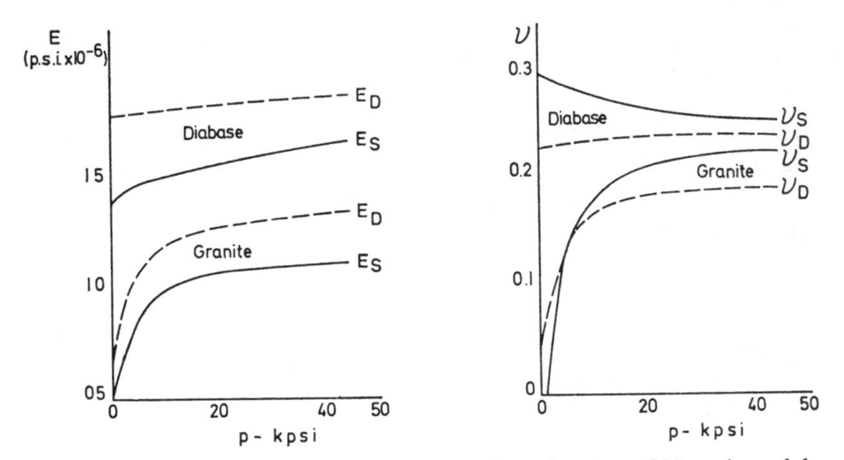

Fig. 6.15.2 A comparison between the static and dynamic values of Young's modulus, and Poisson's ratio, E_S, E_D, ν_S and ν_D, respectively, for two rocks at different confining pressures. (Data from Brace (1964, 1965) and Simmons and Brace (1965).)

lead to early failure. However, the restraint provided by an end-piece during inelastic deformation has the effect of strengthening the end of the specimen. Provided the specimen is of sufficient length in relation to its diameter, failure occurs in the central portion away from the strengthening influence of the end-pieces.

The static and dynamic values of Young's modulus and Poisson's ratio, E_S, E_D, and ν_S, ν_D, respectively, for two different rocks as a function of confining pressure are shown in Fig. 6.15.2. These curves are derived from measurements of static Young's modulus and compressibility on Frederick diabase, *F*, and Westerly Granite, *W*, by Brace (1964b, 1965) and from measurements of the velocities of propagation of waves of dilatation and distortion by Simmons and Brace (1965). They show that both moduli increase with confining pressure and that the dynamic modulus is about 20 per cent greater than the static modulus for both rocks. Both the dynamic and static Poisson's ratio of the relatively porous granite increase with confining pressure. With the exception of the anomalous behaviour of the static Poisson's ratio for granite at low confining pressure, including a negative value for zero pressure, the static value of Poisson's ratio for both rocks is greater than the dynamic ratio for most confining pressures. It should be noted that the determination of the dynamic compressibility from velocities and Poisson's ratio from the compressibility and Young's modulus are essentially inaccurate because they involve the difference of two quantities of similar magnitude, §§ 5.2, 6.14. Nevertheless, the results in Fig. 6.15.2 are consistent with the theoretical predictions of the behaviour of cracked rock, § 12.3. Porosity due to open cracks decreases Young's modulus and Poisson's ratio. Confining pressure closes the open cracks, resulting in increasing modulus and Poisson's ratio with confining pressure. If the surfaces of closed cracks slide past one another the value of Young's modulus decreases, but that of Poisson's ratio increases. Sliding occurs in static compression, resulting in smaller static values of Young's modulus and greater static values of Poisson's ratio than the dynamic values of these properties. The decrease in the static value of Poisson's ratio for the diabase might even be ascribed to a reduction in sliding due to confining pressure.

Chapter Seven

Effects of Size and Stress Gradient

7.1 Introduction

Despite their great importance, the influence of the size of specimens and of stress gradients on the strength of rocks has been little studied.

With regard to size effects, one powerful reason has been the tendency to experiment with specimens of a standard size: further, the effects of the differences in size between ordinary laboratory test specimens are not large, and a careful statistical investigation covering a wide range of sizes is necessary. This can most conveniently be done in uniaxial compression, and most work has been done on coal, being motivated by a desire to extrapolate to large dimensions to estimate the strength of coal pillars. The usual assumption is that effects of size are caused by flaws in the material, but extrapolation to large sizes is not necessarily justified, since there may be several types of flaw on very different scales.

The question of the effect of stress gradient on failure is of equal importance and has been even less studied. This has been partly because of the bias in laboratory testing towards the use of homogeneous stresses. However, in practice, stresses may be highly inhomogeneous, and it is essential to understand the behaviour of rocks under such conditions. As remarked in § 4.4, there is no *a priori* reason why stress gradients should not enter into the criterion for fracture. Quite apart from this, the presence of flaws in the material will contribute effects of stress gradient as well as of size.

Most studies of the effect of stress gradient have been on materials under tensile conditions. It has always been well known that, if the tensile strength is estimated from the extreme fibre strength of a beam in bending, high values are obtained. More recently, similar effects have been observed with diametral compression of hollow cylinders.

Most of these effects can be explained semi-quantitatively on the basis of an elementary statistical theory due to Weibull (1939a), who has made extensive studies (1938, 1939b, 1951, 1952) on phenomena of rupture based, essentially, on the postulate that this is a statistical phenomenon caused by a distribution of flaws in the material. This theory, which will be given in §§ 7.4–7.7, refers only to extension failures. Irwin (1958) gives a review of statistical theories of fracture.

7.2 The influence of size on strength

Most of the experimental information available is for coal and is concerned especially with unconfined compression of cubes of various side lengths. Evans and Pomeroy (1958), and Evans, Pomeroy and Berenbaum (1961) show that cubes of a particular size show a very wide range of strength which is distributed approximately normally. Also, both the mean and modal crushing strengths σ_c decrease with the side length a of the cube according to a power law

$$\sigma_c = ka^{-\alpha}, \tag{1}$$

where α is a constant with values between 0·17 and 0·32 for various coals. Millard, Newman, and Phillips (1955) and Steart (1954) found values of α of the order of 0·5. Gaddy (1956) has found that (1) with $\alpha = 0{\cdot}5$ applies in some cases to cubes of up to 64-in. side length. For non-cubical specimens, (1) is reasonably well generalized into

$$\sigma_c = kd^{\beta}/a^{\alpha}, \tag{2}$$

where a is the thickness of the specimen, d is its least width, and k, β, and α are constants. These and many other measurements on coal have been stimulated by the need to estimate strengths of pillars in coal mines. Greenwald, Howarth, and Hartmann (1941) proposed the formula (2) for this purpose, and it still appears to be adequate.

There seems to be little information available about either the dispersion of compressive strength of rocks or about the effect of specimen size on their strength. Obert *et al.* (1946) concluded that the size effect was small. Lundborg (1967) has recently studied 1 : 1 cylinders of granite of diameters ranging from 1·9 to 5·8 cm and finds a range of from 2,190 to 1,750 kg/cm^2 in crushing strength and reasonable agreement with the Weibull theory of § 7.4.

Evans (1961) found a variation of type (1) of the tensile strength of coal with specimen thickness in the Brazilian test. Durelli and Parks (1962) describe a small variation of the form (1) of tensile strength with size for plastics.

These size effects are usually attributed to the presence of flaws in the material. Suppose that a cube of side a contains m flaws, each of which has a probability P_0 of surviving a given stress, then the probability P_a of a cube of side a surviving this stress is

$$P_a = P_0{}^m. \tag{3}$$

Similarly, if a cube of side b has n flaws the probability P_b of its surviving the same stress will be

$$P_b = P_0{}^n. \tag{4}$$

Also there will be a relation

$$(n/m) = (b/a)^\beta, \tag{5}$$

where β is a constant which indicates the type of distribution of flaws; for a volume distribution $\beta = 3$. From (3)–(5)

$$a^{-\beta} \ln P_a = b^{-\beta} \ln P_b = \text{Constant}. \tag{6}$$

Combining (6) with (1) gives

$$(\sigma_a)^{\beta/\alpha} \ln P_a = \text{Constant},$$

where σ_a is the crushing strength of a cube of side a. The law (6) is reasonably well obeyed in practice.

Many writers have attempted to identify the flaws with Griffith cracks, § 10.13. Gilvarry (1961a, b) considers an extensively cracked material containing cracks of various types. Millard Newman and Phillips (1955) point out that $\alpha = 0{\cdot}5$ is the result to be expected if the lengths of Griffith cracks are proportional to the lengths of the sides of the cubes.

Protodiakonov (1964) has adopted a rather different approach. He regards a mass of rock or coal as being broken up by discontinuities of averaging spacing b. If σ_r is the crushing strength of the rock mass as a whole and σ_d is that of a cylindrical specimen of diameter d he proposes the relation

$$\frac{\sigma_d}{\sigma_r} = 1 + \frac{C - 1}{(d/b) + 1}, \tag{7}$$

where C is a parameter depending on the nature of the rock, which commonly lies between 2 and 10. This empirical relation seems to be reasonably well satisfied.

7.3 The influence of stress gradient on strength

Practically all the available information on this question has been derived from attempts to measure the tensile strength of materials indirectly. It has been well known for many years that if the tensile strength (that is the greatest tensile stress in the specimen at failure calculated on the theory of elasticity) is determined from bending, values considerably higher than those for direct tension are obtained. The same is true for other systems, notably diametral compression of hollow cylinders. Recently, results obtained by various methods have been compared by Berenbaum and Brodie (1959b), Durelli and Parks (1962), and Jaeger and Hoskins (1966a). It is quite clear that different systems employing inhomogeneous stresses lead to different values of the 'tensile strength'. While there are a number of assumptions involved which might affect the results, such as the assumption of linear elasticity up to failure, the most attractive explanation relates them to the effects of size discussed in § 7.2 and, in particular, to the volume of material subjected to the highest stresses. This, of course, is determined by the stress-gradients.

Durelli and Parks (1962) studied the effect of size and stress-gradient on a number of materials and specimens of various shapes, notably dog-bones in tension and bending, and diametral compression of hollow cylinders of various ratios of internal and external diameters. They found that for each material and type of specimen the stress at failure varies regularly with stress-gradient, but in different manners for specimens of different shapes.

They also found that if the failure stress is plotted logarithmically against the volume of the specimen subjected to 95 per cent of the maximum stress, a good straight line covering all specimen shapes is obtained. A physical background for this may be found from Weibull's theory discussed later. If $\bar{\sigma}$ is the breaking stress in pure tension for a specimen of volume V, taking the logarithm of § 7.4 (10) gives

$$\ln \bar{\sigma} = A - (1/m) \ln V, \tag{1}$$

where A and m depend on the material. Similarly, if $\bar{\sigma}_e$ is the breaking stress in pure bending for a specimen of volume V it follows from § 7.5 (3) that

$$\ln \bar{\sigma}_e = A - (1/m) \ln [V/(2m + 2)], \tag{2}$$

where A and m have the same values as in (1). (2) has the same form as (1), except that the volume in the second case is replaced by $V/(2m + 2)$, or $V/20$ if $m = 9$, and this is in fact the volume of material in the beam stressed to 90 per cent of the maximum level.

Experiments using hollow cylinders with external pressure and axial load have been done by Robertson (1955), Bellamy (1960), Hobbs (1962) and Jaeger and Hoskins (1966b). These involve stress-gradients in compression, which have not been considered because inconsistencies such as those described in connection with tension have not appeared.

A comprehensive study of the stamp-loading test, in which a right circular, hard, metal cylinder is loaded against a flat, rock surface, as in a plate-bearing test, § 15.7, has been made by Wagner and Schümann (1971). They found the load-displacement curves from this experiment for five different kinds of rock to be similar to the axial stress–strain curves obtained from triaxial tests for the same rocks. The *stamp bearing strength*, σ_{ST}, given as the peak load divided by the area of the stamp can be described by

$$\sigma_{ST} = Q\,(a/a')^{\alpha}, \tag{3}$$

where $Q =$ the stamp bearing strength for a stamp of radius a'
$a =$ radius of the stamp
$\alpha =$ an exponent which falls in the range from $-0{\cdot}5$ for a completely brittle rock to 0 for a completely ductile rock.

None of these rocks showed a noticeable size-effect in uniaxial compression, so it must be concluded that stress-gradients do have an important effect on the strength of brittle rocks in compression.

7.4 Weibull's discussion of tensile strength

Weibull (1939a) gave a simple discussion of the statistics of failure based on elementary statistical theory and a simple distribution law. This gives a clear idea of the questions involved and allows simple predictions to be made.

Only failure in extension is considered, and for simplicity tension will be reckoned positive in §§ 7.4–7.7.

Consider a set of similar tension test specimens of the same material and unit area, uniformly stressed. All specimens will not break at the same stress, and the experimental results may be expressed by the statement that the probability P_0 of failure at any stress σ is an experimentally determined function, $P_0(\sigma)$, of σ.

The probability that a specimen is not broken by stress σ is $1 - P_0$. Now suppose that n such specimens are arranged in series and stress σ is applied; the probability $1 - P$ that no specimen will break is the product of the probabilities that the individual specimen will not break, that is

$$1 - P = (1 - P_0)^n. \tag{1}$$

This simple result may be expected to be true more generally than for the obvious series arrangement: for example, if the n specimens were arranged in parallel with a total applied load $n\sigma$ equally distributed between them, it remains true.

If the specimens are each of unit length, so that the volume v of them combined is equal to n, (1) becomes

$$1 - P = (1 - P_0)^v, \tag{2}$$

or

$$\ln(1 - P) = v \ln(1 - P_0). \tag{3}$$

It follows from (2) that the probability of failure P increases with the volume of the test piece.

The notation

$$n(\sigma) = -\ln(1 - P_0) \tag{4}$$

will be used, and since $0 < P_0 < 1$, it follows that $0 < n(\sigma) < \infty$ Differentiating (3) with respect to v gives

$$\frac{dP}{1 - P} = -\ln(1 - P_0)\, dv = n(\sigma)\, dv. \tag{5}$$

Inhomogeneous stresses may now be taken into account by assuming that σ depends on position and integrating (5) over the volume of the solid. This gives

$$P = 1 - \exp\{-\int n(\sigma)\, dv\}. \tag{6}$$

The mean stress at failure, $\bar{\sigma}$, is then given by

$$\bar{\sigma} = \int_0^1 \sigma\, dP = -\int_0^\infty \sigma \frac{d(1-P)}{d\sigma}\, d\sigma = -[\sigma(1-P)]_0^\infty + \int_0^\infty (1-P)\, d\sigma$$

$$= \int_0^\infty \exp\{-\int n(\sigma)\, dv\}\, d\sigma, \tag{7}$$

provided that the distribution is such that $(1 - P)\sigma \to 0$ $\sigma \to$ as ∞.

Weibull (1939a) suggested that $n(\sigma)$ is a function of the material, namely,

$$n(\sigma) = [(\sigma - \sigma_u)/\sigma_0]^m, \tag{8}$$

where σ is the tensile stress on an elemental volume, dv, and σ_u, σ_0 and m are constants, σ_u representing the smallest tensile strength which any element can have. Clearly, for rock it is best to assume that σ_u may be zero; Weibull has also made this assumption for the sake of mathematical simplicity. In this case (8) reduces to

$$n(\sigma) = \sigma^m/\sigma_0{}^m = k\sigma^m \tag{9}$$

where $k = \sigma_0{}^{-m}$ and m are now constants depending upon the material. This form has the advantages of involving only two parameters and of being such that the integrals involved can be evaluated in many important cases.

If the stress is homogeneous over a volume V it follows from (6) and (9) that

$$P = 1 - \exp\{-kV\sigma^m\}. \tag{10}$$

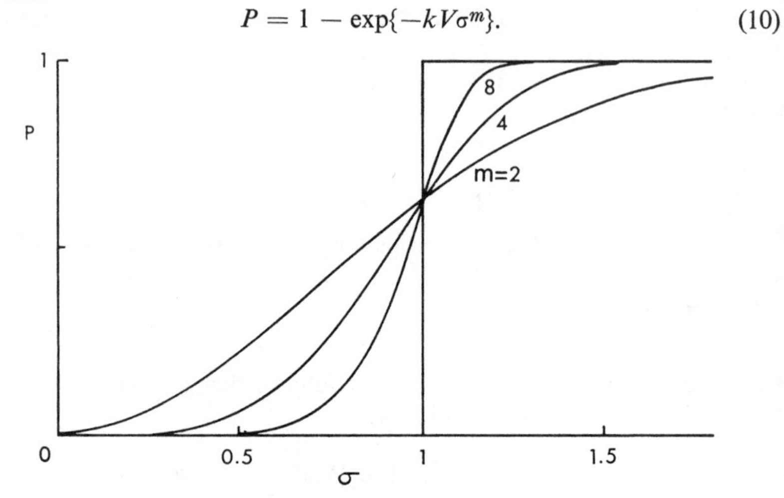

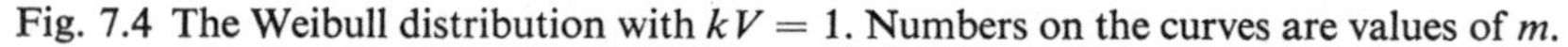

Fig. 7.4 The Weibull distribution with $kV = 1$. Numbers on the curves are values of m.

Fig. 7.4 shows values of P calculated from (10) for $kV = 1$ and $m = 2$, 4, 8, showing the form of the distribution and the way in which it varies with m.

The mean breaking stress $\bar{\sigma}$ calculated from (7) and (9) for the case in which the stress is homogeneous over a volume V is

$$\bar{\sigma} = \int_0^\infty \exp\{-kV\sigma^m\}\, d\sigma = \sigma_0 V^{-1/m} = I_m, \tag{11}$$

where I_m is the integral

$$I_m = \int_0^\infty \exp(-z^m)\, dz. \tag{12}$$

When $m = 2$, I_m has the value $\frac{1}{2}\pi^{\frac{1}{2}}$ and for other values of m is given in Table 7.4

Table 7.4

m	1	2	3	4	8	16	∞
I_m	1	0·886	0·893	0·906	0·942	0·968	1

The factor I_m in (11) is therefore near unity, and its variation is not important. The main result is that the mean breaking strength varies with specimen size like $V^{-1/m}$. This corresponds to the empirical law of § 7.2 (1) with $m = 3/\alpha$. If $\alpha = 0{\cdot}5$, $m = 6$, and if α is smaller, larger values are to be expected.

According to (8) the variation in strengths, σ_1 and σ_2, for volumes V_1 and V_2, would be $(\sigma_1 - \sigma_u)/(\sigma_2 - \sigma_u) = (V_2/V_1)^{1/m}$. The validity of the assumption made in (9) can be checked by studying the distribution (10) or the volume variation (11). Also, if the standard deviation s^2 of the distribution is calculated it is found to be

$$s^2 = \int_0^1 (\sigma - \bar{\sigma})^2\, dP = 2\int_0^\infty \sigma \exp\{-\int n(\sigma)\, dv\}\, d\sigma - \bar{\sigma}^2$$
$$= \sigma_0^2\, V^{-2/m}[I_{\frac{1}{2}m} - I_m], \tag{13}$$

so that $\bar{\sigma}^2/s^2$ should be independent of the specimen volume V. Weibull gives experimental results on plaster and other materials which fit the above formulae very well.

In the following sections some cases of inhomogeneous stresses which lead to elementary integrals will be considered.

7.5 Bending

The simplest case involving inhomogeneous stresses is that of bending. Suppose, first, that a rectangular beam of length l, width b, and depth $2h$

is subjected to pure bending by couples applied to its ends: this case occurs in four-point support, § 5.10. From the theory of that section, the stresses above the neutral axis OO', Fig. 5.10.1 (*b*), are compressive, and the effect of these is neglected. The stress σ at depth y below the neutral axis is

$$\sigma = y\sigma_e/h, \tag{1}$$

where σ_e is the extreme fibre stress at $y = h$.

In this case, using the form § 7.4 (8) of $n(\sigma)$,

$$\int n(\sigma)\,dv = blk\sigma_e{}^m \int_0^h (y/h)^m\,dy = blhk\sigma_e{}^m/(m+1) = kV\sigma_e{}^m/2(m+1), \tag{2}$$

where $V = 2blh$ is the total volume of the beam.

The mean stress $\bar{\sigma}_e$ in the extreme fibre at failure is by § 7.4 (7)

$$\bar{\sigma}_e = \int_0^\infty \exp\{-kV\sigma_e{}^m/2(m+1)\}\,d\sigma_e = \sigma_0[V/2(m+1)]^{-1/m}I_m. \tag{3}$$

Comparing (3) with § 7.4 (11) gives the ratio of the mean extreme fibre stress in pure bending to the mean failure stress $\bar{\sigma}$ in tension. It is

$$\bar{\sigma}_e/\bar{\sigma} = (2m+2)^{1/m}. \tag{4}$$

Some values of this are shown in the first row of Table 7.5.

For a beam of circular cross-section of radius a the corresponding result is

$$\int n(\sigma)\,dv = 2lk\sigma_e{}^m \int_0^a (y/a)^m (a^2 - y^2)^{\frac{1}{2}}\,dy = \frac{2kV\sigma_e{}^m}{\pi}\int_0^{\pi/2} \sin^m\theta\cos^2\theta\,d\theta, \tag{5}$$

where V is the total volume of the beam and, proceeding as before, values of $\bar{\sigma}_e/\bar{\sigma}$ can be found which are given in the second row of Table 7.5.

Another important case is that of three-point bending. For a rectangular beam of length $2l$, width b, and depth $2h$, the stress σ at distance z from the centre of the beam and depth y below the neutral axis is, by § 5.10 (4),

$$\sigma = y(l - z)\sigma_e/hl. \tag{6}$$

It follows that

$$\int n(\sigma)\,dv = 2bk\sigma_e{}^m \int_0^h (y/h)^m\,dy \int_0^l \left(1 - \frac{z}{l}\right)^m dz = kV\sigma_e{}^m/2(m+1)^2, \tag{7}$$

where V is the total volume of the beam.

Therefore, as in (3),

$$\bar{\sigma}_e = \sigma_0[V/2(m+1)^2]^{-1/m}I_m, \tag{8}$$

and

$$\bar{\sigma}_e/\bar{\sigma} = [2(m+1)^2]^{1/m}. \tag{9}$$

Some values of this are shown in Table 7.5 in the third row.

Two conclusions can be drawn from the results of Table 7.5. First, the ratio $\bar{\sigma}_e/\bar{\sigma}$ decreases with increasing m. Secondly, it increases with increasing concentration of stress, being highest for three-point bending where the greatest stresses occur in a restricted region near the central load.

Table 7.5

The ratio $\bar{\sigma}_e/\bar{\sigma}$ of the mean extreme stress at failure in various cases to the mean stress at failure in pure tension

m	2	4	8
Pure bending (rectangle)	2·45	1·78	1·44
Pure bending (circle)	2·83	2·00	1·50
Three-point bending (rectangle)	4·24	2·66	1·89
2:1 hollow cylinder	1·69	1·58	1·41

7.6 Hollow cylinders

The case of a hollow cylinder of internal and external radii R_1 and R_2, respectively, with internal pressure p_1 provides another interesting example. From § 5.11 (10), reckoning tension positive, the stress σ_θ is tensile and has the value

$$\sigma = \frac{p_1\rho^2}{1-\rho^2}\left(1+\frac{R_2{}^2}{r^2}\right) = \frac{\rho^2\sigma_e}{1+\rho^2}\left(1+\frac{R_2{}^2}{r^2}\right), \tag{1}$$

where $\rho = R_1/R_2$ and σ_e is the maximum value of σ_θ which occurs when $r = R_1$.

Using the value § 7.4 (9) of $n(\sigma)$, it follows that if the length of the cylinder is l

$$\int n(\sigma)\,dv = \frac{2\pi lk\rho^{2m}\sigma_e{}^m}{(1+\rho^2)^m}\int_{R_1}^{R_2}\left(1+\frac{R_2{}^2}{r^2}\right)^m r\,dr. \tag{2}$$

The integral can be evaluated for any value of m. For $m = 2$ it gives

$$\int n(\sigma)\,dv = \frac{kV\rho^4\sigma_e{}^2}{(1+\rho^2)^2}\left\{1-\frac{4}{(1-\rho^2)}\ln\rho+\frac{1}{\rho^2}\right\}, \tag{3}$$

where V is the volume of the cylinder. For $\rho = \frac{1}{2}$ this has the value $0{\cdot}348kV\sigma_e{}^2$.

Thus from § 7.4 (7), proceeding as in § 7.5, the mean value $\bar{\sigma}_e$ of σ_e is found to be

$$\bar{\sigma}_e = 0{\cdot}348\sigma_0{}^{m/2}V^{-\frac{1}{2}}I_2, \tag{4}$$

and the ratio of this to the mean breaking stress $\bar{\sigma}$ in pure tension is

$$\bar{\sigma}_e/\bar{\sigma} = (0{\cdot}348)^{-\frac{1}{2}} = 1{\cdot}69. \tag{5}$$

The above result for $m = 2$ and others for $m = 4$, 8 are shown in Table 7.5. It follows from (3) that as ρ increases and the stress becomes more homogeneous, $\bar{\sigma}_e/\bar{\sigma}$ tends towards unity. For $\rho = 0{\cdot}9$, $m = 2$, it is 1·06.

7.7 Biaxial stresses and torsion

The previous discussion refers to tensile stresses in a single direction. The question arises as to whether it can be generalized to stresses in two or three dimensions. Weibull suggests that this can be done by replacing the σ of the previous analysis by a suitable scalar quantity determined by the principal stresses. He does not show how this is to be done in general, but he examines the possibility of using a law involving the magnitude σ_n of the normal stress across any plane.

In place of § 7.4 (9) he assumes the distribution $n_1(\sigma_n)$ of σ_n to be

$$n_1(\sigma_n) = k_1\sigma_n{}^m, \tag{1}$$

where k_1 is a constant of the material, and he evaluates

$$\int n_1(\sigma_n)\, dv \tag{2}$$

over a sphere of radius a.

Consider first the case of uniaxial tension σ_1. In this case the normal stress in a direction at an angle ϕ to σ_1 is

$$\sigma_n = \sigma_1 \cos^2 \phi. \tag{3}$$

The integral (2) over a sphere of radius a is therefore

$$\begin{aligned}\int n_1(\sigma_n)\, dv &= 2\pi k_1\sigma_1{}^m \int_0^a r^2\, dr \int_0^\pi \cos^{2m}\phi \sin\phi\, d\phi \\ &= 4\pi a^3 k_1\sigma_1{}^m/3(2m+1) = Vk_1\sigma_1{}^m/(2m+1), \end{aligned} \tag{4}$$

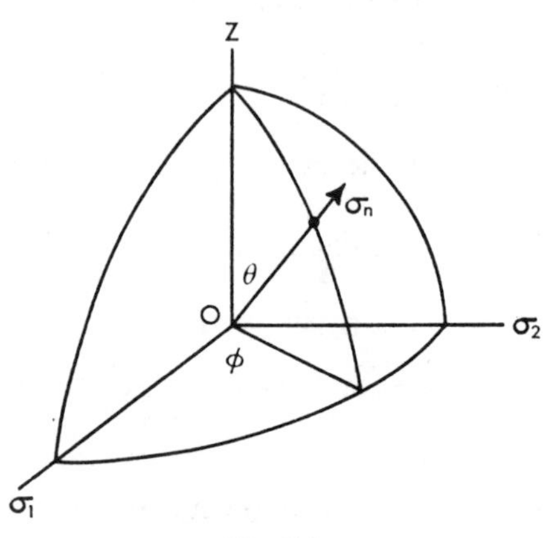

Fig. 7.7

where V is the volume of the sphere. Since this result must be the same as the value $Vk\sigma_1^m$ which follows from § 7.4(9) we must have $k = k_1/(2m+1)$, and the law (1) is replaced by

$$n_1(\sigma_n) = k(2m+1)\sigma_n^m, \tag{5}$$

where k is the constant of the material introduced in § 7.4.

The law will now be applied to plane stresses σ_1, σ_2, $\sigma_1 \geqslant \sigma_2$, the third principal stress being zero. Using the coordinate system of Fig. 7.7, the normal stress over the plane specified by θ, ϕ is by § 2.5 (17)

$$\sigma_n = (\sigma_1 \cos^2 \phi + \sigma_2 \sin^2 \phi) \sin^2 \theta. \tag{6}$$

If both σ_1 and σ_2 are tensile, σ_n is always tensile. If σ_2 is compressive, σ_n is tensile only if $|\phi| < \phi_0$ or $|\pi - \phi| < \phi_0$, where

$$\tan \phi_0 = (-\sigma_1/\sigma_2)^{\frac{1}{2}}. \tag{7}$$

As before, compressive stresses will not be considered, so that if σ_2 is negative only a restricted range of ϕ is in question. The integral (2), taken over the relevant portions of a sphere of radius a, is then

$$\int n_1(\sigma_n)\, dv$$
$$= 4[k(2m+1)]\int_0^a r^2\, dr \int_0^\pi \sin^{2m+1}\theta\, d\theta \int_0^{\phi_0} [\sigma_1 \cos^2 \phi + \sigma_2 \sin^2 \phi]^m\, d\phi$$
$$= \frac{8ka^3(2m+1)\,.\,2m(2m-2)\ldots 2}{3\,.\,(2m+1)(2m-1)\ldots 3} \int_0^{\phi_0} [\sigma_1 \cos^2 \phi + \sigma_2 \sin^2 \phi]^m\, d\phi, \tag{8}$$

where if σ_1 and σ_2 are both tensile ϕ_0 is to be replaced by $\frac{1}{2}\pi$.

The integral in (8) can readily be evaluated for integral values of m. For example, if σ_1 and σ_2 are both tensile and $m = 2$, (8) becomes

$$kV[\sigma_1^2 + \sigma_2^2 + (2/3)\sigma_1\sigma_2] = kV\sigma_1^2[1 + (\sigma_2/\sigma_1)^2 + (2\sigma_2/3\sigma_1)], \tag{9}$$

where V is the volume of the sphere.

Now suppose that the ratio (σ_2/σ_1) is kept constant and σ_1 is increased until failure takes place, then the mean value $\bar{\sigma}_1$ of σ_1 at failure is, as in § 7.4 (10),

$$\bar{\sigma}_1 = \{kV[1 + (\sigma_2/\sigma_1)^2 + (2\sigma_2/3\sigma_1)]\}^{-\frac{1}{2}} I_2, \tag{10}$$

and the ratio of this to the mean breaking strength $\bar{\sigma}$ in tension given by § 7.4 (10) is

$$\bar{\sigma}_1/\bar{\sigma} = [1 + (\sigma_2/\sigma_1)^2 + (2\sigma_2/3\sigma_1)]^{-\frac{1}{2}}. \tag{11}$$

This implies the failure criterion

$$3\sigma_1^2 + 2\sigma_1\sigma_2 + 3\sigma_2^2 = 3\bar{\sigma}^2. \tag{12}$$

This is an example of a criterion for failure deduced from purely statistical considerations.

For the case of biaxial tension, $\sigma_1 = \sigma_2$, with $m = 2$ it follows from (11) that $\bar{\sigma}_1/\bar{\sigma} = 0{\cdot}61$. The corresponding result for any integral value of m follows from (8) and is

$$\bar{\sigma}_1/\bar{\sigma} = \left\{\frac{2m(2m-2)\ldots 2}{(2m-1)\ldots 3}\right\}^{-1/m}. \tag{13}$$

Values of this are shown in Table 7.7 for $m = 2, 4, 8$. It appears that on this theory values of the breaking stress in biaxial tension are less than those in uniaxial tension.

For *pure shear*, $\sigma_2 = -\sigma_1$, $\phi_0 = \pi/4$, (8) becomes

$$\frac{4ka^3 \,.\, 2m(2m-2)\ldots 2}{3 \,.\, (2m-1)(2m-3)\ldots 3}\int_0^{\pi/2} \sigma_1{}^m \cos^m \alpha \, d\alpha. \tag{14}$$

If $m = 2$ this has the value $2kV\sigma_1{}^2/3$ and

$$\bar{\sigma}_1/\bar{\sigma} = (3/2)^{\frac{1}{2}} = 1{\cdot}22. \tag{15}$$

Values of $\bar{\sigma}_1/\bar{\sigma}$ for $m = 4$ and $m = 8$ are given in Table 7.7.

Torsion of a solid cylinder may also be studied. In this case the shear stress τ at radius r is given by

$$\tau = r\tau_1/a, \tag{16}$$

where a is the radius of the cylinder and τ_1 is the shear stress at its surface, corresponding to principal stresses $\sigma_1 = \tau_1$, $\sigma_2 = -\tau_1$. In evaluating $\int n_1(\sigma_n)\, dv$ a term

$$\int_0^a (r/a)^m \,.\, 2\pi r\, dr = 2\pi a^2/(m+2)$$

corresponding to the variation of σ_1 with r appears. Thus the results for $\bar{\sigma}_1$ in torsion may be obtained from those for pure shear by multiplying by $[(m+2)/2]^{1/m}$. Values of $\bar{\sigma}_1/\bar{\sigma}$ are given in Table 7.7.

Table 7.7

The ratio $\bar{\sigma}_e/\bar{\sigma}$ of the mean extreme stress at failure under plane stress to the mean stress at failure in pure tension

m	2	4	8
Biaxial tension	0·61	0·72	0·82
Pure shear	1·22	1·10	1·05
Torsion	1·73	1·45	1·28

It must be emphasized that the calculations of §§ 7.4–7.7, and particularly of this section, are simply the working out of a simple hypothesis and have no absolute quantitative status. Nevertheless, it is useful to have an hypothesis which is applicable to a wide variety of conditions. For example, Tables 7.5, 7.7 suggest that statistical effects of this sort will be most important in three-point bending and much less important in torsion and hollow cylinders with internal pressure. Other cases, such as diametral compression, can be studied by numerical evaluation of the appropriate integrals. It also suggests that a material will be weaker in biaxial tension than uniaxial tension, a possibility which does not emerge from the simple criteria for failure of Chapter 4.

Chapter Eight

Fluid Pressure and Flow in Rocks

8.1 Introduction

The effect of fluid pressure in the pores of a rock and of water movement through the pores is of great practical importance. In general, all rocks at ordinary pressure contain a proportion of empty spaces or voids: some of these may be interconnected to form passages through which fluid can penetrate, and from the present point of view only these are of importance.

While both soil mechanics and rock mechanics are concerned with porous media, there is a considerable difference in the nature of the medium: soils are usually regarded as consisting of discrete particles touching at isolated points, while materials such as sedimentary rocks and concrete are regarded as a solid skeleton traversed by a fine network of capillaries; in rocks of low porosity, such as igneous rocks, the voids probably consist of grain-boundary cracks.

For this reason, most discussions on the effect of pore fluids in soil mechanics, cf. Terzaghi (1943), Scott (1963), and Skempton (1960), concentrate on microscopic behaviour, that is, on discussion of effects at the points of contact of the particles. On the other hand, what is needed in rock mechanics is a macroscopic theory for a solid traversed by a network of pores. In fact, both concepts are applied to both soils and rocks, Skempton (1960) discusses rocks from the microscopic point of view, while Biot (1941a) formulated macroscopic equations with a view to studying problems of soil mechanics. His treatment will be followed here.

In § 8.4 the macroscopic equations will be derived. They are, of course, highly idealized: in particular, they assume that the medium is isotropic (anistropic materials can also be studied) and that the solid material is linearly elastic. It is also assumed here that the pores are filled with water which is treated as incompressible: this assumption can be relaxed and compressible fluid and incomplete saturation treated, but with some difficulty. Finally, an effect which is important for some soils and rocks, namely swelling of some constituents on wetting, is ignored.

8.2 Porosity and permeability

Among the fundamental properties of a porous medium are the amount of space in it which is void or not filled by solid material, and the rate of movement of water or other fluid through the medium.

Various notations are in use. Terzaghi (1943) defines the void ratio e as the ratio of the volume V_v of voids to the volume V_s of solids in any specimen, that is

$$e = V_v/V_s. \tag{1}$$

The porosity n is defined as the proportion of voids per unit total volume, so that

$$n = V_v/(V_s + V_v) = e/(1 + e). \tag{2}$$

Numerical values are frequently given as percentages rather than fractions.

Alternatively, the fractional porosity f is frequently defined as the fraction $f\delta A$ of the area δA of a small surface which contains no solid matter. It is assumed that δA, while small, is large enough to intersect many pores, so that, if the material is homogeneous and isotropic, f will be independent of the position and orientation of the surface, and so is equal to the volume porosity n defined above.

For spheres of equal size in a cubic arrangement the porosity is 47·6 per cent, while in the closely packed rhombohedral arrangement it is 26 per cent. For loose sand, porosity is of the order of 40 per cent, for oil sands it is usually in the range of 10–20 per cent. Tables of values are given by Muskat (1937), and Birch, Schairer, and Spicer (1942). Handin *et al.* (1963) give values of 18·2 per cent for Berea sandstone, 5·6 per cent for Repetto siltstone, and 3·5 per cent for Hasmark dolomite. For massive igneous rocks values are much lower, Brace (1965).

While in soils it may be assumed that the whole of the pore space is interconnected, this may not be true for rocks; and an interconnected system of pores is necessary for transport of fluids. Of different methods of measuring porosity, some measure total porosity and some only the accessible pore space, for example, if ρ_s is the density of the solid material (determined from its mass and volume) and ρ_m is the mineral density (obtained by weighing the crushed rock in toluene or some other inert liquid), then

$$f_t = (\rho_m - \rho_s)/\rho_m \tag{3}$$

measures the total porosity including isolated pores.

On the other hand, if volume V_f of fluid can be injected into volume V of dry solid the ratio

$$f_c = V_f/V \tag{4}$$

measures the interconnected pore space. Comparison of weights dry and saturated with water, Obert *et al.* (1946), also gives this value.

8.3 Flow of fluid through a porous medium

It was established experimentally by Darcy (1856), and has subsequently been confirmed many times, that the rate of flow of fluid V_x, V_y, V_z at a point, defined as the volume of fluid crossing unit area per unit time, is proportional to the gradient of pore pressure p at that point, that is, measuring V_x, V_y, V_z, positively in the negative directions of the axes,

$$V_x = \frac{k'}{\eta}\frac{\partial p}{\partial x}, \quad V_y = \frac{k'}{\eta}\frac{\partial p}{\partial y}, \quad V_z = \frac{k'}{\eta}\frac{\partial p}{\partial z}, \tag{1}$$

where η is the viscosity of the fluid, and k' is a constant which depends on the medium alone and is independent of the fluid. Darcy stated his result in terms of the hydraulic head h, but the alternative form (1) is more suitable in the present context. A very full discussion is given by Muskat (1937). k' in (1) has the dimensions of an area.

For the present purposes the fluid in question is water, whose viscosity may be taken as 0·01 poises, and it is convenient to rewrite (1) for this purpose as

$$V_x = k\frac{\partial p}{\partial x}, \quad V_y = k\frac{\partial p}{\partial y}, \quad V_z = k\frac{\partial p}{\partial z}, \tag{2}$$

where k is a constant called the permeability, and (2) refers explicitly to movement of water: if other fluids are in question (1) must be used where k' may be obtained by taking the viscosity of water as 0·01 (actually 0·01005 at 20° C). The most convenient unit of k to use is the Darcy, which corresponds to a flow of 1 cc/sec cm^2 for a pressure gradient of 1 bar/cm. If all units are cgs, 1 darcy corresponds to a flow of 10^{-6} cc/sec cm^2 for a pressure gradient of 1 (dyne/cm^2)/cm. In fps units it corresponds to a flow of $7{\cdot}32 \times 10^{-5}$ (ft^3/sec ft^2) for a pressure gradient of 1 psi/ft.

If piezometric head is used in place of pressure as a variable, 1 darcy is equivalent to a flow of $9{\cdot}81 \times 10^{-4}$ cc/sec cm^2 for a gradient of 1 cm of water/cm. In these units permeabilities are expressed in cm/sec.

Muskat (1937) gives an extensive table of permeabilities and porosities of oil sands. The range of variation of permeabilities is very large, from 3·4 to 0·00007 darcys, while the porosities vary from 37 to 3 per cent. Handin *et al.* (1963) gave the following values for materials with which they have worked:

	k (millidarcys)	Porosity (%)	Bulk volume change at 2,000 bars (%)
Hasmark dolomite	0·71	3·5	1·8
Marianna limestone	$<$0·05	13·0	1·8
Berea sandstone	217	18·2	3·8
Muddy shale	$<$0·05	4·7	0·6
Repetto siltstone	$<$0·05	5·6	2·9

Theory is also available for flow of fluid through a material with anisotropic permeability, Marcus (1962), Biot (1955).

8.4 The macroscopic stress–strain relations

The stress components for the whole system are defined precisely as in § 2.2. The area δA in Fig. 2.2 (*a*) is supposed to be large enough to be intersected by many pores, and the force $\delta \mathbf{F}$ across it is in fact made up of a component from the solid part and a normal force $np\,\delta A$ across the pores, where p is the fluid pressure in the pores. This breakdown, however, is rarely used. The components of strain and displacement are defined as in § 2.9 by considering displacements in the solid material. To complete the specification of the system, the pressure p of the fluid in the pores at every point is needed as well as a quantity θ to specify the amount of fluid contained in the pores. If v is the volume of pore fluid per unit volume and v_0 is its value in the unstrained condition θ is defined as

$$\theta = v - v_0. \tag{1}$$

If, as will be assumed here, the fluid is incompressible, θ is just $n - n_0$, where n and n_0 are the porosities in the strained and unstrained states.

Just as in § 5.2, linear stress–strain relations will be assumed between θ and the six components of strain and p and the six components of stress.

These will be taken in a form which reduces to the elastic relations of § 5.2 when $p = 0$, so that the suggested generalization of § 5.2 (24), (25) is

$$E\varepsilon_x = \sigma_x - \nu(\sigma_y + \sigma_z) - pE/3H, \tag{2}$$

$$E\varepsilon_y = \sigma_y - \nu(\sigma_x + \sigma_z) - pE/3H, \tag{3}$$

$$E\varepsilon_z = \sigma_z - \nu(\sigma_x + \sigma_y) - pE/3H, \tag{4}$$

$$E\Gamma_{yz} = (1 + \nu)\tau_{yz}, \quad E\Gamma_{zx} = (1 + \nu)\tau_{zx}, \quad E\Gamma_{xy} = (1 + \nu)\tau_{xy}, \tag{5}$$

where H is a constant, and E and ν are Young's modulus and Poisson's ratio for the solid skeleton (these, of course, refer to the skeleton itself containing pores and not to its material). The form (2)–(4) is chosen, since p cannot affect the shear stresses and must affect all components ε_x, ε_y, ε_z in the same way, since the material is supposed to be isotropic. A negative sign is chosen for the term $-pE/3H$, since increase of p will cause expansion of the solid.

A further relation is needed which will connect θ with the stresses: since shear stress will not affect the water content, and normal stresses must affect it equally, this relation may be assumed to be

$$\theta = (p/R) - (\sigma_x + \sigma_y + \sigma_z)/3H_1, \tag{6}$$

where H_1 and R are constants and the signs are chosen from the facts that θ is decreased by σ_x, σ_y, σ_z and increased by p.

It will now be shown from energy considerations that $H_1 = H$. The strain energy per unit volume W, calculated as in § 5.8 (7), is

$$W = \tfrac{1}{2}(\sigma_x\varepsilon_x + \sigma_y\varepsilon_y + \sigma_z\varepsilon_z + 2\Gamma_{yz}\tau_{yz} + 2\Gamma_{zx}\tau_{zx} + 2\Gamma_{xy}\tau_{xy} + p\theta), \quad (7)$$

where the additional term $\frac{1}{2}p\theta$ in (7) is obtained by the same argument as § 5.8 (1). Substituting for the strains in (7) in terms of the stresses by (2)–(6) gives

$$W = \frac{1}{2E}\{\sigma_x^2 + \sigma_y^2 + \sigma_z^2 - 2\nu(\sigma_y\sigma_z + \sigma_z\sigma_x + \sigma_x\sigma_y) +$$
$$2(1+\nu)(\tau_{yz}^2 + \tau_{zx}^2 + \tau_{xy}^2)\} - \tfrac{1}{2}\left[\frac{p}{3H} + \frac{p}{3H_1}\right](\sigma_x + \sigma_y + \sigma_z) + \frac{p^2}{2R}. \quad (8)$$

Therefore

$$\frac{\partial W}{\partial p} = -\tfrac{1}{2}\left[\frac{1}{3H} + \frac{1}{3H_1}\right](\sigma_x + \sigma_y + \sigma_z) + \frac{p}{R}, \quad (9)$$

and since, by the general argument of § 5.8, we must have

$$\frac{\partial W}{\partial p} = \theta,$$

it follows from (6) and (9) that $H = H_1$ and (6) becomes

$$\theta = (p/R) - (\sigma_x + \sigma_y + \sigma_z)/3H. \quad (10)$$

The complete stress–strain relations are (2)–(5) and (10), and contain four constants E and ν (or their equivalents) and H and R. As in § 5.2 (20) (21), they may be solved for stress in terms of strain and give

$$\sigma_x = \lambda\Delta + 2G\varepsilon_x + \alpha p, \quad \sigma_y = \lambda\Delta + 2G\varepsilon_y + \alpha p,$$
$$\sigma_z = \lambda\Delta + 2G\varepsilon_z + \alpha p, \quad (11)$$

$$\tau_{yz} = 2G\Gamma_{yz}, \quad \tau_{zx} = 2G\Gamma_{zx}, \quad \tau_{xy} = 2G\Gamma_{xy}, \quad (12)$$

$$\theta = -\alpha\Delta + (p/Q), \quad (13)$$

where $\Delta = \varepsilon_x + \varepsilon_y + \varepsilon_z,$

$$\alpha = \frac{3\lambda + 2G}{3H} = \frac{2(1+\nu)G}{3(1-2\nu)H} = \frac{E}{3(1-2\nu)H} = \frac{K}{H}, \quad (14)$$

$$\frac{1}{Q} = \frac{1}{R} - \frac{\alpha}{H}, \quad (15)$$

and § 5.2 (13)–(15) have been used.

The principal axes of stress and strain and the principal stresses σ_1, σ_2, σ_3 and the principal strains ε_1, ε_2, ε_3 may be found as usual. The relations (11)–(13) take a particularly simple form when expressed in terms of the stress-and-strain deviations defined in §§ 2.8, 2.12. Adding the three equations (11) and using §§ 2.8 (1), 2.12 (1), and 5.2 (12),

$$s = (3\lambda + 2G)e + \alpha p = 3Ke + \alpha p. \tag{16}$$

Subtracting this from (11) gives in the notations of §§ 2.8, 2.12

$$s_x = 2Ge_x, \quad s_y = 2Ge_y, \quad s_z = 2Ge_z, \tag{17}$$

$$s_{yz} = 2Ge_{yz}, \quad s_{zx} = 2Ge_{zx}, \quad s_{xy} = 2Ge_{xy}, \tag{18}$$

$$\theta = -3\alpha e + (p/Q), \tag{19}$$

so that p and θ occur only in (16) and (19).

As in § 5.2, the physical significance of the new constants H, R, etc., may be found by considering various simple experimental systems.

(i) *Hydrostatic compression drained of fluid,* $\sigma_1 = \sigma_2 = \sigma_3 \neq 0$, $p = 0$
Adding (2) to (4) gives

$$\sigma_1 = \frac{E\Delta}{3(1 - 2\nu)} = K\Delta, \tag{20}$$

and (10) gives

$$\theta = -\sigma_1/H, \tag{21}$$

so $1/H$ measures the change in pore volume caused by external compression.

(ii) *Hydrostatic pressure equal to the pore pressure,* $\sigma_1 = \sigma_2 = \sigma_3 = p$
Adding (2) to (4) gives

$$\Delta = p\left(\frac{1}{K} - \frac{1}{H}\right). \tag{22}$$

The effect of the pore pressure is to decrease the compressibility by $(1/H)$.

(iii) *The material drained and squeezed with no lateral movement,* $\sigma_1 \neq 0$, $\varepsilon_2 = \varepsilon_3 = 0$, $p = 0$
This is a standard test applied in soil mechanics. From (11)

$$\sigma_1 = (\lambda + 2G)\varepsilon_1. \tag{23}$$

This gives the final compression with no pore pressure.

(iv) *Instantaneous compression with no water loss and no lateral movement,* $\sigma_1 \neq 0$, $\varepsilon_2 = \varepsilon_3 = 0$, $\theta = 0$
From (13)

$$p = \alpha Q\Delta = \alpha Q\varepsilon_1, \tag{24}$$

so that the pressure rises instantaneously to this value. Then, using (24) in (11),

$$\sigma_1 = (\lambda + 2G + \alpha^2 Q)\epsilon_1 \tag{25}$$

8.5 The equations of equilibrium, flow, and consolidation

The stress components must satisfy the equations of equilibrium § 5.5 (5)–(7), which are

$$\frac{\partial \sigma_x}{\partial x} + \frac{\partial \tau_{yx}}{\partial y} + \frac{\partial \tau_{zx}}{\partial z} + \rho X = 0, \tag{1}$$

$$\frac{\partial \tau_{xy}}{\partial x} + \frac{\partial \sigma_y}{\partial y} + \frac{\partial \tau_{zy}}{\partial z} + \rho Y = 0, \tag{2}$$

$$\frac{\partial \tau_{xz}}{\partial x} + \frac{\partial \tau_{yz}}{\partial y} + \frac{\partial \sigma_z}{\partial z} + \rho Z = 0. \tag{3}$$

Substituting the stress–strain relation § 8.4 (11) in these gives, precisely as in the derivation of § 5.5 (8)–(10),

$$(\lambda + G)\frac{\partial \Delta}{\partial x} + G\nabla^2 u + \alpha\frac{\partial p}{\partial x} + \rho X = 0, \tag{4}$$

$$(\lambda + G)\frac{\partial \Delta}{\partial y} + G\nabla^2 v + \alpha\frac{\partial p}{\partial y} + \rho Y = 0, \tag{5}$$

$$(\lambda + G)\frac{\partial \Delta}{\partial z} + G\nabla^2 w + \alpha\frac{\partial p}{\partial z} + \rho Z = 0. \tag{6}$$

It appears that a pressure gradient has the same effect on the displacements as a body force of components $(\alpha/\rho)\partial p/\partial x$, etc.

In addition, an equation is needed to describe the flow of pore fluid caused by changes in pore pressure p. The velocity of the fluid may be described by a vector of components V_x, V_y, V_z which satisfies Darcy's law, § 8.3 (2). Using this, and assuming that the fluid is incompressible, the equation of continuity relates the rate of flow of fluid into a small volume to the rate of increase of the amount of fluid in this volume. If the volume is a small cube of side a the rate of flow of fluid into it is

$$-a^3\left\{\frac{\partial V_x}{\partial x} + \frac{\partial V_y}{\partial y} + \frac{\partial V_z}{\partial z}\right\}, \tag{7}$$

while the rate of increase of its fluid content is $a^3\partial\theta/\partial t$. Equating these two quantities and using § 8.3 (2),

$$\frac{\partial \theta}{\partial t} = \frac{\partial V_x}{\partial x} + \frac{\partial V_y}{\partial y} + \frac{\partial V_z}{\partial z} = k\nabla^2 p. \tag{8}$$

Thus, using § 8.4 (13),

$$k\nabla^2 p = \frac{\partial\theta}{\partial t} = -\alpha\frac{\partial\Delta}{\partial t} + \frac{1}{Q}\frac{\partial p}{\partial t}. \tag{9}$$

Next, differentiating (4), (5), (6) with respect to x, y, and z, and adding, gives, for the case of no body forces

$$(\lambda + 2G)\nabla^2\Delta + \alpha\nabla^2 p = 0, \tag{10}$$

and since from § 8.4 (13)

$$\nabla^2\theta = -\alpha\nabla^2\Delta + \frac{1}{Q}\nabla^2 p, \tag{11}$$

it follows that

$$\nabla^2\theta = \left\{\frac{1}{Q} + \frac{\alpha^2}{\lambda + 2G}\right\}\nabla^2 p, \tag{12}$$

and finally from (9)

$$\nabla^2\theta = \frac{1}{C}\frac{\partial\theta}{\partial t}, \tag{13}$$

where

$$\frac{1}{C} = \frac{1}{k}\left\{\frac{1}{Q} + \frac{\alpha^2}{\lambda + 2G}\right\}. \tag{14}$$

C is called the coefficient of consolidation, and (13) shows that θ obeys the equation of diffusion. Problems on stress and water movement in porous media are thus related to, but more complicated than, those of diffusion. Biot (1941, a, b) discusses their solution.

If there are body forces $X = -\partial\phi/\partial x$, etc., derived from a potential ϕ, (13) is replaced by

$$\nabla^2\theta = \frac{1}{C}\frac{\partial\theta}{\partial t} - \frac{\alpha\rho}{\lambda + 2G}\nabla^2\phi. \tag{15}$$

8.6 Consolidation and water movement in one dimension

The problem of movement in one dimension is simple and important. Suppose the z-direction is in question and there is no movement in the x- and y-directions, so that $u = v = 0 = \varepsilon_x = \varepsilon_y$, and

$$\Delta = \varepsilon_1 = \frac{\partial w}{\partial z}. \tag{1}$$

Suppose that the stress in the z-direction is constant, $\sigma_z = S$, then § 8.4 (11) gives

$$(\lambda + 2G)\Delta + \alpha p = S, \tag{2}$$

so that, differentiating,

$$(\lambda + 2G)\frac{\partial\Delta}{\partial t} + \alpha\frac{\partial p}{\partial t} = 0. \tag{3}$$

Using this in § 8.5 (9) gives

$$k\frac{\partial^2 p}{\partial z^2} = \left\{\frac{1}{Q} + \frac{\alpha^2}{\lambda + 2G}\right\}\frac{\partial p}{\partial t}, \tag{4}$$

or

$$\frac{\partial^2 p}{\partial z^2} = \frac{1}{C}\frac{\partial p}{\partial t}, \tag{5}$$

where C is the coefficient of consolidation given by § 8.5 (14). Thus in this case p satisfies the diffusion equation in one dimension which has to be solved with initial and boundary conditions appropriate to the problem. σ_y and σ_z are not constant but depend on p.

It should be noticed that (5) is a special case and does not generalize to $C\nabla^2 p = \partial p/\partial t$ in three dimensions. McNamee and Gibson (1960) state that Terzaghi (1943) incorrectly assumes that this holds. For constant hydrostatic pressure $\sigma_1 = \sigma_2 = \sigma_3$, the arguments used above give in place of (2)

$$(\lambda + 2G/3)\Delta + \alpha p = \sigma_1$$

leading to

$$[kKQ/(K + \alpha^2 Q)]\nabla^2 p = \partial p/\partial t$$

which does have this form but with a constant different from the value C defined in § 8.5 (14).

Problems on consolidation and water movement involve the solution of (5) with the appropriate initial and boundary conditions, and results can be written down immediately from well-known solutions in the theory of conduction of heat.

(i) *Consolidation in the region* $0 < z < h$ *due to an applied stress*

Suppose that the region is initially unstressed with $\theta = 0$ and that constant stress $\sigma_z = S$ is applied at $z = h$ at $t = 0$. Suppose also that the plane $z = 0$ is impervious to fluid, but that fluid can drain freely through a pervious membrane at $z = h$.

The boundary conditions then are, using § 8.3.2

$$\frac{\partial p}{\partial z} = 0, \quad z = 0, \tag{6}$$

$$p = 0, \quad z = h. \tag{7}$$

The initial condition is determined by the fact that pore pressure is required to keep $\theta = 0$ at the instant S is applied. From § 8.4 (13),

$$p = \alpha Q\Delta, \tag{8}$$

and using this in (2) gives

$$p = S[\alpha Q/(\lambda + 2G + \alpha^2 Q)], \tag{9}$$

which is the required value of p when $t = 0$.

The solution of the problem can be written down from Carslaw and Jaeger (1959, § 3.3 (8), (9)) and is

$$p = \frac{4\alpha QS}{\pi[\lambda + 2G + \alpha^2 Q]} \sum_{n=0}^{\infty} \frac{(-1)^n}{(2n+1)} \exp\{-C(2n+1)^2\pi^2 t/4h^2\} \cos\frac{(2n+1)\pi z}{2h}. \quad (10)$$

This solution is most suitable for large values of the time. For smaller values of the time the following alternative form is more suitable:

$$p = \frac{\alpha QS}{(\lambda + 2G + \alpha^2 Q)}\left[1 - \sum_{n=0}^{\infty}(-1)^n\left\{\operatorname{erfc}\frac{(2n+1)h - z}{2(Ct)^{\frac{1}{2}}} + \operatorname{erfc}\frac{(2n+1)h + z}{2(Ct)^{\frac{1}{2}}}\right\}\right], \quad (11)$$

where erfc ξ is the tabulated function

$$\operatorname{erfc}\xi = 1 - \operatorname{erf}\xi = \frac{2}{\pi^{\frac{1}{2}}}\int_0^\infty e^{-u^2}\,du. \quad (12)$$

When p has been found, Δ follows from (2), θ from § 8.4 (13), and the displacement w by integration from (1). For example, when $z = h$ the displacement w is

$$(\lambda + 2G)w = Sh - \frac{8ShQ\alpha^2}{\pi^2(\lambda + 2G + \alpha^2 Q)} \sum_{n=0}^{\infty}(2n+1)^{-2} \exp\{-C(2n+1)^2\pi^2 t/4h^2\}. \quad (13)$$

(ii) *The region* $0 < z < h$ *with zero initial pressure, no flow at* $z = 0$, *constant stress* $\sigma_z = S$ *at* $z = h$, *and constant pressure* p_0 *supplied at* $z = h$ *for* $t > 0$

This is the problem which arises in a test specimen when, as is usually the case, the pore pressure is raised at one end only.

The conditions of the problem are

$$p = 0, \quad 0 < z < h, \quad t = 0 \quad (14)$$

$$p = p_0, \quad z = h, \quad \partial p/\partial z = 0, \quad z = 0, \quad t > 0. \quad (15)$$

The solution is, Carslaw and Jaeger (1959, § 3.4 (2)),

$$p = p_0 - \frac{4p_0}{\pi}\sum_{n=0}^{\infty}\frac{(-1)^n}{(2n+1)}\exp\{-C(2n+1)^2\pi^2 t/4h^2\}\cos\frac{(2n+1)\pi z}{h}. \quad (16)$$

8.7 Simplified equations of consolidation

Two simplifying assumptions are made by Biot (1941a) for the case of a saturated clay. These are that the instantaneous compressive strain

§ 8.4 (25) is negligible compared with the final compressive strain § 8.4 (23). This implies that Q is very large, and as an idealization we take

$$Q = \infty. \tag{1}$$

Secondly, it is approximately true that in these circumstances the change in volume is equal to the volume of water squeezed out, that is $\Delta = -\theta$, so that using (1) in § 8.4 (13)

$$\alpha = 1. \tag{2}$$

It follows from § 8.4 (14) and (15) that

$$R = H = K. \tag{3}$$

Also the coefficient of consolidation C is by § 8.5 (14)

$$C = k(\lambda + 2G). \tag{4}$$

The stress equations § 8.4 (11) become

$$\sigma_x - p = \lambda\Delta + 2G\varepsilon_x, \quad \sigma_y - p = \lambda\Delta + 2G\varepsilon_y, \\ \sigma_z - p = \lambda\Delta + 2G\varepsilon_z \tag{5}$$

so that only the effective stresses

$$\sigma_x - p, \quad \sigma_y - p, \quad \sigma_z - p,$$

which will appear in § 8.8, are involved.

McNamee and Gibson (1960, a, b) have discussed the solution of the consolidation problem for cases of plane strain and axially symmetrical strain on these assumptions.

In this case it follows from § 8.5 (9), (10) that Δ also satisfies the diffusion equation

$$\nabla^2\Delta = \frac{1}{C}\frac{\partial\Delta}{\partial t}, \tag{6}$$

with the value (4) of C, but p does not, except in special circumstances, such as those of § 8.6.

These equations are derived by many other writers. Terzaghi (1943, § 99) takes for one-dimensional consolidation

$$m_{vc}\frac{\partial(\sigma_z - p)}{\partial t} = \frac{\partial\Delta}{\partial t}, \tag{7}$$

so that by (5) his coefficient m_{vc} is our $[\lambda + 2G]^{-1}$ and he finds

$$\frac{\partial^2 p}{\partial z^2} = \frac{1}{C}\frac{\partial p}{\partial t}, \tag{8}$$

with C given by (4).

A rather different situation is considered in the theory of water movement in aquifers. In this case two-dimensional flow is usually considered in the horizontal x,y-plane with σ_z constant. In this case (5) gives

$$\frac{\partial p}{\partial t} = -\lambda \frac{\partial \Delta}{\partial t} \tag{8}$$

and using § 8.5 (9) with $\alpha = 1$, $Q = \infty$, this gives

$$\frac{\partial^2 p}{\partial x^2} + \frac{\partial^2 p}{\partial y^2} = \frac{1}{C}\frac{\partial p}{\partial t} \tag{9}$$

with $C = k\lambda$. In hydrological terms, $1/\lambda$ is the specific storage, which is simply defined as the quantity of water released per unit volume for unit fall in pore pressure, without reference to stresses in the medium.

The discussion above is that usual in soil mechanics: an alternative argument applicable to rocks is that since the compressibility of rock-forming minerals is of the order of 0·1 per cent per kilobar, Birch, Schairer, and Spicer (1942), it is reasonable to neglect changes in mineral volume at laboratory pressures and to take $\theta = -\Delta$, from which it follows that $\alpha = 1$ and $Q \to \infty$, cf. Handin *et al.* (1963).

All the above discussion of consolidation is based on an extension of elastic theory, and thus implies that the behaviour is reversible. In fact, this is far from being the case, particularly in problems such as the consolidation of sediments under high pressure. In such cases there is an equilibrium void ratio corresponding to each value of the pressure. Terzaghi (1923) suggested that there was a linear relation between the equilibrium void ratio n of the sediment and the logarithm of the applied pressure p, that is

$$e = e_1 - c \ln (p/p_1).$$

This relation has been verified by Skempton (1944), and Parasnis (1960) extended experiments to pressures of 1,000 bars without finding important departures from it. The transient process of extruding water while the equilibrium void ratio for any given pressure is being approached shows behaviour similar to that to be expected from the calculations given above.

8.8 Effective stress

The concept of effective stress was introduced for saturated soils by Terzaghi in 1923 on experimental grounds, and has subsequently proved sufficiently accurate for engineering purposes, Skempton (1960). Terzaghi (1943) introduces it with the statements that: (i) if the external hydrostatic stress $\sigma_1 = \sigma_2 = \sigma_3$ and the pore pressure p are increased by the same amount there is negligible change in volume of the material compared to that which would occur if the stress $\sigma_1 = \sigma_2 = \sigma_3$ alone was increased; (ii) in shear failure there is no increase in the shear strength if both the normal stress σ_n and the pore pressure p are increased by the same amount, although if σ_n alone is increased there is a considerable increase in shear strength. From these results he concludes that the '*neutral*' stress p has no influence on deformation or failure and that these are controlled by the *effective stress*

$$\sigma_x' = \sigma_x - p, \quad \sigma_y' = \sigma_y - p, \quad \sigma_z' = \sigma_z - p, \tag{1}$$

$$\tau_{yz}' = \tau_{yz}, \quad \tau_{zx}' = \tau_{zx}, \quad \tau_{xy}' = \tau_{xy}. \tag{2}$$

An important attempt to establish the generality of the effective stress concept has been made by Hubbert and Rubey (1959, 1960, 1961). They consider the equilibrium of portion of a porous solid in which the pore pressure has the form $p = \rho_l gz$ associated with liquid of density ρ_l under gravity, and the solid has total density ρ. Taking the z-axis vertically downwards, the stress is divided into the neutral stress of components $p = \rho_l gz$ and the effective stress defined in (1) and (2). Consider the equilibrium of a volume V of the solid surrounded by surface A. The stress over an element of surface dA of the surface has, by § 2.4 (5)–(7), components

$$p_x\, dA, \quad p_y\, dA, \quad p_z\, dA. \tag{3}$$

The condition of equilibrium is that the surface integral of the normal stress (3) shall be equal to the weight of the body, that is

$$\iint p_x\, dA = 0, \quad \iint p_y\, dA = 0, \tag{4}$$

$$\iint p_z\, dA = \iiint \rho g\, dV. \tag{5}$$

These may be expressed in terms of the corresponding quantities p_x', p_y', p_z' for effective stress and become

$$\iint p_x'\, dA + \iint lp\, dA = 0. \quad \iint p_y'\, dA + \iint mp\, dA = 0, \tag{6}$$

$$\iint p_z'\, dA + \iint np\, dA = \iiint \rho g\, dV, \tag{7}$$

where l, m, n are the direction cosines of the normal to A.

Now by Green's theorem, § 5.8 (21), for vectors along the x-, y-, and z-axes respectively

$$\iint lp\, dA = \iiint \frac{\partial p}{\partial x}\, dV, \quad \iint mp\, dA = \iiint \frac{\partial p}{\partial y}\, dV,$$

$$\iint np\, dA = \iiint \frac{\partial p}{\partial z}\, dV. \tag{8}$$

Using the value $p = \rho_l gz$, these give with (6) and (7)

$$\iint p_x'\, dA = 0, \quad \iint p_y'\, dA = 0, \tag{9}$$

$$\iint p_z'\, dA = \iiint \rho g\, dV - \iiint \rho_l g\, dV. \tag{10}$$

The last integral expresses the fact that the surface integral of the resolved parts in the z-direction of the effective stresses over the surface is equal to the total weight of the material in the volume V less the weight of an equivalent volume of liquid. This may be regarded as an extension of the principle of Archimedes. Hubbert and Rubey interpret it as implying that at any depth z the neutral pressure $p = \rho_l gz$ supports in hydrostatic equilibrium part of the mass of the body, namely a fraction of density ρ_l, leaving stresses set up by the effective stress to support the remainder, and they regard this as justifying the use of effective stress. This resolution is natural if the pore-fluid does completely penetrate the material, establishing a base pressure level p at any point of it. From the mathematical point of view, as pointed out by Moore (1961), the resolution into effective and neutral stresses is merely one possible division of the stress into the sum of two systems: however, for a porous body which is permeated by the fluid pressure p it is the logical one.

Hubbert and Rubey and also Terzaghi (1945) thus regard the neutral stress as a basic hydrostatic stress existing in both the solid and liquid, and the effective stress as arising exclusively from the solid skeleton. In this context it is frequently called the intergranular stress, and this concept is commonly used in soil mechanics.

As introduced by Terzaghi, the concept involves two separate phenomena, deformation and failure, and it is introduced primarily on experimental grounds. It will appear in § 8.9 that there is a considerable body of careful experimental evidence to show that in many cases failure is governed by the effective stresses. However, the question arises as to whether the concept has any fundamental theoretical significance or is merely a good practical approximation.

Considering first the question of deformation and the material specified in § 8.4, the fundamental experiment referred to by Terzaghi is that discussed in § 8.4 (22) in which the volumetric strain is

$$\Delta = p\left(\frac{1}{K} - \frac{1}{H}\right),$$

while in the absence of pore pressure it would be p/K. The statement thus implies $H = K$, and the assumption is essentially that of the simplified theory of § 8.7. In particular, the relations § 8.7 (5) show that only the effective stress is operative in deformation. This result, however, is not true for the more general type of material discussed in § 8.4, in particular the quantities $\sigma_x - \alpha p$, $\sigma_y - \alpha p$, $\sigma_z - \alpha p$, and not the effective stresses (1), appear in the stress–strain relations § 8.4 (11).

In soil mechanics a microscopic approach is more usually adopted, that is, considering the individual grains and their contacts. This is inevitable with partially saturated soils, since air spaces and the menisci separating them from water at the contacts have to be considered. In 1925 Terzaghi suggested that the actual areas of contact were relatively small and that their behaviour was related to the mechanical properties of the grains in a manner similar to that postulated in § 3.2 for the asperities in frictional phenomena.

The discussion of a single contact is relatively easy. Suppose that A_c is the area of the contact between grains of area A, Fig. 8.8, the space outside the area of contact being filled with liquid at pressure p. Then if σ and τ are

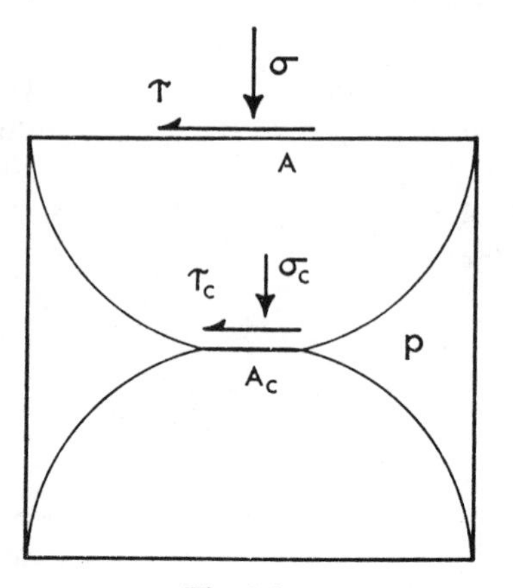

Fig. 8.8

the normal and shear stresses across the area A, and σ_c and τ_c are those across the area of contact A_c,

$$\sigma = a\sigma_c + (1 - a)p, \tag{11}$$

$$\tau = a\tau_c, \tag{12}$$

where

$$a = A_c/A. \tag{13}$$

If a is very small, $a\sigma_c$ tends to a finite limit σ_g, which is sometimes called intergranular stress and is the stress transmitted by the individual grains. By (11) this tends to the effective stress σ' as defined earlier.

Next, considering frictional sliding on the contact, this may be expected to satisfy the relation § 3.3 (1), that is

$$\tau_c = S_0 + \mu\sigma_c, \tag{14}$$

where S_0 and μ are the shear strength and coefficient of friction of the contact. It follows from (11)–(14) that

$$\begin{aligned}\tau &= aS_0 + \mu a\sigma_c \\ &= aS_0 + \mu[\sigma + (1 - a)p],\end{aligned} \tag{15}$$

so that $\sigma - (1 - a)p$ and not $\sigma' = \sigma - p$ enters into the relation for sliding, though if a is small these are indistinguishable. Skempton (1960) has made a careful study of the question, involving, in addition, the mechanical properties of the grains themselves. He concludes that the use of effective stress as determining deformation and failure in saturated soils is valid if the soil grains are incompressible and the yield stress of the grains is independent of pressure. If this is not the case the behaviour is determined by the quantities

$$\sigma_1 - kp, \quad \sigma_2 - kp, \quad \sigma_3 - kp,$$

where k is a constant depending on the area of contact between the grains and their mechanical properties. The case of partially saturated soils is more complex, but the concept of effective stress remains useful for them, Bishop and Blight (1963).

While it is not difficult to discuss the behaviour of single contacts, the fundamental difficulty in setting up a microscopic theory is that the distribution of grain contacts in space is unknown and has to be taken into account. Largely because of this difficulty, the description of soil stresses and the concept of effective stress has always been somewhat controversial, cf. Hubbert and Rubey (1959, 1960, 1961), McHenry (1948), Leliavsky (1958), Harza (1949).

8.9 The effect of pore-pressure on the strength of rocks

Many authors have studied in detail the failure of rocks in the triaxial test when pore-pressure in the rock is included as an additional variable. The pore-fluid is introduced through pores or holes in the platens with which the specimen is in contact. As a preliminary, to ensure that all pores are filled, the specimen is first subjected to a vacuum and subsequently immersed in fluid under pressure. Specimens are jacketed so that pore-pressure and confining pressure are independent variables. The corresponding procedures for soils are fully discussed by Bishop and Henkel (1962).

Major studies on rocks have been made by Handin *et al.* (1963), Heard (1960), Murrell (1965), and Robinson (1959). Their results are in general agreement with the statement that, provided the rocks have connected systems of pores, fracture is controlled by the effective stress

$$\sigma_1' = \sigma_1 - p, \quad \sigma_2' = \sigma_2 - p, \quad \sigma_3' = \sigma_3 - p$$

of § 8.8. Thus, for example, the Coulomb criterion of § 4.6 (20) would be replaced by

$$\sigma_1 - p = C_0 + q(\sigma_3 - p). \tag{1}$$

The way in which pore-pressure affects failure may be seen from Fig. 8.9.1 for Handin's results on sandstone. Here AB is the experimental Mohr envelope for zero pore-pressure. Curve I for $p = 500$ bars corresponds to failure with $\sigma_1 = 5{,}400$, $\sigma_2 = 2{,}000$ bars, so that $\sigma_1' = 4{,}900$, $\sigma_2' = 1{,}500$ and the Mohr circle is seen to touch *AB*. Curve II shows the Mohr circle for $\sigma_1 = 5{,}400$, $\sigma_2 = 2{,}000$ bars with zero pore-pressure which is seen to lie inside the Mohr envelope: as the pore-pressure is increased, this curve is moved to the left until it touches the Mohr envelope and failure can take place. In general, a unique Mohr envelope not depending on p is obtained by working with σ_1' and σ_3'.

Handin *et al.* (1963) show that criteria for failure in terms of effective stress hold reasonably well provided that the permeability is sufficient to

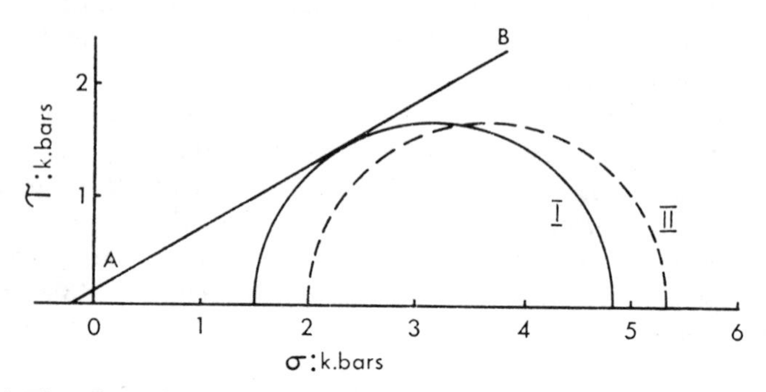

Fig. 8.9.1 The effect of pore-pressure on failure. *AB* is the Mohr envelope. Curve I: Mohr circle for effective stresses. Curve II: Mohr circle for actual stresses.

allow movement of fluid and that connected systems of pores exist, also that the pore-fluid is inert, so that the effects are purely mechanical. If these conditions are not satisfied discrepancies occur. Serdengecti, Boozer, and Hiller (1962) have shown that pore-fluids may affect the cementing material in sedimentary rocks so that additional effects, not attributable to pressure, may occur.

In materials which show a brittle–ductile transformation this is controlled by the effective confining pressure $\sigma_3 - p$, Heard (1960), Robinson (1959). For example, Fig. 8.9.2 shows Robinson's stress–strain curves for limestone at a confining pressure of 10 kpsi and various values of the pore pressure. These curves show a complete transition from ductile to brittle behaviour as p is increased. The curves are also complete stress–strain curves, cf. § 4.2, suggesting that for the same value of the effective confining pressure the effect of pore-pressure is to decrease the brittleness of the material.

An interesting application of the theory is to the effects of the development of pore-pressure by the dehydration of minerals. Raleigh and Paterson (1965) found a loss of strength and reversion from ductile to brittle behaviour in heated serpentinite which they attributed to the pressure of water vapour generated by dehydration. For example, at around 500° C and 3·5 kbars confining pressure there was no loss of strength if the specimen was vented to the atmosphere ($p = 0$), but a very substantial loss if it was sealed ($p = 3{\cdot}5$ kbars). A similar effect was found for gypsum by Heard and Rubey (1963). Pressure of gas in the pores of coal is important in connection with outbursts in coalmines, Hargreaves (1958, 1967).

The effect of fluid pressure p on sliding on faults or joints is obviously of great importance. The criterion for slip, § 3.6 (1), would be replaced by

$$|\tau| = S_0 + \mu(\sigma_n - p),$$

where σ_n is the normal stress across the plane. This has been verified experimentally by Byerlee (1967b) on small specimens. Geological applica-

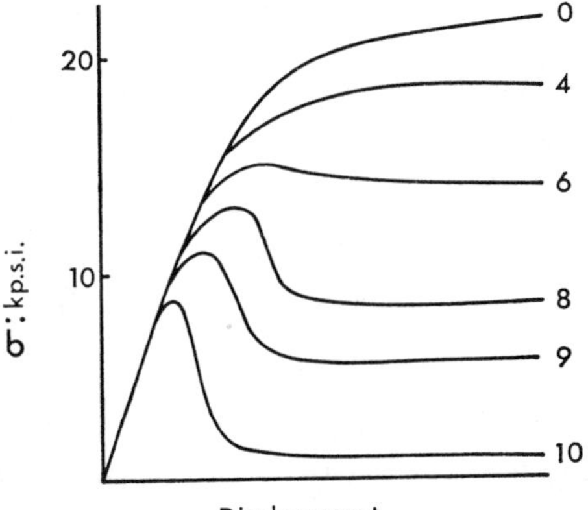

Fig. 8.9.2 The effect of pore-pressure on the brittle–ductile transition in limestone at a confining pressure of 10 kpsi. Numbers on the curves are values of pore pressure in kpsi.

tions are discussed by Secor (1955). The reduction of the shear stress necessary to cause slip is fundamental for the Hubbert–Rubey theory of overthrust faulting which is discussed in § 17.4.

Extension failure under pore-pressure has been discussed theoretically by Murrell (1964b). The effective stress concept would suggest that the criterion for failure would be

$$\sigma_3' = \sigma_3 - p = -T_0, \tag{2}$$

and this is derived in § 10.13 (27) from the Griffith theory for a crack with internal pressure. It has been shown to hold experimentally by Jaeger (1963) and Gurney and Rowe (1945, a, b). This result is fundamental for the theory of hydraulic fracturing, § 8.10.

8.10 Hydraulic fracturing

If liquid is pumped into a borehole to a sufficient pressure, the tangential stresses in the surface of the hole due to the external stresses and the pressure will become tensile, and extension fracture will occur if this stress exceeds the tensile strength T_0 of the material. A joint or fissure intersecting the hole which is held 'closed' by the initial compressive stress field may be 'opened' by tensile stresses when internal pressure is applied, corresponding to the case $T_0 = 0$ above. For example, for a number of underground drill holes into Rand quartzite the holes would not accept water at pressures below 1·5–2·5 times the overburden pressure, but at some pressure in this range would suddenly take water quite freely.

The method is much used in the oil industry to increase yields. In this case a length of the hole is sealed off with packers at its ends so that pressure can be applied only to a restricted range of depths in the hole.

Suppose that the vertical σ_z-direction is a principal direction, and that the horizontal principal stresses are σ_1 and σ_2. Then, neglecting the effect of pore-pressure in the rock, there are two simple possibilities. First, if

$$p > \sigma_z + T_0 \tag{1}$$

there will be a horizontal extension fracture, corresponding to the lifting of the overburden by p. Secondly, the tangential stress σ_t in the surface of the hole corresponding to p, σ_1, and σ_2 is, by § 10.4 (19),

$$\sigma_t = (\sigma_1 + \sigma_2 - p) - 2(\sigma_1 - \sigma_2)\cos 2\theta. \tag{2}$$

It therefore varies from a maximum of $3\sigma_1 - \sigma_2 - p$ when $\theta = \frac{1}{2}\pi$ to a minimum of $3\sigma_2 - \sigma_1 - p$ when $\theta = 0$. If $p > 3\sigma_2 - \sigma_1$ this is tensile, and if

$$p > 3\sigma_2 + T_0 - \sigma_1, \tag{3}$$

tensile failure is possible. This takes place in a radial plane through the direction of the greatest principal stress. This method of determining secondary principal stresses in boreholes was used by Jaeger and Cook (1964).

This theory was set out by Hubbert and Willis (1957), who also discuss other modes of failure. Whether failure will take place according to (1) or (3) depends on the relative values of the horizontal and vertical principal stresses. If the pressure is restricted to portion of a borehole by packers there will be stress concentrations in the neighbourhood of these. The theory is given by Kehle (1964) and Bertrand (1964). Anisotropy of tensile strength of the material may also have a considerable effect. Finally, the effect of pore-pressure in the rock has to be considered and, as shown in § 8.9, this is equivalent to reducing T_0 by the pore-pressure. However, it is difficult to decide what value should be taken for the pore-pressure, since in some circumstances fluid may penetrate from the borehole into the pores of the rock, while in other cases anomalously high fluid pressures are known to exist at depth, cf. Hubbert and Rubey (1959), so that fluid might be expected to move from the rock towards the borehole.

When a fracture has been initiated at the surface of a borehole it is to be expected that it will extend for some distance because of the stress concentration at its end, §§ 10.11, 10.12. The mechanism has been discussed by Odé (1956) and van Poollen (1957).

As remarked earlier, hydraulic fracturing of deep holes should, in principle, give the horizontal tectonic principal stresses and a great deal of discussion, Scheidegger (1960, 1962), Fairhurst (1964a), Gretener (1965), Pulpan and Scheidegger (1965), has been devoted to this possibility.

Rand qtzite need ~ 2x overburden press. to take water freely

if $p > \sigma_z + T_0$ horiz fract.

$\sigma_t = \sigma_1 + \sigma_2 - p - 2(\sigma_1 - \sigma_2)\cos 2\theta$

Chapter Nine

Behaviour of Ductile Materials

9.1 Introduction

In Chapter 4 it was remarked that with increase of confining pressure or temperature or both most rocks exhibit ductile behaviour rather similar to that of metals. A great deal of attention has been paid to ductile behaviour of rocks, partly because of its geophysical importance, and partly because of the interest of geologists in the deformation of materials. In rock mechanics ductility is of much less importance, since relatively few rocks

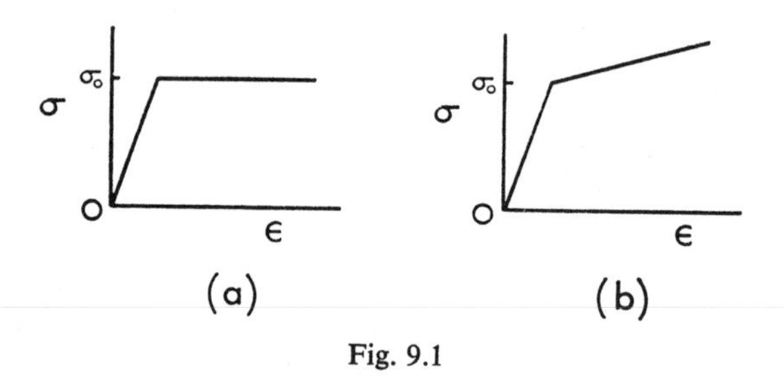

Fig. 9.1

show it at the temperatures and pressures involved. However, in highly stressed regions, either in laboratory experiments or in mining practice, ductile behaviour may be expected in rocks such as limestone, marble, halite, and some sediments.

The theory is also of some interest, since the study of yield and flow has been very highly developed, and the phenomena are essentially simpler than those involved in fracture.

In the study of ductile metals the concept of the perfectly plastic solid has been of the greatest value. This is supposed to deform elastically up to its yield stress σ_0, but to be able to sustain no stress greater than this, so that it will flow indefinitely at this stress unless restricted by some outside agency. Its stress–strain curve is shown in Fig. 9.1 (*a*). It may be generalized to allow for linear strain hardening, Fig. 9.1 (*b*), but this extension will not be used here. Clearly, what has been described as 'fully ductile' behaviour in Chapter 4, Fig. 4.3.2, conforms reasonably closely with the idealization of Fig. 9.1 (*a*). Throughout this chapter this behaviour will be referred to as 'plastic', and the theory is normally referred to as elastic–plastic theory. Full accounts are given in the textbooks of Prager and Hodge (1951), Nadai (1950), Hill (1950).

9.2 Yield criteria

A yield criterion is a relationship between the principal stresses such that, if it is satisfied, the material becomes ductile. Two general requirements for such a criterion appear at once:

(i) Since the process of yield must be independent of the choice of axes in an isotropic material, any yield criterion must be independent of choice of axes, and thus should be expressible in terms of the invariants $I_1, I_2, \ldots$ of stress or $J_2, J_3, \ldots$ of stress deviation defined in §§ 2.4, 2.8. In fact, any function of σ_1, σ_2, σ_3 can be expressed in terms of invariants, but the practice has been to study criteria which have simple expressions in terms of invariants.

(ii) It has been found experimentally that hydrostatic pressure $\sigma_1 = \sigma_2 = \sigma_3$ does not cause yield in metals and crystalline rocks. This suggests that only stress deviation should enter into a criterion for yield. While this is true for yield, it is not true for fracture.

Various simple criteria which have been proposed will now be mentioned briefly.

The *maximum shear stress* or *Tresca's criterion* states that yield will occur when the maximum shear stress attains a value $\frac{1}{2}\sigma_0$ characteristic of the material. This implies that

$$\sigma_1 - \sigma_3 = s_1 - s_3 = \sigma_0, \tag{1}$$

where s_1 and s_3 are the principal stress deviations defined in § 2.8. It implies that the yield stresses in tension and compression both have magnitude σ_0. While this is approximately true for metals, it is not the case for rock. This criterion was proposed by Tresca (1868).

The *von Mises criterion* is obtained by taking J_2 to be a constant: this is the simplest possible expression in terms of invariants. Using the values of § 2.8 (9), (10) of J_2, it may be written in the alternative forms

$$J_2 = \sigma_0^2/3, \tag{2}$$

$$s_1^2 + s_2^2 + s_3^2 = 2\sigma_0^2/3, \tag{3}$$

$$(\sigma_2 - \sigma_3)^2 + (\sigma_3 - \sigma_1)^2 + (\sigma_1 - \sigma_2)^2 = 2\sigma_0^2, \tag{4}$$

$$\tau_{oct} = (2/9)^{\frac{1}{2}}\sigma_0, \tag{5}$$

$$W_d = \sigma_0^2/6G, \tag{6}$$

where in (5), τ_{oct} is the octahedral shear stress defined in § 2.4 (49), and

W_d is the strain energy of distortion defined in § 5.8 (16). (5) and (6) give the physical interpretations that yield occurs when the octahedral shear stress, or the strain energy associated with distortion, reach a value characteristic of the material. The forms (2)–(4) derive from von Mises (1913) and (6) from Hencky (1924).

For uniaxial compression or tension, the magnitude of the yield stress given by (4) is σ_0; for pure shear,

$$\sigma_3 = -\sigma_1, \quad \sigma_2 = 0, \quad \sigma_1 = \sigma_0/3^{\frac{1}{2}}.$$

The von Mises criterion is by far the most commonly used and is adequate for most problems on metals. Its appearance as a criterion for fracture has been noted in § 4.5.

The Coulomb criterion of § 4.6 (20), namely

$$\sigma_1 = C_0 + q\sigma_3 \tag{7}$$

has been used as a criterion of yield by Guest (1940).

The criterion

$$(\sigma_1 - \sigma_2)^2 + (\sigma_2 - \sigma_3)^2 + (\sigma_3 - \sigma_1)^2 = 2(C_0 - T_0)(\sigma_1 + \sigma_2 + \sigma_3) + 2C_0T_0, \tag{8}$$

where C_0 and T_0 are yield strengths in compression and tension has been used by Stassi-D'Alia (1959).

In soil mechanics, the criteria

$$\sigma_1 - \sigma_3 = \alpha(\sigma_1 + \sigma_2 + \sigma_3)/3, \tag{9}$$

and $$(\sigma_1 - \sigma_2)^2 + (\sigma_2 - \sigma_3)^2 + (\sigma_3 - \sigma_1)^2 = \alpha^2(\sigma_1 + \sigma_2 + \sigma_3)^2/9, \tag{10}$$

are described as the *extended Tresca* and *extended von Mises* criteria, Bishop (1966).

Comparisons of the value of the various criteria for representing experimental results under polyaxial stresses have been made by many writers, Taylor and Quinney (1931), Lode (1926), Marin (1935). In doing this, it is frequently convenient to introduce the parameter

$$\mu = (2\sigma_2 - \sigma_1 - \sigma_3)/(\sigma_1 - \sigma_3) \tag{11}$$

which takes values between -1 and 1 as σ_2 runs from σ_3 to σ_1. In terms of this parameter, Tresca's criterion becomes

$$(\sigma_1 - \sigma_3)/\sigma_0 = 1, \tag{12}$$

and von Mises's is

$$(\sigma_1 - \sigma_3)/\sigma_0 = 2(3 + \mu^2)^{-\frac{1}{2}}. \tag{13}$$

For example, Lode used thin-walled tubes with internal pressure and axial tension from which system all values of μ may be obtained. Values of

$\sigma_1 - \sigma_3$ at failure were not constant, and showed better agreement with (13) than (12).

Any criterion of yield may be represented by a relationship between the principal stresses, say

$$\sigma_1 = f(\sigma_2, \sigma_3). \tag{14}$$

This is the equation of a surface called the *yield surface* or limiting surface of yielding, Nadai (1950). If the restriction $\sigma_1 \geqslant \sigma_2 \geqslant \sigma_3$ is maintained, as it will be here, only a portion of this surface is in question. However, in the theories of plasticity and of soil mechanics it is very common to use

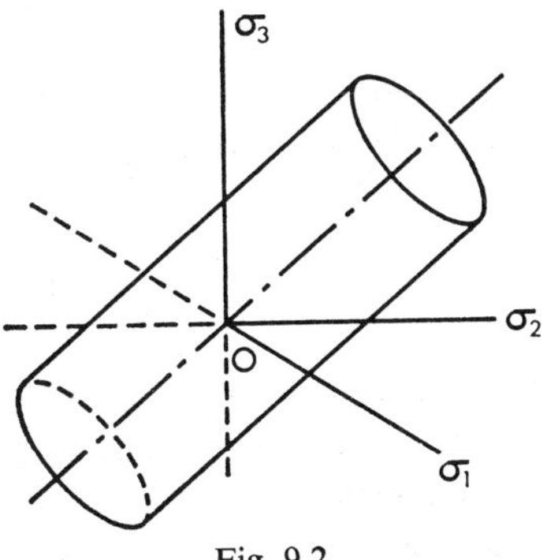

Fig. 9.2

the whole surface for unrestricted values of σ_1, σ_2, σ_3 with a separate notation to indicate which are the greatest and least principal stresses.

It follows from § 2.8 (3) that all points on the line $\sigma_1 = \sigma_2 = \sigma_3$ have zero stress deviation, and that all points on any parallel line $\sigma_1 = a + kr$, $\sigma_2 = b + kr$, $\sigma_3 = c + kr$ have the same stress deviation, so that if the yield criterion can be expressed in terms of stress deviation only, the yield surface will be a cylinder with $\sigma_1 = \sigma_2 = \sigma_3$ as axis. If the criterion involves the mean stress as well this will not be the case.

The yield surface for the criterion (4) is a circular cylinder of radius $(2/3)^{\frac{1}{2}}\sigma_0$ and axis $\sigma_1 = \sigma_2 = \sigma_3$, Fig. 9.2.

The yield surface corresponding to (1) is a hexagonal cylinder whose axis is $\sigma_1 = \sigma_2 = \sigma_3$ and whose edges pass through the points

$$(\sigma_0, 0, 0), \quad (0, \sigma_0, 0), \quad (0, 0, \sigma_0), \quad (0, 2^{\frac{1}{2}}\sigma_0, 2^{\frac{1}{2}}\sigma_0),$$
$$(2^{\frac{1}{2}}\sigma_0, 0, 2^{\frac{1}{2}}\sigma_0), \quad (2^{\frac{1}{2}}\sigma_0, 2^{\frac{1}{2}}\sigma_0, 0).$$

These are, as stated above, the complete geometrical surfaces, and in the present treatment and notation only the portion of them for which $\sigma_1 \geqslant \sigma_2 \geqslant \sigma_3$ is in question.

9.3 The equations of plasticity

When a body is deforming plastically the stresses at every point of it must satisfy the yield criterion at all times. Each yield criterion will give its own set of equations, and here we will usually consider the von Mises criterion of § 9.2 (2), namely,

$$J_2 = [(\sigma_2 - \sigma_3)^2 + (\sigma_3 - \sigma_1)^2 + (\sigma_1 - \sigma_2{}^2)]/6 = \sigma_0{}^2/3, \tag{1}$$

and, since this is the hold at all times in the plastic region, we must have also

$$\dot{J}_2 = 0, \tag{2}$$

where a 'dot' will be used to denote differentiation with respect to the time.

The equations of plasticity are developed in a similar manner to those of viscosity. Considering any point and time, $\dot{u}$, $\dot{v}$, $\dot{w}$ are taken to be the components of the velocity of the particle which happens to be at that point at that time. The quantities

$$\dot{\varepsilon}_x = \frac{\partial \dot{u}}{\partial x}, \quad \dot{\varepsilon}_y = \frac{\partial \dot{v}}{\partial y}, \quad \dot{\varepsilon}_z = \frac{\partial \dot{w}}{\partial z}, \quad \dot{\Gamma}_{xy} = \tfrac{1}{2}\left(\frac{\partial \dot{v}}{\partial x} + \frac{\partial \dot{u}}{\partial y}\right), \text{ etc.} \tag{3}$$

corresponding to § 2.11 (5)–(8) are defined as the components of rate of strain at that point and time. They are supposed to be small, so that squares and products can be neglected. Just as in Chapter 2, principal axes of rate of strain and principal rates of strain $\dot{\varepsilon}_1$, $\dot{\varepsilon}_2$, $\dot{\varepsilon}_3$ can be found. Since plastic deformation is assumed not to be accompanied by any change of volume,

$$\dot{e} = \dot{\varepsilon}_1 + \dot{\varepsilon}_2 + \dot{\varepsilon}_3 = 0, \tag{4}$$

and the principal rates of strain deviation, defined as in § 2.12, are identical with the principal rates of strain.

The fundamental assumption is now made that the principal rates of strain are proportional to the principal stress deviations, that is

$$s_1 = 2\phi\dot{\varepsilon}_1, \quad s_2 = 2\phi\dot{\varepsilon}_2, \quad s_3 = 2\phi\dot{\varepsilon}_3, \tag{5}$$

which is analogous to the stress–strain relations of elasticity, § 5.4, *except* that ϕ is not to be a constant but a function of x, y, z, t determined by the conditions of the problem and the yield criterion. As in Chapter 5, equations (5), referred to any axes become

$$s_x = 2\phi\dot{\varepsilon}_x, \quad s_y = 2\phi\dot{\varepsilon}_y, \quad s_z = 2\phi\dot{\varepsilon}_z, \tag{6}$$

$$s_{yz} = 2\phi\dot{\Gamma}_{yz}, \quad s_{zx} = 2\phi\dot{\Gamma}_{zx}, \quad s_{xy} = 2\phi\dot{\Gamma}_{xy}. \tag{7}$$

These equations, which are due to Saint-Venant (1870), together with the yield criterion, the equations of equilibrium for stresses, § 5.5, and the boundary conditions, determine the solution of the problem.

Here we shall be concerned only with plane problems, say in the *xy*-plane, with the condition that there is no displacement or motion in the *z*-direction, and that all quantities are independent of *z*. This implies that the *z*-direction must be a principal direction, say that of $\dot{\epsilon}_2$, and since $\dot{\epsilon}_2 = 0$, $s_2 = 0$, and so by § 2.8 (3),

$$\sigma_2 = \tfrac{1}{2}(\sigma_1 + \sigma_3). \tag{8}$$

Using (8) in the yield criterion (1), this becomes

$$\sigma_1 - \sigma_3 = 2k, \tag{9}$$

where

$$k = \sigma_0/\sqrt{3}. \tag{10}$$

This form has been chosen, since, taking $k = \sigma_0/2$, (9) becomes Tresca's criterion § 9.2 (1), and so the theory is available for both cases.

In terms of the stress components in the *xy*-plane, (9) becomes, by § 2.3 (14),

$$\tau_{xy}^2 + \tfrac{1}{4}(\sigma_x - \sigma_y)^2 = k^2. \tag{11}$$

The equations of equilibrium in terms of stresses, § 5.5 (19), (20), are

$$\frac{\partial \sigma_x}{\partial x} + \frac{\partial \tau_{xy}}{\partial y} = 0, \tag{12}$$

$$\frac{\partial \tau_{xy}}{\partial x} + \frac{\partial \sigma_y}{\partial y} = 0. \tag{13}$$

These give

$$\frac{\partial^2 \tau_{xy}}{\partial x^2} - \frac{\partial^2 \tau_{xy}}{\partial y^2} = \frac{\partial^2(\sigma_x - \sigma_y)}{\partial x\, \partial y}, \tag{14}$$

and using this in (11) gives

$$\frac{\partial^2 \tau_{xy}}{\partial x^2} - \frac{\partial^2 \tau_{xy}}{\partial y^2} = \pm 2 \frac{\partial^2}{\partial x\, \partial y}(k^2 - \tau_{xy}^2)^{\frac{1}{2}}, \tag{15}$$

which is a partial differential equation for τ_{xy}. If the boundary conditions involve stresses only, the solution is particularly simple, since the stresses are completely determined by (5) and the boundary conditions.

Alternatively, (12) and (13) may be satisfied by introducing the Airy stress function U, § 5.7 (7), and (11) becomes the equation

$$\left(\frac{\partial^2 U}{\partial y^2} - \frac{\partial^2 U}{\partial x^2}\right)^2 + 4\left(\frac{\partial^2 U}{\partial x\, \partial y}\right)^2 = 4k^2, \tag{16}$$

to be solved for U. The complex variable methods of § 10.2 are well suited to problems in plane plastic flow.

In the present context the most usual representation of the stress field is in terms of *slip-lines* or *shear-lines*, which are defined as curves whose direction at every point is that of the maximum rate of shear strian at that point.

Clearly there will be two systems cutting orthogonally, and since the directions of maximum rate of shear strain bisect the angles between the principal stresses, § 2.3 (11), they are readily found.

Suppose that $O\sigma_1$ and $O\sigma_3$ are the directions of the principal stresses inclined at α and $\alpha + \frac{1}{2}\pi$ to Ox, Fig. 9.3, then the shear directions OR and OS will be inclined at β and $\beta + \frac{1}{2}\pi$ to Ox, where $\beta = \alpha + \frac{1}{4}\pi$. Since by (9), $\sigma_1 - \sigma_3 = 2k$, § 2.2 (16), (17), give

$$\sigma_x = \sigma_m + k \cos 2\alpha, \quad \sigma_y = \sigma_m - k \cos 2\alpha, \quad \tau_{xy} = k \sin 2\alpha, \tag{17}$$

where $\sigma_m = (\sigma_1 + \sigma_2)/2$. Written in terms of β these become

$$\sigma_x = \sigma_m + k \sin 2\beta, \quad \sigma_y = \sigma_m - k \sin 2\beta, \quad \tau_{xy} = -k \cos 2\beta. \tag{18}$$

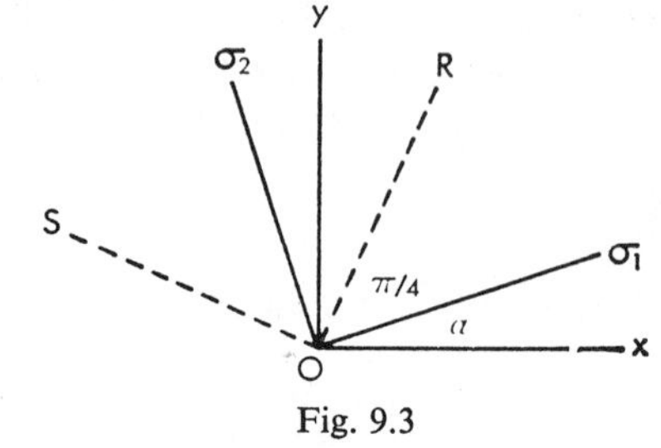

Fig. 9.3

The differential equations of the two systems of slip-lines are therefore

$$\frac{dy}{dx} = \tan \beta = \frac{1 - \cos 2\beta}{\sin 2\beta} = \frac{2(k + \tau_{xy})}{\sigma_x - \sigma_y}, \tag{19}$$

and

$$\frac{dy}{dx} = -\cot \beta = \frac{\sigma_y - \sigma_x}{2(k + \tau_{xy})}. \tag{20}$$

When the stresses are known the slip-lines can be found by solving (19) and (20).

The theory given above is suitable for the study of large plastic flow in which the elastic displacements of Fig. 9.1 can be neglected by comparison with the plastic ones. If this is not the case the Prandtl–Reuss equations have to be used. In these the total rate of strain is assumed to be the sum of elastic and plastic terms. The plastic terms will satisfy (5). The elastic terms, assuming for simplicity that the material is incompressible so that $\varepsilon_1 = e_1$, etc., will satisfy equations obtained by differentiating § 5.4 (2), namely

$$\dot{s}_1 = 2G\dot{\varepsilon}_1, \quad \dot{s}_2 = 2G\dot{\varepsilon}_2, \quad \dot{s}_3 = 2G\dot{\varepsilon}_3. \tag{21}$$

Adding (5) and (21) and writing $\dot{\varepsilon}_1$ for the sum of the elastic and plastic rates of strain in (5) and (21) gives

$$2G\dot{\varepsilon}_1 = \dot{s}_1 + \lambda s_1, \quad 2G\dot{\varepsilon}_2 = \dot{s}_2 + \lambda s_2, \quad 2G\dot{\varepsilon}_3 = \dot{s}_3 + \lambda s_3, \tag{22}$$

where λ is written in place of G/ϕ.

In many cases the yield condition will not be satisfied throughout the body, so that the elastic equations will have to be used in part of it and the equations of plasticity in other parts, the two regions being separated by a surface on which conditions of continuity have to be satisfied.

The stress–strain relations of a viscous incompressible (Newtonian, cf. Chapter 11) fluid are derived in precisely the same way as (6) and (7) and have the same form except that ϕ, which was a function determined by the yield condition, is replaced by a constant, η, the viscosity, so that they are

$$s_x = 2\eta\dot{\varepsilon}_x, \quad s_y = 2\eta\dot{\varepsilon}_y, \quad s_z = 2\eta\dot{\varepsilon}_z, \tag{23}$$

$$s_{yz} = 2\eta\dot{\Gamma}_{yz}, \quad s_{zx} = 2\eta\dot{\Gamma}_{zx}, \quad s_{xy} = 2\eta\dot{\Gamma}_{xy}. \tag{24}$$

These are combined with the equations of motion which, for slow motions, are § 5.5 (21)–(23) with $\rho\,\partial^2 u/\partial t^2$ written in the form $\rho\,\partial U/\partial t$, etc., where U, V, W are the components of the velocity at the point x, y, z and time t.

9.4 Elastic-plastic solutions in cylindrical coordinates

The case to be considered here will be that of the hollow cylinder $R_1 < r < R_2$ in plane strain with pressure p_2 applied at R_2. This arises in the testing of hollow cylinders of material such as marble for which the yield stress may easily be exceeded at the inner surface: it has been studied in this context by Robertson (1955).

The elastic solution has been given in § 5.11 (10), and if the value of the stress difference at the inner surface, namely $\sigma_\theta = 2p_2R_2^2/(R_2^2 - R_1^2)$, satisfies the yield criterion § 9.3 (9) this solution ceases to be valid. Thus there will be a region $R_1 < r < R$ in which plastic theory must be used, while the outer region $R < r < R_2$ remains elastic. The radius R of the elastic–plastic boundary has to be found, and continuity conditions have to be satisfied there.

By symmetry, the radial and tangential stresses will be principal stresses, so that in the region $R_1 < r < R$, $\sigma_1 = \sigma_\theta$, $\sigma_3 = \sigma_r$, and the yield condition § 9.3 (9) becomes

$$\sigma_\theta - \sigma_r = 2k. \tag{1}$$

Also the equation of equilibrium in terms of stresses, § 5.6 (9), is

$$\frac{\partial \sigma_r}{\partial r} + \frac{\sigma_r - \sigma_\theta}{r} = 0, \tag{2}$$

and using (1) this becomes

$$\frac{d\sigma_r}{dr} = \frac{2k}{r}. \tag{3}$$

The solution of this is

$$\sigma_r = 2k \ln r + C, \tag{4}$$

where C is a constant, and the condition $\sigma_r = 0$ when $r = R_1$ gives $C = -2k \ln R_1$, so that

$$\sigma_r = 2k \ln (r/R_1), \quad R_1 < r < R. \tag{5}$$

In the elastic region $R < r < R_2$ it follows from § 5.11 (5), (2), (3) that

$$\sigma_r = 2(\lambda + G)A - \frac{2GB}{r^2}, \quad \sigma_\theta = 2(\lambda + G)A + \frac{2GB}{r^2}, \tag{6}$$

where A and B are constants. The condition $\sigma_r = p_2$ when $r = R_2$ requires

$$2(\lambda + G)A - \frac{2GB}{R_2{}^2} = p_2. \tag{7}$$

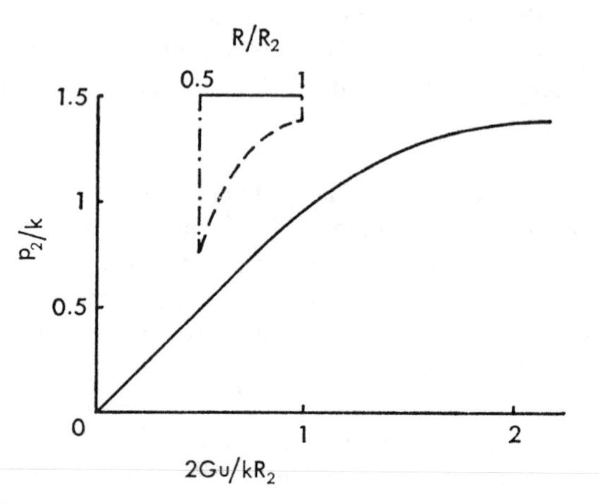

Fig. 9.4.1 Displacement u at the outer surface of a perfectly plastic hollow cylinder of radii R_1 and $R_2 = 2R_1$ as a function of the external pressure p_2. Dotted curve: position of the elastic–plastic boundary R/R_2 as a function of p_2.

At the boundary $r = R$, σ_r and σ_θ given by (6) must satisfy the yield condition (1), and σ_r given by (6) must be equal to the value given by (4). That is

$$\frac{4GB}{R^2} = 2k, \tag{8}$$

$$2(\lambda + G)A - \frac{2GB}{R^2} = 2k \ln \frac{R}{R_1}. \tag{9}$$

It follows from (7), (8), (9) that

$$k\{1 + 2 \ln (R/R_1) - (R/R_2)^2\} = p_2, \tag{10}$$

which is an equation for R. For the von Mises criterion, by § 9.3 (10), k is $\sigma_0/\sqrt{3}$, where σ_0 is the yield stress in compression. This solution is valid only if $R < R_2$, that is, if $p_2 < 2k \ln (R_2/R_1)$.

Fig. 9.4.1 shows the situation for the case $R_2 = 2R_1$, $\lambda = G$. The dotted curve shows the position of the surface of separation R/R_2 as a function of p_2/k. The solid curve shows the displacement u at the outer surface $r = R_2$; this varies linearly with p_2 until yield occurs at the inner surface.

In the plastic region the principal stresses σ_r and σ_θ are radial and transverse, and the slip-lines, PR, PS, Fig. 9.4.2 (a), are inclined at 45° to

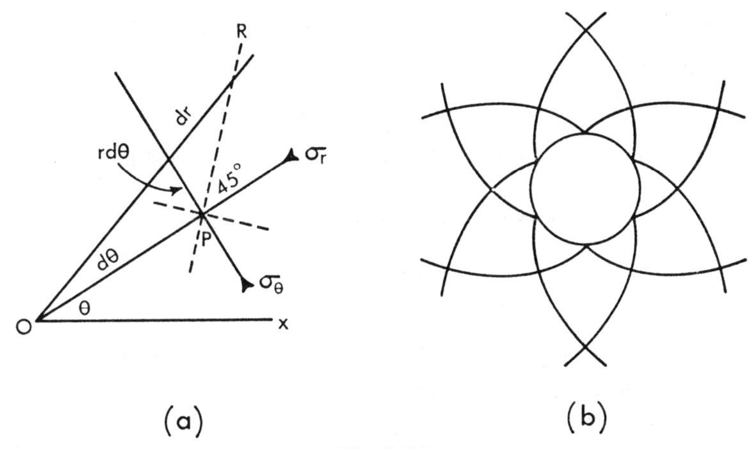

Fig. 9.4.2

them. From the geometry of Fig. 9.4.2 (a) the differential equations of the two systems of slip-lines are

$$\frac{dr}{rd\theta} = \pm \tan \pi/4 = \pm 1, \tag{11}$$

so that their equations are

$$r = ae^{\pm\theta}. \tag{12}$$

These are two families of equiangular spirals, Fig. 9.4.2 (b), and are often seen on the surfaces of radially deformed rocks.

9.5 Flow between flat surfaces

The problem to be considered is that of plane flow of material compressed between the planes $y = \pm a$. If the planes are assumed to be rough and the material is to be slipping over them to the right the boundary conditions will be

$$\tau_{xy} = k, \text{ when } y = a, \text{ and } \tau_{xy} = -k \text{ when } y = -a, \text{ by § 9.3 (9).} \tag{1}$$

The differential equations to be solved are § 9.3 (12), (13), (15), and the form of the boundary condition suggests that the simple expression

$$\tau_{xy} = ky/a \tag{2}$$

be studied. This satisfies (15), and putting it in (12) and (13) gives

$$\sigma_x = f(y) - kx/a, \quad \sigma_y = g(x), \tag{3}$$

where $f(y)$ and $g(x)$ are unknown functions which may be determined by substituting (2) and (3) in § 9.3 (11). This gives

$$k^2y^2/a^2 + \tfrac{1}{4}\{f(y) - g(x) - kx/a\}^2 = k^2,$$

or

$$f(y) - g(x) - kx/a = \pm 2k(1 - y^2/a^2)^{\frac{1}{2}}. \tag{4}$$

It follows that

$$f(y) = P \pm 2k(1 - y^2/a^2)^{\frac{1}{2}}, \quad g(x) = P - kx/a, \tag{5}$$

where P is an arbitrary constant. Choosing the negative sign, the solution is

$$\sigma_x = P - 2k(1 - y^2/a^2)^{\frac{1}{2}} - kx/a, \tag{6}$$

$$\sigma_y = P - kx/a, \tag{7}$$

$$\tau_{xy} = ky/a. \tag{8}$$

The system of stresses on a rectangular region $ABCD$, $0 < x < l$, $-a < y < a$, is shown in Fig. 9.5.1.

On the face AD, $x = 0$,

$$\sigma_x = P - 2k(1 - y^2/a^2)^{\frac{1}{2}}, \tag{9}$$

and so falls parabolically from P at A to $P - 2k$ at $y = 0$. On the face AB, $y = a$, σ_2 falls linearly from P at A to $P - kl/a$ at B. The magnitudes of the normal stresses are indicated by the lengths of the arrows in Fig. 9.5.1.

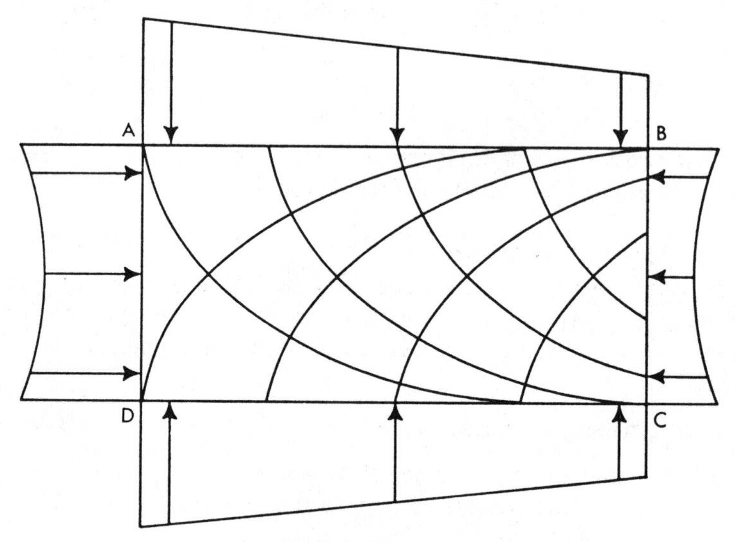

Fig. 9.5.1

The motion corresponding to this system of stresses may be found from § 9.3 (6), (7). It follows from (6) and (7), using § 9.3 (8), that

$$s_x = -s_y = -k(1 - y^2/a^2)^{\frac{1}{2}}. \tag{10}$$

Therefore from § 9.3 (3), (6), writing U and V for the velocities $\dot{u}$ and $\dot{v}$,

$$s_x = -k(1 - y^2/a^2)^{\frac{1}{2}} = 2\phi\dot{\epsilon}_x = 2\phi\frac{\partial U}{\partial x}, \tag{11}$$

$$s_y = k(1 - y^2/a^2)^{\frac{1}{2}} = 2\phi\dot{\epsilon}_y = 2\phi\frac{\partial V}{\partial y}, \tag{12}$$

$$\tau_{xy} = ky/a = 2\phi\dot{\Gamma}_{xy} = \phi\left(\frac{\partial U}{\partial y} + \frac{\partial V}{\partial x}\right), \tag{13}$$

where ϕ is an unknown function involving the strain rate.

Suppose the planes AB and CD are moving towards each other with speed V_0. This condition would be satisfied by

$$V = yV_0/a, \tag{14}$$

and this will satisfy (12) if

$$\phi = k(a^2 - y^2)^{\frac{1}{2}}/2V_0. \tag{15}$$

With this value of ϕ, (11) gives

$$U = -V_0x/a + h(y), \tag{16}$$

where $h(y)$ is an unknown function to be determined from the fact that U, V, and ϕ given by (16), (14), (15) must satisfy (13). This requires

$$h'(y) = -\frac{2yV_0}{a(a^2 - y^2)^{\frac{1}{2}}}$$

so that

$$h(y) = -2V_0(1 - y^2/a^2)^{\frac{1}{2}},$$

neglecting an arbitrary constant. Thus, finally,

$$U = -V_0x/a - 2V_0(1 - y^2/a^2)^{\frac{1}{2}}, \quad V = yV_0/a. \tag{17}$$

This solution corresponds to the plates moving together with speeds V_0 and to the material between them being extruded to the right, its speed past the planes being V_0x/a. The plus sign in (4) gives the solution to the problem of the planes being forced apart by material squeezed in over AD and BC.

It will be noted that the stresses are determined quite independently of, and are not affected by, the speed of motion V_0. This illustrates the difference between plastic flow and viscous flow in which the stresses are determined by V_0.

The equations of the slip-lines for the stresses (6)–(8) may be found from § 9.3 (19), (20). The first of these gives

$$\frac{dx}{dy} = -\frac{a + y}{(a^2 - y^2)^{\frac{1}{2}}}, \tag{18}$$

so that

$$x = -\int \frac{(a^2 - y^2)^{\frac{1}{2}}}{a + y}\, dy.$$

Putting

$$y = a \cos \theta \tag{19}$$

this becomes

$$x = a\int (1 - \cos \theta)\, d\theta = a(\theta - \sin \theta). \tag{20}$$

(18) and (19) are the equations of a cycloid in which, since dy/dx in (18) is negative, only the descending portion must be taken. Similarly, § 9.3 (20)

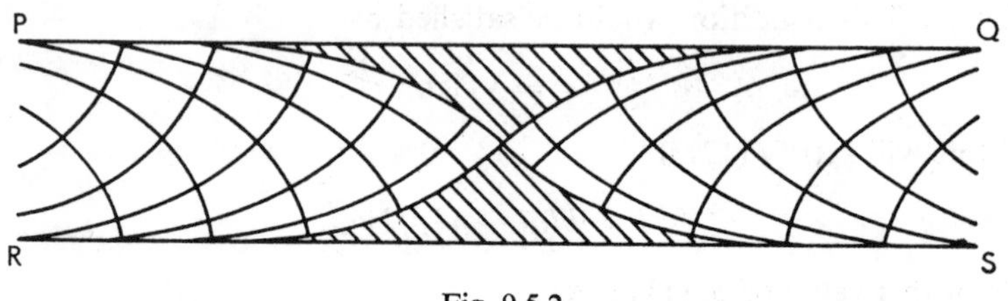

Fig. 9.5.2

gives an orthogonal system of cycloids. These two families are shown in Fig. 9.5.1.

The solution given above is exact, provided that the stresses specified are applied over the planes *ABCD*. In the practical case in which material is compressed between rough parallel plates *PQ*, *RS*, Fig. 9.5.2, fuller analysis is necessary. It is found that there are regions near the centre of the area, shown shaded in Fig. 9.5.2, in which the yield stress is not exceeded, so that they remain undistorted. Material is extruded from the remainder of the region in a similar manner to that discussed earlier, the slip-lines being shown in Fig. 9.5.2.

Nye (1951) extended this solution to cover the flow of glaciers downhill, and Evison (1960) used it to study the growth of continents by plastic extrusion of their deeper layers.

Chapter Ten

Further Problems in Elasticity

10.1 Introduction

In this chapter the solutions of a number of problems which have direct relevance to rock mechanics will be studied. There is a number of different approaches to the solution of problems in elasticity which are described in textbooks on the subject. However, at the present stage of development of rock mechanics the problems which appear are either approximately two-dimensional (such as stresses and displacements around a tunnel) or are very complicated problems in three-dimensions (such as stresses in bord and pillar mining or around a power-house, including excavations for turbines). In these circumstances it seems better to give a fairly full account of two-dimensional problems using the powerful method of the complex variable and a less complete account of three-dimensional problems.

The methods of photoelasticity and finite-element analysis are referred to briefly. These are analogue and numerical methods of solving problems of elasticity in two dimensions which may, with some increase of complexity, be extended to three dimensions. They provide, in effect, the only methods of solving problems for regions of practical, instead of idealized, shapes, and the finite element methods, at least, have the great advantage that they can be applied to inhomogeneous, non-linear, and anisotropic materials. Nevertheless, a fundamental understanding of the mathematical behaviour of the solutions for simple shapes is of the greatest value.

10.2 Complex variable theory in two dimensions

The formulae for change of axes § 2.3 (7)–(9) take a very simple form when the complex variable is introduced. It follows immediately from them that

$$\sigma_{y'} - \sigma_{x'} + 2i\tau_{x'y'} = (\sigma_y - \sigma_x + 2i\tau_{xy})e^{2i\theta}, \tag{1}$$

and these and § 2.3 (10), namely

$$\sigma_{x'} + \sigma_{y'} = \sigma_x + \sigma_y, \tag{2}$$

are the fundamental formulae. Subtracting (1) and (2) gives

$$2(\sigma_{x'} - i\tau_{x'y'}) = \sigma_x + \sigma_y - (\sigma_y - \sigma_x + 2i\tau_{xy})e^{2i\theta}. \tag{3}$$

This may be used to give the boundary conditions at an element of surface whose normal is inclined at θ to the x-axis. If N and T are the normal and shear stresses at this surface the boundary condition is, by (3),

$$2(N - iT) = \sigma_x + \sigma_y - (\sigma_y - \sigma_x + 2i\tau_{xy})e^{2i\theta}. \tag{4}$$

(1) and (2) may also be used to transform to polar coordinates with σ_r inclined at θ to the x-axis. The result is

$$\sigma_r + \sigma_\theta = \sigma_x + \sigma_y, \tag{5}$$

$$\sigma_\theta - \sigma_r + 2i\tau_{r\theta} = (\sigma_y - \sigma_x + 2i\tau_{xy})e^{2i\theta}. \tag{6}$$

The formulae § 2.9 (15) for the transformation of displacements u, v relative to axes Oxy to u', v' for axes rotated through θ also become

$$u' + iv' = (u + iv)e^{-i\theta}. \tag{7}$$

We now show that the Airy stress function, the stresses, and the displacements in plane problems can be expressed in terms of two analytic functions of a complex variable $z = x + iy$. For the theory of analytic functions see, e.g., Muskhelishvili (1953), Churchill (1948), Macrobert (1917). A brief account of the properties needed is given here. A function $\zeta(z)$ of z is said to be analytic in a region if it is finite and single valued and has a definite differential coefficient at all points of the region. The function ζ can be expressed in the form

$$\zeta = \xi + i\eta, \tag{8}$$

where ξ and η are real functions of both x and y. ξ is called the *real part* of ζ, sometimes written $\mathbf{R}(\zeta)$, and η is called the *imaginary part* of ζ, written $\mathbf{I}(\zeta)$. The quantity

$$\bar{\zeta} = \xi - i\eta \tag{9}$$

is called the *conjugate complex* of ζ.

The differential coefficient of $\zeta(z)$ with respect to z is defined in the usual way as

$$\frac{d\zeta}{dz} = \lim_{\delta x \to 0,\, \delta y \to 0} \frac{\delta\zeta}{\delta z} = \lim \frac{\delta\xi + i\,\delta\eta}{\delta x + i\,\delta y}$$

$$= \lim \left[\left\{\left(\frac{\partial\xi}{\partial x} + i\frac{\partial\eta}{\partial x}\right)\delta x + i\left(\frac{\partial\eta}{\partial y} - i\frac{\partial\xi}{\partial y}\right)\delta y\right\}/(\delta x + i\,\delta y)\right]. \tag{10}$$

This limit can only be independent of the ratio $\delta y/\delta x$, and therefore unique, if the coefficients of δy and δx in the numerator are equal, that is, if

$$\frac{\partial\xi}{\partial x} + i\frac{\partial\eta}{\partial x} = \frac{\partial\eta}{\partial y} - i\frac{\partial\xi}{\partial y},$$

or, equating real and imaginary parts,

$$\frac{\partial\xi}{\partial x} = \frac{\partial\eta}{\partial y}, \quad \frac{\partial\eta}{\partial x} = -\frac{\partial\xi}{\partial y}. \tag{11}$$

If this is the case it follows from (10) that

$$\zeta'(z) = \frac{d\zeta}{dz} = \frac{\partial\xi}{\partial x} + i\frac{\partial\eta}{\partial x} = \frac{\partial\xi}{\partial x} - i\frac{\partial\xi}{\partial y}. \tag{12}$$

A 'dash' will be used as in (12) to denote differentiation with respect to z.

Equations (11) are the *Cauchy–Riemann equations* which connect the real and imaginary parts of a function of a complex variable. If they are satisfied it follows that

$$\nabla^2\xi = 0, \quad \nabla^2\eta = 0, \tag{13}$$

where ∇^2 is written for the operator

$$\nabla^2 \equiv \frac{\partial^2}{\partial x^2} + \frac{\partial^2}{\partial y^2}. \tag{14}$$

Thus both the real and imaginary parts of an analytic function of a complex variable satisfy Laplace's equation (13). ξ and η are called conjugate functions, and functions which satisfy Laplace's equation are called harmonic. It can be shown that if ξ is a harmonic function the function η conjugate to it can be found by solving the equations (11), and the function $\xi + i\eta$ so obtained is an analytic function of z.

After these preliminaries a general solution of the biharmonic equation § 5.7 (8)

$$\nabla^2(\nabla^2 U) = 0 \tag{15}$$

can be found in terms of analytic functions. Airy's stress function U satisfies (15), and stresses can be found from it by the formulae § 5.7 (7), which for the case of no body forces are

$$\sigma_x = \frac{\partial^2 U}{\partial y^2}, \quad \sigma_y = \frac{\partial^2 U}{\partial x^2}, \quad \tau_{xy} = -\frac{\partial^2 U}{\partial x\, \partial y}. \tag{16}$$

Let

$$P = \sigma_x + \sigma_y = \frac{\partial^2 U}{\partial x^2} + \frac{\partial^2 U}{\partial y^2} = \nabla^2 U, \tag{17}$$

so that, by (15), P is harmonic. Then, as remarked above, the function Q conjugate to it can, in principle, be found, and the function

$$f(z) = P + iQ \tag{18}$$

will be analytic. So also will its integral, and we define

$$\phi(z) = \tfrac{1}{4}\int f(z)\, dz = p + iq, \tag{19}$$

where p and q are the real and imaginary parts of $\phi(z)$. Then by (12) and (19)

$$\phi'(z) = \frac{\partial p}{\partial x} + i\frac{\partial q}{\partial x} = \tfrac{1}{4}f(z) = \tfrac{1}{4}(P + iQ). \tag{20}$$

Equating real and imaginary parts in (20) and using (11) gives

$$\tfrac{1}{4}P = \frac{\partial p}{\partial x} = \frac{\partial q}{\partial y}, \quad \tfrac{1}{4}Q = \frac{\partial q}{\partial x} = -\frac{\partial p}{\partial y}. \tag{21}$$

Next we show that $U - px - qy$ is harmonic. This follows on differentiating, since

$$\nabla^2(U - px - qy) = \nabla^2 U - x\nabla^2 p - 2\frac{\partial p}{\partial x} - y\nabla^2 q - 2\frac{\partial q}{\partial y} = 0, \tag{22}$$

using (17), (21), and $\nabla^2 p = \nabla^2 q = 0$.

It follows from (22) that

$$U - px - qy = p_1, \tag{23}$$

where p_1 is harmonic, and so may be taken to be the real part of some unknown analytic function, $\chi(z)$, of z. Also $px + qy$ is the real part of $\bar{z}\phi(z) = (x - iy)(p + iq)$. Therefore, finally,

$$U = \mathbf{R}\{\bar{z}\phi(z) + \chi(z)\} \tag{24}$$

$$= \tfrac{1}{2}\{\bar{z}\phi(z) + z\overline{\phi(z)} + \chi(z) + \overline{\chi(z)}\}, \tag{25}$$

and we have obtained a general solution of the biharmonic equation (15) in terms of two analytic functions $\phi(z)$ and $\chi(z)$.

The stress-components may now be obtained from (16) by differentiating (25). This gives

$$\frac{\partial U}{\partial x} = \tfrac{1}{2}\{\phi(z) + \bar{z}\phi'(z) + \overline{\phi(z)} + z\overline{\phi'(z)} + \chi'(z) + \overline{\chi'(z)}\}, \tag{26}$$

$$\frac{\partial U}{\partial y} = \tfrac{1}{2}\{-i\phi(z) + i\bar{z}\phi'(z) + i\overline{\phi(z)} - iz\overline{\phi'(z)} + i\chi'(z) - i\overline{\chi'(z)}\}, \tag{27}$$

$$\frac{\partial^2 U}{\partial x^2} = \tfrac{1}{2}\{2\phi'(z) + \bar{z}\phi''(z) + 2\overline{\phi'(z)} + z\overline{\phi''(z)} + \chi''(z) + \overline{\chi''(z)}\}, \tag{28}$$

$$\frac{\partial^2 U}{\partial y^2} = \tfrac{1}{2}\{2\phi'(z) - \bar{z}\phi''(z) + 2\overline{\phi'(z)} - z\overline{\phi''(z)} - \chi''(z) - \overline{\chi''(z)}\}, \tag{29}$$

$$\frac{\partial^2 U}{\partial x\,\partial y} = \tfrac{1}{2}\{i\bar{z}\phi''(z) - iz\overline{\phi''(z)} + i\chi''(z) - i\overline{\chi''(z)}\}. \tag{30}$$

Adding i times (27) to (26) gives

$$\frac{\partial U}{\partial x} + i\frac{\partial U}{\partial y} = \phi(z) + z\overline{\phi'(z)} + \overline{\chi'(z)}. \tag{31}$$

Also the combination of stresses appearing in (1) is by (16)

$$\sigma_y - \sigma_x + 2i\tau_{xy} = \frac{\partial^2 U}{\partial x^2} - \frac{\partial^2 U}{\partial y^2} - 2i\frac{\partial^2 U}{\partial x\, \partial y}$$
$$= 2\{\bar{z}\phi''(z) + \chi''(z)\}. \quad (32)$$

Also

$$\sigma_x + \sigma_y = \frac{\partial^2 U}{\partial x^2} + \frac{\partial^2 U}{\partial y^2} = 2[\phi'(z) + \overline{\phi'(z)}] = 4\mathbf{R}[\phi'(z)], \quad (33)$$

as could have been seen from (17) and (20).

The displacements for cases of plane stress or strain may be found from § 5.3 (18), (19), which run

$$8G\frac{\partial u}{\partial x} = 8G\varepsilon_x = (\varkappa + 1)\sigma_x + (\varkappa - 3)\sigma_y, \quad (34)$$

$$8G\frac{\partial v}{\partial y} = 8G\varepsilon_y = (\varkappa - 3)\sigma_x + (\varkappa + 1)\sigma_y, \quad (35)$$

where $\varkappa = 3 - 4\nu$ for plane strain, and (36)

$\varkappa = (3 - \nu)/(1 + \nu)$ for plane stress. (37)

These may be written in the form

$$2G\frac{\partial u}{\partial x} = -\sigma_y + \tfrac{1}{4}(\varkappa + 1)(\sigma_x + \sigma_y), \quad (38)$$

$$2G\frac{\partial u}{\partial y} = -\sigma_x + \tfrac{1}{4}(\varkappa + 1)(\sigma_x + \sigma_y). \quad (39)$$

In (38) and (39) we use (16) for σ_x and σ_y and

$$\sigma_x + \sigma_y = P = 4\frac{\partial p}{\partial x} = 4\frac{\partial q}{\partial y}, \quad (40)$$

from (17) and (21). (38) and (39) then become

$$2G\frac{\partial u}{\partial x} = -\frac{\partial^2 U}{\partial x^2} + (\varkappa + 1)\frac{\partial p}{\partial x}, \quad (41)$$

$$2G\frac{\partial v}{\partial y} = -\frac{\partial^2 U}{\partial y^2} + (\varkappa + 1)\frac{\partial q}{\partial y}. \quad (42)$$

Integrating gives

$$2Gu = -\frac{\partial U}{\partial x} + (\varkappa + 1)p + f_1(y), \quad (43)$$

$$2Gv = -\frac{\partial U}{\partial y} + (\varkappa + 1)q + f_2(x), \quad (44)$$

where $f_1(y)$ and $f_2(x)$ are unknown functions of y and x. To study these, the third stress–strain relation, $G\gamma_{xy} = \tau_{xy}$, or

$$G\left(\frac{\partial v}{\partial x} + \frac{\partial u}{\partial y}\right) = -\frac{\partial^2 U}{\partial x\, \partial y} \tag{45}$$

may be used. Substituting (43), (44), and (21) in this gives

$$f_1'(y) + f_2'(x) = 0. \tag{46}$$

The only solution of this is $f_1(y) = \omega y + a$, $f_2(x) = -\omega x + b$, which corresponds to movement as a rigid body and so may be ignored. Putting $f_1(y) = f_2(x) = 0$ in (43) and (44) gives

$$\begin{aligned} 2G(u + iv) &= -\left(\frac{\partial U}{\partial x} + i\frac{\partial U}{\partial y}\right) + (\varkappa + 1)(p + iq) \\ &= \varkappa\phi(z) - z\overline{\phi'(z)} - \overline{\chi'(z)}, \end{aligned} \tag{47}$$

using (19) and (31).

Since $\chi(z)$ does not appear in the fundamental equations (32), (33), (47), these may be simplified slightly by writing

$$\psi(z) = \chi'(z),$$

and they become finally

$$\sigma_x + \sigma_y = 2[\phi'(z) + \overline{\phi'(z)}] = 4\mathbf{R}[\phi'(z)], \tag{48}$$

$$\sigma_y - \sigma_x + 2i\tau_{xy} = 2[\bar{z}\phi''(z) + \psi'(z)], \tag{49}$$

$$2G(u + iv) = \varkappa\phi(z) - z\overline{\phi'(z)} - \overline{\psi(z)}. \tag{50}$$

Finally, the case discussed in § 5.3 (vii), in which there is a uniform strain ε along the z-axis, may be considered. In this case § 5.3 (23), (24) are equivalent to adding terms $-2G\nu\varepsilon$ in the right-hand sides of (38) and (39) and putting $\varkappa = 3 - 4\nu$. The same calculation gives for the displacements in this case

$$2G(u + iv) = (3 - 4\nu)\phi(z) - z\overline{\phi'(z)} - \overline{\psi(z)} - 2\nu\varepsilon Gz. \tag{51}$$

10.3 Simple special cases

The solutions of the previous section were in terms of general analytic functions. Many important problems can be solved by taking $\phi(z)$ and $\psi(z)$ to be polynomials or power series in z or z^{-1}. If the region in question includes the origin only positive powers of z can be used for the functions to remain finite at the origin, while for the region outside a circle only negative powers can be used.

Some important special cases will now be examined.

(i) *Homogeneous stresses*

Suppose

$$\phi(z) = cz, \quad \psi(z) = dz, \tag{1}$$

where c and d are constants which may be complex.

By § 10.2 (48), (49),

$$\sigma_x + \sigma_y = 2\phi'(z) + 2\overline{\phi'(z)} = 2(c + \bar{c}), \tag{2}$$

$$\sigma_y - \sigma_x + 2i\tau_{xy} = 2\{\bar{z}\phi''(z) + \psi'(z)\} = 2d. \tag{3}$$

Since c occurs only in the form $c + \bar{c}$, it may be taken to be real, and (2) may be replaced by

$$\sigma_x + \sigma_y = 4c. \tag{4}$$

(3) and (4) correspond to homogeneous stresses, and if the principal stresses are σ_1 and σ_2 with σ_1 inclined at β to Ox, it follows from § 10.2 (1), (2) that

$$\sigma_2 - \sigma_1 = (\sigma_y - \sigma_x + 2i\tau_{xy})e^{2i\beta} = 2de^{2i\beta}, \tag{5}$$

$$\sigma_1 + \sigma_2 = \sigma_x + \sigma_y = 4c. \tag{6}$$

Therefore this system of stresses is represented by (1) with

$$c = \tfrac{1}{4}(\sigma_1 + \sigma_2), \quad d = \tfrac{1}{2}(\sigma_2 - \sigma_1)e^{-2i\beta}. \tag{7}$$

For hydrostatic stress; $c = \frac{1}{2}\sigma_1$, $d = 0$. For uniaxial stress σ_1 along Ox; $c = \frac{1}{4}\sigma_1$, $d = -\frac{1}{2}\sigma_1$. For pure shear $\sigma_2 = -\sigma_1$, $\beta = \pi/4$; $c = 0$, $d = i\sigma_1$.

The displacements are, by § 10.2 (50),

$$2G(u + iv) = (\varkappa - 1)cz - \bar{d}\bar{z}, \tag{8}$$

since c is real, so that

$$2Gu = (\varkappa - 1)cx - \mathbf{R}(\bar{d}\bar{z}), \quad 2Gv = (\varkappa - 1)cy - \mathbf{I}(\bar{d}\bar{z}). \tag{9}$$

If the axes are principal axes, so that $\beta = 0$, these become

$$8Gu = [(\varkappa + 1)\sigma_1 + (\varkappa - 3)\sigma_2]x, \quad 8Gv = [(\varkappa - 3)\sigma_1 + (\varkappa + 1)\sigma_2]y, \tag{10}$$

in agreement with § 5.3 (14), (15). The point $x = a\cos\theta$, $y = a\sin\theta$ on a circle of radius a is displaced to the point

$$\begin{aligned} &a\{1 - [(\varkappa + 1)\sigma_1 + (\varkappa - 3)\sigma_2]/8G\}\cos\theta, \\ &a\{1 - [(\varkappa - 3)\sigma_1 + (\varkappa + 1)\sigma_2]/8G\}\sin\theta, \end{aligned} \tag{11}$$

which lies on an ellipse.

In polar coordinates, from § 10.2 (7) the displacement is

$$\begin{aligned} 2G(u_r + iu_\theta) &= 2G(u + iv)e^{-i\theta} \\ &= [(\varkappa - 1)cre^{i\theta} - \bar{d}re^{-i\theta}]e^{-i\theta} \\ &= (\varkappa - 1)cr - \bar{d}re^{-2i\theta} \\ &= \tfrac{1}{4}(\varkappa - 1)(\sigma_1 + \sigma_2)r + \tfrac{1}{2}(\sigma_1 - \sigma_2)re^{-2i\theta}, \end{aligned} \tag{12}$$

if the axes are principal axes. Therefore

$$8Gu_r = (\varkappa - 1)(\sigma_1 + \sigma_2)r + 2(\sigma_1 - \sigma_2)r \cos 2\theta, \tag{13}$$

$$8Gu_\theta = -2(\sigma_1 - \sigma_2)r \sin 2\theta. \tag{14}$$

If the principal axes are inclined at β to Ox, as in (7), θ in (13) and (14) is replaced by $\theta - \beta$.

(ii) *The region outside a circle*
Consider

$$\phi(z) = 0, \quad \psi(z) = d/z, \tag{15}$$

with d real and $z = re^{i\theta}$.
Here in polar coordinates, using § 10.2 (5), (6), (48), (49),

$$\sigma_r + \sigma_\theta = 0,$$

$$\sigma_\theta - \sigma_r + 2i\tau_{r\theta} = 2\psi'(z)e^{2i\theta} = -2dr^{-2}. \tag{16}$$

Therefore

$$\sigma_r = d/r^2, \quad \sigma_\theta = -d/r^2. \tag{17}$$

By § 10.2 (7), (50) the displacement in polar coordinates is

$$2G(u_r + iu_\theta) = 2G(u + iv)e^{-i\theta} = -\overline{\psi(z)}e^{-i\theta} = -d/r. \tag{18}$$

If $d = pR^2$ the solution is that for the region outside the cylinder $r = R$ with pressure p on the inner surface which has been found in § 5.11, the results being

$$\sigma_r = pR^2/r^2, \quad \sigma_\theta = -pR^2/r^2, \quad u_r = -pR^2/2Gr. \tag{19}$$

The results for the hollow cylinder found in § 5.11 may be derived by assuming $\phi(z) = cz$, $\psi(z) = d/z$.

(iii) *Rotation*
Suppose

$$\phi(z) = i\omega z, \quad \psi(z) = 0, \tag{20}$$

where ω is real. It follows from § 10.2 (48)–(50) that

$$\sigma_x = \sigma_y = \tau_{xy} = 0, \tag{21}$$

$$2G(u + iv) = (\varkappa + 1)i\omega z. \tag{22}$$

Therefore

$$2Gu = -(\varkappa + 1)\omega y, \quad 2Gv = (\varkappa + 1)\omega x, \tag{23}$$

so that there is no stress, and the displacements correspond to a rigid body rotation through a small angle of $(\varkappa + 1)\omega/2G$.

(iv) *A dislocation*
Consider the case

$$\phi(z) = \ln z, \quad \psi(z) = 0. \tag{24}$$

Since $\ln z$ is not single-valued in a region which permits a complete circuit of the origin, the use of such functions is not always admissible. Their effect may be seen in the following way. By § 10.2 (50)

$$\begin{aligned} 2G(u + iv) &= \varkappa \ln z - z/\bar{z}, \\ &= \varkappa[\ln r + i\theta] - e^{2i\theta}. \end{aligned} \tag{25}$$

It follows that $2Gv = \varkappa\theta - \sin 2\theta$, so that if θ increases by 2π, v increases by $\varkappa\pi/G$. Thus in this case displacements are not single valued, and terms involving $\ln z$ must not appear in $\phi(z)$ or $\psi(z)$. It is possible to study stresses caused by dislocations by including such terms.

10.4 The region outside a circular hole of radius R with given principal stresses at infinity

This is perhaps the most important single problem in rock mechanics. To avoid complicating the discussion, the case of a uniaxial stress p_1 at infinity in the direction of the x-axis will be considered first. The stresses in an infinite region subject to this stress were derived in § 10.3 (7) from $\phi(z) = \frac{1}{4}p_1 z, \psi(z) = -\frac{1}{2}p_1 z$. For the infinite region containing a hole, terms must be added to this to represent the effect of the hole, and these must vanish as $z \to \infty$. We shall assume

$$\phi(z) = \tfrac{1}{4}p_1\left(z + \frac{A}{z}\right), \quad \psi(z) = -\tfrac{1}{2}p_1\left(z + \frac{B}{z} + \frac{C}{z^3}\right), \tag{1}$$

where A, B, C are real constants. The choice of this form will be justified in § 10.5, but it may be regarded at present as including all terms which give rise to terms in σ_r and σ_θ involving functions of the angle 2θ. Differentiating (1) gives

$$\phi'(z) = \tfrac{1}{4}p_1\left(1 - \frac{A}{z^2}\right), \quad \psi'(z) = -\tfrac{1}{2}p_1\left(1 - \frac{B}{z^2} - \frac{3C}{z^4}\right) \tag{2}$$

$$\phi''(z) = Ap_1/2z^3. \tag{3}$$

Working in polar coordinates $z = re^{i\theta}$ it follows from § 10.2 (48) that

$$\begin{aligned} \sigma_r + \sigma_\theta = \sigma_x + \sigma_y &= 2[\phi'(z) + \overline{\phi'(z)}] \\ &= \tfrac{1}{2}p_1[2 - Ar^{-2}e^{-2i\theta} - Ar^{-2}e^{2i\theta}] \\ &= p_1[1 - Ar^{-2}\cos 2\theta]. \end{aligned} \tag{4}$$

Next, from § 10.2 (6), (49),

$$\sigma_\theta - \sigma_r + 2i\tau_{r\theta} = (\sigma_y - \sigma_x + 2i\tau_{xy})e^{2i\theta}$$
$$= p_1\{re^{-i\theta} . Ar^{-3}e^{-3i\theta} - 1 + Br^{-2}e^{-2i\theta} + 3Cr^{-4}e^{-4i\theta}\}e^{2i\theta}$$
$$= p_1\{Br^{-2} - e^{2i\theta} + (Ar^{-2} + 3Cr^{-4})e^{-2i\theta}\}.$$

Therefore

$$\sigma_\theta - \sigma_r = p_1\{Br^{-2} - (1 - Ar^{-2} - 3Cr^{-4}) \cos 2\theta\}, \tag{5}$$

$$\tau_{r\theta} = -\tfrac{1}{2}p_1\{1 + Ar^{-2} + 3Cr^{-4}\} \sin 2\theta. \tag{6}$$

Subtracting (5) from (4),

$$2\sigma_r = p_1\{1 - Br^{-2} + (1 - 2Ar^{-2} - 3Cr^{-4}) \cos 2\theta\}. \tag{7}$$

If there is no stress on the hole $r = R$, the boundary conditions are

$$\sigma_r = \tau_{r\theta} = 0, \quad \text{when} \quad r = R.$$

For this to be satisfied for all values of θ we need

$$1 + AR^{-2} + 3CR^{-4} = 0, \quad 1 - BR^{-2} = 0, \quad 1 - 2AR^{-2} - 3CR^{-4} = 0. \tag{8}$$

The solution of these is

$$B = R^2, \quad A = 2R^2, \quad C = -R^4,$$

and thus finally

$$\sigma_r = \tfrac{1}{2}p_1\left(1 - \frac{R^2}{r^2}\right) + \tfrac{1}{2}p_1\left(1 - \frac{4R^2}{r^2} + \frac{3R^4}{r^4}\right) \cos 2\theta, \tag{9}$$

$$\sigma_\theta = \tfrac{1}{2}p_1\left(1 + \frac{R^2}{r^2}\right) - \tfrac{1}{2}p_1\left(1 + \frac{3R^4}{r^4}\right) \cos 2\theta, \tag{10}$$

$$\tau_{r\theta} = -\tfrac{1}{2}p_1\left(1 + \frac{2R^2}{r^2} - \frac{3R^4}{r^4}\right) \sin 2\theta. \tag{11}$$

These are the fundamental formulae. On the surface $r = R$, σ_θ varies from $-p_1$, tensile, when $\theta = 0$, to $3p_1$ when $\theta = \tfrac{1}{2}\pi$.

Since

$$\sigma_r + \sigma_\theta = p_1\left\{1 - \frac{2R^2}{r^2} \cos 2\theta\right\}, \tag{12}$$

the sum of the principal stresses is negative within the region

$$r^2 = 2R^2 \cos 2\theta \tag{13}$$

which is shown dotted in Fig. 10.4 (*a*).

The principal stresses and their directions at any point can be calculated from (9)–(11). Fig. 10.4 (*b*) show the stress trajectories which are curves whose directions at any point are along the principal axes, cf. § 2.3.

If in addition to p_1 along Ox there was a second principal stress p_2 in the perpendicular direction at infinity, corresponding terms with θ replaced by $\frac{1}{2}\pi + \theta$ would have to be added to (9)–(11) and the stresses are

$$\sigma_r = \tfrac{1}{2}(p_1 + p_2)\left(1 - \frac{R^2}{r^2}\right) + \tfrac{1}{2}(p_1 - p_2)\left(1 - \frac{4R^2}{r^2} + \frac{3R^4}{r^4}\right)\cos 2\theta, \quad (14)$$

$$\sigma_\theta = \tfrac{1}{2}(p_1 + p_2)\left(1 + \frac{R^2}{r^2}\right) - \tfrac{1}{2}(p_1 - p_2)\left(1 + \frac{3R^4}{r^4}\right)\cos 2\theta, \quad (15)$$

$$\tau_{r\theta} = -\tfrac{1}{2}(p_1 - p_2)\left(1 + \frac{2R^2}{r^2} - \frac{3R^4}{r^4}\right)\sin 2\theta. \quad (16)$$

In this case, if $p_1 > p_2$ the tangential stress in the surface varies from

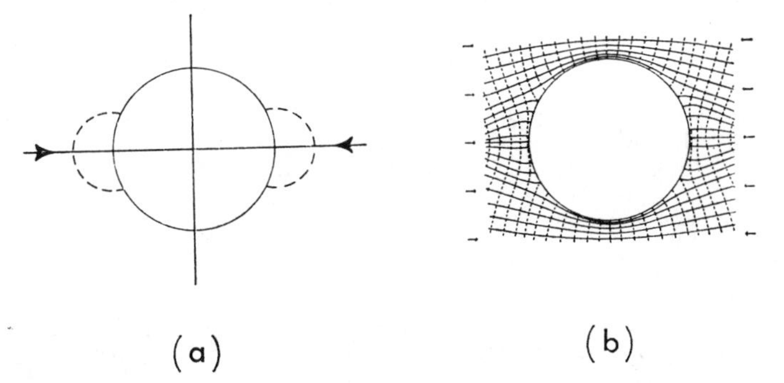

Fig. 10.4 The circular hole in material with uniaxial compressive stress at infinity. (*a*) Region in which the mean stress is negative. (*b*) Stress-trajectories.

$3p_2 - p_1$ when $\theta = 0$ to $3p_1 - p_2$ when $\theta = \frac{1}{2}\pi$. If $3p_2 > p_1$ there are no tensile stresses at any point.

If the boundary condition at $r = R$ had been that of constant pressure p, so that $\sigma_r = p$, when $r = R$, the second of equations (8) would be replaced by

$$2p = p_1(1 - BR^{-2}), \quad (17)$$

and (14) and (15) are replaced by

$$\sigma_r = \tfrac{1}{2}(p_1 + p_2)\left(1 - \frac{R^2}{r^2}\right) + \frac{pR^2}{r^2} + \tfrac{1}{2}(p_1 - p_2)\left(1 - \frac{4R^2}{r^2} + \frac{3R^4}{r^4}\right)\cos 2\theta, \quad (18)$$

$$\sigma_\theta = \tfrac{1}{2}(p_1 + p_2)\left(1 + \frac{R^2}{r^2}\right) - \frac{pR^2}{r^2} - \tfrac{1}{2}(p_1 - p_2)\left(1 + \frac{3R^4}{r^4}\right)\cos 2\theta. \quad (19)$$

These could have been found by superimposing the solution § 10.3 (19) for constant internal pressure on to the previous results (14)–(15).

It appears from (19) that the tangential stress σ_θ in the surface varies from $3p_2 - p_1 - p$ when $\theta = 0$ to $3p_1 - p_2 - p$ when $\theta = \frac{1}{2}\pi$. If

$$p > 3p_2 - p_1$$

there are tensile stresses in the surface, and radial tensile failure becomes possible. This is the simplest case of hydraulic fracturing of a borehole by internal pressure, cf. Hubbert and Willis (1957).

Considering next the displacements derived from (1), working again in polar coordinates, the radial and transverse displacements u_r and u_θ are, by § 10.2 (7), (50),

$$\begin{aligned}2G(u_r + iu_\theta) &= 2G(u + iv)e^{-i\theta} \\ &= \tfrac{1}{4}p_1\{\varkappa(re^{i\theta} + Ar^{-1}e^{-i\theta}) - re^{i\theta}(1 - Ar^{-2}e^{2i\theta}) + \\ &\qquad 2(re^{-i\theta} + Br^{-1}e^{i\theta} + Cr^{-3}e^{3i\theta})\}e^{-i\theta},\end{aligned}$$

or, using the values of A, B, C derived from (8),

$$\frac{8G}{Rp_1}(u_r + iu_\theta) = \varkappa\left(\frac{r}{R} + \frac{2R}{r}e^{-2i\theta}\right) - \frac{r}{R}\left(1 - \frac{2R^2}{r^2}e^{2i\theta}\right) + 2\left(\frac{R}{r} + \frac{r}{R}e^{-2i\theta} - \frac{R^3}{r^3}e^{2i\theta}\right). \quad (20)$$

Hence

$$\frac{8Gu_r}{Rp_1} = (\varkappa - 1 + 2\cos 2\theta)\frac{r}{R} + \frac{2R}{r}\left\{1 + \left(\varkappa + 1 - \frac{R^2}{r^2}\right)\cos 2\theta\right\}, \quad (21)$$

$$\frac{8Gu_\theta}{Rp_1} = \left[-\frac{2r}{R} + \frac{2R}{r}\left\{1 - \varkappa - \frac{R^2}{r^2}\right\}\right]\sin 2\theta. \quad (22)$$

The terms in (r/R) in (21) and (22) are the displacements found in § 10.3 (13), (14) for an infinite solid under uniaxial stress, and the other terms represent the effect of the hole.

(21) and (22) apply to plane stress and plane strain with the appropriate value of $\varkappa$. For the more general case considered in § 5.3 (vii), in which there is an axial strain ε along the hole, § 10.2 (51) has to be used, and in this case $\varkappa = 3 - 4\nu$ and a term

$$-8\nu\varepsilon Gr/Rp_1 \quad (23)$$

has to be added to the right-hand side of (20). In this case

$$\frac{8Gu_r}{Rp_1} = 2[1 - 2\nu + \cos 2\theta]\frac{r}{R} + \frac{2R}{r}\left\{1 + \left(4 - 4\nu - \frac{R^2}{r^2}\right)\cos 2\theta\right\} - \frac{8\nu\varepsilon Gr}{Rp_1}. \quad (24)$$

Since u_r and u_θ are both small, the change in radial distance is u_r and, in particular, the decrease in radius ΔR of the hole is the value of u_r when $r = R$, namely

$$\frac{u_r}{R} = \frac{\Delta R}{R} = \frac{(1-\nu)p_1}{2G}\{1 + 2\cos 2\theta\} - \nu\varepsilon$$
$$= \frac{(1-\nu^2)p_1}{E}\{1 + 2\cos 2\theta\} - \nu\varepsilon. \quad (25)$$

If there is an additional principal stress p_2 perpendicular to p_1,

$$\frac{u_r}{R} = \frac{\Delta R}{R} = \frac{(1-\nu^2)}{E}\{(p_1 + p_2) + 2(p_1 - p_2)\cos 2\theta\} - \nu\varepsilon.$$

Also, by § 5.3 (22), $E\varepsilon = p_3 - \nu(p_1 + p_2)$, where p_3 is the stress in the direction perpendicular to p_1 and p_2, so that finally

$$\frac{u_r}{R} = \frac{\Delta R}{R} = \frac{1}{E}\{p_1 + p_2 - \nu p_3 + 2(1-\nu^2)(p_1 - p_2)\cos 2\theta\}. \quad (26)$$

These formulae are fundamental both for the calculation of stresses about a circular opening and for the theory of many stress-measuring devices operated from boreholes. Since they are two-dimensional only, the question arises as to how far they are applicable to more general states of stress. The general case of the region outside a circular cylinder may be studied by using the elastic equations in cylindrical coordinates § 5.6 (11)–(13), with the z-axis along the axis of the cylinder. For a long tunnel with prescribed stresses at infinity it may be assumed that the displacements u, v are independent of z and the strain $\varepsilon_z = \partial w/\partial z$ is a constant ε so that § 5.3 (21) (22) hold. Making this assumption in § 5.6 (11)–(13) with zero body forces, § 5.6 (13) becomes

$$\frac{\partial^2 w}{\partial r^2} + \frac{1}{r}\frac{\partial w}{\partial r} + \frac{1}{r^2}\frac{\partial^2 w}{\partial \theta^2} = 0, \quad (27)$$

while § 5.6 (11), (12) are independent of w. The previous discussion of this section thus holds for displacements and stresses in the r, θ plane, and the only new feature is the variation of w derived from (27) which gives the shear stresses

$$\tau_{rz} = G\gamma_{rz} = G\frac{\partial w}{\partial r}, \quad \tau_{\theta z} = G\gamma_{\theta z} = G\frac{\partial w}{r\partial\theta}. \quad (28)$$

A general solution of (27) is

$$w = (Ar + Br^{-1})\cos\theta + (Dr + Er^{-1})\sin\theta + \varepsilon z, \quad (29)$$

where A, B, D, E are constants. It follows that

$$\tau_{rz} = G(A - Br^{-2})\cos\theta + G(D - Er^{-2})\sin\theta. \quad (30)$$

Since $\tau_{rz} = 0$ at the free surface $r = R$ we must have $B = AR^2$, $E = DR^2$, and the complete solution is

$$w = A(r + R^2r^{-1}) \cos \theta + D(r + R^2r^{-1}) \sin \theta + \varepsilon z, \tag{31}$$

$$\tau_{rz} = GA(1 - R^2r^{-2}) \cos \theta + GD(1 - R^2r^{-2}) \sin \theta, \tag{32}$$

$$\tau_{\theta z} = -GA(1 + R^2r^{-2}) \sin \theta + GD(1 + R^2r^{-2}) \cos \theta. \tag{33}$$

If the shear stresses at infinity are known, A and D can be found from them and the solution is complete.

If the axes in the r, θ plane are those of subsidiary principal stresses p_1 and p_2, σ_r and σ_θ are given by (18) and (19), u_r and u_θ by (21) and (22) with $\varkappa = 3 - 4\upsilon$, and, using (14), (15) and § 5.3 (22),

$$\sigma_z = \upsilon(p_1 + p_2) - 2\upsilon(p_1 - p_2)R^2r^{-2} \cos 2\theta + E\varepsilon. \tag{34}$$

10.5 Solutions in the form of infinite series

In problems involving circular boundaries, the normal and tangential stresses applied to the boundary can be represented by Fourier series in the angle θ, that is,

$$f(\theta) = a_0 + \sum_{n=1}^{\infty} (a_n \cos n\theta + b_n \sin n\theta), \tag{1}$$

where $-\pi < \theta < \pi$, and

$$a_0 = \frac{1}{2\pi}\int_{-\pi}^{\pi} f(\theta)\, d\theta, \quad a_n = \frac{1}{\pi}\int_{-\pi}^{\pi} f(\theta) \cos n\theta\, d\theta, \quad b_n = \frac{1}{\pi}\int_{-\pi}^{\pi} f(\theta) \sin n\theta\, d\theta. \tag{2}$$

Using the exponential forms of the sine and cosine, (1) may be written in the form

$$f(\theta) = \sum_{n=-\infty}^{\infty} \alpha_n e^{in\theta}, \tag{3}$$

where $\alpha_n = \frac{1}{2}(a_n - ib_n)$ if $n > 0$, $\alpha_0 = a_0$, and $\alpha_n = \frac{1}{2}(a_n + ib_n)$ if $n < 0$, so the α_n are complex.

For problems on the region $r > R$ with prescribed normal and tangential stresses N and T at the surface $r = R$, it may be assumed that the combination $N - iT$ of these which appears in the boundary condition § 10.2 (4) may be represented by the complex Fourier series

$$N - iT = \sum_{n=-\infty}^{\infty} A_n e^{in\theta}. \tag{4}$$

If the calculation of stresses only is in question, only $\phi'(z)$, $\phi''(z)$ and $\psi'(z)$ appear in the formulae § 10.2 (48), (49), and the most general forms which can be assumed for them which give stresses finite at infinity are the power series in z^{-1}

$$\phi'(z) = \sum_{n=0}^{\infty} c_n z^{-n}, \quad \psi'(z) = \sum_{n=0}^{\infty} d_n z^{-n} \tag{5}$$

The boundary condition at $r = R$, § 10.2 (4), now gives, using § 10.2 (48), (49),

$$\sum_{n=-\infty}^{\infty} A_n e^{in\theta} = \phi'(z) + \overline{\phi'(z)} - [\bar{z}\phi''(z) + \psi'(z)]e^{2i\theta}$$

$$= \sum_{n=0}^{\infty} c_n R^{-n} e^{-in\theta} + \sum_{n=0}^{\infty} \bar{c}_n R^{-n} e^{in\theta} + \sum_{n=1}^{\infty} n c_n R^{-n} e^{-in\theta} - \sum_{n=0}^{\infty} d_n R^{-n} e^{-i(n-2)\theta}$$

$$= \sum_{n=0}^{\infty} (n+1) c_n R^{-n} e^{-in\theta} + \sum_{n=0}^{\infty} \bar{c}_n R^{-n} e^{in\theta} - \sum_{n=0}^{\infty} d_n R^{-n} e^{(2-n)i\theta}. \quad (6)$$

Equating coefficients of $e^{in\theta}$ for all n gives

$$A_n = \bar{c}_n R^{-n}, \quad n \geqslant 3, \quad (7)$$

$$A_2 = \bar{c}_2 R^{-2} - d_0, \quad A_1 = \bar{c}_1 R^{-1} - d_1 R^{-1}, \quad A_0 = c_0 + \bar{c}_0 - d_2 R^{-2}, \quad (8)$$

$$A_{-n} = (n+1) c_n R^{-n} - d_{n+2} R^{-n-2}, \quad n \geqslant 1. \quad (9)$$

In addition to this, there is the general requirement discussed in § 10.3 (iv) which requires that displacements be single valued. This requires

$$c_1 = d_1 = 0. \quad (10)$$

If the problem of § 10.4 is treated in this way, the fact that there is no stress on the boundary $r = R$ gives $A_n = 0$ for all n. Next, uniaxial stress p_1 at infinity requires as in § 10.3 (7)

$$c_0 = \tfrac{1}{4} p_1, \quad d_0 = -\tfrac{1}{2} p_1. \quad (11)$$

Also, since c_0 only appears in the form $c_0 + \bar{c}_0$ we take $\bar{c}_0 = c_0 = \frac{1}{4} p_1$. (7) to (9) then give

$$\bar{c}_2 = -\tfrac{1}{2} R^2 p_1, \quad d_2 = \tfrac{1}{2} R^2 p_1, \quad d_4 = -\tfrac{3}{2} R^4 p_1, \quad (12)$$

and all the other c_n and d_n vanish. This is the form assumed in § 10.4 (2), and we have justified this assumption and derived § 10.4 (9)–(11) in a more general way.

10.6 Stress applied to the surface of a circular hole

The general solution can be written down from the equations § 10.5 (5)–(10) derived in the previous section, and then important special cases can be discussed.

The solution of § 10.5 (7)–(9) is

$$\left. \begin{aligned} &c_n = \bar{A}_n R^n, \ n \geqslant 3; \quad c_2 = (\bar{A}_2 + \bar{d}_0) R^2; \quad \bar{c}_1 - d_1 = A_1 R, \\ &d_2 = (c_0 + \bar{c}_0 - A_0) R^2; \quad d_{n+2} = (n+1) c_n R^2 - A_{-n} R^{n+2}, \ n \geqslant 1. \end{aligned} \right\} \quad (1)$$

There will also be other conditions, for example, c_0 and d_0 are given from the conditions at infinity as in § 10.3 (7), and c_1 and d_1 vanish because of single-valuedness of displacements as in § 10.3 (iv).

Consider the case of uniform radial pressure p applied over the arcs $-\alpha < \theta < \alpha$ and $\pi - \alpha < \theta < \pi + \alpha$ of Fig. 10.6. The Fourier coefficients § 10.5 (2) for this distribution of stress are

$$a_0 = 2\alpha p/\pi; \quad a_n = (2p/n\pi)[1 + (-1)^n] \sin n\alpha, \; n > 0; \quad b_n = 0. \quad (2)$$

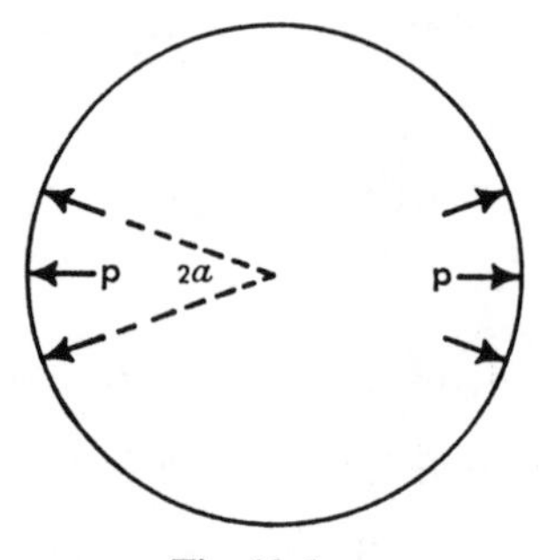

Fig. 10.6

Therefore in § 10.5 (4)

$$A_0 = \frac{2\alpha p}{\pi}; \quad A_{2m} = A_{-2m} = \frac{p}{m\pi} \sin 2m\alpha, \; m = 1, 2, \ldots,$$

$$A_{2m+1} = A_{-(2m+1)} = 0. \quad (3)$$

Also, if there is no stress at infinity, it follows from § 10.3 (7) that

$$c_0 = d_0 = 0. \quad (4)$$

Using the values (3) in (1) gives

$$c_{2m} = (pR^{2m}/m\pi) \sin 2m\alpha, \; m \geqslant 1, \quad (5)$$

$$d_2 = -2\alpha pR^2/\pi; \quad d_{2m+2} = (2pR^{2m+2}/\pi) \sin 2m\alpha, \; m \geqslant 1. \quad (6)$$

Using these results in § 10.5 (5) gives finally

$$\phi'(z) = \frac{p}{\pi} \sum_{m=1}^{\infty} \frac{\sin 2m\alpha}{m} \left(\frac{R}{z}\right)^{2m}, \quad (7)$$

$$\psi'(z) = -\frac{2\alpha pR^2}{\pi z^2} + \frac{2p}{\pi} \sum_{m=1}^{\infty} \left(\frac{R}{z}\right)^{2m+2} \sin 2m\alpha. \quad (8)$$

Then from § 10.2 (48)

$$\sigma_r + \sigma_\theta = \frac{4p}{\pi} \sum_{m=1}^{\infty} \frac{1}{m} \left(\frac{R}{r}\right)^{2m} \sin 2m\alpha \cos 2m\theta. \quad (9)$$

And from § 10.2 (49) and § 10.2 (6),

$$\sigma_\theta - \sigma_r + 2i\tau_{r\theta} = -\frac{4p\alpha R^2}{\pi r^2} + \frac{4p}{\pi}\sum_{m=1}^{\infty}\left(\frac{R}{r}\right)^{2m}\left[\frac{R^2}{r^2} - 1\right]e^{-2mi\theta}\sin 2m\alpha. \quad (10)$$

Hence the stresses are determined. It follows from (9) and (10) that at the surface $r = R$

$$\sigma_r = \frac{2p\alpha}{\pi} + \frac{2p}{\pi}\sum_{m=1}^{\infty}\frac{1}{m}\sin 2m\alpha\cos 2m\theta, \quad (11)$$

$$\sigma_\theta = -\frac{2p\alpha}{\pi} + \frac{2p}{\pi}\sum_{m=1}^{\infty}\frac{1}{m}\sin 2m\alpha\cos 2m\theta. \quad (12)$$

Now the Fourier series (2) for the load implies that

$$\left.\begin{aligned} &\frac{2p\alpha}{\pi} + \frac{2p}{\pi}\sum_{m=1}^{\infty}\frac{1}{m}\sin 2m\alpha\cos 2m\theta \\ &\qquad = p, \quad \text{if} \quad -\alpha < \theta < \alpha \quad \text{or} \quad \pi + \alpha > \theta > \pi - \alpha, \\ &\qquad = 0, \quad \text{if} \quad \alpha < \theta < \pi - \alpha \quad \text{or} \quad \pi + \alpha < \theta < 2\pi - \alpha. \end{aligned}\right\} \quad (13)$$

Using this result in (11) verifies that σ_r has the value p in the loaded areas, and is zero in the unloaded areas. Also for σ_θ, from (12),

$$\left.\begin{aligned} \sigma_\theta &= p - (4p\alpha/\pi), \quad -\alpha < \theta < \alpha, \quad \pi - \alpha < \theta < \pi + \alpha, \\ &= -4p\alpha/\pi, \quad \alpha < \theta < \pi - \alpha, \quad \pi + \alpha < \theta < 2\pi - \alpha. \end{aligned}\right\} \quad (14)$$

It follows that there is a constant tensile stress $(-4p\alpha/\pi)$ in the surface outside the loaded region.

The displacements can be calculated from § 10.2 (7), (50). These give, for the radial displacement of the surface $r = R$,

$$\frac{2Gu_r}{R} = -\frac{2\alpha p}{\pi} - \frac{p}{\pi}\sum_{m=1}^{\infty}\frac{1}{m}\left\{\frac{\varkappa}{2m-1} + \frac{1}{2m+1}\right\}\sin 2m\alpha\cos 2m\theta. \quad (15)$$

If $\theta = 0$ the series can be summed using results in Bromwich (1926, § 121) and the result is

$$\begin{aligned} \frac{2\pi G u_r}{Rp} = -2\alpha - \varkappa[\tfrac{1}{2}\pi\cos\alpha + \sin\alpha\ln\cot\tfrac{1}{2}\alpha - \tfrac{1}{2}\pi + \alpha] - \\ [\tfrac{1}{2}\pi - \alpha - \tfrac{1}{2}\pi\cos\alpha + \sin\alpha\ln\cot\tfrac{1}{2}\alpha]. \quad (16) \end{aligned}$$

If $\theta = \frac{1}{2}\pi$, in the same way,

$$\frac{2\pi G u_r}{Rp} = (\varkappa + 1)\{\cos\alpha\ln(\sec\alpha + \tan\alpha) - \alpha\} + \tfrac{1}{2}\pi(\varkappa - 1)\sin\alpha. \quad (17)$$

These results are of interest. (16) and (17) imply that if a borehole is stressed in this way measurement of the displacement either along or perpendicular to the direction of the load will give information about the elastic moduli. (14) implies that if a borehole is stressed in this way failure will take place in the direction of the greatest principal stress at infinity, provided this direction does not intersect the stressed portion of the surface. More elaborate applications are given by Jaeger and Cook (1964).

10.7 Stress applied to the surface of a solid cylinder

Problems on the region $r < R$ with prescribed stresses N and T applied at the surface $r = R$ may be treated in the same way. Since the region includes the origin, only positive powers of z may be included in $\phi'(z)$ and $\psi'(z)$, so that, corresponding to § 10.5 (5), the assumption now is

$$\phi'(z) = \sum_{n=0}^{\infty} c_n z^n, \quad \psi'(z) = \sum_{n=0}^{\infty} d_n z^n. \tag{1}$$

Consider the case, Fig. 10.7 (a), of radial pressure p applied to the surface over the region $-\alpha < \theta < \alpha$, $\pi - \alpha < \theta < \pi + \alpha$, so that

$$N - iT = A_0 + \sum_{m=-\infty}^{\infty} A_{2m} e^{2im\theta}, \tag{2}$$

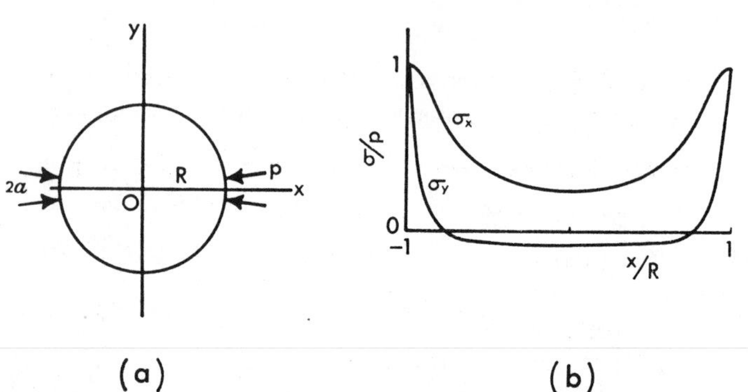

Fig. 10.7 (a) Diametral compression of a circular cylinder over an angular width of 2α. (b) Variation of the stress σ_x and σ_y on the loaded diameter for $\alpha = 7{\cdot}5°$.

with A_0 and A_{2m} given by § 10.6 (3). Proceeding as in §§ 10.5, 10.6 it is found that

$$\phi'(z) = \frac{\alpha p}{\pi} + \frac{p}{\pi} \sum_{m=1}^{\infty} \frac{1}{m} \left(\frac{z}{R}\right)^{2m} \sin 2m\alpha, \tag{3}$$

$$\psi'(z) = -\frac{2p}{\pi} \sum_{m=0}^{\infty} \left(\frac{z}{R}\right)^{2m} \sin 2(m+1)\alpha. \tag{4}$$

It follows in the usual way from § 10.2 (6), (48), (49) that the stresses in polar coordinates are

$$\sigma_r = \frac{2\alpha p}{\pi} + \frac{2p}{\pi} \sum_{m=1}^{\infty} \left(\frac{r}{R}\right)^{2m-2} \left\{1 - \left(1 - \frac{1}{m}\right)\left(\frac{r}{R}\right)^2\right\} \sin 2m\alpha \cos 2m\theta, \tag{5}$$

$$\sigma_\theta = \frac{2\alpha p}{\pi} - \frac{2p}{\pi} \sum_{m=1}^{\infty} \left(\frac{r}{R}\right)^{2m-2} \left\{1 - \left(1 + \frac{1}{m}\right)\left(\frac{r}{R}\right)^2\right\} \sin 2m\alpha \cos 2m\theta, \tag{6}$$

$$\tau_{r\theta} = \frac{2p}{\pi} \sum_{m=1}^{\infty} \left\{\left(\frac{r}{R}\right)^{2m} - \left(\frac{r}{R}\right)^{2m-2}\right\} \sin 2m\alpha \sin 2m\theta. \tag{7}$$

These results were derived by Hondros (1959) by another method. If $\theta = 0$, the series can be summed, and he finds that

$$(\sigma_r)_{\theta=0} = \frac{2p}{\pi}\left\{\frac{(1-\rho^2)\sin 2\alpha}{(1-2\rho^2\cos 2\alpha+\rho^4)} + \tan^{-1}\left[\frac{(1+\rho^2)}{(1-\rho^2)}\tan\alpha\right]\right\}, \quad (8)$$

$$(\sigma_\theta)_{\theta=0} = -\frac{2p}{\pi}\left\{\frac{(1-\rho^2)\sin 2\alpha}{(1-2\rho^2\cos 2\alpha+\rho^4)} - \tan^{-1}\left[\frac{(1+\rho^2)}{(1-\rho^2)}\tan\alpha\right]\right\}, \quad (9)$$

where $\rho = r/R$. Also on the vertical diameter $\theta = \pi/2$ (10)

$$(\sigma_r)_{\theta=\frac{1}{2}\pi} = -\frac{2p}{\pi}\left\{\frac{(1-\rho^2)\sin 2\alpha}{(1+2\rho^2\cos 2\alpha+\rho^4)} - \tan^{-1}\left[\frac{(1-\rho^2)}{(1+\rho^2)}\tan\alpha\right]\right\}, \quad (11)$$

$$(\sigma_\theta)_{\theta=\frac{1}{2}\pi} = \frac{2p}{\pi}\left\{\frac{(1-\rho^2)\sin 2\alpha}{(1+2\rho^2\cos 2\alpha+\rho^4)} + \tan^{-1}\left[\frac{(1-\rho^2)}{(1+\rho^2)}\tan\alpha\right]\right\}. \quad (12)$$

If α is small, that is, for line loads $W = 2p\alpha R$ per unit length of the cylinder applied at $\theta = 0$ and $\theta = \pi$, (8) and (9) become

$$(\sigma_r)_{\theta=0} = \frac{W(3+\rho^2)}{\pi R(1-\rho^2)}, \quad (13)$$

$$(\sigma_\theta)_{\theta=0} = -\frac{W}{\pi R}. \quad (14)$$

It appears that in this case there is a constant tensile stress $-W/\pi R$ across the loaded diameter, while the stress in this diametral plane increases from $3W/\pi$ at the centre to infinity at the loaded points. For finite values of α the infinite stress-concentration is removed, the stresses for $2\alpha = 15°$ being shown in Fig. 10.7 (*b*). The stress-distribution in general is discussed by Jaeger and Hoskins (1966b). This theory is that of the well-known Brazilian or indirect tensile test which will be discussed further in § 6.11.

Stresses of this type may readily be applied by curved jacks, Jaeger and Cook (1964). If $2\alpha = 45°$, the variation of stress with radius at the centre is slow and any combination of stresses is readily obtained by the use of two pairs of such 'quadrantal' jacks. This system may be used with strain gauges for the measurement of elastic moduli of cores.

Displacements can be found from (3) and (4) and § 10.2 (50), and the radial displacement u_r at $r = R$ is found to be

$$\frac{2\pi G u_r}{Rp} = \alpha(\varkappa - 1) + \sum_{m=1}^{\infty}\frac{1}{m}\left\{\frac{\varkappa}{2m+1} + \frac{1}{2m-1}\right\}\cos 2m\theta \sin 2m\alpha. \quad (15)$$

As in § 10.6, this series can be summed if $\theta = 0$ or $\theta = \frac{1}{2}\pi$. If $\theta = \frac{1}{2}\pi$ and α is small, the displacement u_r is

$$u_r = -[2(\varkappa+1) - \pi(\varkappa-1)]\alpha p R/4\pi G. \quad (16)$$

This result, also, may be used for measurement of the elastic modulus.

The problem of the circular cylinder with several line loadings has been considered by many authors, Timoshenko and Goodier (1951), Michell (1900, 1902), Jaeger (1967a).

10.8 The circular ring stressed over parts of its outer or inner surfaces

In this case we consider the cylindrical region $R_1 < r < R_2$ and suppose that the stresses at its inner and outer surfaces are given by Fourier series

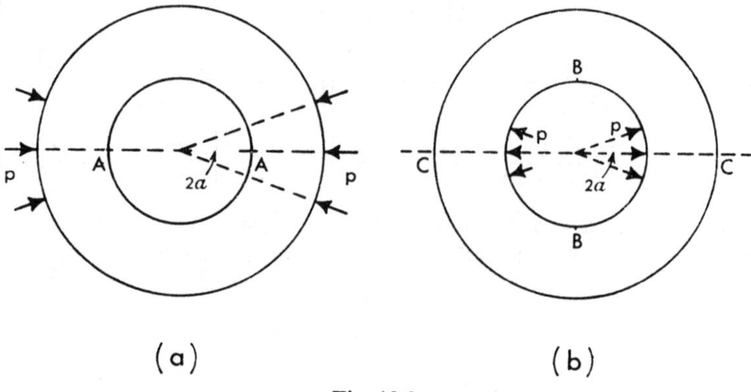

Fig. 10.8

of type § 10.5 (4). Since the region in question now extends neither to infinity nor to the origin, both positive and negative powers of z may be included in $\phi'(z)$ and $\psi'(z)$.

The problem has been discussed, using the present methods, by Muskhelishvili (1953), and also by Filon (1924), Ripperger and Davids (1947), and Jaeger and Hoskins (1966a).

The most interesting cases are the generalization to this system of the problems of §§ 10.6, 10.7, namely, radial pressure p applied over the arcs

Table 10.8
Stress concentration factors for $\alpha = 7{\cdot}5°$

ρ	0	0·1	0·2	0·3	0·4	0·5	0·6	0·7	0·8	0·9
$f(\rho)$	0·492	0·524	0·622	0·804	1·112	1·643	2·636	4·799	10·98	44·3
$g(\rho)$		0·182	0·231	0·323	0·482	0·761	1·296	2·504	6·10	26·3
$h(\rho)$		0·009	0·039	0·105	0·241	0·522	1·136	2·671	7·62	37·4

$-\alpha < \theta < \alpha$, $\pi - \alpha < \theta < \pi + \alpha$, either externally as in Fig. 10.8 (*a*) or internally as in Fig. 10.8 (*b*). Formulae and numerical values for these two cases are given by Jaeger and Hoskins (1966a). For the external case, Fig. 10.8 (*a*), the maximum tensile stress is attained at the points A, A on the loaded diameter and may be written $pf(\rho)$, where $\rho = R_1/R_2$, and $f(\rho)$ is given in Table 10.8. For the internal case, Fig. 10.8 (*b*), the maximum tensile stress is usually attained at points B, B on a diameter perpendicular to the loaded one and is $pg(\rho)$ where $g(\rho)$ is given in Table 10.8. However, there are also tensile stresses on portion of the loaded diameter which have a maximum value $ph(\rho)$ at the points CC. $h(\rho)$ becomes greater than $g(\rho)$ for thin rings for which $\rho > 0{\cdot}6$. $f(\rho)$, $g(\rho)$, and $h(\rho)$ in the table are calculated for the case $\alpha = 7{\cdot}5°$.

The effects of eccentricity of the central hole are discussed by Hobbs (1965). The use of hollow cylinders stressed as in Fig. 10.8 (*a*) for measurement of tensile strength has been proposed by Hobbs (1965) and Ripperger and Davids (1947) and is discussed further in § 6.11.

10.9 Inclusions in an infinite region

Suppose that the region $r > R$ consists of one material of elastic properties G, ν, and that the region $r < R$ contains a different material of properties G_0, ν_0. The materials are supposed to be firmly attached ('welded') at the boundary $r = R$, so that the boundary conditions there are continuity of normal and shear stress and of displacement, that is, using a suffix 0 for the interior,

$$\sigma_r = \sigma_{r0}, \quad \tau_{r\theta} = \tau_{r\theta 0}, \quad u_r = u_{r0}, \quad v_r = v_{r0}, \quad \text{when} \quad r = R. \tag{1}$$

The stress at infinity will be taken to be p_1 in the x-direction.

Following the discussion of § 10.4 for the region outside a circular hole, we assume for the region $r > R$ as in § 10.4 (1)

$$\phi(z) = \tfrac{1}{4}p_1\left(z + \frac{AR^2}{z}\right), \quad \psi(z) = -\tfrac{1}{2}p_1\left(z + \frac{BR^2}{z} + \frac{CR^4}{z^3}\right), \tag{2}$$

where A, B, C are real constants to be determined, and for large values of z the stress is p_1 in the x-direction, as in § 10.3 (7).

In the region $r < R$ only positive powers of z can appear, since the solution must be finite at $r = 0$. We assume for this region

$$\phi_0(z) = \tfrac{1}{4}p_1\left(A_0 z + \frac{B_0 z^3}{R^2}\right), \quad \psi_0(z) = -\tfrac{1}{2}p_1 C_0 z, \tag{3}$$

since these will be found to be the only terms which give rise to cos 2θ and sin 2θ in the solution. As in §§ 10.4, 10.5, if complete infinite series were assumed, all other terms would be found to vanish.

The quantities A, B, C, A_0, B_0, C_0 are now found from (1).

The displacements at $r = R$ derived from (2) are

$$4G(u_r + iu_\theta)/p_1R = \tfrac{1}{2}(\varkappa - 1) + B + (\tfrac{1}{2}\varkappa A + 1)e^{-2i\theta} + (\tfrac{1}{2}A + C)e^{2i\theta}, \tag{4}$$

while those derived from (3) are

$$4G_0(u_{r0} + iu_{\theta 0})/p_1R = \tfrac{1}{2}A_0(\varkappa_0 - 1) + \tfrac{1}{2}\varkappa_0 B_0 e^{2i\theta} + (C_0 - \tfrac{3}{2}B_0)e^{-2i\theta}. \tag{5}$$

Equating coefficients and writing

$$k = G_0/G, \tag{6}$$

gives

$$k[\varkappa - 1 + 2B] = A_0(\varkappa_0 - 1); \quad k(\varkappa A + 2) = 2C_0 - 3B_0;$$
$$k(A + 2C) = \varkappa_0 B_0. \tag{7}$$

The normal and tangential stresses across $r = R$ are given by § 10.2 (4), (48), (49),

$$(N - iT) = \phi'(z) + \overline{\phi'(z)} - [\bar{z}\phi''(z) + \psi'(z)]e^{2i\theta}.$$

Calculated from (2) they are

$$2(N - iT)/p_1 = 1 - B - (\tfrac{3}{2}A + 3C)e^{-2i\theta} + (1 - \tfrac{1}{2}A)e^{2i\theta}, \tag{8}$$

and calculated from (3)

$$2(N_0 - iT_0)/p_1 = A_0 + (C_0 - \tfrac{3}{2}B_0)e^{2i\theta} + \tfrac{3}{2}B_0 e^{-2i\theta}. \tag{9}$$

To secure continuity of normal and tangential stresses at $r = R$ we equate coefficients between (8) and (9), which gives

$$A_0 = 1 - B, \quad B_0 = -A - 2C, \quad 2C_0 - 3B_0 = 2 - A. \tag{10}$$

It follows from the determinental solution of (7) and (10) that

$$B_0 = 0, \tag{11}$$

and using this result the other constants are easily found. They are

$$A_0 = k(\varkappa + 1)/(2k + \varkappa_0 - 1), \quad C_0 = k(\varkappa + 1)/(k\varkappa + 1), \tag{12}$$

$$A = 2(1 - k)/(k\varkappa + 1), \quad B = [\varkappa_0 - 1 - k(\varkappa - 1)]/(2k + \varkappa_0 - 1),$$
$$C = (k - 1)/(k\varkappa + 1). \tag{13}$$

For the region $r < R$ we have now

$$\phi_0(z) = \tfrac{1}{4}p_1 A_0 z, \quad \psi_0(z) = -\tfrac{1}{2}p_1 C_0 z. \tag{14}$$

Therefore by § 10.2 (48), (49),

$$\left.\begin{aligned} \sigma_x + \sigma_y &= A_0 p_1 \\ \sigma_y - \sigma_x + 2i\tau_{xy} &= -C_0 p_1. \end{aligned}\right\}$$

Thus the stress in the interior $r < R$ is homogeneous and

$$\sigma_x = \tfrac{1}{2}(A_0 + C_0)p_1, \quad \sigma_y = \tfrac{1}{2}(A_0 - C_0)p_1. \tag{15}$$

If the principal stresses at infinity are σ_1 along Ox and σ_2 along Oy the principal stresses σ_{10}, σ_{20} in the region $r < R$ are, from (12) and (13),

$$\sigma_{10} = \frac{\{k(\varkappa + 2) + \varkappa_0\}k(\varkappa + 1)}{2(2k + \varkappa_0 - 1)(k\varkappa + 1)}\sigma_1 + \frac{[k(\varkappa - 2) - \varkappa_0 + 2]k(\varkappa + 1)}{2(2k + \varkappa_0 - 1)(k\varkappa + 1)}\sigma_2, \tag{16}$$

$$\sigma_{20} = \frac{[k(\varkappa - 2) - \varkappa_0 + 2]k(\varkappa + 1)}{2(2k + \varkappa_0 - 1)(k\varkappa + 1)}\sigma_1 + \frac{[k(\varkappa + 2) + \varkappa_0]k(\varkappa + 1)}{2(2k + \varkappa_0 - 1)(k\varkappa + 1)}\sigma_2. \quad (17)$$

Solving for σ_1 and σ_2 they become

$$\sigma_1 = \frac{k(\varkappa + 2) + \varkappa_0}{2k(\varkappa + 1)}\sigma_{10} + \frac{\varkappa_0 - 2 - k(\varkappa - 2)}{2k(\varkappa + 1)}\sigma_{20}, \quad (18)$$

$$\sigma_2 = \frac{\varkappa_0 - 2 - k(\varkappa - 2)}{2k(\varkappa + 1)}\sigma_{10} + \frac{k(\varkappa + 2) + \varkappa_0}{2k(\varkappa + 1)}\sigma_{20}. \quad (19)$$

In the region $r > R$ with principal stresses σ_1, σ_2 at infinity, we find as in § 10.4

$$\sigma_r = \tfrac{1}{2}(\sigma_1 + \sigma_2)\left(1 - \frac{BR^2}{r^2}\right) + \tfrac{1}{2}(\sigma_1 - \sigma_2)\left\{1 - \frac{2AR^2}{r^2} - \frac{3CR^4}{r^4}\right\}\cos 2\theta, \quad (20)$$

$$\sigma_\theta = \tfrac{1}{2}(\sigma_1 + \sigma_2)\left(1 + \frac{BR^2}{r^2}\right) - \tfrac{1}{2}(\sigma_1 - \sigma_2)\left\{1 - \frac{3CR^4}{r^4}\right\}\cos 2\theta, \quad (21)$$

$$\tau_{r\theta} = -\tfrac{1}{2}(\sigma_1 - \sigma_2)\left\{1 + \frac{AR^2}{r^2} + \frac{3CR^4}{r^4}\right\}\sin 2\theta, \quad (22)$$

where A, B, C are given by (13).

These equations are derived by Muskhelishvili (1953). They have also been derived or quoted by other authors, Goodier (1933), Sezawa and

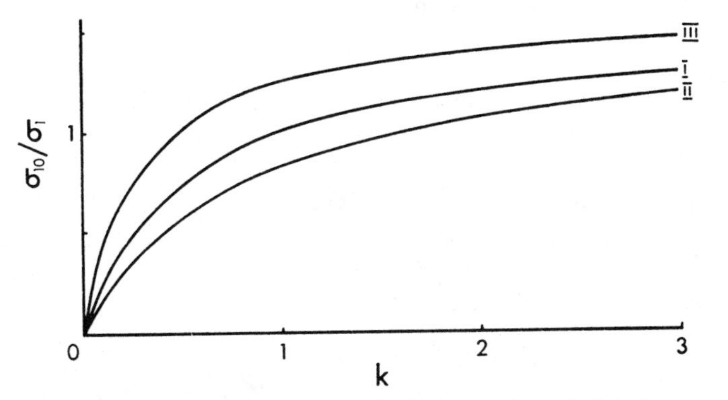

Fig. 10.9 Ratio σ_{10}/σ_1 of the stress σ_{10} in an inclusion to the uniaxial stress σ_1 at a distance as a function of $k = G_0/G$. Curve I: $\nu = \nu_0 = 0{\cdot}25$. Curve II: $\nu_0 = 0{\cdot}1$, $\nu = 0{\cdot}4$. Curve III: $\nu_0 = 0{\cdot}4$, $\nu = 0{\cdot}1$.

Nishimura (1931), whose results have been used by Continho (1949) and Wilson (1961). Sezawa (1933) gave simplified formulae for plane strain and the case $\nu = \nu_0 = \frac{1}{4}$. Muskhelishvili (1953) also discusses the case in which the material in the region $r < R$ is merely inserted into the hole $r = R$ with normal stress but no friction across the boundary. He also considers the case in which the insertion is originally slightly larger than the hole.

The results (16), (17) that the stress in the inclusion is homogeneous are of great practical interest and enable σ_1 and σ_2 at infinity to be determined from measurements of σ_{10} and σ_{20}. For the simplest case, $\sigma_2 = 0$ and plane strain with $\nu = \nu_0 = \frac{1}{4}$, so that $\varkappa = \varkappa_0 = 2$, (16) gives

$$\frac{\sigma_{10}}{\sigma_1} = \frac{3k}{2k+1}, \tag{23}$$

which tends to 3/2 as $k = G_0/G \to \infty$ and to zero as $k \to 0$. Values of σ_{10}/σ_1 for plane strain and various values of ν are shown in Fig. 10.9.

The theory of this case has been given in detail to illustrate the principles involved. Many calculations on the effects of rigid and elastic circular and elliptical inclusions will be found in Muskhelishvili (1953) and Savin (1961). The elliptic inclusion is discussed by Donnell (1941). Taylor (1934) studies the interesting case of an elliptical inclusion with zero rigidity. Edwards (1951) gives numerical results for spheroidal inclusions which are useful in the present context. Goodier (1933) gives formulae for spherical and cylindrical inclusions, and Eshelby (1957) for ellipsoidal ones. The effect of two rigid spherical inclusions is studied by Shelley and Yi-Yuan Yu (1966) and the combination of a circular inclusion and a circular hole by Bharvaga and Kapoor (1966).

10.10 Curvilinear coordinates

Suppose that the transformation

$$z = \omega(\zeta), \tag{1}$$

where $\omega(\zeta)$ is an analytic function of ζ connects the position (ξ, η) of a point $\zeta = \xi + i\eta$ in the ζ-plane with the position (x, y) of a point $z = x + iy$ in the z-plane.

The ζ-plane is shown in Fig. 10.10 (*a*) and the z-plane in Fig. 10.10 (*b*). The point P', (ξ_0, η_0) in the ζ-plane will correspond to P in the z-plane. The line $\eta = \eta_0$, constant, in the ζ-plane will correspond to a curve PA in the z-plane: the slope of this curve is obtained from

$$dx + i\,dy = dz = \omega'(\zeta)\,d\zeta = \omega'(\zeta)[d\xi + i\,d\eta] \tag{2}$$

$$= Me^{i\delta}[d\xi + i\,d\eta], \tag{3}$$

where M and δ are the modulus and argument of $\omega'(\zeta)$. δ is most conveniently expressed in the form

$$e^{2i\delta} = \omega'(\zeta)/\overline{\omega'(\zeta)}. \tag{4}$$

If $d\eta = 0$ in (3), the slope of the tangent $P\xi$ to the curve in the z-plane corresponding to $P'A'$ in the direction of ξ increasing is by (3)

$$\frac{dy}{dx} = \tan\delta. \tag{5}$$

Similarly, $\xi = \xi_0$, constant, corresponds to a curve in the z-plane and the tangent $P\eta$ to this at P has slope

$$\frac{dy}{dx} = -\cot \delta, \tag{6}$$

which is found from (3) with $d\xi = 0$. Thus the curves in the z-plane corresponding to $\xi = \xi_0$, constant, and $\eta = \eta_0$, constant, cut orthogonally at all points. ξ and η are thus called orthogonal curvilinear coordinates.

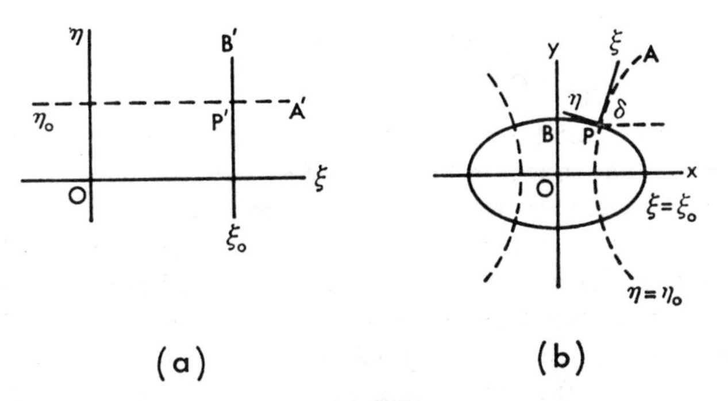

Fig. 10.10

The formulae § 10.2 (1), (2), (7), (48), (49), (50) may now be used to write down results relative to curvilinear coordinates. They are

$$\sigma_\xi + \sigma_\eta = \sigma_x + \sigma_y = 2[\phi'(z) + \overline{\phi'(z)}], \tag{7}$$

$$\begin{aligned} \sigma_\eta - \sigma_\xi + 2i\tau_{\xi\eta} &= (\sigma_y - \sigma_x + 2i\tau_{xy})e^{2i\delta}, \\ &= 2[\bar{z}\phi''(z) + \psi'(z)][\omega'(\zeta)/\overline{\omega'(\zeta)}], \end{aligned} \tag{8}$$

$$\begin{aligned} 2G(u_\xi + iu_\eta) &= 2G(u + iv)e^{-i\delta} \\ &= [\varkappa\phi(z) - z\overline{\phi'(z)} - \overline{\psi(z)}][\overline{\omega'(\zeta)}/\omega'(\zeta)]^{\frac{1}{2}}, \end{aligned} \tag{9}$$

using (4). In these $\phi(z)$, $\phi'(z)$, etc., will be expressed as functions of ζ using (1), but it has to be remembered that the differentiations are with respect to z, so that, for example,

$$\phi'(z) = \frac{d\phi}{d\zeta}\frac{d\zeta}{dz} = \frac{1}{\omega'(\zeta)}\frac{d\phi}{d\zeta}. \tag{10}$$

As an example, consider the case of elliptic coordinates. In this case the transformation (1) is

$$z = x + iy = c \cosh \zeta = c \cosh (\xi + i\eta), \tag{11}$$

so that

$$x = c \cosh \xi \cos \eta, \quad y = c \sinh \xi \sin \eta. \tag{12}$$

If ξ has the constant value ξ_0 the corresponding curve in the x, y plane is the ellipse

$$\frac{x^2}{c^2 \cosh^2 \xi_0} + \frac{y^2}{c^2 \sinh^2 \xi_0} = 1 \tag{13}$$

of semi-axes

$$a = c \cosh \xi_0, \quad b = c \sinh \xi_0. \tag{14}$$

If $\xi_0 = 0$ this becomes the slit from $x = -c$ to $x = +c$.

Similarly, the curves $\eta = \eta_0$, constant, are the hyperbolae

$$\frac{x^2}{c^2 \cos^2 \eta_0} - \frac{y^2}{c^2 \sin^2 \eta_0} = 1. \tag{15}$$

The ellipses (13) and hyperbolae (15) are confocal with foci at $x = \pm c$. The system is that shown in Fig. 10.10 (*b*).

The other system of orthogonal curvilinear coordinates of great importance in the present context is that of bipolar coordinates which arise in the solution of problems relating to two circles, Jeffery (1921).

There are now two possible methods of approach to the solution of problems involving regions of simple shapes. The first is to regard the problem as one in the appropriate curvilinear coordinates and to work directly in these: this method will be used for the elliptic hole in § 10.11. The second method is to use the transformation (1) to transform the given region in, say, the ζ-plane to a simpler one in the z-plane for which a solution is known. This method will be illustrated in § 10.15.

10.11 The infinite region with an elliptic hole

This problem has been solved in elliptic coordinates by Stevenson (1945), and a great many results can be found from his values of $\phi(z)$ and $\psi(z)$. For the elliptic hole $\xi = \xi_0$ of semi-axes

$$a = c \cosh \xi_0, \quad b = c \sinh \xi_0 \tag{1}$$

in an infinite region with uniaxial stress p_2 at an infinity in a direction inclined at β to the major axis Ox of the ellipse, Fig. 10.11.1 (*a*), the functions $\phi(z)$ and $\psi(z)$ are found to be

$$\phi(z) = \tfrac{1}{4}p_2ce^{2\xi_0} \cos 2\beta \cosh \zeta + \tfrac{1}{4}p_2c(1 - e^{2\xi_0+2i\beta}) \sinh \zeta, \tag{2}$$

$$\psi(z) = -\tfrac{1}{4}p_2c[\cosh 2\xi_0 - \cos 2\beta + e^{2\xi_0} \sinh 2(\zeta - \xi_0 - i\beta)] \operatorname{cosech} \zeta, \tag{3}$$

where, as in § 10.10 (11), z and ζ are connected by

$$z = \omega(\zeta) = c \cosh \zeta, \tag{4}$$

so

$$\frac{dz}{d\zeta} = \omega'(\zeta) = c \sinh \zeta. \tag{5}$$

It can be shown that (2) and (3) satisfy the required conditions at infinity and at the surface $\xi = \xi_0$ of the hole.

Using (5), it follows that

$$\phi'(z) = \frac{d\phi}{d\zeta}\frac{d\zeta}{dz}$$

$$= \tfrac{1}{4}p_2 e^{2\xi_0} \cos 2\beta + \tfrac{1}{4}p_2(1 - e^{2\xi_0 + 2i\beta}) \coth \zeta. \tag{6}$$

The most important quantity, the tangential stress σ_t in the surface

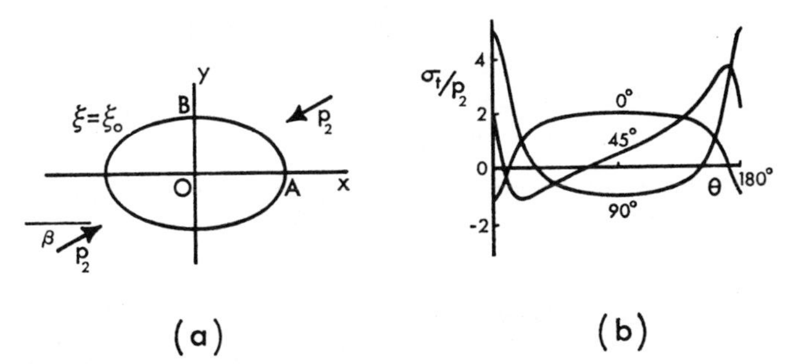

Fig. 10.11.1 (*a*) Elliptic hole with stress p_2 at infinity. (*b*) The variation of the tangential surface stress σ_t with polar angle θ: numbers on the curves are the values of the angle β between p_2 and the major axis. $a = 2b$.

$\xi = \xi_0$, can be found very simply, since here $\sigma_\xi = 0$ and $\sigma_t = \sigma_\eta$, so that by (6) and § 10.10 (7)

$$\begin{aligned}
\sigma_t &= 2[\phi'(\xi_0 + i\eta) + \phi'(\xi_0 - i\eta)] \\
&= p_2 e^{2\xi_0} \cos 2\beta + \tfrac{1}{2}p_2(1 - e^{2\xi_0 + 2i\beta}) \coth (\xi_0 + i\eta) + \\
&\qquad \tfrac{1}{2}p_2(1 - e^{2\xi_0 - 2i\beta}) \coth (\xi_0 - i\eta) \\
&= p_2 e^{2\xi_0} \cos 2\beta + p_2[\cosh 2\xi_0 - \cos 2\eta]^{-1} \sinh 2\xi_0 - \\
&\qquad p_2 e^{2\xi_0} [\cos 2\beta \sinh 2\xi_0 + \sin 2\beta \sin 2\eta][\cosh 2\xi_0 - \cos 2\eta]^{-1}
\end{aligned}$$

$$= p_2 \frac{\sinh 2\xi_0 + \cos 2\beta - \exp (2\xi_0) \cos 2(\beta - \eta)}{\cosh 2\xi_0 - \cos 2\eta}, \tag{7}$$

$$= p_2 \frac{2ab + (a^2 - b^2) \cos 2\beta - (a + b)^2 \cos 2(\beta - \eta)}{a^2 + b^2 - (a^2 - b^2) \cos 2\eta}, \tag{8}$$

using (1). The elliptic coordinate η in (8) is related to the angle θ of polar coordinates $x = r \cos \theta$, $y = r \sin \theta$, since by § 10.10 (12), (14)

$$\tan \theta = y/x = \tanh \xi_0 \tan \eta = (b/a) \tan \eta, \tag{9}$$

so that η is the 'eccentric angle' of the theory of conic sections.

The results (8) and (9) enable the variation of tangential stress around the surface to be studied.

If $\beta = 0$, $\sigma_t = -p_2$ at the end A of the major axis, $\theta = \eta = 0$, and $\sigma_t = [1 + (2b/a)]p_2$ at the end B of the minor axis, $\theta = \eta = \tfrac{1}{2}\pi$.

If $\beta = \frac{1}{2}\pi$, $\sigma_t = [1 + (2a/b)]p_2$ at the end A of the major axis, $\theta = \eta = 0$, and $\sigma_t = -p_2$ at the end B of the minor axis. It follows that, if b/a is small, there is a considerable stress concentration at the point A.

Some values of σ_t for the case of inclined stress are shown in Fig. 10.11.1 (*b*) for the case $a = 2b$.

If there are two principal stresses at infinity, p_2 inclined at β to Ox, and p_1 at $\frac{1}{2}\pi + \beta$, (8) is replaced by

$$\sigma_t = \frac{2ab(p_1 + p_2) + (p_1 - p_2)[(a + b)^2 \cos 2(\beta - \eta) - (a^2 - b^2) \cos 2\beta]}{a^2 + b^2 - (a^2 - b^2) \cos 2\eta}. \tag{10}$$

There is no difficulty in finding the stresses at any point, but the algebra involved is heavy and the formulae are cumbersome. The most important

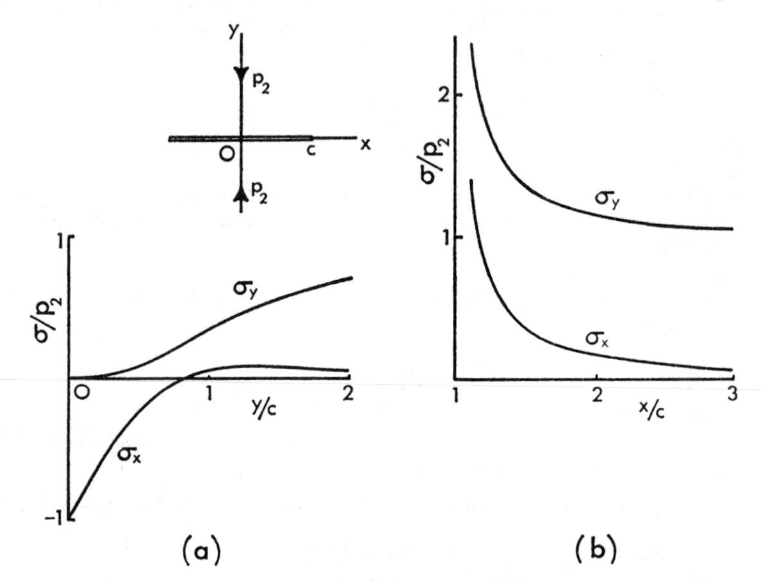

Fig. 10.11.2 Stresses for an elliptic crack of length $2c$ with stress p_2 at infinity perpendicular to its plane. (*a*) Stresses on the y-axis. (*b*) Stresses on the x-axis.

case is that of the *flat elliptic crack* $\xi_0 = 0$. Here, for uniaxial stress p_2 at infinity inclined at β to the plane of the crack

$$\sigma_\xi + \sigma_\eta = p_2 \cos 2\beta + \alpha p_2[(1 - \cos 2\beta) \sinh 2\xi - \sin 2\beta \sin 2\eta], \tag{11}$$

$$\begin{aligned}\sigma_\xi - \sigma_\eta = {} & \alpha p_2 \cosh 2\xi \cos 2(\eta - \beta) + \\ & \alpha^2 p_2\{(1 - \cos 2\beta)(\cos 2\eta - 1) \sinh 2\xi - \cosh 2\xi \cos 2\beta + \\ & \cos 2(\eta - \beta) - \cosh 2\xi \sin 2\beta \sin 2\eta\},\end{aligned} \tag{12}$$

$$\begin{aligned}\tau_{\xi\eta} = {} & \tfrac{1}{2}p_2\alpha \sinh 2\xi \sin 2(\beta - \eta) + \tfrac{1}{2}p_2\alpha^2\{\sinh 2\xi \sin 2\beta\,(\cos 2\eta - 1) + \\ & (1 - \cos 2\beta)(\cosh 2\xi - 1) \sin 2\eta\},\end{aligned} \tag{13}$$

where $\alpha = (\cosh 2\xi - \cos 2\eta)^{-1}$. (14)

The stresses on the axes Ox and Oy for the case $\beta = \frac{1}{2}\pi$ are shown in Fig. 10.11.2.

For the case of the crack in pure shear, that is, with principal stresses at infinity of p_2 at $\beta = \pi/4$ and $-p_2$ at $3\pi/4$, there is a considerable simplification and the results are

$$\sigma_\xi = p_2 (\cosh 2\xi - 1)(\alpha - \alpha^2) \sin 2\eta, \tag{15}$$

$$\sigma_\eta = p_2[\alpha^2 (\cosh 2\xi - 1) - \alpha (\cosh 2\xi + 1)] \sin 2\eta, \tag{16}$$

$$\tau_{\xi\eta} = p_2[\alpha \cos 2\eta - \alpha^2(1 - \cos 2\eta)] \sinh 2\xi, \tag{17}$$

where α is defined in (14).

The other simple case is that in which the stress is hydrostatic at infinity, corresponding to $p_1 = p_2 = p$ in (10), which becomes

$$\sigma_t = \frac{4ab\, p}{a^2 + b^2 - (a^2 - b^2) \cos 2\eta}. \tag{18}$$

The case of internal pressure p in an elliptic hole may be obtained by superposing this solution with negative p on the solution for constant pressure p, for which, by § 10.3 (7), $\phi(z) = \frac{1}{2}pz$, $\psi(z) = 0$. The tangential stress at the surface is, using (18),

$$\sigma_t = p - \frac{4ab\, p}{a^2 + b^2 - (a^2 - b^2) \cos 2\eta}. \tag{19}$$

Results for the slit with internal pressure may be written down in this way from (11)–(14).

For the calculation of displacements we may either find u_ξ and u_η in elliptic coordinates, using § 10.10 (9), or determine them in x, y coordinates from § 10.2 (50), namely

$$2G(u + iv) = \varkappa\phi(z) - z\overline{\phi'(z)} - \overline{\psi(z)}, \tag{20}$$

for many applications this procedure is simpler.

The most important case is that of a flat crack for which ξ_0 is small, so that ξ_0^2 may be neglected, with stress p_2 at infinity perpendicular to it. In this case (2) and (3) become

$$\phi(z) = -\tfrac{1}{4}p_2 c[e^{2\xi_0} \cosh \zeta - (1 + e^{2\xi_0}) \sinh \zeta], \tag{21}$$

$$\psi(z) = -\tfrac{1}{4}p_2 c[1 + \cosh 2\xi_0 - e^{2\xi_0} \sinh 2(\zeta - \xi_0)] \operatorname{cosech} \zeta, \tag{22}$$

$$\phi'(z) = -\tfrac{1}{4}p_2[e^{2\xi_0} - (1 + e^{2\xi_0}) \coth \zeta]. \tag{23}$$

It follows that

$$\begin{aligned} 8G(u + iv)/cp_2 = {} & -\varkappa e^{2\xi_0} \cosh \zeta + \varkappa(1 + e^{2\xi_0}) \sinh \zeta + \\ & [e^{2\xi_0} - (1 + e^{2\xi_0}) \coth \bar{\zeta}] \cosh \zeta + \\ & [1 + \cosh 2\xi_0 - e^{2\xi_0} \sinh 2(\bar{\zeta} - \xi_0)] \operatorname{cosech} \bar{\zeta}. \end{aligned} \tag{24}$$

The surface displacement of a flat crack is found by putting $\xi = \xi_0 = 0$, $\zeta = i\eta$, $\cos\eta = x/c$ in (24) and is

$$v = [(\varkappa + 1)p_2/4G](c^2 - x^2)^{\frac{1}{2}}. \tag{24a}$$

On the axis Oy, $\eta = \frac{1}{2}\pi$, $\zeta = \xi + \frac{1}{2}\pi i$, $\sinh\zeta = i\cosh\xi$, $\cosh\zeta = i\sinh\xi$, and (24) gives $u = 0$,

$$\begin{aligned} 8Gv/cp_2 = -\varkappa e^{2\xi_0}\sinh\xi + \varkappa(1 + e^{2\xi_0})\cosh\xi &+ \\ [e^{2\xi_0} - (1 + e^{2\xi_0})\tanh\xi]\sinh\xi &+ \\ [1 + \cosh 2\xi_0 + e^{2\xi_0}\sinh 2(\xi - \xi_0)]\,\mathrm{sech}\,\xi&. \end{aligned} \tag{25}$$

So far, ξ_0 has not been assumed to be small, and if, now, ξ_0^2 is neglected, (25) becomes

$$\begin{aligned} 8Gv/cp_2 = \varkappa(2\cosh\xi - \sinh\xi) + 3\sinh\xi - 2\cosh\xi + 4\,\mathrm{sech}\,\xi &+ \\ 2\xi_0[\varkappa(\cosh\xi - \sinh\xi) + 3\sinh\xi - 3\cosh\xi + 2\,\mathrm{sech}\,\xi]&. \end{aligned} \quad (26$$

Here, by § 10.10 (12) with $\eta = \frac{1}{2}\pi$,

$$\sinh\xi = y/c, \quad \cosh\xi = [1 + (y^2/c^2)]^{\frac{1}{2}}, \tag{27}$$

and using these values in (26) gives

$$\begin{aligned} 8Gv/cp_2 = (3 - \varkappa)(y/c) + 2(\varkappa - 1)[1 + (y^2/c^2)]^{\frac{1}{2}} + 4[1 + (y^2/c^2)]^{-\frac{1}{2}} &+ \\ 2\xi_0\{(3 - \varkappa)(y/c) + (\varkappa - 3)[1 + (y^2/c^2)]^{\frac{1}{2}} + 2[1 + (y^2/c^2)]^{-\frac{1}{2}}\}&. \end{aligned} \tag{28}$$

Important results follow immediately from this. For the case of plane stress, $\varkappa = (3 - \nu)/(1 + \nu)$, it becomes

$$\begin{aligned} v = \frac{cp_2}{E}\left\{\frac{\nu y}{c} + (1 - \nu)\left[1 + \frac{y^2}{c^2}\right]^{\frac{1}{2}} + (1 + \nu)\left[1 + \frac{y^2}{c^2}\right]^{-\frac{1}{2}}\right\} &+ \\ \frac{c\xi_0 p_2}{E}\left\{-2\nu\left[\left(1 + \frac{y^2}{c^2}\right)^{\frac{1}{2}} - \frac{y}{c}\right] + (1 + \nu)\left[1 + \frac{y^2}{c^2}\right]^{-\frac{1}{2}}\right\}&. \end{aligned} \tag{29}$$

This is the displacement caused by applying stress p_2 at infinity to the material containing the slot under conditions of plane stress. In the absence of the slot, the displacement would be just yp_2/E. Subtracting this gives for the displacement caused by cutting the slot

$$\frac{cp_2}{E}\left\{(1 - \nu)\left[\left(1 + \frac{y^2}{c^2}\right)^{\frac{1}{2}} - \frac{y}{c}\right] + (1 + \nu)\left[1 + \frac{y^2}{c^2}\right]^{-\frac{1}{2}}\right\}, \tag{30}$$

neglecting the term in ξ_0 for shortness. This gives the displacement when a slot is cut for a flat-jack; it gives the variation in displacement for measuring points at various distances and may be used to determine E. These formulae are discussed from this point of view by Alexander (1960).

Another result of great importance follows from (26). If p_2 is positive the crack will close when the displacement v of the point $y = b\ c\xi_0 =$ on its surface is equal to $c\xi_0$, that is, when, neglecting squares of ξ_0,

$$p_2 = 4G\xi_0/(\varkappa + 1). \tag{31}$$

If the principal stresses at infinity are σ_2 at β to Ox and σ_1 at $\frac{1}{2}\pi + \beta$ to Ox this result becomes

$$\sigma_1 \cos^2 \beta + \sigma_2 \sin^2 \beta = 4G\xi_0/(\varkappa + 1). \tag{32}$$

The above discussion has referred to the case $\beta = \frac{1}{2}\pi$ with the stress p_2 normal to the crack. If $\beta = 0$, so that the stress is along the crack, we find in the same way that, neglecting $\xi_0{}^2$,

$$8Gv/cp_2 = (\varkappa - 3) \sinh \xi + 2\xi_0\{(\varkappa - 3)(\sinh \xi - \cosh \xi) - 2 \operatorname{sech} \xi\} \tag{33}$$

$$= (\varkappa - 3)(y/c) + 2\xi_0(\varkappa - 3)(y/c) - 2\xi_0(\varkappa - 3)[1 + (y^2/c^2)]^{\frac{1}{2}} - 4\xi_0[1 + (y^2/c^2)]^{-\frac{1}{2}}. \tag{34}$$

For the case of plane stress in which $\varkappa - 3 = -4\nu/(1 + \nu)$, (34) gives

$$v = -\frac{\nu p_2 y}{E} + \frac{c\xi_0 p_2}{E}\left\{-\frac{2\nu y}{E} + 2\nu\left[1 + \frac{y^2}{c^2}\right]^{\frac{1}{2}} - (1 + \nu)\left[1 + \frac{y^2}{c^2}\right]^{-\frac{1}{2}}\right\}. \tag{35}$$

The first term, $-\nu p_2 y/E$, in (35) is the displacement in an infinite medium with no slot, and the second term represents the effect of the slot. (35) represents a correction to (29) which allows for stress in the plane of the slot.

Finally, the case of the crack in pure shear is of considerable interest. In this case the stresses at infinity are p at $\beta = \pi/4$ and $-p$ at $\beta = 3\pi/4$, and with these values and taking $\xi_0 = 0$ for simplicity, (2) and (3) become

$$\phi(z) = -\tfrac{1}{2}\, ipc \sinh \zeta, \quad \psi(z) = \tfrac{1}{2} ipc \cosh 2\zeta \operatorname{cosech} \zeta. \tag{36}$$

Hence from (20)

$$4G(u + iv)/pc = -i\varkappa \sinh \zeta - i \cosh \zeta \coth \bar{\zeta} + i \cosh 2\bar{\zeta} \operatorname{cosech} \bar{\zeta}, \tag{37}$$

and on the y-axis, $\zeta = \xi + i\pi/2$, this becomes

$$4G(u + iv)/pc = \{\varkappa \cosh^2 \xi + \sinh^2 \xi + \cosh 2\xi\}/\cosh \xi \tag{38}$$

$$= (\varkappa + 3)\left[1 + \frac{y^2}{c^2}\right]^{\frac{1}{2}} - 2\left[1 + \frac{y^2}{c^2}\right]^{-\frac{1}{2}}, \tag{39}$$

using (27). This allows shear stress to be measured by measuring displacements parallel to the crack. The relative displacement of the crack surfaces at the centre of the crack is $[(\varkappa + 1)/2G]cp$.

In elliptic coordinates the components of displacement u_ξ and u_η are by (37) and § 10.10 (9)

$$4G(u_\xi + iu_\eta)/pc = \frac{-i[\varkappa \sinh \zeta \sinh \bar{\zeta} + \cosh \zeta \cosh \bar{\zeta} - \cosh 2\bar{\zeta}]}{[\sinh \zeta \sinh \bar{\zeta}]^{\frac{1}{2}}}. \qquad (40)$$

It follows that

$$4Gu_\xi/pc = (2\alpha)^{\frac{1}{2}} \sinh 2\xi \sin 2\eta, \qquad (41)$$

$$4Gu_\eta/pc = -\varkappa(2\alpha)^{-\frac{1}{2}} - (\alpha/2)^{\frac{1}{2}}[\cosh 2\xi + \cos 2\eta] + (2\alpha)^{\frac{1}{2}} \cosh 2\xi \cos 2\eta, \qquad (42)$$

where α is defined in (14). In this case the matter is complicated by the fact that there is a rotation, and as shown in §§ 10.2 and 10.3 (iii), a rigid body rotation may be added without affecting the stresses. The problem is discussed by Starr (1928).

The mathematics of the stresses around an elliptic hole have been discussed by many authors, cf. Timoshenko and Goodier (1951). Inglis (1913) determined stresses and displacements for a variety of conditions, and his formulae have been used by Griffith (1921, 1924), McClintock and Walsh (1962), and Anderson (1938, 1951); the latter author uses them to study the dynamics of intrusion and stresses around a fault. Inglis's analysis is not very convenient, so Pöschl (1921–22) used the Airy stress function, and his results have been used by Murrell (1964a). Starr (1928) gives a very full discussion of the crack in pure shear based on the Airy stress function. Murrell (1964b) discusses the crack with internal pressure, and the combination of a crack with a dislocation. Stroh (1954, 1955) also treats the latter problem. Symonds (1946) and Green (1947) study the elliptic hole stressed by equal and opposite forces at the ends of the minor axis. Stevenson (1945) develops the method used here and is followed by Walsh (1965a).

All the references given above refer to the use of elliptic coordinates, which is the classical method of applied mathematics for treating the slit or the elliptic hole. However, other methods are available – transformation on to a circle is used by Muskhelishvili (1953) and Savin (1961). They take

$$z = R(\rho e^{i\theta} + m\rho^{-1}e^{-i\theta}) \qquad (43)$$

in place of (4), so that solutions in the two notations are related by

$$\rho = m^{\frac{1}{2}}e^{\xi}, \quad c = 2Rm^{\frac{1}{2}}, \quad \theta = \eta, \qquad (44)$$

and $m = 1$ corresponds to the case of a slit. The case of a slit with zero displacement and zero shear stress across its surface if it is closed is discussed by Mossakovskii and Rybka (1965).

An entirely different mathematical approach for the flat crack has been adopted by other authors. Instead of considering it as the limiting case of an ellipse, they consider the ideal mathematical crack of zero thickness in the region $-c < x < c$. In this case the solutions have singularities at the crack tips $x = \pm c$. Sneddon and Elliott (1946) have applied Fourier transform theory to this problem. Westergaard (1939) developed a simple type of stress function applicable to a restricted but important class of problems, including the crack in tension, a row of cracks, a crack with undulating surface, and the opening of cracks by internal forces. Sneddon (1946) uses this method for the crack with internal pressure.

As a rough generalization it may be said that most workers on rock mechanics have used the flat elliptic crack, while most workers on fracture mechanics have used the mathematical crack. The reason is probably partly historical and partly because of the relative importance of compressive conditions in rock mechanics.

10.12 Stresses near the tip of a crack

To study in detail the stresses near the tip of the crack $\xi = \xi_0$ at the point P,

$$x = c \cosh \xi \cos \eta, \quad y = c \sinh \xi \sin \eta, \tag{1}$$

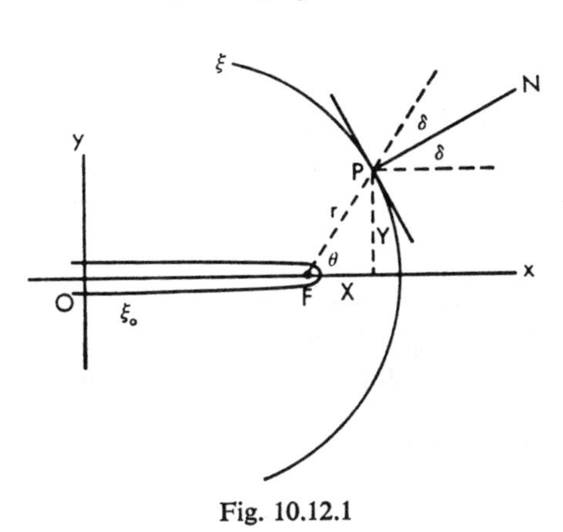

Fig. 10.12.1

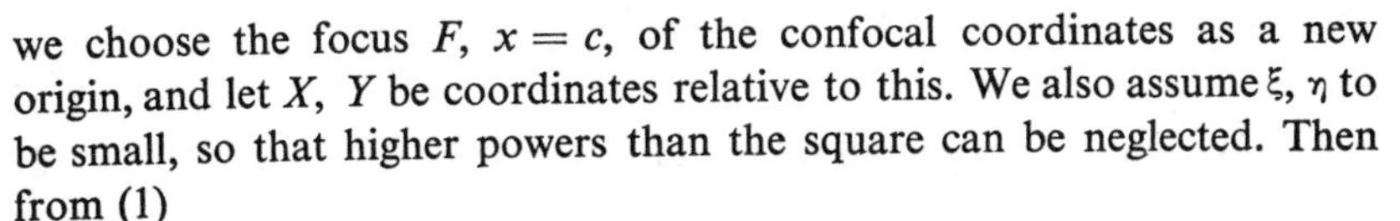

we choose the focus F, $x = c$, of the confocal coordinates as a new origin, and let X, Y be coordinates relative to this. We also assume ξ, η to be small, so that higher powers than the square can be neglected. Then from (1)

$$X = x - c = c(\xi^2 - \eta^2)/2, \tag{2}$$

$$Y = y = c\xi\eta. \tag{3}$$

In assuming ξ small we are tacitly assuming that the ξ_0 of the crack is still smaller, and we shall take it to be zero here, though, in a fuller treatment, powers of ξ_0 can also be included.

ξ and η can now be expressed in terms of the Cartesian coordinates X, Y of P, or, more conveniently, in terms of its polar coordinates

$$r = (X^2 + Y^2)^{\frac{1}{2}}, \quad X = r \cos \theta, \quad Y = r \sin \theta. \tag{4}$$

From (2) and (3)

$$(\xi^2 + \eta^2) = 2(X^2 + Y^2)^{\frac{1}{2}}/c = 2r/c. \tag{5}$$

And from (2) and (5)

$$\xi = (r/c)^{\frac{1}{2}}(1 + \cos \theta)^{\frac{1}{2}} = (2r/c)^{\frac{1}{2}} \cos \tfrac{1}{2}\theta, \tag{6}$$

$$\eta = (r/c)^{\frac{1}{2}}(1 - \cos \theta)^{\frac{1}{2}} = (2r/c)^{\frac{1}{2}} \sin \tfrac{1}{2}\theta. \tag{7}$$

Also, the quantity α defined in § 10.11 (14) becomes

$$\alpha = [\cosh 2\xi - \cos 2\eta]^{-1} = 1/2(\xi^2 + \eta^2) = c/4r. \tag{8}$$

The angle δ between the normal PN to the ellipse ξ and the x-axis is by § 10.10 (4)

$$e^{2i\delta} = \frac{\sinh (\xi + i\eta)}{\sinh (\xi - i\eta)} = \frac{\xi + i\eta}{\xi - i\eta} = e^{i\theta},$$

so that

$$\delta = \tfrac{1}{2}\theta. \tag{9}$$

Using (6)–(9) in § 10.11 (11)–(14) the stresses in the neighbourhood of the tip of the flat elliptic crack $\xi_0 = 0$ can be written down. Thus, for the case of stress p_2 at infinity at an angle β to the crack,

$$\begin{aligned}
\sigma_r + \sigma_\theta &= \sigma_\xi + \sigma_\eta \\
&= p_2 \cos 2\beta + \alpha p_2[(1 - \cos 2\beta) \sinh 2\xi - \sin 2\beta \sin 2\eta] \\
&= p_2 \cos 2\beta + \alpha p_2[(1 - \cos 2\beta) \,.\, 2\xi - 2\eta \sin 2\beta] \\
&= p_2 \cos 2\beta + p_2(c/2r)^{\frac{1}{2}}[(1 - \cos 2\beta) \cos \tfrac{1}{2}\theta - \sin 2\beta \sin \tfrac{1}{2}\theta]. \quad (10)
\end{aligned}$$

The term $p_2 \cos 2\beta$ is a constant which may be neglected by comparison with the term in $r^{-\frac{1}{2}}$.

Proceeding in this way we get finally

$$(8r/cp_2{}^2)^{\frac{1}{2}}\sigma_r = \sin \tfrac{1}{2}\theta(1 - 3 \sin^2 \tfrac{1}{2}\theta) \sin 2\beta + 2 \cos \tfrac{1}{2}\theta(1 + \sin^2 \tfrac{1}{2}\theta) \sin^2 \beta, \tag{11}$$

$$(8r/cp_2{}^2)^{\frac{1}{2}}\sigma_\theta = -3 \sin \tfrac{1}{2}\theta \cos^2 \tfrac{1}{2}\theta \sin 2\beta + 2 \cos^3 \tfrac{1}{2}\theta \sin^2 \beta, \tag{12}$$

$$(8r/cp_2{}^2)^{\frac{1}{2}}\tau_{r\theta} = \cos \tfrac{1}{2}\theta(3 \cos^2 \tfrac{1}{2}\theta - 2) \sin 2\beta + 2 \cos^2 \tfrac{1}{2}\theta \sin \tfrac{1}{2}\theta \sin^2 \beta. \tag{13}$$

These show that all stresses increase like $r^{-\frac{1}{2}}$ as the tip of the crack is approached.

In the theory of fracture mechanics these or equivalent results are usually deduced from the theory of the mathematical flat crack, Erdogan and Sih (1963), Irwin (1958, 1957), Westergaard (1939). Attention also is usually concentrated on the crack with normal tension or in pure shear. If (11)–(13) are taken as fundamental it follows by differentiating (12) with respect to θ that σ_θ is a maximum when $\theta = \pi$ or when θ is given by

$$\tan\beta \sin\theta + 3\cos\theta = 1, \tag{14}$$

and it follows from (13) that if (14) is satisfied $\tau_{r\theta} = 0$. In this case, then, σ_r and σ_θ are principal stresses and σ_θ has a maximum value, so that this value of θ might be expected to be the direction of failure. Results obtained in this way are not altogether compatible with those obtained from the Griffith theory of § 10.13, but the models are essentially different, cf. Erdogan and Sih (1963).

The case of the crack with internal pressure p is of special interest, and, in

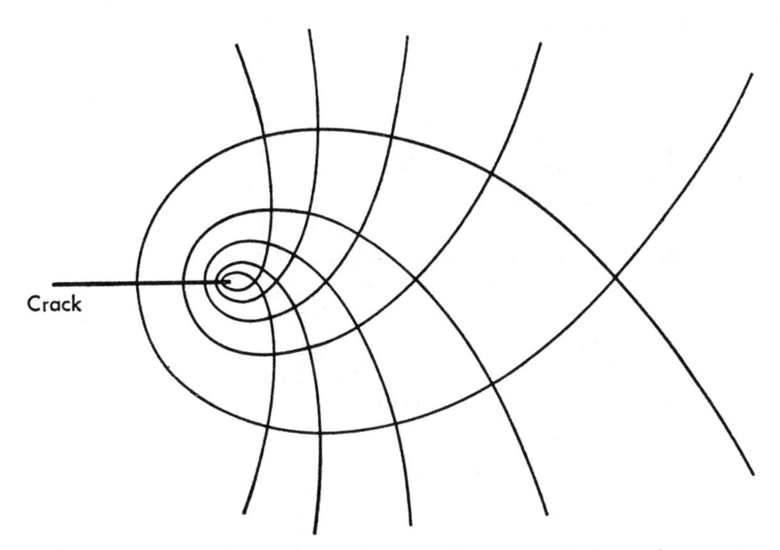

Fig. 10.12.2 Stress trajectories near the end of a crack with internal pressure.

addition, a uniaxial stress, p_2, in a direction perpendicular to the crack may be superimposed without loss of generality. For this system it follows from (11)–(13), again only retaining terms in $r^{-\frac{1}{2}}$, that

$$(8r/c)^{\frac{1}{2}}\sigma_r = 2(p_2 - p)\cos\tfrac{1}{2}\theta(1 + \sin^2\tfrac{1}{2}\theta), \tag{15}$$

$$(8r/c)^{\frac{1}{2}}\sigma_\theta = 2(p_2 - p)\cos^3\tfrac{1}{2}\theta, \tag{16}$$

$$(8r/c)^{\frac{1}{2}}\tau_{r\theta} = 2(p_2 - p)\cos^2\tfrac{1}{2}\theta\sin\tfrac{1}{2}\theta. \tag{17}$$

It follows from § 2.3 (14) that the principal stresses are

$$(c/2r)^{\frac{1}{2}}(p_2 - p)\cos\tfrac{1}{2}\theta(1 \pm \sin\tfrac{1}{2}\theta). \tag{18}$$

This has its greatest magnitude when $\cos\theta = \sin\tfrac{1}{2}\theta$, so that $\theta = 60°$.

For example, if $p > p_2$ both principal stresses are tensile, and the greatest tensile stress of

$$1.3(p - p_2)(c/2r)^{\frac{1}{2}} \tag{19}$$

is attained when $\theta = 60°$.

The inclination ϕ of the principal axes to the σ_r direction is by § 2.3 (11)

$$\tan 2\phi = \frac{\sin \theta}{1 - \cos \theta} = \cot \tfrac{1}{2}\theta, \tag{20}$$

so that

$$\phi = (\pi/4) - (\theta/4) \quad \text{or} \quad \phi = (3\pi/4) - (\theta/4). \tag{21}$$

The differential equation of the first set of stress trajectories is

$$\frac{rd\theta}{dr} = \tan \phi = \tan\left(\frac{\pi}{4} - \frac{\theta}{4}\right),$$

so that their equation is

$$r \cos^4\left(\frac{\pi}{4} + \frac{\theta}{4}\right) = \text{constant}. \tag{22}$$

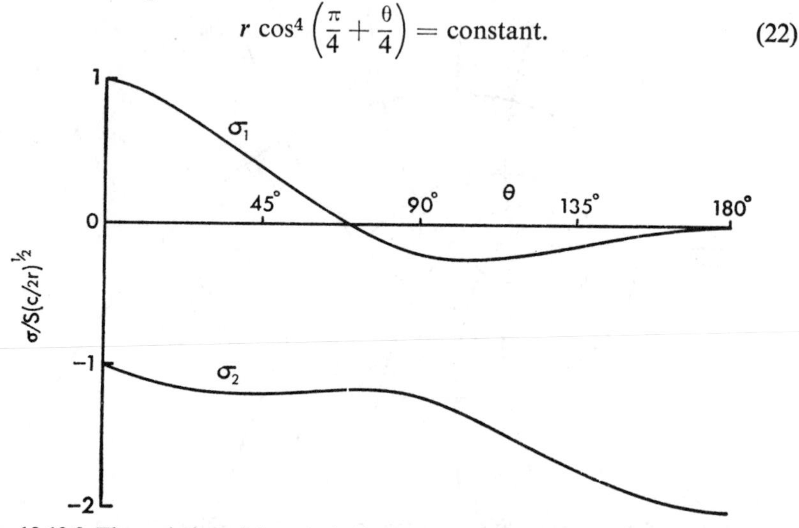

Fig. 10.12.3 The variation of the principal stresses with θ at constant distance r from one end of a long crack under pure shear.

Similarly, the equation of the second set is

$$r \cos^4\left(\frac{3\pi}{4} + \frac{\theta}{4}\right) = \text{constant}. \tag{23}$$

The stress-trajectories are shown in Fig. 10.12.2.

Another case of practical importance is that of *pure shear S*, obtained from (11)–(13) by combining $p_2 = S$ at $\beta = \pi/4$ with $p_2 = -S$ at $\beta = 3\pi/4$. The stresses are

$$(2r/cS^2)^{\frac{1}{2}}\sigma_r = \sin \tfrac{1}{2}\theta(1 - 3 \sin^2 \tfrac{1}{2}\theta), \tag{24}$$

$$(2r/cS^2)^{\frac{1}{2}}\sigma_\theta = -3 \sin \tfrac{1}{2}\theta \cos^2 \tfrac{1}{2}\theta, \tag{25}$$

$$(2r/cS^2)^{\frac{1}{2}}\tau_{r\theta} = \cos \tfrac{1}{2}\theta(3 \cos^2 \tfrac{1}{2}\theta - 2). \tag{26}$$

It follows that

$$(2r/cS^2)^{\frac{1}{2}}(\sigma_r + \sigma_\theta) = -2 \sin \tfrac{1}{2}\theta,$$
$$(2r/cS^2)[(\sigma_r - \sigma_\theta)^2 + 4\tau_{r\theta}^2] = 1 + 3 \cos^2 \theta.$$

Thus, by § 2.3 (14), the principal stresses are given by

$$(2r/cS^2)^{\frac{1}{2}}\sigma = -\sin \tfrac{1}{2}\theta \pm \tfrac{1}{2}(1 + 3 \cos^2 \theta)^{\frac{1}{2}} \tag{27}$$

The variation of σ_1 and σ_2 with θ is shown in Fig. 10.12.3.

10.13 Griffith theory in two dimensions

Griffith (1924) derived his criterion for failure under two-dimensional stresses σ_1, σ_2 at infinity by studying in detail the variation of the tangential stress σ_t in the surface of a flat elliptical crack of semi-axes

$$a = c \cosh \xi_0, \quad b = c \sinh \xi_0,$$

where ξ_0 is small. The principal stresses σ_1, σ_2 are taken, as in Fig. 10.13 (*a*),

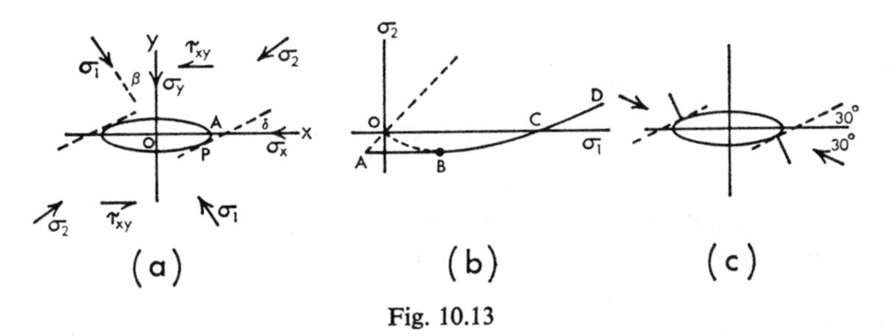

Fig. 10.13

to be inclined at $\frac{1}{2}\pi + \beta$ and β to Ox, but it is also convenient to use their components in crack coordinates, namely by § 2.3 (7)–(9),

$$\sigma_x = \sigma_1 \sin^2 \beta + \sigma_2 \cos^2 \beta, \quad \sigma_y = \sigma_1 \cos^2 \beta + \sigma_2 \sin^2 \beta, \tag{1}$$

$$\tau_{xy} = -\tfrac{1}{2}(\sigma_1 - \sigma_2) \sin 2\beta. \tag{2}$$

The tangential stress σ_t in the surface of the crack is, by § 10.11 (7),

$$\sigma_t = \frac{(\sigma_1 + \sigma_2) \sinh 2\xi_0 + (\sigma_1 - \sigma_2)[\exp(2\xi_0) \cos 2(\beta - \eta) - \cos 2\beta]}{\cosh 2\xi_0 - \cos 2\beta} \tag{3}$$

or, using the stresses (1), (2) in crack coordinates,

$$\sigma_t = \frac{2\sigma_y \sinh 2\xi_0 + 2\tau_{xy}[(1 + \sinh 2\xi_0) \cot 2\beta - \exp(2\xi_0) \cos 2(\beta - \eta) \operatorname{cosec} 2\beta]}{\cosh 2\xi_0 - \cos 2\beta}. \tag{4}$$

If ξ_0 is small (a flat crack) and also η is small, corresponding to points near the end A, (4) becomes

$$\sigma_t = \frac{2(\xi_0 \sigma_y - \eta \tau_{xy})}{\xi_0^2 + \eta^2}. \tag{5}$$

Next we find the maximum and minimum values of the tangential stress as a function of η. Differentiating (5), $d\sigma_t/d\eta = 0$ when

$$\eta^2\tau_{xy} - 2\eta\xi_0\sigma_y - \xi_0^2\tau_{xy} = 0, \tag{6}$$

or

$$\eta = \xi_0[\sigma_y \pm (\sigma_y^2 + \tau_{xy}^2)^{\frac{1}{2}}]/\tau_{xy}. \tag{7}$$

Substituting (7) in (5) gives for the values of σ_t at these points

$$\xi_0\sigma_t = \sigma_y \mp [\tau_{xy}^2 + \sigma_y^2]^{\frac{1}{2}}. \tag{8}$$

Of the values (8), the negative sign gives σ_t negative, tensile, and it is this tensile stress which is assumed to cause failure. The result thus is that the maximum tensile stress σ_e in the crack surface is given by

$$\xi_0\sigma_e = \sigma_y - [\tau_{xy}^2 + \sigma_y^2]^{\frac{1}{2}} \tag{9}$$

$$= (\sigma_1 \cos^2\beta + \sigma_2 \sin^2\beta) - [\sigma_1^2 \cos^2\beta + \sigma_2^2 \sin^2\beta]^{\frac{1}{2}}, \tag{10}$$

and occurs when

$$\eta/\xi_0 = [\sigma_y + (\sigma_y^2 + \tau_{xy}^2)^{\frac{1}{2}}]/\tau_{xy} \tag{11}$$

$$= 2[(\sigma_1 \cos^2\beta + \sigma_2 \sin^2\beta) + (\sigma_1^2 \cos^2\beta + \sigma_2^2 \sin^2\beta)^{\frac{1}{2}}]/(\sigma_2 - \sigma_1)\sin 2\beta. \tag{12}$$

A more refined calculation, Odé (1960), which does not assume the approximation (5) gives the same result.

Next, we determine the orientation β of the stresses σ_1, σ_2 which makes σ_e in (10) a maximum. Differentiating (10) gives

$$\xi_0\frac{d\sigma_e}{d\beta} = \left\{2\sigma_2 - 2\sigma_1 + \frac{\sigma_1^2 - \sigma_2^2}{(\sigma_1^2 \cos^2\beta + \sigma_2^2 \sin^2\beta)^{\frac{1}{2}}}\right\} \sin\beta\cos\beta. \tag{13}$$

This is zero if either $\beta = 0$, $\beta = \frac{1}{2}\pi$, or

$$\cos 2\beta = -\tfrac{1}{2}(\sigma_1 - \sigma_2)/(\sigma_1 + \sigma_2). \tag{14}$$

The inclined position (14) exists only if $|\cos 2\beta| < 1$, which requires that

$$\sigma_1 + 3\sigma_2 > 0. \tag{15}$$

Using (14) in (10) gives as the value of σ_e for this case

$$\sigma_e = -\frac{(\sigma_1 - \sigma_2)^2}{4(\sigma_1 + \sigma_2)\xi_0}. \tag{16}$$

If the inequality (15) is not satisfied σ_2 must be negative, and from (10) and (12), or directly from (3), the greatest tensile stress in the crack surface occurs when $\beta = \frac{1}{2}\pi$ and has the value

$$\sigma_e = 2\sigma_2/\xi_0. \tag{17}$$

It is now assumed that failure takes place when the maximum tensile stress in the crack of most dangerous orientation reaches some value characteristic of the material. This can be found from (17) for the case of uniaxial tension normal to the plane of the crack, and if T_0 is the uniaxial tensile strength, so that $\sigma_2 = -T_0$, (17) gives $\sigma_e = -2T_0/\xi_0$, and using this value in (16) gives

$$(\sigma_1 - \sigma_2)^2 - 8T_0(\sigma_1 + \sigma_2) = 0, \quad \text{if} \quad \sigma_1 + 3\sigma_2 > 0, \tag{18}$$

which with

$$\sigma_2 = -T_0, \quad \text{if} \quad \sigma_1 + 3\sigma_2 < 0, \tag{19}$$

constitutes the Griffith criterion for failure. This relation has already been discussed in § 4.8. It is shown in Fig. 10.13 (*b*).

For the case of the inclined crack the position (12) of maximum tensile stress corresponds to a negative value of η at the point P, Fig. 10.13. The slope δ of the tangent to the ellipse at this point is given by

$$\tan \delta = \frac{dy}{dx} = -\frac{\xi_0}{\eta}$$
$$= \frac{(\sigma_1 - \sigma_2) \sin 2\beta}{2[\sigma_1 \cos^2 \beta + \sigma_2 \sin^2 \beta + (\sigma_1^2 \cos^2 \beta + \sigma_2^2 \sin^2 \beta)^{\frac{1}{2}}]}. \tag{20}$$

Using the value (14) of β in this, it follows that at failure

$$\delta = 2\beta - \tfrac{1}{2}\pi, \tag{21}$$

and the direction of the actual fracture, which may be assumed to be normal to this, always runs towards the greatest principal stress. For example, for uniaxial compression, $\beta = 60°$ by (14) and $\delta = 30°$ by (21). The situation is shown in Fig. 10.13 (*c*). The theory thus predicts that actual failure will occur near but not at the end of the original crack, and that the direction in which fracture is initiated is not in the direction of the original crack. This effect has been observed with artificial cracks in glass and plastic, Brace and Bombalakis (1963), Hoek and Bieniawski (1965). For a set of cracks *en echelon* interaction occurs, and failure at a lower stress is possible. Fairhurst and Cook (1966) use the fact that the directions of fracture are directed towards the direction of maximum principal stress to explain the phenomenon of longitudinal splitting under uniaxial compressive stress.

This simple theory is very attractive, since it is an accurate calculation based on no assumptions except that failure will take place normal to the surface of the crack when the tangential stress in the crack reaches a value characteristic of the material. It predicts the direction of the crack and of the initial failure in it.

If there is fluid pressure p in the interior of the crack in addition to stresses σ_1, σ_2 at infinity as before, the solution § 10.11 (19) for internal pressure p

has to be added to (3) to give the tangential stress σ_t in the crack surface, which then becomes

$$\sigma_t = p + \frac{(\sigma_1 + \sigma_2 - 2p)\sinh 2\xi_0 + (\sigma_1 - \sigma_2)[\exp(2\xi_0)\cos 2(\beta - \eta) - \cos 2\beta]}{\cosh 2\xi_0 - \cos 2\eta}. \tag{22}$$

The essential change is that σ_1 and σ_2 are replaced by the effective stresses $\sigma_1 - p$ and $\sigma_2 - p$. The calculation then proceeds as before; (16) is replaced by

$$\sigma_e = p - \frac{(\sigma_1 - \sigma_2)^2}{4(\sigma_1 + \sigma_2 - 2p)\xi_0}, \quad \text{if} \quad \sigma_1 + 3\sigma_2 > 4p. \tag{23}$$

Combining this with (17) gives

$$(\sigma_1 - \sigma_2)^2 - 8T_0(\sigma_1 + \sigma_2 - 2p) - 4p\xi_0(\sigma_1 + \sigma_2 - 2p) = 0.$$

Since ξ_0 is small, the last term may be neglected and the criterion for failure becomes

$$(\sigma_1 - \sigma_2)^2 - 8T_0(\sigma_1 + \sigma_2 - 2p) = 0, \quad \sigma_1 + 3\sigma_2 > 4p, \tag{24}$$

$$\sigma_2 - p + T_0 = 0, \quad \sigma_1 + 3\sigma_2 < 4p, \tag{25}$$

or, in terms of the effective stresses σ_1', σ_2' defined in § 8.8,

$$(\sigma_1' - \sigma_2')^2 - 8T_0(\sigma_1' + \sigma_2') = 0, \quad \sigma_1' + 3\sigma_2' > 0, \tag{26}$$

$$\sigma_2' + T_0 = 0, \quad \sigma_1' + 3\sigma_2' < 0. \tag{27}$$

10.14 The effect of friction between crack surfaces

The theory of § 10.13 applies to tensile conditions, but under compressive conditions the crack may close. The condition for this has been found in § 10.11 (32) to be

$$\sigma_y = \sigma_1 \cos^2\beta + \sigma_2 \sin^2\beta = 4G\xi_0/(\varkappa + 1), \tag{1}$$

where the system is that of Fig. 10.13 and σ_y, given by § 10.13 (1), is the normal stress on the crack. When the crack has closed, further displacement can be achieved only by sliding across the closed surface, and this will be resisted by sliding friction across this surface. McClintock and Walsh (1962) first modified the Griffith theory to take this effect into account, and the theory was developed further by Brace (1960b), Murrell (1964b), and Hoek and Bieniawski (1965).

McClintock and Walsh (1962) assume that a normal stress σ_c at infinity is necessary to close the crack. They then assume that the normal stress across the surface of the closed crack of Fig. 10.14 (*a*) is $\sigma_n = \sigma_y - \sigma_c$ and that a frictional force $\tau_f = \mu\sigma_n$ resists sliding across this surface.

The actual direction of τ_f must be chosen to be opposite to the direction of slipping. Now with the coordinate system chosen, τ_{xy} is given by § 10.13 (*a*) is negative, that is, movement is in the direction of the arrows at *B*, *B'*, and so τ_f must be in the direction shown in Fig. 10.14 (*a*). If the stress systems of uniform tension $-\sigma_n$, and uniform shear τ_f, Figs. 10.14 (*b*), (*c*) are superimposed on this, the result is that of a crack with normal stress $\sigma_y - \sigma_n = \sigma_c$ and shear stress $\tau_{xy} + \tau_f = \tau_{xy} + \mu(\sigma_y - \sigma_c)$ at infinity and zero internal stress, Fig. 10.14 (*d*). Since the stresses due to the uniform systems of Fig. 10.14 (*b*), (*c*) are small compared with those at the crack tip, they may be neglected by comparison, and the stresses near the crack

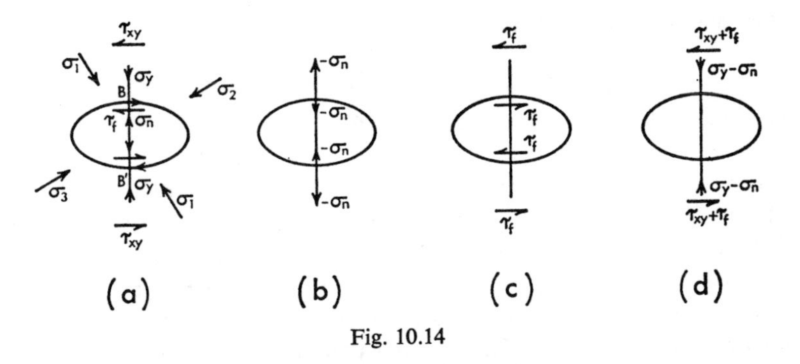

Fig. 10.14

tip for the system of Fig. 10.14 (*a*) are the same as those of Fig. 10.14 (*d*) and may be obtained from § 10.13 (5) which gives for the tangential stress σ_t in the crack surface.

$$\sigma_t = \frac{2\xi_0\sigma_c - 2\eta[\tau_{xy} + \mu(\sigma_y - \sigma_c)]}{\xi_0^2 + \eta^2} \tag{2}$$

$$= [2\xi_0\sigma_c + \eta\sigma^*]/(\xi_0^2 + \eta^2), \tag{3}$$

where, using 10.13 (1), (2),

$$\sigma^* = (\sigma_1 - \sigma_2)[\sin 2\beta - \mu \cos 2\beta] - \mu(\sigma_1 + \sigma_2 - 2\sigma_c). \tag{4}$$

Differentiating (3) with respect to η, σ_t is found to have maximum or minimum values when

$$\eta/\xi_0 = [-2\sigma_c \pm (4\sigma_c^2 + \sigma^{*2})^{\frac{1}{2}}]/\sigma^*, \tag{5}$$

and inserting this value of η in (3) gives for the extreme values σ_e of σ_t,

$$\sigma_e = [2\sigma_c \pm (4\sigma_c^2 + \sigma^{*2})^{\frac{1}{2}}]/2\xi_0. \tag{6}$$

In (6) the negative sign is to be taken, since we are interested in tensile values. The maximum value of σ_e as a function of β occurs when $d\sigma^*/d\beta = 0$, or from (4) when

$$\tan 2\beta = -1/\mu, \tag{7}$$

so that

$$\sin 2\beta = (\mu^2 + 1)^{-\frac{1}{2}}, \quad \cos 2\beta = -\mu(\mu^2 + 1)^{-\frac{1}{2}}. \tag{8}$$

As in § 10.13 (17), we equate the maximum negative value (6) of σ_e to $-2T_0/\xi_0$, where T_0 is the uniaxial tensile strength, and find

$$(4\sigma_c^2 + \sigma^{*2})^{\frac{1}{2}} - 2\sigma_c = 2T_0,$$

or

$$\sigma^* = 4T_0(1 + \sigma_c/T_0)^{\frac{1}{2}},$$

or, using (8) in (4),

$$\sigma_1[(\mu^2 + 1)^{\frac{1}{2}} - \mu] - \sigma_2[(\mu^2 + 1)^{\frac{1}{2}} + \mu] = 4T_0[1 + \sigma_c/T_0]^{\frac{1}{2}} - 2\mu\sigma_c \tag{9}$$

If σ_c may be neglected, this reduces to

$$\sigma_1[(\mu^2 + 1)^{\frac{1}{2}} - \mu] - \sigma_2[(\mu^2 + 1)^{\frac{1}{2}} + \mu] = 4T_0, \tag{10}$$

which is the old Coulomb criterion, § 4.6.

It follows from (9) that if all cracks are closed there is a linear relation between σ_1 and σ_2; if all cracks are open the Griffith theory of § 10.13 applies; there will be an intermediate region which is discussed by McClintock and Walsh (1962). The case $\mu = 0$ has been discussed by Mossakovskii and Rybka (1965) and leads to different results.

The present theory is known as the modified Griffith theory. Its status has to be noted carefully. The original theory of § 10.13 involved no assumptions except the single one stated. The modified theory makes various approximations and ends by maximizing the quantity $\sin 2\beta - \mu \cos 2\beta$, which is the same as that which appears in the Coulomb theory of § 4.6. In fact, for the case $\sigma_c = 0$ the analysis is identical, and the only change is that μ is now a coefficient of ordinary sliding friction and the cohesion S_0 is replaced by $2T_0$.

10.15 The nearly rectangular hole and openings of other shapes

By far the most powerful method for the solution of two-dimensional problems is the detailed use of complex variable theory and conformal representation as developed in the books of Muskhelishvili (1953) and Savin (1961). In particular, solutions for the region outside a hole of specified shape may be obtained by a transformation which transforms this region into the region outside or inside a circle.

The simplest and most important case is that of the transformation

$$z = x + iy = \omega(\zeta) = a(\zeta^{-1} - \tfrac{1}{6}\zeta^3). \tag{1}$$

This transforms the unit circle $\zeta = \exp i\alpha$ into the curve

$$x = a\,(\cos\alpha - \tfrac{1}{6}\cos 3\alpha), \quad y = -a\,(\sin\alpha + \tfrac{1}{6}\sin 3\alpha), \tag{2}$$

and maps the region outside the curve (2) on the inside of the unit circle.

The shape of the curve (2) is shown in Fig. 10.15: it appears that it is nearly a square with rounded corners. Since from (2), $x = 5a/6$ when $\alpha = 0$, the 'side' of the square is of length $5a/3$. The radius of curvature at the corners is approximately $a/10$ or 0·06 times the side length.

Savin (1961), Chapter II, discusses in detail the determination of the fundamental functions $\phi(z)$, $\psi(z)$ of § 10.2 for this problem. For stress p_1 at infinity at an angle β to Ox the results are

$$\phi(\zeta) = p_1 a\left[\frac{1}{4\zeta} + (\tfrac{3}{7}\cos 2\beta + \tfrac{3}{5}i\sin 2\beta)\zeta + \tfrac{1}{24}\zeta^3\right], \tag{3}$$

$$\psi(\zeta) = -p_1 a\left[\frac{1}{2\zeta}e^{-2i\beta} + \{13\zeta - 26(\tfrac{3}{7}\cos 2\beta + \tfrac{3}{5}i\sin 2\beta)\zeta^3\}/12(2+\zeta^4)\right], \tag{4}$$

and the stresses and displacements can be calculated from these by

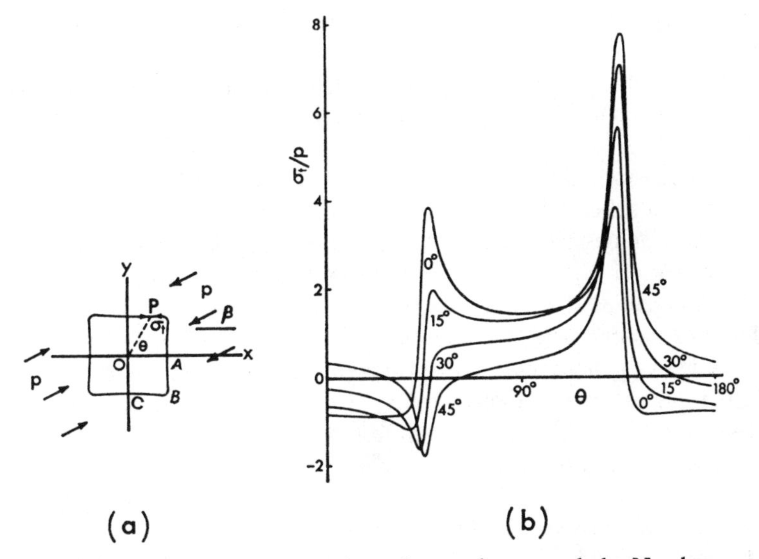

Fig. 10.15 Tangential stress in the surface of a nearly square hole. Numbers on the curves are the inclination β of the uniaxial principal stress at infinity to a side of the square.

§ 10.2 (48)–(50). In particular, the tangential stress in the surface at the point corresponding to $\zeta = \exp i\alpha$ is

$$\begin{aligned}\sigma_t &= 4\mathbf{R}[\phi'(z)] = 4[\mathbf{R}\{\phi'(\zeta)/\omega'(\zeta)\}]_{\zeta=i\alpha}\\ &= p_1\mathbf{R}[\{70\zeta^{-2} - (120\cos 2\beta + 168i\sin 2\beta) - 35\zeta^2\}/35(2\zeta^{-2}+\zeta^2)]_{\zeta=i\alpha}\\ &= p_1\{105 - 360\cos 2\beta\cos 2\alpha + 168\sin 2\beta\sin 2\alpha\}/35(5+4\cos 4\alpha).\end{aligned} \tag{5}$$

When $\beta = 0$, $\sigma_t = -51p_1/63$ when $\alpha = 0$, that is at the point A; $\sigma_t = 93p_1/63$, at the point C, $\alpha = 90°$; $\sigma_t = 3p_1$ at the point B, $\alpha = 45°$. These may be compared with the values of $-p_1$ and $3p_1$ for the circle. The maximum stress in the present case is not attained at the 'corner' B but near it and is in fact almost $4p_1$. If β is not zero greater stresses are attained near the corners.

Fig. 10.15 (*b*) shows the variation of σ_t with the polar angle $\theta = \tan^{-1}(y/x)$ for various values of β. It appears that the stress concentration near the corners reaches a maximum value of about $8p_1$ when $\beta = 45°$. Savin gives results for more sharply rounded corners – if the radius is 0·025 times the side length the tangential stress rises to $11{\cdot}5p_1$ for $\beta = 45°$.

Similar results are obtained for the nearly rectangular hole. For an aspect ratio of 5 : 1 with unfavourable directions of stress, tensile stresses as high as $-2p_1$ and compressive stresses as high as $12{\cdot}5p_1$ occur. Savin (1961) gives very full information for various cases, including plots for principal stresses and stress trajections.

Other cross-sections, for example, the equilateral triangle, are also treated.

Finally, a brief reference should be made to a number of important problems which will not be treated here. These are:

(i) *The spherical hole*

The case of a spherical hole in an infinite medium with hydrostatic stress at infinity will be described in § 10.17. For a uniaxial stress p_1 at infinity the solution is given by Timoshenko and Goodier (1951), and the results for surface tangential stresses are as follows:

There is tension – $(3 + 15\nu)p_1/2(7 - 5\nu)$ at the points of the sphere in the direction of the stress; that is, approximately $-\frac{1}{2}p_1$.

In the equator perpendicular to the direction of p_1 there is a hoop stress $(15\nu - 3)p_1/2(7 - 5\nu)$ and a compressive stress $(27 - 15\nu)p_1/2(7 - 5\nu)$, approximately $2p_1$, in the perpendicular direction.

(ii) *Ellipsoidal and spheroidal holes*

These have been discussed with reference to Griffith cracks by Sack (1946) and Elliott (1947). Numerical information is given by Sadowsky and Sternberg (1947, 1949), Terzaghi and Richart (1952), and Luré (1964).

(iii) *The circular tunnel under gravity*

This is discussed by Mindlin (1939) and Yi-Yuan Yu (1952).

(iv) *Parallel circular openings*

Jeffery (1921) has given a full account of problems in bipolar coordinates, see also Green (1945). The most interesting case of two equal parallel circular holes is discussed by Savin (1961) and Ling (1948), who give numerical results.

If the radii of the holes are a and the distance between their centres is $d + 2a$ the greatest tangential surface stress (at the nearest points of the holes) for uniaxial stress p_1 normal to the plane of their axes increases from $3p_1$ for large d to $3{\cdot}26p_1$ for $d = a$. Thus even at this close approach these stresses are not greatly affected, although for smaller values of d/a they increase rapidly.

The case of an infinite row of equally spaced holes with their centres in a plane is discussed by Howland (1935).

(v) *Notches*

The hyperbolic notch is considered by Neuber (1933, 1958) and the circular notch by Ling (1947).

(vi) *Pressure applied over portion of a cylindrical boundary*

Stresses for the case of pressure, p, applied to a finite length of a solid circular cylinder are calculated by Tranter and Craggs (1945). The case of pressure, p, applied to a finite length of a cylindrical hole in an infinite medium is studied by Kehle (1964) and Bertrand (1964).

(vii) *Displacements around mining excavations*

A number of writers, Berry (1960), Berry and Sales (1961, 1962), Salamon (1963, 1964), and Ryder and Officer (1964), have made calculations more nearly related to actual situations than the idealizations given above. Many of these refer to anisotropic ground, with particular reference to coal mines.

10.16 Surface loads on a semi-infinite region: two-dimensional theory

We shall take the region to be $x > 0$, Fig. 10.16.1 (a), and suppose that normal and tangential stresses N and T are applied to the surface. The total load applied (per unit length perpendicular to the plane Oxy) to the surface between points P_1 and P_2 may be found from

$$\int_{P_1}^{P_2} (N + iT)\,dy = \int_{P_1}^{P_2} \left[\frac{\partial^2 U}{\partial y^2} - i\frac{\partial^2 U}{\partial x\,\partial y}\right] dy = -i\left[\frac{\partial U}{\partial x} + i\frac{\partial U}{\partial y}\right]_{P_1}^{P_2}$$

$$= -i\left[\phi(z) + z\overline{\phi'(z)} + \overline{\psi(z)}\right]_{P_1}^{P_2}, \tag{1}$$

where § 10.2 (16), (31) have been used. This result will be needed later.

Now consider the functions

$$\phi(z) = A \ln z, \quad \psi(z) = B \ln z \tag{2}$$

which are single valued in the region concerned but have a singularity at

$z = 0$ and might be expected to correspond to a concentrated load at this point. Suppose that in (1), P_2 is the point $r_2 e^{i\pi/2}$ and that P_1 is $r_1 e^{-i\pi/2}$, then using (2) in (1) gives

$$\int_{P_1}^{P_2} (N + iT)\, dy = -i[A \ln (r_2/r_1) + i\pi A + \bar{B} \ln (r_2/r_1) - i\pi \bar{B}]. \quad (3)$$

If this is to correspond to a concentrated load of components X, Y at O the term in $\ln (r_2/r_1)$ must vanish, and so (3) gives

$$A + \bar{B} = 0, \quad \pi(A - \bar{B}) = X + iY. \quad (4)$$

It follows that

$$\phi(z) = \frac{X + iY}{2\pi} \ln z, \quad \psi(z) = -\frac{X - iY}{2\pi} \ln z, \quad (5)$$

are the functions appropriate to a line load of components X, Y, per unit length perpendicular to the plane of the paper applied at O. It is convenient to consider the normal and tangential components separately.

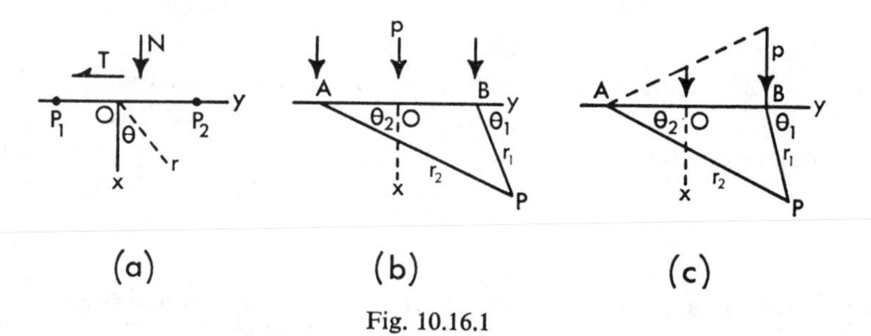

Fig. 10.16.1

(i) *Normal line load*

From (5) with $Y = 0$ and § 10.2 (5), (6), (48)–(50), the stresses in polar coordinates are

$$\sigma_r + \sigma_\theta = \frac{2X}{\pi} \mathbf{R}\left(\frac{1}{z}\right) = \frac{2X \cos \theta}{r\pi}, \quad (6)$$

$$\sigma_\theta - \sigma_r + 2i\tau_{r\theta} = \frac{X}{\pi}\left\{-\frac{\bar{z}}{z^2} - \frac{1}{z}\right\} e^{2i\theta} = -\frac{2X \cos \theta}{r\pi}. \quad (7)$$

It follows that

$$\sigma_r = (2X/r\pi) \cos \theta, \quad \sigma_\theta = \tau_{r\theta} = 0. \quad (8)$$

By considering the stresses over a small semicircle around O it is easy to verify that they are statically equivalent to a normal force X at O.

Despite the simplicity of (8), it is more convenient to work in Cartesian coordinates. These give

$$\sigma_x + \sigma_y = 2Xx/\pi r^2, \tag{9}$$

where

$$r^2 = x^2 + y^2.$$

$$\sigma_y - \sigma_x + 2i\tau_{xy} = \frac{X}{\pi}\left\{-\frac{(x-iy)^3}{r^4} - \frac{(x-iy)}{r^2}\right\}. \tag{10}$$

And so finally

$$\sigma_x = \frac{2Xx^3}{\pi r^4}, \quad \sigma_y = \frac{2Xxy^2}{\pi r^4}, \quad \tau_{xy} = \frac{2Xx^2y}{\pi r^4}, \tag{11}$$

The displacement is, by § 10.2 (50),

$$2G(u + iv) = (X/2\pi)\{(\varkappa + 1)\ln r - \cos 2\theta + i[(\varkappa - 1)\theta - \sin 2\theta]\} \tag{12}$$

so that the normal displacement of the surface is

$$[(\varkappa + 1)X/4\pi G]\ln r. \tag{13}$$

An arbitrary constant displacement may be added to (12) and (13).

(ii) *Tangential line load*
Proceeding in the same way with $X = 0$ in (5), we find

$$\sigma_x = \frac{2Yx^2y}{\pi r^4}, \quad \sigma_y = \frac{2Yy^3}{\pi r^4}, \quad \tau_{xy} = \frac{2Yxy^2}{\pi r^4}, \tag{14}$$

$$2G(u + iv) = (Y/2\pi)\{(1 - \varkappa)\theta - \sin 2\theta + i[(\varkappa + 1)\ln r + \cos 2\theta]\}. \tag{15}$$

Results for many cases of distributed loading may either be written down by developing appropriate values of $\phi(z)$ and $\psi(z)$, cf. Muskhelishvili (1953), or by integration of the elementary solutions (12)–(15). Some important results will be given here using the latter method.

(iii) *Normal load p per unit length over a strip AB of width 2a*
The system and coordinates are shown in Fig. 10.16.1 (*b*). Using (11), the stress σ_x at P, (x, y) is

$$\sigma_x = \frac{2px^3}{\pi}\int_{-a}^{a}\frac{dy'}{[x^2 + (y - y')^2]^2}$$

$$= (p/\pi)\{\theta_1 - \theta_2 - x(y - a)/r_1^2 + x(y + a)/r_2^2\} \tag{16}$$

$$= (p/\pi)\{\theta_1 - \theta_2 - \sin(\theta_1 - \theta_2)\cos(\theta_1 + \theta_2)\}. \tag{17}$$

Similarly

$$\sigma_y = (p/\pi)\{\theta_1 - \theta_2 + x(y - a)/r_1^2 - x(y + a)/r_2^2\}, \tag{18}$$

$$= (p/\pi)\{\theta_1 - \theta_2 + \sin(\theta_1 - \theta_2)\cos(\theta_1 + \theta_2)\}. \tag{19}$$

$$\tau_{xy} = px^2(r_2^2 - r_1^2)/r_1^2 r_2^2, \tag{20}$$

$$= (p/\pi)\sin(\theta_1 - \theta_2)\sin(\theta_1 + \theta_2). \tag{21}$$

The displacement of the surface in the direction of Ox is, except for an additive constant,

$$[(\varkappa + 1)p/4\pi G]\{2a + (y - a)\ln|y - a| - (y + a)\ln|y + a|\}, \quad (22)$$

which holds both inside and outside the loaded area. It is shown in Fig. 10.16.2.

By § 2.3 (14) the principal stresses at (x, y) are

$$(p/\pi)\{\theta_1 - \theta_2 \pm 2ax/r_1r_2\} = (p/\pi)\{\theta_1 - \theta_2 \pm \sin(\theta_1 - \theta_2)\}. \quad (23)$$

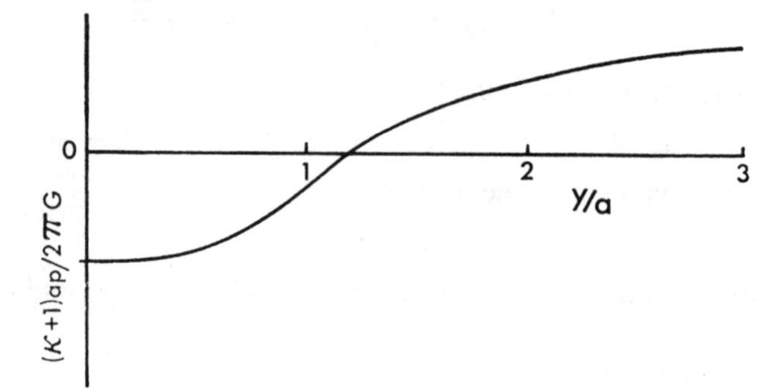

Fig. 10.16.2 Deflection below a loaded strip of width $2a$ on the surface of a semi-infinite region.

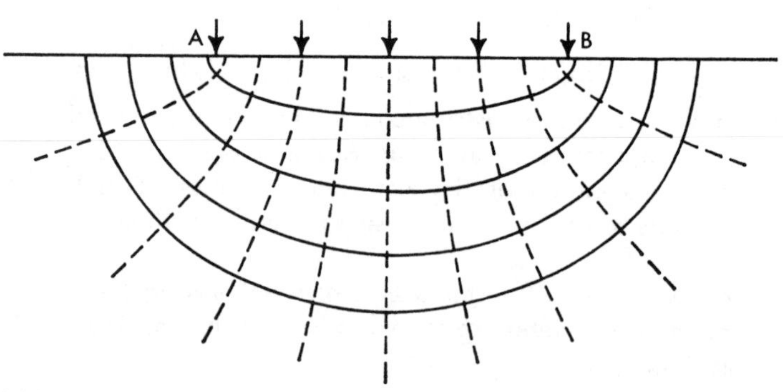

Fig. 10.16.3 Stress-trajectories for uniform loading over a strip AB on the surface of a semi-infinite region.

It follows that the maximum shear stress at any point is

$$2axp/\pi r_1r_2 = (p/\pi)\sin(\theta_1 - \theta_2), \quad (24)$$

and this has a constant value on any circle through AB. The absolute maximum value of p/π is obtained if $\theta_1 - \theta_2 = \frac{1}{2}\pi$, that is, on a semicircle described on AB.

From (17), (19), (21) the angle θ which the principal axes at any point P make with Ox is given by $\tan 2\theta = -\tan(\theta_1 + \theta_2)$, so that

$$\theta = \tfrac{1}{2}(\pi - \theta_1 - \theta_2) \quad \text{or} \quad \theta = -\tfrac{1}{2}(\theta_1 + \theta_2).$$

This implies that the principal axes bisect the angle BPA, so that, by a well-known property of the ellipse, the stress trajectories are ellipses and hyperbolae with foci at A and B. They are shown in Fig. 10.16.3.

(iv) *A linearly increasing load*

Suppose the load increases linearly from zero at A to p at B, as in Fig. 10.16.1 (*c*), in which the coordinates and notation are as for Fig. 10.16.1 (*b*). Then the results are

$$\sigma_x = (p/2\pi)\{[1 + (y/a)](\theta_1 - \theta_2) - \sin 2\theta_1\}, \tag{25}$$

$$\sigma_y = (p/2\pi)\{[1 + (y/a)](\theta_1 - \theta_2) + \sin 2\theta_1 - (x/a)\ln (r_2^2/r_1^2)\}, \tag{26}$$

$$\tau_{xy} = (p/2\pi)\{1 - (x/a)(\theta_1 - \theta_2) - \cos 2\theta_1\}. \tag{27}$$

The results for combinations of these two conditions may be superimposed to give the effect of a wide variety of loading conditions. Jeffreys (1952) uses (25)–(27) to estimate stresses in the crust below mountain ranges.

(v) *A harmonic load*

If the load is $p \cos \omega y$ it follows from (10) that

$$\sigma_x = \frac{2px^3}{\pi}\int_{-\infty}^{\infty} \frac{\cos \omega y' \, dy'}{[x^2 + (y - y')^2]^2}$$

$$= p(1 + \omega x)e^{-\omega x} \cos \omega y, \tag{28}$$

on evaluating the integral. Similarly,

$$\sigma_y = p(1 - \omega x)e^{-\omega x} \cos \omega y, \tag{29}$$

$$\tau_{xy} = \omega p x e^{-\omega x} \sin \omega y. \tag{30}$$

These show the way in which disturbances at the surface die off in depth.

(vi) *Tangential load q per unit length over a strip of width* $2a$

Here, with the coordinates of Fig. 10.16.1 (*b*),

$$\sigma_x = (q/\pi) \sin (\theta_1 - \theta_2) \sin (\theta_1 + \theta_2), \tag{31}$$

$$\sigma_y = (q/\pi)[\ln (r_2^2/r_1^2) - \sin (\theta_1 - \theta_2) \sin (\theta_1 + \theta_2)], \tag{32}$$

$$\tau_{xy} = (q/\pi)[\theta_1 - \theta_2 + \sin (\theta_1 - \theta_2) \cos (\theta_1 + \theta_2)]. \tag{33}$$

Results of this type are fundamental for any problems of surface loading and, in particular, for soil mechanics. Further formulae and references are given by Scott (1963). The problem is one ideally suited for the application of the theory of Fourier transforms and is treated by this method of Sneddon (1951), who also discusses the problem of the strip with stresses applied at both surfaces.

(vii) *Other systems of loading*
Since the previous formulae (17), (19), (21), (25), (26), (27), (31), (32), (33) are expressed almost entirely in terms of angles subtended at the point (x, y) and its distances from the ends of the loaded region, it follows that combinations of these types of loading may readily be expressed in the same way. For example, Voight (1966a) and Jürgensen (1934) give for the loading

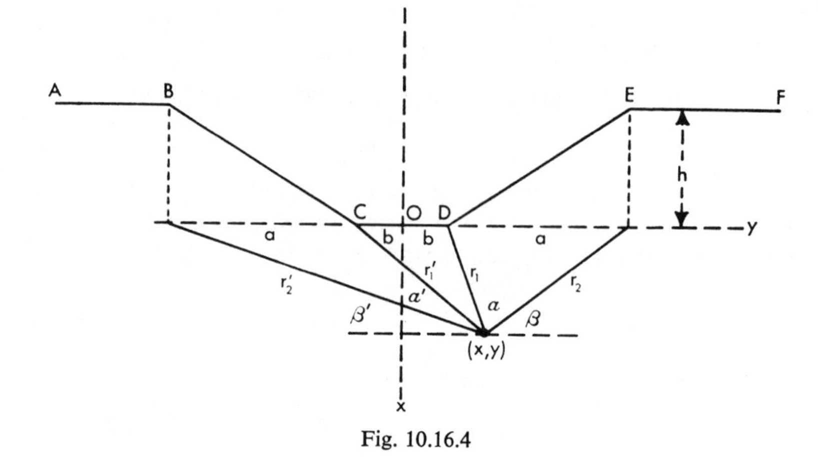

Fig. 10.16.4

represented by the line $ABCDEF$ of material of density ρ on the plane Oy, Fig. 10.16.4,

$$\sigma_x = (\rho gh/\pi)[\beta + \beta' - (b/a)(\alpha + \alpha') + (y/a)(\alpha - \alpha')], \tag{34}$$

$$\sigma_y = (\rho gh/\pi)[\beta + \beta' - (b/a)(\alpha + \alpha') + (y/a)(\alpha - \alpha') + (2x/a)\ln(r_2 r_2'/r_1 r_1')], \tag{35}$$

$$\tau_{xy} = (\rho gh/\pi)(x/a)(\alpha - \alpha'). \tag{36}$$

10.17 Surface loads on a semi-infinite region: three-dimensional theory

There is a number of simple solutions of the equations of elasticity corresponding to various types of 'nuclei of strain'. By combining these, results for more complicated problems may be obtained. Further results of this type may be found in Love (1927) and Mindlin (1936).

The equations of equilibrium in terms of displacements, § 5.5 (10)–(13), are

$$(\lambda + G)\frac{\partial \Delta}{\partial x} + G\nabla^2 u = 0, \quad (\lambda + G)\frac{\partial \Delta}{\partial y} + G\nabla^2 v = 0,$$

$$(\lambda + G)\frac{\partial \Delta}{\partial z} + G\nabla^2 w = 0. \tag{1}$$

(i) *Purely radial displacements or point compression*
The displacements

$$u = Ax/r^3, \quad v = Ay/r^3, \quad w = Az/r^3, \tag{2}$$

where

$$r = (x^2 + y^2 + z^2)^{\frac{1}{2}}, \tag{3}$$

are easily shown to satisfy (1). From them

$$\varepsilon_x = \frac{A}{r^3} - \frac{3Ax^2}{r^5}, \quad \varepsilon_y = \frac{A}{r^3} - \frac{3Ay^2}{r^5}, \quad \varepsilon_z = \frac{A}{r^3} - \frac{3Az^2}{r^5}, \tag{4}$$

$$\Gamma_{xy} = -3Axyr^{-5}, \quad \Gamma_{yz} = -3Ayzr^{-5}, \quad \Gamma_{zx} = -3Azxr^{-5}, \tag{5}$$

$$\Delta = 0. \tag{6}$$

The stress–strain relations, § 5.2 (20), (21), give

$$\sigma_x = 2GA(r^{-3} - 3x^2r^{-5}), \quad \sigma_y = 2GA(r^{-3} - 3y^2r^{-5}),$$
$$\sigma_z = 2GA(r^{-3} - 3z^2r^{-5}), \tag{7}$$

$$\tau_{xy} = -6GAxyr^{-5}, \quad \tau_{yz} = -6GAyzr^{-5}, \quad \tau_{zx} = -6GAzxr^{-5}. \tag{8}$$

Using these and § 2.5 (1), the radial stress in the direction whose direction cosines are x/r, y/r, z/r is

$$\sigma_r = 2GAr^{-7}\{x^2(y^2 + z^2 - 2x^2) + y^2(z^2 + x^2 - 2y^2) +$$
$$z^2(x^2 + y^2 - 2z^2) - 6x^2y^2 - 6y^2z^2 - 6z^2x^2\}$$
$$= -4GAr^{-3}. \tag{9}$$

Similarly, the tangential stress σ_θ in any direction perpendicular to this is found to be

$$\sigma_\theta = 2GAr^{-3}. \tag{10}$$

This solution is appropriate for the case of radial pressure p inside a spherical cavity of radius R. Here, $A = -pR^3/4G$, and

$$\sigma_r = pR^3/r^3, \quad \sigma_\theta = -pR^3/2r^3. \tag{11}$$

If $R \to 0$ and pR^3 is finite it is described as a 'point pressure'.

The solution for the case of a spherical cavity with zero internal pressure in an infinite medium with hydrostatic pressure p at infinity follows from (11) and is

$$\sigma_r = p(1 - R^3r^{-3}), \quad \sigma_\theta = p(1 + \tfrac{1}{2}R^3r^{-3}), \tag{12}$$

so that there is a biaxial tangential compression of $3p/2$ at the surface.

(ii) *The 'point-push' or first type of simple solution*
The displacements in this case are

$$u = \frac{Bxz}{r^3}, \quad v = \frac{Byz}{r^3}, \quad w = B\left(\frac{z^2}{r^3} + \frac{\lambda + 3G}{r(\lambda + G)}\right) \tag{13}$$

$$\Delta = -2GBz/(\lambda + G)r^3, \tag{14}$$

and may be shown to satisfy (1). The stresses follow from § 5.2 (20), (21), and are

$$\left.\begin{aligned}
\sigma_x &= \frac{2GBz}{r^3}\left\{\frac{G}{\lambda+G}-\frac{3x^2}{r^2}\right\}, & \sigma_y &= \frac{2GBz}{r^3}\left\{\frac{G}{\lambda+G}-\frac{3y^2}{r^2}\right\},\\
\sigma_z &= -\frac{2GBz}{r^3}\left\{\frac{G}{\lambda+G}+\frac{3z^2}{r^2}\right\}, & \tau_{yz} &= -\frac{2GBy}{r^3}\left\{\frac{G}{\lambda+G}+\frac{3z^2}{r^2}\right\},\\
\tau_{zx} &= -\frac{2GBx}{r^3}\left\{\frac{G}{\lambda+G}+\frac{3z^2}{r^2}\right\}, & \tau_{xy} &= -\frac{6GBxyz}{r^5}.
\end{aligned}\right\} \quad (15)$$

These become infinite as $r \to 0$, and to interpret them we consider the stresses on a small sphere surrounding the origin. If p_z is the component of stress in the z-direction over an element of surface of this sphere whose normal has direction cosines x/r, y/r, z/r it follows from § 2.4 (7) that

$$\begin{aligned}
p_z &= -\frac{2GBx^2}{r^4}\left\{\frac{G}{\lambda+G}+\frac{3z^2}{r^2}\right\} - \frac{2GBy^2}{r^4}\left\{\frac{G}{\lambda+G}+\frac{3z^2}{r^2}\right\} - \frac{2GBz^2}{r^4}\left\{\frac{G}{\lambda+G}+\frac{3z^2}{r^2}\right\}\\
&= -\frac{2GB}{r^2}\left\{\frac{G}{\lambda+G}+\frac{3z^2}{r^2}\right\}.
\end{aligned}$$

Integrating this over a sphere of radius r gives

$$-8\pi GB\int_0^{\frac{1}{2}\pi}\left\{\frac{G}{\lambda+G}+3\cos^2\theta\right\}\sin\theta\, d\theta = -\frac{8\pi GB(\lambda+2G)}{\lambda+G}. \quad (16)$$

In the same way the resultant forces in the perpendicular directions are

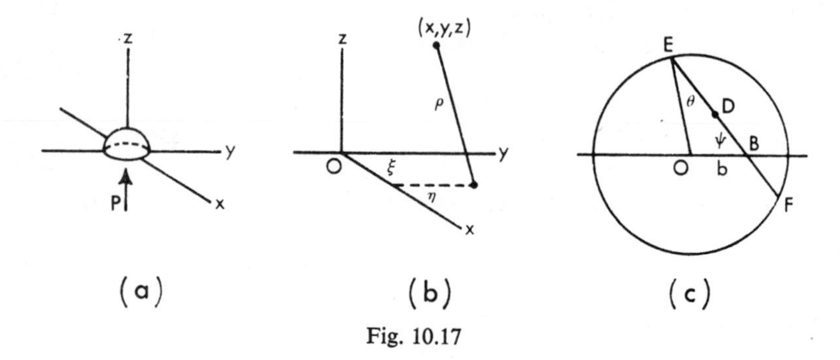

Fig. 10.17

found to be zero. Thus the solution (13)–(15) corresponds to a localized force

$$-8\pi GB(\lambda+2G)/(\lambda+G) \quad (17)$$

applied at the origin in an infinite medium in the direction of the positive z-axis.

For the semi-infinite region $z > 0$ the displacements and stresses (13) and (15) could be produced by stresses

$$\sigma_z = 0, \quad \tau_{yz} = -2G^2By/(\lambda + G)(x^2 + y^2)^{3/2},$$
$$\tau_{zx} = -2G^2Bx/(\lambda + G)(x^2 + y^2)^{3/2}, \tag{18}$$

in the plane $z = 0$, together with a resultant force Q in the z-direction applied over a vanishingly small hemisphere surrounding the origin, Fig. 10.17 (*a*), where

$$Q = -4\pi GB(\lambda + 2G)/(\lambda + G). \tag{19}$$

(iii) *The 'second type' of simple solution*

The displacements in this case are

$$u = \frac{Cx}{r(z + r)}, \quad v = \frac{Cy}{r(z + r)}, \quad w = \frac{C}{r}, \tag{20}$$

where r is defined in (3). It may be verified that they satisfy (1) except on the negative z-axis, $z \leqslant 0$, where they become infinite. They correspond to a line of point compressions along the negative z-axis, and so may be obtained by integration from (2).

The corresponding stress-components are

$$\left.\begin{aligned}
\sigma_x &= 2GC\left\{\frac{y^2 + z^2}{r^3(z + r)} - \frac{x^2}{r^2(z + r)^2}\right\}, \\
\sigma_y &= 2GC\left\{\frac{z^2 + x^2}{r^3(z + r)} - \frac{y^2}{r^2(z + r)^2}\right\}, \\
\sigma_z &= -2GCzr^{-3}, \quad \tau_{yz} = -2GCyr^{-3}, \\
\tau_{zx} &= -2GCxr^{-3}, \quad \tau_{xy} = -2GCxy(z + 2r)r^{-3}(z + r)^{-2}.
\end{aligned}\right\} \tag{21}$$

On the plane $z = 0$ these give, except at the origin,

$$\sigma_z = 0, \quad \tau_{yz} = -2GCy(x^2 + y^2)^{-3/2},$$
$$\tau_{zx} = -2GCx(x^2 + y^2)^{-3/2}. \tag{22}$$

Also the type of argument leading to (17) shows that the resultant force Q' in the z-direction across a small hemisphere in the region $z > 0$ with centre at the origin is

$$Q' = -4\pi GC. \tag{23}$$

(iv) *The semi-infinite region $z > 0$ with a concentrated normal force P at the origin in the z-direction*

The situation is that of Fig. 10.17 (*a*) and may be obtained by adding the solutions (13) and (20) for this region. It follows from (15) and (21) that

$$\sigma_z = \tau_{yz} = \tau_{zx} = 0 \text{ on the plane } z = 0 \quad \text{if} \qquad GB = -(\lambda + G)C, \tag{24}$$

except at the origin. Also, in the region of the origin the total force P over the small hemisphere is by (19) and (23)

$$P = Q + Q' = -4\pi GC - 4\pi GB(\lambda + 2G)/(\lambda + G). \tag{25}$$

It follows from (24) and (25) that

$$B = -P/4\pi G, \quad C = P/4\pi(\lambda + G).$$

Combining (13) with (20), and (15) with (21), with these values gives

$$u = \frac{P}{4\pi}\left\{\frac{x}{(\lambda + G)r(z + r)} - \frac{xz}{Gr^3}\right\}, \tag{26}$$

$$v = \frac{P}{4\pi}\left\{\frac{y}{(\lambda + G)r(z + r)} - \frac{yz}{Gr^3}\right\}, \tag{27}$$

$$w = -\frac{P}{4\pi}\left\{\frac{z^2}{Gr^3} + \frac{\lambda + 2G}{G(\lambda + G)r}\right\}, \tag{28}$$

$$\sigma_x = \frac{P}{2\pi}\left\{\frac{3x^2z}{r^5} + \frac{G(y^2 + z^2)}{(\lambda + G)r^3(z + r)} - \frac{Gz}{(\lambda + G)r^3} - \frac{Gx^2}{(\lambda + G)r^2(z + r)^2}\right\}, \tag{29}$$

$$\sigma_y = \frac{P}{2\pi}\left\{\frac{3y^2z}{r^5} + \frac{G(z^2 + x^2)}{(\lambda + G)r^3(z + r)} - \frac{Gz}{(\lambda + G)r^3} - \frac{Gy^2}{(\lambda + G)r^2(z + r)^2}\right\}, \tag{30}$$

$$\sigma_z = 3Pz^3/2\pi r^5, \quad \tau_{yz} = 3Pyz^2/2\pi r^5, \quad \tau_{zx} = 3Pxz^2/2\pi r^5, \tag{31}$$

$$\tau_{xy} = \frac{P}{2\pi}\left\{\frac{3xyz}{r^5} - \frac{Gxy(z + 2r)}{(\lambda + G)r^3(z + r)^2}\right\}. \tag{32}$$

(v) *Distributed normal loading $p(x, y)$ over a region of the surface $z = 0$ of the semi-infinite solid $z > 0$*

The displacements and stresses may be obtained by integration from (26) to (32). For example, from (26) the displacement u at x, y, z due to normal force $p(\xi, \eta)\, d\xi\, d\eta$ at $(\xi, \eta, 0)$ is

$$\frac{1}{4\pi}\left\{\frac{x - \xi}{(\lambda + G)\rho(z + \rho)} - \frac{z(x - \xi)}{G\rho^3}\right\}p(\xi, \eta)\, d\xi\, d\eta, \tag{33}$$

where

$$\rho = [(x - \xi)^2 + (y - \eta)^2 + z^2]^{\frac{1}{2}}. \tag{34}$$

The coordinate system is shown in Fig. 10.17 (*b*). Integrating (33) over the loaded portion of the plane gives

$$u = \frac{1}{4\pi G}\iint\left\{\frac{G(x-\xi)}{(\lambda+G)\rho(z+\rho)} - \frac{z(x-\xi)}{\rho^3}\right\}p(\xi,\eta)\,d\xi\,d\eta, \tag{35}$$

$$v = \frac{1}{4\pi G}\iint\left\{\frac{G(y-\eta)}{(\lambda+G)\rho(z+\rho)} - \frac{z(y-\eta)}{\rho^3}\right\}p(\xi,\eta)\,d\xi\,d\eta, \tag{36}$$

$$w = -\frac{1}{4\pi G}\iint\left\{\frac{z^2}{\rho^3} + \frac{\lambda+2G}{(\lambda+G)\rho}\right\}p(\xi,\eta)\,d\xi\,d\eta, \tag{37}$$

$$\sigma_x = \frac{1}{2\pi}\iint\left\{\frac{3(x-\xi)^2 z}{\rho^5} + \frac{G[(y-\eta)^2+z^2]}{(\lambda+G)\rho^3(z+\rho)} - \frac{Gz}{(\lambda+G)\rho^3} - \frac{G(x-\xi)^2}{(\lambda+G)\rho^2(z+\rho)^2}\right\}p(\xi,\eta)\,d\xi\,d\eta, \tag{38}$$

$$\sigma_y = \frac{1}{2\pi}\iint\left\{\frac{3(y-\eta)^2 z}{\rho^5} + \frac{G[(x-\xi)^2+z^2]}{(\lambda+G)\rho^3(z+\rho)} - \frac{Gz}{(\lambda+G)\rho^3} - \frac{G(y-\eta)^2}{(\lambda+G)\rho^2(z+\rho)^2}\right\}p(\xi,\eta)\,d\xi\,d\eta, \tag{39}$$

$$\sigma_z = \frac{3z^3}{2\pi}\iint \rho^{-5}p(\xi,\eta)\,d\xi\,d\eta, \tag{40}$$

$$\tau_{yz} = \frac{3z^2}{2\pi}\iint (y-\eta)\rho^{-5}p(\xi,\eta)\,d\xi\,d\eta, \tag{41}$$

$$\tau_{zx} = \frac{3z^2}{2\pi}\iint (x-\xi)\rho^{-5}p(\xi,\eta)\,d\xi\,d\eta, \tag{42}$$

$$\tau_{xy} = \frac{1}{2\pi}\iint\left\{\frac{3z(x-\xi)(y-\eta)}{\rho^5} - \frac{G(x-\xi)(y-\eta)(z+2\rho)}{(\lambda+G)\rho^3(z+\rho)^2}\right\}p(\xi,\eta)\,d\xi\,d\eta. \tag{43}$$

Since $$G/(\lambda+G) = 1-2\nu, \quad G = E/2(1+\nu),$$

these are readily written in terms of E and ν.

An important result is that the normal displacement w on the plane $z = 0$ is

$$w = -\frac{(\lambda+2G)}{4\pi G(\lambda+G)}\iint\frac{p(\xi,\eta)\,d\xi\,d\eta}{[(x-\xi)^2+(y-\eta)^2]^{\frac{1}{2}}}. \tag{44}$$

The above relations can be expressed simply in terms of the quantity

$$\chi = \iint \ln(\rho+z)p(\xi,\eta)\,d\xi\,d\eta, \tag{45}$$

for which

$$\frac{\partial \chi}{\partial x} = \iint \frac{(x-\xi)}{\rho(z+\rho)} p(\xi, \eta)\, d\xi\, d\eta, \tag{46}$$

$$\frac{\partial \chi}{\partial y} = \iint \frac{(y-\eta)}{\rho(z+\rho)} p(\xi, \eta)\, d\xi\, d\eta, \tag{47}$$

$$\frac{\partial \chi}{\partial z} = \iint \rho^{-1} p(\xi, \eta)\, d\xi\, d\eta = \Phi, \tag{48}$$

where Φ is the Newtonian potential of a distribution $p(x, y)$ on the surface $z = 0$. Writing

$$4\pi\Omega = \frac{z\Phi}{G} + \frac{\chi}{\lambda + G}, \tag{49}$$

(35) to (37) become

$$u = \frac{\partial \Omega}{\partial x}, \quad v = \frac{\partial \Omega}{\partial y}, \quad w = \frac{\partial \Omega}{\partial z} - \frac{(\lambda + 2G)}{2\pi G(\lambda + G)}\Phi. \tag{50}$$

In many interesting cases, Φ can be written down from known results in potential theory, and so the tedious integrations involved in (35)–(37) are avoided.

(vi) *Constant normal pressure p over a circle of radius a*

The most interesting question is that of the displacement w in the plane $z = 0$. It will be calculated for the point B, $(b, 0, 0)$, inside the circle, Fig. 10.17 (*c*). The result (44) then is

$$w = -\frac{(\lambda + 2G)}{4\pi G(\lambda + G)} \iint \rho^{-1}\, d\xi\, d\eta,$$

where $\rho^2 = (\xi - b)^2 + \eta^2$, and the integral is taken over the circle. Introducing polar coordinates with B as origin this becomes

$$w = -\frac{(\lambda + 2G)p}{4\pi G(\lambda + G)} \cdot 2\int_0^{\pi/2} d\psi \int d\rho, \tag{51}$$

where the integral in ρ is taken over the chord EF and so is $EF = 2a \cos \theta$. Since $b \sin \psi = a \sin \theta$, this becomes

$$w = -\frac{(\lambda + 2G)pa}{\pi G(\lambda + G)} \int_0^{\pi/2} \{1 - (b^2/a^2) \sin^2 \psi\}^{\frac{1}{2}}\, d\psi \tag{52}$$

$$= -\frac{4pa(1 - \nu^2)}{\pi E} \int_0^{\pi/2} \{1 - (b^2/a^2) \sin^2 \psi\}^{\frac{1}{2}}\, d\psi. \tag{53}$$

This is an elliptic integral. When $b = 0$ its value is $-2pa(1 - \nu^2)/E$. On the circumference, $b = a$, its value is $-4pa(1 - \nu^2)/\pi E$.

The average value w_{av} of w over the circle is

$$w_{av} = -0{\cdot}54\pi pa(1 - \nu^2)/E. \tag{54}$$

This situation occurs in plate bearing tests, § 15.7. The problem of uniform displacement $w = w_0$ by a rigid stamp of radius a can also be solved, and in this case σ_z varies across the circle and the mean value σ_{av} is given by

$$w_0 = -\pi a\sigma_{av}(1 - \nu^2)/2E. \tag{55}$$

These problems are the celebrated problems of Boussinesq, and they, and problems of indentation by cones, may readily be solved by the use of the Hankel transform (Sneddon, 1951).

10.18 Anisotropic materials

Two-dimensional problems for orthotropic materials may be solved by the methods developed in § 5.7.

Introducing the Airy function § 5.7 (7), the elastic equations § 5.12 (33) become

$$\varepsilon_x = s_{11}\frac{\partial^2 U}{\partial y^2} + s_{12}\frac{\partial^2 U}{\partial x^2}, \quad \varepsilon_y = s_{12}\frac{\partial^2 U}{\partial y^2} + s_{22}\frac{\partial^2 U}{\partial x^2}, \quad \nu_{xy} = -s_{66}\frac{\partial^2 U}{\partial x\,\partial y}. \tag{1}$$

Substituting these in the compatibility condition for strains, § 5.7 (1) gives

$$s_{22}\frac{\partial^4 U}{\partial x^4} + (s_{66} + 2s_{12})\frac{\partial^4 U}{\partial x^2\,\partial y^2} + s_{11}\frac{\partial^4 U}{\partial y^4} = 0. \tag{2}$$

This may be written in the form

$$\left(\frac{\partial^2}{\partial x^2} + \alpha_1\frac{\partial^2}{\partial y^2}\right)\left(\frac{\partial^2}{\partial x^2} + \alpha_2\frac{\partial^2}{\partial y^2}\right)U = 0, \tag{3}$$

where
$$\alpha_1\alpha_2 = s_{11}/s_{22} = E_2/E_1, \tag{4}$$

$$\alpha_1 + \alpha_2 = (s_{66} + 2s_{12})/s_{22} = (E_2/G) - 2\nu_{21}. \tag{5}$$

α_1 and α_2 may be real or complex. They are real and positive for many materials, and this will be assumed to be the case here. (3) is a generalization of the biharmonic equation. It may be solved either by an extension of the complex variable methods of § 10.2, Savin (1961), Lekhnitskii (1963), or by more elementary methods.

Introducing three sets of polar coordinates

$$x + iy = re^{i\theta}, \quad x + iy/\alpha_1^{\frac{1}{2}} = r_1e^{i\theta_1}, \quad x + iy/\alpha_2^{\frac{1}{2}} = r_2e^{i\theta_2}, \tag{6}$$

it may be verified that (3) has elementary solutions of the forms

$$r_1^{\pm n}\cos n\theta_1, \quad r_1^{\pm n}\sin n\theta_1, \quad r_2^{\pm n}\cos n\theta_1, \quad r_2^{\pm n}\sin n\theta_2, \tag{7}$$

as well as terms involving $\ln r_1$, $\ln r_2$.

Green and Taylor (1939, 1945) develop solutions in the form of series of this type. In particular, in the second paper they study the problem of the circular hole in an infinite region with uniaxial compression p_1 at infinity in the direction of Ox, which is an axis of symmetry. The tangential stress σ_θ at the surface of the hole at a point whose polar angle relative to Ox is θ is found to be

$$\sigma_\theta = \frac{p_1(1+\gamma_1)(1+\gamma_2)(1+\gamma_1+\gamma_2-\gamma_1\gamma_2-2\cos 2\theta)}{(1+\gamma_1^2-2\gamma_1\cos 2\theta)(1+\gamma_2^2-2\gamma_2\cos 2\theta)}, \tag{8}$$

where $\gamma_1 = (\alpha_1^{\frac{1}{2}}-1)/(\alpha_1^{\frac{1}{2}}+1), \quad \gamma_2 = (\alpha_2^{\frac{1}{2}}-1)/(\alpha_2^{\frac{1}{2}}+1).$ (9)

Green (1942, 1945) develops a solution by the use of the complex variable and applies it to the elliptic hole, to two circular holes, and to the circular hole with various internal conditions.

For the circular hole with pure shear S at infinity referred to the axes Ox, Oy the tangential stress σ_θ in the surface of the hole is (Green, 1939)

$$\sigma_\theta = \frac{4S(\gamma_1\gamma_2-1)\sin 2\theta}{(1+\gamma_1^2-2\gamma_1\cos 2\theta)(1+\gamma_2^2-2\gamma_2\cos 2\theta)}. \tag{10}$$

For the circular hole with radial pressure p at its surface and no stress at infinity, the tangential stress σ_θ at the surface is (Green, 1939)

$$\sigma_\theta = p.\frac{1+(\gamma_1+\gamma_2)^2-3\gamma_1^2\gamma_2^2-2(\gamma_1+\gamma_2)(1-\gamma_1\gamma_2)\cos 2\theta-2\gamma_1\gamma_2\cos 4\theta}{(1+\gamma_1^2-2\gamma_1\cos 2\theta)(1+\gamma_2^2-2\gamma_2\cos 2\theta)}. \tag{11}$$

The difficulty in applying these and similar solutions is lack of information about the four parameters involved. Green and Taylor (loc. cit.) give numerical calculations for wood, but these are not of great value in the present context, since the ratio of the moduli E_1/E_2 is very large (greater than 6), while for rocks it is rarely as high as 2. To illustrate the effects to be expected, calculations from (8), (10), (11) are shown in Fig. 10.18 (*a*), (*b*), (*c*) for the case

$$E_1 = 5 \times 10^6 \text{ psi}, \quad E_2 = 2{\cdot}5 \times 10^6, \quad G = 1{\cdot}25 \times 10^6, \quad \nu_{21} = 0{\cdot}15, \quad \nu_{12} = 0{\cdot}3.$$

Fig. 10.18 (*a*), Curve I, shows the tangential tension σ_θ in the surface of the hole as a function of θ for the case of uniaxial compression p_1 at infinity parallel to E_1. Curve II is the corresponding result for p_1 parallel to E_2, and Curve III is the value for the isotropic case. It is seen that the difference between the curves is not large, and the principal effect observed is that the tangential tension which occurs in the direction of the stress p_1 is greater by 50 per cent if p_1 is in the direction of the greatest Young's modulus.

Fig. 10.18 (*b*), Curve I, shows the value of σ_θ/S for a uniform shear S at infinity, while Curve II is the result for the isotropic case. It appears that the direction of maximum tension is displaced by about 10° towards the direction of lower Young's modulus.

Fig. 10.18 (*c*), Curve I, shows the effect of internal pressure p. The change from the value $-p$ for isotropic material is small, being about a 7 per cent increase in the direction of both greatest and least moduli.

As remarked above, the 2 : 1 ratio of $E_1 : E_2$ used above is rather extreme, so it is probably true that the effects of anistropy of strength of rocks are much more important in failure under inhomogeneous stresses than effects of anisotropy of elasticity.

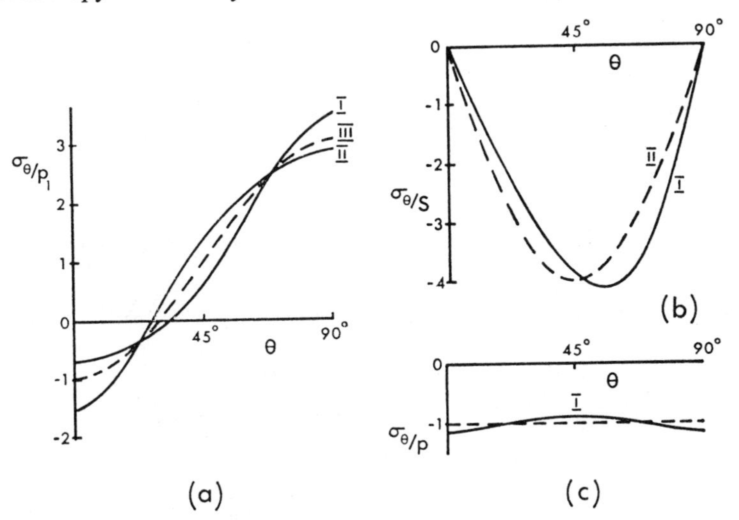

Fig. 10.18 Variation of tangential tension in the surface of a circular hole in an orthotropic material. (*a*) Compression p_1 at infinity: Curve I, parallel to maximum Young's modulus: Curve II, parallel to least Young's modulus: Curve III, isotropic material. (*b*) Pure shear S at infinity: Curve I, orthotropic case: Curve II, isotropic case. (*c*) Internal pressure p.

Many other solutions of importance are available in the literature. Green and Taylor (1939) discuss the point load in an infinite orthotropic plate, and Green (1942) point loads in the semi-infinite plate. See also Hearmon (1961), Savin (1961), and Lekhnitskii (1963).

10.19 Photoelasticity

Photoelastic measurements are one of the most important tools in rock mechanics. They enable the stresses around any two-dimensional opening or combination of openings to be measured and have been extensively employed for this purpose. More recently they have been extended to handle problems in three dimensions, and have been used for other purposes, such as *in situ* stress measurement.

The theory and practice of photoelasticity is fully discussed in the standard works, Coker and Filon (1957), Hetényi (1950), Durelli, Phillips and Tsao (1958), and only a brief account will be given here to indicate the type of information which can be obtained.

The principle behind the method is that certain materials become doubly refracting under stress. If σ_1 and σ_2 are the principal stresses at a point in a plane sheet of such material it is found experimentally that there are two principal directions of optical vibration which coincide with the principal axes of stress, and the indices of refraction n_1 and n_2 in these directions are connected with the principal stresses by the relation

$$n_1 - n_2 = C(\sigma_1 - \sigma_2), \tag{1}$$

where C is a constant of the material.

Suppose, now, that Ox and Oy, Fig. 10.19 (*a*), are the directions of the

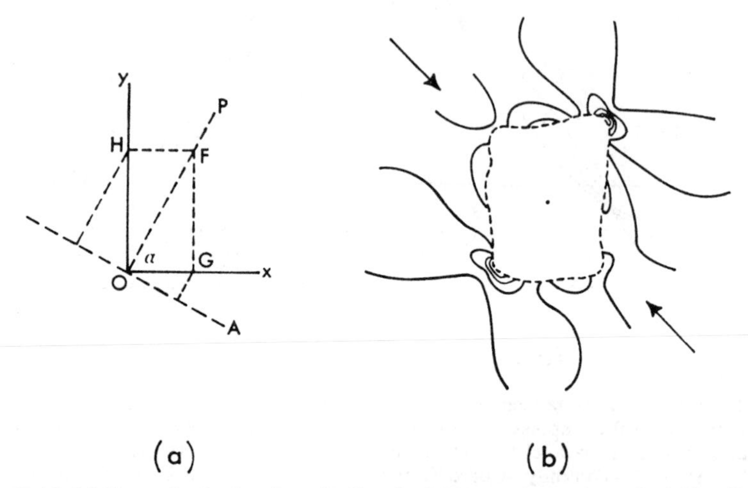

Fig. 10.19 (*a*) Photoelastic situation. (*b*) Sketch of fringe patterns for uniaxial loading in the direction of the arrow applied to an actual tunnel cross-section.

principal axes and that a ray of monochromatic plane polarized light is incident normally on the sheet, its vibration being $OF = a \sin \omega t$ in the plane OP inclined at α to Ox. This vibration will be resolved into components $OG = a \cos \alpha \sin \omega t$ and $OH = a \sin \alpha \sin \omega t$ along the principal directions Ox and Oy, and these will be propagated with refractive indices n_1 and n_2. If the emergent beam is viewed through an analyser which only accepts vibrations in the plane OA perpendicular to OP ('crossed Nicols') the vibration ξ observed through this will be

$$\begin{aligned} \xi &= a \cos \alpha \sin \omega(t - n_1 l/c) \sin \alpha - a \sin \alpha \sin \omega(t - n_2 l/c) \cos \alpha \\ &= -a \sin 2\alpha \sin [\omega l(n_1 - n_2)/2c] \cos [\omega(t - [n_1 + n_2]l/2c)], \end{aligned} \tag{2}$$

where l is the thickness of the plate and c is the speed of light *in vacuo*.

If the material is unstressed, $n_1 = n_2$, and $\xi = 0$. For stressed material, ξ vanishes if either

$$\omega l(n_1 - n_2)/c = 2m\pi, \quad m = 1, 2, \ldots \tag{3}$$

or

$$\sin 2\alpha = 0. \tag{4}$$

Using (1), the first of these conditions becomes

$$\sigma_1 - \sigma_2 = m\lambda/lC, \quad m = 1, 2, \ldots \tag{5}$$

where λ is the wavelength of the light. Thus curves on which the stress-difference is constant and an integral multiple of λ/lC will appear dark. The constant λ/lC may be determined by a calibration experiment. The stress difference $\sigma_1 - \sigma_2$ can then be determined absolutely by counting the number of curves or 'fringes'. If white light is used instead of monochromatic, different wavelengths will be extinguished in different positions and the curves will exhibit a regular gradation of colours, hence the name isochromatics used for such curves in § 2.3.

Photoelastic observations thus immediately determine the stress-difference $\sigma_1 - \sigma_2$ at any point. To determine σ_1 and σ_2, use is made of the fact that $\sigma_1 + \sigma_2$ satisfies Laplace's equation in two dimensions, § 5.5 (16). A determination of $\sigma_1 + \sigma_2$ can therefore be made by an independent analogue method, such as the use of an electrolytic tank or conducting paper, and hence σ_1 and σ_2 are found.

The other condition (4) for no light to be transmitted implies that $\alpha = 0$ or $\alpha = \frac{1}{2}\pi$, corresponding to all points at which the directions of the principal axes coincide with the directions of the polarizer and analyser. Such points lie on an isoclinic, § 2.3, and by rotating the polarizer (with the analyser remaining perpendicular to it) all isoclinics may be found. From the isoclinics, or directly, the directions of the principal stresses at every point may be found. With the simple arrangement described above, points corresponding to both isoclinics and isochromatics are observed simultaneously. A simple modification, using circular polarization, allows the isochromatics to be observed alone. If it is desired, for example, to study the stresses around a tunnel, an opening of the required shape is cut in a sheet of photoelastic material and stresses in two chosen directions are applied to the sheet mechanically. An example of the isochromatics obtained in this way is shown in Fig. 10.19 (*b*).

It should be added that the method is not limited to homogeneous materials, but may be used to study effects in layered, jointed, and granular materials. Such studies, however, must remain largely qualitative, since if much complexity is introduced the individual elements become too small for precise measurements of conditions within them to be made.

Three-dimensional photoelastic measurements depend on the fact that certain materials, such as some resins, behave to a first approximation as

Kelvin materials, § 11.3, in which the viscosity of the materials in the dash-pot depends strongly on temperature. If the material is heated to a temperature (of the order of 100–150° C) at which the viscosity is negligible the springs alone behave as an elastic body in equilibrium: if the material is now cooled, the stresses in the spring system are 'frozen' and the viscosity in the dash-pots in fact becomes so high that the material may be sectioned (and the stresses in the sections determined photoelastically) without appreciable creep. Effects due to gravity may be simulated by centrifuging. This method has the disadvantage that for the resins at present in use Poisson's ratio is high, of the order of 0·5, which complicates comparison with the behaviour of rocks. One important application has been to the study of the stress-concentration near the end of a flat-ended borehole or drive, Galle and Wilhoit (1962).

10.20 Finite element methods

There are many numerical methods of solving the equations of elasticity. Most of these involve the use of a rectangular grid, and therefore are not very suitable for the study of irregular boundaries. Recently, a new method, known as finite element analysis, was developed for engineering problems and has come into use very rapidly. It has the advantage that the discussion throughout is in terms of matrices, so that it is very well adapted to modern computers. To date, the problems studied have been mostly two-dimensional or axially symmetrical. The method has the great advantage that it is capable of handling anisotropic, inhomogeneous, and non-linear or bilinear materials of irregular shapes without additional complication. It does, however, need a large computer.

Essentially, in plane problems, the region considered is divided up into a number of 'finite elements', usually triangles. These elements are supposed to be connected only at *nodal points*, which are the vertices of triangles, the sides of triangles between nodal points being free. The forces acting on the body, both surface forces and body forces, are replaced by a statically equivalent system of forces acting at the nodal points.

The fundamental assumption is that stress and strain are homogeneous in each element. This implies that straight lines remain straight lines, so that the edges of two elements which are in contact remain in contact during straining. The components of stress and strain in each element are constant over the element. The forces exerted on each element by the stresses across its faces are replaced by a statically equivalent set of forces acting at its vertices, which are nodal points.

The fundamental quantities for each triangular element are the components of displacement u_x, u_y and force s_x, s_y in the x- and y-directions at each of its vertices 1, 2, 3. The components of displacement

$$u_{x1},\quad u_{y1},\quad u_{x2},\quad u_{y2},\quad u_{x3},\quad u_{y3}, \tag{1}$$

may be represented by a matrix $[u]$ of six rows and one column. The components of strain ε_x, ε_y, γ_{xy} may be represented by a matrix $[\varepsilon]$ of three rows and one column. The components of strain may be calculated from the displacements (1), and the relationship may be expressed by the matrix equation

$$[\varepsilon] = [A] \times [u], \tag{2}$$

where $[A]$ is a matrix of three rows and six columns depending on the shape and orientation of the triangle.

Similarly, the forces

$$s_{x1}, \quad s_{y1}, \quad s_{x2}, \quad s_{y2}, \quad s_{x3}, \quad s_{y3}, \tag{3}$$

acting at the vertices of the triangle may be represented by a 6×1 matrix $[s]$ and the stress-components σ_x, σ_y, τ_{xy} by a 3×1 matrix $[\sigma]$. The forces may be calculated from the stress-components, and the result takes the form

$$[s] = [B] \times [\sigma], \tag{4}$$

where $[B]$ is a matrix of six rows and three columns depending on the shape and orientation of the triangle.

Finally, the stress–strain relations for the material and the conditions postulated (e.g. plane strain) can be expressed by a matrix equation

$$[\sigma] = [C] \times [\varepsilon]. \tag{5}$$

Combining (2), (4), (5) gives

$$[s] = [B] \times [C] \times [A] \times [u], \tag{6}$$

or

$$[s] = [k] \times [u], \tag{7}$$

where

$$[k] = [B] \times [C] \times [A] \tag{8}$$

is a 6×6 matrix called the stiffness matrix of the triangular element.

If $\mathbf{s}_1$, $\mathbf{s}_2$, $\mathbf{s}_3$ are the vector forces at the vertices and $\mathbf{u}_1$, $\mathbf{u}_2$, $\mathbf{u}_3$ are the vector displacements (so that $\mathbf{s}_1$ has components $\mathbf{s}_{x1}$, $\mathbf{s}_{y1}$, and so on), (7) can be reduced to the form

$$[\mathbf{s}] = [K] \times [\mathbf{u}], \tag{9}$$

where K is a 3×3 matrix.

So far, only one triangle has been considered. Suppose, now, that two contiguous triangles A, with vertices 1, 2, 3, and B, with vertices 2, 3, 4, have vertices 2 and 3 in common. An equation relating the force at each nodal point with the displacements at the four nodal points can be written down from the stiffness matrices for the two triangles. The four such relations so obtained can be combined in the matrix relation

$$[\mathbf{s}] = [K_4] \times [\mathbf{u}], \tag{10}$$

where now **s** and **u** are 4×1 matrices referring to the four nodal points and K_4 is a 4×4 matrix which can be built up from the stiffness matrices of the two triangles.

In the same way, if there are n nodal points a relation

$$[\mathbf{s}] = [K_n] \times [\mathbf{u}], \tag{11}$$

where now $[K_n]$ is an $n \times n$ matrix, will hold. While this analysis may appear complicated, the individual operations are well adapted for computer work and systems with several hundred nodal points can readily be handled.

The sketch given above is based on the 'engineering' approach in much of the early work: a more sophisticated analysis can be based on energy considerations, Goodman and Taylor (1967). The method has been applied to a number of problems with known solutions to check its accuracy, Clough (1960, 1965). Recently it has been applied to a wide variety of problems: dams and foundations, Clough and Wilson (1963) and Zienkiewicz and Cheung (1966); tunnels and openings, Goodman (1966) and Reyes and Deere (1966); slope problems, Anderson and Dodd (1966) and Goodman and Taylor (1967).

Clearly the method has many advantages for use in problems in rock mechanics:

(i) The elements can be chosen in any way and of any size so that irregular boundaries can be fitted without trouble, and a greater density of nodal points can be chosen in regions of greater stress concentration.

(ii) Variable surface and body forces are easily handled.

(iii) Each element has its own stiffness matrix, so that the material need not be homogeneous or isotropic. Also non-linear and bilinear materials can easily be studied.

(iv) The method is not restricted to elastic behaviour, and other rheological models can be introduced.

(v) Friction on joint surfaces can be allowed for.

(vi) The final output can be processed in the computer in any desired fashion.

Chapter Eleven

Time-dependent Effects

11.1 Introduction

The study of time-dependent effects, usually spoken of under the general title of 'creep', is of the greatest importance in rock mechanics and geophysics.

In underground work the movements which occur after excavation are of creep type and the requirement is to find laws by which future behaviour

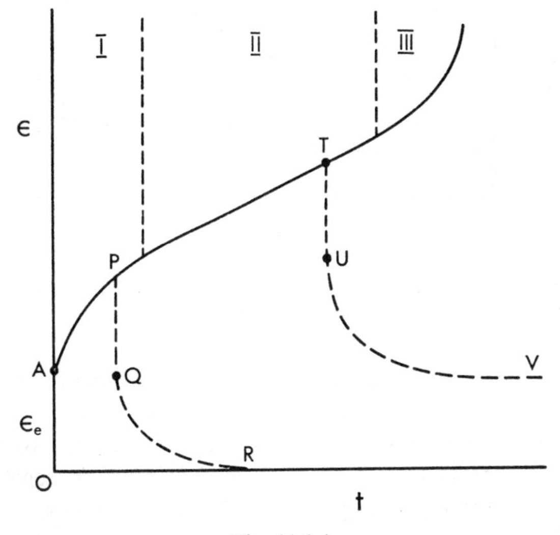

Fig. 11.1.1

may be predicted. On a smaller scale, creep movements occur in measuring systems such as flat-jacks, and it is necessary to be able to correct for them. Creep of pillars is of the greatest importance in mining such materials as halite and some shales.

Time-dependent effects are observed in much laboratory testing, and for this reason it is usual to increase stress at a standard rate, Obert *et al.* (1946). The next step is to study these effects under controlled conditions, in particular at high temperatures and confining pressures, which may be expected to provide fundamental information about the possibility of flow of material in the earth's interior.

There are many useful reviews of creep of rocks, Robertson (1964) and Murrell and Misra (1961–62). Creep of metals at elevated temperatures has been much studied because of its economic importance, Sully (1956) and Rotherham (1951), and the general behaviour of rocks is very similar. The typical behaviour is shown in Fig. 11.1.1. If constant stress is applied to the material an instantaneous elastic strain ε_e first appears. This is followed by a region I in which the strain–time curve is concave downwards: creep in this region is called *primary* or *transient* creep. This is followed by

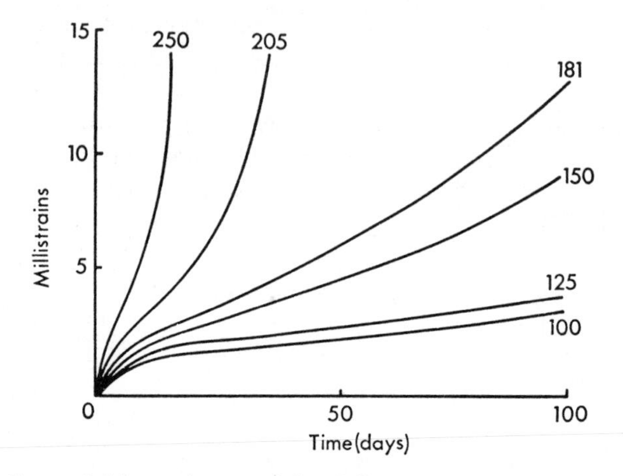

Fig. 11.1.2 Creep of alabaster in water (after Griggs, 1940). Numbers on the curves are values of the uniaxial compressive stress in bars.

region II, in which the curve has approximately constant slope: creep in the region is secondary or *steady-state* creep. Finally, region III corresponds to accelerating or *tertiary* creep and leads rapidly to failure.

If the applied stress is suddenly reduced to zero in region I the ε, t curve takes the form PQR, Fig. 11.1.1, in which $PQ = \varepsilon_e$ and QR tends asymptotically to zero. There is thus no permanent set, so that the material remains elastic in the sense of § 4.2; this behaviour is described as time-dependent elasticity or anelasticity, and is studied in detail by Zener (1948). If the applied stress is suddenly reduced to zero in the steady-creep region II a curve TUV is traversed which leads asymptotically to permanent set.

An actual set of creep curves in uniaxial compression is shown in Fig. 11.1.2, the numbers on the curves being the values of the stress. This illustrates one of the difficulties of creep experiments – if the stress is too low, little effect is produced: if it is too high, accelerating creep and failure occur too soon: the choice of stress, therefore, is a matter of difficulty and importance, since each experiment may occupy an apparatus for a considerable time. This difficulty may be partly avoided by incremental creep measurements (Hardy, 1966a), in which the stress is increased by fixed amounts at regular intervals, but these experiments are rather more difficult to interpret. Partly because of these difficulties, and partly because of the difficulty of maintaining the load constant over long periods of time, an alternative approach to time-dependent behaviour has been made by some workers by using constant, slow rates of strain. Serdengecti and Boozer (1961) have used strain rates from 10^{-2} to 10^{3} millistrains/sec, and Heard (1963) has studied Yule marble in triaxial extension with strain rates of 3×10^{-4} to 400 millistrains/sec. It is usual to distinguish between 'slow straining' apparatus of this type and 'creep' apparatus in which the load is maintained constant. In modern versions of the latter the load is maintained constant by a servo-control system.

Two other difficulties of creep measurements should be mentioned. In most cases the strains are small so that temperature effects on the specimen and measuring apparatus are important. Secondly, with many sediments, absorption of water produces effects similar to, and of the same order as, those of creep, so that humidity, also, has to be very carefully controlled.

While direct tension is very commonly used with metals, it is rarely used for rocks. Bending is the simplest method for studying creep and the effects of moisture on rocks. It was extensively used by Phillips (1931, 1932, 1948) with particular reference to sediments in coal measures. Price (1964) has also used bending to study mine rock, and Pomeroy (1956) has used it to study creep of coal.

Important pioneer experiments on igneous rocks using torsion were made by Michelson (1917, 1920), and further work with this method was done by Lomnitz (1956).

By far the greatest number of workers have used uniaxial compression, for example, Evans and Wood (1937) on granite, marble, and slate; Matsushima (1960) on granite; Nishihara (1958), Hardy (1959, 1966), Price (1964) on sediments. Other workers, notably Griggs (1936, 1939, 1940), Robertson (1960), and le Comte (1965) on rock salt, have used triaxial compression. Griggs (1940) demonstrated that pressure of pore water had a very marked effect on creep of some materials. Griggs's most recent apparatus is designed for year-long runs at temperatures of up to 1,000° C and confining pressures of up to 20 kbars. Load is applied and can be maintained constant by a servo-controlled screw mechanism which can provide strain rates as low as 10^{-10} sec^{-1}.

11.2 Empirical laws of creep

The creep strain shown in Fig. 11.1.1 may be represented by

$$\varepsilon = \varepsilon_e + \varepsilon_1(t) + Vt + \varepsilon_3(t), \tag{1}$$

where ε_e is the instantaneous elastic strain, $\varepsilon_1(t)$ is the transient creep, Vt is the steady-state creep, and $\varepsilon_3(t)$ the accelerating creep. We shall confine our attention to transient and steady-state creep, so that only the magnitude of V and the form of $\varepsilon_1(t)$ are in question.

Steady-state creep is sometimes called pseudo-viscous flow and expressed in terms of a viscosity η by the relations $\dot{\varepsilon} = \sigma/3\eta$ for a Maxwell substance in uniaxial compression or $\dot{\varepsilon} = (\sigma - \sigma_0)/3\eta$ for a Bingham substance, § 11.3, Griggs (1939).

Two empirical laws have been much used to represent transient creep of metals and have also been used for rocks, namely the power law, Cottrell (1952),

$$\varepsilon_1(t) = At^n, \quad 0 < n < 1, \tag{2}$$

in particular Andrade's (1910, 1914) one-third power law

$$\varepsilon_1(t) = At^{1/3}, \tag{3}$$

and the logarithmic law

$$\varepsilon_1(t) = A \ln t. \tag{4}$$

In these, the 'constants' A refer to the particular conditions of the experiment and depend on temperature and confining pressure in a manner which will be discussed later.

None of (2)–(4) conform strictly to the idea of transient creep as stated above, since they increase continually with time instead of tending to a constant value. Also they suffer from the mathematical disadvantage that the creep velocities $\dot{\varepsilon}_1(t)$ become infinite as $t \to 0$. To avoid this, Lomnitz (1956, 1957) proposed the law

$$\varepsilon_1(t) = A \ln [1 + \alpha t] \tag{5}$$

and interpreted his experiments in terms of it. Jeffreys (1958) proposed the 'modified Lomnitz' law

$$\varepsilon_1(t) = A[(1 + at)^\alpha - 1], \tag{6}$$

which has considerable mathematical advantages, and used it in geophysical work.

The logarithmic law (4) has been much used for interpretation of experimental results, in particular by Griggs (1939), who found, for example, for Solenhofen limestone at a uniaxial compressive stress of 1,400 bars

$$\varepsilon = (6{\cdot}1 + 5{\cdot}2 \log T) \times 10^{-5},$$

where T is the time in days. This gives a very good fit for times of up to one year

The various linear rheological models which will be discussed in detail in § 11.3 give laws of which the simplest is

$$\varepsilon_1(t) = A[1 - e^{-t/t_1}], \tag{7}$$

where t_1 is a constant, and more complicated systems lead to expressions of type

$$\varepsilon_1(t) = A[1 - e^{-t/t_1}] + B[1 - e^{-t/t_2}] + \ldots \tag{8}$$

There is a complete difference in principle between the laws (2)–(6),

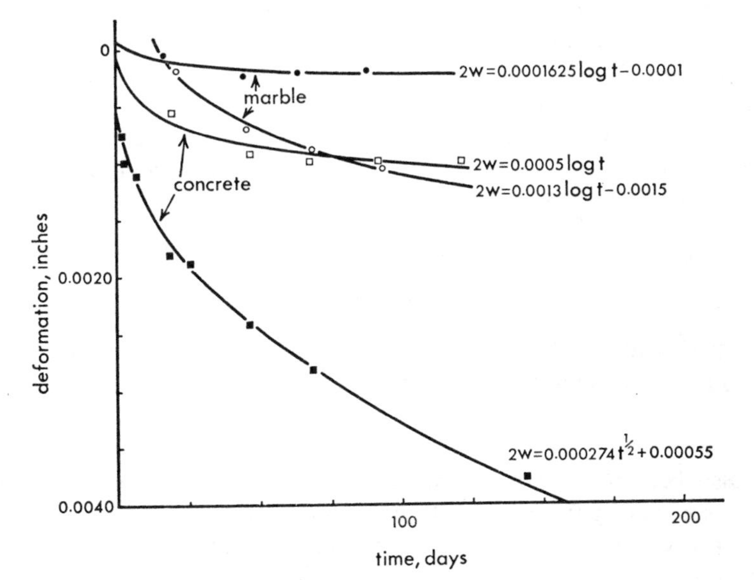

Fig. 11.2 Creep in a flat-jack with time after slot-cutting. The lowest curve is for very weak concrete.

which are essentially empirical, and (7) and (8) derived from linear rheological theory. The former provide useful representations of experimental data and can be used for extrapolation for short times, but they cannot be used as the basis of a theory which will predict the behaviour of more complicated systems. The linear rheological models discussed in § 11.3, which lead to one-dimensional solutions of types (7) or (8), can be generalized to give a theory applicable to time-dependent problems in three dimensions, and therefore, although they are based on very simple physical idealizations, they have a very great advantage.

Nishihara (1958) has discussed the representation of experimental results in terms of (7) and (8), and Hardy (1966) has shown that many experimental results can be expressed in terms of the Burgers model § 11.3 (25). However, in many practical cases the information available does not permit discrimination between, say, (4) and (7). This is of considerable importance in attempting to make long-term predictions, since (4) increases steadily with time, while (7) tends to a constant value.

As a specific practical example, Fig. 11.2 shows the creep between the measuring pins of a flat-jack, § 15.2, after cutting the slot. It appears that three of the four curves can be well represented by a logarithmic law of type (4). This implies that the logarithmic law can be used for extrapolation to estimate the effect of slot-cutting on later measurements, and this is usually done. The curve of Fig. 11.2 representing a very weak concrete, did not fit the logarithmic law well, and the exponential law (7) was rather inferior in all cases.

As remarked above, the 'constants' V in (1) and A in (3)–(8) refer only to the conditions of the particular experiment and depend strongly on temperature and pressure. Most attention has been concentrated on the steady state creep velocity $\dot{\epsilon} = V$. In metal theory, empirical laws of dependence with the stress σ

$$V = V_0 e^{\sigma/\sigma_0}, \tag{9}$$

and

$$V = V_0 \sinh(\sigma/\sigma_0), \tag{10}$$

were suggested by Ludwik (1909) and Nadai (1938), respectively. Nadai (1963) gives an extensive discussion. The power law

$$V = V_0 \sigma^n \tag{11}$$

has also been used to represent experimental results on rocks with exponents n varying from 1 to 5, Robertson (1964). Creep is in fact a molecular process, Cottrell (1953), and should be controlled by the Eyring equation

$$V = V_0 \exp[-E/RT] \sinh(\sigma/\sigma_0), \tag{12}$$

where E is an activation energy, R is the gas constant, and T is the absolute temperature. Thus (12) gives a theoretically based generalization of (10) to include the effects of temperature, Ree *et al.* (1960). Equation (12) has been used by Heard (1963) to interpret results on Yule marble in extension. For example, he finds at 5,000 bars confining pressure and 10 per cent strain normal to the foliation

$$V = 2{\cdot}9 \times 10^8 \exp[-62{,}400/RT] \sinh 10\sigma, \tag{13}$$

where σ is the value in kbars of the differential stress.

Information of this sort is fundamental for the study of processes in the interior of the earth, Stacey (1963), for example, Griggs and Handin (1960) suggested that thermal instability of creep was responsible for deep focus earthquakes.

Finally, it should be emphasized that it is not clear that the constants deduced from laboratory measurements on intact specimens of rock are applicable to practical situations in which extensively fractured rock may occur: the essential point is that empirical laws of the types discussed above appear to apply in both cases with the appropriate constants.

11.3 Simple rheological models

Rheology, which is the study of flow in general, has made much use of simple models to define fundamental types of behaviour, Reiner (1947, 1949), Eirich (1956). These models are built up as combinations of simple ideas, such as linear elasticity and viscosity, and may conveniently be represented by simple mechanical models. The equations in one dimension obtained in this way may be generalized to give three-dimensional theory analogous to the theory of elasticity.

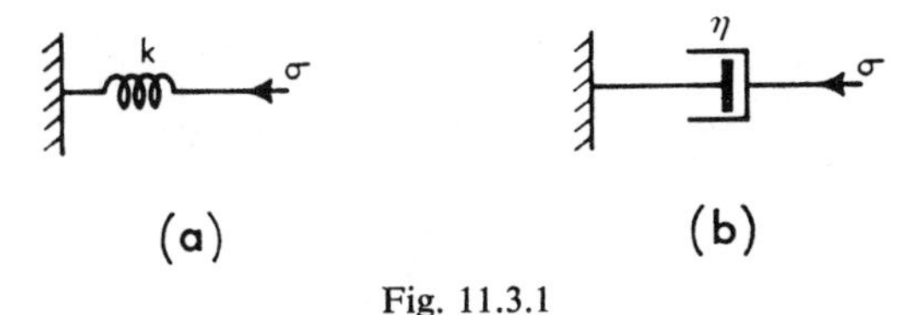

Fig. 11.3.1

The simple models will be stated in terms of stress and strain in uniaxial compression, but they apply equally well to other systems, such as shear. For this reason the non-committal symbol k will be used for the modulus in place of E.

The two basic models are the linearly elastic or *Hookean* substance, which has the stress–strain relation

$$\sigma = k\varepsilon \tag{1}$$

and may be represented by a spring of stiffness k, Fig. 11.3.1 (*a*), and the perfectly viscous or *Newtonian* substance which may be represented by a dash-pot, Fig. 11.3.1 (*b*), and has the stress–strain relation

$$\sigma = \eta\dot{\varepsilon}, \tag{2}$$

where, as usual, dots denote differentiation with respect to the time. In (1) and (2) k and η are constants.

It follows from (2) that if the strain ε is zero when $t = 0$,

$$\varepsilon = \sigma t/\eta.$$

More complicated systems may be built up from combinations of the simple elements of Fig. 11.3.1.

(i) *The Maxwell or elastico-viscous substance*

This consists of a spring k and a dash-pot η in series, Fig. 11.3.2 (*a*), so that the same stress σ is applied to both. If ε_1 and ε_2 are the strains in them it follows from (1) and (2) that

$$\varepsilon_1 = \sigma/k, \quad \dot{\varepsilon}_2 = \sigma/\eta, \tag{3}$$

so that the total strain $\varepsilon = \varepsilon_1 + \varepsilon_2$ satisfies the differential equation

$$\dot{\varepsilon} = (\dot{\sigma}/k) + (\sigma/\eta). \tag{4}$$

If the system is unstrained at $t = 0$ and constant stress S is applied at that time it follows from (3) that

$$\varepsilon = (S/k) + St/\eta, \tag{5}$$

so that there is an instantaneous strain S/k followed by linearly increasing strain St/η, Fig. 11.3.2 (*b*).

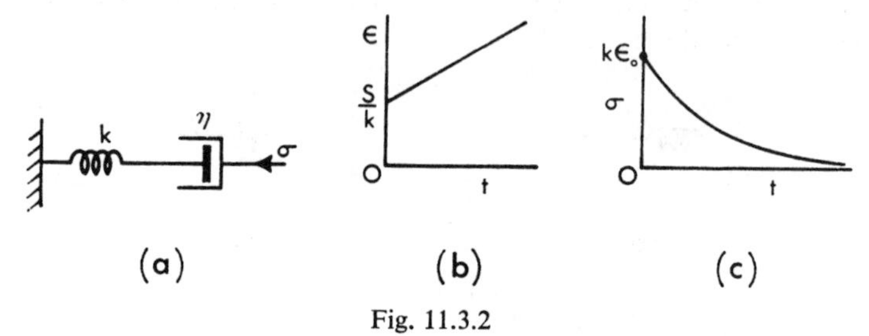

Fig. 11.3.2

If a fixed strain ε_0 is suddenly imposed on the system at $t = 0$ the stress rises suddenly to $k\varepsilon_0$, and subsequently is given by

$$\sigma = k\varepsilon_0 e^{-t/t_1}, \tag{6}$$

where

$$t_1 = \eta/k. \tag{7}$$

The stress thus falls off exponentially or *relaxes* with time, Fig. 11.3.2 (*c*). The time t_1 at which it reaches $1/e$ times its initial value is called the relaxation time.

This model was introduced by Maxwell to describe the behaviour of substances such as pitch which show instantaneous elasticity but flow under small stresses if they are applied for a sufficiently long time. It has been applied to the study of the earth's mantle, which must behave elastically for short times, since it transmits shear waves, but is assumed to be able to flow under long-term stresses. It is essentially the rheid model introduced by Carey (1953).

(ii) *The Kelvin, Voigt, or firmo-viscous substance*

This may be represented by the combination of a spring k and dash-pot η in parallel, Fig. 11.3.3 (*a*). In this case, if ε is in the strain, the stresses across the spring and dash-pot will be given by (1) and (2), respectively, so that σ, which is the sum of these individual stresses, will be given by

$$\sigma = \eta\dot{\varepsilon} + k\varepsilon. \tag{8}$$

This differential equation has to be solved with specified initial conditions. Here and subsequently the method of solution will be assumed known and solutions written down: they may most simply be obtained by the use of the Laplace transformation method, which is sketched briefly in § 11.4.

Suppose, first, that the system is compressed so that the strain is ε_0 when

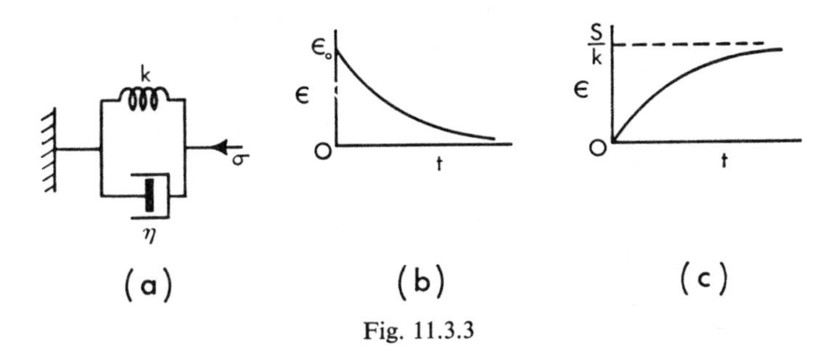

Fig. 11.3.3

$t = 0$ and that the stress is then released. The differential equation (8) in this case becomes

$$\eta\dot{\varepsilon} + k\varepsilon = 0, \quad t > 0, \tag{9}$$

to be solved with $\varepsilon = \varepsilon_0$ when $t = 0$. The solution is

$$\varepsilon = \varepsilon_0 e^{-t/t_1}, \tag{10}$$

where

$$t_1 = \eta/k. \tag{11}$$

The strain thus decays exponentially from its initial value ε_0 with relaxation time t_1, Fig. 11.3.3 (*b*).

If constant stress S is suddenly applied at $t = 0$ when the system is unstrained, (8) becomes

$$\eta\dot{\varepsilon} + k\varepsilon = S, \tag{12}$$

to be solved with $\varepsilon = 0$ when $t = 0$. The solution is

$$\varepsilon = (S/k)[1 - e^{-t/t_1}] \tag{13}$$

where t_1 is given by (11). The elastic strain S/k is now approached exponentially with time constant t_1.

This is the formula § 11.2 (7) for transient creep.

This model does not adequately represent the typical behaviour of Fig. 11.1 (*a*), since it does not give an instantaneous elastic strain ε_e. This deficiency is remedied in the generalization of it which follows.

(iii) *The generalized Kelvin substance*
This consists of a Kelvin element k_1, η_1 arranged in series with a spring k_2, Fig. 11.3.4 (*a*). If ε_1 and ε_2 are the strains in the two elements, respectively, it follows from (1) and (8) that

$$\left.\begin{aligned} \sigma = \eta_1\dot{\varepsilon}_1 + k_1\varepsilon_1, \quad \sigma = k_2\varepsilon_2, \\ \text{and} \qquad \varepsilon = \varepsilon_1 + \varepsilon_2. \end{aligned}\right\} \tag{14}$$

Eliminating ε_1 and ε_2 from (14) gives

$$\sigma = \eta_1(\dot{\varepsilon} - \dot{\sigma}/k_2) + k_1(\varepsilon - \sigma/k_2),$$

or

$$\eta_1\dot{\sigma} + (k_1 + k_2)\sigma = k_2(\eta_1\dot{\varepsilon} + k_1\varepsilon). \tag{15}$$

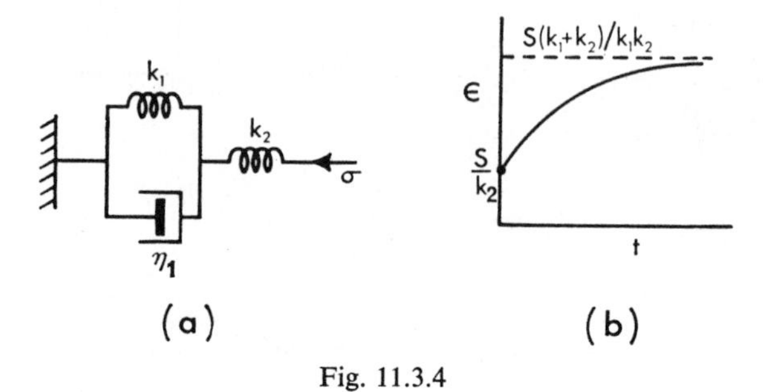

Fig. 11.3.4

If constant stress S is suddenly applied at $t = 0$ when the strain is zero, there is an instantaneous strain ε_e given by

$$\varepsilon_e = S/k_2, \tag{16}$$

and the value of the strain subsequently is

$$\varepsilon = \frac{S}{k_2} + \frac{S}{k_1}(1 - e^{-t/t_1}), \tag{17}$$

where

$$t_1 = \eta_1/k_1. \tag{18}$$

The strain thus tends finally to the value $S(k_1 + k_2)/k_1k_2$ with time constant t_1, Fig. 11.3.4 (*b*).

This model, and the simple Kelvin model, have been used to represent the damping of oscillations by internal friction. The term 'internal friction' is used in this context to denote dissipation of energy in the dash-pot and not, of course, in the sense of § 4.6. Suppose that there are steady periodic oscillations of period $2\pi/\omega$ in the system, so that

$$\varepsilon = \xi e^{i\omega t}, \quad \sigma = Se^{i\omega t}. \tag{19}$$

Substituting (19) in (15) gives

$$\frac{S}{\xi} = \frac{k_2(k_1 + i\omega\eta_1)}{(k_1 + k_2) + i\omega\eta_1} = Ke^{i\delta} \tag{20}$$

where $\tan \delta = \omega(t_1 - t_2)/(1 + \omega^2 t_1 t_2),$ (21)

and $t_1 = \eta_1/k_1, \quad t_2 = \eta_1/(k_1 + k_2).$

δ measures the lag of strain behind stress and provides a measure of the damping. The internal friction of solids is frequently measured by studying the variation of δ with ω, Zener (1948).

(iv) *The Burgers substance*

This consists of a Kelvin element, k_1, η_1, in series with a Maxwell element, k_2, η_2, Fig. 11.3.5 (*a*). If ε_1 and ε_2 are the strains in these two elements, (8) and (4) give

$$\sigma = \eta_1\dot{\varepsilon}_1 + k_1\varepsilon_1, \quad \dot{\varepsilon}_2 = (\dot{\sigma}/k_2) + (\sigma/\eta_2), \tag{22}$$

with

$$\varepsilon = \varepsilon_1 + \varepsilon_2. \tag{23}$$

Eliminating ε_1 and ε_2 from (22) and (23) gives

$$\eta_1\ddot{\varepsilon} + k_1\dot{\varepsilon} = (\eta_1/k_2)\ddot{\sigma} + [1 + (k_1/k_2) + (\eta_1/\eta_2)]\dot{\sigma} + (k_1/\eta_2)\sigma. \tag{24}$$

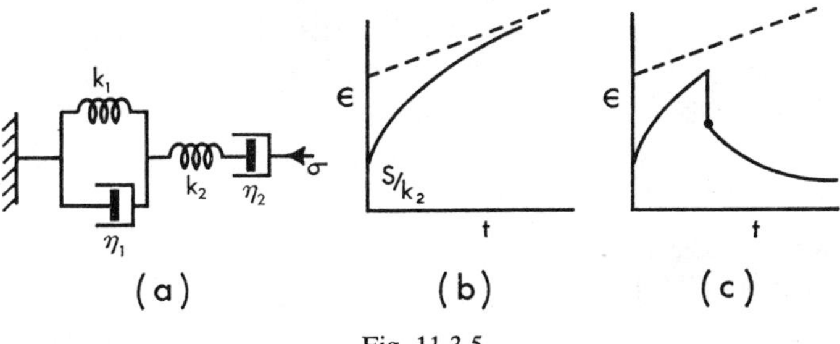

Fig. 11.3.5

If constant stress S is applied to the system at $t = 0$ when it is unstrained there is an instantaneous strain S/k_2, and subsequently

$$\varepsilon = \frac{S}{k_2} + \frac{S}{k_1}[1 - e^{-t/t_1}] + \frac{S}{\eta_2}t, \tag{25}$$

where $t_1 = \eta_1/k_1$.

This model gives the simplest representation of a material which shows an instantaneous strain, transient creep, and steady-state creep, Fig. 11.3.5 (*b*). Nishihara (1958) and Hardy (1966) have applied it to experimental results. The latter used curve fitting by a computer to determine the four parameters k_1, k_2, η_1, η_2. If the stress is released after some time permanent displacement results, Fig. 11.3.5 (*c*).

(v) *Linear viscoelastic substances and the principle of superposition*
Clearly, any more complicated arrangement of the elements previously used will lead to a linear differential equation connecting σ and ε. This will take the general form

$$f(D)\sigma = g(D)\varepsilon, \tag{26}$$

where $f(D)$ and $g(D)$ are polynomials in D, and D denotes differentiation with respect to the time. Such substances will be called linear viscoelastic substances. The solution of (26) will lead to a relation between stress and strain of the form

$$\varepsilon/S = A + \sum_{n} B_n(1 - e^{-t/t_n}), \tag{27}$$

if constant stress S is applied when the system is unstrained.

One important property of such systems is the principle of superposition, which may be stated as follows. Suppose that $f(t)$ is the strain produced by unit stress applied suddenly at $t = 0$ when the system is unstrained. Then the strain $\varepsilon(t)$ at time t caused by stresses S applied at $t = 0$, S_1 at $t = T_1$ (that is, a change from S to S_1 at T_1), S_2 at $t = T_2$, and so on is

$$\varepsilon(t) = Sf(t) + (S_1 - S)f(t - T_1) + (S_2 - S_1)f(t - T_2) + \ldots \tag{28}$$

that is, the effect at time t is the sum of the changes of stress multiplied by the value of $f(t)$ for the time for which each changed stress is in operation. If the stress varies continuously so that it may be written $S(t)$, (28) becomes the integral

$$\varepsilon(t) = \int_0^t S(u)f'(t - u)\,du. \tag{29}$$

As a simple example from (25), suppose that the stress is reduced from S to zero at time T. Then for $t > T$

$$\begin{aligned}\varepsilon(t) &= \frac{S}{k_2} + \frac{S}{k_1}[1 - e^{-t/t_1}] + \frac{S}{\eta_2}t - \frac{S}{k_2} - \frac{S}{k_1}[1 - e^{-(t-T)/t_1}] - \frac{S\,(t - T)}{\eta_2} \\ &= \frac{S}{k_1}[e^{-(t-T)/t_1} - e^{-t/t_1}] + \frac{ST}{\eta_2}.\end{aligned} \tag{30}$$

This is shown in Fig. 11.3.5 (c) and is of the general type described in connection with Fig. 11.1 (a). In principle, all viscoelastic materials 'remember' their previous stress-history.

(vi) *Non-linear models. The Bingham substance*
Non-linear models offer much greater mathematical difficulty. They are of two types: the first of these, which are much studied in rheology, replace the constants k and η used above by functions of stress or strain.

The second, which are of more interest in the present context, introduce the simple mechanical model of a frictional contact to represent a yield point. If this contact can supply frictional stress σ_0 we have the *Saint Venant* model of Fig. 11.3.6 (*a*), in which there is no strain if $\sigma < \sigma_0$ and indefinite strain if $\sigma > \sigma_0$.

The most important generalization of this is the Bingham substance, Fig. 11.3.6 (*b*), in which stress is applied to a Saint Venant element through a spring k, and the motion of the element is resisted by a dash-pot η. In this case

$$\varepsilon = \sigma/k, \quad \text{if} \quad \sigma < \sigma_0, \tag{31}$$

$$\varepsilon = (\sigma - \sigma_0)t/\eta + \sigma/k, \quad \text{if} \quad \sigma < \sigma_0. \tag{32}$$

The theory of the Bingham substance has a close relationship to the

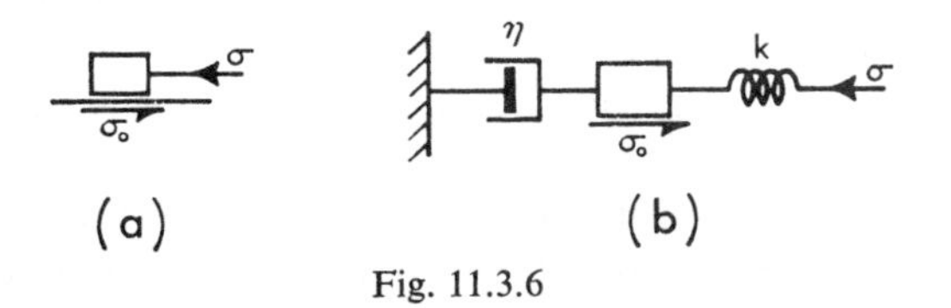

Fig. 11.3.6

theory of plasticity and may be extended to study problems in two dimensions. Many such solutions contain regions in which the yield condition is not satisfied, and these regions move as if they were solid.

Other non-linear mechanical models have been suggested. For example, Attewell (1962) has proposed a representation for porous rock which contains a dash-pot of variable viscosity. Price (1964) has proposed a *Bingham–Voigt* model in which the spring of the Bingham model is replaced by a generalized Kelvin element: this behaves as a generalized Kelvin substance for stresses below the yield stress in creep, σ_0, and for greater stresses and long times as a Maxwell substance. This gives a reasonable representation of laboratory experiments on many rocks.

11.4 Behaviour of more complicated systems

Just as the three-dimensional theory of elasticity was obtained by assuming linear stress–strain relations suggested by the simple Hooke's law in uniaxial compression, so three-dimensional equations of viscoelasticity may be set up involving substances with the properties described in § 11.3.

The stress–strain relations of § 5.4 (4), (1) may be written

$$s_{rs} = 2Ge_{rs}, \tag{1}$$

$$s = 3Ke, \tag{2}$$

where (1) comprises the six equations § 5.4 (2), (3), r and s being supposed to run through the letters, x, y, z and s_{xx}, s_{yy}, s_{zz}, being written for s_x, s_y, s_z. This is, essentially, the tensor notation which is especially convenient here. Of these equations, the set (1) refers to distortion and (2) to hydrostatic compression, each having its own elastic modulus.

The obvious generalization of § 11.3 (26) to three dimensions is to assume in place of (1)

$$f(D)s_{rs} = 2g(D)e_{rs}, \tag{3}$$

where $f(D)$ and $g(D)$ have forms suggested by the models of § 11.3, but now relating to distortion. Similarly, in place of (2) we take

$$f_1(D)s = 3g_1(D)e, \tag{4}$$

where, again, $f_1(D)$ and $g_1(D)$ have forms suggested by the models § 11.3, but relating to hydrostatic compression. $f_1(D)$ and $g_1(D)$ need not be the same as $f(D)$ and $g(D)$, and, for simplicity, it will usually be assumed here that the material behaves elastically in hydrostatic compression, so that (4) reduces to (2).

(3) and (4), or (3) and (1), are a system of ordinary linear differential equations which have to be solved in conjunction with the stress-equations of equilibrium and the boundary conditions. Lee (1955) pointed out that the solution of these equations is greatly simplified by the use of the Laplace transformation, which enables the solutions of many problems in elasticity to be taken over for viscoelastic problems.

The theory of the Laplace transform is given by many authors, e.g. Jaeger (1961a), Churchill (1958), and Gardner and Barnes (1942), and will be summarized very briefly.

If $u(t)$ is a function of the time (and of the space variables) its Laplace transform $\bar{u}$ with respect to the time is defined as

$$\bar{u} = \int_0^\infty e^{-pt}u(t)\,dt, \tag{5}$$

where p is a real positive number sufficiently large to make the integral converge. $\bar{u}$ is, of course, a function of p and the space variables.

In particular if

$$u(t) = 1, \quad \bar{u} = 1/p, \tag{6}$$

$$u(t) = t, \quad \bar{u} = 1/p^2, \tag{7}$$

$$u(t) = e^{-at}, \quad \bar{u} = 1/(p+a). \tag{8}$$

If $u(t) = u_0$ when $t = 0$, it follows by integration by parts that

$$\int_0^\infty e^{-pt}(Du)\,dt = p\bar{u} - u_0, \tag{9}$$

and so if $u_0 = 0$ the Laplace transform of Du is $p\bar{u}$, that is, is obtained by replacing D by p and u by $\bar{u}$. Similarly, if the values of u, Du, D^2u, . . . are all zero when $t = 0$, and $f(D)$ is a polynomial in D,

$$\int_0^\infty e^{-pt}[f(D)u]\,dt = f(p)\bar{u}, \tag{10}$$

and so is obtained simply by replacing $f(D)u$ by $f(p)\bar{u}$.

Differential equations are then solved by writing down their Laplace transforms, finding $\bar{u}$ algebraically, and then writing down $u(t)$ from it by using a table of Laplace transforms such as (6)–(8). For example, the Laplace transform of § 11.3 (8) with $\varepsilon = 0$, when $t = 0$ is, using (9),

$$\eta p\bar{\varepsilon} + k\bar{\varepsilon} = \bar{\sigma}, \tag{11}$$

and if $\sigma = S_0$, constant, $\bar{\sigma} = S_0/p$ by (6), so that

$$\bar{\varepsilon} = \frac{S_0}{p(\eta p + k)} = \frac{S_0}{k}\left\{\frac{1}{p} - \frac{1}{p + k/\eta}\right\}. \tag{12}$$

It then follows from (6) and (8) that

$$\varepsilon = (S_0/k)[1 - e^{-kt/\eta}],$$

which is the previous solution § 11.3 (13).

As a further example, if the stress is linearly increasing, so that $\sigma = S_1 t$, $\bar{\sigma} = S_1/p^2$ by (7) and we get in place of (12)

$$\bar{\varepsilon} = \frac{S_1}{k}\left\{\frac{1}{p^2} - \frac{1}{p(p + k/\eta)}\right\} = \frac{S_1}{k}\left\{\frac{1}{p^2} - \frac{\eta}{k}\left[\frac{1}{p} - \frac{1}{p + k/\eta}\right]\right\},$$

so that from (6)–(8)

$$\varepsilon = \frac{S_1 t}{k} - \frac{S_1 \eta}{k^2}[1 - e^{-kt/\eta}]. \tag{13}$$

All the differential equations given in § 11.3 may easily be solved in this way.

For the solution of three-dimensional problems which start at $t = 0$ from undisturbed conditions, the Laplace transforms of the stress–strain relations (3), (4) are

$$f(p)\bar{s}_{rs} = 2g(p)\bar{e}_{rs}, \tag{14}$$

$$f_1(p)\bar{s} = 3g_1(p)\bar{e}, \tag{15}$$

which are the same as (1) and (2) with G replaced by $g(p)/f(p)$ and K replaced by $g_1(p)/f_1(p)$. Suppose, now, that we know the solution of a problem in elasticity in which stresses are applied to the body at time $t = 0$ when it is undisturbed. If, in this solution, the stress is replaced by its Laplace transform, G is replaced by $g(p)/f(p)$, and K is replaced by

$g_1(p)/f_1(p)$, the Laplace transform of the solution of the corresponding viscoelastic problem is obtained. The solution itself can then be written down from a table of Laplace transforms.

Some examples of this simple procedure will now be given. It should be understood that it applies only to cases in which the system is initially unstrained and at rest, but it can easily be extended to other cases. More complicated solutions with specific reference to mining problems are available, Gnirk and Johnson (1964) and Serata (1964).

(i) Constant uniaxial stress S is applied at $t = 0$ to a substance which is elastic in hydrostatic compression and behaves as a Kelvin substance in shear.

For the Kelvin substance, from § 11.3 (8), $f(p) = 1$, $g(p) = \eta p + k$. (14) and (15) then become

$$\bar{s}_{rs} = 2(\eta p + k)\bar{e}_{rs}, \quad \bar{s} = 3K\bar{e}. \tag{16}$$

Suppose that S is in the x-direction, so that $\sigma_y = \sigma_z = 0$, then

$$\bar{s} = S/3p, \quad \bar{s}_{xx} = 2S/3p. \tag{17}$$

It follows from (16) that $\bar{e} = S/9Kp$ and

$$\bar{s}_{xx} = 2S/3p = 2(\eta p + k)[\bar{\varepsilon}_x - \bar{e}] = 2(\eta p + k)[\bar{\varepsilon}_x - S/9Kp].$$

Therefore

$$(\eta p + k)\bar{\varepsilon}_x = S[\eta p + k + 3K]/9Kp,$$

$$\bar{\varepsilon}_x = \frac{S}{9K}\left\{\frac{k + 3K}{kp} - \frac{3K/k}{p + k/\eta}\right\},$$

$$\varepsilon_x = S\{k + 3K - 3Ke^{-kt/\eta}\}/9Kk. \tag{18}$$

This example has been given to illustrate the relationship of the constants in the various laws. k here is a modulus of rigidity. It is seen that a substance which is elastic in hydrostatic compression and Kelvin in distortion follows a generalized Kelvin law in uniaxial compression.

(ii) Pressure P is applied at $t = 0$ to the interior of a circular hole of radius R_1 in an infinite medium. The material is supposed to be elastic in hydrostatic compression and to behave as a generalized Kelvin substance in distortion.

From § 11.3 (15) for this case

$$f(p) = (\eta_1 p + k_1 + k_2), \quad g(p) = k_2(\eta_1 p + k_1). \tag{19}$$

Now from § 5.11 (11) the displacement u at radius r in the elastic case is

$$u = -PR_1^2/2Gr. \tag{20}$$

For the viscoelastic problem we replace P by P/p and G by $g(p)/f(p)$, so that

$$\bar{u} = -\frac{PR_1^2(\eta_1 p + k_1 + k_2)}{2rpk_2(\eta_1 p + k_1)} = -\frac{PR_1^2}{2rk_1k_2}\left\{\frac{k_1 + k_2}{p} - \frac{k_2}{(p + k_1/\eta_1)}\right\}.$$

Therefore, by (6) and (8),

$$u = -\frac{PR_1^2(k_1 + k_2)}{2rk_1k_2} + \frac{PR_1^2}{2rk_1}e^{-k_1t/\eta_1}. \tag{21}$$

There is an initial displacement of $-PR_1^2/2k_2r$ and the final displacement of $-PR_1^2(k_1 + k_2)/2rk_1k_2$ is approached with time constant η_1/k_1.

If an infinite solid material is subject to uniform all-round pressure P and a hole of radius R_1 is suddenly opened in it, the change of displacement at radius r is obtained by changing the sign of P in (21), and so is

$$\frac{PR_1^2(k_1 + k_2)}{2rk_1k_2} - \frac{PR_1^2}{2rk_1}e^{-k_1t/\eta_1}. \tag{22}$$

This gives the change in displacement in a measuring bolt, § 15.9, when a tunnel cross-section is opened.

(iii) Constant normal stress P applied at $t = 0$ to the surface of a semi-infinite solid over a circular area of radius R. The material is supposed to be elastic in hydrostatic compression and to behave as a generalized Kelvin substance in distortion, so that $f(p)$ and $g(p)$ are given by (19).

In the elastic case, the displacement w at the centre of the circle is by § 10.17 (53)

$$w = \frac{2PR(1 - \nu^2)}{E} = \frac{PR}{2}\left\{\frac{1}{G} + \frac{3}{3K + G}\right\}. \tag{23}$$

Replacing P by P/p and G by $g(p)/f(p)$ gives for $\bar{w}$ in the viscoelastic case

$$\bar{w} = \frac{PR}{2p}\left\{\frac{\eta_1 p + k_1k_2}{k_2(\eta_1 p + k_1)} + \frac{3(\eta_1 p + k + k_2)}{3K(\eta_1 p + k_1 + k_2) + k_2(\eta_1 p + k_1)}\right\}$$

$$= \frac{PR}{2k_1k_2}\left\{\frac{k_1 + k_2}{p} - \frac{k_2}{(p + k_1/\eta_1)}\right\} + \frac{3PR}{2\alpha\eta_1(3K + k_2)}\left\{\frac{k_1 + k_2}{p} - \frac{k_2^2}{(3K + k_2)(p + \alpha)}\right\},$$

where $\alpha = [3K(k_1 + k_2) + k_1k_2]/(3K + k_2)\eta_1.$

It follows from (6) and (8) that

$$w = \frac{PR(k_1 + k_2)}{2k_1k_2} + \frac{3PR(k_1 + k_2)}{2[3K(k_1 + k_2) + k_1k_2]} - \frac{PR}{2k_1}e^{-k_1t/\eta_1} - \frac{3PRk_2^2}{2(3K + k_2)[3K(k_1 + k_2) + k_1k_2]}e^{-\alpha t}. \tag{24}$$

This gives the way in which the displacement in a plate-bearing test, § 15.7, varies with time on the present hypothesis.

(iv) Stress S applied to a material at $t = 0$ along the z-axis with no lateral displacement. The material behaves elastically in hydrostatic compression and as a Maxwell substance in distortion.

In this case the elastic solution for the stress is by § 5.3 (2)

$$\sigma_z = S, \quad \sigma_x = \sigma_y = \frac{\nu}{1-\nu}S = \frac{3K - 2G}{3K + 4G}S. \tag{25}$$

For the Maxwell substance, § 11.3 (4), $f(p) = (p/k) + (1/\eta)$, $g(p) = p$, and so for the viscoelastic problem

$$\bar{\sigma}_z = \frac{S}{p}, \quad \bar{\sigma}_x = \bar{\sigma}_y = \frac{S\{3K(\eta p + k) - 2pk\eta\}}{p\{3K(\eta p + k) + 4pk\eta\}}.$$

It follows that

$$\bar{\sigma}_x = \bar{\sigma}_y = S\left\{\frac{1}{p} - \frac{6k\eta}{(3K + 4k)\eta p + 3Kk}\right\},$$

therefore

$$\sigma_z = S, \quad \sigma_x = \sigma_y = S - \frac{6k\eta S}{(3K + 4k)\eta}e^{-t/t_1}, \tag{26}$$

where

$$t_1 = (3K + 4k)\eta/3Kk. \tag{27}$$

It follows that the stress tends to become hydrostatic with time constant t_1.

Chapter Twelve

Crack Phenomena and the Mechanism of Fracture

12.1 Introduction

Cracks in rock range in size from faults measuring hundreds of thousands of feet in length to intra-granular cracks with lengths of as little as a thousandth of a foot. In this sense, cracks must be defined as confined regions within a body of rock across which the displacements, usually shear but sometimes normal, can be discontinuous under certain stress conditions.

Griffith (1921) suggested that the low tensile strength of glass was due to failure caused by stress concentrations at the ends of minute internal and surface flaws. These flaws have become known as *Griffith cracks.* Their existence on glass surfaces has been demonstrated, although they are too narrow to be observed optically. Gordon *et al.* (1959) consider their widths to be of the order of 100 Å and their depths to be about 1,000 Å, and Levengood (1959) describes methods of observing them. The existence of internal cracks is still a matter of argument, Ernsberger (1965), but Gilvarry (1961) and Drucker (1963) regard the fact that glass spheres shatter into a large number of fragments when compressed between parallel plates as evidence for the existence of internal cracks. In metals in the unstrained state, cracks are exceptional, Stroh (1954), although grey cast iron may be regarded as an extensively cracked material, Grassi and Cornet (1949). Hsu *et al.* (1963) find that in unstressed concrete, cracks cover about 10 per cent of the surface of the aggregate. In coal, cracks of the order of a micron are visible under a microscope, and there is evidence of a large number of cracks, Pomeroy and Morgans (1956) and Terry (1959). Most rocks are granular and porous, indicating the existence of cracks with the dimensions of small fractions of an inch. Brace (1961) and Hoek (1965) give evidence that the failure of rocks takes place from grain boundaries and conclude that the lengths of Griffith cracks in rock are about equal to the maximum grain diameter. Small inter-granular cracks in rock have assumed major significance in connection with the plane and modified Griffith criteria for the strength of rock, §§ 4.8, 10.14.

The applicability of the Griffith concept of failure, based on grain-sized cracks, to a heterogeneous aggregate such as rock is questionable. However, the body of theory which has grown up around this concept is very useful for studying the effect of these cracks on the elastic properties of rock, and there seems to be no reason why it should not be applied to the behaviour of much larger cracks using a scale on which rock can be regarded as homogeneous, thereby providing an approach to the growth of fractures in rock.

12.2 The strain energy associated with a crack

The elastic strain energy stored in a body containing an elliptical crack can be calculated by applying the integral § 5.8 (18) to the formulae for the stresses around an elliptical hole, § 10.11. For an extended body the total strain energy is, of course, infinite, and it is necessary to consider the difference between the strain energy, W, of the body containing the crack under specified stress and the strain energy, W_0, of the same body without the crack under the same stresses. The strain energy due to the crack, W_c, is defined as the difference between these two quantities, that is,

$$W_c = W - W_0. \tag{1}$$

The quantity, W_c, may be expressed as an integral over the volume of the body or as an integral over the external surface of the body and the surface of the crack, § 5.8 (23). The most convenient method for calculating W_c is to make use of the reciprocal theorem, Walsh (1965b, c), § 5.8 (24).

It is sufficient to determine the strain energy due to a crack for two cases: a normal stress, σ_n, perpendicular to a flat elliptical crack and a shear stress, τ, parallel to such a crack. Consider a rectangular plate of length l, width b, and thickness t, containing a flat elliptical crack of length $2c$, lying parallel to the sides of width b. The reciprocal theorem states that for a body acted upon separately by two sets of applied stresses the work done by the first set acting over the displacements produced by the second set equals the work done by the second set acting over the displacements produced by the first set. Let the first set of stresses be a uniaxial compressive stress, σ_c, applied only to the external sides of width b. Let the second set of stresses, σ_a, be a uniform compressive stress applied in the same direction as σ_c on both these sides and the surfaces of the crack. The second set of stresses produces a uniaxial compressive strain of σ_a/E throughout the plate, where E is the Young's modulus of the material of the plate. The work done by the first set of stresses acting over the uniform displacements of the second set is

$$W_{12} = (\sigma_c \sigma_a/E)(btl). \tag{2}$$

The work done by the second set of stresses acting over the displacements of the first set comprises the work done by these stresses over the displacements of the external sides of the plate and over the displacements of the surfaces of the crack. The displacement of the external sides can be written in terms of the effective Young's modulus of the plate containing the crack, E_{eff}, as $(\sigma_c/E_{\text{eff}})(l)$. Let the displacement of the crack surface in the direction of σ_a be v and the components of crack surface normal to this direction be tda. The work done by the second set of stresses acting over the displacement of the external sides is $(\sigma_a\sigma_c/E_{\text{eff}})(lbt)$ and over the displacements of the crack surfaces is

$$W_{21} = -2\int_{-c}^{+c} \sigma_a v t da. \tag{3}$$

Adding these two quantities and using the reciprocal theorem gives

$$\frac{\sigma_c\sigma_a}{E}(blt) = \frac{\sigma_a\sigma_c}{E_{\text{eff}}}(blt) - 2t\sigma_a\int_{-c}^{+c} vda, \tag{4}$$

for the case of a plate in plane stress. The values of σ_c and σ_a are quite arbitrary provided that σ_c does not cause the crack to close, and they can be chosen so that $\sigma_c = \sigma_a = \sigma$. In this case (4) can be written as

$$\frac{\sigma^2}{2E_{\text{eff}}}(blt) = \frac{\sigma^2}{2E}(blt) + t\sigma\int_{-c}^{+c} vda. \tag{5}$$

The term on the left-hand side is the strain energy, W, in the plate containing the crack, and the first term on the right-hand side is the strain energy of the plate without the crack, W_0. Thus (5) becomes

$$W = W_0 + t\sigma\int_{-c}^{+c} vda, \tag{6}$$

and, by the definition (1),

$$W_c = t\sigma\int_{-c}^{+c} vda. \tag{7}$$

The displacement, v, around the crack can be calculated from the theory developed in § 10.11. For a very flat crack the shape of the displacement profile across the length of the crack differs insignificantly from an ellipse. The length of the minor semi-axis of the displacement ellipse is given by § 10.11 (28) with $y = c\xi_0$ as ξ_0 tends to zero, and the length of the major semi-axis is c. For a very flat crack, therefore, the integral (7) is equal to the volume of an elliptical semi-cylinder with a major semi-axis of c and minor semi-axis of

$$v = 2\sigma c/E \tag{8}$$

$$v = 2\sigma c(1 - \nu^2)/E, \tag{9}$$

for conditions of plane stress and plane strain, respectively. Therefore, the strain energy due to a flat elliptical crack in uniaxial compression, or tension, is

$$W_c = \pi\sigma^2 c^2 t/E, \tag{10}$$

$$W_c = \pi\sigma^2 c^2 t(1 - \nu^2)/E, \tag{11}$$

for plane stress and plane strain, respectively. A similar argument for the case of a very flat crack with a shear stress, τ, parallel to it yields expressions identical to (10) and (11) with σ^2 replaced by τ^2. The strain energy due to a very flat crack subject to normal and shear stresses, σ_n and τ, arising from any system of applied stresses is given by superposition as

$$W_c = \pi c^2 t(\sigma_n^2 + \tau^2)/E = \pi c^2 t(\sigma_1^2 \cos^2 \beta + \sigma_2^2 \sin^2 \beta)/E, \tag{12}$$

$$W_c = \pi c^2 t(1 - \nu^2)(\sigma_n^2 + \tau^2)/E = \pi c^2 t(1 - \nu^2)(\sigma_1^2 \cos^2 \beta + \sigma_2^2 \sin^2 \beta)/E, \tag{13}$$

for conditions of plane stress and plane strain, respectively, where σ_1 and σ_2 are the principal stresses at infinity and β is the angle between the major axis of the crack and the direction of the minor principal stress, Fig. 10.11.1 (*a*).

Starr (1928) calculated W_c directly for the case of pure shear $\sigma_1 = -\sigma_2$, $\beta = \pi/4$ and obtained the value given by (12). The value for W_c used by Griffith (1921) was incorrect and was corrected by Griffith (1924). The same results as those obtained above for the flat elliptical crack are obtained for the mathematical flat crack, § 10.11. The three-dimensional or 'penny-shaped' crack has been treated as a flat spheroid with major diameter $2c$ by Sack (1946), who finds the strain energy due to such a crack for a stress σ normal to the diametral plane to be

$$W_c = 8(1 - \nu^2)c^3\sigma^2/\pi E. \tag{14}$$

This shape of crack is conveniently treated using the Hankel transform, Sneddon (1946, 1951). It is interesting to note that the strain energies for two-dimensional flat cracks in plane stress and plane strain and the three-dimensional flat crack given by (10), (11), and (14) differ only by numerical factors near unity if $t = c$.

12.3 The effect of cracks on elastic properties

For many years phenomena in rock such as the concavity of the early part of the stress–strain curve, Adams and Williamson (1923) and Nishihara (1958), and the difference between static and dynamic elastic moduli, Zisman (1933) and Ide (1936), have been attributed to the existence of cracks or pores. The effect of cracks on the properties of rocks has been the subject of many recent studies, both theoretical, Walsh (1965a, b, c) and experimental, Brace (1965), Cook and Hodgson (1965), and Walsh and Brace (1966).

The effect of voids, in the form of cavities, very flat open cracks, and closed cracks, on the elastic properties of a body is most easily studied with the aid of the reciprocal theorem. The elastic constants of a body containing cracks and voids are defined to be the *effective elastic constants* of that body, while the elastic constants of the material of the body, uninfluenced by the presence of these imperfections, are referred to as the *intrinsic elastic constants* of the material.

Consider a volume of rock, V_r, containing a void of any shape with a volume, V_v. Let this volume be subject to two separate sets of stresses; the first being uniform hydrostatic compression, p, on the external surfaces of the volume, V_r, and on the internal surface of the void, V_v, the second being uniform hydrostatic tension, $-p$, on the internal surface of the void only. The first set of stresses produces hydrostatic compression throughout the volumes, V_r and V_v, so that the changes in these volumes are

$$\Delta_1 V_r = V_r p/K \tag{1}$$

$$\Delta_1 V_v = V_v p/K \tag{2}$$

where K is the intrinsic bulk modulus of the rock. The second set of stresses produces compression of the volumes, V_r and V_v, of $\Delta_2 V_r$ and $\Delta_2 V_v$, respectively. The work done by the first set of stresses acting over the displacements of the second set is

$$W_{12} = p\Delta_2 V_r - p\Delta_2 V_v \tag{3}$$

and that done by the second set acting over the displacements of the first set is

$$W_{21} = p\Delta_1 V_v.$$

Using the reciprocal theorem,

$$p\Delta_2 V_r = p(\Delta_1 V_v + \Delta_2 V_v). \tag{4}$$

Superposition of the second set of stresses on the first set produces the conditions obtaining when a volume of rock, V_r, containing a void, V_v, is subjected to an external hydrostatic pressure, p, only. Let the volumetric compression of the volume, V_r, containing a void, V_v, be ΔV_r when subject to an external hydrostatic pressure, p, then

$$\Delta V_r = \Delta_1 V_r + \Delta_2 V_r. \tag{5}$$

Substituting from (4) into (5) yields

$$\Delta V_r = \Delta_1 V_r + \Delta_1 V_v + \Delta_2 V_v. \tag{6}$$

The increase in volumetric compression of V_r due to the void is

$$\Delta V_r - \Delta_1 V_r = \Delta_1 V_v + \Delta_2 V_v. \tag{7}$$

The effective bulk modulus of a body containing a void is

$$1/K_{\text{eff}} = \Delta V_r/pV_r = (\Delta_1 V_r + \Delta_1 V_v + \Delta_1 V_v)/pV_r$$
$$= 1/K + (\Delta_1 V_v + \Delta_2 V_v)/pV_r \tag{8}$$

Consider a plate of length l, width b, and thickness, t, containing a flat, open elliptical crack as in § 12.2. The volumetric compression of a very flat, open crack due to a hydrostatic pressure, p, applied to the external surfaces of the plate is virtually the same as the volumetric compression of the same crack by a uniaxial compressive stress, $\sigma_c = p$, normal to the plane of this crack. Using the notation of §§ 10.11, 12.2, this volumetric compression is

$$\Delta_1 V_v + \Delta_2 V_v = 2\int_{-c}^{+c} tv da. \tag{9}$$

Making use of § 12.2 (7), (10) and putting $p = \sigma$ this becomes

$$\Delta_1 V_v + \Delta_2 V_v = 2W_c/p = 2\pi pc^2 t/E. \tag{10}$$

The effective bulk modulus of a body containing a very flat open crack is

$$1/K_{\text{flat}} = 1/K + (2\pi pc^2 t)/pEV_r. \tag{11}$$

Since $V_r = blt$ (11) becomes

$$1/K_{\text{flat}} = 1/K + 2\pi c^2 L/E, \tag{12}$$

where $L = 1/bl$ can be regarded as the crack length per unit volume. Using the relation § 5.2 (14) between the intrinsic Young's modulus, bulk modulus and Poisson's ratio, (12) can be written as

$$1/K_{\text{flat}} = (1/K)[1 + 2\pi c^2 L/3(1 - 2\nu)]. \tag{13}$$

The effective Young's modulus of a body containing a very flat, open crack lying normal to the direction of compression is found, using § 12.2 (5), (7), (10) to be

$$1/E_{\text{flat}} = 1/E + 2\pi c^2 L/E \tag{14}$$

that is

$$E_{\text{flat}} = E/(1 + 2\pi c^2 L). \tag{15}$$

Flat, open cracks of all orientations are compressed equally by a hydrostatic pressure, p. This is not so for a uniaxial compressive stress, σ_c, which compresses a flat, open crack lying normal to it by the same amount as does a pressure $p = \sigma_c$, but compresses cracks of all other orientations to a lesser extent. The effective Young's modulus of a body containing a large number of randomly oriented, flat, open cracks with an average length L per unit volume can be found by simple spherical averaging.

If the plane of the crack is at $\pi/2 - \beta$ to the direction of uniaxial stress the only change in § 12.2 (5) is a factor $\cos^2 \beta$ in the last term, so that for flat, open cracks inclined at β

$$1/E_{\text{flat}} = 1/E + 2\pi c^2 L \cos^2 \beta / E. \tag{16}$$

If the cracks are randomly distributed a spherical average may be taken over all directions and, since

$$\left(\frac{1}{2\pi}\right) \int_0^{\pi/2} 2\pi \sin \beta \cos^2 \beta d\beta = 1/3,$$

this gives for randomly orientated cracks

$$1/E_{\text{flat}} = 1/E + 2\pi c^2 L / 3E. \tag{17}$$

$$E_{\text{flat}} = E/(1 + 2\pi c^2 L/3). \tag{18}$$

It can be expected and shown, Walsh (1965c), that the usual relations between elastic constants apply to the effective constants of an elastic body with open voids. § 5.2 (14) can be written in the form

$$\nu/E = \tfrac{1}{2}(1/E - 1/3K). \tag{19}$$

Substituting from (12) and (17) into the right-hand side of (19) yields, for the case of flat, open cracks

$$\nu_{\text{flat}}/E_{\text{flat}} = \nu/E. \tag{20}$$

This shows that flat, open cracks have the same effect on the Poisson's ratio of a body as they have on its Young's modulus, that is, the rate of lateral strain for a body with open cracks is the same as that of the solid body without cracks.

Using calculations for a spherical cavity in arguments similar to those leading to (12), (14), and (20), it can be shown that the values of Young's modulus, E_{equi}, the bulk modulus, K_{equi}, and the ratio $\nu_{\text{equi}}/E_{\text{equi}}$ for a body containing equidimensional voids are all less than the corresponding intrinsic values for the material of the body, where Poisson's ratio ν_{equi} is given by

$$\nu_{\text{equi}} = \tfrac{1}{2}(1 - E_{\text{equi}}/3K_{\text{equi}}). \tag{21}$$

In the case of a body with closed cracks, it is obvious that $\Delta_1 V_v + \Delta_2 V_v = 0$. Therefore, from (8), the effective bulk modulus of a body containing closed cracks must be equal to its intrinsic bulk modulus.

Between the surfaces of any closed crack in a body subject to uniaxial compression there are shear, τ, and normal, σ_n, components of stress. For a crack of any orientation with respect to the direction of this stress, these components can be found using the methods of § 2.3. Let the coefficient of sliding friction between the crack surfaces be μ. If $|\tau| > \mu\sigma_n$ these surfaces slide past one another due to an effective shear stress,

$$\tau_{\text{eff}} = |\tau| - \mu\sigma_n. \tag{22}$$

The effective shear stress is, by § 4.6 (6), a maximum for cracks oriented so that

$$\tan 2\beta = -1/\mu. \tag{23}$$

The effective Young's modulus of a plate of length l, width b, and thickness t containing a closed crack of length $2c$ can be found by considering the work done by a uniaxial compressive stress σ_c applied to the sides of width b. If this stress produces a strain ε_c the work done by σ_c is

$$W = blt\sigma_c\varepsilon_c/2. \tag{24}$$

This can be equated to the sum of the strain energy in the plate without the crack, W_0, the additional strain energy due to the crack, W_c, and the work done against friction as the crack surfaces slide past one another, W_f. As in § 12.2 (10),

$$W_0 = \sigma_c^2 blt/2E, \tag{25}$$

$$W_c = \pi c^2 t \tau_{eff}^2/E. \tag{26}$$

The work done against friction is obviously the integral around the crack of the product of the shear displacement produced by τ_{eff} and the mean frictional force against which the displacement worked. From § 12.2 (3), (7) it can be seen that the strain energy around a crack due to a stress is equal to half the integral around the crack of the product of the displacement produced by the stress and that stress. Let the displacement produced by τ_{eff} around the crack be $f(\eta)$ then

$$W_c = \tfrac{1}{2}\tau_{eff}\oint f(\eta)\, d\eta, \tag{27}$$

and

$$W_f = \tfrac{1}{2}\mu\sigma_n\oint f(\eta)\, d\eta, \tag{28}$$

so that

$$W_f = W_c\mu\sigma_n/\tau_{eff}. \tag{29}$$

Substituting (26) into (29) and using (22) yields

$$W_f = \pi c^2 t\mu\sigma_n\tau_{eff}/E. \tag{30}$$

Multiplying

$$W = W_0 + W_c + W_f \tag{31}$$

by $2/(blt\sigma_c)$ yields

$$\varepsilon_c = \sigma_c/E + 2\pi c^2 L(\tau_{eff}^2 + \mu\sigma_n\tau_{eff})/\sigma_c E, \tag{32}$$

which, using (22), can be written

$$\varepsilon_c = \sigma_c/E + 2\pi c^2 L(\tau^2 - \mu\sigma_n \mid \tau \mid)/\sigma_c E. \tag{33}$$

If the crack is oriented so as to satisfy (23), (33), using § 4.6 (4), (5), (7), becomes

$$\varepsilon_c = \frac{\sigma_c}{E}\left[1 + \pi c^2 L\frac{(\mu^2+1)^{\frac{1}{2}} - \mu}{2(\mu^2+1)^{\frac{1}{2}}}\right]. \tag{34}$$

The effective Young's modulus of a body containing such a closed crack is

$$E_{\text{eff}} = E/\{1 + \pi c^2 L[(\mu^2 + 1)^{\frac{1}{2}} - \mu]/2(\mu^2 + 1)^{\frac{1}{2}}\}. \tag{35}$$

Again, the effective Young's modulus of a body containing a large number of randomly oriented closed cracks of average length L per unit volume can be found by averaging over all cracks for which $|\tau| > \mu\sigma_n$, that is, $\tan^{-1}\mu < \beta < \pi/2$, Walsh (1965b). This yields

$$E_{\text{eff}} = E/\left\{1 + \frac{2\pi L}{15}\left[\frac{2 + 3\mu^2 + 2\mu^4}{(1 + \mu^2)^{2/3}} - 2\mu\right]\right\}. \tag{36}$$

As the effective and intrinsic bulk moduli of a body containing closed cracks are equal,

$$(1 - 2\nu_{\text{eff}})/E_{\text{eff}} = 1/3K = (1 - 2\nu)/E \tag{37}$$

and

$$(1 - 2\nu_{\text{eff}})/(1 - 2\nu) = E_{\text{eff}}/E. \tag{38}$$

Relative motion between the surfaces of a closed crack is possible only while $|\tau| > \mu\sigma_n$. Assume that at the end of a loading cycle the uniaxial compressive stress in the body, the shear stress, the frictional resistance, and the effective shear stress on the crack have values of σ_c', τ', $\mu\sigma_n'$, τ_{eff}', respectively. As soon as unloading begins the direction of the frictional resistance reverses, so that the shear stress resulting from the deformation of the crack has to overcome both the frictional resistance and the shear stress due to the load before reverse sliding commences. The shear stress due to the deformation of the crack is equal in magnitude but opposite in sign to the effective shear stress which produced this deformation. The condition for reverse sliding can be expressed as

$$\tau_{\text{eff}}' \geqslant |\tau| + \mu\sigma_n. \tag{39}$$

At the conclusion of the loading cycle the conditions were

$$|\tau'| = \tau_{\text{eff}}' + \mu\sigma_n', \tag{40}$$

so that (39) becomes

$$|\tau'| - \mu\sigma_n' \geqslant |\tau| + \mu\sigma_n, \tag{41}$$

or

$$\Delta\tau + \mu\Delta\sigma_n \geqslant 2\mu\sigma_n', \tag{42}$$

where $\Delta\tau$ and $\Delta\sigma_n$ are the reductions in shear and normal stresses on the crack surface which are necessary to start reverse sliding. Using § 2.3 (15), (17), (42) becomes

$$\Delta\sigma_c[1 + (1/\mu)\tan\beta] \geqslant 2\sigma_c, \tag{43}$$

where $\Delta\sigma_c$ is the reduction in the applied compressive stress necessary to start reverse sliding. The factor in parenthesis on the left-hand side of (43) is finite, except for $\mu = 0$, or β approaching $\pi/2$, that is, for cracks parallel to the direction of compression. These cracks have no effect on Young's modulus, however, so that a finite reduction in the applied stress, $\Delta\sigma_c$, is necessary to cause reverse sliding. The initial Young's modulus for unloading a body containing closed cracks is therefore equal to its intrinsic Young's modulus. As unloading proceeds, more cracks at progressively smaller values of β than $\pi/2$ begin to slide in reverse and the unloading Young's modulus decreases progressively.

It is interesting to study the distribution and dissipation of energy in a body containing a closed crack during loading and unloading, Fig. 12.3

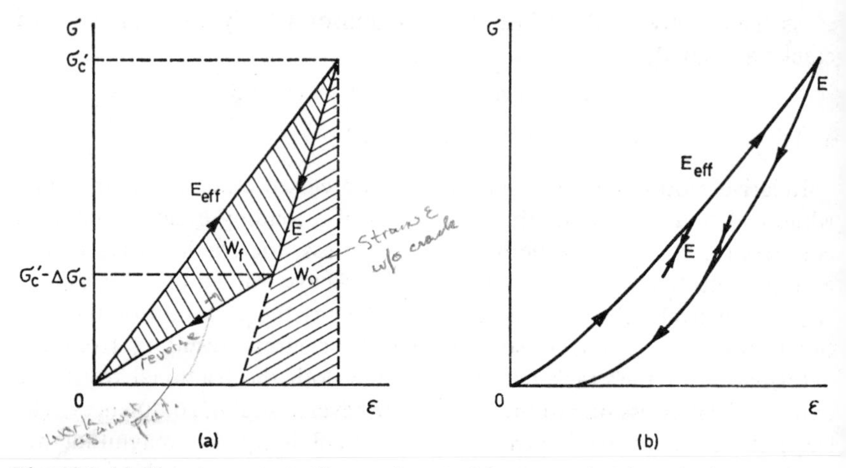

Fig. 12.3 (*a*) The stress–strain diagram in uniaxial compression for a body containing a single, inclined, closed crack. (*b*) The first two regions of a complete stress–strain curve for rock during loading and unloading in uniaxial compression.

(*a*). The loading part of the stress–strain diagram is linear with an effective Young's modulus, E_{eff}, given by (35), which is lower than the intrinsic Young's modulus of the body without a crack. At the start of unloading the stress–strain curve is linear, with a slope equal to that of the intrinsic Young's modulus, E. If this line is extended to intersect the strain axis the area beneath it represents the strain energy, W_0, that would be stored at maximum load in the body without the crack. The area between this line and the loading line represents the strain energy due to the crack and the work done against friction between the crack surfaces during loading and unloading. Part of the strain energy due to the crack is recovered during unloading. Reverse sliding of the crack surfaces commences when the load has been decreased by the amount given by (43). When reverse sliding starts, the unloading modulus decreases to a value even lower than the effective Young's modulus. The area between the loading and unloading lines represents the work done against friction between the crack surfaces during the complete cycle. Sliding proceeds continuously up to the maximum stress, σ_c', during loading, but begins only at the stress, $\sigma_c' - \Delta\sigma_c$, during unloading. In each case the relative displacement between the crack surfaces is identical. Therefore, the work done against friction during a complete cycle is divided between the loading and unloading portions in the ratio of $\sigma_c'/(\sigma_c' - \Delta\sigma_c)$, respectively. This discussion applies to the case of uniaxial compression with no other normal stress across the crack. If a normal stress exists across a crack, due either to a confining pressure or stresses within the body, forward sliding commences only when the load has increased by an amount sufficient to overcome the frictional resistance to this sliding, and the stress–strain curve starts with a slope equal to the intrinsic Young's modulus. Furthermore, the frictional resistance due to this additional stress prevents the surfaces of the crack from returning to their original relative positions by reverse sliding during unloading, and some strain can be expected to remain in the body at zero load after a complete cycle.

It should be noted that very flat, open cracks are likely to close under compressive stress when the normal stress across such cracks is of the order of $G\xi_0$, § 10.11 (31), which amounts to a few kilobars for cracks with aspect ratios of the order of 0·001, Birch (1961). One of the most sensitive methods for studying the closing of cracks is by measurement of the electrical resistivity of rock. Brace, Orange, and Madden (1965) show that for rocks such as granite and quartzite a change in confining pressure of 10 kilobars causes a change in resistivity of two to three orders of magnitude. A similar but less pronounced effect occurs with thermal conductivity, Walsh and Decker (1966).

The effect of cracks on the elastic properties of a body can be summarized as follows: (i) The effective Young's modulus of a body containing open cracks and cavities is less than the intrinsic Young's modulus of a solid body. (ii) The effective Young's modulus of a body containing closed cracks is less than its intrinsic Young's modulus if the surfaces of the cracks slide past one another. (iii) Closed cracks which have undergone sliding do not slide in the opposite sense immediately the load is reversed. (iv) The effective Poisson's ratio of a body containing equidimensional cavities is less than the intrinsic Poisson's ratio of a solid body. (v) The effective Poisson's ratio of a body containing very flat, open cracks is less than its intrinsic Poisson's ratio. (vi) The effective Poisson's ratio of a body containing closed cracks is greater than its intrinsic Poisson's ratio.

In most rocks, voids, open cracks, and closed cracks of different and often random orientations are present. The behaviour of rock in the first two regions of the complete stress–strain curve, § 4.2, can be attributed to the effects due to these. The increasing slope of the stress–strain curve with load in the first region, Fig. 12.3 (*b*), is probably caused by the closing of very flat open cracks; the linear part of the second region with a slope defining the effective Young's modulus is probably due to the surfaces of these and other closed cracks sliding past one another; the initial slope of the stress–strain curve at the start of unloading provides a measure of the intrinsic Young's modulus of the rock, unless open voids are still present; the modulus decreases progressively as unloading proceeds and increasing numbers of cracks slide in reverse; the strain at zero load provides an indication of the amount of inherent compressive stress across crack surfaces in the rock; small stress changes superposed on the major loading and unloading cycle at any point are generally insufficient to cause sliding of crack surfaces and provide a measure of the intrinsic elastic properties. Corresponding changes in Poisson's ratio can be expected. Thus, the slope of the early part of the stress–strain curve for rock, the hysteresis observed in loading and unloading cycles, and the differences between static and dynamic moduli can be explained in terms of phenomena associated with the existence of cracks and cavities.

12.4 Energy theories of failure

Griffith (1921) attempted to estimate the tensile strength of a cracked material from energy considerations. The surfaces of the crack are supposed to possess surface energy α per unit area (associated with the rupturing of atomic bonds when the crack is formed) and the energy balance between the surface energy of the crack, the strain energy associated with it, and the work done by the external forces provides a criterion to determine whether the crack will grow. The surface energy of cracks in silicate crystals has been measured for this purpose by Brace and Walsh (1962) using a method of Gilman (1959). Elliott (1947) gives a discussion involving intratomic forces, and Cottrell (1965) gives one based on dislocation theory. Griffith's theory was set up originally to explain fracture of glass; a review of its present status in this context is given by Anderson (1959). The application of Griffith theory to the brittle fracture of ductile metals is questionable, Drucker (1962). Certainly it is necessary to modify it to take account of the energy of plastic deformation; Orowan (1952) proposed that this should be added to the surface energy. Further discussion of the fracture of metals is given by Allen (1959) and Orowan (1959). For rocks, also, it seems likely that the energy of plastic deformation should be allowed for and α should be called the *apparent surface energy of fracture* to distinguish it from the true surface energy, which may have a significantly smaller value.

Two distinct and important cases arise. The first is that in which there is no displacement of the surface of the body and the surface energy of the crack has to be supplied from the strain energy of the body, which will diminish. This case is discussed by Timoshenko and Goodier (1959). The second case, and the one which will be considered here, is that in which the external stresses are held constant while the crack extends; in this case, by § 5.8, the work done by these stresses is equal to $2\Delta W$ or twice the increase in strain energy of the body, and therefore the amount ΔW is available for increasing the surface energy of the crack. Thus the condition for unstable equilibrium of a crack of length $2c$ in material of thickness t is

$$\frac{\partial}{\partial c}(W - 4\alpha ct) = 0. \tag{1}$$

Most of the discussion which follows will refer to the case of uniaxial tension, $-\sigma_2$, normal to the plane of a flat crack. For this case, using § 12.2 (5), (7), (10) for W in plane stress (1) becomes

$$\frac{\partial}{\partial c}\left[\frac{blt\sigma_2{}^2}{2E}\left(1 + \frac{2\pi c^2}{bl}\right) - 4\alpha ct\right] = 0,$$

or

$$\pi\sigma_2{}^2 c = 2\alpha E. \tag{2}$$

The tensile stress, T_g, at which a crack of length c will begin to grow is

$$T_g = (2\alpha E/\pi c)^{\frac{1}{2}}, \tag{3}$$

for plane stress.

Similarly, from (11) and (14) of § 12.2

$$T_g = [2\alpha E/\pi(1 - \nu^2)c]^{\frac{1}{2}} \tag{4}$$

for plane strain, and for the penny-shaped crack,

$$T_g = [\pi\alpha E/4(1 - \nu^2)c]^{\frac{1}{2}}. \tag{5}$$

These three estimates of the tensile strength of a body containing a crack of linear dimension $2c$ differ only in the following numerical factors $2/\pi$, $2/\pi(1 - \nu^2)^{\frac{1}{2}}$, $\pi/4(1 - \nu^2)$, which are all just less than unity, showing that the strength is not critically dependent upon the geometry of the crack.

The importance of this formulation of the Griffith theory is that it analyses failure in terms of stability and requires no precise knowledge of the nature of the flaws. When Griffith extended his theory to conditions of compressive stress he resorted to consideration of the stress concentration at crack tips, which implicitly involves important assumptions concerning the shape of the crack. Part of the difficulty created by this is overcome by expressing the criterion in terms of the tensile strength. This renders consideration of the tensile strength in terms of the stress concentration at the tip of a crack useful. The tangential stress at the tip of a crack, such as that considered above, is shown in § 10.11 to be

$$\sigma_t = 2\sigma(c/r)^{\frac{1}{2}}, \tag{6}$$

where r is the radius of curvature of the crack tip. Failure occurs when σ_t reaches a value equal to the inherent tensile strength of the material.

Orowan (1949) proposed the following simple 'order of magnitude' argument to estimate the inherent tensile strength, σ_m. If the crack is to extend, the additional surface energy must be provided by the strain energy near the tip of the crack, where the stress must be σ_m. Assuming that the volume from which the strain energy is derived has dimensions of the order of the intermolecular spacing, a, and that this is also the value of the radius of curvature, r, the strain and surface energies can be equated to yield

$$\sigma_m = (4\alpha E/a)^{\frac{1}{2}}. \tag{7}$$

Equating (6) and (7) gives

$$T_g = (\alpha E/c)^{\frac{1}{2}} \tag{8}$$

which is not very different from (3).

The process of straining a body containing a crack may now be considered more fully. Suppose the rectangular body described in §§ 12.2, 12.3 of length l, width b, and thickness t containing a crack is strained by tension normal to the plane of the crack. As in the case of compression normal to the crack, § 12.3 (15), the effective Young's modulus of the body containing the crack is

$$E_{\text{eff}} = E/(1 + 2\pi L c^2) \tag{9}$$

so that the cracked material behaves as if it had a lower Young's modulus than the solid.

Deformation of the body proceeds linearly according to (9) until the tensile stress, T, reaches the critical value T_g given by (3) at which the crack of length c can grow. Using (3) in (9) and eliminating c gives a relation between the stress T_g at which the crack becomes unstable and the strain ε_g at that time. This is

$$\varepsilon_g = (T_g/E) + (8L\alpha^2 E/\pi T_g^3). \tag{10}$$

This locus APB, Fig. 12.4, in the stress–strain plane is called the Griffith locus, Berry (1960a) and Cook (1965). For large values of T_g it approaches the line $T = E\varepsilon$ and $T_g \to 0$ as $\varepsilon_g \to \infty$. It has a vertical tangent $d\varepsilon_g/dT_g = 0$ at the point V, and this occurs for a crack of length $c_c = (6L\pi)^{-\frac{1}{2}}$.

For any crack length the tensile stress–strain curve is linear elastic according to (9) until it intersects the Griffith locus at some point P', when the crack begins to extend. If crack extension is halted at some increased length, represented by a point farther along the Griffith locus, P'', the stress–strain curve resulting from decreasing the load will again be linear elastic, but with a diminished modulus. The area bounded by the Griffith locus and the two lines OP' and $P''O$ is a measure of the energy absorbed by the extension of the crack. Extension is stable or unstable, depending upon whether it is possible or not to follow the Griffith locus.

The initial extension of cracks shorter than c_c is necessarily unstable, as it is not possible for an applied stress to follow the Griffith locus while its slope is positive. A similar discussion applies to the case of plane strain and to penny-shaped cracks. These questions will be discussed further in connection with the dynamics of crack propagation in § 12.6.

This discussion has referred only to the case of uniaxial tension normal to the plane of the crack. Clearly the same arguments can be applied to general two-dimensional stresses, for which (3) would be replaced by

$$\sigma_1^2 \cos^2 \beta + \sigma_2^2 \sin^2 \beta = \sigma_y^2 + \tau_{xy}^2 = 2\alpha E/\pi c. \tag{11}$$

Here a fundamental difficulty arises, since this takes no account of the signs of the stresses. The crack will obviously not propagate under some

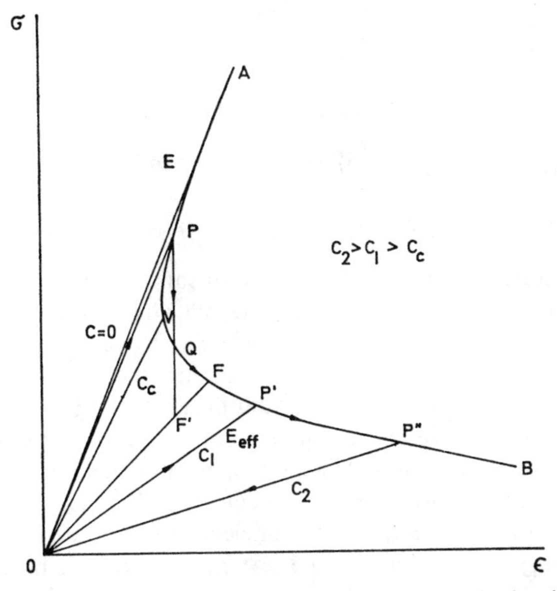

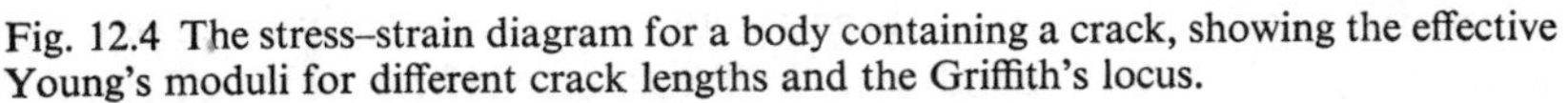
Fig. 12.4 The stress–strain diagram for a body containing a crack, showing the effective Young's moduli for different crack lengths and the Griffith's locus.

compressive conditions, and where it does propagate it may not do so in its own plane. Nevertheless, in view of the fact that failure in triaxial tests so frequently manifests itself in the form of a plane shear fracture, it seems worth while to examine the stability of such a fracture in a manner analogous to Griffith's analysis of the stability of a tensile fracture.

The condition for unstable equilibrium of a crack in shear can be written, in a manner analogous to (1), as

$$\frac{\partial}{\partial c}[W - 4\alpha ct] = 0. \tag{12}$$

The only term in the strain energy, W, which depends upon the crack length is the strain energy due to the crack. In the case of a closed crack, the only component of stress free to generate additional strain energy around the crack is the effective shear stress, $(|\tau| - \mu\sigma_n)$. Substituting this into § 12.2 (12) yields

$$W_c = \pi c^2 t(|\tau| - \mu\sigma_n)^2/E. \tag{13}$$

Substituting (13) in (12) gives

$$(|\tau| - \mu\sigma_n) = (2\alpha E/\pi c)^{\frac{1}{2}} \tag{14}$$

and comparing this with (3), the criterion for unstable shear can be expressed as

$$(|\tau| - \mu\sigma_n) = T_g. \tag{15}$$

The effective shear stress, $(|\tau| - \mu\sigma_n)$, is maximum when

$$\tan 2\beta = -1/\mu \tag{16}$$

so that using § 2.3 (16), (17), (15) can be expressed in terms of the principal stresses as

$$\sigma_1[(\mu^2 + 1)^{\frac{1}{2}} - \mu] - \sigma_3[(\mu^2 + 1)^{\frac{1}{2}} + \mu] = 2T_g. \tag{17}$$

A comparison between (17) and § 10.14 (10) shows that, whereas in terms of the criterion based on the maximum tensile stress, failure occurs when the effective shear stress is twice the tensile strength; in terms of the criterion based on stability, failure occurs when the effective shear stress is equal to the tensile strength. The latter criterion suggests that the ratio between the compressive and tensile strengths of rock is much less than it is observed to be, and must therefore be regarded as unacceptable. The reason why shear failure does not occur according to (17) is that the stresses around the fracture are not sufficient to propagate it, even though sufficient energy would be available if extension were possible in the plane of the fracture. Nevertheless, shear fractures are frequently observed and must be governed by a condition similar to (12). If the value of the apparent surface energy, α, applicable to the final stage of failure by shear fracture is much greater than the value applicable to the original cracks, and if smaller stresses are required to propagate the shear fracture than are required to extend the original cracks, then the final stage of failure can be described by (12) and the considerations that follow from it.

The falling portion of the stress–strain curve during failure by shear fracture can be studied by deriving the Griffith locus for compressive stress in the same way as the Griffith locus for tensile stress is derived, Cook (1965).

12.5 Elementary theory of crack propagation

The modified and extended Griffith criteria compare well with the strengths of rock measured in triaxial, § 6.6, and other tests. More detailed investigations show that crack propagation in rock initiates at much smaller values of the major stress than the strength, corresponding to the beginning of the third, ductile, region of the complete stress–strain curve, Cook (1965), Hoek (1965), and Brace and Byerlee (1967). The basic assumption of all forms of the Griffith criterion is that failure starts when some condition around the most dangerous crack reaches a critical condition.

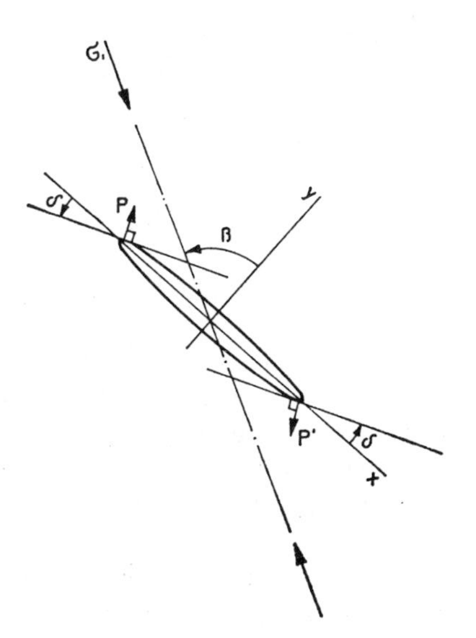

Fig. 12.5 An open crack in compression showing the positions (P, P') of maximum tangential tension and the directions of initial crack extension.

If this were to result in the extension of this crack in its own plane, as is the case in tension, the condition of the crack would become progressively worse, and the ability of a body containing such a crack to resist load would diminish with continued crack growth. This situation is unstable, and extension of the crack results in failure of the body. It has been suggested that the stresses at which crack propagation initiates in rock are correctly described by one or other form of the Griffith criteria, but that the extension of the most dangerous crack is not catastrophic, Hoek and Bieniawski (1966). It has been proposed that the interaction of a number of such cracks, perhaps in the form of an array, are responsible for ultimate failure, Brace and Bombolakis (1963) and Hoek and Bieniawski (1965). These difficulties raise important questions concerning the propagation of Griffith cracks and an even more important question concerning the coincidence between the observed strength of brittle rock, marking the transition from ductile to brittle behaviour, and the predictions of the Griffith criteria.

The basic assumption of the plane and modified Griffith criterion as it applies in compression is that crack propagation starts where the maximum tensile stress around the most dangerous crack reaches some critical value §§ 10.13, 10.14. The position of the points of maximum tensile stress around a flat elliptical crack is derived in § 10.12. These positions are the points P and P' in Fig. 12.5 where the tangents to the elliptical boundary of the crack make angles δ with its major axis. For flat, open cracks inclined so that their normals make angles, β, with the direction of the major principal stress satisfying § 10.13 (14), the tangents are inclined so that

$$\delta = 2\beta - \pi/2 \tag{1}$$

and for closed cracks inclined so that β satisfies § 10.14 (7) the tangents are inclined so that

$$\delta = \pi/4. \tag{2}$$

The assumption made above implies that the extension of a crack begins in a direction perpendicular to the maximum tensile stress, that is, perpendicular to the tangents inclined at δ to the major axis of the crack. For open cracks in uniaxial compression, the initial extension of the crack runs towards the axis of the major principal stress at an angle of $\frac{1}{2}\pi - \beta$, and for closed cracks in compression it runs towards this axis at an angle of $\beta - \pi/4$. Experiments indicate that, with further crack extension, the cracks tend to turn parallel to the axis of the major principal stress, Brace and Bombolakis (1963).

Crack propagation in this direction becomes very stable, and progressively greater stress increments are necessary to produce a given crack extension as the crack lengthens.

This phenomenon has been used by Fairhurst and Cook (1966) to explain splitting of rock in the vicinity of a surface parallel to the direction of the major compressive stress. For cracks that have extended much beyond their original length they propose that sliding of the original inclined crack surfaces past one another generates the displacement necessary to produce crack extension and reduces the stresses in the vicinity of the original inclined crack. Sliding and crack extension cease when the shear stresses across the original crack become insufficient to generate tensile stresses at the tip of the extended crack sufficient to overcome the inherent tensile strength of the material and any compressive stresses at the crack tip, due to concentrations of the minor principal stress in that vicinity. Analysis reveals that the value of the major principal stress required to extend a crack in the direction of the major stress axis is given, approximately, by an expression in the form

$$\sigma_1 = k_1\sigma_2 + k_2[\sigma_2 c_1 + k_3(c_1)^{\frac{1}{2}}], \tag{3}$$

where the k's are positive constants and c_1 is the length of the extended crack. This relationship shows that crack extension is an essentially stable process, where an increase in length always requires an increase in the major stress, as has been observed by Hoek (1964) in experiments. The minor stress has two effects. The value of σ_1 necessary to start crack extension and that necessary to produce a given extension both increase with σ_2. At any stress in excess of the Griffith stress necessary to initiate crack extension, rock is partially cleaved in a direction parallel to the direction of the major principal stress.

Thus, in compression Griffith crack propagation appears to be a stable process which does not lead to catastrophic failure as does the unstable propagation of Griffith cracks in tension.

The effect of cracks on the first two regions of the complete stress–strain curve for rock is discussed in § 12.3. This discussion neglects any effects due to crack propagation which are considered in § 12.4 and lead to the Griffith locus in the stress–strain plane which probably corresponds to the fourth, or brittle, region of the complete stress–strain curve. These approaches can be generalized to allow for the effects of stable crack propagation in the third, ductile, region of the stress–strain curve. The strain, ε, developed by a stress, σ, in rock can be expressed in the form

$$\varepsilon = (\sigma/E)[1 + \Sigma f(c)], \tag{4}$$

where $\Sigma f(c)$ is the sum through a unit volume of rock of the contribution of each crack to the strain in excess of that due to the stress and intrinsic Young's modulus. Failure will initiate at some of these cracks when the stress attains a value

$$\sigma_g = F(c), \tag{5}$$

where $F(c)$ describes the stress at which the crack extends, in terms of its size, inclination, coefficient of friction, and the properties of the rock. The extension of a crack is stable or unstable depending upon whether the derivative $\partial\sigma_g/\partial c = F'$ is positive or negative, respectively.

The effect of crack extension on the stress–strain curve can be explained using (4) and (5).

Differentiating (4) with respect to σ yields

$$\frac{d\varepsilon}{d\sigma} = \frac{1}{E}[1 + \Sigma f(c)] + \frac{\sigma}{E}\Sigma\left[\frac{df(c)}{dc}\frac{dc}{d\sigma}\right]. \tag{6}$$

While $\sigma < \sigma_g$ of the weakest crack its length is independent of the stress and $dc/d\sigma = 0$, but this becomes equal to $1/F'$ when σ exceeds σ_g. Equation (6) may be written as

$$\frac{d\sigma}{d\varepsilon} = \frac{E}{1 + \Sigma f(c) + \sigma \Sigma (f'/F')}, \tag{7}$$

which shows that, prior to crack extension, the slope of the stress–strain curve is linear with a Young's modulus.

$$E_{\text{eff}} = E/[1 + \Sigma f(c)]. \tag{8}$$

As soon as extension of the weakest crack begins, the slope of the stress–strain curve changes due to the effect of the last term in the denominator of (8). The charge will be different for stable and unstable modes of crack extension. As the value of $f(c)$ always increases with c, f' is always positive and, for stable extension, F' is also positive. Provided that the magnitude of the last term in the denominator of (7) increases with increasing stress, the slope of the stress–strain curve decreases with increasing stress. As soon as a crack extends unstably F' becomes negative and the ability of the rock to resist load decreases. If the magnitude of the last term in the denominator of (7) is small the slope of the stress–strain curve during failure may remain positive, as is shown by the Griffith locus for cracks shorter than C_c, § 12.4. Otherwise, the slope of the stress–strain curve changes from positive through zero, when $F' = 0$, to negative, as F' becomes negative.

12.6 The dynamics of crack propagation

The theory of § 12.4 can very simply be carried further by a discussion due to Berry (1960a, b) which gives an insight into the behaviour of a propagating crack. As before, we consider the case of a plate of length l, breadth b, and unit thickness with tension normal to the crack. Suppose that $2c_i$ is the width of the crack and that the applied tensile stress is T_i.

Assuming plane stress, the work done by the external forces in establishing this state of stress is, by § 12.2 (5),

$$W_i = blT_i^2[1 + 2\pi c_i^2/bl]/2E. \tag{1}$$

The total energy V_i of the system, including the surface energy $4\alpha c_i$ of the crack, is

$$V_i = blT_i^2[1 + 2\pi c_i^2/bl]/2E + 4\alpha c_i. \tag{2}$$

Suppose, now, that the crack extends while the applied tension is maintained constant at T_i. Then when the length of the crack is c the total energy, V, of the system will be

$$V = blT_i^2[1 + 2\pi c^2/bl]/2E + 4\alpha c + K, \tag{3}$$

where K is the kinetic energy stored in the solid. The work $W - W_i$ done by T_i during this extension is

$$W - W_i = blT_i(\varepsilon - \varepsilon_i), \tag{4}$$

where ε and ε_i are, by § 12.2 (5), the strains corresponding to crack lengths c and c_i, that is

$$W - W_i = 2\pi T_i^2(c^2 - c_i^2)/E. \tag{5}$$

Equating this to $V - V_i$ given by (2) and (3) gives

$$\pi T_i^2(c^2 - c_i^2)/E = 4\alpha(c - c_i) + K. \tag{6}$$

Writing

$$n = 4\alpha E/\pi c_i T_i^2, \tag{7}$$

this becomes

$$K = \pi c^2 T_i^2(1 - c_i/c)[1 - (n - 1)c_i/c]/E. \tag{8}$$

Mott (1948) has shown that on dimensional grounds

$$K = k\rho T_i^2 c^2 v_c^2/2E^2, \tag{9}$$

where $v_c = dc/dt$ is the velocity at which the crack extends, ρ is the density of the solid, and k is a numerical constant.

Using (9) in (8) gives

$$v_c^2 = \frac{2\pi E}{k\rho}\left(1 - \frac{c_i}{c}\right)\left[1 - (n - 1)\frac{c_i}{c}\right] \tag{10}$$

$$= v_m^2\left(1 - \frac{c_i}{c}\right)\left[1 - (n - 1)\frac{c_i}{c}\right], \tag{11}$$

where

$$v_m = (2\pi E/k\rho)^{\frac{1}{2}}, \tag{12}$$

is the maximum velocity of crack extension which is attained for large c. The quantity k was evaluated by Roberts and Wells (1954) and leads to $v_m = 0{\cdot}38(E/\rho)^{\frac{1}{2}}$.

Differentiating (11) gives for the acceleration of the moving crack tip

$$\frac{dv_c}{dt} = \frac{\pi E c_i}{k\rho c^2}\left[n - \frac{2(n - 1)c_i}{c}\right]. \tag{13}$$

It follows from (11) that the velocity of the crack is zero when $c = c_i$ and also from (13) that if $n = 2$ the acceleration is zero also. Now from (7), $n = 2$ implies $\pi c_i T_i^2 = 2\alpha E$, which is just the Griffith relation § 12.4 (2) between c_i and T_i. Thus, the crack of Griffith length is in unstable equilibrium but does not propagate. The whole of the preceding argument essentially extends the Griffith discussion using dynamical considerations.

If T_g is the tensile stress corresponding on Griffith theory to crack length c_i, (7) may be written

$$n = 2T_g^2/T_i^2. \tag{14}$$

If $n < 2$, or $T_i > T_g$, the crack propagates and the equation of motion (11) can be solved to give its length at any time. This has been done by Berry (1960a), who also treats the case of a cleavage crack held open by constant force.

Berry (1960b) has considered the case of the opening of a crack under constant strain. In this case the stress is increased until a point ε_g, T_g on the Griffith locus is attained and the strain is then held constant. The equation of the Griffith locus is again found to be § 12.4 (10), but two cases arise corresponding to short and long cracks. A short crack with $c_i < (6\pi L)^{-\frac{1}{2}}$ will correspond to a stress–strain curve OP, Fig. 12.4.1, which meets the Griffith locus at a point P above the point of minimum strain, V. Deformation at constant strain then corresponds to a path PQF which passes through the unstable region and intersects the locus again at Q. It can be shown that the crack will continue to grow to an equilibrium length corresponding to the point F for which the area PVQ is equal to the area QFF'. On the other hand, a long crack with $c_i > (6L\pi)^{-\frac{1}{2}}$ will correspond to an initial stress–strain curve which intersects the Griffith locus below V and will propagate only if $T_i > T_g$.

12.7 Disruption of rock by crack propagation

In the Coulomb, Mohr, and Griffith theories failure is presumed to start when the stresses on the plane, or crack, with the most dangerous orientation reach certain critical values. It is implied that, once failure has begun at certain values of the applied stresses, it will continue at the same, or even lower, values of these stresses. However, more detailed studies of failure, § 12.5, suggest that the primary processes of failure following initiation are essentially stable. In terms of the concept of cracks, this implies that as stress is applied to the rock, failure starts first from the crack with the most dangerous orientation. An increase in the applied stresses is required to extend this crack further. As a result of this increase, failures start from other cracks with less dangerous orientations. While the crack with the most dangerous orientation must be parallel to the axis of the intermediate principal stress, this is not necessarily the case for other cracks, which subsequently extend.

The importance of the effective shear stress in connection with the strength and failure of rock is demonstrated in §§ 4.6, 10.14, 12.4. Consider a volume of rock with randomly oriented and homogeneously distributed closed cracks. Suppose that each crack which can slide contributes in the primary process of stable failure to the disruption of the rock, and that its contribution is proportional to the strain energy around the crack. The effect of the intermediate principal stress is to change the effective shear stress on all those cracks which are not parallel to its axis. When the minor and intermediate principal stresses are equal all cracks within a range of angles for which $(|\tau| - \mu\sigma_n) > 0$ symmetrical about the axis of the major principal stress contribute to the disruption of the rock. At values of the intermediate stress between those of the minor and major stresses some of these cracks are prevented from sliding, but other cracks inclined to the direction of the intermediate stress begin to slide.

Postulate that the strength of rock is determined by some maximum value of the *effective shear strain energy*, obtained by summing the strain energy due to the effective shear stress on all those cracks for which

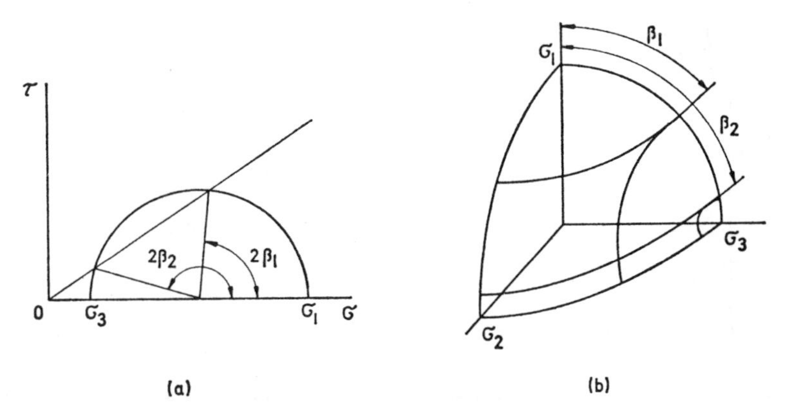

Fig. 12.7.1 (*a*) A Mohr diagram for triaxial and biaxial compression showing that cracks with their normals in the interval $2\beta_1$ to $2\beta_2$ of this diagram can slide. (*b*) The solid angle subtended by the normals of cracks in the interval β_1 to β_2 for triaxial and biaxial compression.

$(|\tau| - \mu\sigma_n) > 0$. Then it can be surmised that the effect of increasing the intermediate principal stress is to increase the strength from that obtained in triaxial stress conditions to a higher value. An analytical formulation of this transition is so complex that its meaning is not obvious. However, by definition, the value of the intermediate principal stress lies between those of the minor and major principal stresses. Several interesting corollaries of the postulate follow from a consideration of these two extreme cases when the value of the intermediate stress is equal to that of the minor and major principal stress respectively. The Mohr diagram, Fig. 12.7.1 (*a*) shows the stresses at failure on cracks of all orientations for these two cases and shows the condition $\tau = \mu\sigma_n$. In each case cracks with orientations in the interval $2\beta_1$ to $2\beta_2$ slide and contribute effective strain energy proportional to $(|\tau| - \mu\sigma_n)^2$. Assume that the number of cracks with normals lying in a unit solid angle is N. The solid angle subtended by the interval $2\beta_1$ to

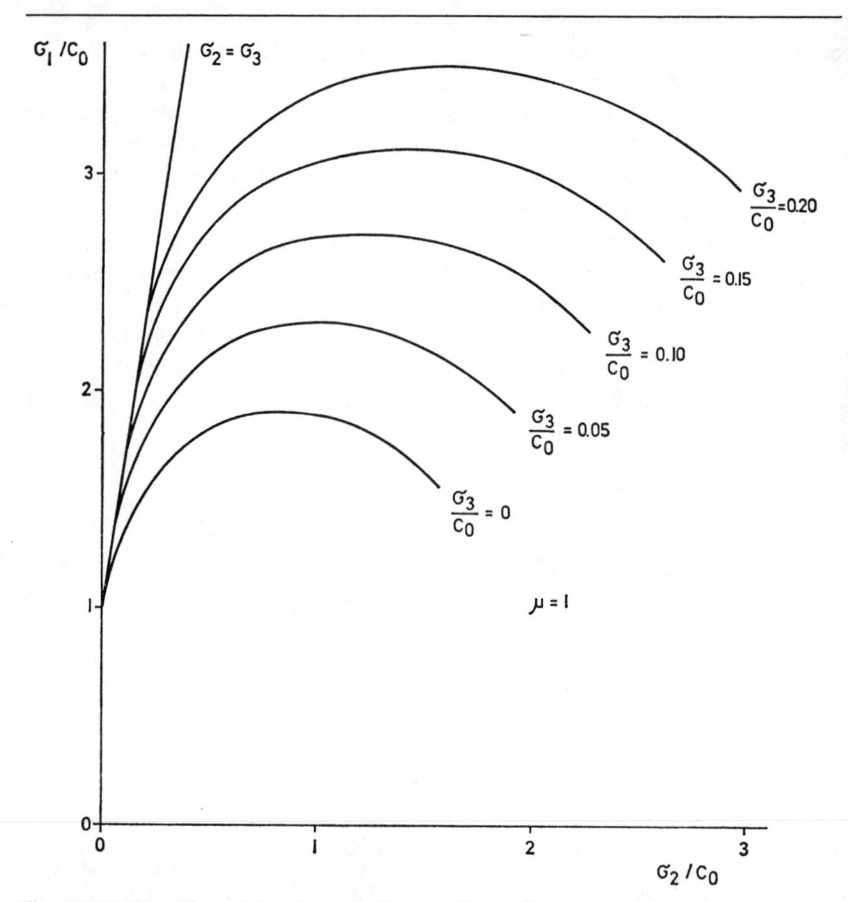

Fig. 12.7.2 The effect of the minor and intermediate principal stresses on the strength of rock according to the effective shear strain energy criterion.

$2\beta_2$ in one octant of a unit sphere is shown in Fig. 12.7.1 (*b*) for each of the two cases being considered. The effective shear strain energy for these cases of stress is given by:

$$W_{\text{eff}} = 2\pi N\bar{c}^2 \int_{\beta_1}^{\beta_2} (|\,\tau\,| - \mu\sigma_n)^2 \sin\beta \, d\beta, \quad \text{for } \sigma_1 > \sigma_2 = \sigma_3, \tag{1}$$

$$W_{\text{eff}} = 2\pi N\bar{c}^2 \int_{\beta_1}^{\beta_2} (|\,\tau\,| - \mu\sigma_n)^2 \cos\beta \, d\beta, \quad \text{for } \sigma_1 = \sigma_2 > \sigma_3, \tag{2}$$

where $\bar{c}$ is a mean factor determined by the size and shape of the cracks.

In the case of a uniaxial stress and the biaxial case when the minor stress is zero, $\beta_2 = \pi$ and (1) and (2) reduce to

$$W_{\text{eff}} = \frac{\pi N \sigma_1^2}{2}\left[\frac{7\mu^6 + 15\mu^4 + 9\mu^2 + 1}{30(\mu^2+1)^{\frac{5}{2}}} + \frac{5\mu^2+3}{6(\mu^2+1)^{\frac{1}{2}}} - \frac{16\mu}{15}\right] \tag{3}$$

$$W_{\text{eff}} = \frac{\pi N \sigma_2^2}{2}\left[\frac{\mu^7 + 9\mu^5 + 15\mu^3 + 7\mu}{30(\mu^2+1)^{\frac{5}{2}}} - \frac{2\mu^5 + 6\mu^3 + 4\mu}{3(\mu^2+1)^{\frac{3}{2}}} - \frac{3\mu^3+\mu}{(\mu^2+1)^{\frac{1}{2}}} + \frac{32\mu^2+8}{15}\right]. \tag{4}$$

Three significant characteristics of the effective strain energy failure criterion emerge from a full study of its properties, as has been done numerically by Wiebols and Cook (1968).

(i) In triaxial compression, $\sigma_1 > \sigma_2 = \sigma_3$, the strength increases linearly with confining stress, as is the case with the Coulomb criterion. However, the rate of increase corresponds to a value of the Coulomb coefficient of internal friction greater than the value of the coefficient of sliding friction between the surfaces of the cracks.

(ii) In compression, with $\sigma_1 = \sigma_2 > \sigma_3$, the strength increases linearly with confining stress at the same rate as in triaxial compression, but the biaxial compressive strength, $\sigma_3 = 0$, is greater than the uniaxial compressive strength.

(iii) For any constant value of the minor principal stress, the strength passes through some maximum value as the value of the intermediate principal stress varies between $\sigma_2 = \sigma_3$ and $\sigma_2 = \sigma_1$.

A family of curves showing these properties for a rock with a coefficient of sliding friction $\mu = 1$ is shown in Fig. 12.7.2.

Chapter Thirteen

Strain Waves

13.1 Introduction

It is usually assumed that rock is at rest under the action of static stresses, and most problems in rock mechanics are treated as problems of statics. There are a number of important situations in which the stresses are of a dynamic nature, and the propagation of these stresses through the rock as waves must be considered. Such situations arise in connection with explosive blasting, rockbursts in mining, and earthquakes. The amplitudes of these dynamic stresses, except in the immediate vicinity of their source, are generally small in relation to the compressive strength of rock, and the time of their application is short. It is to such problems that the theory of elasticity is most applicable in rock mechanics. Most of the ensuing discussion is concerned with the propagation of elastic waves in rock, or *seismic waves*.

Much of the mathematical theory concerning elastic wave propagation preceded meaningful seismic field measurements. As early as 1828, Cauchy and Poisson determined the equations of motion of a disturbance in a perfectly elastic body. Rayleigh (1885) described the propagation of waves over the surface of a solid body, and Love (1911) explained the occurrence of surface waves with particle motion parallel to the boundary surface.

Notable developments in seismic instrumentation occurred in 1892, when Milne developed a compact seismograph which was superseded by new types developed by Omori (1901) and Wiechert (1903). Galitzin (1914) introduced galvanometric recording seismographs, and elegant improvements to seismographs were made by Benioff (1932). Recent developments in electronics and instrumentation make it relatively simple to measure seismic waves with periods ranging from a millisecond to many minutes, having amplitudes from millimicrons to microns, respectively.

Seismic waves originating from earthquakes are used to locate the foci and study the mechanism at the source of earthquakes, Hirasawa and Stauder (1965), and to deduce the structure of the earth's interior, Gutenberg (1959). Seismic waves from explosions are used to study the near surface structure of the earth's crust in exploration for minerals, especially oil, Dix (1952). Recently a great deal of effort has been devoted to detailed seismic measurements to detect and locate the origin of large, nuclear explosions, Vesiac Report (1962). There are a large number of texts on seismology and wave propagation, such as Macelwane and Sohon (1936), Officer, (1958), Brekhovskikh (1960), Kolsky (1963), Bullen (1963), and White (1965).

In rock mechanics seismic waves are important because they allow information to be gained concerning rock failures and the properties of rock in regions of a rock mass to which there is no direct access, Cook (1963). Strain waves play an important part in blasting, Duvall and Atchison (1957). The strain waves from a spherical explosive source have been analysed by Sharpe (1942), and Starfield and Pugliese (1967) have extended this work by numerical procedures to the case of a long cylindrical explosive detonating with finite velocity. Small chemical explosions radiate as little as 10^3 ft-lb of seismic energy per lb weight of explosive, large rockbursts as much as 10^9 ft-lb of seismic energy, and great earthquakes up to 10^{17} ft-lb of seismic energy.

13.2 Elastic waves

The *equations of motion* for an elastic solid, § 5.5 (24), (25), (26), are

$$\rho\, \partial^2 u/\partial t^2 = (\lambda + G)\, \partial\Delta/\partial x + G\, \nabla^2 u + \rho X, \tag{1}$$

$$\rho\, \partial^2 v/\partial t^2 = (\lambda + G)\, \partial\Delta/\partial y + G\, \nabla^2 v + \rho Y, \tag{2}$$

$$\rho\, \partial^2 w/\partial t^2 = (\lambda + G)\, \partial\Delta/\partial z + G\, \nabla^2 w + \rho Z. \tag{3}$$

Differentiating (1), (2), (3) partially with respect to x, y, z, respectively, and adding the results for the case where the density and body forces are constant, gives

$$\partial^2 \Delta/\partial t^2 = ([\lambda + 2G]/\rho)\, \nabla^2\Delta, \tag{4}$$

which is the equation of a *wave of dilatation* with a velocity of propagation

$$C_P = [(\lambda + 2G)/\rho]^{\frac{1}{2}} = [E(1 - \nu)/(1 + \nu)(1 - 2\nu)\rho]^{\frac{1}{2}}. \tag{5}$$

Differentiating (2) and (3) partially with respect to z and y respectively, subtracting the results for the case of constant density and body forces, and dividing by 2 yields

$$\rho\, \partial^2[\tfrac{1}{2}(\partial w/\partial y - \partial v/\partial z)]/\partial t^2 = G\, \nabla^2[\tfrac{1}{2}(\partial w/\partial y - \partial v/\partial z)]. \tag{6}$$

The terms in parentheses are defined in § 2.11 (9) as the components of rotation, so that

$$\partial^2\omega_x/\partial t^2 = (G/\rho)\, \nabla^2(\omega_x), \tag{7}$$

and

$$\partial^2\omega_y/\partial t^2 = (G/\rho)\, \nabla^2(\omega_y), \tag{8}$$

$$\partial^2\omega_z/\partial t^2 = (G/\rho)\, \nabla^2(\omega_z), \tag{9}$$

which are wave equations of *waves of rotation* with velocity of propagation along the x, y, z axes, respectively, of

$$C_S = (G/\rho)^{\frac{1}{2}}. \tag{10}$$

If the dilatation is zero, and the density and body forces are constant, (1), (2), and (3) reduce to wave equations in terms of the displacements

$$\partial^2 u/\partial t^2 = C_S{}^2 \nabla^2 u, \tag{11}$$

$$\partial^2 u/\partial t^2 = C_S{}^2 \nabla^2 v, \tag{12}$$

$$\partial^2 w/\partial t^2 = C_S{}^2 \nabla^2 w. \tag{13}$$

In the case when the rotations are zero, $\omega_x = \omega_y = \omega_z = 0$ and

$$\partial\Delta/\partial x = \partial^2 u/\partial x^2 + \partial^2 v/\partial y\,\partial x + \partial^2 w/\partial z\,\partial x = \nabla^2 u, \tag{14}$$

$$\partial\Delta/\partial y = \partial^2 u/\partial x\,\partial y + \partial^2 v/\partial y^2 + \partial^2 w/\partial z\,\partial y = \nabla^2 v, \tag{15}$$

$$\partial\Delta/\partial z = \partial^2 u/\partial x\,\partial z + \partial^2 v/\partial y\,\partial z + \partial^2 w/\partial z^2 = \nabla^2 w, \tag{16}$$

and if the density and body forces are constant (1) (2), (3) yield

$$\partial^2 u/\partial t^2 = C_P{}^2 \nabla^2 u, \tag{17}$$

$$\partial^2 v/\partial t^2 = C_P{}^2 \nabla^2 v, \tag{18}$$

$$\partial^2 w/\partial t^2 = C_P{}^2 \nabla^2 w. \tag{19}$$

The equations of motion can be solved in a general way by means of displacement potentials, Love (1927). It can be verified by direct substitution in (1), (2), and (3) that there is a *scalar potential*, Φ, such that the components of displacement are given by $u = \partial\Phi/\partial x$, $v = \partial\Phi/\partial y$, $w = \partial\Phi/\partial z$. Likewise, there is a *vector potential* with components ψ_x, ψ_y, ψ_z such that $u = \partial\psi_z/\partial y - \partial\psi_y/\partial z$, $v = \partial\psi_x/\partial z - \partial\psi_z/\partial x$, $w = \partial\psi_y/\partial x - \partial\psi_x/\partial y$. The potential Φ is associated with dilatation and the potential ψ with rotation. The introduction of these two potentials allows the effects of dilatation and rotation to be separated, which proves to be convenient in the treatment of many problems involving boundaries.

The preceding wave equations are all of the form

$$\partial^2\alpha/\partial t^2 = C^2 \nabla^2 \alpha. \tag{20}$$

When the deformation is a function of one coordinate only, say a plane wave travelling along the x direction, (20) becomes

$$\partial^2\alpha/\partial t^2 = C^2\, \partial^2\alpha/\partial x^2. \tag{21}$$

The general solution of this is

$$\alpha = f(x - Ct) + F(x + Ct), \tag{22}$$

where f and F are functions depending upon the initial conditions, and represent waves travelling in the positive and negative x-directions, respectively, with a velocity of propagation, C. At any time t_1 the wave $\alpha = f(x - Ct)$ travelling in the x-direction has a given shape in the medium which is the same as the shape a distance $C(t_2 - t_1)$ along the x-axis at a later time, t_2.

A spherical wave is also a function of one coordinate, the radius, r, from its source. In this case (20) becomes

$$\partial^2(r\alpha)/\partial t^2 = C^2 \, \partial^2(r\alpha)/\partial r^2 \tag{23}$$

which has a solution

$$r\alpha = f(r - Ct) + F(r + Ct) \tag{24}$$

showing that the amplitude of a spherical wave is inversely proportional to the radius r.

In an unbounded elastic material energy is transmitted by *waves of dilatation,* often referred to as *primary* or *compressional waves,* which propagate with a velocity $C_P = [(\lambda + 2G)/\rho]^{\frac{1}{2}}$ or by *waves* of *distortion,* often referred to as *secondary* or *shear waves,* which propagate with a velocity $C_S = [G/\rho]^{\frac{1}{2}}$. Two further important modes of wave propagation occur in thin bars and along the surface of a semi-infinite solid.

Where a solid is bounded by extensive surfaces, elastic surface waves, of which the most important are *Rayleigh waves* and *Love waves,* may occur, Bullen (1963) and Kolsky (1963). Rayleigh waves are polarized so that their particle motion consists of a retrograde ellipse perpendicular to the free surface with an amplitude which decreases exponentially with increasing distance beneath the surface. In a homogeneous solid Rayleigh waves are not dispersive, and propagate with a velocity $C_R = \gamma C_s$, where γ is a factor with a maximum value of 0·9553 when $\nu = 0{\cdot}5$ and a value of 0·9194 when Poisson's relation holds, that is, for $\nu = 0{\cdot}25$. In a stratified solid, Love waves appear if the velocity of propagation of waves of distortion is greater in the lower stratum. Love waves have a particle motion parallel to the free surface and perpendicular to the direction of propagation and propagate with a velocity lying between the velocities of propagation of waves of distortion in the surface and lower layer. Surface waves play an important part in the study of earthquakes, where, diverging parallel to the surface only, they acquire relatively increasing importance with increasing distance from their source.

In thin bars, waves of distortion, corresponding to torsional disturbances, propagate with the same velocity as waves of distortion in an unbounded body, that is, $C_{BS} = C_S = (G/\rho)^{\frac{1}{2}}$. Waves of dilatation in thin bars, corresponding to tensile or compressive disturbances, can be shown to propagate with a velocity $C_{BP} = (E/\rho)^{\frac{1}{2}}$ if it is assumed that the disturbance is accompanied by free Poisson expansion or contraction of the bar and that the inertial forces due to this radial motion can be neglected, Timoshenko and Goodier (1951). This is a reasonable approximation only if the wavelength of the disturbance is several times greater than the cross-sectional dimensions of the bar. The problem of wave propagation in cylindrical bars was investigated by Pochammer (1876), and accounts of his treatment are found in Love (1927) and Kolsky (1963). A full analysis shows that the propagation of waves of dilatation in thin bars is dispersive and that the velocity of propagation of waves of short length in relation to the diameter of the bar tends towards the velocity of propagation of Rayleigh waves near the surface of a semi-infinite solid.

13.3 Energy, particle velocity, and stress

In most instances waves take the form of transient pulses. The linearity of the wave equation allows solutions to the propagation of complicated wave shapes or pulses to be obtained by superposition using a Fourier series or transform, Papoulis (1962). Much can be learnt about the nature of wave propagation by studying the propagation of a simple harmonic wave. Any function $f(x - Ct)$ represents a wave propagating in the x-direction, and the function

$$A = A_0 \sin 2\pi(x - Ct)/L, \tag{1}$$

represents a simple harmonic wave with an amplitude A_0 and a wavelength L propagating in this direction.

For mathematical convenience, many texts write (1) in the form

$$A = A_0 \exp [ik(x - Ct)], \tag{2}$$

where $k = 2\pi/L$ is known as the *wave-number*.

If A is a displacement, the particle velocity of this wave is

$$\partial A/\partial t = -(2\pi C A_0/L) \cos 2\pi(x - Ct)/L, \tag{3}$$

and the strain due to it is

$$\partial A/\partial x = (2\pi A_0/L) \cos 2\pi(x - Ct)/L. \tag{4}$$

The energy propagated by a wave is part potential and part kinetic. Consider an element, dx, of a filament of unit cross-section extending in the direction of wave propagation. The kinetic energy of this element is

$$\tfrac{1}{2}\rho\,(\partial A/\partial t)^2\,dx = (2\pi^2 C^2 A_0^2 \rho/L^2) \cos^2 [2\pi(x - Ct)/L]\,dx. \tag{5}$$

The strain energy of the same element depends upon whether the wave is one of dilatation or distortion, that is, whether the particle motion, A, is parallel or perpendicular to the direction of propagation, respectively. In the former case the strain energy is

$$\tfrac{1}{2}(\lambda + 2G)\,(\partial A/\partial x)^2\,dx = [2\pi^2 A_0^2\,(\lambda + 2G)/L^2] \cos^2 [2\pi(x - Ct)/L]\,dx, \tag{6}$$

whereas in the latter case it is

$$\tfrac{1}{2}G\,(\partial A/\partial x)^2\,dx = [2\pi^2 G A_0^2/L^2]\cos^2[2\pi(x - Ct)/L]\,dx \tag{7}$$

Substituting the velocities of propagation, § 13.2 (5) (10), into (6) and (7) and comparing the result with (5) shows that, at any instant, the kinetic and potential energies of an element are equal, from which it is concluded that the energy propagated by a wave is divided equally between the kinetic and potential energies.

A comparison of (3) and (4) shows that the particle velocity and strain are related by

$$\partial A/\partial t = -C\,\partial A/\partial x, \tag{8}$$

so that

$$\varepsilon = -\dot{A}/C, \tag{9}$$

where $\dot{A}$ is the particle velocity.

In terms of stress, this yields

$$\sigma = (\lambda + 2G)\varepsilon = -\dot{A}E(1 - \nu)/[(1 + \nu)(1 - 2\nu)C_P], \tag{10}$$

and

$$\tau = -\dot{A}G/C_S, \tag{11}$$

for body waves, depending upon whether the particle motion, A, is parallel or transverse to the direction of propagation, respectively, and

$$\sigma = -\dot{A}E/C_B, \tag{12}$$

for compressional waves in thin bars. The ratio between the stress and particle velocity at a point is $-\rho C$, which is often referred to as the *acoustic impedence*.

The energy flux, p, through a unit area perpendicular to the direction of propagation in unit time is found by integrating the energy in a unit cross-section of length C along the wave. For a single harmonic component this yields

$$\begin{aligned} p &= \int_{x-C}^{x} [4\pi^2 C^2 \rho A_0^2/L^2]\cos^2[2\pi(x - Ct)/L]\,dx \\ &= 2\pi^2 \rho C^3 A_0^2/L^2. \end{aligned} \tag{13}$$

In the case of a spherical wave the total flux, P, in unit time through an envelope of large radius r is

$$P = (8\pi^3 \rho C^3 A_0^2 r^2)/L^2, \tag{14}$$

where A_0 is the displacement amplitude at r.

For a simple harmonic wave the peak particle velocity and strain are

$$\dot{A}_0 = 2\pi A_0 C/L, \tag{15}$$

$$\varepsilon_0 = 2\pi A_0/L. \tag{16}$$

As no energy is lost in a perfectly elastic medium, P in (14) is a constant. Substituting (15) and (16) in (14) shows that the amplitude of the particle velocity, the strain, and hence the stress must decrease inversely with the distance, r, from the source.

13.4 Reflection and refraction at an interface

In general, when an elastic wave intercepts a plane border between two different media four new waves are generated. A wave of dilatation and one of distortion are refracted into the medium away from the wave source, and two similar waves are reflected back into the medium near the wave source. The normal and tangential displacements and the normal and tangential stresses across the interface must be equal. Consider a plane wave in the x–y plane incident on an interface, $x = 0$, and designate the medium near the source by subscript 1 and that away from the source by subscript 2, so that the boundary conditions become

$$\text{(i) } \Sigma u_1 = \Sigma u_2 \tag{1}$$

$$\text{(ii) } \Sigma v_1 = \Sigma v_2; \quad \Sigma w_1 = \Sigma w_2 \tag{2}$$

$$\text{(iii) } \Sigma(\sigma_x)_1 = \Sigma(\sigma_x)_2$$

$$\text{that is } \Sigma(\lambda\Delta + 2G\ \partial u/\partial x)_1 = \Sigma(\lambda\Delta + 2G\ \partial u/\partial x)_2 \tag{3}$$

$$\text{(iv) } \Sigma(\tau_{xy})_1 = \Sigma(\tau_{xy})_2; \quad \Sigma(\tau_{xz})_1 = \Sigma(\tau_{xz})_2$$

$$\begin{aligned} \text{that is } \Sigma[G\,(\partial v/\partial x + \partial u/\partial y)]_1 &= \Sigma[G(\partial v/\partial x + \partial u/\partial y)]_2 \\ \Sigma[G\,(\partial w/\partial x + \partial u/\partial z)]_1 &= \Sigma[G(\partial w/\partial x + \partial u/\partial z)]_2. \end{aligned} \tag{4}$$

The boundary conditions are satisfied, Macelwane and Sohon (1936), if Huygens' principle is applied to the incident and generated waves. For an incident wave of dilatation this gives

$$\sin\alpha_i/C_{P1} = \sin\alpha_1/C_{P1} = \sin\beta_1/C_{S1} = \sin\alpha_2/C_{P2} = \sin\beta_2/C_{S2} \tag{5}$$

and for one of distortion it gives

$$\sin\beta_i/C_{S1} = \sin\alpha_1/C_{P1} = \sin\beta_1/C_{S1} = \sin\alpha_2/C_{P2} = \sin\beta_2/C_{S2}, \tag{6}$$

where α and β are the angles between the normal to the interface and the incident, reflected, and refracted waves of dilatation and distortion, respectively, Fig. 13.4.1.

When the velocity of propagation of a reflected or refracted wave is greater than that of the incident wave there exists a critical angle of incidence which makes the angle of reflection or refraction $\pi/2$. For angles of incidence greater than this critical value, (5) and (6) break down, and it is found that a disturbance, the amplitude of which decays exponentially with distance from the interface, is set up. Geometrically, a critically refracted ray appears to propagate along the interface with a velocity equal to the velocity of propagation in the faster of the two media. Plane-wave theory predicts that no energy can be transmitted along this path. However, such refracted waves are observed daily and form the basis of seismic refraction surveys, § 13.6. The explanation of this paradox is found in the complicated theory of curved wavefronts, Grant and West (1965).

The case of a wave of dilatation with an amplitude, A_i, incident normal to the interface is satisfied when the amplitude, A_1 of the reflected wave and that, A_2 of the refracted wave are given by

$$A_1 = A_i(\rho_2 C_{P2} - \rho_1 C_{P1})/(\rho_2 C_{P2} + \rho_1 C_{P1}) \tag{7}$$

$$A_2 = 2A_i\rho_1 C_{P1}/(\rho_2 C_{P2} + \rho_1 C_{P1}). \tag{8}$$

If the impedance of the second medium is zero, that is, the interface is a free surface, the amplitude of the reflected wave is equal to that of the incident wave but is of opposite sign, so that the amplitude of vibration at

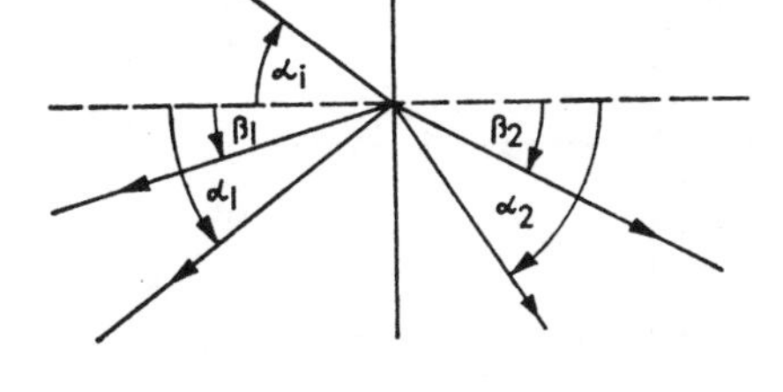

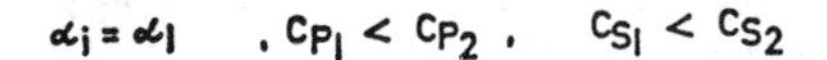

Fig. 13.4.1 Reflected and refracted waves of dilatation and distortion due to the incidence of a plane wave of dilatation on the boundary between 1 and 2.

the boundary is twice that of the original incident wave, as is indicated by (8), although no energy is transmitted by a refracted wave in a medium with zero impedence. The resultant amplitude and stress in the first medium are the differences between those of the incident and reflected waves. The latter can be considered as an inverted image of the former, propagating in the opposite direction. The situation of a compressional pulse incident upon a free surface and its reflected tensile image is shown in Fig. 13.4.2 immediately before incidence (*a*), at incidence (*b*), just after incidence (*c*), and immediately after complete reflection (*d*). The resultant of the incident and reflected pulses is the heavy line in each case. It is seen that the resultant stress changes from compression at incidence to tension at complete reflection. If the resultant tension at any stage exceeds the tensile strength of the material an extension fracture occurs at that point. The material between the free surface and this fracture flies off with a momentum equal to that in a length of the original pulse corresponding to twice the distance between the free surface and the fracture. This phenomenon is the basis of fracture by 'scabbing' when plates are impacted by projectiles or explosives, Hopkinson (1912) and Rinehart (1960), and fracture by 'slabbing' in rock blasting, Duvall and Atchison (1957).

Conversely, when the impedence of the second medium is infinite the

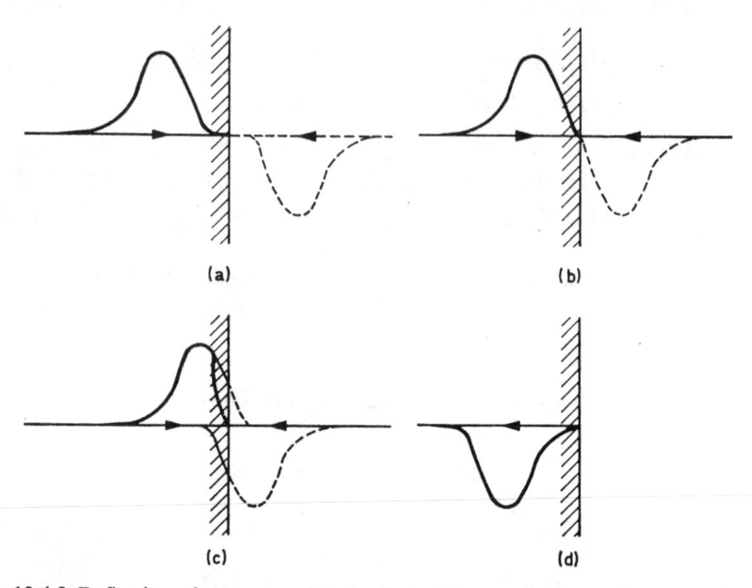

Fig. 13.4.2 Reflection of a compressional pulse incident on the free surface, showing the resultant pulse, heavy curve, and the incident and reflected components, light and dashed curves, respectively.

reflected wave can be considered an image of the incident wave propagating in the opposite direction.

13.5 Seismic location of failures

One of the most important uses of seismic waves in rock mechanics has been the study of violent rock failures by analysis of the waves radiated from them, Gane *et al.* (1952), Cook (1963), and Joughin (1966). Seismic techniques provide one of the few means by which changes in the rock mass around an excavation can be studied. An important aspect of this is the location of failure foci using a network of seismic transducers, known as seismometers or geophones, placed in the rock mass within which the failures are to be studied.

The accuracy of seismic location is determined by the precision with which the arrival, at the seismometers, of the seismic wave radiating from the failure can be measured and by the accuracy with which the velocities of propagation are known. The precision with which the time of arrival can be determined is set by the rise-time of the initial seismic pulse. The rise-time can be kept small if the attenuation of the high-frequency components of the seismic wave is limited by keeping the lengths of the seismic paths between the failures and the seismometers short. Uncertainties in the velocities of propagation are also minimized when these paths are short. For accurate location a network of seismometers should closely surround the region in which the failures are to be located. Experience, Cook (1962) and Joughin (1966), has shown that the accuracy of location is usually about 1 per cent of the path length.

If a system of rectangular coordinates x, y, z defines any point in the rock and seismometers are situated at points x_n, y_n, z_n the length of a seismic path in homogeneous rock between a failure at x, y, z and a seismometer is

$$r_n = [(x_n - x)^2 + (y_n - y)^2 + (z_n - z)^2]^{\frac{1}{2}}. \tag{1}$$

Where the times of arrival t_{nP} and t_{nS} of the waves of dilatation and distortion, respectively, can be determined,

$$r_n = (t_{nS} - t_{nP})C_P C_S/(C_P - C_S). \tag{2}$$

The times of arrival at three seismometers are required to solve (1) and (2) to find the focus of the failure but yield two possible solutions symmetric about the plane of the seismometers. A fourth seismometer outside this plane is necessary to resolve this ambiguity.

The time of arrival of the distortion wave cannot be determined with the same accuracy as that of the dilatation wave, as the coda of the latter wave impairs the signal-to-noise ratio at the time when the former wave arrives. Locations with short-range seismic networks are usually made on the basis of the time of arrival of the wave of dilatation only. If t_0 is the time of origin of the failure

$$r_n = (t_{nP} - t_0)C_P. \tag{3}$$

The times of arrival at four seismometers are required to solve (1) and (3) to find the focus of the failure, but again two solutions result on account of the quadratic nature of (1), although one of these is often imaginary. A fifth seismometer is required to resolve the ambiguity when it is not, and the seismometers must not be coplanar.

In practice, variations in the velocities of propagation and uncertainties in defining the time of arrival make it desirable to use additional seismometers and obtain a least-squares solution from them for the location of the focus of the failure. The coordinates of a focus can be found by direct computation or by the use of an analogue computer, Cook (1963).

Fractures extend over an area of finite dimensions, which raises a question concerning the position of the located focus in relation to the fracture surface. New fractures propagate at velocities less than the velocity of propagation of waves of distortion, § 12.6. The first pulse to radiate from a failure emanates from the position at which the fracture originates and propagates with the velocity of waves of dilatation, that is, faster than the velocity of fracture propagation. If the first pulse is detected at all the seismometers the position of the focus is the position of the origin of the fracture. Where the first pulse is not sufficiently large to be detected at all the seismometers, the position of the focus is some point along the fracture.

An existing fracture may slide as a result of the disturbance due to a stress wave passing across it. If the fracture is inclined to the direction of propagation of such a wave the apparent velocity of the disturbance along the fracture is greater than the velocity of propagation. Such a fracture might appear to propagate with a velocity in excess of the velocity of wave propagation, but in this case the question of the location of the focus is also complicated by the existence of the wave which triggered the new fracture process.

Many rocks, especially hard rocks with a coarse and heterogeneous structure, have been observed to produce microseismic vibrations when subjected to an increasing stress, § 6.14. Microseismic activity in solid rocks usually starts when the stress exceeds three-quarters of the ultimate strength of the rock and increases in intensity with increasing stress. Stick–slip motion on a fracture surface might also be expected to give rise to microseismic activity. These observations suggest that warning of an impending rock failure might be gained by monitoring microseismic activity, Obert and Duvall (1957, 1967).

Where rock failure is the exception rather than the rule and phenomena associated with failure, such as microseismic activity, are not ordinarily expected, then the observation of such phenomena does provide a means for monitoring conditions approaching failure. In such cases, measurements of microseismic activity might provide some indication of an impending failure if the rock is of a type prone to produce microseismic noise before failure. The use of microseismic activity to predict the occurrence of a failure requires that the pattern of microseismic activity leading up to failure be known and that it have a clearly recognizable pattern. Furthermore, the microseismic detector system must have the capacity to locate the origin of the activity it is detecting, otherwise the intensity of each burst of activity cannot be determined and no distinction can be made between a local source of intense activity and distributed sources each of lesser activity.

Where rock failure is an inevitable and an almost continuous process, as it is around deep mine excavations, all the phenomena leading up to failure exist most of the time. In this case virtually continuous microseismic activity must be expected. Only if the form of activity preceding a dangerous failure is different from the continuous background can microseismicity be expected to be useful as a predictive technique. Its effectiveness depends primarily upon the ratio between the microseismic signal pattern preceding a dangerous failure and the background microseismic noise. Capacity to locate the origin of each burst of activity becomes even more important in these circumstances, and can present a formidable problem.

13.6 Seismic measurements

The velocities of seismic wave propagation depend upon the rigidity, compressibility, and density of the rock. The values of the elastic properties can therefore be deduced from measurements of the velocities of wave propagation if the density is known, § 6.14. All three properties depend on the composition and the condition of the rock, such as the extent to which it is weathered or fractured, and on the state of stress to which it is subject. Measurements of seismic velocities provide a means of assessing these factors which is important, because it does not require direct access to the region of rock where the assessment has to be made.

In general, three techniques are used for making these measurements. First, the time of propagation of a pulse direct through the rock between source and receiver, often situated in long boreholes, is measured, Obert and Duvall (1961). Second, subsurface structure is studied by reflection or refraction seismic surveys, much as those used in prospecting, Dix (1952). Third, direct reflection of ultrasonic pulses by cracks in the rock is measured using a combined transmitting and receiving transducer, much as in inspectoscopes used in examining mechanical components for flaws, Mason (1958). The first method is straightforward and small explosive charges or mechanical hammers, McDonal *et al.* (1958) and Swain (1962), are used as sources. These generate sufficient seismic energy with frequencies of the order of hundreds of cycles per second, so that attenuation is not a serious problem and the method can be used over distances of a few thousand feet. The second method is the most flexible of all, and can be applied on a small scale using a hammer source, Lutsch and Szendrei (1958), or on a crustal scale using rockbursts or very large explosive sources, Willmore *et al.* (1952) and Steinhart and Meyer (1961). The third method requires very high frequencies, of the order of many kilocycles per second, to obtain good reflections, and its range is limited to tens of feet on account of high-frequency attenuation.

The attenuation of waves in solids due to dissipative phenomena is complex and varies with the nature of the solid. In most rocks dissipative attenuation assumes significance when the propagation distances exceed the wavelength by more than a few orders of magnitude. Dissipation can be defined in terms of the *specific energy loss*, which is the ratio between the amount of energy, ΔW, lost in taking a body through a stress cycle and the elastic strain energy in the body at the maximum stress during that cycle, W, that is, the ratio $\Delta W/W$. The specific energy loss is frequently expressed in terms of the parameter Q, used also to describe the *quality* of a resonant electrical circuit, where $Q = 2\pi W/\Delta W$. Its reciprocal, $1/Q$, is the tangent to the phase angle by which strain lags behind applied stress in the case of sinusoidal excitation of a linear system. If rheological properties, § 11.3, are incorporated in the equations of motion, § 5.5 (24), (25), (26), wave propagation in a body possessing these time-dependent properties is seen to be both dissipative and dispersive, that is, the velocities of propagation are functions of frequency. There is increasing evidence to suggest that the specific energy loss for many materials ranging from polymers to rocks is independent of frequency over a range of frequencies covering many orders of magnitude, Kolsky (1965) and Bradley and Newman (1966), and that the velocities of wave propagation in rocks are virtually independent of frequency over a similar range, White (1965). It would appear that wave propagation in rock is dissipative but only slightly

Table 13.6

Typical velocities of propagation and the dissipation constants for seismic waves in common materials*

Material	C_P (ft/millisecond)	C_S (ft/millisecond)	Q (approx)
Water (room temp.)	4·9	0	10^5
Steel	19·5	10·6	3,000
Glass	22·3	10·7	600
Taconite, Minnesota	17·5	—	300
Sandstone, Pennsylvania	9·5	—	50
Chalk, Texas	9·2	3·6	80
Granite, Westerly	19·0	10·6	40
Quincy	19·3	9·7	70
Limestone, Solenhofen	19·6	9·5	150
Norite, Sudbury	20·4	1·1	125

* Based on data from Memoir 97 of the Geological Society of America, Clark (1966).

dispersive. Knopoff (1956) proposed that seismic dissipation is due to solid friction, and Walsh (1966) has discussed the attenuation of seismic waves caused by friction on closed cracks.

The geometrical and dissipative attenuation of seismic waves spreading from a point source can be determined from

$$A = (A'/r)e^{-\alpha(r-1)}, \tag{1}$$

where r = the distance from the source;
A' = the amplitude at unit distance from the source;
A = the amplitude at the distance r from the source;
$\alpha = \pi f/QC$;
f = the frequency of the wave;
C = the velocity of propagation.

Typical values of C_P, C_S, and Q for various materials and rocks, based on data in Memoir 97 of the Geological Society of America, Clark (1966), are given in Table 13.6.

In reflection or refraction surveys the times of arrival of seismic waves from a surface or near surface source, S, at a line of surface or near surface receivers, R_1, R_2, . . ., Fig. 13.6.1 (a), are measured.

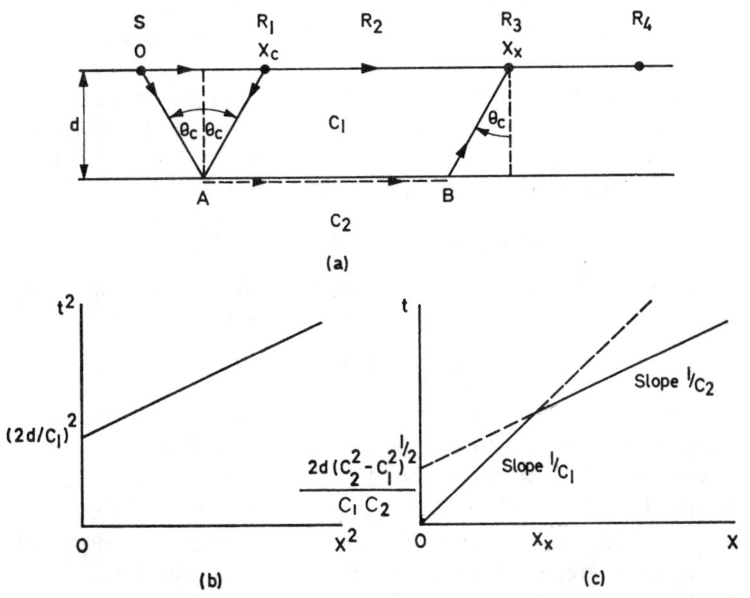

Fig. 13.6.1 (a) The geometry of reflected and refracted seismic paths in a surface layer. (b) A travel–time plot for reflected pulses. (c) A travel–time plot for direct and refracted pulses.

Consider the simple case of a surface layer with uniform thickness, d, having a velocity of propagation, C_1, and an impedance different from that in the underlying material. A wave from S reaching R after reflection from the interface at A will have travelled the path SAR, which has a length $(4d^2 + x^2)^{\frac{1}{2}}$, where x is the distance along the surface between S and R. The time taken by this wave to travel the path SAR is

$$t = (4d^2 + x^2)^{\frac{1}{2}}/C_1. \tag{2}$$

Plotting t^2 against x^2, Fig. 13.6.1 (*b*), gives a line having a slope $1/C_1^2$ with an intercept $(2d/C_1)^2$ on the t-axis, from which d and C_1 can be determined. The reflection method has the disadvantage that the direct wave from S to R travels with the same velocity C_1 and always arrives at R before the reflected wave, making it difficult to determine the time at which the latter arrives. For this reason it is desirable to use source pulses of very short duration in reflection surveys and to maintain the distance S–R small so that the direct wave has passed before the reflected wave arrives. It is also desirable to minimize the generation and detection of surface waves, which is often the reason for using subsurface sources and transducers. The case of a layer with non-uniform thickness, that of several layers, and the question of multiple reflectors can be analysed similarly.

If the velocity of propagation in the surface layer, C_1, is less than the velocity of propagation in the underlying material, C_2, as is usually the case, a wave from S reaches R by the refracted path $SABR$ for values of x greater than a distance, x_c, corresponding to the critical angle of reflection, θ_c. If x is sufficiently great the refracted wave arrives before the direct wave SR. From § 13.4 (5), (6), the critical angle, θ_c, of the paths SA and BR is given by $\sin\theta_c = C_1/C_2$, which makes $x_c = 2dC_1/(C_2^2 - C_1^2)^{\frac{1}{2}}$. The time taken by the wave to travel the path $SABR$ is

$$t = 2d(C_2^2 - C_1^2)^{\frac{1}{2}}/C_1C_2 + x/C_2. \tag{3}$$

The direct wave SR arrives at a time $t_d = x/C_1$. Beyond the cross-over distance, x_x, the refracted wave arrives first, where x_x is given by

$$x_x = 2d[(C_2 + C_1)/(C_2 - C_1)]^{\frac{1}{2}}. \tag{4}$$

If t is plotted against x, Fig. 13.6.1 (*c*), two straight lines are obtained. The first, corresponding to the direct wave, passes through the origin and has a slope $1/C_1$. The second has a slope $1/C_2$ for $x > x_x$ and intercepts the t-axis at $2d(C_2^2 - C_1^2)^{\frac{1}{2}}/C_1C_2$. It follows that C_1, C_2, and d can be determined from such results. The case of a layer with non-uniform thickness can be treated similarly, but is more complicated and requires a reversed refraction profile, S, on the opposite side of R, to be made before a solution can be obtained. These and other complications are discussed more fully by Grant and West (1965).

When the velocities of propagation have been measured it is usually necessary to relate these observations to other properties of the rock which are of more direct concern. In Fig. 6.15.1 (*a*) the relation between the static and dynamic values of Young's modulus for many different rocks is shown.

The effects of stress on Young's modulus are discussed in § 12.3, and some examples are shown in Figs. 6.14.1 and 6.15.2. This effect is most marked at relatively low values of stress, where changes in seismic velocities hold some promise of providing a means for measuring changes in stress level. Stress measurements are more likely to be of interest at high values of stress approaching the strength. Unfortunately, seismic velocities become relatively insensitive to changes in stress at high stresses. At high stresses irreversible changes in the structure of the rock are brought about. The effect of these changes on the dynamic modulus is different from that on the static modulus. The resulting changes are orthotropic with respect to the principal stress axes, and can be expected to decrease the seismic velocity in the direction of the minor principal stress while having little effect in the direction of the major principal stress. Attempts to determine the approach of failure from measurements of seismic velocities must take cognizance of these irreversible and anisotropic changes.

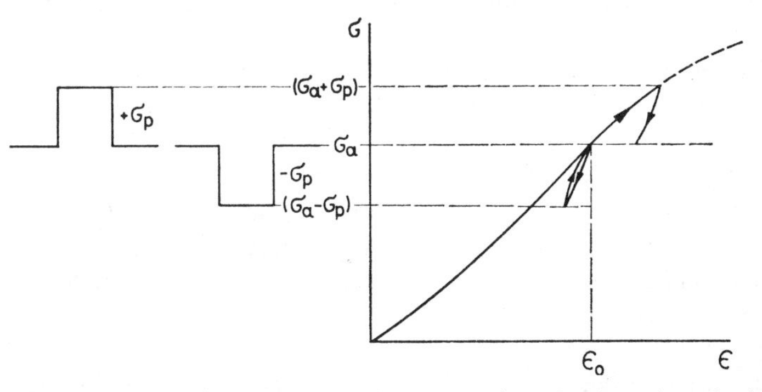

Fig. 13.6.2 The stress–strain paths due to the propagation of pulses of rarefaction and compression through a rock subject to a compressive stress σ_a.

Fig. 13.6.2 shows a portion of a stress–strain curve for rock in which the compressive stress has increased to a value, σ_a, where the slope of the curve is D and the strain is ε_a. Consider the effect of the irreversible nature of the stress–strain curve on the propagation of a rectangular pulse of rarefaction and one of compression. A rarefaction pulse with an amplitude $-\sigma_p$ first decreases the stress to a value $\sigma_a - \sigma_p$ along a line with a slope, E, close to that of the intrinsic Young's modulus of the rock and then returns it to σ_a along almost the same line. This occurs with little dissipation, and the pulse propagates with a velocity corresponding to that of the intrinsic Young's modulus. On the other hand, a compression pulse of the same amplitude first increases the stress to a value, $\sigma_a + \sigma_p$ along the stress–strain curve with the slope, D, and then returns it to σ_a along a different line with a greater slope, E. The area beneath the first half of the stress cycle of a compression pulse is greater than that beneath the second half, so that energy is dissipated due to the irreversible changes brought about by the stress increment due to the pulse. This dissipation does not affect the amplitude of the pulse but shortens its duration. The front of the pulse propagates with a velocity corresponding to the slope D, while the back of the pulse propagates faster, with a velocity corresponding to the slope E. A similar argument applies to rock in which the compressive stress has decreased to σ_a from some greater value, but it is the compressive pulse which propagates with little dissipation in this case.

Seismic wave propagation is affected by fractures in the rock. Information concerning fractures is often of great significance.

A fracture surface in rock may be thought of as a plane containing an irregular pattern of areas of contact and flat voids. The reduction of the elastic modulus of rock by voids has already been discussed, § 12.3. Where the length of the seismic waves is long in relation to the dimensions of these voids, the seismic wave passes through the fracture with negligible reflection, but its velocity is reduced by an amount corresponding to the reduction in the modulus of the rock on account of the fracture. Measurements in Witwatersrand quartzite around a mine 8,000 ft below surface, Cook (1962), showed that the velocities of waves of dilatation ranged from 21 ft/millisecond along paths in solid rock to 12 ft/millisecond along a path in the same rock which was extensively fractured by the high stresses near the face. The reflection of a wave by a discontinuity increases rapidly as the wavelength becomes smaller than the dimension of the discontinuity. If the wavelength of ultrasonic seismic waves is shorter than the dimensions of the voids in the fracture plane good reflections from these voids can be expected.

13.7 Inelastic waves

Reference is sometimes made to two other types of waves which may occasionally be of significance in rock, namely, *plastic waves* and *shock waves*. The treatment of the former type of inelastic wave in this section follows that of a plastic wave in a wire by von Kármán and Duwez (1950), but uses the notation, § 2.9, of displacement being positive when in the negative direction of the axis.

If rock is subjected to a sufficiently large strain its deformation ceases to be elastic, and the slope of the stress–strain curve for a continuously increasing stress decreases, § 4.2. Consider the effect of a strain wave with a very large amplitude, travelling along the x-direction, on an element of rock with unit cross-section at a distance x from the origin in the unstrained state. If the displacement of this element at a time, t, is u its equation of motion is

$$\rho\ \partial^2 u/\partial t^2 = \partial\sigma/\partial x, \tag{1}$$

which can be written as

$$\rho\ \partial^2 u/\partial t^2 = \partial\varepsilon/\partial x\ .\ d\sigma/d\varepsilon, \tag{2}$$

provided that the strain is a single-valued function of an increasing stress. Let $d\sigma/d\varepsilon = D(\varepsilon)$, so that (2) becomes

$$\partial^2 u/\partial t^2 = (D/\rho)\partial^2 u/\partial x^2. \tag{3}$$

The non-linear nature of $D(\varepsilon)$ makes (3) difficult to solve in the general case. However, by assuming that the strain wave is the result of a constant velocity, $-V$, at $x = 0$, for all positive values of t one can show that three different solutions of (3) together describe the resulting wave of deformation. One solution to (3) which satisfies the boundary condition is

$$u = -Vt + \varepsilon_I x, \tag{4}$$

which corresponds to a constant strain ε_I in the interval 0 to x. By analogy with the elastic wave equation another solution can be expected to exist when

$$D/\rho = \bar{C}^2. \tag{5}$$

Since D is a function of ε, this makes

$$\varepsilon = f(\bar{C}). \tag{6}$$

Assume that $\bar{C} = x/t$

$$u = \int_{\infty}^{x} \partial u/\partial x \, dx = \int_{\infty}^{x} \varepsilon \, dx = t\int_{\infty}^{\bar{C}} f(\bar{C}) \, d\bar{C}. \tag{7}$$

Differentiating (7) twice with respect to t yields

$$\partial^2 u/\partial t^2 = (\bar{C}^2/t) df/d\bar{C} \tag{8}$$

also

$$\partial^2 u/\partial x^2 = (1/t) df/d\bar{C}. \tag{9}$$

Substituting (8) and (9) in (3) shows that either

$$df/d\bar{C} = 0 \tag{10}$$

or

$$D/\rho = \bar{C}^2. \tag{11}$$

The solution given by (10) corresponds to (4), and that given by (11) corresponds to an inelastic wave propagating with a velocity $\bar{C}$.

There are three regions in an inelastic wave. First, at $x = C_P t$ there is a discontinuity in strain which propagates with the velocity of an elastic wave of dilatation because, initially, $D = E(1 - \nu)/(1 + \nu)(1 - 2\nu)$. Second, between $x = C_P t$ and $x = C_I t$, where C_I is the velocity of propagation of the inelastic wave, the relation $\bar{C}^2 = D/\rho$ holds. Third, behind $x = C_I t$ the strain is constant and equal to ε_I. The velocity of propagation behind the elastic wave front is $\bar{C}$, which depends upon the strain, ε. The strain in the interval $C_I t < x < C_P t$, therefore, changes as the wave propagates. The velocity of the inelastic wave part, where the strain, ε_I, is constant, depends upon the value of D corresponding to that strain.

When D becomes zero, that is, the slope of the stress–strain curve is zero, the velocity of inelastic wave propagation is zero. Thus, waves cannot propagate in an ideally plastic material. Assume that the slope of the stress–strain curve becomes zero at a critical value of the stress, σ_a, or strain

ε_a. If the amplitude of the source of some disturbance is in excess of these critical values only a part of the disturbance, with a maximum amplitude equal to the critical values, is propagated as a wave.

For most solids, D decreases with increasing strain, and plastic waves with velocities of propagation less than those of the elastic wave are usually set up as a result of disturbances with a large amplitude. However, under conditions of exceptionally great compression the rigidity of solids increases with compression, Bridgman (1931). When D increases with increasing strain large strains propagate faster than smaller ones, and any pulse with a large amplitude tends to acquire a steep front. This phenomenon gives rise to a shock wave which propagates with a velocity greater than the velocity of propagation of elastic waves of small amplitude.

Consider a filament of unit cross-section through which a plane shock wave is travelling with a constant velocity C_K. Ahead of the shock front the rock is undisturbed, and behind it the particle velocity, $-\dot{u}$, pressure, p, and specific volume, s, are assumed to be constant. Choose a system of reference so that the shock front and the transition zone immediately behind it appear to be at rest, and designate the region ahead of the shock by subscript 1 and that behind it by subscript 2. From the conservation of mass the amounts of material, m, entering and leaving the shock zone are equal, so that

$$m = \dot{u}_1/s_1 = \dot{u}_2/s_2. \tag{12}$$

Equating the change of momentum across the shock zone in unit time to the force across that zone yields

$$m(\dot{u}_1 - \dot{u}_2) = p_2 - p_1. \tag{13}$$

Let the change in internal energy per unit mass across the shock zone be ΔW, then the equation for the conservation of energy across the shock zone can be written as

$$p_2\dot{u}_2 - p_1\dot{u}_1 = m[(\dot{u}_1^2 - \dot{u}_2^2)/2 + \Delta W]. \tag{14}$$

The velocity of propagation of the shock front through the rock is equal to the velocity of the undisturbed region relative to this front, that is, $\dot{u}_1$, but it is in the opposite direction. From (12) and (13) the velocity of propagation of the shock front is

$$C_K = -\dot{u}_1 = -s_1[(p_2 - p_1)/(s_1 - s_2)]^{\frac{1}{2}}. \tag{15}$$

The particle velocity behind the shock zone is

$$-\dot{u} = \dot{u}_2 - \dot{u}_1 = -[(p_2 - p_1)(s_1 - s_2)]^{\frac{1}{2}} \tag{16}$$

and the change in internal energy across the shock zone is

$$\Delta W = \tfrac{1}{2}(s_1 - s_2)(p_2 + p_1). \tag{17}$$

Equations (12), (13), and (14) are the Rankine conditions for a shock in a perfect gas, and (17) is known as the Hugoniot relation implying a change of entropy across a shock. For a small pressure difference (15) reduces to

$$C_K = [s^2 \Delta p / \Delta s]^{\frac{1}{2}}, \tag{18}$$

which is the same as the velocity of propagation of an elastic wave of dilatation.

Chapter Fourteen

The State of Stress Underground

14.1 Introduction

In order to calculate the stresses close to an excavation, it is first necessary to know the rock stresses which would have existed in its absence. The stress concentration around the excavation can then, in principle, be calculated by the methods of Chapters 10 and 11.

Methods of measuring stress underground will be discussed in Chapter 15, and unfortunately practically all of these are themselves made from excavations, so that, while an attempt to correct for the disturbance of the original stress field by the excavation is made, some doubt about this correction always remains, particularly if, as in the case of old mines, there has been some disturbance over a very large volume.

It will be seen in § 14.3 that the results of measurements to date show very wide variations. Also, unfortunately, there is no standard condition with which they can be compared. It might at first sight be expected that the stresses due to gravity would be determinate, but, as will be seen in § 14.2, they are not, because of the uncertainty about conditions at infinity. Nevertheless, it is useful first to study the stresses which might be expected if gravity alone was acting, and to attempt to attribute causes to large deviations from them.

The term *hydrostatic stress* will as before be used for a stress in which $\sigma_1 = \sigma_2 = \sigma_3$, independent of its origin. It is, unfortunately, a little confusing in the present context, in which water may be present.

If the density of rock is ρ, stresses which at depth z are hydrostatic and have the value $\rho g z$ will be described as *lithostatic*. The stress is just the weight of a column of the overburden of unit area, and if the density is variable it is

$$\int_0^z \rho g \, dz. \tag{1}$$

The assumption that the stresses at depth are lithostatic is described as *Heim's rule* by Talobre (1957) and has been used by many workers on rock mechanics. Heim stated it in the form that stresses in rock tend to become lithostatic because of creep. It is also frequently argued that since the stresses in sediments are lithostatic throughout the process of formation, the final state of stress in them should also be lithostatic.

These various effects are analysed by Denkhaus (1966), who points out that the terminology is not yet settled – he describes all stresses other than simple gravitational ones as 'latent'.

In practical cases it is useful to distinguish between the *virgin rock stresses*, which are those which would exist in the absence of any excavation at all, and the *field stresses* for any part of the excavation, which are those which would exist in the neighbourhood of this part of the excavation before it was made. The field stresses are thus determined by the virgin rock stresses and the remainder of the excavation. It is usually these which are of importance in mining practice.

14.2 Stresses due to gravity

The first problem to be considered is that of the flat-lying horizontal region of density ρ under gravity. The simplest assumption to make is that there is no horizontal displacement anywhere. Then, taking the z-axis vertically downwards, the components of the body force are $X = Y = 0, Z = -g$. The equations of equilibrium, § 5.5 (5)–(7), are then satisfied by

$$\tau_{xy} = \tau_{yz} = \tau_{zx} = 0, \quad \sigma_z = \rho g z, \tag{1}$$

and with σ_x and σ_y functions of z only.

Since, by hypothesis, $\varepsilon_x = \varepsilon_y = 0$, $\Delta = \varepsilon_z$, and the stress–strain relations, § 5.2 (20), give

$$\sigma_x = \sigma_y = \lambda\varepsilon_z, \quad \sigma_z = (\lambda + 2G)\varepsilon_z. \tag{2}$$

It follows that

$$\sigma_x = \sigma_y = \frac{\lambda}{\lambda + 2G}\sigma_z = \frac{\nu}{1 - \nu}\sigma_z = \frac{\nu}{1 - \nu}\rho g z. \tag{3}$$

Thus the horizontal stresses are $\nu/(1 - \nu)$ times the vertical stress, or of the order of one-third of it if ν is approximately $\frac{1}{4}$. Since the density of water is approximately one-third of that of rock, the horizontal stresses are of the same order as the water pressure in a water-filled joint.

A second problem capable of exact solution is that of a slab $0 < z < l$ supported by normal stress $\sigma_z = \rho g l$ at $z = l$ and with no lateral stress. The stresses

$$\sigma_x = \sigma_y = \tau_{xy} = \tau_{yz} = \tau_{zx} = 0, \quad \sigma_z = \rho g z \tag{4}$$

then satisfy the stress-equations of equilibrium, and the stress–strain relations give

$$\varepsilon_x = \varepsilon_y = \frac{\partial u}{\partial x} = \frac{\partial v}{\partial y} = -\nu\rho g z/E, \quad \varepsilon_z = \frac{\partial w}{\partial z} = \rho g z/E, \tag{5}$$

together with

$$\Gamma_{yz} = \Gamma_{zx} = \Gamma_{zy} = 0.$$

Integrating these gives the displacements, and when $z = l$,

$$u = -\rho g l \nu x/E, \quad v = -\rho g l \nu y/E, \quad w = \rho g[\nu(x^2 + y^2) + l^2]/2E. \tag{6}$$

That is, if there is no lateral restraint there is a lateral expansion determined by Poisson's ratio.

There is thus no indication from the theory of elasticity as to what the horizontal stresses should be. Many writers take

$$\sigma_x = \sigma_y = k\sigma_z, \quad 0 < k < 1, \tag{7}$$

where k varies from zero for no lateral restraint to 1 for lithostatic stresses. It was shown in § 11.4 that for a viscoelastic material of Maxwell (or Burgers) type the stresses tend to the lithostatic values if no lateral movement is permitted: this is Heim's rule as stated in § 14.1.

If the surface is not flat the problem becomes much more complicated. Considering, for simplicity, the two-dimensional case, suppose that the form of the surface is $z = -h(x)$, where h is the (positive) height above the plane $z = 0$, then the stresses for $z > 0$ can be estimated very roughly from formulae for stresses in the region $z > 0$ with a surface stress of $\rho g h(x)$.

For example, if $h(x) = h$ in the region $|x| < a$, and $h(x) = 0$ for $|x| > a$, the stresses may be written down from § 10.16 (16)–(21). In particular, when $x = 0$

$$\sigma_z = \rho g h\{1 - (2/\pi) \tan^{-1}(z/a) + 2az/\pi(z^2 + a^2)\}, \tag{8}$$

$$\sigma_x = \rho g h\{1 - (2/\pi) \tan^{-1}(z/a) - 2az/\pi(z^2 + a^2)\}. \tag{9}$$

When $z = 0$, $\sigma_z = \sigma_x$, but as z increases the relative importance of σ_x decreases and when $z = a$ and $x = 0$

$$\sigma_z = \rho g h\{\tfrac{1}{2} + (1/\pi)\}, \quad \sigma_x = \rho g h\{\tfrac{1}{2} - (1/\pi)\}. \tag{10}$$

Perhaps the most useful solution of this type is that of § 10.16 (34)–(36), which gives an estimate of the stresses below a V-notch valley, Voight (1966).

Similarly, the loading $\rho g h \cos \omega x$ corresponds to a periodic hill and valley system, and by § 10.16 (28)–(30) the stresses are

$$\sigma_z = \rho g h(1 + \omega z)e^{-\omega z} \cos \omega x, \tag{11}$$

$$\sigma_x = \rho g h(1 - \omega z)e^{-\omega z} \cos \omega x, \tag{12}$$

$$\tau_{xz} = -\rho g h \omega z e^{-\omega z} \sin \omega x. \tag{13}$$

This shows the way in which stresses due to surface irregularities die off with depth.

Accurate analytical solutions for the elastic medium under gravity with an irregular surface are not available. The problem may be studied by photoelastic or finite element methods, but even with these a difficulty appears which is related to those previously discussed. Both these methods study the stresses in a region of interest, Fig. 14.2, but boundary conditions have to be applied over arbitrary surfaces bounding this region, and it is not clear what these should be. Fig. 14.2 shows stress trajectories for a hill

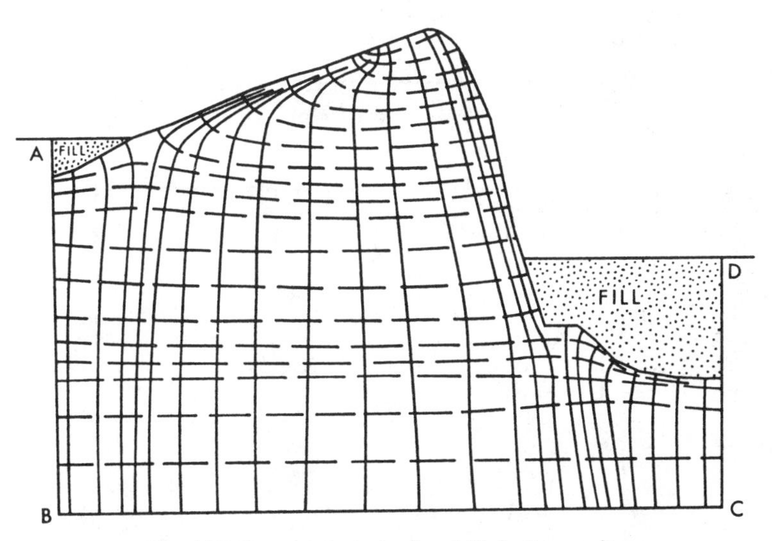

Fig. 14.2 Stress-trajectories in a hill due to gravity.

of the shape shown calculated by finite-element analysis on the assumptions that the stresses on *AB* and *CD* are normal and in equilibrium and that there is no displacement on *BC*.

14.3 Results of stress measurements

While a great many measurements of stress have been made underground, a relatively small amount of published information is available. This is partly due to the fact that until recently many of the methods of measurement were in a phase of development.

Few of the results are explicable on the simple gravity theories of § 14.2, and only a selection is given here to indicate the effects observed. Everling (1964) states that the stresses in the tectonically deformed Carboniferous sediments of Central Europe are in exceptionally good agreement with Heim's rule. Hast (1958) found relatively high horizontal stresses in the iron mines of Sweden in a region undergoing tectonic uplift. In the Snowy Mountains region of Australia, vertical and horizontal stresses of 1,800, 1,500, 1,500 psi, respectively, were measured at T1 power station, which had 1,100 feet of cover; while at T2 power station, with 650 feet of cover, the values were 1,500, 1,900, 1,800 psi. Both these stations are near the bottoms of V-notch valleys. Horizontal stresses greater than the vertical have also been measured in tunnels in the Snowy Mountains region and in coal mines in the coastal escarpment of New South Wales. At Poatina power station, Tasmania, with 500 feet of cover, the stresses measured

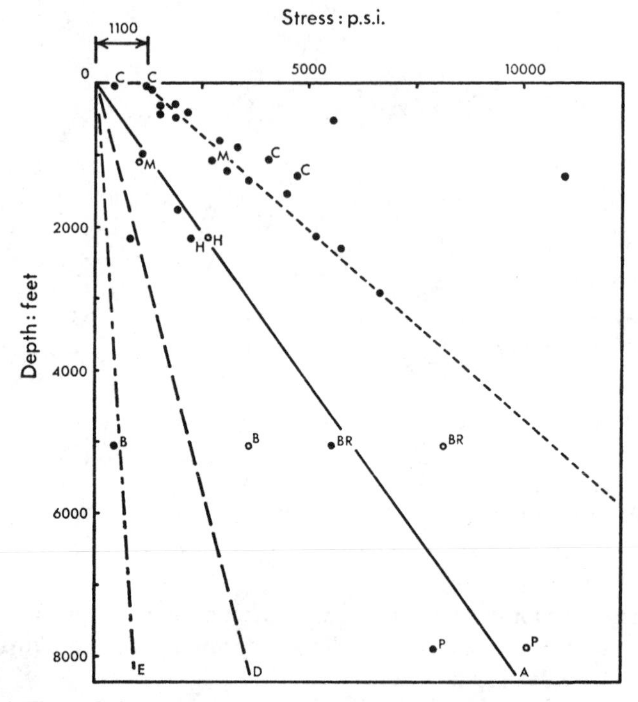

Fig. 14.3.1. The variation of vertical (σ_v) and horizontal (σ_{h1}, σ_{h2}) virgin rock stresses with depth. $\sigma_{o \cdot b.}$ is the overburden stress. a and b are the transverse and axial stress concentration factors for a flat-ended borehole, cf. § 15.5.

OA: $\sigma_v = \sigma_{h1} = \sigma_{h2} = \sigma_{o.b.}$
OP: Pore-pressure of water.
OE: $\sigma_{h1} = \sigma_{h2} = [\nu/(1-\nu)]\sigma_v = [\nu/(1-\nu)]\sigma_{o.b.}$
●: $\frac{1}{2}(\sigma_{h1} + \sigma_{h2})$ measured by Hast in Scandinavia.
●C: $\frac{1}{2}(\sigma_{h1} + \sigma_{h2})$ measured by Coates and Grant in Canada.
●H: $\frac{1}{2}(\sigma_{h1} + \sigma_{h2})$ measured by Hoskins in Australia.
○H: σ_v measured by Hoskins in Australia.
●M: σ_{h1} measured by Moye in Australia.
○M: σ_v measured by Moye in Australia.
●B: $\frac{1}{2}(\sigma_{h1} + \sigma_{h2})$ measured by Barron in South Africa using $a = 1{\cdot}38$, $b = 0$.
○B: σ_v measured by Barron in South Africa using $a = 1{\cdot}38$, $b = 0$.
●BR: $\frac{1}{2}(\sigma_{h1} + \sigma_{h2})$ measured by Barron in South Africa using $a = 1{\cdot}1$, $b = -0{\cdot}75$.
○BR: σ_v measured by Barron in South Africa using $a = 1{\cdot}1$, $b = -0{\cdot}75$.
●P: $\frac{1}{2}(\sigma_{h1} + \sigma_{h2})$ measured by Pallister in South Africa using $a = 1{\cdot}1$, $b = -0{\cdot}75$.
○P: σ_v measured by Pallister in South Africa using $a = 1{\cdot}1$, $b = -0{\cdot}75$.

were 1,240 psi vertically, and 2,400 and 1,800 psi horizontally, Endersbee and Hofto (1963). Some European measurements are described by Talobre (1958). At Picote power station, Portugal, vertical stresses of 500 and 2,850 psi were measured on opposite sides of the station, which is under only 300 feet of cover but in the neighbourhood of a fault, Serafim (1961).

It appears from the results quoted that, while horizontal stresses are frequently of the same order as the vertical as would be expected from Heim's rule, they are frequently higher. Also, and more difficult to explain, the vertical stresses are occasionally considerably higher than the weight of the overburden. The most extreme case of high horizontal stresses occurs in the measurements of Hooker and Duvall (1966) at depths of only a few tens of feet in a rock outcrop in which they observed horizontal stresses of between 500 and 3,000 psi.

As remarked in § 8.10, it seems likely that stresses can be determined from measurements of pressure taken during the hydraulic fracturing of wells. Scheidegger (1960) discussed this using a spherical source of pressure as a model, and Kehle (1964) using the cylindrical model. Kehle studied five wells of depth of up to 2,100 m and concluded that the stresses were approximately lithostatic and that the results from Scheidegger's spherical model were too high.

The results of most stress measurements to date are shown in Fig. 14.3, which illustrates the large variation of the measured values of horizontal stress in relation to depth and the importance of the value used for the stress-concentration factors at the flat end of a borehole. The line OA represents the rate at which the vertical stress can be expected to increase with depth due to the weight of the overlying rock, § 14.1 (1). Measured values of the vertical stress agree reasonably well with this, as do a few values of the mean horizontal stress; corresponding to $k = 1$ in § 14.2 (7). The lines OE and OD correspond, respectively, to $k = \upsilon/(1 - \upsilon)$ and $k = 1/\rho$, where ρ is an average specific gravity for the overlying rock. Most of the measured values of the mean horizontal stress are greater than those suggested by these lines. Many of them lie close to the dashed line intersecting the surface at a stress of 1,100 psi, which may be a result of residual stress, § 14.4.

One simple explanation for the presence of high values for the horizontal components of stress near surface, where relatively extensive denudation has taken place, can be given in terms of the theory of elasticity. Assume that at some time when the present surface was buried at considerable depth the stresses were in accordance with Heim's rule so that the vertical and horizontal components were equal and their values were those generated by the weight of the overburden. If relatively rapid denudation then took place, the value of the vertical stress would decrease elastically according to the reduction in the weight of the overburden. The elastic relaxation of the values of the horizontal components, however, would be only $\nu/(1 - \nu)$ times that of the vertical stress, as given by equation (3), § 14.2. As a result, values of horizontal stress at surface would be only a little less than those which prevailed at depth before denudation.

An important feature of rock stresses emerged from detailed measurements made at a depth of 2,5 km below surface in the quartzitic sediments of the Witwatersrand System. Geological surveys of these sediments have provided a quantitative measure of their degree of continuity. In Fig. 14.3.2 is plotted the frequency with which sections of different lengths,

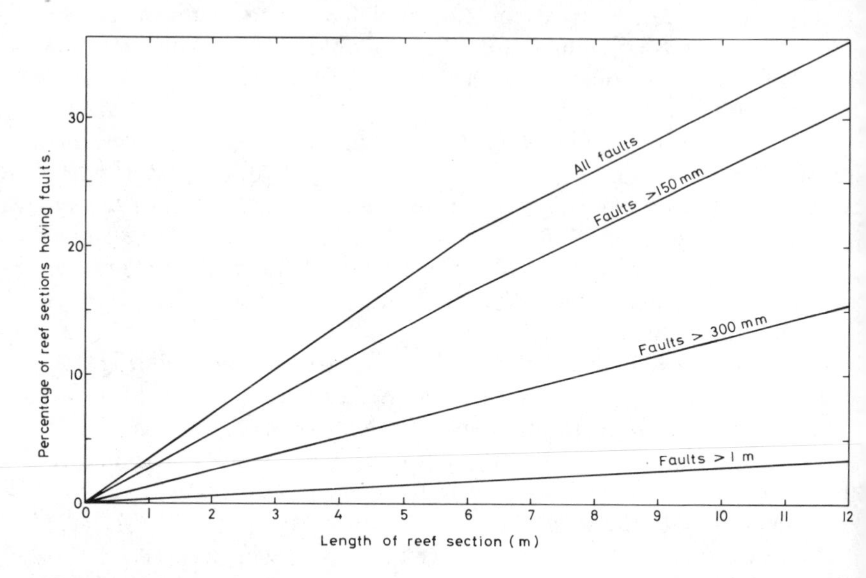

Fig. 14.3.2 Graphs showing the frequency with which different lengths of reef are disturbed by faults having various sizes of throw.

measured in the plane of one of the gold reefs in these sediments, are disturbed by faults having different sizes of throw. It can be seen that the spacing of faults or discontinuities is of the order of 10 m. The magnitudes and directions of the principal strains as deduced from overcoring electric resistance strain gauge rosettes cemented to the ends of a borehole at various depths during drilling to determine rock stresses, § 15.5, are shown in Fig. 14.3.3. The fracture zone around the tunnel from which the hole was drilled and the stress concentration ahead of this can be seen, but of much greater interest is the enormous variation in the magnitudes and directions of the principal strains and, presumably, the stresses along the length of the hole, especially where the core showed the presence of pre-existing slippage planes or minor faults. This evidence suggests that

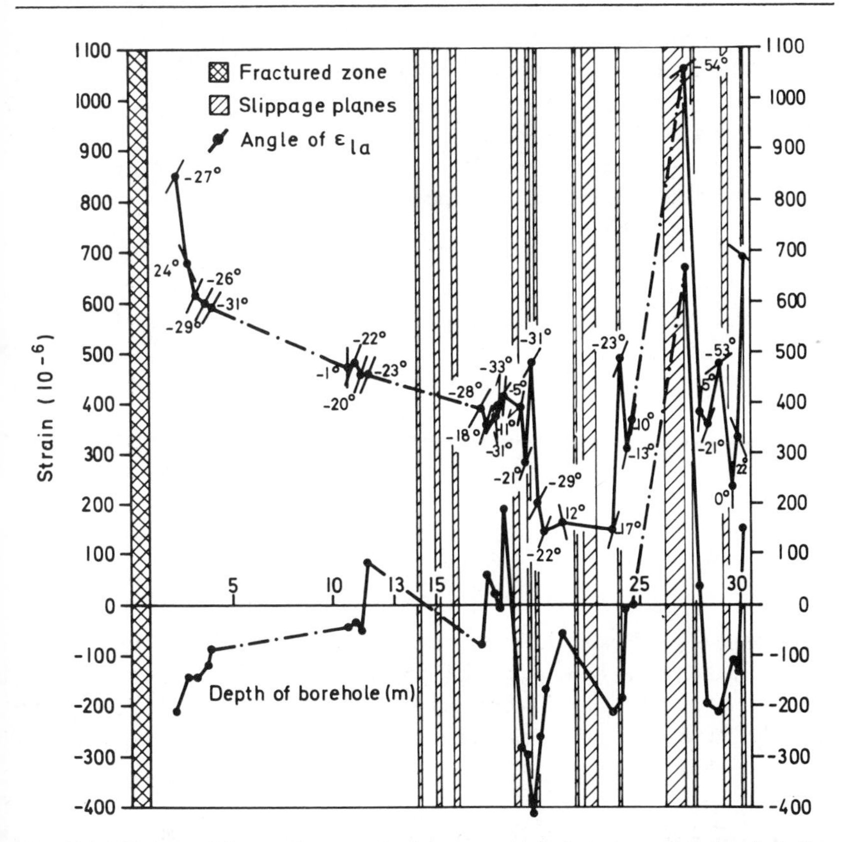

Fig. 14.3.3 Values of the maximum and minimum principal strains and their orientation in a plane normal to the axis of a borehole, as deduced from overcoring electric resistance strain gauge rosettes cemented to the ends of the borehole at various depths as it was deepened. The fracture zone around the tunnel from which the hole was drilled and geological planes on which slippage had occurred are shown.

geological discontinuities on a scale of metres give rise to variations of the stress in the rock which may be likened to those which must exist on an almost microscopic scale between one grain in the rock and another.

14.4 Tectonic, structural, and residual stresses

It is clear from the results given in § 14.3 that in many cases the measured stresses are anomalous in the sense that they cannot be attributed to gravity, even allowing for variations in surface level. A number of explanations may be invoked to account for this.

Tectonic stresses are known to occur and to be responsible for shallow-focus earthquakes. Information may be obtained about their general directions from fault-plane studies. Even in a seismically inactive region such as south-eastern Australia it is possible to infer a roughly North-West direction of a principal stress, and it is reasonable to assume that regions which show no seismic activity may nevertheless be subject to some tectonic stress.

If a horizontal tectonic stress σ_x occurs in the x-direction it will be accompanied by stress in other directions. For example, if it is assumed that there is no displacement in the perpendicular y-direction and zero stress in the vertical z-direction the stresses are, by § 5.3 (3),

$$\sigma_y = \nu\sigma_x, \quad \sigma_z = 0 \tag{1}$$

which have to be added to § 14.2 (3).

Tectonic stresses may arise from other causes, for example, Price (1959) has suggested that if a rock mass is uplifted by an amount u with no dis-

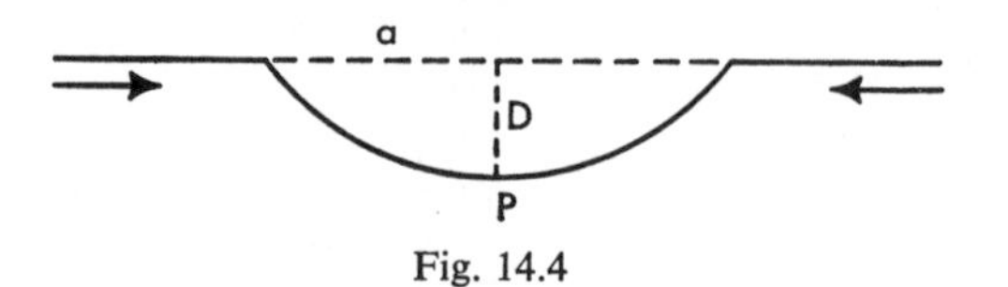

Fig. 14.4

placement in the perpendicular direction, so that $v = 0$ in cylindrical co-ordinates, there will be a tangential strain $-u/R$ by § 2.15 (12) and hence a tangential tension $-Eu/R$, where R is the radius of the earth. He suggested that this tension might be responsible for jointing. Hast (1958) has measured anomalous stresses in Fennoscandia, which is known to be undergoing uplift.

Detailed calculations of the stresses in a region for simple conditions corresponding to uplift, horizontal tectonic stresses, or faulting are given in § 17.3.

If horizontal tectonic stresses occur, they will be greatly modified by the stress-concentrations caused by surface irregularities. This is particularly important in civil engineering practice, since hydro-electric works are commonly in regions of rugged topography. The effect of simple model shapes, such as a V-shaped notch, or of actual topography is usually studied photoelastically.

A simple estimate of the order of magnitude of the stress-concentration can be found from a result of Chih Bing Ling (1947). He calculates the stress-concentration factor C at the point P at the bottom of a notch in the form of an arc of a circle of depth D and chord $2a$ in the surface of a semi-infinite region under uniaxial compressive stress, Fig. 14.4. Values of C in terms of D/a are given in Table 14.4.

The negative values of D correspond to a hill instead of a valley, and it is noteworthy that tension may be developed at the top of a semicircular hill.

Structural stresses are caused by inhomogeneity of the region under consideration. For example, it was shown in § 10.9 (23) that the stress concentration in a cylindrical region of modulus of rigidity G_0 within material of modulus G is of the order

$$3G_0/(2G_0 + G). \tag{2}$$

Since mines commonly involve rock of very different properties to country rock, such effects are to be expected. Faults will also induce a major alteration of stresses.

Table 14.4

D/a	3·73	1·73	1	0·58	0·27	0	−0·27	−0·58	−1	−1·73	−3·73
C	3·88	3·57	3·06	2·42	1·71	1	0·41	0·04	−0·1	−0·08	−0·02

The fact that isallostress patterns, Chapter 17, Lensen (1958) and Scheidegger (1963, 1966), tend to correlate with geophysical anomalies suggests a correlation between crustal inhomogeneity and the stress pattern.

Residual stresses are defined in the sense of the theory of elasticity as 'locked-in' stresses associated with the previous history of the rock. They form a set of stresses in equilibrium on which external stresses may be superimposed. Such stresses are known to exist, both on a small and a large scale, Voight (1966b) and Emery (1964), and they may well be of very great importance, but at present it is not known how widely distributed they are. Related to them are stresses of thermal type produced by a change in volume of part of the material. This might be caused by swelling due to absorption of water or contraction on drying, or by generation of heat during metamorphism.

Residual stresses are usually postulated when anomalously high vertical stresses are measured, since it is difficult to account for these by any of the mechanisms previously discussed. One likely possibility is that of visco-elastic effects caused by erosion. It was shown in § 11.4 (27) in a simple case that the stresses in a material which behaved as a Maxwell substance tended to become lithostatic with a time constant t_1. In many cases, erosion of river valleys takes place relatively rapidly, so that the higher stresses which existed previously may not have died away completely. A viscoelastic substance cannot retain residual stresses indefinitely, but some materials possessing a yield stress can do so.

Chapter Fifteen

Underground Measurements

15.1 Introduction

Essentially, underground measurements are made for two purposes. First, rock stresses and large-scale rock properties are measured to provide the basis for design calculations. Second, the response of the rock mass to the changes brought about by excavation are measured to provide experimental information concerning the behaviour of the rock around underground excavations.

Generally, underground measurements can be divided into three different types: measurements of rock stresses, measurements of the displacements induced in the rock by an excavation, and measurements of the strength of large blocks of rock in place. All three types of measurement are used both to provide design data and to study the behaviour of the rock around excavations.

The difficulties of making refined measurements in rock underground and the limitations of physical access are such that none of the various methods in use can be regarded as entirely satisfactory.

To date, much of the work on underground measurements has been concerned with the development and testing of techniques and equipment; especially that on rock stress measurement, Leeman (1964), Merrill (1964), and Hoskins (1966, 1967). Careful measurements of the displacements induced in the rock by an excavation have made major contributions to knowledge concerning the behaviour of rock masses in this respect, Potts (1964), Ortlepp and Cook (1964), and Ortlepp and Nicoll (1964). Up to the present, most underground measurements have been made with a view to furthering understanding of the behaviour of rock around excavations, but stress measurements and large-scale tests are coming into more extensive use to provide basic design data.

15.2 Flat-jack measurements

In principle, the flat-jack method of measuring stress, Mayer *et al.* (1951) and Tincelin (1951), is simple and attractive. A slot is cut into a rock surface, such as the wall of a tunnel, by drilling a series of overlapping holes or with a saw, Fig. 15.2.1 (*a*). This completely relieves the surfaces of the slot of the stresses originally across them and frees the rock on either side of the slot of some of the stresses which existed parallel to the wall prior to cutting. This relief brings about an expansion of the rock into the slot, which can be measured by the convergence across the slot or the convergence between points on opposite sides of the slot. A flat, hydraulic, or Freysinnet jack is embedded in the slot, with grout in the case of a thick slot, and the pressure in this jack is increased until the displacements brought about by cutting the slot are cancelled, Fig. 15.2.1 (*b*). The cancellation pressure is essentially equal to the original stress normal to the plane of the slot which existed in the rock before cutting the slot. Thus, the flat-jack method is a null

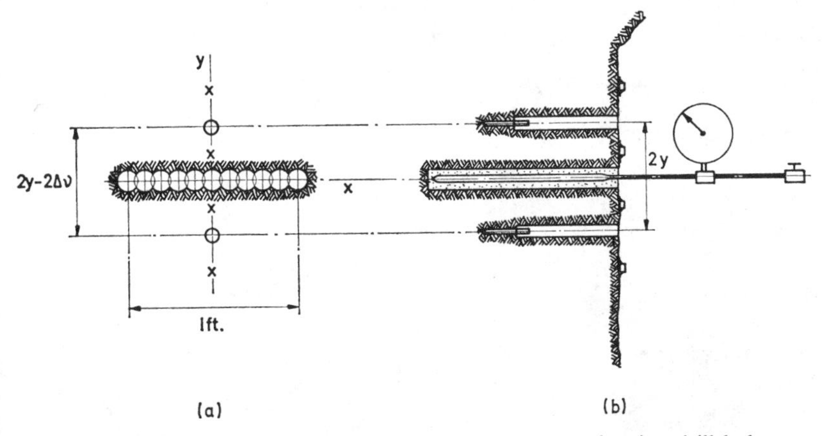

Fig. 15.2.1 (*a*) A flat-jack slot cut into a rock surface by overlapping drill holes producing a convergence $2\Delta v$ between measuring pins. (*b*) A section through the same slot after the flat-jack has been installed and the cancellation pressure applied.

technique and one of the few in which rock stresses can be measured direct. It does, however, suffer from a number of disadvantages. First, as usually applied, it is limited to the measurement of stresses near the surface. Second, it is complicated by the effects of creep in the rock during cutting, grouting, and cancellation.

Flat-jacks are usually made from thin metal plates about 1 ft square welded together around their perimeter and provided with a communicating tube to allow oil to enter the space between them. Pins for measuring displacements across the slot are placed on each side of the slot, preferably along the centre line normal to the plane of the slot. These pins are installed before cutting starts, and may be cemented to the surface or grouted into holes drilled into the rock. They are usually spaced from an inch to a foot away from the slot. The displacements between the pins are measured with a Huggenberger deformeter, precision dial gauges, or other suitable measuring devices. Panek (1961) has used an embedded strain gauge to detect the deformation resulting from slot cutting and cancellation.

If the slot is regarded as a very flat ellipse in plane stress, normal displacement between measuring pins across the slot is caused only by changes in the normal stress across the slot and is unaffected by changes in the shear stress parallel to it, § 10.11 (30), (39). In this case the pressure required on the whole surface of the slot to cancel the normal displacements brought about by slot cutting should be identical to the original normal stress across the slot. In practice, a region, d, near the edge of a flat-jack is inoperative and the width of the jack, $2c_j$, is less than that of the slot, $2c$. With such a system the relation between the original normal stress, σ_n, across the slot and the cancellation jack pressure, p_c, might be approximated by

$$\sigma_n = p_c(c_j - d)/c. \tag{1}$$

The convergence caused by slot cutting, $2\Delta v$, between measuring pins

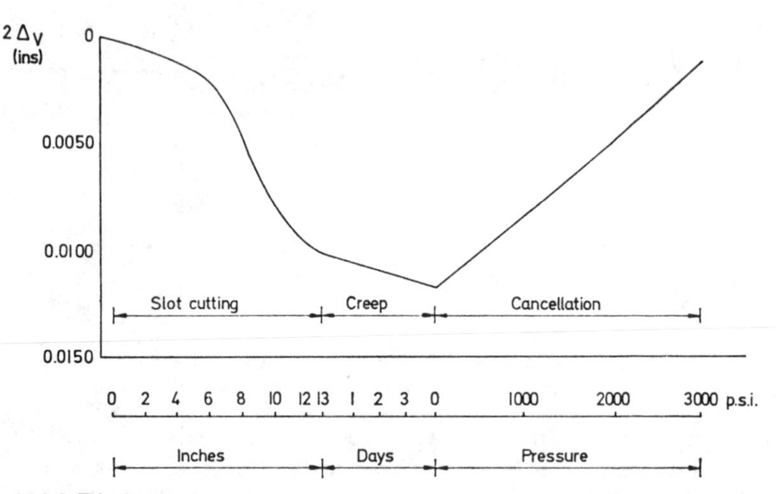

Fig. 15.2.2 Displacements between measuring pins during slot-cutting, grouting, and the application of flat-jack pressure (from Alexander, 1960).

spaced at equal distances y from the plane of the slot along the centre line normal to its plane is, from § 10.11 (30)

$$2\Delta v = 2c\sigma_n/E\{(1 - \nu)[(1 + y^2/c^2)^{\frac{1}{2}} - y/c] + (1 + \nu)(1 + y^2/c^2)^{-\frac{1}{2}}\}. \tag{2}$$

From measurements at two or more different distances, y, both Young's modulus and Poisson's ratio can be calculated. Alexander (1960) proposed that the slot be regarded as an ellipse of finite width and showed that two small corrections should be applied to (2). The first, due to the effect of a finite width on the displacements caused by the stress normal to the plane of the slot, is included in § 10.11 (29). The second, due to any normal stress parallel to the plane of the slot, is given by § 10.11 (35). These two corrections tend to cancel one another. Measurements of the displacements parallel to the slot caused by slot-cutting can be used in § 10.11 (39) to provide an estimate of the shear stress in the plane of the slot.

Alexander (1960) has studied the effects of creep and jointing on flat-jack measurements in the field, and Hoskins (1966) has made similar investigations in laboratory experiments. Typical displacements between measuring pins during slot-cutting, grouting, and pressure restoration are shown in Fig. 15.2.2. Panek and Stock (1964) proposed two procedures to allow for the effects of creep on flat-jack tests. The first involves holding the flat-jack near the cancellation pressure for a time comparable with that for which the slot stood relieved, to permit reverse creep to take place before the cancellation pressure is determined. In the second method they deduct all the time-dependent deformation from the total deformation and cancel only the difference with the flat-jack.

It is customary to make flat-jack measurements in several mutually perpendicular slots, Alexander (1960) and Merrill (1964). In a tunnel, for example, horizontal and vertical slots may be cut in the sidewall to measure the vertical and axial tangential components of stress at the surface, and a vertical longitudinal slot might be cut in the roof to measure the horizontal tangential component of stress on the surface. Measurements can also be made at the end, or face, of a tunnel. If these measurements are to provide data concerning the field stresses away from the influence of the excavation the results must be corrected for the stress concentrations around the excavation. For simple shapes of excavation, such as a circular tunnel, the stress concentrations can be estimated from analytical solutions, §§ 10.4, 10.11, 10.15. More complicated situations, such as measurements of the end of a tunnel, can be interpreted in the light of results, such as the 'frozen stress' photoelastic analysis of the stresses at the flat end of a blind hole by Galle and Wilhoit (1962), § 15.5. Sometimes, photoelastic analysis of the stress concentrations around a particular excavation is necessary.

Flat-jack measurements must be made at a surface where the rock is almost invariably damaged as a result of weathering, stress relaxation, and the processes of excavation. Correction of these measurements for the stress concentrations around the excavation is a difficult and uncertain procedure. Jaeger and Cook (1964) proposed that the flat-jack technique be adapted for use in drill holes up to 20 ft in length to make measurements in sound rock farther from the influence of the excavation. For this purpose they developed thin elongated curved jacks to fit in the annular space cut by a diamond core drill. As in the flat-jack method, curved jacks, usually four quadrantal jacks, are used to restore the stresses in the core relieved by drilling. The stresses in the core can be measured with another set of jacks or any other suitable device. The complete theory of the elastic displacements and stresses generated by concentric jacks is set out by

Jaeger and Cook (1964), and the essential basis of this theory is given in §§ 10.4–10.7.

15.3 Stressing of the interior of a borehole

Stresses are applied to the interior of boreholes either to produce deformations in order to determine the modulus of the rock or to induce fractures from which the direction and magnitude of the minor principal rock stress may be deduced.

When the modulus of the rock is to be found, uniform fluid pressure is usually applied through an impermeable membrane to a length of the borehole between two packers. The diameter of the holes range from a few inches, in the case of boreholes, to several feet, in the case of pressure-chamber tests, Oberti (1960). From § 5.11 (13), the radial displacement u_R, induced by an internal pressure p in a circular hole of radius R is

$$u_R = -pR/2G \tag{1}$$

at the circumference.

Measurements of the diametral expansion, $2u_R$, and p provide a means for determining the modulus of rigidity, G, of the rock around a borehole or tunnel, Talobre (1957). The displacements induced by a uniform radial pressure over opposite arcs of the hole are derived in § 10.6 (15), (16), (17). These depend upon Poisson's ratio, so that two different measurements should provide a means for determining both this ratio and G. If pressure is applied over opposite quadrants the radial displacement measured at the edge of these quadrants depends only on G, as in the case in (1). With quadrantal pressure the effect of Poisson's ratio on the displacements is greatest along and normal to the diameter through the middle of the quadrants.

The drilling of a hole or the excavation of a tunnel constitutes a de-stressing process which is the inverse of the stressing process discussed above. Assume that the cross-section of a borehole or tunnel is measured at its end, or face, and that the same cross-section is measured again after the end has been advanced. The resulting deformations can be regarded, crudely, as borehole deformations induced by the removal of the field stresses from the circumference of the hole. The induced deformation at the circumference of a hole of radius R is found from § 10.4 (21) to be

$$u_R = R(1 + \nu)[\sigma_x + \sigma_y + (\sigma_x - \sigma_y)(3 - 4\nu) \cos 2\theta]/2E, \tag{2}$$

for plane strain conditions where σ_x and σ_y are the field stresses normal to the axis of the hole and θ is the direction between the radius R and the direction of σ_x. If the induced deformations are measured across three or more diameters of different orientation, and Young's modulus is known or found from a subsequent stressing experiment, (2) can be used to estimate the field stresses around a tunnel or borehole.

In the hydraulic fracturing of a borehole, originally a technique used to stimulate production from an oil well, a section of the borehole is sealed off with packers and fluid pressure is applied to the bare walls of the hole between the packers. The pressure is increased until a fracture develops in the wall and extends as the pressure fluid flows into it. It is generally believed that the fracture propagates normal to the direction of the minor principal stress, Scheidegger (1962) and that the fluid pressure required to propagate the fracture is a little greater than the value of that stress.

The stresses induced in the walls of a borehole in an impervious rock by fluid pressure in the region between two packers have been calculated by Kehle (1964). In most of the region between the packers the fluid pressure, p, induces a tangential tension, $\sigma_\theta = -p$, and a radial compression, $\sigma_r = p$, and zero axial stress, as suggested by § 5.11 (12). Near the packers the tangential stress decreases to zero and the axial stress becomes tensile with a value very near $-p$. Assume that a fracture initiates from the surface of the hole as soon as the pressure, p, creates a net tension, T, sufficient to cause failure. Fracture initiation appears to be possible either along an axial plane in the central region between the packers or along a plane normal to the axis of the hole in the end regions near the packers. If the axis of the borehole is parallel to one of the principal stresses in the rock, σ_z, and the principal stresses normal to the axis of the hole are $\sigma_x > \sigma_y$, the fractures initiate when, by § 10.4,

$$p \geqslant T + (3\sigma_y - \sigma_x), \tag{3}$$

or

$$p \geqslant T + \sigma_z + \nu(3\sigma_y - \sigma_x). \tag{4}$$

When $(3\sigma_y - \sigma_x) < \sigma_z + \nu(3\sigma_y - \sigma_x)$ fracture initiates in the axial plane parallel to the direction of σ_x and propagates in the same plane provided that $\sigma_z > \sigma_y$. Fracture initiates normal to the axis of the hole when $\sigma_z + \nu(3\sigma_y - \sigma_x) < (3\sigma_y - \sigma_x)$ and propagates in the same plane provided that $\sigma_z < \sigma_y$. While this discussion has been concerned with the initiation of fracture around a hole in elastic rock, it must be noted that the conditions for fracture propagation are only that p be equal to or slightly greater than the minor principal stress, whether the rock is elastic or not. The fluid pressure required to propagate a fracture provides a direct measure of the value of the minor principal stress in the rock whatever its properties.

Jaeger and Cook (1964) have proposed stressing the interior of a borehole so as to generate tangential tension around its circumference, § 10.6 (14), and fracture the rock normal to the direction of the minor principal stress. By overcoring such a hole and observing the direction of the resultant fracture it is possible to find the orientation of the minor principal stress.

15.4 Borehole deformation

Any opening in an elastic body is deformed by stresses in that body. If a circular hole is drilled into rock the hole deforms when the stresses in the rock change. Thus, deformation of a borehole can be used to measure changes of stress in an elastic rock around the hole. Alternatively, the stresses in the rock around the borehole can be relieved by trepanning or overcoring the hole, and the resultant deformations of the hole can be

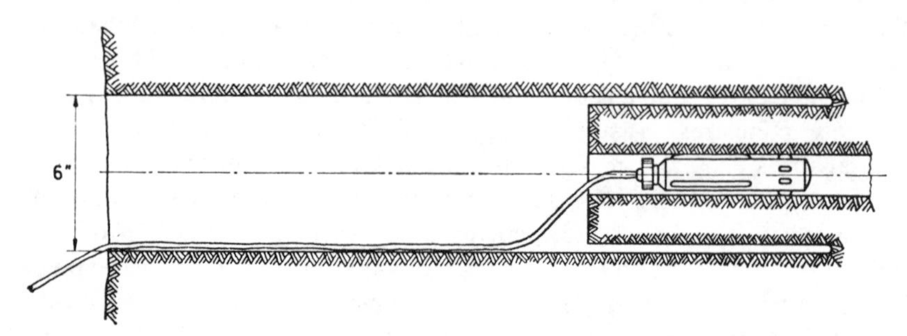

Fig. 15.4 A section through a borehole with a deformation gauge which has just been overcored.

used to calculate the absolute stress in the rock, provided that the properties of the rock are known.

Several investigators, Jacobi and Brändle (1956), Obert *et al.* (1962), Leeman (1964), and Merrill (1967), have developed borehole deformation gauges which measure the deformation across one or more diameters of a hole about $1\frac{1}{2}$ in. in diameter. Borehole deformation gauges are essentially different from stress meters, § 15.6, in that they are designed to be resilient, or soft, and to exert only a negligible load on the interior of the borehole. They are essentially remote-reading borehole calipers and should be capable of measuring 0·001 in. of deformation and have a range of about 0·1 in. The designs of borehole gauges have been reviewed by Leeman (1964). By overcoring with an annulus having an outside diameter of 6 in. it is possible to make measurements twenty or more feet deep into the rock, Fig. 15.4.

Assuming that the axis of the hole is in the direction of a principal stress, σ_z, the increase in length of a diameter inclined at θ to the direction of the greater of the other two principal stresses, $\sigma_x > \sigma_y$, due to the relief of the principal stresses by overcoring is, from § 10.4 (26),

$$\delta = (2R/E)[(\sigma_x + \sigma_y) + 2(\sigma_x - \sigma_y)(1 - \nu^2) \cos 2\theta - \nu\sigma_z]. \qquad (1)$$

If the change in length of three diameters with different orientations is measured three equations such as (1) can be solved to determine σ_x and σ_y in terms of σ_z, and θ. It is necessary to know the value of Young's modulus for the rock with the same accuracy as is required for the principal stresses. The accuracy with which the value of Poisson's ratio is required to be known is less. Fairhurst (1965) has considered the case when a borehole is not parallel to one of the principal stresses. He has shown that the component of shear stress parallel to the axis of the hole has no effect on the length of the diameter and that the normal components of stress perpendicular to such a hole can be found in terms of the component parallel to it from measurements of the deformation of three diameters of different orientations, as is the case when the axis of the hole is parallel to a principal stress. The complete state of stress can be found from measurements in three mutually inclined holes, § 2.13. The walls of holes inclined to the direction of a principal stress are deformed tangentially and longitudinally. Borehole deformation gauges should therefore be made insensitive to deformation of the walls of the hole normal to its diameter.

Borehole deformation gauges are unsuitable for measuring stresses approaching the strength of the rock because the surfaces of the borehole fail, giving rise to inelastic deformations, and overcoring becomes impossible due to discing, § 6.12.

The question of the influence of anisotropy and time-dependent deformation on borehole deformation has been discussed by Berry and Fairhurst (1965), who concluded that significant errors in the determination of stress can arise where allowance is not made for pronounced anisotropy in the rock. The case where a borehole is parallel to transversely isotropic rock, such as bedded strata, can be analysed using the relations of § 10.18.

15.5 Measurements at the end of a borehole

Several investigators, Mohr (1956), Leeman (1964), and Hawkes and Moxon (1965), have sought to determine the stresses in rock by fixing strain gauges to the smoothed end of a borehole and measuring the relief brought about by overcoring the gauges with an annulus having the same outside diameter as the original borehole, Fig. 15.5. This method has the advantage of avoiding the problems associated with overcoring a borehole using a large-diameter annulus, and measurements have been made at the bottom of boreholes more than a hundred feet in length. The major difficulty with this method is that it is necessary to relate the strain relief at the end of the borehole to the stresses in the rock.

The stress concentrations at the end of a borehole have not been determined theoretically. Galle and Wilhoit (1962) investigated the stresses at the flat end of a borehole using 'frozen stress' photoelasticity. They concluded that, when the axis of the hole is coincident with the direction of one of the principal stresses, radial field stresses are concentrated at the centre of the flattened end by 1·56 times in their own direction, zero times normal to this direction, and that the axial field stress is concentrated −1·04 times in all radial directions. These stress-concentration factors are essentially constant over a concentric circular region of the end with a diameter half that of the borehole. In laboratory tests using large blocks of rock Hoskins (1967) verified these factors but Bonnechere (1967) and Pallister (1967) obtained values for the concentration of a radial field stress along its own direction of 1·25 and 1·1, respectively. Nevertheless, all experiments show the concentration to be zero in a direction normal to that of the field stress. Hoskins (1967) also investigated the stress concentration factors at the bottom of a borehole with a hemispherical end. He concluded that the

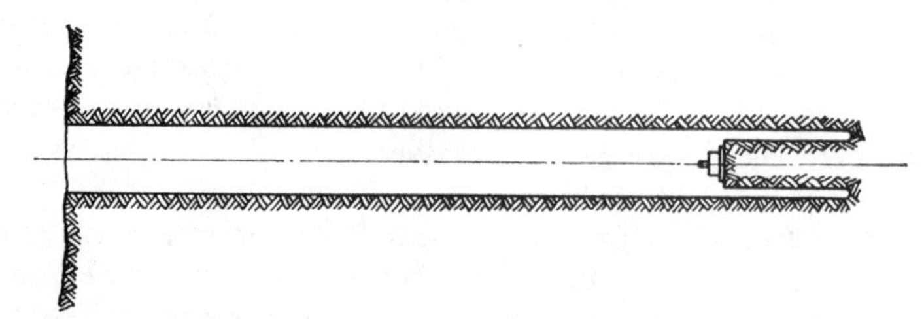

Fig. 15.5 A section through a borehole with a flat end to which a strain gauge has been cemented and overcored.

stress-concentration factors are 2·0 for the radial field stress and −0·5 for the axial field stress. These results are consistent with the stress concentrations around a spherical cavity, § 10.15, with $\nu = 0{\cdot}2$.

If x, y, z are the directions of the principal stresses in the rock and the axis of the borehole is in the z-direction, the strain in the x-direction on the end of the borehole, which is in plane strain, is

$$\varepsilon_x = [a(\sigma_x - \nu\sigma_y) - b(1 - \nu)\sigma_z]/E. \tag{1}$$

Experimental values for a range from 1·1 to 1·6 and for b from 0·07 to 1·04 for a borehole with a flat end, while a is 2·0 and b is 0·5 for a borehole with a hemispherical end. Since the factor in a direction normal to the field stress is zero for a flat-ended borehole the problem can be analysed using an axi-symmetric finite element procedure, which gives $a = 1{\cdot}13$ and $b = 0{\cdot}29$ for $\nu = 0{\cdot}25$ (Wiebols, personal communication). The effect of shear stresses parallel to the borehole is thought to be negligible.

The strains at the end of a borehole can be measured in three directions, OP, OQ, OR, inclined at known angles, α and β, to one another using a strain rosette. Suppose that OP is inclined at some unknown angle θ to the direction of the principal stress σ_x and that the strains resulting from overcoring the rosette are ε'_P, ε'_Q, ε'_R. The usual relations for finding the principal strains and their orientation in two dimensions from three strain measurements, § 2.13, can be applied to find the orientation of ε_x, ε_y and the magnitudes of ε_x', ε_y', where $\varepsilon_x' = \varepsilon_x + b(1 - \nu)\sigma_z/E$, $\varepsilon_y' = \varepsilon_y + b(1 - \nu)\sigma_z/E$. In the important practical case when $\alpha = \beta = \pi/4$

$$\left.\begin{aligned} \varepsilon_x' + \varepsilon_y' &= \varepsilon_P' + \varepsilon_R' \\ \varepsilon_x' - \varepsilon_y' &= [(\varepsilon_P' - 2\varepsilon_Q' + \varepsilon_R')^2 + (\varepsilon_P - \varepsilon_R)^2]^{\frac{1}{2}} \\ 2\theta &= \tan^{-1}(\varepsilon_P' - 2\varepsilon_Q' + \varepsilon_R')/(\varepsilon_P' - \varepsilon_R'). \end{aligned}\right\} \quad (2)$$

The complete state of stress can be found either by making sets of measurements at the end of two perpendicular holes (assuming that they are drilled in the direction of a principal stress) or by making two sets of measurements close together in the same hole using a flat end for one set and a hemispherical end for the other. The different constants in (1) for flat and hemispherical ends allow σ_x, σ_y, σ_z to be found from measurements in the same hole. If the directions of the principal stresses are not known measurements must be made in an additional borehole which is neither normal nor parallel to the holes used in the previous cases.

15.6 Inclusion stressmeters

Borehole inclusion stressmeters differ from borehole deformation gauges in that they are stiff devices having an elastic modulus greater than that of the rock. They are used to measure stress direct rather than deformations from which the stresses can be computed. The essential theory of inclusion stressmeters is developed in § 10.9. If a cylindrical inclusion stressmeter is cemented into a borehole so that the stresses and displacements across the boundary between the stressmeter and the surrounding rock are always continuous, any change in the values of the field stresses in the rock produces a change of the stresses in the stressmeter, governed by § 10.9 (16), (17). The relationship between the stresses in the stressmeter and those in the rock is mainly dependent upon the ratio between the moduli of rigidity of the material of the stressmeter, G_0, and of the rock, G, and mildly dependent upon the relative values of Poisson's ratio. Assume that the Poisson's ratios for the rock and the material of the stressmeter are both 0·25. For plane strain conditions a change in the value of any stress normal to the axis of the borehole, σ, produces a corresponding change of stress, σ_0, in the same direction in the inclusion. These changes are governed by

$$\sigma_0/\sigma = 3k/(2k + 1), \quad (1)$$

where $k = G_0/G$. This tends to a constant value 3/2 as k tends to infinity. Provided that the modulus of the stressmeter is many times that of the rock, changes of stress in the rock are reflected as changes of stress in the stressmeter, which are more or less independent of the modulus of the rock.

Coutinho (1949) first suggested using a rigid inclusion for determining stresses in concrete. Subsequently, inclusion stressmeters have been built for measuring stresses in rock by Potts (1954), Hiramatsu *et al.* (1957), Hast (1958), May (1959), Wilson (1961), and others. Stresses in the inclusion are measured with hydraulic capsules, birefringent glass plugs, magnetostrictive effects, and electric resistance strain gauges. The theory on which the operation of inclusion stressmeters is based supposes that displacements and stresses across the whole of the boundary between the stressmeter and the borehole always remain continuous. Leeman (1964) pointed out that, where a stressmeter is an exact fit in a circular hole, a uniaxial stress change results in loss of contact between the stressmeter and the rock over part of the boundary. For this reason stressmeters are generally installed with a high pre-stress and are cemented into the boreholes.

Inclusion stressmeters have generally been used to monitor changes in stress. Berry and Fairhurst (1966) have analysed the problem of stressmeters in a viscoelastic material and show that the stresses in the inclusion change appreciably with time due to viscous effects, even though the field stresses in the rock remain constant. In rock which exhibits appreciable creep this might provide a means for determining the field stresses in the rock.

Hast's work combined a high-modulus stressmeter with stress relief by overcoring to determine the absolute values of the stresses in the rock. His gauge constitutes a rectangular inclusion diametrically across the borehole. The results of stress-relief measurements with this inclusion are less sensitive to the modulus of the rock than are stress measurements deduced from deformation gauges and the high gauge pre-stress ensures that it remains in intimate contact with the rock during overcoring. Four measurements with different orientations allow the stresses normal and parallel to any borehole to be calculated, and measurements in two boreholes which are neither perpendicular nor parallel to one another allow the complete state of stress to be deduced, as in § 15.4.

15.7 Plate-bearing tests

The rock mass forms an integral part of large structures founded upon or constructed within it, such as dams and hydro-electric power-station halls. The deformation of the rock mass under the loads imposed by these structures has important implications for their design. No reliable method exists for predicting the stiffness or strength of the rock mass from the results of tests on laboratory specimens. The plate-bearing test provides a convenient means for determining the modulus of deformation of rock in the field, Rocha *et al.* (1955). It can also be used to assess the strength of rock in the field, Serafim (1964) and Coates and Gyenge (1966).

In the plate-bearing test a part of an exposed rock surface is loaded by a normal force and the resulting displacements of the rock surface are measured. The stresses and displacements are related to the applied load by the Boussinesq solution, § 10.16, for the distribution of stress and strain in a semi-infinite elastic medium subject to surface loads. The load is conveniently applied by hydraulic jacks, and tests are generally carried out in tunnels or galleries; the jack reaction being taken by the opposite wall. If the results are to be representative of the rock mass it is essential that the volume of rock affected by the test be large enough to embrace sufficient of the structural features which differentiate the rock mass from a specimen. The capacity of jacks used for plate-bearing tests is, in practice, limited to several hundred tons, and the size of the loaded area is limited to a fraction of the radius of the tunnel in which the test is made. The dimensions of the loaded area generally range from several inches to a few feet, and bearing pressures seldom exceed a few thousand pounds per square inch.

In practice, the boundary conditions beneath the bearing plate must be between the extremes represented by a constant displacement and a uniform stress, respectively. The stress at the perimeter of a completely rigid bearing plate would be infinite, and the rock must in fact fail. Because of this difficulty and others arising from the flexibility of actual bearing plates, most investigators seek to achieve conditions of uniform stress under the bearing plate either by inserting a pad of flexible material, such as rubber, or an hydraulic cushion, much like a sealed flat-jack, between the plate and the rock. With this refinement, the relationship between the average displacement of the bearing area, W_{av}, and the pressure, p, on it is given by § 10.17 (54)

$$W_{av} = 0{\cdot}54\pi pa(1 - \nu^2)/E, \tag{1}$$

where a is the radius of a circular bearing plate. Timoshenko and Goodier (1951) show that the average deflections of rectangular plates can be put in the form

$$W_{av} = mP(1 - \nu^2)A^{\frac{1}{2}}/E, \tag{2}$$

where A is the area of the plate and m is a numerical factor ranging from 0·96 for a circle through 0·95 for a square to 0·71 for a rectangle with a ratio of 10 between its length and breadth.

The question of failure of the rock in a plate-bearing test is complicated by the inhomogeneous stress conditions which prevail. Failure which allows indentation of the plate must include failure of the rock somewhere along the axis normal to the rock surface through the centre of the plate. The stresses parallel and radial to this axis can be found by the methods of § 10.17. They are given by Timoshenko and Goodier (1951) as

$$\sigma_z = p(1 - b^3), \tag{3}$$

$$\sigma_r = \sigma_\theta = p[(1 + 2\nu) + b^3 - 2(1 + \nu)b]/2, \tag{4}$$

where $b = z/(a^2 + z^2)^{\frac{1}{2}}$. Using a relation for the strength of the form § 4.6 (20), failure can occur when

$$\sigma_z = C_0 + q\sigma_r, \tag{5}$$

that is,

$$p = 2C_0/[2(1 - b^3) - q(1 + 2\nu) - qb^3 + 2q(1 + \nu)b]. \tag{6}$$

The value of the denominator in (6) is a maximum for

$$b = [2(1 + \nu)q/3(q + 2)]^{\frac{1}{2}},$$

which makes the minimum value of p fall in the range from one to two times the uniaxial compressive strength. Furthermore, the maximum value of b is likely to range between a half and unity, indicating that failure occurs at a depth of the order of the radius of the bearing plate.

To enable plate-bearing tests to be made on flat, open surfaces, Zienkiewicz and Stagg (1967) propose that the reaction of the jacks be taken by cables anchored in boreholes drilled into the rock beneath the bearing plate. The depth of the anchor hole should be about ten times the lateral dimension of the bearing plate. If two proximate bearing plates are used, additional loads parallel to the surface of the rock can be applied between them, and the resulting normal and parallel displacements can be used to estimate the moduli of the rock in these two directions.

15.8 Large-scale compression and shear tests

The strength of rock depends in a complicated way on its structure, size, and the stress conditions. The influence of these factors is discussed in Chapters 4, 6, and 7. However, it is difficult if not impossible to predict the strength of large blocks of rock in the field with the nicety demanded by the economic context of large projects such as collieries and dams. In these situations it is often necessary to measure the strength of large pillars and blocks on site, to refine other estimates of their strength, and to provide an additional check on the safety of the structure.

First, consider tests to determine the compressive strength of pillars. The strength of a pillar depends upon the rock, the geometry of the pillar, and its size; usually in that order of importance. The geometry and geological structure determine the constraints applied to the ends of the pillar by the strata between which it is compressed. These constraints have an overriding influence on the compressive strength of short pillars. If test results are intended to define the strength of a pillar it is important that the end constraints be disturbed as little as possible by the test. Consider the compression of one of many pillars by convergence across the excavation. Convergence is often near uniform across the width of the pillar, and a rectangular or circular pillar is deformed symmetrically about its axis.

If the strata at each end of the pillar are similar the deformation of the pillar is also symmetrical about a plane through the middle of the pillar and normal to its axis. Such a plane remains flat during compression of the pillar by convergence, and is subject only to normal stress. If the pillar is compressed a distance Δh the distance between this plane and the strata at each end of the pillar decreases by a uniform distance, $\Delta h/2$, Fig. 15.8.1 (*a*), (*b*). In a test the compression of this pillar can most readily be simulated by forcing the two surfaces of the plane of symmetry apart a uniform distance $\Delta h/2$ through the application of uniform normal forces between them, Fig. 15.8.1 (*c*). As only normal forces are required, relatively simple jacks can be used for such an experiment. These jacks must allow free lateral

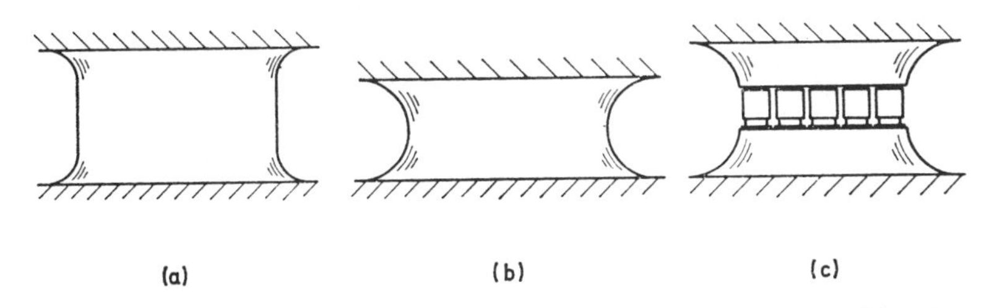

Fig. 15.8.1 (*a*) A pillar before compression. (*b*) A pillar which has been compressed by convergence across an excavation. (*c*) A pillar each half of which has been compressed by jacks installed across its mid-plane.

expansion of the pillar during compression and must be arranged so that they expand by a fixed amount irrespective of the resistance of any part of the pillar to a particular jack.

If the jacks are installed in a slot cut through the plane of symmetry this constitutes two experiments (one on each half of the pillar), and facilitates measurement of the compression of the pillar, which must be identical with the expansion of the jacks. The experiment could be made on one half of the pillar only. This configuration emphasizes the fact that, using normal loads only, the height of an experimental pillar is only half that of the actual pillar which it simulates, due to the changed constraints at the jack end.

The load capacity of jacks for compression testing should be such that they can generate a stress at least twice the average strength of the pillar as a whole. The expansion of most jacks is about a third. This can be regarded as the ultimate strain capacity of a jack. The width necessary to accommodate test jacks is simply related to the height of a pillar and its ultimate strain, ε_u. Let w_j be the width of the jacks and h_t be the height of the test pillar; then $w_j/h_t = 3\varepsilon_u$.

Triaxial tests have been made on large specimens using flat-jacks to apply lateral confinement and piston jacks to apply the major principal stress along the axis of the specimen, Nose (1964). The results of such tests are affected by end-effects similar to those affecting laboratory tests, §§ 6.2, 6.4, but they are probably of less significance at the end of the specimen in contact with the jacks than at the end in contact with the rock.

Large, on-site shear tests are usually made in connection with the strength of dam foundations, especially where the shear strength of geological joints is in question. The size of the specimen should be as large as is reasonably possible so that the area of the plane to be sheared incorporates

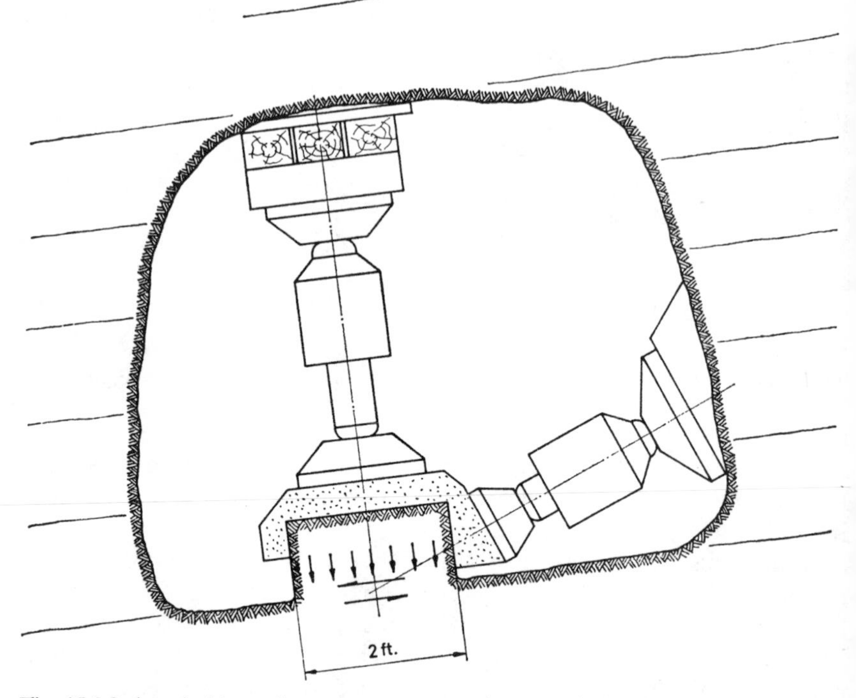

Fig. 15.8.2 A typical large shear test arrangement in a tunnel, showing the inclined shear jack and the normal jack.

a sufficient number of flaws and heterogenieties to be representative of the properties of the rock as a whole. The side dimensions of shear specimens range from a foot to several feet, being limited by the capacity of the available jacks, which is usually a few hundred tons. A typical arrangement for a shear test is shown in Fig. 15.8.2. The shear-force jack is inclined at some small angle, which must be less than the angle of friction, to the shear plane, so that its line of thrust passes through the centre of the shear plane. The shear jack, therefore, contributes a component of normal force which is supplemented by the force of the normal jack. Flat-jacks have been used for applying normal force, but the arrangement shown in the figure is to be preferred because the spherical seats on the normal jack allow the block to slide with negligible restraint from that jack. Loads are usually increased in steps, normal force increments preceding shear-force increments, and time is allowed for the block to come to rest before the next increment, Rocha (1964). Nose (1964) recommends that specimens should be loaded and unloaded during tests to simulate load variations due to changes in the water level in dams.

Shear tests are also made on concrete blocks cast on to the rock surface to determine the strength of foundation contacts.

The cost of large-scale tests makes it desirable that the sites for them be chosen carefully in the light of a full geological map of the site and that they be accompanied by a programme of laboratory tests on specimens of the same rock. Large-scale specimens must be prepared with care so as to disturb them as little as possible. It is desirable to cut them from the surrounding rock with coal cutters, rope saws, or overlapping drill holes.

15.9 Measuring bolts

Measuring bolts are simply rockbolts used to make measurements of elongations in the rock along the length of the bolt between its anchor and collar. They are generally used in four configurations: untensioned bolts anchored into the rock at their far ends but otherwise lying free in their holes, tensioned load bolts anchored at their far ends and tightened at their collars against a bolt load-cell, tensioned hollow displacement bolts anchored at their far ends and at their collars with a thin reference rod lying free in the bore of the bolt between the collar and the anchor to which it is fixed, hollow grouted displacement bolts of similar design.

Untensioned measuring bolts are usually installed in groups of holes of different lengths drilled into the side-walls or roof of an excavation; the length of each bolt being just less than that of the hole, Fig. 15.9 (*a*). The lengths of the bolts range from 2 to 20 ft or more. Measurements of the displacements between the end of the longest bolts, the ends of the other bolts, and the surface of the excavation provide a means for determining the elongation of the rock, normal to the surface of the excavation, in the intervals between bolt anchors, to an accuracy of a hundredth of an inch or so. If groups of bolts are installed in diametrically opposite pairs across a symmetrical excavation the absolute displacement of the rock at the position of the bolt anchors can be found by measuring the diametral distance between corresponding bolts, Fig. 15.9 (*b*). The absolute displacements of the anchor positions of vertical bolts can be followed with precise levelling surveys, § 15.10. The ends of measuring bolts are usually fitted with stainless-steel sockets or hooks, to facilitate attaching tapes or extensometer rods.

The rock in the vicinity of a group of untensioned bolts might behave differently from the rest of the rock, if that is supported by a pattern of tensioned or grouted rockbolts. In the case of tensioned support bolts, some of them can be fitted with load-cells to study the elongation of the rock along the length of the bolt. Changes in bolt load indicated by the cell can be related to elongation of the bolt through its length, section, and stress–strain curve. This method is generally applicable only within the

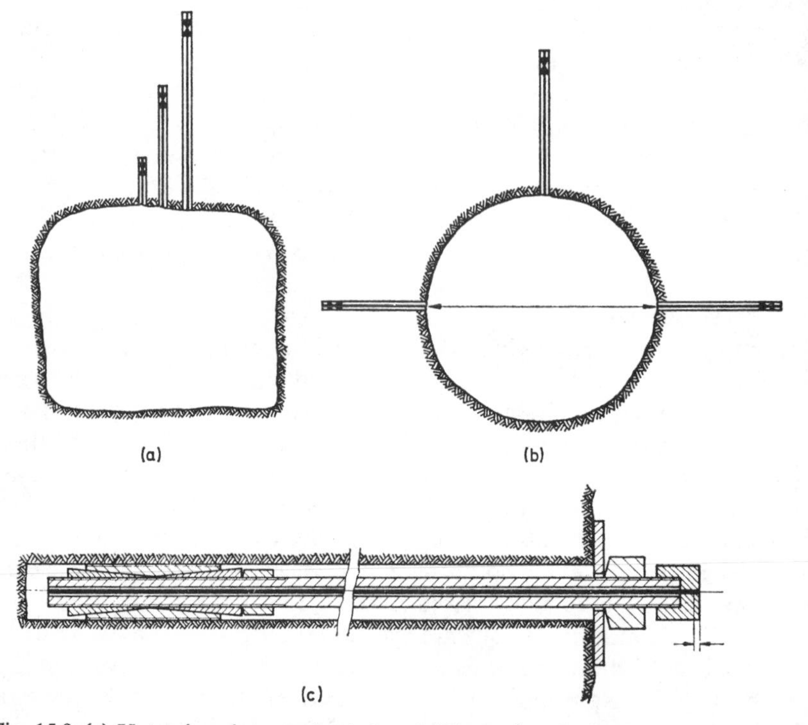

Fig. 15.9 (*a*) Untensioned measuring bolts of different lengths in the roof of a tunnel. (*b*) Diametrically opposed horizontal measuring bolts in a circular tunnel. (*c*) A section through a borehole in which a hollow measuring bolt tensioned by a nut and washer has been installed.

range of elastic deformation of the bolt, and is therefore limited to relatively small elongations.

Hollow rockbolts overcome the limited elongation range of rockbolts with load-cells. A thin reference rod lying freely in the bore of the hollow bolt between the collar and anchor ends of the bolt is welded to the bolt at the latter end. The reference rod is often made of stainless steel about a quarter of an inch or less in diameter, and the bolt is usually fitted with a stainless-steel reference cap and rod guide at its collar end, Fig. 15.9 (*c*). Elongation of the bolt between its anchor and collar is detected by measuring the displacement between the collar reference cap and the end of the reference rod with a vernier caliper, dial gauge, or an electrical displacement tranducer if remote recording is required. Hollow measuring bolts may be tensioned with a nut and washer at their collars, as in Fig. 15.9 (*c*), or they may be grouted into the rock. If the bolt is of ductile metal elongations beyond the elastic limit of the bolt can be followed because these are measured direct, and the bolt tension can be deduced with adequate accuracy from a knowledge of its length, cross-section, and stress–strain curve.

Measurements of the elongation of rock normal to the surface of an excavation are made for a number of purposes. It is often useful to have some knowledge of the movement of the rock towards an excavation. If measuring bolts are installed near the advancing end of a tunnel or in the early stages of making a large excavation, such as a power-station hall, subsequent extension of these excavations causes movement of the rock towards the excavation and elongation normal to the walls of the excavation. Movements and elongations in excess of those likely to occur as a result of the elasticity of the rock, and the field stresses around the excavation, provide a measure of the extent to which the rock around the excavation is damaged by blasting. Alexander (1960) installed bolts 8 ft in length around a section 1 ft from a tunnel end and measured the elongation of these bolts after the tunnel had been advanced. The radial displacements induced in the rock around a circular tunnel of radius R by excavation are, from § 10.4 (21), given by

$$u_r = R^2(1 + \nu)[\sigma_x + \sigma_y + (\sigma_x - \sigma_y)(4 - 4\nu - R^2/r^2)\cos 2\theta]/2Er, \quad (1)$$

where σ_x, σ_y are the field stresses normal to the axis of the tunnel and θ is the direction between the radius r and the direction of σ_x. From elongation measurements and a knowledge of the field-stress values, Alexander was able to calculate the elastic modulus of the rock around the tunnel in a direction normal to its surface. The value of this modulus was found to be five times less than the modulus tangential to the surface of the tunnel obtained from flat-jack measurements. Measuring bolts can also be used to follow the movement of the rock around an excavation with time. Such elongations may be caused by increasing field stresses, rock creep, time-dependent fracturing, or weathering. Measuring bolts anchored at different depths into the sidewall and roof of large excavations provide an excellent means of monitoring the behaviour of the rock around them and the effectiveness of its support. If one of the bolts is anchored at sufficient depth to be in solid rock virtually unaffected by the presence of the excavation, the displacements of other shorter bolts will show the position of any separation which occurs and provide an indication of the adequacy of the support to retain the separated rock.

15.10 Gross displacements in the rock mass from precise levelling

Underground levelling provides a convenient means of following the vertical displacements of benchmarks in excavations. The excavations resulting from mining do not always lend themselves to continued levelling, but tunnels running through the rock in the vicinity of such excavations provide useful access for studying the displacements induced by the main mine excavation in the rock through which the tunnel runs. Several factors combine to make levelling underground more accurate than the equivalent operation on surface. Benchmarks are usually short rockbolts anchored

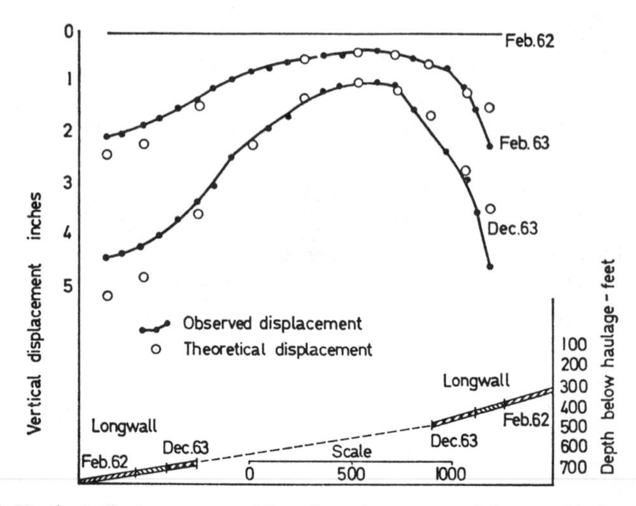
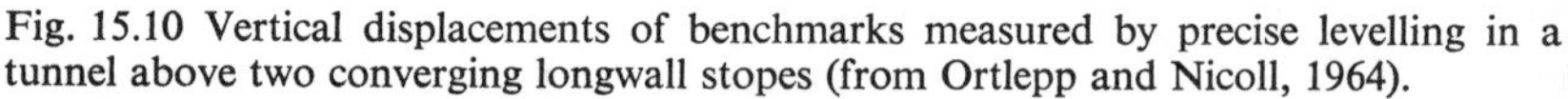

Fig. 15.10 Vertical displacements of benchmarks measured by precise levelling in a tunnel above two converging longwall stopes (from Ortlepp and Nicoll, 1964).

in the roof and equipped to take suspended staves, so that damage to benchmarks is minimized and carefully balanced staves hang vertically. Temperature conditions underground are generally more constant than on surface, and errors due to differential expansion in the level and refraction of the line of sight are less. High-speed air currents can be troublesome in tunnels used as ventilation airways. Using sight distances of about a 100 ft, mean square errors of $\frac{1}{10}$ in. per mile can be expected, if lines are observed forward and back.

The results of levelling along a haulage situated at about 8,000 ft below surface above two converging longwall stopes, Ortlepp and Nicoll (1964), are shown in Fig. 15.10, together with the vertical displacements calculated from the theory of elasticity with the aid of an analogue computer, § 18.5. The measured displacements induced by the extension of the longwalls compare well with the displacements calculated from the theory of elasticity using a value of Young's modulus of 11×10^6 psi determined from tests on laboratory specimens of rock and a vertical virgin rock stress equal to the weight of the overburden.

15.11 Longitudinal displacements in long boreholes

The measurement of the elongation along the axis of a borehole provides a precise means for studying small displacements in a large volume of rock. If a borehole is drilled in the undisturbed rock ahead of an advancing excavation and wires are anchored at intervals along the length of the borehole and are brought out together at the collar of the hole, the exposed ends of the wires will move relative to one another as the advancing excavation induces displacements in the rock around the borehole. Provided the anchors are adequate and the effects of temperature, creep,

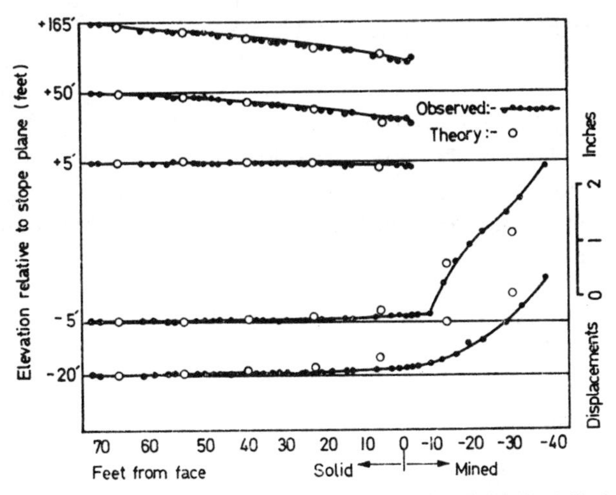

Fig. 15.11 Longitudinal displacements in a vertical borehole initially drilled ahead of an advancing face (from Ortlepp and Cook, 1964).

tension, and friction on the wires are made insignificant, the displacements of the wires correspond to the longitudinal displacements of the anchor points along the borehole.

The borehole ends of individual wires are generally fixed to mechanical anchors which lock to the sides of the borehole by friction, or fit into circumferential grooves specially cut around the hole during drilling. Individual anchors are equipped with low-friction bushes to allow wires anchored farther along the hole to pass through them. At the collar the individual wires are usually kept separated by passing each of them over a pulley and tensioning them with light weights or springs. When observations are made intermittently it is good practice to measure the displacement of each wire while the tension is increased and then decreased. The hysteresis in such displacement measurements provides an indication of the magnitude of any frictional effects and a check on the stability of the anchor. Where continuous measurements are made by means of transducers fixed to the end of the wire, it is necessary to maintain the tension in the wire at a greater value than that needed merely to keep them separated. Care must be taken that this tension is not so great as to produce significant creep in the wire. For precise continuous measurements it is desirable to reduce the number of wires per hole and increase their diameter.

Measurements of longitudinal displacements in boreholes have been used extensively by Potts (1964), who has devised a precise tensioning device and micrometer known as an 'extensiometer' for measuring the displacements of the wires. Ortlepp and Cook (1964) have used up to six wires in vertical and horizontal boreholes more than 150 ft long drilled in the rock around deep-level mine excavations. The measured displacements agreed remarkably well with those calculated on the basis of the theory of elasticity using a value of Young's modulus equal to that of laboratory specimens of the surrounding rock, Fig. 15.11. Salamon and Oravecz (1966) have applied the same technique in boreholes drilled from surface through a coal seam. The measured displacements compared well with those calculated on the basis of the theory of elasticity when a value for Young's modulus is used which is several times less than that of small specimens of coal and the surrounding strata.

Chapter Sixteen

Granular Materials

16.1 Introduction

Much of the preceding discussion on the properties of rock was concerned with the behaviour of unbroken rock. In many practical situations rock is so fractured and jointed that its granular nature becomes important. The properties of extensively weathered rock approach those of soils. Most of the information relevant to the behaviour of granular materials is to be found in the literature on soil mechanics, Terzaghi (1943), Bishop and Henkel (1951), and Sokolovskii (1965). Most rocks and soils have in common the Coulomb criterion, § 4.6,

$$\tau = S_0 + \mu\sigma_n, \tag{1}$$

which defines the limiting state of stress for static equilibrium within the material at which inelastic deformation begins. Soil-mechanics analysis is applied to problems of slope stability and rock-fill dams. However, there are important differences between the behaviour of soils and intensely broken rock, due mainly to interlocking effects in the latter which are ignored in the usual soil-mechanics analysis. As a result, the strength of closely jointed rock is often underestimated. It is the purpose of this chapter to discuss the differences between the behaviour of soils, aggregates, jointed rock, and unbroken rock.

16.2 The stress–strain curve for weathered rock and aggregate

An ideal elastic-Coulomb material, which will be described for shortness as a *Coulomb material*, is one which has two modes of deformation: bodily elastic deformation while the shear stress is everywhere less than that defined by the Coulomb criterion, § 16.1 (1), and shear along slip surfaces whenever the stresses satisfy that condition for limiting static equilibrium. Consider a specimen of Coulomb material subject to hydrostatic stress, σ_a, and let one principal stress increase. Ideally, the deformation is linearly elastic along the path *PQ*, Fig. 16.2.1 (*a*), until this principal stress reaches the value σ_b for which the Mohr circle on σ_a and σ_b touches the line representing the Coulomb criterion at *A*, Fig. 16.2.1 (*b*). The material can then deform in a fully ductile manner along the path *QR*, Fig. 16.2.1 (*a*).

If this principal stress is then gradually decreased the deformation is linearly elastic along the path *RS* until the stress has decreased to the point where it is now the minor principal stress and has a value σ_c such that the Mohr circle on σ_a and σ_c touches the line representing the Coulomb criterion at *B*. The material then deforms in a fully ductile manner in the reverse direction along the path *ST*. These concepts are capable of generalization to tests such as plate-bearing and pressure-chamber tests, from which it is theoretically possible to obtain Mohr circles by cycling the load, Talobre (1957). Ideal behaviour of this sort is observed in laboratory tests on triaxial specimens of rock with a single shear plane, but in practice the

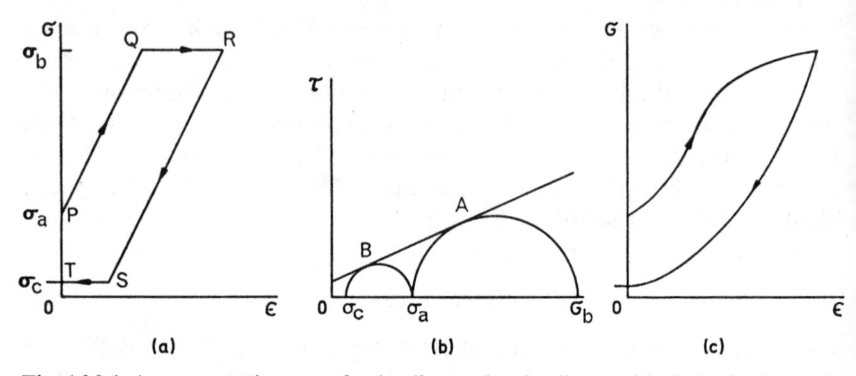

Fig. 16.2.1 A stress–strain curve for loading and unloading an ideal elastic-Coulomb material: (*a*) together with the corresponding Mohr circles; (*b*) and the curve for a real Coulomb material (*c*).

elastic deformation of real materials tends to be curved, and the transition from elastic deformation to Coulomb slip occurs gradually along a curved path, making it difficult to recognize the points of transition, Fig. 12.2.1 (*c*).

Non-linear elastic deformation in aggregates and jointed rock arises partially from the closing under compression of open cracks within the aggregate in much the same way as cracks affect the modulus of unbroken rock, § 12.3, but also as a result of the indentation of one particle of rock by another and the resulting crushing and compaction. Non-linearity from the closing and opening of cracks is largely reversible, whereas much of the non-linearity observed in loading and unloading aggregate or extensively jointed rock is accompanied by considerable hysteresis. The deformation of such materials must therefore depend upon their strain history. However, a clearly recognizable but not fully understood pattern of deformation emerges from almost any loading test on these materials. If the material is completely contained, the loading modulus increases continuously with compression. Otherwise it increases until the stress conditions allow of Coulomb slip, when it starts to decrease, goes through zero at maximum load, reaches a maximum negative slope, and flattens once again. If the load is cycled at any stage the initial unloading modulus is found to be considerably greater than the loading modulus at the same stress and its slope decreases continuously with unloading until reverse Coulomb slip occurs. Increasing the load traces a curve more or less parallel with the unloading curve, which crosses the latter curve at about two-thirds the original maximum load with a modulus decreasing as the load increases, until it rejoins the continuation of the original loading curve at a somewhat greater deformation than that at which the unloading started, Fig. 16.2.2.

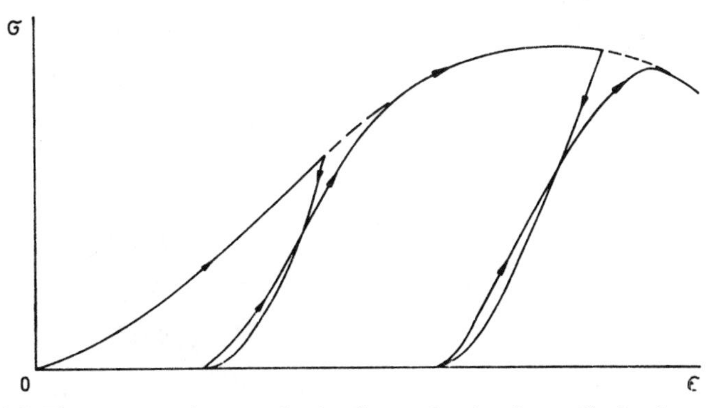

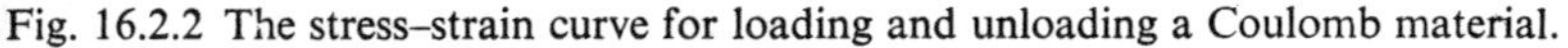

Fig. 16.2.2 The stress–strain curve for loading and unloading a Coulomb material.

16.3 Triaxial tests on soil and aggregate

The triaxial test has been very highly developed as a method of studying the mechanical properties of soil. The essential difference in practice between triaxial tests on soils and solid rock is that, while rocks are usually tested dry and pore-pressure is occasionally added as an additional parameter, in soils the pore-pressure is a vital element and must always be considered; specimens are tested either drained, in which escape of pore-water through the platens is allowed, or at a prescribed pore-pressure. Further, the movement of water through the pores is slow, so that time must be allowed for equilibrium of pressure to be attained. Finally, because of compaction and for other reasons, the stress-path by which failure is approached may have an important effect. The procedures are fully described by Bishop and Henkel (1962).

The confining pressures used in ordinary testing of soils are relatively low, and higher values are used only in studies of compaction and geophysically orientated work, Parasnis (1960).

The testing of loose rock and aggregate has assumed considerable importance in connection with rock-fill dams and problems of slope stability. In these cases it is essential to know the angle of friction for the material as a whole, and since the material usually contains relatively large rock fragments, it is necessary to make tests on specimens which are as large as possible. Since the confining pressures needed in practice are not high, the use of large specimens is quite practicable, and several large triaxial cells have been built.

The triaxial cell built by the Snowy Mountains Authority takes specimens 22·5 in. in diameter and 45 in. long. Confining pressures up to 200 psi may be used, and the chamber has Perspex windows through which the specimen can be viewed. Aggregate up to a size which will pass a 4·5-in. screen

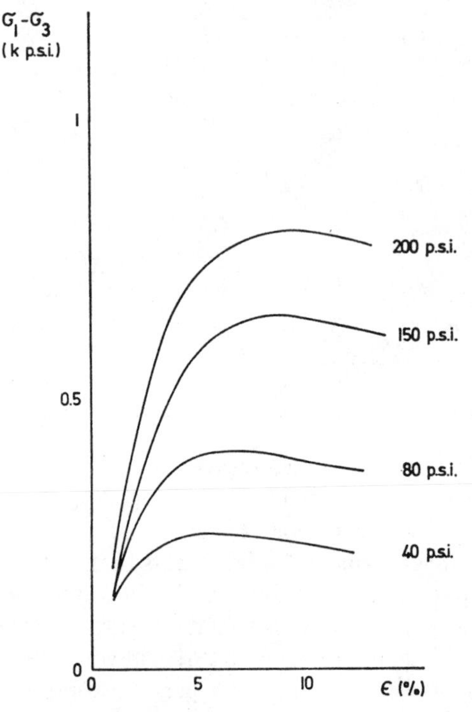

Fig. 16.3.1 Triaxial stress–strain curves at different confining pressures for coarse saturated aggregate (from Snowy Mountains Authority).

may be used, and it is pre-compacted inside the jacket in a steel former to the prescribed density and grading. When this has been done, the sample in its jacket is transferred to the triaxial cell. Arrangements are made for measuring changes in volume of the specimen as well as changes of length, and the pressure and volume of interstitial fluid are measured.

Fig. 16.3.1 shows the stress–strain curves at various confining pressures for aggregate compacted to a density of 125 lb/ft^3, saturated, and tested fully drained. Fig. 16.3.2 shows the corresponding Mohr envelopes for the maximum stresses attained, which are consistent with very low cohesion and an angle of friction of about 40°.

It appears from Fig. 16.3.2 and many similar tests on soils and aggregate that their behaviour can be very well represented by the Coulomb criterion of § 4.6, the intrinsic shear stress or cohesion S_0 being small. One of the most important differences between the behaviour of aggregates and that of unbroken rock is that the cohesion, S_0, of the former changes only slightly when the deformation ceases to be elastic, whereas the cohesion of unbroken rock changes by several orders of magnitude when it fails. It is the magnitude of this change which determines whether a material behaves in a brittle or ductile manner; the smaller and more gradual the change, the more ductile is the material.

An alternative method of studying the behaviour of aggregate, which is

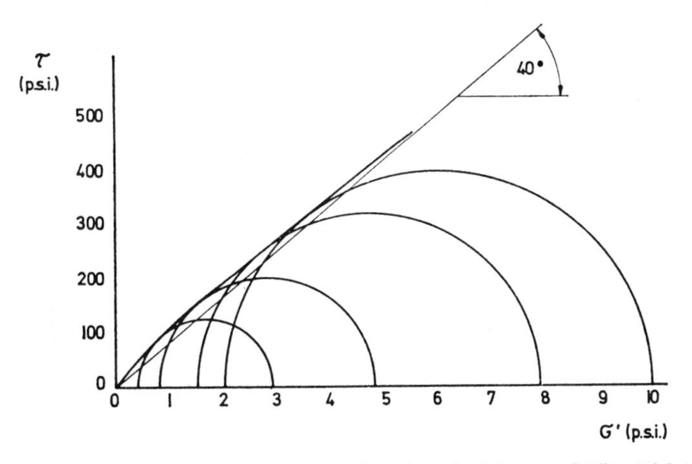

Fig. 16.3.2 The Mohr circles and envelope for the triaxial test of Fig. 16.3.1 (from Snowy Mountains Authority).

very simple, is to confine the material within a steel tube of large diameter and to compress it axially by circular platens within the tube. Radial stress is applied to the aggregate by the confining tube, and its amount can be measured by strain gauges attached to the tube. The stress–strain curve of $-\frac{1}{8}$-in. granite aggregate tested in such a cylinder to a maximum stress of about 40,000 psi is shown in Fig. 16.3.3. These results include unloading cycles at a number of different stresses, which are shown on an enlarged scale. The unloading modulus is always considerably greater than the loading modulus, and its value increases with increasing compression of the aggregate. It should also be noted that the unloading modulus does not comprise two linear parts corresponding to elastic relaxation and reverse sliding as predicted for an ideal Coulomb material but decreases continuously as unloading proceeds in a fashion very similar to that predicted for cracked rock, § 12.3, Fig. 12.3 (*b*). This would seem to suggest that a significant amount of interlocking occurs between the particles of aggre-

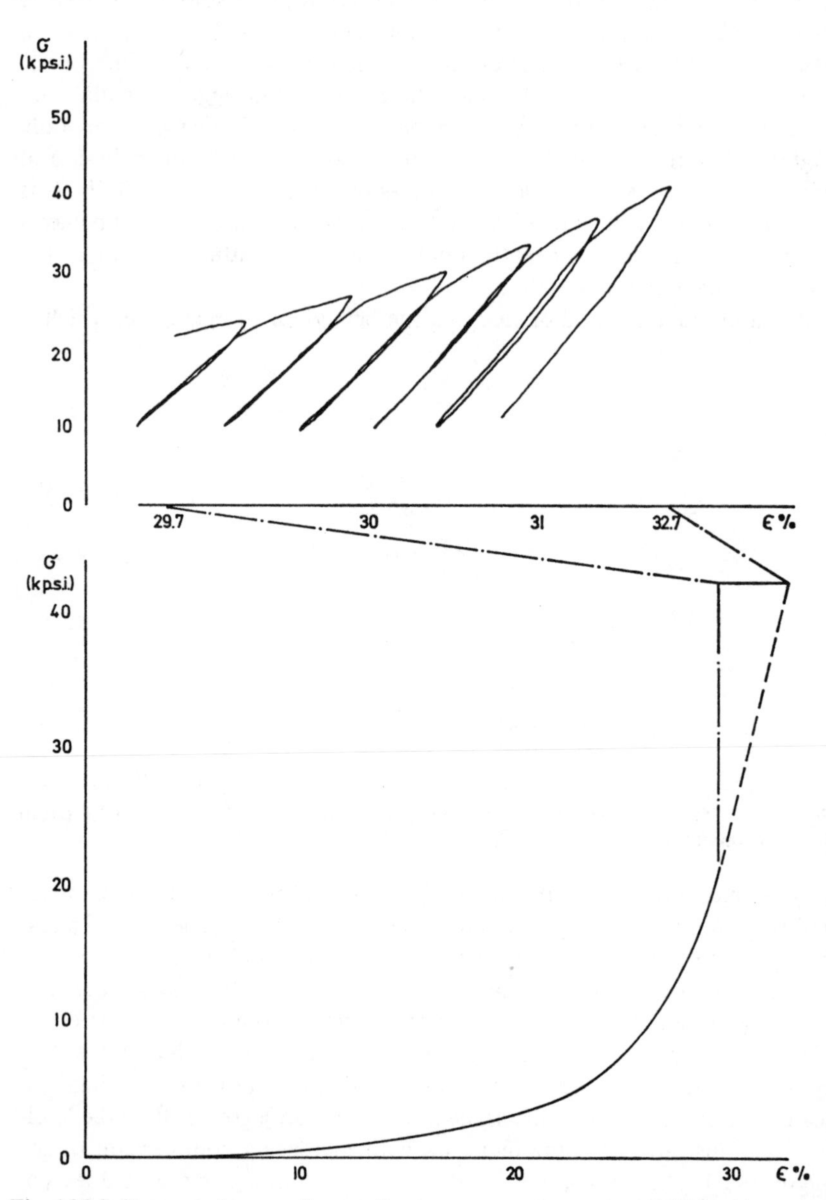

Fig. 16.3.3 Stress–strain curve for granite aggregate compressed to 40,000 psi in a steel cylinder.

gate, allowing strain energy to be stored in them which is released by reverse sliding during unloading, much as the strain energy around cracks is released.

It appears from the typical behaviour shown in Fig. 16.3.3 that under any specified conditions of void ratio and stress, the behaviour of aggregate in a restricted portion of a cycle of loading and unloading is approximately linear so that approximate 'elastic' constants covering this region may be assigned and used for calculations of its response to loading under these conditions.

For the case of confined compression, $\varepsilon_2 = \varepsilon_3 = 0$, it follows from § 5.3 (2) that,

$$\sigma_1 = (\lambda + 2G)\varepsilon_1 = \frac{(1 - \nu)E}{(1 + \nu)(1 - \nu)}\,\varepsilon, \tag{1}$$

For the triaxial test, $\sigma_2 = \sigma_3$,

$$\sigma_1 = E\varepsilon_1 + 2\nu\sigma_2, \tag{2}$$

so that, in principle, E and ν may be determined.

A plate-bearing test on the surface of aggregate which had been compressed to 40,000 psi showed it to have a loading modulus much less than its unloading modulus and a bearing strength of about 12,600 psi, corresponding to a uniaxial compressive strength of about 6,000 psi.

The irreversible nature of load–compression curves for soils and aggregate must be attributed to the work done during compaction. Direct evidence for this is found in a comparison of particle-size distributions for virgin and compressed aggregates, which show an increased proportion of small particles in the compressed aggregate produced by crushing during loading.

16.4 Finely cracked rock

Aggregate, the behaviour of which is discussed in § 16.3, consists of randomly orientated, irregularly shaped blocks, and behaves very much in the manner of soils. Much rock encountered in the field consists of small pieces of rock with dimensions of the order of an inch or two but closely packed with a regular pattern of joints. This close packing has several important effects which result in behaviour under confining pressure which is much more nearly that of rock than that of aggregate or soil. Unless the joints in finely cracked rock are continuous, plane surfaces inclined to the direction of the major principal stress at an angle close to the Coulomb surface given by $\beta = \tan^{-1}(-1/\mu)$, § 4.6 (6), any fracture surface with an average inclination close to the Coulomb angle must consist of a multitude of smaller surfaces, the inclination of each of which deviates from the average. Whether these surfaces are inclined at greater or smaller angles to the direction of the major principal stress than the Coulomb fracture plane, the ratio between the shear and normal stresses on them is less advantageous for sliding than it is on a plane of the Coulomb inclination, Fig. 16.4.1 (*a*). Any shear displacement which occurs on a Coulomb fracture plane separates all those parts of the fracture surface which are inclined at a smaller angle to the direction of the major principal stress than that plane and increases both the shear and normal stresses on those parts inclined at greater angles. The ratio between the average shear and normal stresses required to cause sliding on a Coulomb fracture plane with an irregular surface is greater than it would be on a smooth plane with the same inclination and coefficient of sliding friction. The coefficient of internal friction of an irregular Coulomb fracture plane appears to be greater than that of a smooth plane, and a couple is applied to each of the irregularities. At low confining stresses sliding takes place on the irregular surface with an apparently high coefficient of internal friction. As the confining stress increases, a stage is reached at which the stresses on the irregularities cause them to fail either by shear or rotation, and gross shear of the fracture plane occurs with a coefficient of internal friction equal to that of a smooth Coulomb fracture plane. The Mohr envelope of a finely cracked rock can be expected to have a greater coefficient of internal friction at low confining stress than it has at higher confining stress. At high confining stresses the effect of this is the same as that of an increased cohesion, Fig. 16.4.1 (*b*).

Unfortunately it is difficult to test a finely jointed rock in the laboratory, but an effect of this sort may be observed in marble whose grain boundaries have been broken by heating. Calcite has a considerable anistropy of thermal expansion, and Lord Rayleigh (1934) showed that heating of it caused grain boundary cracking and a considerable decrease of strength. A similar effect is observed with some alloys, Boas and Honeycombe (1947).

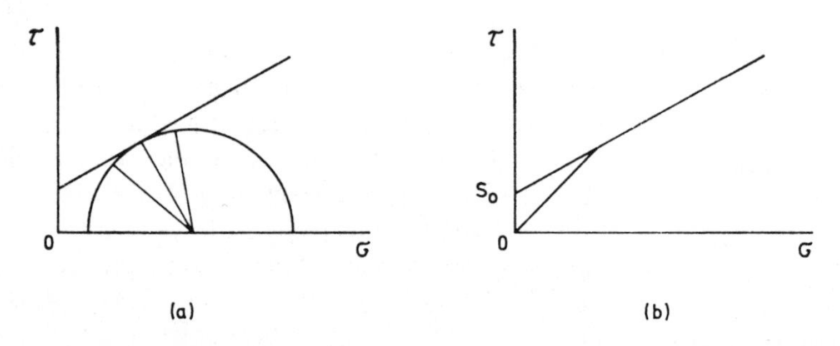

Fig. 16.4.1 (*a*) The shear and normal stresses on parts of a Coulomb fracture plane inclined at different directions from the average direction of that plane. (*b*) The Mohr envelope for an irregular Coulomb fracture plane.

If Wombeyan marble, which is fairly coarse grained with a grain size of around 2 mm, is heated to 500° C in an inert atmosphere its tensile strength is reduced to about 10 psi from about 500 psi, and fracture is entirely intergranular. Its uniaxial compressive strength is not reduced so drastically; from 12,000 to 2,000 psi, and with increase in confining pressure, σ_1 at failure rises rapidly to a value not far short of that of the original rock. This effect is shown in Fig. 16.4.2. It is undoubtedly due to this close contact of the grains and suggests that, with moderate confining pressure, closely jointed rock may retain considerable strength, Rosengren and Jaeger (1968).

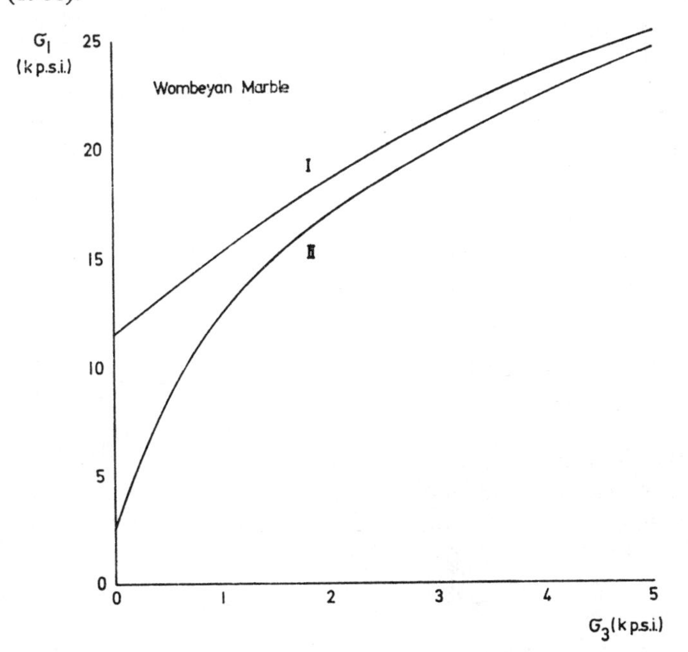

Fig. 16.4.2 The triaxial strength of Wombeyan Marble, (I) original marble, (II) marble with grain boundaries disintegrated by heating.

The effect on Young's modulus is shown in the stress–strain curves at a confining pressure of 500 psi for the original marble and one which has been heated to 500° C, Fig. 16.4.3. Both curves start from a base axial stress of 1,900 psi, at which load considerable axial strain of the heated sample had taken place. The curve for the cracked marble, II, shows further compaction, but a tangent modulus of about a third that of the original material is attained.

It is frequently suggested that soil mechanics theory should be used for broken rock. These results suggest that for closely and regularly jointed rock this procedure may underestimate the strength.

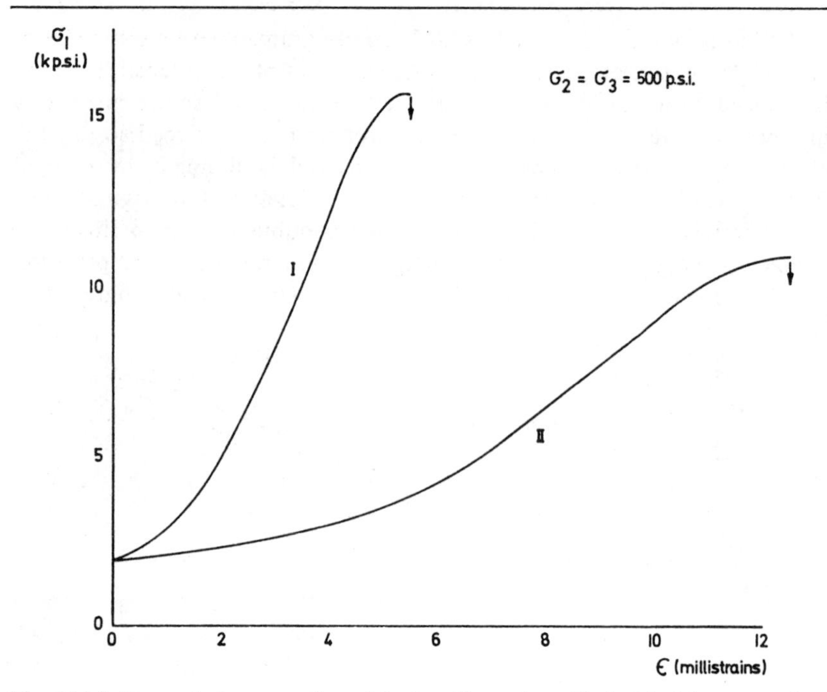

Fig. 16.4.3 Stress–strain curves for original and heated marble in triaxial compression.

16.5 The Coulomb material: theoretical solutions

A *Coulomb material* will be defined as one in which the principal stresses satisfy the Coulomb criterion of failure of § 4.6. In soil mechanics this is usually taken in the form § 4.6 (18),

$$\sigma_1 = 2S_0 \tan \alpha + \sigma_3 \tan^2 \alpha, \tag{1}$$

where

$$\alpha = (\pi/4) + (\phi/2), \tag{2}$$

and ϕ is the angle of friction. The direction of shear failure is inclined at an angle

$$\psi = \tfrac{1}{2}\pi - \alpha = (\pi/4) - (\phi/2) \tag{3}$$

to the direction of σ_1.

For shortness, (1) will frequently be used in the alternative form, § 4.6 (20),

$$\sigma_1 = C_0 + q\sigma_3, \tag{4}$$

where $C_0 = 2S_0 \tan \alpha$ and $q = \tan^2 \alpha$.

As remarked in § 16.3, the behaviour of soils and aggregate is described very well by (1), S_0 usually being small, or equal to zero for a cohesionless material such as sand.

The study of stress and motion in a body composed of Coulomb material is a generalization of the theory of the perfectly plastic solid, Chapter 10; in fact the Tresca criterion corresponds to taking $q = 1$ in (4). In general, the body will be divided into two regions, in one of which the stresses are not sufficiently high to satisfy (1), so that the material behaves elastically in this region, while in the other region the criterion (1) is satisfied at every point. An extensive discussion is given in the book of Sokolovskii (1965). Some simple problems of importance in soil mechanics are discussed in (i) and (iii) below and an application to rock mechanics in (iv).

(i) *The normal pressure necessary to support a vertical surface*

Suppose that Coulomb material of density ρ occupies the region AOB, $x > 0$, $y > 0$ of Fig. 16.5.1 (a) and (b) with the x-axis vertically downwards and that it is required to support the surface OB by normal stress σ_y distributed over it. The vertical stress at the surface will be $\rho g x$, and since this will be the greater principal stress in (1), it follows that

$$\sigma_y \tan^2 \alpha = \rho g x - 2S_0 \tan \alpha. \tag{5}$$

If normal stress, given as a function of x by (5) is applied to the surface, all points in the material will be in a state of incipient slip down planes PQ inclined at α to the horizontal. This is known as the *active Rankine state of equilibrium*.

Another solution arises if the horizontal stress σ_y is greater than $\rho g x$. In this case from (1)

$$\sigma_y = 2S_0 \tan \alpha + \rho g x \tan^2 \alpha, \tag{6}$$

and the material will be in a state of incipient slip up planes PQ inclined at α to the vertical, Fig. 16.5.1 (b). This is known as the *passive Rankine state of equilibrium*.

These values are shown on the Mohr diagram of Fig. 16.5.1 (c), in which the point C corresponds to the vertical stress $\rho g x$ at depth x, the Mohr circle on AC to the active state with σ_y given by (5) and the circle on CP to the passive state with σ_y given by (6). It appears that failure will not take place for intermediate values of the stress σ_y.

If there is pressure p of pore-water in the material the criterion (1) is to be replaced by

$$(\sigma_1 - p) = 2S_0 \tan \alpha + (\sigma_3 - p) \tan^2 \alpha \tag{7}$$

as in § 8.9 (1), and thus (5) is replaced by

$$\sigma_y \tan^2 \alpha = \rho g x + p(\tan^2 \alpha - 1) - 2S_0 \tan \alpha. \tag{8}$$

(ii) *The stresses around a circular hole in a Coulomb material with hydrostatic stress at a great distance*

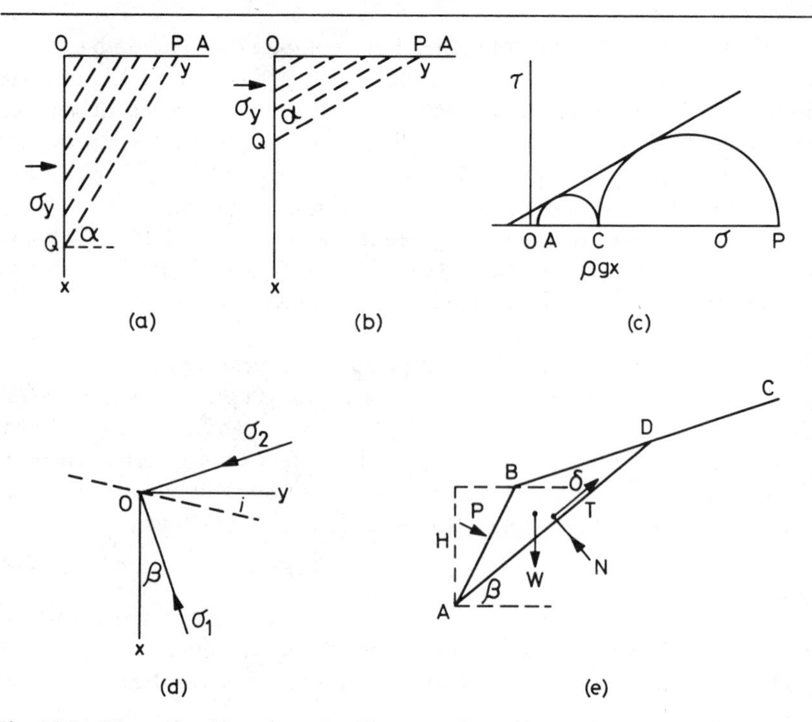

Fig. 16.5.1 The active (*a*) and passive (*b*) states of Rankine equilibrium for a Coulomb material subject to gravity, the corresponding Mohr circles (*c*), and the axes of principal stress for an inclined slope (*d*) and Coulomb's approximate solution for the load on a retaining wall (*e*).

Suppose that the opening is of radius R_1 and that the stress at a great distance is p_0. Suppose that the material is held in equilibrium by radial stress p_i applied at $r = R_1$. There will be a region $R_1 < r < R$, in which the material satisfies the failure criterion (4) at every point; outside this there will be a region $r > R$ in which it behaves elastically. The radius R of the boundary surface between the two regions has to be found, and continuity conditions have to be satisfied on it.

In the region $R_1 < r < R$ the stress-equation of equilibrium § 5.6 (9) holds and is

$$\frac{d\sigma_r}{dr} + \frac{\sigma_r - \sigma_\theta}{r} = 0. \tag{9}$$

Assuming that $\sigma_\theta > \sigma_r$, (4) becomes

$$\sigma_\theta = C_0 + q\sigma_r \tag{10}$$

and substituting this in (9) gives the differential equation

$$\frac{d\sigma_r}{dr} = \frac{C_0 + (q-1)\sigma_r}{r}. \tag{11}$$

The solution of this is

$$\sigma_r = \frac{C_0}{1-q} + Ar^{q-1}, \tag{12}$$

where A is a constant to be determined from the condition $\sigma_r = p_i$ when $r = R_1$. Using this condition gives

$$\sigma_r = \frac{C_0}{1-q} + \left(p_i - \frac{C_0}{1-q}\right)\left(\frac{r}{R_1}\right)^{q-1}. \tag{13}$$

Then from (10)

$$\sigma_\theta = \frac{C_0}{1-q} + q\left(p_i - \frac{C_0}{1-q}\right)\left(\frac{r}{R_1}\right)^{q-1}. \tag{14}$$

In the region $r > R$ it follows from § 5.11 (12) that the solution has the form

$$\sigma_r = p_0 - Br^{-2}, \quad \sigma_\theta = p_0 + Br^{-2}, \tag{15}$$

where B is an unknown constant.

At the boundary $r = R$ two conditions must be satisfied. First, continuity of radial stress requires that the values of the radial stress σ_r derived from (13) and (15) must be equal. Secondly, when $r = R$ the stresses given by (15) must just satisfy the criterion (10). Therefore

$$\frac{C_0}{1-q} + \left(p_i - \frac{C_0}{1-q}\right)\left(\frac{R}{R_1}\right)^{q-1} = p_0 - BR^{-2}, \tag{16}$$

$$p_0 + BR^{-2} = C_0 + q(p_0 - BR^{-2}). \tag{17}$$

These equations also imply that σ_θ is continuous at $r = R$.

Solving (16) and (17) gives

$$\frac{R}{R_1} = \left\{\frac{2[p_0(q-1) + C_0]}{[p_i(q-1) + C_0](q+1)}\right\}^{1/(q-1)}, \tag{18}$$

$$B = R^2[p_0(q-1) + C_0]/(q+1), \tag{19}$$

and the solution is completely determined.

In the region $R_1 < r < R$ the directions of slip are inclined at α to the direction of least compressive stress, which is radial, so that they are given by

$$\frac{dr}{r\,d\theta} = \pm \cot \alpha. \tag{20}$$

The solution of this, which gives the slip-lines, is the two families of equiangular spirals

$$r = r_0 \exp(\pm\theta \cot \alpha), \tag{21}$$

where r_0 is a constant.

The discussion above has been set out on the assumption that $\sigma_\theta > \sigma_r$, which corresponds to the practical case. If it is assumed instead that $\sigma_r > \sigma_\theta$ an alternative solution arises; the two cases correspond to the active and passive Rankine states in Fig. 16.5.1.

As a simple example, suppose that $\phi = 60°$, so that $\alpha = 60°$ and $q = 3$. Then (18) gives

$$R/R_1 = [\tfrac{1}{2}(C_0 + 2p_0)/(C_0 + 2p_i)]^{\frac{1}{2}}, \tag{22}$$

and the stresses are

$$\sigma_r = (p_i + \tfrac{1}{2}C_0)(r/R_1)^2 - \tfrac{1}{2}C_0, \quad \sigma_\theta = 3(p_i + \tfrac{1}{2}C_0)(r/R_1)^2 - \tfrac{1}{2}C_0, \quad R_1 < r < R, \tag{23}$$

$$\sigma_r = p_0 - \tfrac{1}{2}(p_0 + \tfrac{1}{2}C_0)(R/r)^2, \quad \sigma_\theta = p_0 + \tfrac{1}{2}(p_0 + \tfrac{1}{2}C_0)(R/r)^2 \quad r > R. \tag{24}$$

Two simple cases of (22) through (24) are of interest. If $C_0 = 0$, zero cohesion, they become

$$R/R_1 = (p_0/2p_i)^{\frac{1}{2}}; \quad \sigma_r = p_i(r/R_1)^2, \sigma_\theta = 3p_i(r/R_1)^2, R_1 < r < R;$$
$$\sigma_r = p_0 - (p_0^2R_1^2/4r^2p_i), \sigma_\theta = p_0 + (p_0^2R_1^2/4r^2p_i); \quad r > R. \tag{25}$$

Also if $C_0 \neq 0$ but $p_i = 0$ they become

$$R/R_1 = [(C_0 + 2p_0)/2C_0]^{\frac{1}{2}}, \tag{26}$$

$$\sigma_r = \tfrac{1}{2}C_0[(r/R_1)^2 - 1], \quad \sigma_\theta = \tfrac{1}{2}C_0[3(r/R_1)^2 - 1], \quad R_1 < r < R, \tag{27}$$

and for $r > R_1$, σ_r and σ_θ are given by (24).

These results are well known from soil mechanics; Terzaghi (1943), Morrison and Coates (1955), and Talobre (1957) have suggested that they may be applicable to corresponding problems in rock mechanics. The validity of this assumption will be discussed later.

Hobbs (1966) has suggested the use of the criterion

$$\sigma_\theta - \sigma_r = D\sigma_r^b, \tag{28}$$

where D and b are constants, in place of the Coulomb criterion (1) and (4). Proceeding in the same way this gives

$$\frac{d\sigma_r}{\sigma_r^\theta} = D\frac{dr}{r}, \tag{29}$$

and

$$\frac{1}{1-b}\sigma_r^{1-b} = D \ln r + A, \tag{30}$$

in place of (11) and (12). The condition $\sigma_r = p_i$ when $r = R_1$ gives from (30)

$$\sigma_r = [D(1 - b) \ln (r/R_1) + p_i^{1-b}]^{1/(1-b)} \tag{31}$$

and from (28)

$$\sigma_\theta = [D(1-b)\ln(r/R_1) + p_i^{1-b}]^{1/(1-b)} + D[D(1-b)\ln(r/R_1) + p_i^{1-b}]^{b/(1-b)}. \quad (32)$$

For $r > R$ the stresses are given by (15) as before, and the continuity conditions at $r = R$ lead to

$$D[D(1-b)\ln(R/R_1) + p_i^{1-b}]^{b/(1-b)} + 2[D(1-b)\ln(R/R_1) + p_i^{1-b}]^{1/(1-b)} = 2p_0, \quad (33)$$

$$2BR^{-2} = D[(1-b)D\ln(R/R_1) + p_i^{1-b}]^{b/(1-b)}. \quad (34)$$

R can be found by solving (33) numerically and the solution completed. Hobbs (1966) has used this theory to study stresses around tunnels.

(iii) *The circular region $R_1 < r < R_2$ of Coulomb material with radial body force*

If there is constant radial body force P per unit mass the stress-equation of equilibrium, § 5.6 (9), is

$$\frac{d\sigma_r}{dr} + \frac{\sigma_r - \sigma_\theta}{r} + \rho P = 0, \quad (35)$$

in place of (9). Using the Coulomb relation, (10) thus becomes

$$\frac{d\sigma_r}{dr} - \frac{(q-1)}{r}\sigma_r = \frac{C_0}{r} - \rho P. \quad (36)$$

The solution of this is

$$\sigma_r = Ar^{q-1} - \frac{C_0}{q-1} + \frac{\rho P r}{q-2}, \quad (37)$$

where A is an unknown constant.

Now suppose that the boundary condition at $r = R_2$ is $\sigma_r = 0$. Using this, (37) becomes

$$\sigma_r = \left(\frac{C_0}{q-1} - \frac{\rho P R_2}{q-2}\right)\left(\frac{r}{R_2}\right)^{q-1} - \frac{C_0}{q-1} + \frac{\rho P r}{q-2}. \quad (38)$$

This gives the radial stress σ_r at any radius, r, necessary to maintain equilibrium. The assumed boundary condition $\sigma_r = 0$ when $r = R_2$ would arise if Coulomb material tended to separate from sound rock at $r = R_2$. This solution has been used as a very crude approximation to the effect of gravity on the roof of a horizontal tunnel surrounded by loose material.

(iv) *The combination of Coulomb material and solid rock around a circular opening of radius* R_1

The disadvantage of the preceding theory from the point of view of application to rock mechanics is that the whole of the material outside the opening is assumed to have the same properties specified by (4). A much more realistic assumption is that the region $r > R$, where, as before, R has to be determined, is solid rock in which the stresses at $r = R$ are sufficient to cause failure. The region $R_1 < r < R$ is supposed to contain broken rock, which behaves as a Coulomb material according to (4).

The criterion for failure of the solid rock may also be taken to be of the form § 4.6 (20), that is

$$\sigma_\theta = C_0' + q'\sigma_r, \tag{39}$$

where C_0' and q' are constants appropriate to the failure of solid rock and, in particular, C_0' is its uniaxial compressive strength of the solid rock. It will be assumed again that there is radial pressure p_i at $r = R_1$ and that the stresses at a great distance are hydrostatic and have the value p_0. The whole of the previous calculation, (9) through (16), remains valid, and the only change is that (17) is replaced by

$$p_0 + BR^{-2} = C_0' + q'(p_0 - BR^{-2}). \tag{40}$$

Solving (16) and (40) gives

$$\frac{R}{R_1} = \left\{\frac{(2p_0 - C_0')(1 - q) - C_0(1 + q')}{(1 + q')[p_i(1 - q) - C_0]}\right\}^{1/(q-1)} \tag{41}$$

$$B = R^2[p_0(q' - 1) + C_0']/(1 + q'). \tag{42}$$

This solution reduces to (18) and (19) if $C_0' = C_0$, $q' = q$, but the essential change is that $C_0' \gg C_0$. As a simple example suppose that $q' = q = 3$, $C_0 = 0$, then (41) and (42) become

$$R = R_1\{(2p_0 - C_0')/4p_i\}^{\frac{1}{2}}, \quad B = R^2(2p_0 + C_0')/4. \tag{43}$$

The stresses are

$$\sigma_r = p_i(r/R_1)^2, \quad \sigma_\theta = 3p_i(r/R_1)^2, \quad R_1 < r < R, \tag{44}$$

$$\sigma_r = p_0 - R^2(2p_0 + C_0')/4r^2, \quad \sigma_\theta = p_0 + R^2(2p_0 + C_0')/4r^2, \quad r > R. \tag{45}$$

In the absence of the Coulomb layer the rock would fail when $p_0 = C_0'/2$. Suppose that $p_0 = C_0'$, that is, twice this value. Then (43) gives

$$R = R_1(C_0'/4p_i)^{\frac{1}{2}} \tag{46}$$

so that if $p_i = C_0'/100$, $R = 5R_1$. Fig. 16.5.2 shows the variation of σ_r and σ_θ with r for this case. It shows the way in which a small radial stress at the surface is enhanced by the Coulomb material and also the discontinuity in σ_θ at the radius R at which the solid rock fails.

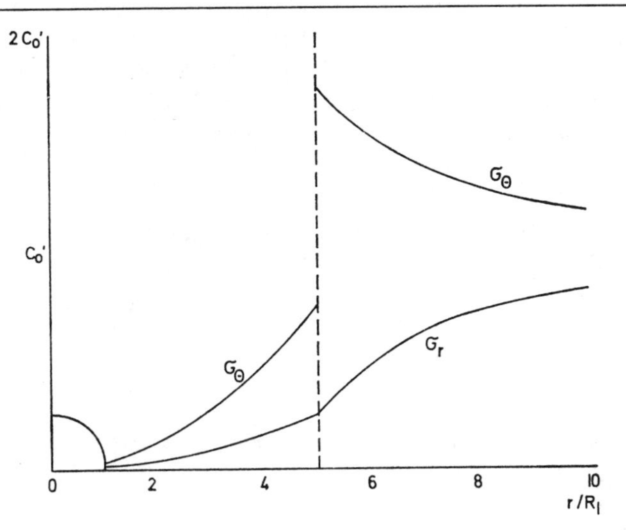

Fig. 16.5.2 Radial and tangential stresses around an annular opening of radius R_1 with Coulomb material in the interval $R_1 < r < R$ and elastic material with a uniaxial compressive strength C_0' for $r > R$.

(v) *Stress equations in two dimensions*

It is convenient to take the criterion (1) in the alternative form § 4.6 (23)

$$\tau_m = \sigma_m \sin \phi + S_0 \cos \phi = \bar{\sigma} \sin \phi, \tag{47}$$

where

$$\bar{\sigma} = \sigma_m + S_0 \cot \phi. \tag{48}$$

Then if σ_1 is inclined at β to O_x, Fig. 16.5.1 (*d*), is an x, y-coordinate system with O_x vertically downwards, by, § 2.3 (16), (17),

$$\sigma_x = \sigma_m + \tau_m \cos 2\beta = (\bar{\sigma} - S_0 \cot \phi) + \bar{\sigma} \sin \phi \cos 2\beta, \tag{49}$$

$$\sigma_y = \sigma_m - \tau_m \cos 2\beta = (\bar{\sigma} - S_0 \cot \phi) - \bar{\sigma} \sin \phi \cos 2\beta, \tag{50}$$

$$\tau_{x,y} = \tau_m \sin 2\beta \qquad = \bar{\sigma} \sin \phi \sin 2\beta. \tag{51}$$

Expressed in terms of $\bar{\sigma}$ and β, the equations of equilibrium § 5.5 (19), (20) become

$$(1 + \sin \phi \cos 2\beta) \frac{\partial \bar{\sigma}}{\partial x} + \sin \phi \sin 2\beta \frac{\partial \bar{\sigma}}{\partial y} - 2\bar{\sigma} \sin \phi \left[\frac{\partial \beta}{\partial x} \sin 2\beta - \frac{\partial \beta}{\partial y} \cos 2\beta\right] = \rho g, \tag{52}$$

$$\sin \phi \sin 2\beta \frac{\partial \bar{\sigma}}{\partial x} + (1 - \sin \phi \cos 2\beta) \frac{\partial \bar{\sigma}}{\partial y} + 2\bar{\sigma} \sin \phi \left[\frac{\partial \beta}{\partial x} \cos 2\beta + \frac{\partial \beta}{\partial y} \sin 2\beta\right] = 0, \tag{53}$$

where gravity, along O_x, is the only body force.

These equations give the directions and magnitudes of the principal stresses at any point. The slip-lines are in directions inclined at $\pm(\frac{1}{4}\pi - \frac{1}{2}\phi)$ to $O\sigma_1$, that is, they are inclined at $\beta + \frac{1}{4}\pi - \frac{1}{2}\phi = \theta$ and

$$\beta - \tfrac{1}{4}\pi - \tfrac{1}{2}\phi = \theta + \phi - \tfrac{1}{2}\pi \text{ to } O_x.$$

Rewriting (52) and (53) in terms of θ and the arc-lengths of the slip-lines leads to differential equations for the slip-lines.

(vi) *The infinite uniform slope in a cohesionless material*

Suppose that the slope (dotted in Fig. 16.5.1 (*d*)) is inclined at i to the horizontal and the origin is taken in it so that the equation of the surface is $x = y \tan i$. Since the slope is infinite and the material must be in limiting equilibrium at all points, the orientation of the principal axes must be the same at all points, that is, β must be constant. Using this and $S_0 = 0$ so that $\bar{\sigma} = \sigma_m$, (52) and (53) give

$$\frac{\partial \sigma_m}{\partial x} = \rho g(1 - \sin\phi \cos 2\beta) \sec^2\phi, \tag{54}$$

$$\frac{\partial \sigma_m}{\partial y} = -\rho g \sin\phi \sin 2\beta \sec^2\phi. \tag{55}$$

Interpreting gives:

$$\sigma_m = \rho g[x(1 - \sin\phi \cos 2\beta) - y \sin\phi \sin 2\beta] \sec^2\phi. \tag{56}$$

The vertical depth d of the point x, y below the surface is $d = x - y \tan i$, so that (56) becomes

$$\sigma_m = \rho g[d(1 - \sin\phi \cos 2\beta) + y \sec i\{\sin i - \sin\phi \sin(2\beta + i)\}] \sec^2\phi \tag{57}$$

For an infinite slope, σ_m must be independent of friction along the slope so that (57) must be independent of y which requires

$$\sin(2\beta + i) = \sin i \operatorname{cosec}\phi. \tag{58}$$

This determines β, and σ_m is given by

$$\sigma_m = gd(1 - \sin\phi \cos 2\beta) \sec^2\phi. \tag{59}$$

For a horizontal surface, $i = 0$, by (58) either $\beta = 0$ or $\beta = \frac{1}{2}\pi$. If $\beta = 0$

$$\sigma_m = \rho g d/(1 + \sin\phi), \quad \tau_m = \rho g d \sin\phi/(1 + \sin\phi),$$
$$\sigma_1 = \sigma_x = \rho g d, \quad \sigma_2 = \sigma_y = \cot^2(\pi/4 + \tfrac{1}{2}\phi). \tag{60}$$

This is the active Rankine state already found in (5). Similarly, $\beta = \frac{1}{2}\pi$ leads to the passive Rankine state.

For inclined slopes it follows from (58) that no solution exists if $i > \phi$, the angle of repose.

If $i = \phi$, (58) gives $\beta = \frac{1}{4}\pi - \frac{1}{2}\phi$, and by (59), (49), (50), (51),

$$\sigma_m = \rho g d, \quad \sigma_x = \rho g d(1 + \sin^2 \phi), \quad \sigma_y = \rho g d \cos^2 \phi, \quad \tau_{xy} = \rho g d \sin \phi \cos \phi. \tag{61}$$

The slip-lines, which are inclined at $\pm(\frac{1}{4}\pi - \frac{1}{2}\phi)$ to $0\sigma_1$ are vertical and parallel to the free surface.

In the general case $0 < i < \phi$, (58) has two solutions which may be written explicitly as

$$\cos 2\beta = [\sin^2 i \pm (\sin^2 \phi - \sin^2 i)^{\frac{1}{2}} \cos i] \operatorname{cosec} \phi. \tag{62}$$

The resultant horizontal force over a vertical strip of unit width and height H is

$$\int_0^H \sigma_y \, dx = \tfrac{1}{2}\rho g H^2 (1 - \sin \phi \cos 2\beta)^2 \sec^2 \phi. \tag{63}$$

(vii) *Coulomb's wedge approximation. Retaining walls*

In the active Rankine state for cohesionless material (5) the resultant horizontal force across a vertical strip OQ, Fig. 16.5.1 (*a*) of unit width and depth H below the surface is

$$\int_0^H \sigma_y \, dx = \tfrac{1}{2}\rho g H^2 \tan^2(\tfrac{1}{4}\pi - \tfrac{1}{2}\phi) \tag{64}$$

and its line of action is at depth $2H/3$ below the surface. This is the force which would have to be applied to a smooth vertical retaining wall OQ to maintain the region POQ in equilibrium. This solution is an accurate one in which the Coulomb criterion is satisfied at all points of the region POQ. In actual problems on retaining walls, the wall may not be vertical, may be rough, and the upper surface may not be horizontal. In such cases the solution of the equations of equilibrium with the Coulomb condition is much more complicated.

Coulomb (1773) proposed a simple approximate method which has been much used, particularly in soil mechanics, Lambe and Whitman (1969), and is applicable to many problems. Instead of assuming that there is limiting equilibrium at all points, he assumes that there is incipient slip, satisfying the Coulomb criterion, of a single (unknown) plane, the rest of the material moving as if solid. The portion of the plane is determined to make the force needed to prevent motion a maximum. As a specific example consider a smooth wall AB, Fig. 16.5.1 (*e*), of vertical height H and inclined at i to the horizontal supporting material ABC whose upper surface BC is inclined at δ to the horizontal. The assumption is that a wedge of material ABD will slide down the plane AD at an (as yet unknown) angle β to the horizontal. If T and N are the resultant tangential and normal forces across a unit width of this plane they will be connected by the law of friction.

$$T = lS_0 + N \tan \phi, \tag{65}$$

where S_0 is the cohesion, ϕ the angle of friction, and l is the length AD. If P is the normal force on the plane AB and W is the weight (per unit thickness) of the wedge ABD, resolving horizontally and vertically gives

$$P \sin i = N \sin \beta - T \cos \beta = N(\sin \beta - \cos \beta \tan \phi) - lS_0 \cos \beta, \tag{66}$$
$$P \cos i + W = N \cos \beta + T \sin \beta = N(\cos \beta + \sin \beta \tan \phi) + lS_0 \sin \beta. \tag{67}$$

Solving gives

$$P \sin(i - \beta + \phi) = W \sin(\beta - \phi) - lS_0 \cos \phi, \tag{68}$$

where $\quad l = H \operatorname{cosec} i \operatorname{cosec}(\beta - \delta) \sin(i - \delta), \quad (69)$

$$W = \tfrac{1}{2}\rho g H l \operatorname{cosec} i \sin(i - \beta). \tag{70}$$

Thus, finally,

$$P = \frac{\rho g H^2 \sin(i - \beta) \sin(i - \delta) \sin(\beta - \phi)}{2 \sin^2 i \sin(i - \beta + \phi) \sin(\beta - \delta)} - \frac{HS_0 \cos \phi \sin(i - \delta)}{\sin(i - \beta + \phi) \sin(\beta - \delta) \sin i}. \tag{71}$$

Plotting P as a function of β gives the maximum value of P and the corresponding value of β.

For the case $S_0 = 0$, $\delta = 0$, $i = \frac{1}{2}\pi$, (71) becomes

$$P = \tfrac{1}{2}\rho g H^2 \tan(\beta - \phi) \cot \beta. \tag{72}$$

This has a maximum value of $\frac{1}{2}\rho g H^2 \tan^2(\pi/4 - \frac{1}{2}\phi)$ when $\beta = \frac{1}{4}\pi + \frac{1}{2}\phi$ which is identical with the result (64). Thus, in this particular case, the method gives the accurate result, but in the general case this is not so.

This calculation may be extended to the case of a rough wall (Terzaghi, 1943) and to many other problems.

Chapter Seventeen

Geological and Geophysical Applications

17.1 Introduction

As remarked in Chapter 1, many of the problems of structural geology are similar to those of rock mechanics, except that they are on a larger scale. Over the years a number of theories of faulting, geological stress systems, the mechanics of intrusion, and so on have been developed for geological purposes, and all these might equally well have been developed in the context of rock mechanics. In particular, the theories of the mechanics of faulting relate back directly to the laboratory studies of shear fracture discussed in Chapter 4.

These questions will be discussed in this chapter. Also, a brief discussion of finite strain will be given to supplement the treatment of Chapter 2, which was confined to infinitesimal strain.

17.2 Faulting

All the phenomena of brittle fracture studied in Chapter 4 on a laboratory scale appear to be reproduced on a geological scale. Faults are geological fractures of rock in which there is relative displacement in the plane of fracture. They are thus shear fractures in the sense of § 4.5, and Griggs and Handin (1960) and others have used the term 'fault' for shear fractures both on the laboratory and geological scale.

The surface of fracture is the *fault plane*, and is specified by its dip and strike. If the relative motion in the fault plane is in the direction of the strike the fault is described as a *strike–slip* fault; if it is in the direction of the dip the fault is *dip–slip*; if in any other direction, *oblique–slip*.

In strike–slip faulting the observed fault planes are commonly nearly vertical. If the movement relative to an observer facing the fault plane is to the right it is described as *right-handed* or dextral; if it is to the left the fault is *left-handed* or sinistral. The terms 'wrench fault' or 'transcurrent fault' are also used for strike–slip faults.

A dip–slip fault is referred to as a *normal* fault if its dip is greater than 45°, and if the upper surface or *hanging wall* moves downwards relative to the lower surface or *foot wall*; if the hanging wall moves upwards the fault is described as a reverse fault. If the dip of the fault plane is less than 45°

and the hanging wall moves upwards the fault is a *thrust fault*. Thrust faults with very shallow dips of the order of 10° or less are called *overthrusts*.

These types of faulting were discussed by Anderson (1951) on the basis of the Coulomb and Mohr theories of shear fracture and classified according to the relative magnitudes of the principal stresses. On these theories, §§ 4.6, 4.7, fracture takes place in one or both of a pair of conjugate planes which pass through the direction of the intermediate principal stress and are equally inclined at angles of less than 45° to the direction of the greatest principal stress. Since the surface of the earth is a free surface, one of the principal stresses at the surface must be normal to it, so it is reasonable to assume that one principal stress is vertical at moderate depths. There are then three cases.

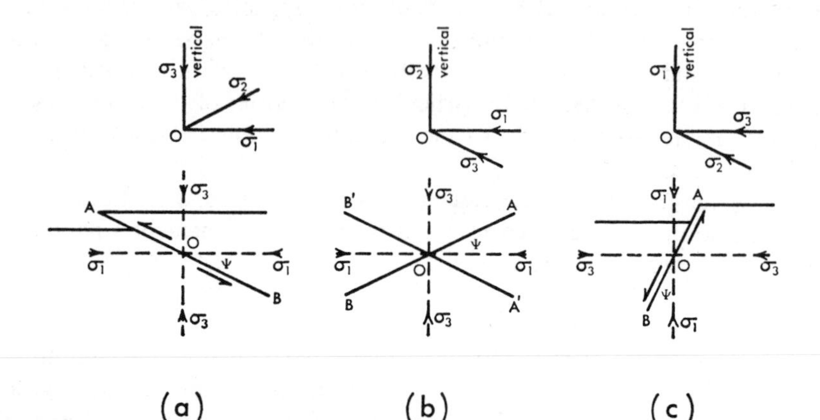

Fig. 17.2.1 (*a*) Thrust faulting. (*b*) Strike–slip faulting. (*c*) Normal faulting.

(i) *Thrust faulting*
If the least principal stress σ_3 is vertical, Fig. 17.2.1 (*a*), failure may take place on either of two planes such as *AOB* inclined at $\psi < 45°$ to the horizontal.

(ii) *Strike–slip faulting*
If the intermediate principal stress σ_2 is vertical, Fig. 17.2.1 (*b*), failure may take place on either of two vertical planes *AOB*, *A'OB'*, passing through $O\sigma_2$ and inclined at angles $\psi < 45°$ to $O\sigma_1$.

(iii) *Normal faulting*
If the greatest principal stress σ_1 is vertical, Fig. 17.2.1 (*c*), failure may take place on either of two planes such as *AOB* inclined at $\psi < 45°$ to the vertical.

The angle ψ between the fault plane and the direction of maximum principal stress is, by § 4.6 (17),

$$\psi = (\pi/2) - \alpha = (\pi/4) - (\phi/2), \tag{1}$$

where ϕ is the angle of internal friction. Since values of ϕ from 30° to 55° are commonly measured in laboratory experiments, values of ψ between 17° and 30° might be expected in the field. Measurements of ψ by Sax (1946) quoted by Hubbert (1951) give 25°–30° for normal faults and 20°–25° for thrust faults. Price (1962) obtained similar values. For strike–slip faults, values of ψ of the order of 30° are commonly measured, Moody and Hill (1956) and Williams (1959). There is thus quite good agreement on this simple theory between the laboratory and geological measurements.

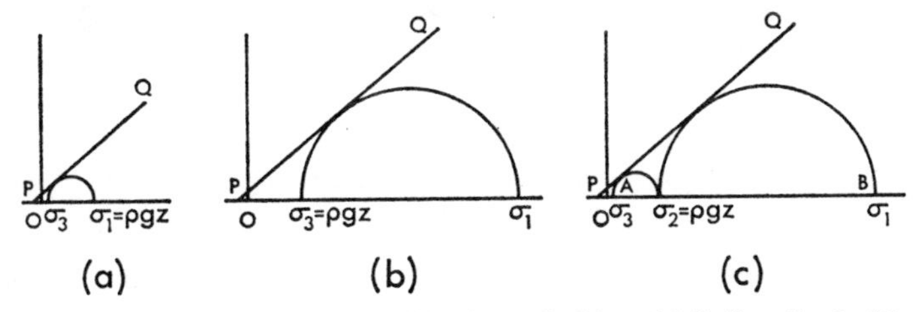

Fig. 17.2.2 (*a*) Normal faulting. (*b*) Thrust faulting. (*c*) Strike–slip faulting.

The above considerations depend on the relative values of the principal stresses, and are therefore much affected by the depth at which fracture is initiated. In the absence of other information the only simple assumption that can be made is that at depth z one principal stress is vertical and equal to ρgz. An informative discussion based on this assumption and the Mohr diagram, was given by Hubbert (1951). To take a definite example, suppose that the material is such that $S_0 = 150$ bars and the angle of internal friction ϕ is 40°. The Mohr envelope is then PQ of Fig. 17.2.2 and the criterion of failure is, by § 4.6 (18),

$$\sigma_1 = 640 + 4{\cdot}6\sigma_3. \tag{2}$$

For normal faulting to occur at depth z, $\sigma_1 = \rho gz$ is fixed and σ_3 has to be reduced until the Mohr circle on σ_1 and σ_3 touches PQ. Or, from (2) $\sigma_3 = (\rho gz - 640)/4{\cdot}6$. If ρgz is less than the uniaxial compressive strength, σ_3 must be negative (tensile). This suggests that horizontal stresses will be small in regions in which normal faulting occurs.

For thrust faulting, Fig. 17.2.2 (*b*), $\sigma_3 = \rho gz$ is fixed and σ_1 must be increased until the Mohr circle on σ_1 and σ_3 touches PQ, or, from (1) $\sigma_1 = 640 + 4{\cdot}6\rho gz$.

For strike–slip faulting ρgz is to be the intermediate principal stress, and the other principal stresses must lie between the extreme values A

and B in Fig. 17.2.2 (c). Here A corresponds to $\sigma_3 = (\rho gz - 640)/4{\cdot}6$, $\sigma_1 = \sigma_2 = \rho gz$ and B to $\sigma_2 = \sigma_3 = \rho gz$, $\sigma_1 = 640 + 4{\cdot}6\rho gz$.

Fig. 17.2.2 illustrates very clearly the variations in the principal stresses involved in the various types of faulting. Hubbert (1951) has shown that the phenomena of normal and thrust faulting may be illustrated by experiments on sand.

Anderson (1951) regards the standard state of stress at depth z as consisting of lithostatic stresses $\sigma_x = \sigma_y = \sigma_z = \rho gz$ and assumes that the stresses causing faulting are superposed on this state. Price (1959) has examined the effect of the assumption of no horizontal displacement, § 14.2 (3), in which

$$\sigma_x = \sigma_y = [\nu/(1 - \nu)]\rho gz, \quad \sigma_z = \rho gz. \tag{3}$$

If there is an additional tectonic stress σ_t along the x-axis the effect of this is approximately that of the system

$$\sigma_x = \sigma_t, \quad \varepsilon_y = 0, \quad \sigma_z = 0, \tag{4}$$

for which, by § 5.3 (3), $\sigma_y = \nu\sigma_t$. Adding these to (3) gives for the total stress system

$$\sigma_x = [\nu/(1 - \nu)]\rho gz + \sigma_t, \quad \sigma_y = [\nu/(1 - \nu)]\rho gz + \nu\sigma_t, \quad \sigma_z = \rho gz. \tag{5}$$

Certain deductions follow from this. For example, for strike–slip faulting it is necessary that $\sigma_x > \sigma_z > \sigma_y$, and this requires, by (5),

$$(1 - 2\nu)\rho gz > \nu(1 - \nu)\sigma_t > \nu(1 - 2\nu)\rho gz,$$

and using this in (5), it follows that

$$\sigma_x/\sigma_y < (1 - \nu)/\nu. \tag{6}$$

Since σ_x and σ_y are also connected by the criterion of failure § 4.6 (18) and from this $\sigma_x < \sigma_y \tan^2 \alpha$, it follows that σ_x and σ_y must lie in a rather restricted region for strike–slip faulting to occur. This may be a reason why strike–slip faulting is relatively rare. Alternatively, as pointed out in § 14.2, the assumption (3) has little status.

These simple considerations have proved remarkably successful in providing a simple physical background for geological faulting. They may be supplemented in various ways. First, on a global scale, stresses may be symmetrical about a point in the earth's crust, that is, vertical, radial, and tangential; in this case a similar theory can be stated, leading, in some cases to cones of fracture. Secondly, it has been suggested that in some cases stresses may be tensile, and in this case extension fractures are possible; some cases leading to these are studied in § 17.5. Thirdly, the effects of anisotropy have been neglected and these may be very important; it was remarked in § 4.10 that the effect of anisotropy on shear failure was to replace the two conjugate planes of fracture possible on the Coulomb and

Mohr theories by a single plane lying between the plane of minimum shear strength and the nearest of the two conjugate Coulomb–Mohr planes – this effect may be expected to appear on the geological scale also.

Next, some rather more complicated systems of faulting will be considered.

(iv) *Oblique–slip Faulting*

In this case it is necessary to study the direction of the shear stress in an arbitrary plane in greater detail. Suppose that OP, Fig. 17.2.3, is the normal to the plane and has direction cosines l, m, n relative to principal axes

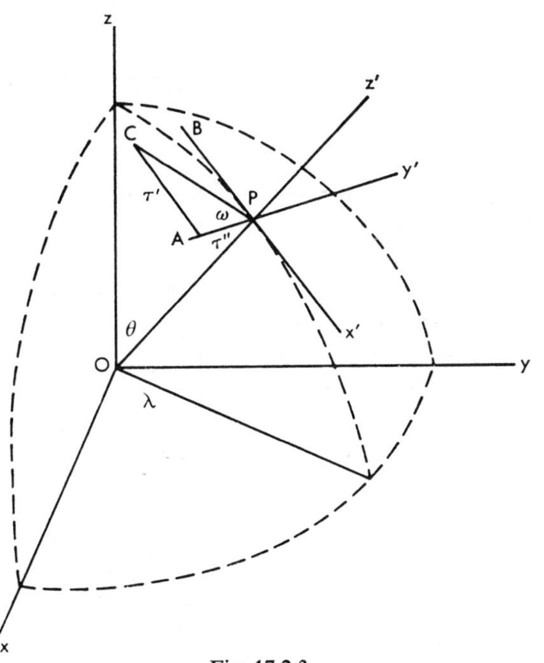

Fig. 17.2.3

$Oxyz$, and suppose that OP is also specified by its colatitude θ and longitude λ. The principal stresses will be σ_x, σ_y, σ_z, but their relative magnitudes will not be specified at this stage.

Taking new axes Pz' along OP, Px' in the plane zOP and Py' perpendicular, as in Fig. 17.2.3, the components of the stress across the plane $Px'y'$ have been found in § 2.5 (17)–(19). They are:

$$\sigma_{z'} = (\sigma_x \cos^2 \lambda + \sigma_y \sin^2 \lambda) \sin^2 \theta + \sigma_z \cos^2 \theta, \tag{7}$$

$$\tau' = \tau_{x'z'} = \tfrac{1}{2}(\sigma_x \cos^2 \lambda + \sigma_y \sin^2 \lambda - \sigma_z) \sin 2\theta, \tag{8}$$

$$\tau'' = \tau_{y'z'} = -\tfrac{1}{2}(\sigma_x - \sigma_y) \sin \theta \sin 2\lambda, \tag{9}$$

where, for shortness, τ' and τ'' have been written for $\tau_{x'z'}$ and $\tau_{y'z'}$, respectively. Introducing the direction cosines l, m, n of Oz' from § 2.5 (14), namely,

$$l = \sin\theta\cos\lambda, \quad m = \sin\theta\sin\lambda, \quad n = \cos\theta, \tag{10}$$

(9) and (8) become

$$\tau'' = (\sigma_y - \sigma_x)lm(1 - n^2)^{-\frac{1}{2}}, \tag{11}$$

$$\tau' = [m^2(\sigma_y - \sigma_x) - (1 - n^2)(\sigma_z - \sigma_x)]n(1 - n^2)^{-\frac{1}{2}}. \tag{12}$$

With the present convention of sign, τ'' and τ' are positive in the directions PA and PB, Fig. 17.2.3, and the angle ω which the resultant shear stress makes with PA is given by

$$\tan\omega = \frac{n}{lm}\left\{m^2 - (1 - n^2)\frac{\sigma_z - \sigma_x}{\sigma_y - \sigma_x}\right\}. \tag{13}$$

Suppose, now that σ_x, σ_y, σ_z are principal stresses in the crust, σ_z being vertical, and consider positive values of l, m, n. If faulting takes place under this system of stresses the direction of slip will be parallel to PC, and the system can be characterized by this direction. If $\sigma_y > \sigma_x$, $\tau'' > 0$, and the direction of slip is to the right, so the system is called *dextral*. If $\sigma_y < \sigma_x$, $\tau'' < 0$, and the system is *sinistral*.

If $\sigma_y > \sigma_x > \sigma_z$ it follows from (13) that $\tan\omega \to \infty$ if $\sigma_y \to \sigma_x$, and $\tan\omega \to nm/l$ if $\sigma_x \to \sigma_z$, so that

$$90° > \omega > \tan^{-1}(nm/l) \quad \text{if} \quad \sigma_y > \sigma_x > \sigma_z, \tag{14}$$

and the system is called *dextral thrust*.

Similarly,

$$\tan^{-1}(nm/l) > \omega > -\tan^{-1}(ln/m) \quad \text{if} \quad \sigma_y > \sigma_z > \sigma_x, \tag{15}$$

and the system is called *dextral wrench*.

Finally,

$$-\tan^{-1}(ln/m) > \omega > -90° \quad \text{if} \quad \sigma_z > \sigma_y > \sigma_x, \tag{16}$$

and the system is called *dextral gravity*.

Similarly, there are three sinistral cases and, allowing for the six cases in which two stresses are equal, Bott (1959) distinguishes twelve 'tectonic regimes' corresponding to various combinations of the tectonic stresses and the orientation of the plane.

Another useful concept, particularly relevant to strike–slip faulting, has been introduced by Lensen (1958). For pure strike–slip faulting the two conjugate fault planes intersect the earth's surface in traces which make an angle 2ψ with one another, Fig. 17.2.1 (*b*). If the σ_2-axis is not vertical and two traces are observed they will make an angle γ with one another which depends on the direction of the principal axes of stress. Here we shall only consider the symmetrical case in which the σ_1 and σ_2 axes lie in a vertical plane and $O\sigma_2$ dips at θ, Fig. 17.2.4. Here PA and PC will be the conjugate fault planes, and in the spherical triangle PAB, $AB = \frac{1}{2}\gamma$, $BP = \theta$, $APB = \psi$, $PBA = \frac{1}{2}\pi$. Writing δ for AP it follows from spherical trigonometry that

$$\cos \delta = \cos \theta \cos \tfrac{1}{2}\gamma, \tag{17}$$

$$\sin \delta = \operatorname{cosec} \psi \sin \tfrac{1}{2}\gamma. \tag{18}$$

Therefore

$$1 - \operatorname{cosec}^2 \psi \sin^2 \tfrac{1}{2}\gamma = \cos^2 \theta \cos^2 \tfrac{1}{2}\gamma,$$

$$\sin^2 \theta \cos^2 \tfrac{1}{2}\gamma = \sin^2 \tfrac{1}{2}\gamma \,[\operatorname{cosec}^2 \psi - 1],$$

or

$$\sin \theta = \pm\cot \psi \tan \tfrac{1}{2}\gamma. \tag{19}$$

If $\theta = 90°$, $\gamma = 2\psi$ for pure strike–slip faulting, and as $\theta \to 0$, $\gamma \to 0$. Thus for a given material γ is a measure of the dip of the σ_2-axis. The

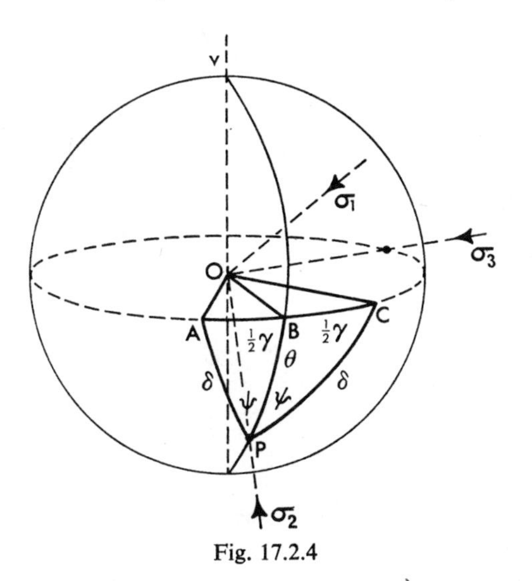

Fig. 17.2.4

general case in which the σ_3-axis is not horizontal is considered by Lensen (1958) and Scheidegger (1963). Lensen (1958) introduced the idea of measuring γ for all fault traces in a region and plotting contours of equal values of γ which he called isallostress curves. These curves show some correlation with geophysical anomalies, suggesting that the structure which gives rise to the anomalies also influences the stresses, Scheidegger (1966).

(v) *Secondary Faulting*

It is well known in the case of strike–slip faulting that off-shoot faults which make a small angle with the main fault are observed, Price (1966). It is frequently suggested that they are caused by failure under the altered system of stresses which will be present after the redistribution of stresses, which must take place after fracture on the main fault plane.

McKinstry (1953) considers the case of a slab of rock which is subjected only to normal stress p_n and a frictional shear stress μp_n, Fig. 17.2.5 (*a*), where μ is a coefficient of ordinary sliding friction on the plane Ox. In this case the principal axes are inclined at θ and $\frac{1}{2}\pi + \theta$ to Ox where, by § 2.3 (11),

$$\tan 2\theta = -2\mu, \tag{20}$$

and, by § 2.3 (14), the principal stresses are

$$\tfrac{1}{2}\{1 \pm (1 + 4\mu^2)^{\frac{1}{2}}\}p_n. \tag{21}$$

For example, if $\mu = 0{\cdot}5$, the inclinations of the principal axes are

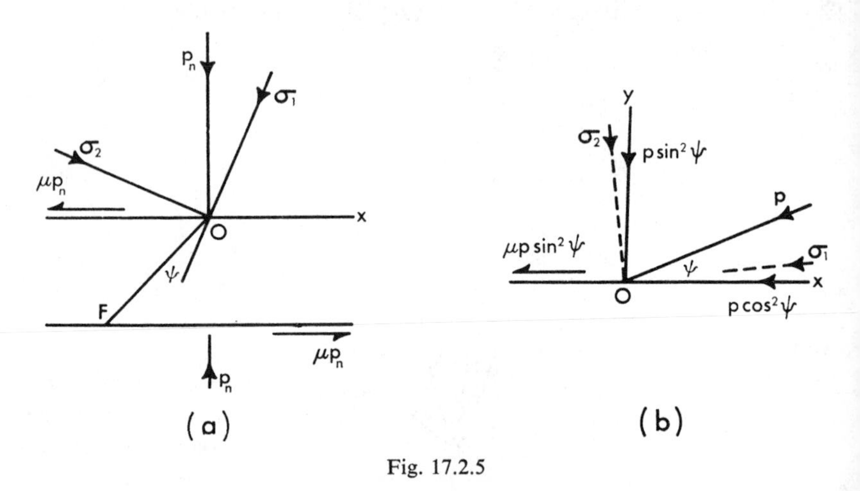

Fig. 17.2.5

$3\pi/8$ and $7\pi/8$ and the principal stresses are $1{\cdot}21p_n$ and $-0{\cdot}21p_n$. These directions are shown in Fig. 17.2.5 (*a*).

Since σ_2 is negative, the possibility of extension failure in a plane perpendicular to it cannot be ruled out. However, assuming that shear failure takes place according to the Coulomb criterion, it will, by § 4.6 (14), occur in a plane inclined at

$$\tfrac{1}{2} \tan^{-1} (1/\mu_s) \tag{22}$$

to $O\sigma_1$, where μ_s is the coefficient of internal friction of the solid material. If $\mu_s = 1$, this angle is $\pi/8$ and the direction of fracture is OF, Fig. 17.2.5.

Effects of this sort undoubtedly may occur. However, McKinstry (1953) and Moody and Hill (1956) have made further assumptions about the relief of stress caused by faulting. They suppose that the original fault is produced by uniaxial compression p at infinity, which must therefore be inclined at an angle $\psi = \frac{1}{2}\tan^{-1}(1/\mu_s)$ to it, so that before failure

$$\sigma_x = p\cos^2\psi, \quad \sigma_y = p\sin^2\psi, \quad \tau_{xy} = p\sin\psi\cos\psi, \tag{23}$$

Fig. 17.2.5 (*b*). After failure they assume that σ_x is reduced to zero and that τ_{xy} is replaced by μp_y, leading to the theory given earlier. This has been criticized by Chinnery (1966), who states that failure will only relieve the shear stress on the plane of fracture and possibly replace it by $\mu\sigma_y$, so that the stresses after failure will be

$$\sigma_x = p\cos^2\psi, \quad \sigma_y = p\sin^2\psi, \quad \tau_{xy} = \mu p\sin^2\psi. \tag{24}$$

The directions of the principal axes are now given by

$$\tan 2\theta = \frac{2\mu\sin^2\psi}{\cos 2\psi} = \mu[(1+\mu_s^2)^{\frac{1}{2}} - \mu_s]/\mu_s. \tag{25}$$

If the previous values of $\mu = 0{\cdot}5$, $\mu_s = 1$ are used this gives $\theta = 6°$ or $96°$, and the principal axes are $O\sigma_1$ and $O\sigma_2$, shown dotted in Fig. 17.2.5 (*b*), and are quite different from those shown in Fig. 17.2.5 (*a*). It is clear that the discrepancy arises from lack of knowledge of the way in which tectonic stress redistributes itself after fracture.

(vi) *The modification of stress produced by faulting*

As remarked in (v), many attempts have been made to explain the secondary faulting associated, in particular, with strike–slip faulting in terms of a redistribution of stresses. Clearly, for a satisfactory explanation, the stress-distribution around the region containing the master fault has to be studied in detail. The simplest representation of a master fault is the two-dimensional one of a flat elliptic crack, and the use of this implies that the fault extends downwards to a level of 'detachment' such that the material below exerts no influence on it.

The problem then becomes that of the flat elliptic crack in a two-dimensional stress-field, which is treated in § 10.11. For the simplest case of the crack subjected to pure shear S the stresses are given by § 10.11 (15)–(17), and in the surface $\xi = 0$ they are

$$\sigma_\xi = \tau_{\xi\eta} = 0, \quad \sigma_\eta = -\frac{2S\sin 2\eta}{1-\cos 2\eta} = -2S\cot\eta. \tag{26}$$

It follows that σ_η is compressive in the second and fourth quadrants and its magnitude is

$$2S\,|\,x\,|\,/(c^2 - x^2)^{\frac{1}{2}}, \tag{27}$$

and in the first and third quadrants it is tensile and of this magnitude. In both cases (27) becomes infinite as $x \to \pm c$, so that large stresses are attained.

The compressive stresses in the second and fourth quadrants can cause shear failure along planes AB, $A'B'$, Fig. 17.2.6, inclined at ψ given by (1) to the direction of the crack and initiated at points A, A' near the highly stressed ends, P, P' of the crack. This explanation of 'splay' fracturing has been given by Anderson (1951). He discusses the stress-trajectories of the system more fully.

Chinnery (1966) gives an entirely different discussion based on the dislocation model described in § 17.3, Fig. 17.3.3. He calculates stress trajectories for the case $d = 0$, $D = 10$ km, $L = 100$ km, and a relative

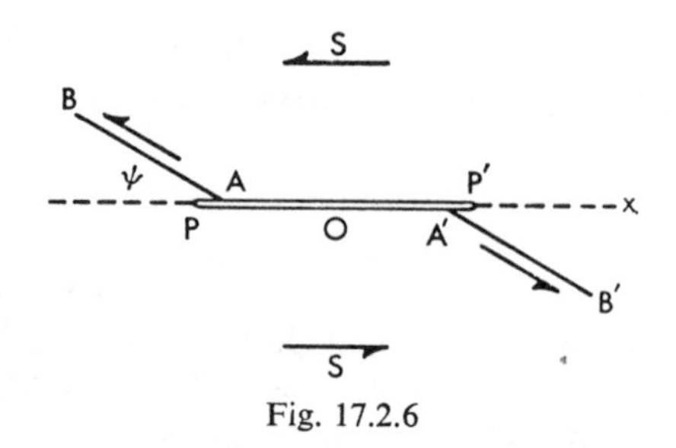

Fig. 17.2.6

surface displacement of 5m corresponding to a large fault. From these, he determines possible directions of faulting.

The contrast between these two discussions indicates the difficulties of the subject. Anderson's discussion assumes 'detachment' of the faulted region from the material below it, while in the dislocation theory stresses transmitted elastically from this region have an important effect.

17.3 Geological stress systems

Frequent mention has been made earlier of tectonic stresses, both those actually measured and those postulated to explain geological phenomena. It is therefore natural to study the variation of stress in simple systems and to see how the various criteria of failure would operate. Because of our fundamental lack of knowledge of actual conditions, there is no point in considering other than very simple situations for which mathematical solutions can easily be found. The general procedure is to specify a system of stresses or displacements on the boundaries of a (usually rectangular) region and to calculate the stresses and stress trajectories on elastic theory and also the directions in which failure would take place on some simple criterion of failure.

(i) *Hafner's discussion*

Hafner (1951) used simple polynomial solutions of the stress-equations of equilibrium § 5.5 (5)–(7). For example, the stresses

$$\sigma_x = \rho g x, \quad \sigma_y = p - ky + \rho g x, \quad \tau_{xy} = kx, \tag{1}$$

satisfy them with body force $X = -g$ corresponding to gravity. Consider this system applied to the rectangle $OABC$, Fig. 17.3.1, of length $OA = l$ and depth $OC = h$. The x-axis is vertically downwards and the y-axis horizontal. On OC, $y = 0$, $\sigma_x = \rho g x$, $\sigma_y = p + \rho g x$, $\tau_{xy} = kx$; apart from the shear stress kx, this corresponds to the lithostatic stress $\rho g x$ augmented by a uniform horizontal stress p. The length l of the block is chosen to be $l = p/k$, so that on AB, $y = l$, $\sigma_x = \sigma_y = \rho g x$, $\tau_{xy} = kx$, and the horizontal stress p has disappeared. On CB, $x = h$, $\sigma_x = \rho g h$, $\sigma_y = p + \rho g h - ky$, $\tau_{xy} = kh$. The stresses on OC, CB, and BA keep the block in static equilibrium.

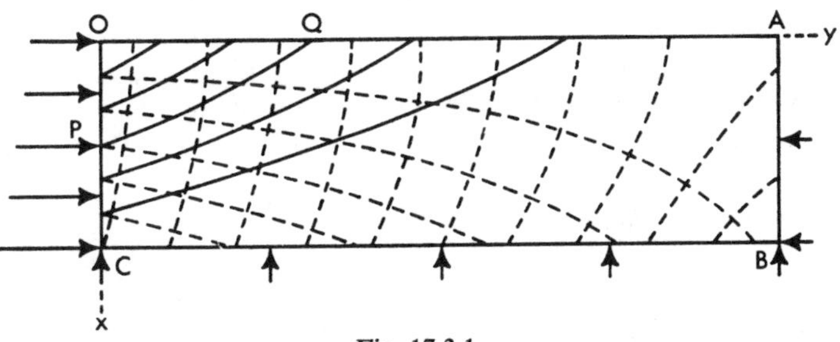

Fig. 17.3.1

By § 2.3 (11) the principal axes at x, y are inclined at θ to Ox, where

$$\tan 2\theta = \frac{2kx}{ky - p} = \frac{2x}{y - l}, \tag{2}$$

using the value $p = lk$. On $x = 0$, $\theta = 0$, so that the principal axes are horizontal and vertical. On $y = 0$, $\tan 2\theta = -2x/l$, so that they are rotated by an increasing amount in a clock-wise direction. The stress-trajectories are shown by the dotted lines in Fig. 17.3.1.

To determine whether and where failure will occur, the stresses (1) must be substituted in a failure criterion such as the Coulomb criterion § 4.6 (23), which becomes

$$[4k^2x^2 + (p - ky)^2]^{\frac{1}{2}} = (p - ky + 2\rho g x) \sin \phi + 2S_0 \cos \phi, \tag{3}$$

and failure will take place in a direction inclined at $(\pi/4) - (\phi/2)$ to the greatest principal stress. Once the criterion (3) is satisfied at any point, fracture will take place at this point, and the solution given above will cease to be valid. Nevertheless, it is interesting to proceed as if it did not and to draw a family of curves intersecting the trajectories of the greater principal stress at $(\pi/4) - (\phi/2)$. These are drawn in Fig. 17.3.1 for $\phi = 30°$, and may be regarded as directions of faulting, a typical curve being PQ. Despite the fact that these curves have no theoretical status, they do correspond quite well with observation, both on the geological scale and in experimental observations. Hubbert (1951) has made an extensive series of experiments, in which the system of Fig. 17.3.1 (without the gravity field) has been reproduced by compressing sand or other materials in a 'sand box'. Thrusting along curves such as PQ is observed.

A full theoretical discussion of Hubbert's experiments would involve studying the corresponding problem for a granular medium, but the qualitative similarity between the results may be expected from the facts that in both cases the stress-equations of equilibrium form part of the formulation of the problem and that the slip lines are related to the stress trajectories. A careful discussion of the matter is given by Odé (1960).

In conclusion, it may be added that Hafner gives several more complicated examples, and that model experiments of the type referred to briefly are a very useful qualitative method of studying tectonic effects. The force of gravity may be simulated by the use of a centrifuge, Ramberg (1963). The theory of scaling in models is discussed by Hubbert (1937, 1945).

(ii) *Sanford's discussion*

Sanford (1959) has studied a rather different class of problem, namely the stresses in a uniform upper layer caused by movements of the basement. He, again, obtains an exact solution on elastic theory for his specified boundary conditions and determines stresses and stress trajectories. He also makes 'sand-box' experiments and obtains a good correlation between theory and experiment.

He also chooses, in most cases, to specify displacements on some boundaries, and this makes the calculation more sophisticated than Hafner's. If only stresses are in question, a wide variety of problems can be solved very easily by studying simple solutions (e.g. polynomials) of the biharmonic equation § 5.7 (9). If displacements are needed it is better to use the complex variable method of § 10.2.

Sanford considers the slab $-\infty < x < \infty$, $0 < y < h$ with one free surface and seeks solutions which are periodic in x with wavelength $2\pi/\omega$. In particular, if the displacements are of this form, namely,

$$u = f(y) \cos \omega x, \quad v = g(y) \sin \omega x, \tag{4}$$

the solution will be suitable for the study of stresses and displacements in a layer caused by uplifting or sinking of the basement. Any form of displacement may be studied by the use of a Fourier series.

We now discuss the choice of the functions $\phi(z)$, $\psi(z)$ of § 10.2 to give solutions of the required type. Clearly, choosing

$$\phi(z) = Ae^{i\omega z} + Be^{-i\omega z}, \tag{5}$$

where A and B are real constants, will do so. Also, since $z\overline{\phi'(z)}$ occurs in § 10.2 (50), it is to be expected that $\psi(z)$ will contain terms of type

$z \exp(\pm i\omega z)$. However, there is a restriction on these, since § 10.2 (50) involves the conjugate of

$$\bar{z}\phi'(z) + \psi(z) = (x - iy)[i\omega A e^{i\omega z} - i\omega B e^{-i\omega z}] + \psi(z), \tag{6}$$

and if all quantities are to have the form (4) the terms in x in (6) must disappear. This requires that $\psi(z)$ should contain a term $-z\phi'(z)$, and since, for generality, terms in $\exp(\pm i\omega z)$ may be added it follows that the required form of $\psi(z)$ is

$$\psi(z) = -i\omega z A e^{i\omega z} + i\omega B z e^{-i\omega z} + C e^{i\omega z} + D e^{-i\omega z}, \tag{7}$$

where C and D are real constants.

The displacements and stresses then follow from § 10.2 (48)–(50). They are

$$2Gu = \{[A(\varkappa - 2\omega y) - C]e^{-\omega y} + [B(\varkappa + 2\omega y) - D]e^{\omega y}\} \cos \omega x, \tag{8}$$

$$2Gv = \{[A(\varkappa + 2\omega y) + C]e^{-\omega y} + [B(-\varkappa + 2\omega y) - D]e^{\omega y}\} \sin \omega x, \tag{9}$$

$$\sigma_x + \sigma_y = -4\omega[Ae^{-\omega y} + Be^{\omega y}] \sin \omega x, \tag{10}$$

$$\sigma_x = \{\omega[(2\omega y - 3)A + C]e^{-\omega y} + \omega[D - (2\omega y + 3)B]e^{\omega y}\} \sin \omega x, \tag{11}$$

$$\sigma_y = \{-\omega[(2\omega y + 1)A + C]e^{-\omega y} + \omega[(2\omega y - 1)B - D]e^{\omega y}\} \sin \omega x, \tag{12}$$

$$\tau_{xy} = \{\omega[(2\omega y - 1)A + C]e^{-\omega y} + \omega[(2\omega y + 1)B - D]e^{\omega y}\} \cos \omega x. \tag{13}$$

These are the required general solutions of this type.

Suppose, for example, that $y = 0$ is a free surface and that the displacements are prescribed on the surface $y = h$ so that the boundary conditions are

$$\sigma_y = \tau_{xy} = 0, \quad y = 0, \tag{14}$$

$$u = u_0 \cos \omega x, \quad v = v_0 \sin \omega x, \quad y = h. \tag{15}$$

Then (12), (13), (14) give

$$D = -A, \quad C = -B, \tag{16}$$

and (8), (9), (15), (16) give finally

$$2Gu_0 = [A(\varkappa - 2\omega h) + B]e^{-\omega h} + [A + B(\varkappa + 2\omega h)]e^{\omega h}, \tag{17}$$

$$2Gv_0 = [A(\varkappa + 2\omega h) - B]e^{-\omega h} + [A - B(\varkappa - 2\omega h)]e^{\omega h}, \tag{18}$$

so that A, B, C, D are determined and the solution is complete.

This solution corresponds to a 'welded' contact at the basement $y = h$.

Sanford also considered the case in which only the vertical displacements are prescribed and movement is free in the horizontal direction, that is, (15) is replaced by

$$\tau_{xy} = 0, \quad v = v_0 \sin \omega x. \tag{19}$$

In this case (16) and (18) still hold, and (17) is replaced by

$$[(2\omega h - 1)A - B]e^{-\omega h} + [(2\omega h + 1)B + A]e^{\omega h} = 0. \tag{20}$$

Sanford considers a number of cases numerically. For example, for a half wavelength $\pi/\omega = 25$ km, $h = 5$ km, $u_0 = 0$, $v_0 = 8{\cdot}2$ m, $\nu = 0{\cdot}25$, $G = 2 \times 10^{-11}$ dynes/cm^2, he finds $\sigma_x = 1{,}260$, $\sigma_y = 1{,}280$ kg/cm^2 at the point B, Fig. 17.3.2. In this figure the displacements v on the surface

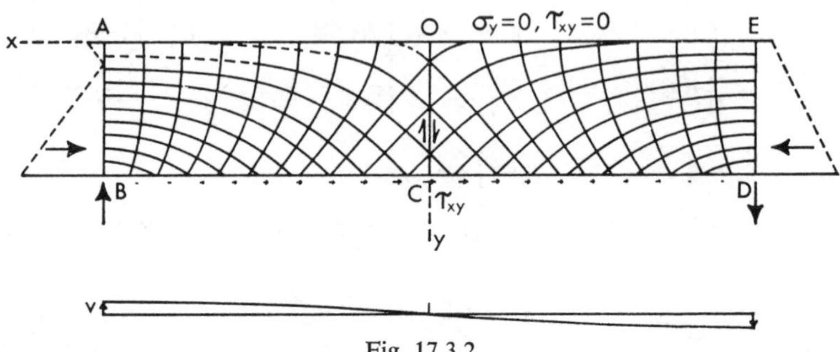

Fig. 17.3.2

BCD are shown below, and the stress-trajectories are also shown. In the region near A one principal stress becomes tensile and the appropriate portion of the stress-trajectory is shown dotted. He also gives numerical results for other cases, including prescribed stress on $y = h$. By the use of a Fourier series he obtains a solution for a step-function displacement at $y = h$. Instead of plotting directions of fracture as Hafner does, he gives contours of the strain energy of distortion to indicate the region in which failure is most likely to occur.

(iii) *Stresses around faults*

In a Volterra dislocation in an elastic solid a cut is made across a surface in the solid, a shearing displacement is made in this surface, and the two sides of it are welded together. The displacements and stresses in the region can then be calculated.

Chinnery (1961, 1966) following Steketee (1958) considers the result of strike–slip faulting to be the introduction of a dislocation in the fault plane. He takes the dislocation surface to be the region, Fig. 17.3.3 (*a*),

$$-L < x < L, \quad -(D + d) < z < -d, \quad y = 0 \tag{21}$$

of the semi-infinite solid $z < 0$ with $z = 0$ as a free surface. The amount of the dislocation, that is the relative displacement of the two surfaces, is U, constant over the area.

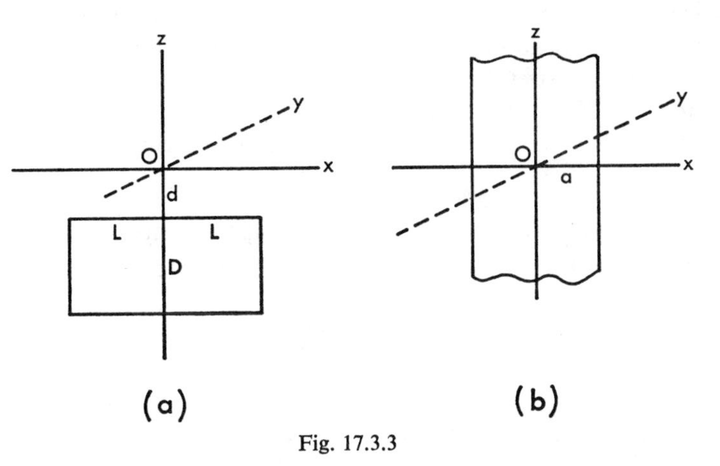

Fig. 17.3.3

For the general case, computer solution is necessary, but for a very long dislocation, $L = \infty$, the result takes the simple form

$$u = \frac{U}{2\pi}\left\{\tan^{-1}\frac{D+z}{y} + \tan^{-1}\frac{D-z}{y} - \tan^{-1}\frac{d+z}{y} - \tan^{-1}\frac{d-z}{y}\right\}, \tag{22}$$

$$v = w = 0. \tag{23}$$

It follows that

$$\sigma_x = \sigma_y = \sigma_z = \tau_{yz} = 0, \tag{24}$$

$$\tau_{xy} = \frac{GU}{2\pi}\left\{\frac{d+z}{y^2+(d+z)^2} + \frac{d-z}{y^2+(d-z)^2} - \frac{D+z}{y^2+(D+z)^2} - \frac{D-z}{y^2+(D-z)^2}\right\}, \tag{25}$$

$$\tau_{xz} = \frac{GUy}{2\pi}\left\{\frac{1}{y^2+(D+z)^2} - \frac{1}{y^2+(D-z)^2} - \frac{1}{y^2+(d+z)^2} + \frac{1}{y^2+(d-z)^2}\right\}. \tag{26}$$

When $z = 0$, $\tau_{xz} = 0$, as it should be for a free surface, and

$$u = \frac{U}{\pi}\left\{\tan^{-1}\frac{D}{y} - \tan^{-1}\frac{d}{y}\right\}, \tag{27}$$

$$\tau_{xy} = \frac{GU}{\pi}\left\{\frac{d}{d^2+y^2} - \frac{D}{D^2+y^2}\right\}. \tag{28}$$

These show the way in which effects die away with distance. Chinnery gives many figures calculated for the case of finite length L. These results may be important in connection with the stresses around faults, since any movement which is subsequently locked, for example, by friction, corresponds to a dislocation and will produce a stress field independent of the externally applied stresses.

Knopoff (1958) has used another system for which a very simple theory is available. He considers an infinite medium with a cut in the region

$$y = 0, \quad |x| < a, \quad -\infty < z < \infty, \tag{29}$$

giving an infinitely long slit, Fig. 17.3.3 (b), which may be taken as a model of a strike–slip fault. His solution, which corresponds to displacements along the fault, is

$$u = v = 0, \quad w = A_0\mathbf{I}[(x + iy)^2 - a^2]^{\frac{1}{2}}. \tag{30}$$

In the plane of the fault, $y = 0$, this gives

$$w = 0, \; |x| > a; \quad w = \pm A_0(a^2 - x^2)^{\frac{1}{2}}, \; |x| < a, \tag{31}$$

where A_0 is a constant, so that the relative movement across the fault varies from 0 at $x = a$ to $2A_0a$ when $x = 0$. In the plane $x = 0$ the displacement is $\pm A_0(y^2 + a^2)^{\frac{1}{2}}$.

The non-vanishing components of strain are

$$\Gamma_{yz} = \tfrac{1}{2}A_0\mathbf{R}\{(x + iy)[(x + iy)^2 - a^2]^{-\frac{1}{2}}\}, \tag{32}$$

$$\Gamma_{zx} = \tfrac{1}{2}A_0\mathbf{I}\{(x + iy)[(x + iy)^2 - a^2]^{-\frac{1}{2}}\}. \tag{33}$$

It appears that $\varepsilon_{zx} = 0$ when $x = 0$, that is, there is no stress on the surface $x = 0$, which may thus be taken to be a free surface. The solution is therefore applicable to a vertical strike–slip fault of depth a with relative displacement of $2A_0a$ at the surface.

17.4 Overthrust faulting and sliding under gravity

Hubbert and Rubey (1959) have stressed the importance of pore-fluid pressure in the mechanics of sliding of large masses of rock with particular reference to low-angle overthrust faulting.

Consider the sliding of the block $OABC$ of height h and length l on the horizontal plane CB at which the coefficient of sliding friction is μ. The origin is taken at O, Fig. 17.4.1 (a), and the x-axis vertically downwards. The bulk density of the fluid-filled rock is supposed to be ρ, so that the vertical stress σ_x at depth x is $\sigma_x = \rho g x$. The pore-pressure p at depth x is taken to be $p = \lambda\sigma_x = \lambda\rho g x$, where λ may be a function of x, though it will usually be assumed to be constant. If $\sigma_x' = \sigma_x - p$ is the effective stress, we thus have

$$\sigma_x = \rho g x, \quad \sigma_x' = (1 - \lambda)\rho g x, \quad p = \lambda\rho g x. \tag{1}$$

The block is now supposed to be pushed to the right by stresses σ_y applied on OC, which are taken to be the maximum possible, so that the material is on the point of failure at every point of OC and the problem is to determine the greatest length l which can be moved in this way. The criterion of failure will be taken to be § 4.6 (20), except that, since pore pressure is present, it must be written in terms of effective stresses, as in § 8.9 (1), so that it becomes

$$\sigma_y' = C_0 + q\sigma_x'. \tag{2}$$

It follows that

$$\sigma_y = C_0 + \rho g x\{q + \lambda(1 - q)\}. \tag{3}$$

The condition for sliding on BC, again involving effective stresses, is § 8.9 (2), namely

$$\tau_{xy} = \mu(1 - \lambda_1)\rho g h + S_0, \tag{4}$$

where λ_1 is the value of λ at $x = h$, S_0 is the inherent shear strength of the surface, and μ its coefficient of friction.

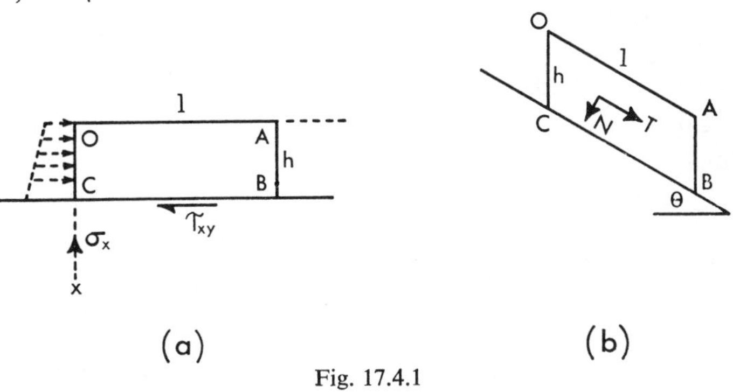

Fig. 17.4.1

For the system to be in equilibrium, the resultant force on the block in the y-direction must be zero, that is

$$\int_0^h \sigma_y \, dx = \int_0^l \tau_{xy} \, dy. \tag{5}$$

Using the values (3) and (4), this becomes

$$hC_0 + \tfrac{1}{2}\rho g q h^2 + \rho g(1 - q)\int_0^h \lambda x \, dx = \mu(1 - \lambda_1)\rho g h l + S_0 l. \tag{6}$$

Taking λ constant for simplicity, this gives

$$l = \frac{2hC_0 + \rho g h^2[q + \lambda(1 - q)]}{2S_0 + 2\mu(1 - \lambda_1)\rho g h}. \tag{7}$$

λ should be at least 1/3, and Hubbert and Rubey quote measurements to show that it may be of the order of 0·85 or more. They show numerically that values of l calculated from (7) are an order of magnitude larger than those in the absence of pore-pressure.

A second case considered is that of the sliding of a block $OABC$, Fig. 17.4.1 (b), of length l and vertical depth h under gravity down a surface CB of slope θ, coefficient of sliding friction μ, and inherent shear strength S_0. If ρ is the bulk density of the water-saturated material, the resultant force T on the block down the slope is

$$T = \rho glh \sin\theta \cos\theta, \tag{8}$$

and the normal force is

$$N = \rho glh \cos^2\theta. \tag{9}$$

The criterion for sliding, including the effects of pore-pressure, is taken to be

$$T = (1 - \lambda_1)\mu N + S_0 l, \tag{10}$$

or

$$\rho glh \sin\theta \cos\theta = (1 - \lambda_1)\rho ghl\mu \cos^2\theta + S_0 l. \tag{11}$$

If $S_0 = 0$, this becomes

$$\tan\theta = (1 - \lambda_1)\mu = (1 - \lambda_1)\tan\phi, \tag{12}$$

where ϕ is the angle of friction. It follows that if $\lambda_1 \neq 0$ sliding is possible at angles smaller than the angle of friction.

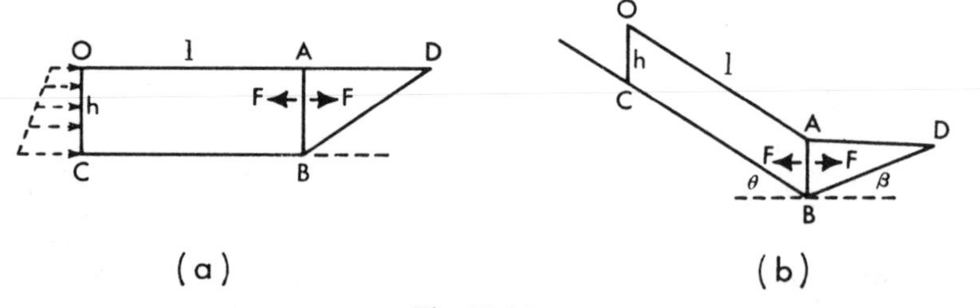

Fig. 17.4.2

This simple discussion, while clearly correct in principle, has been criticized in matters of detail by Birch (1961). First, Hubbert and Rubey take $S_0 = 0$, although its finite value (which may be of the order of 200 bars) may be of importance in this context. This objection is to some extent offset by the fact that sliding may well be on a layer of low S_0; also this layer may well be impervious, so that the effects of pore-pressure are confined to the equation of sliding (4) and need not be considered in the criterion of failure (2). The other obvious oversimplification is the assumed vertical face AB at the free end of the slab. Raleigh and Griggs (1963) have considered two slightly more complicated models to overcome this.

In the first, Fig. 17.4.2 (a), they assume that the end of the slab, after sliding on CB, is pushed up a face BD inclined at β to the horizontal and that material which moves upwards is removed by erosion, so that the surface *OAD* remains flat. The two regions *OABC* and *ABD* are then considered separately. Let the resultant horizontal force across the boundary *AB* be *F*. Then including this, and taking λ to be constant, (6) for the region *OABC* is replaced by

$$hC_0 + \tfrac{1}{2}\rho g h^2[q + \lambda(1 - q)] = \mu(1 - \lambda_1)\rho g h l + S_0 l + F. \qquad (13)$$

Next, considering the sliding of the block *ABD* up the plane *BD*, the total force *T* on the block parallel to the plane is

$$T = F \cos \beta - \tfrac{1}{2}\rho g h^2 \cot \beta \sin \beta,$$

and the force *N* normal to the plane is

$$N = F \sin \beta + \tfrac{1}{2}\rho g h^2 \cot \beta \cos \beta.$$

The condition for sliding, including the effects of pore-pressure, is taken to be

$$T = (1 - \lambda_2)\mu N + S_0 h \operatorname{cosec} \beta, \qquad (14)$$

which gives

$$2F[\cos \beta - (1 - \lambda_2)\mu \sin \beta] = \{\rho g h^2[\tan \beta + (1 - \lambda_2)\mu] \cos^2 \beta + 2S_0 h\} \operatorname{cosec} \beta. \qquad (15)$$

Eliminating *F* from (13) and (15) gives *l*. Using typical values, in particular $\phi = \beta = 30°$, Raleigh and Griggs find that the effect of the toe is to reduce the total length of the thrust block from the simple value (7) by an amount of 10–20 per cent.

Raleigh and Griggs also consider the case in which a block *OABC*, Fig. 17.4.2 (*b*), slides down a slope *CB* inclined at θ to the horizontal to meet a slope *BD* inclined at β. As before, rising material is supposed to be removed by erosion. Assuming, again, a horizontal resultant force *F* acting over *AB*, the condition for sliding of the block *ABD* is again (15). Also for the block *ABCD* the conditions are found by replacing *T* and *N* in (8) and (9) by $T - F \cos \theta$ and $N + F \sin \theta$ so that (11) is replaced by

$$\rho g l h \sin \theta \cos \theta - F \cos \theta = (1 - \lambda_1)\mu[\rho g h l \cos^2 \theta + F \sin \theta] + l S_0. \qquad (16)$$

Eliminating *F* from (15) and (16) gives the required solution.

Raleigh and Griggs neglect S_0 in the above discussion, also a further simplification is possible if it is desired to find the least angle θ down which sliding is possible; since this is expected to be small, $\sin \theta$ may be replaced by θ, and (15) and (16) give with this approximation and $S_0 = 0$

$$[1 + (1 - \lambda_1{}^2)\mu]\theta = (1 - \lambda_1)\mu + \frac{h[\tan \beta + (1 - \lambda_2)\mu] \tan \beta}{2l[1 - (1 - \lambda_2)\mu \tan \beta]}. \qquad (17)$$

For example, suppose that $\beta = \phi = 30°$, $\lambda_1 = \lambda_2 = 0{\cdot}85$, then

$$\theta = 4{\cdot}9[1 + 7(h/l)] \text{ degrees.} \tag{18}$$

Thus under these conditions, even allowing for the effect of the toe, a long block can slide down a slope of 5° in place of the 30° in the absence of pore-pressure.

17.5 The mechanics of intrusion

The primary mechanism of igneous intrusion is thought to be tensile failure under stresses caused by the pressure of magma, and several models for this process have been studied.

(i) *Sheet intrusions*

The obvious model for this case is the flat elliptical crack filled with magma at pressure p. This has been discussed in connection with the Griffith criterion for a crack with internal pressure, and the condition for failure, § 10.13 (25), is

$$\sigma_3 - p + T_0 = 0, \quad \text{provided} \quad \sigma_1 + 3\sigma_3 < 4p. \tag{1}$$

Vertical dykes are frequently associated with normal faulting, so that it may be assumed that they have been formed under conditions in which the horizontal principal stress σ_h is less than the vertical stress ρgz. In this case (1) becomes

$$p = \sigma_h + T_0, \tag{2}$$

provided $3\sigma_h + \rho gz < 4p$, and this condition is likely to be satisfied, since p may be expected to be of the order of ρgz.

For the formation of a horizontal sheet on the present assumptions it is necessary that $\sigma_h > \rho gz$, and (1) gives

$$p = \rho gz + T_0, \tag{3}$$

provided that $\sigma_h + 3\rho gz < 4p$.

Anderson (1937) has given a discussion based on the formulae § 10.12 (15)–(17) for the stresses near the tip of a flat horizontal elliptic crack with internal pressure p and overburden pressure p_2. These formulae show that if $p > p_2$ tensile stresses proportional to $r^{-\frac{1}{2}}$ are developed near the tip of the crack, where r is the distance of the point considered from the tip of the crack. The stress trajectories near the tip of the crack are shown in Fig. 10.12.2. It has been shown in § 10.12 (19) that the tensile stresses are greatest in planes inclined at 60° to the direction of the crack, so that it is not clear from this discussion that the crack will remain plane.

(ii) *Radial dyke swarms*

Pressure of magma in a vertical volcanic neck may give rise to fractures in precisely the manner described in connection with hydraulic fracturing in

§ 8.10. According to conditions, these may be either horizontal sheets or a radial system of dykes. Odé (1957) has studied the theory for the case of dykes running from a vertical neck in the presence of a parallel impervious barrier.

(iii) *Ring dykes and cone sheets*
Anderson (1935) discussed the effects caused by pressure in a magma chamber near the earth's surface.

His first model was a centre of pressure at depth a below the earth's surface, for which the displacements are given by § 10.17 (2). Taking the

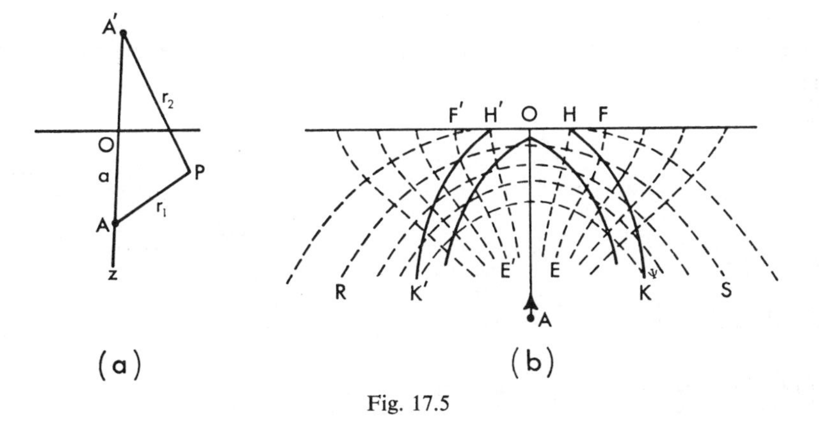

Fig. 17.5

positive z-axis vertically downwards, this would be at A, Fig. 17.5 (a). As a first step towards allowing for the effect of the surface he added an equal centre at A' (0, 0, $-a$). The combined effect of these two in an infinite medium is to give displacements, § 10.17 (2),

$$u_1 = Ax(r_1^{-3} + r_2^{-3}), \quad v_1 = Ay(r_1^{-3} + r_2^{-3}),$$
$$w_1 = A[(z - a)r_1^{-3} + (z + a)r_2^{-3}], \tag{4}$$

where r_1 and r_3 are AP and $A'P$ in Fig. 17.5 (a), so that

$$r_1 = [x^2 + y^2 + (z - a)^2]^{\frac{1}{2}}, \quad r_2 = [x^2 + y^2 + (z + a)^2]^{\frac{1}{2}}. \tag{5}$$

It follows from § 10.17 (7), (8) that on $z = 0$, $\tau_{yz} = \tau_{zx} = 0$, and

$$\sigma_z = 4GA[r_1^{-3} - 3a^2 r_1^{-5}]. \tag{6}$$

The solution for the region $z > 0$ with the centre of pressure at $z = a$ may be found by combining (4) with the displacements u_2, v_2, w_2 found as in § 10.17 (50) for the region with surface loading $-\sigma_z$ given by (6).

These are found to be

$$u_2 = -2A\left\{\frac{3xz(z+a)}{r_2^5} - \frac{Gx}{(\lambda+G)r_2^3}\right\}, \tag{7}$$

$$v_2 = -2A\left\{\frac{3yz(z+a)}{r_2^5} - \frac{Gy}{(\lambda+G)r_2^3}\right\}, \tag{8}$$

$$w_2 = -2A\left\{\frac{3z(z+a)^2}{r_2^5} + \frac{Gz}{(\lambda+G)r_2^3} + \frac{(\lambda+2G)a}{(\lambda+G)r_2^3}\right\}. \tag{9}$$

The complete solution of the problem is $u = u_1 + u_2$, $v = v_1 + v_2$, $w = w_1 + w_2$, and the stresses are then found as in § 10.17.

As a second example, Anderson treated the point-push along the z-axis acting at a point A at depth a below the free surface of a semi-infinite solid. He finds

$$u = -Bx(z-a)r_1^{-3} + Bx(z+a)r_2^{-3} + 2GBax/(\lambda+G)r_2^3 - \frac{2B(\lambda+2G)xz}{(\lambda+G)r_2^3} + \frac{2BG(\lambda+2G)x}{(\lambda+G)^2 r_2(r_2+z+a)} - \frac{6Baxz(z+a)}{r_2^5}, \tag{10}$$

$$w = -B(z-a)^2 r_1^{-3} + B(z+a)^2 r_2^{-3} - [B(\lambda+3G)/(\lambda+G)r_1] + [B(\lambda+3G)/(\lambda+G)r_2] - [2B(\lambda+2G)^2/(\lambda+G)^2 r_2] - 6Baz(z+a)^2 r_2^{-5} - \frac{2B[(\lambda+2G)z^2 + (\lambda+3G)az + (\lambda+2G)a^2]}{(\lambda+G)r_2^3}, \tag{11}$$

and v has the same form as u, with x replaced by y.

This problem has some interest as representing the application of a force in a semi-infinite material. It has also been discussed by Mindlin (1936). Anderson gives numerical values for the principal stresses and drawings of the stress-trajectories for both cases.

For the geological applications it is sufficient to consider the vertical point-push at A, whose stress-trajectories in a vertical plane are shown in Fig. 17.5 (*b*). Of these, the set of type EF, $E'F'$, corresponds to a tensional principal stress and intersects the surface vertically in circles of centre O and diameter FF'. Thus a force at A due to upward-thrusting magma would be expected to cause tensional failure on these surfaces and to give rise to a circular system of dykes, dipping approximately vertically. These are the commonly observed *ring-dykes*. The theory for a centre of pressure, derived from (4)–(9), gives a similar result.

Associated with ring dykes, another circular system, *cone-sheets*, usually occurs. These dip outwards from the centre. The mechanism suggested for these involves a retreat of magma from the magma chamber. Since this region will now have a density less than the normal value, this effect may be simulated by a negative (i.e. downwards) push at A. The stress-trajectories will be as before, but σ_3 will now lie along the set EF and σ_1

is the orthogonal direction. Shear failure may then be expected to take place, as in § 17.3, along curves such as HK inclined at an angle ψ to the trajectories, such as RS, of the greatest principal stress, where ψ is of the order of 30°. As seen in Fig. 17.5 (*b*), these directions, shown in heavy lines, dip outwards in the required fashion. Further discussion is given by Robson and Barr (1964).

17.6 Finite strain

Hitherto, only infinitesimal strains and small displacements have been considered. Quite large strains, however, can be imposed in some laboratory experiments, and in field geology very large strains are frequently seen to have occurred. New methods of analysis have to be developed for the study of these.

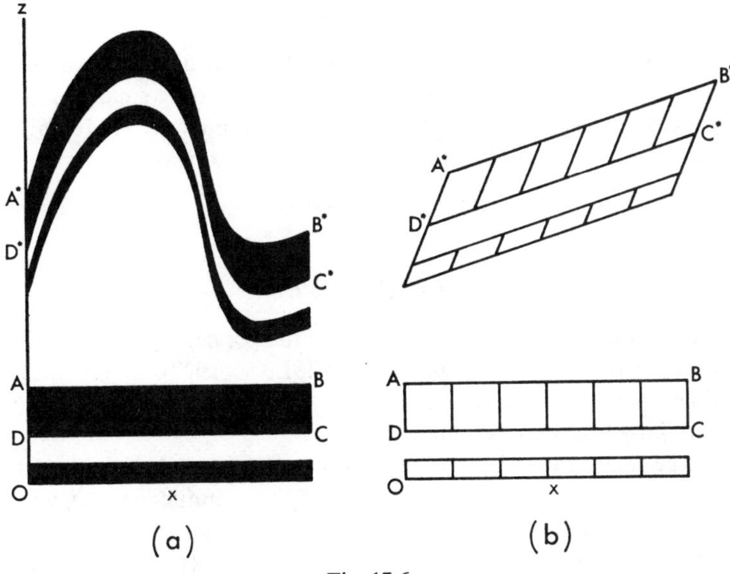

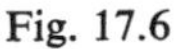

Fig. 17.6

Essentially, as described in § 2.9, the final position x^*, y^*, z^* of the point initially at x, y, z is assumed to be known and to be specified by a numerical or mathematical relation

$$x^* = x^*(x, y, z), \quad y^* = y^*(x, y, z), \quad z^* = z^*(x, y, z). \tag{1}$$

For example, a simple situation frequently postulated in structural geology is that of Fig. 17.6 (*a*), namely

$$z^* = z + f(x), \quad y^* = y, \quad x^* = x, \tag{2}$$

where $f(x)$ may be a very complicated function. This displaces the bed $ABCD$ to the position $A^*B^*C^*D^*$.

It is essential to notice that (1) is simply a relationship between the initial and final positions of the particles with no reference to the way in which the movement was made. In general, the movement may comprise both relative movement of the particles and translation and rotation as a rigid body, Fig. 17.6 (*b*).

The geological problem is complicated by the fact that probably several different sets of complicated displacements may have been imposed on the particles at various times, and only the final position of the particles is known. Clearly this situation is beyond mathematical analysis, but the study of superimposed displacements of the type of Fig. 17.6 (*a*) may be studied by numerical or analogue methods, for example, O'Driscoll (1962, 1964a, b) uses card models. The unravelling of situations of this type is discussed in works on structural geology, Turner and Weiss (1963) and Ramsay (1967).

A mathematical theory of finite strain based on the transformation (1) may be set up. In this the lengths of two corresponding small elements l^* and l in the strained and unstrained states are compared. As remarked in § 2.9, it is more convenient to use the quadratic elongation

$$\lambda = (l^*/l)^2 \tag{3}$$

than the elongation

$$\varepsilon = (l - l^*)/l. \tag{4}$$

A calculation on the lines of § 2.10 (15) can then be carried through without the assumption that $(\partial u/\partial x)$, etc., are small, Love (1927). Shear strain may be defined by considering the changes in the angles between three directions which initially are perpendicular.

Both the definitions (3), (4) have the disadvantage that they relate the final length l^* or the change in length $l - l^*$ to the initial length l, while the final length l^* might equally well be regarded as fundamental, especially as, in many cases, it is the only observable quantity. It will be seen in §§ 17.7, 17.8 that in fact it is useful to work in terms of both the strained and unstrained states. However, apart from this, it is more natural to have a definition which relates the change in length $\delta l'$ to the length l' at any stage of the deformation from l to l^*. Since $\delta l'$ is negative for positive strains, we define the increment of *natural strain* $\delta\bar{\varepsilon}$ as

$$\delta\bar{\varepsilon} = -\delta l'/l', \tag{5}$$

and the natural strain $\bar{\varepsilon}$ is

$$\bar{\varepsilon} = -\int_l^{l^*} \frac{dl'}{l'} = -\ln(l^*/l) = -\ln(1 - \varepsilon) = -\tfrac{1}{2}\ln\lambda. \tag{6}$$

Because of the occurrence of the logarithm in (6), $\bar{\varepsilon}$ is frequently called *logarithmic strain.*

The stress–strain relations may now be regarded as connecting increments of stress and natural strain, so that, referred to instantaneous principal axes of stress and natural strain, § 5.2 (1)–(3) are replaced by

$$\delta\sigma_1 = (\lambda + 2G)\delta\bar{\varepsilon}_1 + \lambda\delta\bar{\varepsilon}_2 + \lambda\bar{\varepsilon}_3, \tag{7}$$

$$\delta\sigma_2 = \lambda\delta\bar{\varepsilon}_1 + (\lambda + 2G)\delta\bar{\varepsilon}_2 + \lambda\delta\bar{\varepsilon}_3, \tag{8}$$

$$\delta\sigma_3 = \lambda\delta\bar{\varepsilon}_1 + \lambda\delta\bar{\varepsilon}_2 + (\lambda + 2G)\delta\bar{\varepsilon}_3. \tag{9}$$

For example, for uniaxial stress $\delta\sigma_2 = \delta\sigma_3 = 0$, these give

$$\delta\sigma_1 = [(3\lambda + 2G)G/(\lambda + G)]\delta\bar{\varepsilon}_1. \tag{10}$$

Now for purely uniaxial stress the principal axes of stress and strain always coincide, so that (10) may be integrated, and if λ and G remain constant throughout the straining it gives

$$\sigma_1 = [(3\lambda + 2G)G/(\lambda + G)]\bar{\varepsilon}_1 = \bar{\varepsilon}_1 E. \tag{11}$$

Similarly, if the principal axes do not change their directions during straining (7)–(9) become

$$\sigma_1 = (\lambda + 2G)\bar{\varepsilon}_1 + \lambda\bar{\varepsilon}_2 + \lambda\bar{\varepsilon}_3, \text{ etc.} \tag{12}$$

In general, however, the principal axes rotate during straining, and the situation becomes much more complicated. Clearly, to complete the discussion of natural strain, natural shears, corresponding to the change in angle between lines which are instantaneously at right angles, should also be defined: this is done by Nadai (1950).

In studying finite strain the logical procedure is to study progressively more complicated forms of the functional relationship (1) beginning with the simplest case, that of the linear transformation

$$x^* = a_1x + b_1y + c_1z, \quad y^* = a_2x + b_2y + c_2z, \quad z^* = a_3x + b_3y + c_3z, \tag{13}$$

or, in the matrix notation of § 2.14

$$[x^*] = [A] \times [x], \tag{14}$$

where

$$x^* = \begin{bmatrix} x^* \\ y^* \\ z^* \end{bmatrix}, \quad [x] = \begin{bmatrix} x \\ y \\ z \end{bmatrix}, \quad [A] = \begin{bmatrix} a_1 & b_1 & c_1 \\ a_2 & b_2 & c_2 \\ a_3 & b_3 & c_3 \end{bmatrix}. \tag{15}$$

It follows from (13), and will be demonstrated in more detail in § 17.7, that planes which are parallel before straining remain parallel planes after straining, as in Fig. 17.6 (*b*), so that the geometrical nature of the strain is the same at all points. For this reason the strain is called *homogeneous*. This study of finite homogeneous strain involves, in principle, merely elementary algebra and trigonometry. Much of the theory was set out by Thomson and Tait (1962), but Becker (1893) developed it further. Becker and others have attempted, unsuccessfully, to develop criteria for failure based on it. The understanding of finite homogeneous strain may be regarded as an essential preliminary to the study of more complicated cases; however, even this relatively simple theory runs into great algebraical complexity and has not yet been fully worked out; modern discussions are given by Ramsay (1967) and Flinn (1962).

Successive displacements may be handled by matrix multiplication, for example, if $[x^*]$ of (14) is strained to a new position $[x^{**}]$ given by

$$[x^{**}] = [B] \times [x^*], \tag{16}$$

it follows from (14) and (16) that

$$[x^{**}] = [B] \times [A] \times [x]. \tag{17}$$

Since matrix multiplication is not in general commutative, it follows that the effect of a strain $[A]$ followed by a strain $[B]$ is not necessarily the same as that of a strain $[B]$ followed by a strain $[A]$.

A more detailed discussion of the analysis of finite homogeneous strain in two and three dimensions will be given in §§ 17.7, 17.8.

17.7 Finite homogeneous strain in two dimensions

After making a displacement so that the origins of coordinates in the strained and unstrained states coincide, the general linear transformation in two dimensions may be written

$$x^* = ax + by, \quad y^* = cx + dy, \tag{1}$$

where a, b, c, d are constants, and a and d may be assumed to be positive. Solving (1) gives

$$h^2x = dx^* - by^*, \quad h^2y = -cx^* + ay^*, \tag{2}$$

where

$$h^2 = ad - bc, \tag{3}$$

and it will be assumed that $h^2 > 0$. The significance of this will appear later.

Two special cases which are frequently referred to are *pure shear*,

$$x^* = kx, \quad y^* = k^{-1}y, \tag{4}$$

and *simple shear*,

$$x^* = x + 2sy, \quad y^* = y. \tag{5}$$

The classical method of procedure, Becker (1893), is the study of the geometrical effect of the transformation (1). For example, the straight line

$$y = mx + f \tag{6}$$

becomes after straining

$$(a + mb)y^* = (c + md)x^* + h^2 f, \tag{7}$$

so that straight lines which are parallel before straining remain parallel after straining, though their inclination and spacing may be changed. This is the fundamental property of homogeneous strain, § 17.6.

A circle of unit radius

$$x^2 + y^2 = 1 \tag{8}$$

in the unstrained state becomes the ellipse

$$(d^2 + c^2)x^{*2} - 2(ac + bd)x^*y^* + (a^2 + b^2)y^{*2} = h^4. \tag{9}$$

This is the *strain ellipse*, which gives an immediate picture of the deformation and has been made fundamental in many works on structural geology. Also from (1), the ellipse

$$(a^2 + c^2)x^2 + 2(ab + cd)xy + (b^2 + d^2)y^2 = 1, \tag{10}$$

in the unstrained state, becomes the circle

$$x^{*2} + y^{*2} = 1 \tag{11}$$

in the strained state. (10) is known as the *reciprocal strain ellipse*. Further discussion of the effects of the transformation could be based on the detailed geometry of these ellipses, but it is rather more satisfactory to study the variation of quadratic elongation and shear strain with direction in a manner analogous to that used for infinitesimal strain in Chapter 2.

To determine the quadratic elongation λ, § 17.6 (3), corresponding to a direction θ, we take $x = l \cos\theta$, $y = l \sin\theta$ in (1), and it follows that

$$\begin{aligned}\lambda &= (x^{*2} + y^{*2})/l^2 \\ &= (a\cos\theta + b\sin\theta)^2 + (c\cos\theta + d\sin\theta)^2 \\ &= (a^2 + c^2)\cos^2\theta + 2(ab + cd)\sin\theta\cos\theta + (b^2 + d^2)\sin^2\theta, \qquad (12)\\ &= \tfrac{1}{2}(a^2 + b^2 + c^2 + d^2) + \tfrac{1}{2}(a^2 + c^2 - b^2 - d^2)\cos 2\theta + \\ &\qquad (ab + cd)\sin 2\theta. \qquad (13)\end{aligned}$$

Differentiating (13), it follows that λ has a maximum or minimum value when $\theta = \alpha$ or $\theta = \alpha + \frac{1}{2}\pi$, where

$$(a^2 + c^2 - b^2 - d^2)\sin 2\alpha - 2(ab + cd)\cos 2\alpha = 0. \tag{14}$$

The values λ_1 and λ_2 of λ corresponding to these directions will be called the *principal quadratic elongations.* They may be found as follows: adding equation (13) for α to the same equation for $\frac{1}{2}\pi + \alpha$ gives

$$\lambda_1 + \lambda_2 = a^2 + b^2 + c^2 + d^2, \tag{15}$$

and subtracting them gives

$$\lambda_1 - \lambda_2 = (a^2 + c^2 - b^2 - d^2) \cos 2\alpha + 2(ab + cd) \sin 2\alpha. \tag{16}$$

Squaring and adding (14) and (16), it follows that

$$(\lambda_1 - \lambda_2)^2 = (a^2 + c^2 - b^2 - d^2)^2 + 4(ab + cd)^2. \tag{17}$$

Then from (15) and (17)

$$\lambda_1\lambda_2 = (ad - bc)^2 = h^4, \tag{18}$$

where h^2 is defined in (3).

λ_1 and λ_2 may be found numerically from a, b, c, d by using (15), (17) (18). It follows from (14), (15), and (17) that (13) may be written

$$\lambda = \tfrac{1}{2}(\lambda_1 + \lambda_2) + \tfrac{1}{2}(\lambda_1 - \lambda_2) \cos 2(\theta - \alpha). \tag{19}$$

The significance of the angle α determined by (14) will become clearer after the question of the shear strain associated with a direction θ has been discussed. To do this we consider the line

$$y = mx, \tag{20}$$

where

$$m = \tan \theta, \tag{21}$$

which is inclined at θ to Ox. By (2) this transforms into the line

$$y^* = [(c + dm)/(a + bm)]x^*. \tag{22}$$

Similarly, the line perpendicular to (20), namely

$$y = -x/m \tag{23}$$

transforms into

$$y^* = [(c - dm^{-1})/(a - bm^{-1})]x^*. \tag{24}$$

The angle $\frac{1}{2}\pi + \psi$ between the lines (22) and (24) is given by

$$\begin{aligned} \tan(\tfrac{1}{2}\pi + \psi) &= \frac{(c - dm^{-1})(a + bm) - (a - bm^{-1})(c + dm)}{(a - bm^{-1})(a + bm) + (c - dm^{-1})(c + dm)} \\ &= \frac{2(bc - ad)}{(a^2 + c^2 - b^2 - d^2) \sin 2\theta - 2(ab + cd) \cos 2\theta}, \end{aligned} \tag{25}$$

using $m = \tan \theta$. Now the shear strain γ associated with the direction θ was defined in § 2.9 (13) as $\gamma = \tan \psi$, so it follows from (25) that

$$\gamma = [2(ab + cd) \cos 2\theta - (a^2 + c^2 - b^2 - d^2) \sin 2\theta]/2(bc - ad). \tag{26}$$

Thus $\gamma = 0$ if θ has the values α or $\alpha + \frac{1}{2}\pi$ defined by (14). The directions $\theta = \alpha$ and $\theta = \alpha + \frac{1}{2}\pi$ remain perpendicular after straining, and are called the *initial directions of the principal axes of strain.* To find their positions in the strained state suppose that these are

$$y^* = m'x^* \quad \text{and} \quad y^* = -x^*/m' \tag{27}$$

corresponding to (22) and (24), where $m' = \tan \alpha'$. Then (22), (24), and (27) give

$$bmm' + am' - dm - c = 0, \tag{28}$$

$$cmm' - dm' + am - b = 0. \tag{29}$$

Adding and subtracting (28) and (29) gives

$$\tan (\alpha' + \alpha) = \frac{m' + m}{1 - mm'} = \frac{b + c}{a - d}, \tag{30}$$

$$\tan (\alpha' - \alpha) = \frac{m' - m}{1 + mm'} = \frac{c - b}{a + d}, \tag{31}$$

and hence

$$\tan 2\alpha' = \frac{2(ac + bd)}{a^2 + b^2 - c^2 - d^2}. \tag{32}$$

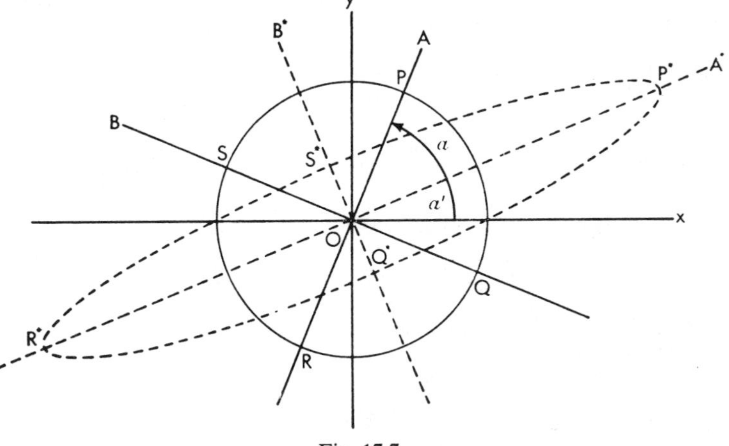

Fig. 17.7

The *rotation* of the principal axes between their initial and final states, $\alpha' - \alpha$, is given by (31) and vanishes if

$$b = c. \tag{33}$$

In this case the strain is called *irrotational.*

The complete situation may now be seen from Fig. 17.7. OA and OB are the initial positions of the principal axes inclined at α and $\frac{1}{2}\pi + \alpha$ to

Ox. Their final positions are OA^* and OB^* inclined at α' and $\frac{1}{2}\pi + \alpha'$. The circle $PQRS$ of unit radius becomes the strain ellipse $P^*Q^*R^*S^*$ with axes along OA^* and OB^*, and its semi-axes are $\lambda_1^{\frac{1}{2}}$ and $\lambda_2^{\frac{1}{2}}$. The area of the strain ellipse is

$$\pi\lambda_1^{\frac{1}{2}}\lambda_2^{\frac{1}{2}} = \pi h^2, \tag{34}$$

by (18), so that h^2 represents the fractional change in area due to the strain and the significance of the assumption that $h^2 > 0$ now becomes clear.

Using the value (14) of α in (26) together with (17) and (18) gives

$$\gamma = \frac{\lambda_1 - \lambda_2}{2\lambda_1^{\frac{1}{2}}\lambda_2^{\frac{1}{2}}} \sin 2(\theta - \alpha). \tag{35}$$

(19) and (35) give the variation of quadratic elongation and shear strain with direction.

The discussion above follows the method which is most natural mathematically, namely, starting from certain configurations in the unstrained state and studying how they are modified in the final state. Alternatively, the strained state might have been taken as a starting-point. This approach, which is perhaps more logical from the geological point of view, in which the strained state is accessible and the unstrained state is not, does in fact lead to a certain simplification of formulae and will be briefly indicated.

Beginning with the point $x^* = l^* \cos \theta^*$, $y^* = l^* \sin \theta^*$, and using (2)

$$\frac{1}{\lambda} = \frac{l^2}{l^{*2}} = \frac{x^2 + y^2}{l^{*2}}$$
$$= [\tfrac{1}{2}(a^2 + b^2 + c^2 + d^2) + \tfrac{1}{2}(c^2 + d^2 - a^2 - b^2) \cos 2\theta^* - (bd + ac) \sin 2\theta^*] \times h^{-4}. \tag{36}$$

This has stationary values when $\theta^* = \alpha'$ or $\alpha' + \frac{1}{2}\pi$ given by

$$(a^2 + b^2 - c^2 - d^2) \sin 2\alpha' - 2(bd + ac) \cos 2\alpha' = 0, \tag{37}$$

which is just (32) as it should be.

Proceeding as in the derivation of (15)–(18) gives

$$\lambda_1^{-1} + \lambda_2^{-1} = (a^2 + b^2 + c^2 + d^2)h^{-4}, \tag{38}$$

$$[\lambda_1^{-1} - \lambda_2^{-1}]^2 = [(a^2 + b^2 - c^2 - d^2)^2 + 4(ac + bd)]^2 h^{-8}, \tag{39}$$

$$\lambda_1^{-1}\lambda_2^{-1} = h^{-4}. \tag{40}$$

Using these, (36) becomes

$$\lambda^{-1} = \tfrac{1}{2}(\lambda_1^{-1} + \lambda_2^{-1}) + \tfrac{1}{2}(\lambda_1^{-1} - \lambda_2^{-1}) \cos 2(\theta^* - \alpha'). \tag{41}$$

To study the shear strain corresponding to the direction θ^* it is necessary to write down the condition that the lines in the initial state corresponding to θ^* and $\theta^* + \frac{1}{2}\pi + \tan^{-1}\gamma$ are perpendicular. Doing this, and using (36) and (39), gives finally

$$\gamma/\lambda = \tfrac{1}{2}(\lambda_2^{-1} - \lambda_1^{-1}) \sin 2(\theta^* - \alpha'). \tag{42}$$

It follows from (41) and (42) that

$$[\lambda^{-1} - \tfrac{1}{2}(\lambda_1^{-1} + \lambda_2^{-1})]^2 + (\gamma/\lambda)^2 = \tfrac{1}{4}(\lambda_2^{-1} - \lambda_1^{-1})^2, \tag{43}$$

and that λ^{-1} and γ/λ can be found from λ_2^{-1}, λ_1^{-1}, and $\theta^* - \alpha'$ by precisely the Mohr circle construction of Fig. 2.3.4. The previous relations (19) and (35) in terms of directions in the unstrained condition lead to

$$[\lambda - \tfrac{1}{2}(\lambda_1 - \lambda_2)]^2 + \lambda_1\lambda_2\gamma^2 = \tfrac{1}{4}(\lambda_1 - \lambda_2)^2, \tag{44}$$

which is an ellipse and gives a slightly more complicated representation.

Many interesting properties of the transformation may be deduced from the preceding formulae. For example, there are two directions in which lengths are unaltered, so that $\lambda = 1$. By (19) these are

$$\cos 2(\theta - \alpha) = (2 - \lambda_1 - \lambda_2)/(\lambda_1 - \lambda_2) \tag{45}$$

and are real only if $(\lambda_1 - 1)(\lambda_2 - 1) < 0$. From (41) their final positions are given by

$$\cos 2(\theta^* - \alpha') = (2\lambda_1\lambda_2 - \lambda_1 - \lambda_2)/(\lambda_2 - \lambda_1). \tag{46}$$

It follows from (35) that the directions in the unstrained state for which the shear strain is a maximum are given by

$$\theta = \alpha \pm \tfrac{1}{4}\pi, \tag{47}$$

while on differentiating (42) and using (41) it follows that in the final state these are given by

$$\cos 2(\theta^* - \alpha') = (\lambda_1 - \lambda_2)/(\lambda_1 + \lambda_2). \tag{48}$$

It must be emphasized that all discussion of this type is simply a comparison of the initial and final positions of a set of particles determined by the transformation (1). The question of the paths by which the particles move between these positions is not in question. This point is discussed further in § 17.8.

As an example, the important case of simple shear (5), for which $a = d = 1$, $b = 2s$, $c = 0$ will be considered.

It follows from (15) and (18) that

$$\lambda_1 + \lambda_2 = 2(1 + 2s^2), \quad \lambda_1\lambda_2 = h^4 = 1, \tag{49}$$

so that there is no change in area on straining. From (49)

$$\lambda_1 = 1 + 2s^2 + 2s(1 + s^2)^{\frac{1}{2}} = [(s^2 + 1)^{\frac{1}{2}} + s]^2, \tag{50}$$

$$\lambda_2 = 1 + 2s^2 - 2s(1 + s^2)^{\frac{1}{2}} = [(s^2 + 1)^{\frac{1}{2}} - s]^2. \tag{51}$$

It follows that the semi-axes $\lambda_1^{\frac{1}{2}}$, $\lambda_2^{\frac{1}{2}}$ of the strain ellipse are $(s^2 + 1)^{\frac{1}{2}} \pm s$. From (30) and (31),

$$\alpha + \alpha' = \tfrac{1}{2}\pi, \quad \alpha - \alpha' = \tan^{-1} s,$$

so that

$$\alpha = \tfrac{1}{4}\pi + \tfrac{1}{2}\tan^{-1} s, \quad \alpha' = \tfrac{1}{4}\pi - \tfrac{1}{2}\tan^{-1} s, \tag{52}$$

which gives the initial and final positions of the principal axes of strain. Fig. 17.7 has been drawn for this situation with $s = 1$.

The situation is that the final position may also be attained by a strain with $\lambda_1^{\frac{1}{2}} = (s^2 + 1)^{\frac{1}{2}} + s$, $\lambda_2^{\frac{1}{2}} = (s^2 + 1)^{\frac{1}{2}} - s$, which is a pure shear as in (4), relative to the axes OA, OB, followed by a rotation of the axes through $\alpha' - \alpha$ to the position OA^*, OB^*.

The results of this section show that the theory of finite homogeneous strain in two dimensions gives rise to formulae of the same type as those developed in § 2.10 for infinitesimal strain (which, of course, is homogeneous). All the results of § 2.10 may be deduced as special cases of the present theory. The elongation ε is connected with the quadratic elongation λ by

$$\lambda = (1 - \varepsilon)^2 = 1 - 2\varepsilon,$$

if ε is so small that its square can be neglected. In the present section there has been no need to consider the relative magnitudes of λ_1 and λ_2, so we may take $\lambda_1 = 1 - 2\varepsilon_1$ corresponding to $\theta = \alpha$ and $\lambda_2 = 1 - 2\varepsilon_2$ corresponding to the perpendicular direction, and (19) and (35) become

$$\varepsilon = \tfrac{1}{2}(\varepsilon_1 + \varepsilon_2) + \tfrac{1}{2}(\varepsilon_1 - \varepsilon_2)\cos 2(\theta - \alpha),$$
$$\gamma = (\varepsilon_2 - \varepsilon_1)\sin 2(\theta - \alpha),$$

which are § 2.10 (26), (27).

17.8 Finite homogeneous strain in three dimensions

The transformation is § 17.6 (13), and the general theory is complicated. However, it follows the same lines as § 17.7. Planes transform into planes, and a sphere becomes an ellipsoid, the *strain ellipsoid*, and a particular ellipsoid, the *reciprocal strain ellipsoid* becomes a sphere. The quadratic elongation corresponding to any direction in the initial state can be calculated, and it is found that this is stationary for three mutually perpendicular directions, the initial directions of the principal axes of strain, which remain mutually perpendicular in the strained state and become the axes of the strain ellipsoid.

The situation is greatly complicated by rotation, and only the irrotational case, in which the initial and final positions of the principal axes coincide, will be considered here. Taking these as axes, the transformation becomes

$$x^* = \lambda_1^{\frac{1}{2}}x, \quad y^* = \lambda_2^{\frac{1}{2}}y, \quad z^* = \lambda_2^{\frac{1}{2}}z, \tag{1}$$

where λ_1, λ_2, λ_3 are the principal quadratic elongations. The quadratic elongation for the direction of direction cosines l, m, n, obtained by considering the point whose initial position is $x = lr$, $y = mr$, $z = nr$, is

$$\lambda = (x^{*2} + y^{*2} + z^{*2})/r^2 = \lambda_1 l^2 + \lambda_2 m^2 + \lambda_3 n^2. \tag{2}$$

To determine the shear strain corresponding to the direction l, m, n, we consider this line and the plane

$$lx + my + nz = 0 \tag{3}$$

to which it is normal. The line becomes

$$x^*/l\lambda_1^{\frac{1}{2}} = y^*/m\lambda_2^{\frac{1}{2}} = z^*/n\lambda_3^{\frac{1}{2}}, \tag{4}$$

and the plane becomes

$$\frac{lx^*}{\lambda_1^{\frac{1}{2}}} + \frac{my^*}{\lambda_2^{\frac{1}{2}}} + \frac{nz^*}{\lambda_3^{\frac{1}{2}}} = 0. \tag{5}$$

The shear strain γ corresponding to the direction (l, m, n) is given by $\gamma = \tan\psi$, where ψ is the angle between the line (4) and the normal to the plane (5). That is

$$\cos\psi = [(l^2\lambda_1 + m^2\lambda_2 + n_2\lambda_3)(l^2\lambda_1^{-1} + m^2\lambda_2^{-1} + n^2\lambda_3^{-1})]^{-\frac{1}{2}}. \tag{6}$$

It follows that

$$\gamma^2 = \tan^2\psi = (l^2\lambda_1 + m^2\lambda_2 + n^2\lambda_3)(l^2\lambda_1^{-1} + m^2\lambda_2^{-1} + n^2\lambda_3^{-1}) - 1.$$

Multiplying out and using $(l^2 + m^2 + n^2) = 1$ gives

$$\lambda_1\lambda_2\lambda_3\gamma^2 = \lambda_1(\lambda_2 - \lambda_3)^2 m^2 n^2 + \lambda_2(\lambda_1 - \lambda_3)^2 n^2 l^2 + \lambda_3(\lambda_1 - \lambda_2)^2 l^2 m^2. \tag{7}$$

(2) and (7) give the quadratic elongation and shear corresponding to a direction l, m, n in the unstrained state. As in two dimensions, slightly simpler relations are obtained by working in terms of the direction l^*, m^*, n^* in the strained state corresponding to l, m, n in the unstrained state. From (4) and (2) it follows that

$$l^* = l\lambda_1^{\frac{1}{2}}\lambda^{-\frac{1}{2}}, \quad m^* = m\lambda_2^{\frac{1}{2}}\lambda^{-\frac{1}{2}}, \quad n^* = n\lambda_3^{\frac{1}{2}}\lambda^{-\frac{1}{2}}. \tag{8}$$

Using (8) in $l^2 + m^2 + n^2 = 1$ gives

$$\lambda^{-1} = \lambda_1^{-1}l^{*2} + \lambda_2^{-1}m^{*2} + \lambda_3^{-1}n^{*2}, \tag{9}$$

and using (8) in (7) gives

$$(\gamma\lambda^{-1})^2 = (\lambda_2^{-1} - \lambda_3^{-1})^2 m^{*2} n^{2*} + (\lambda_3^{-1} - \lambda_1^{-1})^2 n^{*2} l^{*2} + (\lambda_1^{-1} - \lambda_2^{-1})^2 l^{*2} m^{*2}. \tag{10}$$

These have precisely the same form as the relations § 2.4 (25), (26) for normal and shear stress with σ, τ, σ_1, σ_2, σ_3 replaced by λ^{-1}, $\gamma\lambda^{-1}$, λ_1^{-1},

λ_2^{-1}, λ_3^{-1}, respectively, so that the Mohr circle construction of § 2.6 can be used to determine λ^{-1} and $\gamma\lambda^{-1}$. A full discussion of the application of the Mohr construction to problems in finite homogeneous strain is given by Brace (1961). A similar construction, but based on ellipses in place of circles, follows from (2) and (7), Nadai (1950).

In the past a great deal of attention has been paid to the geometrical aspects of the transformation, in particular to the strain ellipsoid, whose equation is, by (1),

$$\frac{x^{*2}}{\lambda_1} + \frac{y^{*2}}{\lambda_2} + \frac{z^{*2}}{\lambda_3} = 1, \tag{11}$$

and the reciprocal strain ellipsoid

$$\lambda_1 x^2 + \lambda_2 y^2 + \lambda_3 z^2 = 1. \tag{12}$$

The strain ellipsoid gives a very simple representation of the transformation and is of direct practical importance, since objects of known shape, such as some fossils, are frequently found to be deformed, Cloos (1947), and by measuring this deformation the strain ellipsoid may be determined. The problem of determining the parameters of the ellipsoid by measurements made of various sections of it is discussed in detail by Ramsay (1967). A great deal of attention has been paid to the circular sections of the strain ellipsoid. These are defined by its intersection with the sphere of radius λ_2, namely

$$x^{*2} + y^{*2} + z^{*2} = \lambda_2. \tag{13}$$

Subtracting (11) and (13) shows that this lies on the pair of planes

$$(\lambda_1^{-1} - \lambda_2^{-1})x^{*2} + (\lambda_3^{-1} - \lambda_2^{-1})z^{*2} = 0. \tag{14}$$

If $\lambda_1 > \lambda_2 > \lambda_3$ these planes are real and inclined to the x^*-axis at angles

$$\pm\tan^{-1} [\lambda_3(\lambda_1 - \lambda_2)/\lambda_1(\lambda_2 - \lambda_3)]^{\frac{1}{2}}, \tag{15}$$

and they, and planes parallel to them, intersect the ellipsoid in circles. If $\lambda_2 = 1$, $\lambda_3 = 1/\lambda_1$, so that there is no change in volume on straining, (15) becomes

$$\pm\tan^{-1} (\lambda_1^{-\frac{1}{2}}). \tag{16}$$

One of the old strain theories of failure associated this with the circular sections and assumed it to take place in the planes (16), Becker (1893); the validity of this was discussed by Griggs (1935).

The whole of this discussion has referred to the comparison of a single initial and final state. In structural geology the main concern is with continuous or progressive deformation. In this the state of strain is supposed to be specified at any time by the strain ellipsoid or the principal strains and axes, and the variation of these as the strain proceeds defines a strain path or deformation path. Clearly, because of the number of parameters involved, the representation of this is a matter of considerable complexity even in the irrotational case. There are various ways in which the instantaneous state of strain can be represented:

(i) Plotting λ_1, λ_2, λ_3 in rectangular coordinates.
(ii) Plotting logarithmic principal strains $\bar{\varepsilon}_1$, $\bar{\varepsilon}_2$, $\bar{\varepsilon}_3$.
(iii) Plotting on a triangular diagram.
(iv) The use of other parameters. For example, Flinn (1962) plots $a = (1 + \varepsilon_1)/(1 + \varepsilon_2)$ against $b = (1 + \varepsilon_2)/(1 + \varepsilon_3)$ and uses

$$k = (a - 1)/(b - 1) = (1 + \varepsilon_3)(\varepsilon_1 - \varepsilon_2)/(1 + \varepsilon_2)(\varepsilon_2 - \bar{\varepsilon}_3)$$

to describe the nature of the strain. Alternatively, ln *a* may be plotted against ln *b*.

These questions are disccussed in detail by Flinn (1962) and Ramsay (1967).

The general problem of successive strains, which may be specified relative to different axes, may be handled by matrix multiplication.

17.9 Joint and fault sets in sedimentary basins

Large sedimentary basins are of considerable geological and, often economic, interest. Accordingly, they have been the subject of several investigations, and the sets of joints and faults in them are distinctive because of the absence of tectonic deformation (Raistrick and Marshall, 1939; Price, 1966; Hancock, 1969).

Price (1974) has provided an explanation for the origin of these sets of joints and faults in terms of the stresses resulting from downwarp, water pressures and uplift in the history of such basins. Essentially he shows that, as a result of deposition and associated downwarping of large, elliptical sedimentary basins, the sediments in a large central portion of the basin are subject to a vertical stress equal to the weight of the overburden and elastic horizontal stresses given by equation (3), § 14.2, modified by the strain caused by the depression of the basin into a spherical earth. The maximum value of this horizontal strain in a direction parallel to that of the long axis of the basis is given by

$$\varepsilon = 1 - \sin \theta/\theta, \qquad (1)$$

where θ = the angle subtended by the basin at the centre of the earth. The horizontal strain in a direction parallel to that of the short axis is tensile as a result of the small radius of curvature of the bottom of the basin in a vertical section through this axis.

Water is likely to be trapped in such sediments and the values of its pressure can easily be from 0,5 to 0,8 times the lithostatic stress, § 14.1,

(Hubbert and Rubey, 1959). The resulting effective stresses, § 8.8, are likely to give rise to normal faults parallel to the long axis of the basin in the harder strata, and vertical hydraulic fractures normal to this axis in the harder strata and parallel to it in the weaker strata. He concludes, also, that the value of the horizontal stress can never be less than the greatest pore fluid pressure minus the tensile strength of the rock.

During subsequent uplift two situations need to be considered. First, where there is no change in the dip of the strata, the value of the horizontal components of stress decreases elastically in both directions as a result of the reduction in the value of the vertical component, § 14.3, but also because of a lateral strain resulting from the curvature of the earth, which is equal to

$$\varepsilon = \Delta Z/R, \tag{2}$$

where ΔZ = the uplift
R = the radius of the earth, or 6400 km.

Consequently, the values of the horizontal stresses are likely to decrease more rapidly during uplift than is the value of the pressure of trapped water. A second set of vertical fractures parallel to both the long and short axes of the basin may then develop, as a result of which the horizontal stress will increase to equal the pore fluid pressure.

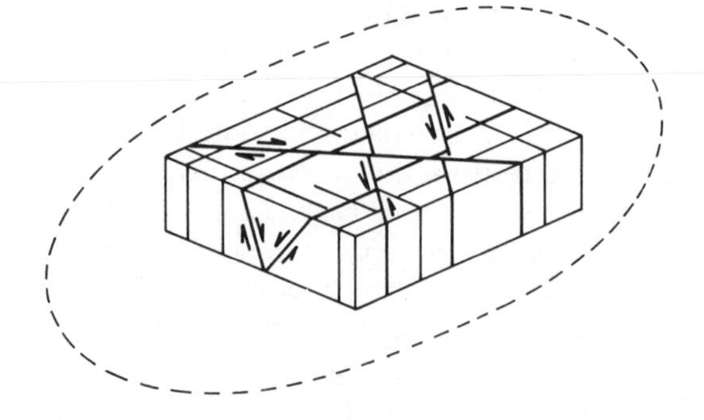

Fig. 17.9.1 A composite diagram showing the normal faulting, vertical jointing and strike–slip faults which develop during downwarping and uplift of large, elliptical sedimentary basins, and their orientation relative to the shape of the basin (after Price, 1974).

If, on the other hand, the dip of the strata undergoes a progressive reduction as uplift takes place, the horizontal strains are not symmetrical; the strain in the direction of the short axis is compressive and that in the direction of the long axis tensile. Vertical hydraulic fractures develop parallel to the short axis of the basin and the value of the minimum principal stress becomes equal to the pore fluid pressure. The value of the horizontal stress parallel to the short axis increases so that it becomes the maximum principal stress which can be of sufficient magnitude to cause strike–slip faulting.

A composite diagram showing the orientation of the sets of joints and faults described above is shown in Fig. 17.9.1.

17.10 Tectonophysics

Tectonophysics may be defined as the study of the relation between deformation and forces in the earth. The mechanical properties of the earth change greatly with depth and it is convenient to make a distinction between the upper crust where brittle phenomena are important, the lower crust where there is a transition from brittle to ductile deformation, and the mantle where creep is predominant. The crust of the earth varies in thickness from 20 km to 70 km under the continents, and between 5 km and 10 km under the oceans.

Faulting due to earthquakes is one of the most striking surface phenomena associated with tectonophysics. There are two outstanding features of earthquakes. First, earthquakes are concentrated into narrow zones forming a network around the earth and, second, those which occur at depths of less than 60 km below surface account for three-quarters of all the energy released by earthquakes. These observations appear to be consistent with the theory of plate tectonics, proposed originally by Wegener (1915) as 'continental drift' (Hallam, 1973). According to this theory, the crust comprises a number of discrete plates moving across the surface of the earth in response to viscous flow in the mantle (Fig. 17.10.1). Deep earthquakes are believed to be caused by slabs of crustal material sinking under deep trenches such as those in Alaska, Japan and South America. A striking feature of both trenches and ridges is their offset in a number of places by transform faults. Wilson (1965) has shown how the large shear displacements of these faults can be accommodated by the creation and destruction of crustal material along ridges and trenches. Fault plane solutions of earthquakes from ocean ridges have been obtained by Sykes (1967) which verify Wilson's hypothesis. Where relative motion of these plates is resisted, mechanical instabilities in the form of earthquakes occur.

Many theories have been put forward to account for these instabilities. Reid (1910) analysed the deformations which occurred across the San Andreas fault in central California as a result of the 1906 San Francisco earthquake of magnitude 8·3. He concluded that shear strain had been accumulating across the fault prior to the earthquake at a rate of the order of 10^{-6} per year. When the resulting stress across the fault reached a value equal to the strength of the fault, sudden slippage occurred releasing much of the accumulated strain. Clearly, in tectonophysics instabilities in com-

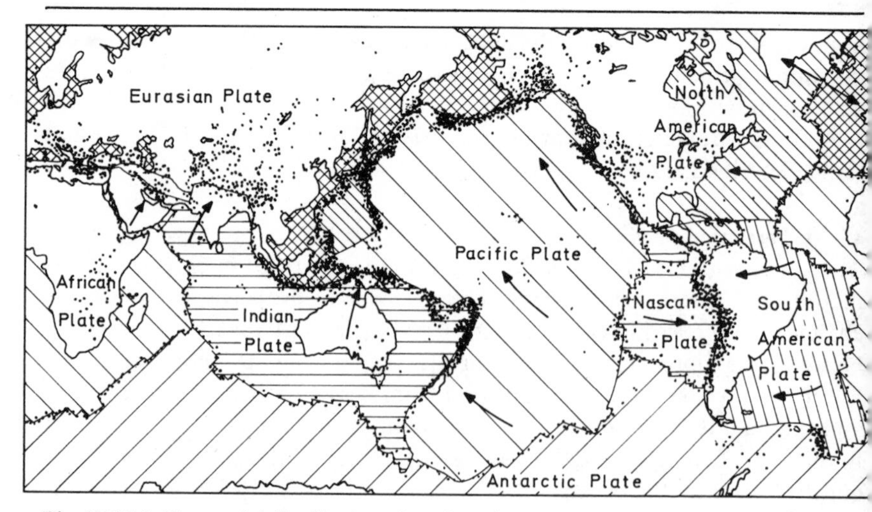

Fig. 17.10.1 The spatial distribution of earthquakes over the earth and their relationship to tectonic plates (after Nur, 1974).

pression are of special interest. Phase changes (Bridgmann, 1945; Griggs and Handin, 1960; Benioff, 1964; Raleigh and Paterson, 1965), shear melting (Griggs and Baker, 1969), and creep instability (Orowan, 1960) have been proposed as potential mechanisms for such instabilities. Sudden fracture and 'stick–slip' motion during sliding friction, § 3.5, are equally plausible and attractive mechanisms on account of their simplicity.

Relative slip between the two sides of earthquake faults on land has been observed to vary from a few centimetres to about 10 metres for the largest earthquakes (Nur, 1974). The horizontal length of fault slip zones varies with the magnitude of the earthquake from a few metres to 1000 kilometres (Iida, 1965). Surface displacements from slip on faults that intersect the surface have been measured for a number of earthquakes. Horizontal near-field displacements from predominantly strike–slip earthquakes show that the relative displacement between the two sides varies from 2 m to 5 m (Nur, 1974) and that these displacements diminish rapidly to negligible values within 20 km to 30 km on each side of the fault. This suggests that the displacement is controlled by the minimum dimension of the fault surface, namely, its depth which, probably, does not exceed about 20 km.

Theoretical studies of the far-field static displacement (Press, 1965) show that this should decrease inversely as the distance from the focus, that is, the strain should decrease inversely as the square of the distance. Static strain changes of the order of 10^{-8} to 10^{-9} have been measured, but such few measurements as have been made suggest that the strain decreases less rapidly with distance from the focus than is suggested above (Wideman and

Major, 1967); this discrepancy can be ascribed, possibly, to physical properties of the crust and mantle different from those assumed in simple theoretical models.

Models of earthquake faults involving simple boundary conditions invariably give rise to stresses at the edges of the fault with infinite values, which is physically unreasonable (Nur, 1974). However, only a somewhat more complicated geometrical model (Fairhurst and Cook, 1966) or more comprehensive boundary conditions, such as increasing friction with depth (Weertman, 1969), are necessary to remove this objection.

The stress drop that accompanies faulting can be calculated, approximately, from

$$\Delta\tau = \bar{u}G/L, \tag{1}$$

where $\bar{u}$ = the average relative displacement between the fault surfaces;
G = the modulus of rigidity;
L = the smallest dimension of a buried fault or twice the depth of a vertical strike–slip fault intersecting the surface.

Stress drops calculated in this way (Chinnery, 1969) for earthquakes with magnitudes in the range from 6·8 to 8·3 lie between 20 bar and 100 bar, which are considerably less than would be expected from the fracture of solid rock, § 17.11. These data suggest strongly that these earthquakes are the result of an incremental extension of a pre-existing fault (Berg, 1967) or from stick–slip motion in such a fault (Dieterich, 1969).

The seismic moment, § 17.11, is given by $M_0 = \bar{u}GA$ where $\bar{u}$ is the average displacement between the two sides of a fault, G is the modulus of rigidity and A is the area of the mobilized portion of the fault. The seismic moment can be estimated from the shear wave amplitude spectrum level at long periods (Keilis-Borok, 1960; Brune, 1970) from

$$M_0 \approx [\overline{\Omega^2(\omega)}]^{\frac{1}{2}}\, \pi\rho R C_S{}^3 10, \tag{1}$$

where $[\overline{\Omega^2(\omega)}]^{\frac{1}{2}}$ = the average far-field root mean square shear wave amplitude spectrum level;
ρ = the density of the rock;
R = the distance from the earthquake to the measuring point;
C_S = seismic shear wave velocity.

The minimum fault dimension can be estimated from the effect of the rise-time of the fault dislocation on the far-field radiation but the result is not unambiguous. Using the simple stick–slip model, § 3.5, Nur (1974) shows that the far-field spectrum is bounded by a low frequency trend with zero slope and a high frequency trend with a slope of -2. The intersection of these two trends defines a 'corner frequency', $\alpha = C_S/L$ where L is the fault dimension; or Brune's (1970) model $\alpha = 1{,}17\ C_S/L$.

In addition to unstable motion, fault creep, a slow process by which the offset across a fault increases steadily with time, is observed. The rate of offset is very slow compared with seismic movements, being of the order of a few centimetres a year for steady creep, and a few millimetres a day for episodic creep. The greatest seismic creep rate on the San Andreas fault is about 50 mm per year at Monarch Peak north of Parkfield. Such a rate of creep is consistent with the rate of ocean floor spreading but it diminishes rapidly north and south of Monarch Peak. Logarithmically decaying post-earthquake creep is common (Scholtz, 1972) and has been observed after rockbursts, § 18.12, by Hodgson and Cook (1971).

The phenomenon of dilatancy, §§ 4.2, 4.5 and 6.5 and the triggering of earthquakes by increasing pore fluid pressures have given added impetus to explanations of earthquake instabilities based on simple mechanisms such as 'stick–slip' motion.

In the early 1960's a well for the disposal of liquid wastes was drilled through 3700 m of sedimentary rocks and completed in Precambrian granite gneiss near Denver, Colorado. In 1966 Evans drew attention to the remarkable coincidence between a series of earthquakes in Denver and the rate of injection of liquid into this well. Since then this has been the subject of intensive study. An updated plot of the frequency of earthquakes and the volume and well-head pressure is shown in Fig. 17.10.2. In 1969 a pro-

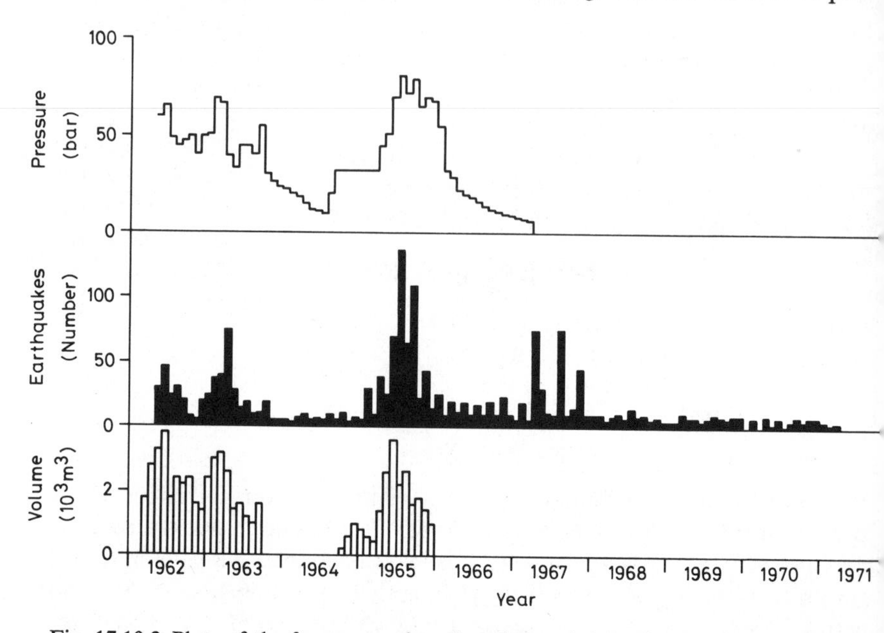

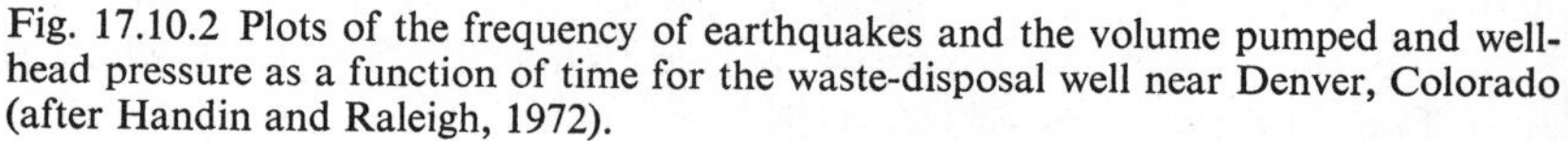
Fig. 17.10.2 Plots of the frequency of earthquakes and the volume pumped and well-head pressure as a function of time for the waste-disposal well near Denver, Colorado (after Handin and Raleigh, 1972).

gramme of research into the relationship between earthquakes and pore fluid pressures in the Rangely oil field was initiated (Handin and Raleigh, 1972). These experiments have shown that seismic activity in the Rangely field can be predicted remarkably well from a knowledge of the virgin state of stress, coefficients of sliding friction for the Weber sandstone as determined from laboratory studies, and the effective stresses, § 8.8, in the strata, taking into account the fluid pore pressure.

World-wide efforts are being made to predict or control earthquakes (Brace, 1975). Much of this work is based on the phenomena which have been observed in the laboratory to precede fracture of rock specimens, § 6.6. Two basic models have been advanced to explain precursors of an earthquake. Both involve changes in rock properties brought about by microcracking. The model proposed by scientists in the U.S.S.R. follows virtually the processes described in § 4.5 for a specimen of hard rock tested in a stiff triaxial cell. The other model, supported by most scientists in the U.S.A., also envisages microcracking and dilatancy of the rock but involves the flow of pore fluid into the fault area to reduce the effective normal stress so that an earthquake occurs. Both models predict precursive changes, in the vicinity of a pending earthquake, of the ratio between the velocities of compressional and shear waves, ground movements, electrical resistivity, seismic foreshocks and radon emission, as a result of dilatant microcracking (Nur, 1974; Brace, 1975).

17.11 Earthquake mechanics

It seems probable that most earthquakes are mechanical instabilities which result from the sudden failure of rock to sustain the shear stresses acting across a surface; the surface may be a pre-existing fault or a new fracture caused by the failure. If it is assumed as a first approximation that most of the rock around such a surface responds in a linear, elastic way to the sudden change in stress associated with the earthquake, much of the mechanics of an earthquake can be explained, in simple terms.

From such an assumption, it follows that a linear relationship exists between any change in the value of the shear stress across such a surface and the relative displacement of the two sides past one another, § 10.11. It is convenient to think in terms of an average value of this displacement across the entire surface as a function of the shear stress. The slope of the straight line representing this function in a shear stress–average displacement diagram is determined by the shape of the surface, its size and the elastic moduli of the rock. Apart from the moduli, the most important factor is the minimum dimension of the mobilized portion of the surface. In Fig. 17.11.1 a shear stress–average displacement diagram is plotted for a surface with a minimum dimension of the order of 10 km in solid rock with a modulus of rigidity of the order of 50 MPa. If the shear stress across

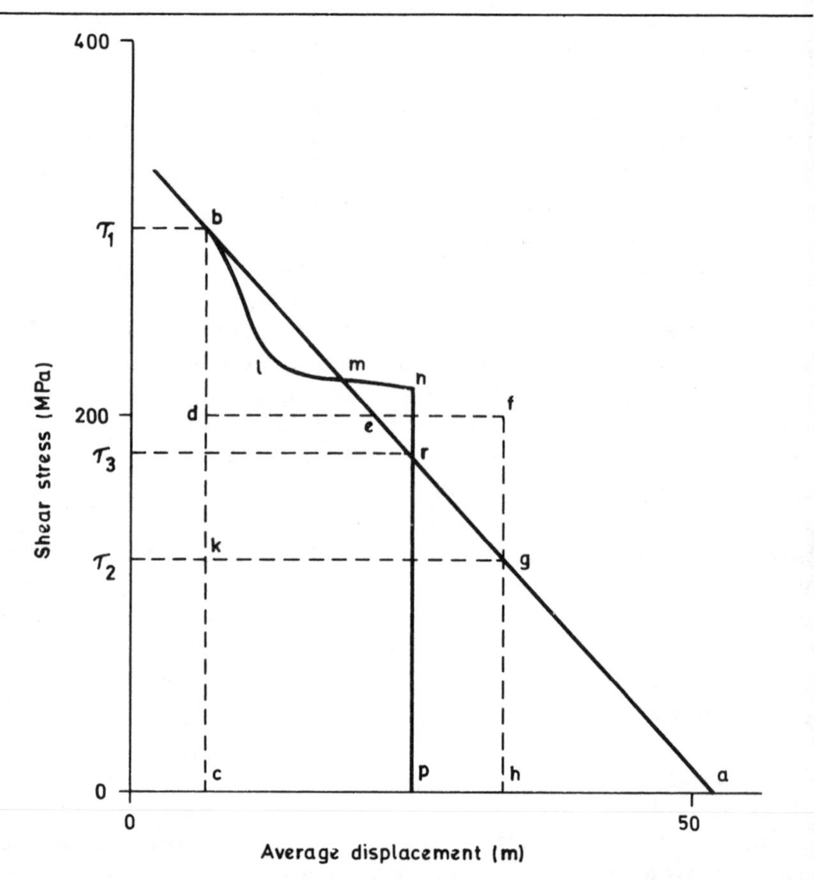

Fig. 17.11.1 A diagram showing the average shear stress drop as a function of the average displacement across a fault surface for a variety of conditions.

the earthquake surface were τ_1 before failure and this stress were to drop to zero after failure, an amount of energy proportional to the area *abc* would be released. The total amount of energy release, E_{T1}, is given by

$$E_{T1} = A\tau_1\bar{u}_1/2, \qquad (1)$$

where A = the area of the surface;
τ = the shear stress drop across the surface;
$\bar{u}_1$ = the mean displacement between two sides of the surface.

The seismic moment is given by

$$M_{01} = A\bar{u}_1 G, \qquad (2)$$

where G is the modulus of rigidity of the rock around the focus.

The situation discussed above is grossly oversimplified since it assumes that there is no resistance to motion between the two sides. Such could be the case if, for example, the rock adjacent to each side were to melt. However, it seems more realistic to assume that a significant normal stress exists across the surface and that this gives rise to some frictional resistance to shearing. If this were the case and the shear stress across the surface were still τ_1 before failure the shear stress might drop to a value τ_2, given by the point *g*, after the earthquake. For this to happen, the frictional shear stress would have had to be less than $(\tau_1 + \tau_2)/2$, given by the value of point *e*. The kinetic energy acquired by the rock on each side of the surface causes these sides to overshoot one another from *e* to *g*. In this situation a total amount of energy proportional to the area *hgebdc* would be released by the earthquake. This released energy, E_{T2} is given by

$$E_{T2} = A(\tau_1 + \tau_2)\bar{u}_2/2, \tag{3}$$

where $\bar{u}_2 = \bar{u}_1(\tau_1 - \tau_2)/\tau_1$,
and the seismic moment is now

$$M_{02} = A\bar{u}_2 G = A\bar{u}_1(\tau_1 - \tau_2)G/\tau_1. \tag{4}$$

In sliding from *b* to *e* the amount of potential energy released exceeds that dissipated in friction by an amount proportional to the area *ebd*. As a result of the kinetic overshoot a further amount of energy proportional to the area *gfe* is dissipated mostly in friction. The area *efg* must be less than the area *ebd* by an amount proportional to the energy radiated seismically plus any other losses except friction across the surfaces. It follows that the seismic radiation efficiency given by [(*ebd*) — (*gfe*)]/*hgebdc* has a small value. The seismic stress drop is of the order of $(\tau_1 - \tau_2)/2$.

In fact, the real situation is probably even more complicated than this; the earthquake has been assumed to give rise to a new fracture surface. In this case, the loss of strength on the fracture with displacement is likely to follow a shape similar to that of a complete stress–strain curve, §§ 4.2, 4.5 and 6.6. The resistance of this fracture to sliding would then be given by a curve similar to that shown by *blmn* in the figure. The total amount of energy released would be proportional to the area *prmbdc* and given by

$$E_{T3} = A(\tau_1 + \tau_3)\bar{u}_3/2, \tag{5}$$

where $\bar{u}_3 = \bar{u}_1(\tau_1 - \tau_3)/\tau_1$,
and the seismic moment becomes

$$M_{03} = A\bar{u}_3 G = A\bar{u}_1(\tau_1 - \tau_3)G/\tau_1. \tag{6}$$

The seismic radiation efficiency is given by the area

[(*mbl*) — (*rnm*)]/*prmbdc*,

which is even smaller than before. The seismic stress drop is of the order of the vertical separation between the line *mb* and the curve *mlb*.

There is no reason to suppose that there is in fact a qualitative difference between the resistance to shearing along a new fracture surface or along a pre-existing one. Qualitatively, the last of the three arguments presented above, therefore, probably provides a realistic description of both types of earthquake. Quantitatively, the stress differences are likely to be an order of magnitude less when sliding on a pre-existing surface is involved than when a new fracture surface is created.

From this discussion it follows that seismic radiation efficiencies should be low, of the order of one per cent, and seismic stress drops small in relation to the ambient values of stress near an earthquake focus.

By taking the ratio of the total energy release to the seismic moment the unknown quantities of the area of the mobilized surface and the average displacement between its two sides (Brune, 1968; Wyss, 1970) can be eliminated, thus

$$E_T/M_0 = \frac{A(\tau_1 + \tau_3)\bar{u}_1(\tau_1 - \tau_3)\tau_1}{2\tau_1 A\bar{u}_1(\tau_1 - \tau_3)G}, \\ = \frac{(\tau_1 + \tau_3)}{2G}. \tag{7}$$

The average ambient stress at the focus of an earthquake $(\tau_1 + \tau_3)/2$ can be found from the seismic energy, if the seismic radiation efficiency and G are known or can be estimated.

A common criticism of this kind of earthquake model is that it gives rise to very high, if not infinite, values of stress near the edges of the fault. Such values seem to be physically unreasonable and raise a question as to how the fault can become stable, even temporarily. However, it is important to note that the directions of these very high stresses are usually such as not to extend the fault in its own plane, § 12.15. In fact, these high stresses provide a reasonable explanation for the origin of sympathetic fractures breaking off the main fault, which are observed in practice and vary considerably in size and significance. These sympathetic fractures may themselves play a major part, by relieving stresses in the vicinity of the edges of the main fault, in determining its stability.

This simple earthquake model can account for yet another form of instability, arising from an incremental enlargement of a pre-existing fault. The average displacement, $\bar{u}$, across a fault surface is given by,

$$\bar{u} = KL(\tau_1 - \tau_n), \tag{8}$$

where K = a constant determined by the shape of the fault and the elastic constants of the rock;

L = the smallest dimension of the fault;

$(\tau_1 - \tau_n)$ = the stress drop across the surface.

If the fault were now to enlarge suddenly by a small amount, ΔL, then the average incremental displacement, $\Delta\bar{u}$, across the fault is

$$\Delta\bar{u} = K\Delta L(\tau_1 - \tau_n). \tag{9}$$

By comparing equations (8) and (9) it can be seen that the displacement due to an incremental enlargement is proportional to, ΔL, whereas that due to a new fault is proportional to L.

This provides a very satisfactory explanation of the small co-seismic displacements observed across many faults of known length, § 17.10, which otherwise yield surprisingly low values of stress drops and low coefficients of sliding friction inconsistent with the known properties of the rock, when L is used instead of ΔL.

Chapter Eighteen

Mining and Other Engineering Applications

18.1 Introduction

The purpose in applying knowledge of rock mechanics to engineering problems involving structures in, or founded on, rock is to predict and allow for the response of the rock to the loads imposed on it by these structures. The response of the rock can have important implications in the design of structures ranging from buildings to underground mines. Loads may be *imposed loads*, such as the additional loads due to the weight and pressure of water in a dam or *induced loads*, such as those due to the redistribution of stresses around an underground excavation.

In any case, it is necessary to start with some knowledge of the magnitudes of these loads and the properties of the rocks to which they are applied. The magnitude of imposed loads can usually be calculated with little difficulty. The magnitudes of induced loads are dependent upon the magnitudes of the virgin stresses in the undisturbed rock. The techniques for measuring these stresses and the difficulties associated with such measurements are discussed in §§ 15.2, 15.4, 15.5, 15.6. Some knowledge of the properties of the rock can be gained from measurements on laboratory specimens, Chapter 6. It is of equal and often greater importance to determine the properties of the body of rock affected by the structure as a whole. These properties are markedly dependent upon large-scale structural features in the rock, such as bedding, schistosity, and jointing, §§ 3.6, 3.8, 3.9. Furthermore, pore- and particularly joint-water pressures have a major influence on the strength of the body of rock, § 8.9. These factors must be evaluated by field investigations, such as geological mapping, boring, and seismic surveys, §§ 13.6, 18.9. It may even be necessary to make field measurements of the properties using plate-bearing tests, § 15.7, and large-scale compression and shear tests, § 15.8.

Once the loads and properties have been determined, the response of the rock to these loads can be calculated. In the simplest case, where the loads are modest in relation to the strength of the rock, the calculation yields the deformations of the rock, for which allowance may have to be made in the design of the structure. If the loads produce stresses in the rock approaching its strength it may become necessary to alter the design of the structure so as to reduce these stresses to safer values or to enhance the strength of the rock by techniques such as rockbolting or grouting. In extreme cases, as often occurs in mining, it may not be possible to design the excavation or strengthen the rock, so as to maintain stresses less than the strength of the rock. Hence the structural stability of failed rock has to be considered.

18.2 Criteria for the design and support of underground excavations

In mechanical or civil engineering practice it is customary to ensure the stability of a structure by designing it so that the stresses in each element of the structure are always less than the strength of that element, defined in some appropriate way. This situation is achieved by a suitable choice of the material, section, and disposition of the elements of the structure. Despite the highly developed state of design in many fields of engineering, success depends as much on experience and empirical results as it does on any fundamental understanding of the behaviour of the material of the structure. Thus, the strength that is used may be the yield point, the ultimate strength, or the fatigue strength, depending upon the nature of the load to which the structure is subjected. Also, 'safety factors' are used to allow for uncertainties concerning the applicability of the analysis which is used and to allow for the variability in the properties of the materials which are used.

The criterion that the stresses should always be less than the strength would certainly appear to be adequate to ensure the stability of a structure and to be applicable to those structures in the form of underground excavations. However, the options which the designer of an excavation has available to him are far fewer than those available to his mechanical or civil engineering counterpart. In the first place, other than by moving the site of the excavation, there is no choice of an alternative material, though, to a limited extent, it may be possible to reinforce the rock surrounding the excavation. Second, the rock around an excavation usually extends from the boundary of the excavation to the surface of the earth, so that the only options available in respect of the section and disposition of the structural elements lie in the shape of the excavation and the relative positions of neighbouring excavations. These options are often circumscribed by other considerations, especially in mining, where the major excavation results from the extraction of the ore body. Situations, such as an isolated tunnel in hard rock, exist where it is possible to satisfy the usual design criterion despite these limitations. However, in many excavations, especially those resulting from mining, it is not possible to keep the stresses everywhere in the rock less than its strength. Neither does experience indicate that this is necessary to ensure the stability of the excavation. The rock fails in parts surrounding many underground excavations, but only occasionally does this fact impair the stability or safety of the excavation. The criterion that the stresses must always be less than the strength may therefore be sufficient to ensure the stability of a structure, but it is certainly not necessary. In fact, it is often unacceptable, as it would preclude mining in many situations where it is practised with comparative safety. One is therefore forced to examine the conditions which are necessary for structural stability if rock mechanics is to be applied usefully to the design of many underground excavations. Observation of underground excavations surrounded by failed rock often shows that this rock, though failed, is still subject to stress. Turning to the ideas developed in connection with the complete stress–strain curve for rock, § 4.2, and the stiffness of testing machines, § 6.13, they suggest that, while the stresses in the failed rock must have exceeded its strength, the stresses applied to it are in stable equilibrium

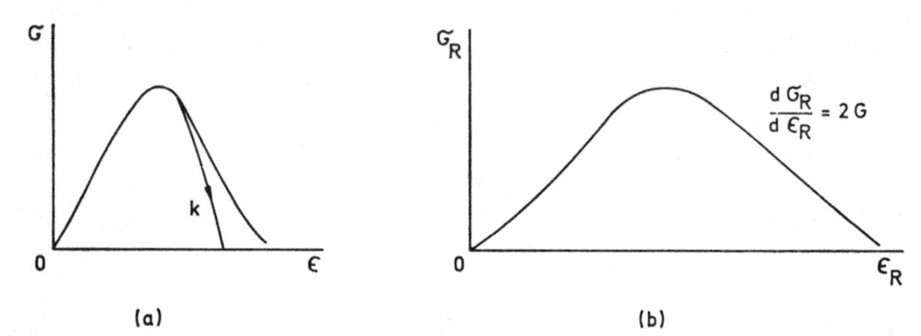

Fig. 18.2 (*a*) The complete tangential stress–strain curve for an annulus of failed rock in stable equilibrium with the tangential stress applied to it by its surround. (*b*) The complete radial stress–elongation curve for an annulus of failed rock in critical equilibrium with the radial stress applied to it by the surrounding elastic rock.

with its resistance to them. In § 12.4 it is shown that failure does not occur when sufficient energy is available but the stresses are too small, and it now seems that unstable failure cannot occur even when the stresses are adequate if insufficient energy is available.

This concept is best illustrated by a simple example of a circular tunnel subject to a hydrostatic field stress, p, such that the stresses in a thin annulus of rock of thickness t around the interior of a tunnel of radius R have just exceeded the strength of this rock. Examination of the complete tangential stress–strain curve for the failed annulus, Fig. 18.2 (*a*), shows that it is in stable equilibrium with the stresses applied to it by the surround of unfailed rock, provided that

$$\left|\frac{d\sigma}{d\varepsilon}\right| < |k|, \tag{1}$$

where σ and ε are the tangential stress and strain in the annulus and k is the stiffness in terms of annular tangential stress and strain of the stress applied by the surrounding rock to the annulus.

Instead of using (1), it is more convenient to consider the criterion for stability of the complete annulus with respect to the radial stress applied to it by the rock mass, in terms of this stress, σ_R, the radial elongation of the annulus, ε_R, and the displacement, u_R, at the outer circumference of the annulus, that is

$$\left| \frac{d\sigma_R}{d\varepsilon_R} \right| < | k_R |. \tag{2}$$

Static equilibrium between the tangential forces in the annulus and the radial forces on its circumference, R, requires that

$$R\sigma_R = t\sigma. \tag{3}$$

The diametral elongation of the annulus is

$$\varepsilon_R = u_R/R, \tag{4}$$

and the tangential strain in it is

$$\varepsilon = u_R/R, \tag{5}$$

so that

$$\frac{d\sigma_R}{d\varepsilon_R} = \frac{t}{R}\frac{d\sigma}{d\varepsilon}. \tag{6}$$

Assuming that for $r > R$ the rock around the tunnel behaves linearly and elastically, the stiffness of the radial stress applied to the annulus by its surround is given by § 5.11 (13) as

$$k_R = \sigma_R/\varepsilon_R = 2G. \tag{7}$$

Substituting (6) and (7) in (2) yields

$$\left| \frac{t}{R}\frac{d\sigma}{d\varepsilon} \right| < | 2G |, \tag{8}$$

as the criterion for the stability of a failed annulus with respect to the stresses applied to it by the surrounding rock. This shows that a thin annulus is 'soft' in relation to the stiffness of the stresses applied to it by the surrounding rock, and is therefore stable with respect to them.

It is interesting and relevant to consider the effects of a relatively greater field stress, resulting in a thicker annulus of failed rock. Assume that the failed rock is incompressible, so that the radial displacement at any circumference is governed by

$$ur = u_1R_1, \tag{9}$$

where R_1 is the radius of the inner circumference of the failed annulus. If the slope of the stress–strain curve of the failed rock in the brittle state is approximated by a constant, $-B$, then any radial compression, Δu_r, of the failed annulus leads to a reduction in the tangential stress in it given by

$$\Delta\sigma = -B\,\Delta u_r/r. \tag{10}$$

Consideration of the static equilibrium of forces across a diameter of the annulus due to the tangential stresses in the annulus and the radial stresses at its circumference yields

$$\Delta\sigma_R = -B\,\Delta u_R\left[\frac{1}{R_1} - \frac{1}{R}\right]. \tag{11}$$

The slope of the radial stress–strain curve of the annulus in terms of the stress and strain at R is therefore

$$\frac{d\sigma_R}{d\varepsilon_R} = -B\left[\frac{R}{R_1} - 1\right], \tag{12}$$

the modulus of which increases as the thickness of the annulus increases. As before, the annulus is stable with respect to the stress applied to it by its surround until

$$\left|\frac{d\sigma_R}{d\varepsilon_R}\right| \geqslant k_R, \tag{13}$$

that is,

$$\left[\frac{R}{R_1} - 1\right] \geqslant \frac{2G}{B}, \tag{14}$$

as shown in Fig. 18.2 (*b*). The preceding analysis can be expressed in terms of the energy absorbed by the radial compression of a failed annulus and that supplied by the surrounding rock, by multiplying the radial elongation by a factor $2\pi R^2$, to convert radial elongation to the change in volume. This shows that the failed rock is stable when insufficient energy is available from the surrounding elastic rock to complete the process of failure. Stability breaks down and instant failure ensues when sufficient energy is available from the surrounding elastic rock to destroy all the cohesion of the failed rock.

If the ratio $2G/B$ is relatively large, conditions for spontaneous instability as defined by (14) may never arise because the inner surface of the failed annulus can become so strained by stable deformation that its cohesion is completely destroyed before R is sufficiently large to satisfy the requirements of (14). In this case it is necessary to protect the tunnel not from violent failure but from closing due to the collapse and continued growth of the completely destroyed inner surface of the failed annulus.

At this stage it is worth considering these two situations for the Coulomb material, § 16.5, which fails according to § 4.6 (20). From § 16.5 (15), (19) the radial stress at R, the boundary between the failed and elastic rock, is

$$\sigma_r = p[2 - C_0/p]/(q + 1), \tag{15}$$

and from § 5.11 (11) the radial displacement of this boundary is

$$u = pR(1 - 2\nu)/2G + (p - \sigma_r)R/2G. \tag{16}$$

The maximum radial elongation of the inner circumference of the failed annulus before it becomes completely failed is

$$\varepsilon_{1(\max)} = C_0\left(\frac{1}{2(1+\nu)G} + \frac{1}{|B|}\right) = u_1/R_1. \tag{17}$$

From (9)

$$\varepsilon_1 = uR/R_1^2, \tag{18}$$

so that (16) can be written as

$$2G\varepsilon_1 = pR^2(1-2\nu)/R_1^2 + pR^2[1-(2-C_0/p)/(q+1)]/R_1^2 \tag{19}$$

using (15) and (18), from which

$$R^2/R_1^2 = C_0/p[2G/B + 1/(1+\nu)]/[2(1-\nu)-(2-C_0/p)/(q+1)]. \tag{20}$$

According to (14),

$$R^2/R_1^2 \geqslant (2G/B+1)^2, \tag{21}$$

for failure of the annulus to be unstable, so that instability in a Coulomb material occurs only if

$$(C_0/p)/[2(1-\nu)-(2-C_0/p)/(q+1)] \geqslant (2G/B+1)^2/[2G/B+1/(1+\nu)]. \tag{22}$$

Failure cannot occur at all unless C_0/p is less than 2, which shows that $2G/B$ must be less than unity or q must be very small for unstable failure to occur.

The influence of support must now be considered. If the annulus is not completely failed, a radial stress, σ_S, at its inner circumference due to support has the effect of diminishing the field stress, p, by σ_S, which is often negligible in comparison with p, and increasing the strength of the rock by $q\sigma_S$, which is also often negligible in comparison with C_0. A more important situation is that where an annulus, $R_S < r < R_1$, the cohesion of which has been completely destroyed, has to be supported. The radial stresses at R_S and R_1 in such an annulus are related by § 16.5 (13), which yields

$$\sigma_1 = \sigma_S(R_1/R_S)^{(q-1)}. \tag{23}$$

A properly supported annulus of completely failed rock provides greatly enhanced support at R_1 to the partially failed rock, which is equivalent to diminishing the field stresses by σ_1 and enhancing the strength of the rock in accordance with the appropriate failure criterion, § 16.5. If the support stress remains constant with radial compression the stability of the tunnel is enhanced only as a result of these two factors. Support which generates an increasing stress with compression also enhances the stability of the tunnel by 'softening' the combined behaviour of the partially and completely failed annuli.

The behaviour of the rock around a tunnel and its support can be summarized as follows: (1) Failure of the rock around a tunnel can occur without instability and violence and without complete loss of cohesion if the field stress is sufficient to start failure. (2) As the field stress increases, instability of the failed rock or complete loss of cohesion occurs, depending upon the properties of the rock. (3) If the fractured rock resulting from an unstable failure is retained by radial support stresses it can be used to enhance the effect of these stresses sufficiently to ensure that the partially failed annulus is kept to a stable thickness. (4) If the cohesion of the inner part of the annulus is completely destroyed by stable failure radial support of this can be used both to retain the cohesionless material in place and to prevent further growth of the failed annuli.

Support is generally provided by linings, arches, or rockbolts. Provided that the rockbolts are sufficiently long to be anchored in the solid elastic rock and to disperse their loads, the support generated by all these techniques can be equated to σ_S. If shorter rockbolts are used which are anchored in the failed rock, as is often the case, some modification to (23) becomes necessary. Consider the case of radial rockbolts generating an average support stress σ_S at R_S and anchored at R_A in the completely failed rock. The average inward radial stress generated by these bolts at R_A is $\sigma_A = \sigma_S R_S / R_A$. The effective outward radial support stress at R_A then becomes

$$\sigma_{AS} = \sigma_S (R_A/R_S)^{q-1} - \sigma_S (R_A/R_S)^{-1} = \sigma_S R_S [(R_A/R_S)^q - 1]/R_A. \quad (24)$$

The preceding discussion is also of direct relevance when considering the results of tests to determine the strength of rock from the unstable failure of specimens in the shape of hollow cylinders with inhomogeneous stresses, § 6.8. It can also be generalized to other cases involving inhomogeneous stresses.

18.3 The energy released by making an underground excavation

Having argued that the criterion for structural stability based on maintaining stresses in a structure always less than the strength is neither necessary nor acceptable when designing an underground excavation, and having demonstrated the fundamental importance of energy in this connection, it is necessary to analyse the energy changes brought about by making an excavation underground.

The changes which result from making an excavation deep in an elastic solid subject to gravity are readily studied with the aid of the reciprocal theorem, §§ 5.8, 12.2, 12.3. The theorem states that, for an elastic body in equilibrium under the action of two separate sets of stresses, the work done by the first set acting over the displacements produced by the second set is equal to the work done by the second set acting over the displacements produced by the first set.

Consider a large rectangular prism of rock surrounding the excavation and bounded by the surface of the earth and planes perpendicular to the axes of the principal virgin rock stresses, σ_1, σ_2, σ_3. Let the first set of stresses, designated by single superscripts, be stresses applied to the surface of the excavation only and let the value of these stresses be equal in magnitude but opposite in sign to those of the virgin rock stresses on these surfaces prior to excavation. The first set of stresses produces:

Displacements of the excavation surface having components u'_1, u'_2, u'_3 in the directions of the principal virgin rock stresses.

Displacements v'_2 and v'_3 of each of the vertical surfaces of the prism.

A displacement v'_1 of the surface of the earth. It is assumed that the bottom of the prism does not move relative to the centre of the earth.

The first set of stresses superposed on the virgin rock stresses yields the stresses around the excavation. The first set of stresses therefore corresponds to the *induced stresses*, and the displacements produced by them are identical with those that result from making the excavation.

Let the second set of stresses, designated by double superscripts, be the same as the first set within the excavation. In addition to this, however, let stresses be applied uniformly to the outer surfaces of the prism and let the value of the stress on each surface be equal to the average value over the surface of the excavation of the principal virgin rock stress, which is perpendicular to that surface of the prism. The second set of stresses produced:

(i) Displacements of the excavation surface u''_1, u''_2, u''_3, corresponding to the expansion of a volume of rock with the shape of the excavation when subjected to the same stresses.

(ii) Displacements of the surfaces of the prism which are of no concern as the values of the first set of stresses on these surfaces is zero.

Let da_1, da_2, da_3 be components of the surface of the excavation perpendicular to the axes of the principal virgin rock stresses and let dA_1, dA_2, dA_3 be components of the surfaces of the prism. The work done by the first set of stresses acting over the displacements of the second set is equal to the work done by the second set acting over the displacements of the first set. The former and latter amounts of work are given by the left- and right-hand sides, respectively, of the following equation

$$
\begin{aligned}
-\int_a \sigma'_1 u''_1\, da_1 - \int_a \sigma'_2 u''_2\, da_2 - \int_a \sigma'_3 u''_3\, da_3 \\
= \int_a \sigma''_1 u'_1\, da_1 + \int_a \sigma''_2 u'_2\, da_2 + \int_a \sigma''_3 u'_3\, da_3 - \\
\int_{A_1} \sigma''_1 v'_1\, dA_1 - 2\int_{A_2} \sigma''_2 v'_2\, dA_2 - 2\int_{A_3} \sigma''_3 v'_3\, dA_3. \quad (1)
\end{aligned}
$$

If stresses similar to those of the first set but equal in sign to the virgin rock stresses are applied to the surface of an excavation as it is mined, the rock mass is not disturbed by that excavation. The transition from this hypothetical undisturbed state of the rock mass to the practical case, disturbed by an excavation without surface stresses, is made by reducing the surface stresses to zero. In an elastic rock mass these stresses decrease linearly with displacement, as a result of which work is done on the surface stresses given by

$$W_R = \frac{1}{2}\int_a \sigma_1 u'_1 \, da_1 + \frac{1}{2}\int_a \sigma_2 u'_2 \, da_2 + \frac{1}{2}\int_a \sigma_3 u'_3 \, da_3. \tag{2}$$

This is half the sum of the first three terms on the right-hand side of (1) and is the minimum amount of energy which is necessarily released when an unsupported excavation is made underground. If the rock mass does not behave elastically the displacements are greater and the amount of energy which is necessarily released is greater than in the elastic case. The maximum amount of energy which can be released by any underground excavation is

$$W_R = \sigma_v V, \tag{3}$$

where V is the volume of the excavation and σ_v is the greatest value of the virgin rock stress around the excavation. The validity of (3) can be verified by considering the case of an excavation in a fluid subject to gravity. The amount of energy released by an excavation in an elastic rock mass can approach this maximum value closely when the opposite surfaces of the excavation make contact with one another, generating contact stresses with values close to those of the virgin rock stresses.

In addition to the energy which is released when an excavation is made underground, changes in potential energy occur due to the displacement of the surfaces of the prism. The displacements of the vertical surfaces of the prism abstract work from the horizontal components of the virgin rock stresses, the amount of which is

$$W_H = \sigma_2 V_2 + \sigma_3 V_3, \tag{4}$$

where V_2 and V_3 are the volumetric displacements given by the last two integrals in (1). The displacement of the horizontal surface of the prism represents the amount by which the rock mass above the excavation moves towards the centre of the earth, resulting in a loss of gravitational potential given by

$$W_G = V_1 \sigma_1, \tag{5}$$

where V_1 is the volumetric displacement given by the fourth integral on the right-hand side of (1). The sum of the integrals on the left-hand side of (1) is $-2W_v$, where W_v is the strain energy due to the virgin rock stresses in the volume of rock removed by mining.

It is convenient to express (1) in terms of these energy changes as

$$W_G + W_H = 2W_R + 2W_v = W_R + W_I + 2W_v. \tag{6}$$

Defining the stored energy, W_S, as the difference between the total energy change and the energy which is released it becomes

$$W_S = W_I + 2W_v, \tag{7}$$

where $W_I = W_R$ is the strain energy around the excavation generated by the induced stresses, § 5.8 (23). When the virgin rock stresses are hydrostatic, the integrals in (1) represent volumes multiplied by a constant stress and (6) can be written as

$$V_f = (V_1 + V_2 + V_3) - V_v, \tag{8}$$

where V_f is the volumetric convergence of the excavation and V_v is the volumetric compression by the virgin rock stress of the volume of rock originally in the excavation. This shows that the volumetric compression of the prism exceeds the volumetric convergence of the excavation by V_v.

It is informative to illustrate the preceding analysis by applying it to the simple example of a deep, circular tunnel of radius R_1 in rock where the field stress, p, is hydrostatic. Consider a portion of the rock mass in the annulus $R_2 \geqslant r \geqslant R_1$, where $R_2 \gg R_1$, so that the virgin rock stress at R_2 is virtually unaffected by the presence of the tunnel. Formulae for the stresses, displacements, and strain energy in such an annulus are given in § 5.11. Prior to driving the tunnel the strain energy per unit length of the annulus due to the virgin rock stress, p, is, by § 5.11 (21),

$$W_i = \pi(1 + \nu)(1 - 2\nu)p^2(R_2^2 - R_1^2)/E. \tag{9}$$

After the tunnel is driven it becomes

$$W_f = \pi(1 + \nu)p^2[(1 - 2\nu)R_2^4 + R_2^2R_1^2]/E(R_2^2 - R_1^2) \tag{10}$$

so that the additional stored energy due to the tunnel, for $R_2 \gg R_1$ so that R_1^2 is negligible in comparison with R_2^2, is

$$W_S = W_f - W_i = \pi p^2R_1^2(1 + \nu)(3 - 4\nu)/E. \tag{11}$$

From § 5.11 (21)

$$W_I = \pi p^2R_1^2(1 + \nu)/E, \tag{12}$$

and

$$W_v = \pi(1 + \nu)(1 - 2\nu)p^2R_1^2/E. \tag{13}$$

Substituting (12) and (13) in (7) yields

$$W_S = \pi p^2R_1^2(1 + \nu)(3 - 4\nu)/E, \tag{14}$$

which is the same as (11).

The displacements induced at R_2 by driving the tunnel are found, using § 5.11 (8), to be

$$u_{21} = 2(1 + \nu)(1 - \nu)pR_1^2R_2/(R_2^2 - R_1^2)E. \tag{15}$$

The corresponding volumetric compression of the annulus by the virgin rock stress, p, when $R_2 \gg R_1$, so that R_1^2 is negligible compared with R_2^2, is

$$V = 4\pi(1 + \nu)(1 - \nu)pR_1^2/E, \tag{16}$$

and the work done by the rock mass is

$$(W_G + W_H) = 4\pi(1 + \nu)(1 - \nu)p^2R_1^2/E. \tag{17}$$

Using (12) and (13),

$$2(W_I + W_v) = 4\pi(1 + \nu)(1 - \nu)p^2R_1^2/E, \tag{18}$$

so that (17) and (18) are in accordance with (6).

The preceding analysis has shown that the energy which is released when an excavation is made depends upon the volumetric convergence of the excavation and that the energy which is stored in stress concentrations around the excavation depends upon this and the volume of rock removed in making the excavation. By definition, the stored energy is theoretically recoverable and cannot contribute to the destruction of the rock around the excavation. However, the stored energy does determine the stresses around an excavation, and therefore affects any other excavations which are made in this stress field. The released energy is the quantity most directly related to the failure of rock around an underground excavation. It is conveniently studied by means of stress–volume diagrams showing the volumetric convergence of an excavation, as the stresses on its surface are diminished from values equal to the virgin rock stress, σ_v, to values equal to the stresses generated by support in the excavation, σ_s.

In the case of an open excavation in an elastic rock mass, Fig. 18.3.1 (*a*), the released energy, W_R, and the contribution of convergence to the stored energy, W_I, are equal if failure does not occur. If the surfaces of an excavation make contact with one another or with support stresses W_R is greater than W_I, Fig. 18.3.1 (*b*). When failure occurs the volumetric convergence and the released energy increase, Fig. 18.3.1 (*c*).

The amount and degree of rock failure around an excavation can be reduced by decreasing the volumetric convergence of an excavation. This can be accomplished by minimizing the volume of the excavation, introducing artificial support, filling the excavation, or leaving pillars in the excavation.

If the surfaces of an excavation are supported by evenly distributed loads generating a mean stress, σ_s, the volumetric convergence of the excavation, V, is less than the volumetric convergence, V_f, when the surfaces are free from stress. However, the mean stress during convergence increases, Fig. 18.3.2 (*a*), so that the difference between the energy released by an unsupported and a supported excavation is

$$\Delta W_R = V_f \sigma_s^2 / 2\sigma_v, \tag{19}$$

which is less than the difference between the contributions to the stored energy due to convergence,

$$\Delta W_I = V_f \sigma_s (1 - \sigma_s / 2\sigma_v). \tag{20}$$

It should be noted that if the support stress is applied during convergence a fraction of the released energy is used in compressing the support which

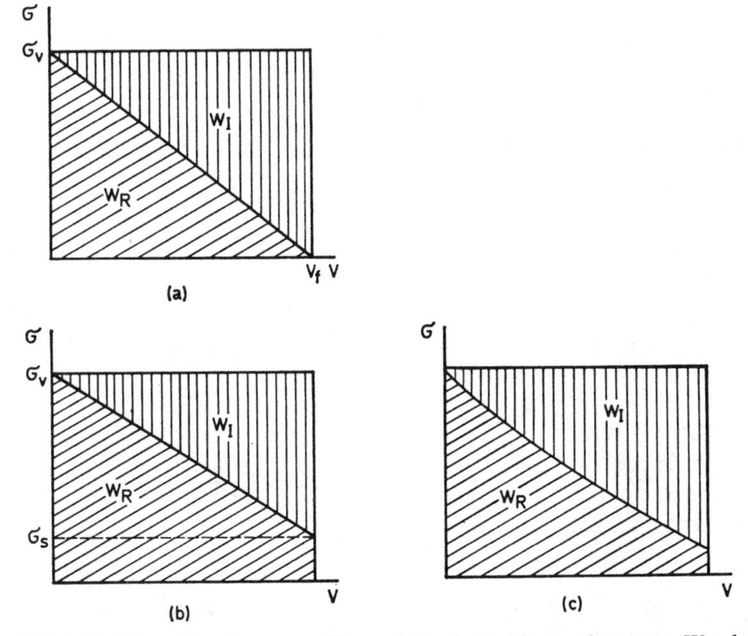

Fig. 18.3.1 (*a*) The released energy, W_R, and the induced stored energy, W_I, due to convergence of an open excavation in a completely elastic rock mass. (*b*) The released and stored energies due to an excavation with surface loads in an elastic rock mass. (*c*) The released and stored energies when failure of part of the rock around an excavation occurs.

makes $\Delta W_R = \Delta W_I$. Equations (19) and (20) show that the support stress must be a significant fraction of the virgin rock stress if it is to be effective in reducing the amount of energy released and hence the damage to the rock surrounding an excavation.

Filling an excavation with sand or rock is effective in reducing the amount of energy released by the excavation only if the fill reduces the volumetric convergence of the excavation. To do this, the volume of fill, when compacted, must exceed the volume of the void which would be left in the same excavation without support. Provided that this condition is satisfied, the effects of filling are shown in Fig. 18.3.2 (*b*) for the case where little convergence is required to compress the fill, that is, where the volumetric convergence in the absence of support is a large fraction of the volume of rock removed from the excavation. Filling is less effective where the volumetric convergence of the unsupported excavation is a small fraction of the volume of rock extracted from the excavation and a

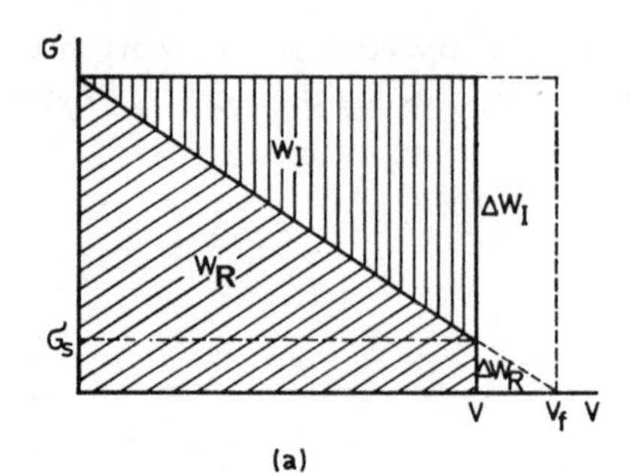

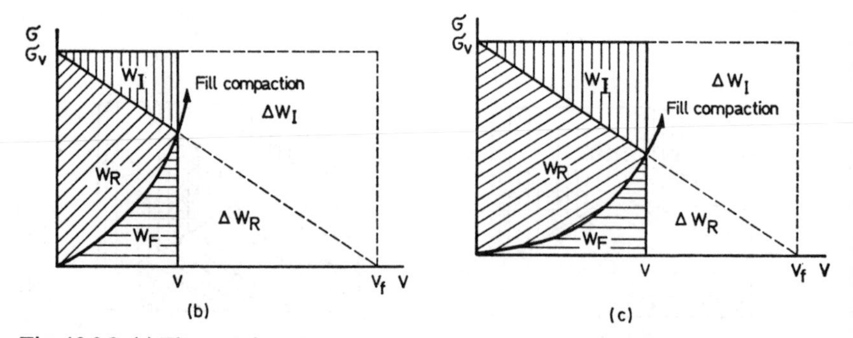

Fig. 18.3.2 (*a*) The energies released and stored due to the convergence of an excavation with a support stress, σ_s. The released, W_R, stored, W_I, and absorbed, W_F, energies in an excavation with relatively incompressible fill (*b*) and relatively compressible fill (*c*).

relatively large volume of fill has to be compressed, Fig. 18.3.2 (*c*). The difference between the energy released by an unsupported and a filled excavation is

$$\Delta W_R = \sigma_v (V_f - V)^2 / 2V_f. \tag{21}$$

Some of the released energy, W_F, is absorbed in compaction of the fill. The difference between the contribution to the stored energy due to convergence of unsupported and filled excavations is

$$\Delta W_I = \sigma_v (V_f^2 - V^2) / 2V_f. \tag{22}$$

18.4 Tabular excavations

Many mineral deposits, such as most coal seams and the Witwatersrand gold reefs, occur in the form of thin strata extending over very great distances compared with their thickness. The planes of these strata are often almost horizontal, though they may be inclined at any angle up to the vertical. Mining of these deposits gives rise to *tabular excavations* which have a negligible thickness in comparison with their lateral extent and depth below surface. This type of excavation warrants special discussion on account of its importance and frequency of occurrence and because its geometry makes it amenable to relatively simple analysis.

To avoid having to consider gravitational body forces it is convenient to work in terms of induced stresses and displacements, obtaining the total stresses and displacements by superposing the induced components on the virgin components. In the case of deep excavations, where the induced components are negligible at a distance from the excavation equal to the depth below surface, solutions for the stresses around openings in infinite media can be used to calculate the induced components of stress and displacement. If allowance must be made for the proximity of the surface this can be done by iteration; superposing on the surface, stresses equal in magnitude but opposite in sign to the induced stresses at the surface.

Many tabular excavations have their major axis so much longer than their intermediate axis that they can be regarded as two-dimensional and can be analysed accordingly with fair approximation. Furthermore, two-dimensional analysis provides a maximum estimate of the induced changes, as the effect of truncation along the third dimension can only provide additional support which is assumed to be absent in the two-dimensional analysis. It should also be noted that a distinct similarity exists between the analysis of tabular excavations in the form of very flat ellipses or slits and the analysis of Griffith cracks, §§ 10.12, 10.13, 10.14, 12.4.

First, consider a tabular excavation approximating an isolated slit with a normal component of the virgin rock stress, σ_v, perpendicular to the plane of the excavation and negligible shear components of stress in this plane. In this situation the strain energy in the volume of rock removed by mining, W_v of § 18.3, is negligible and the amount of energy released by the excavation, W_R, and the stored strain energy, W_S, are very nearly equal, until the slit closes. For a unit length of open slit of width $2c$ the convergence between the hangingwall and footwall at any distance x from its centre is by § 10.11 (24a),

$$2v = 4(1 - \nu^2)\sigma_v(c^2 - x^2)^{\frac{1}{2}}/E, \tag{1}$$

and the released and stored energies are, by § 12.2 (11),

$$W_R = W_S = \pi(1 - \nu^2)c^2\sigma_v^2/E. \tag{2}$$

The convergence, $2v$, becomes equal to the thickness, v_0, of the slit and contact between the hangingwall and footwalls occurs at a width $2c_0$ given by

$$c_0 = v_0 E/4(1 - \nu^2)\sigma_v. \tag{3}$$

For widths greater than $2c_0$ the stored energy approaches a maximum value $W_{S\,\max} = \pi v_0 c_0 \sigma_v/2$ and the released energy approaches a value $W_R = 2c_0 v_0 \sigma_v$, Cook *et al.* (1966).

Second, consider the important case of a number of tabular excavations in one plane. Salamon (1964) has calculated the volumetric convergence of a series of slits on centres of $2S$. From this result the energy released and stored by a unit length of each slit is found to be

$$W_R = W_S = 8(\nu^2 - 1)S^2\sigma_v^2 \ln(\cos \pi c/2S)/\pi E. \tag{4}$$

While the total stored strain energy provides a measure of the extent and magnitude of the stress concentrations around an excavation, in practice it is the amount of energy released per unit extension to the width of an excavation which determines the amount of damage done to the rock in the vicinity of the working face. The spatial rates of energy release for an isolated slit-like excavation and a series of such excavations are

$$\frac{\partial W_R}{\partial c} = 2\pi c(1 - \nu^2)\sigma_v^2/E, \tag{5}$$

$$\frac{\partial W_R}{\partial c} = 4S(\nu^2 - 1)\sigma_v^2 \tan\left(\frac{\pi c}{2S}\right)/E, \tag{6}$$

respectively. These have the characteristics illustrated in Fig. 18.4.1, which show that significant interaction between a series of co-planar excavations occurs only after about three-quarters of the area has been mined.

The extent of the interaction between excavations in two different planes can be gauged from § 10.16 (28). This shows that the stress variation due to a harmonic load on a plane decreases exponentially with distance from that plane expressed in terms of the wavelength of the load. Taking the fundamental wavelength $2S$ for the pillar loads between slits, it is seen that this variation is significantly attenuated at a distance S from the plane of the slits. The interaction between two planes of such excavations becomes negligible if their separation is $2S$ or more.

The relationship between rock failures, strata control problems, and the spatial rate of energy release has been discussed by Cook (1967) and investigated by Hodgson and Joughin (1967). Observations of the incidence of violent rock failures were made using continuous recording seismic networks at two mines. On the one mine the workings at an average depth of 9,000 ft below surface were surrounded by hard, massive quartzites, and at the other mine the average depth was 5,000 ft below surface, and the

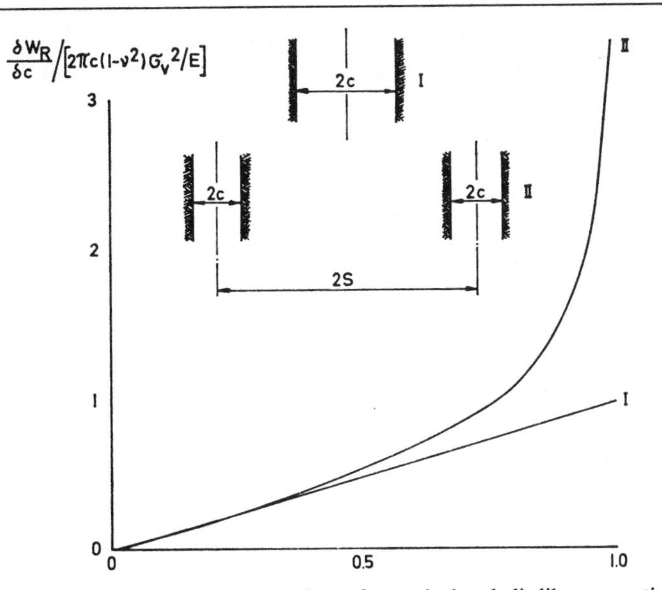

Fig. 18.4.1 The spatial rates of energy release for an isolated slit-like excavation in an elastic rock mass and a co-planar series of such excavations.

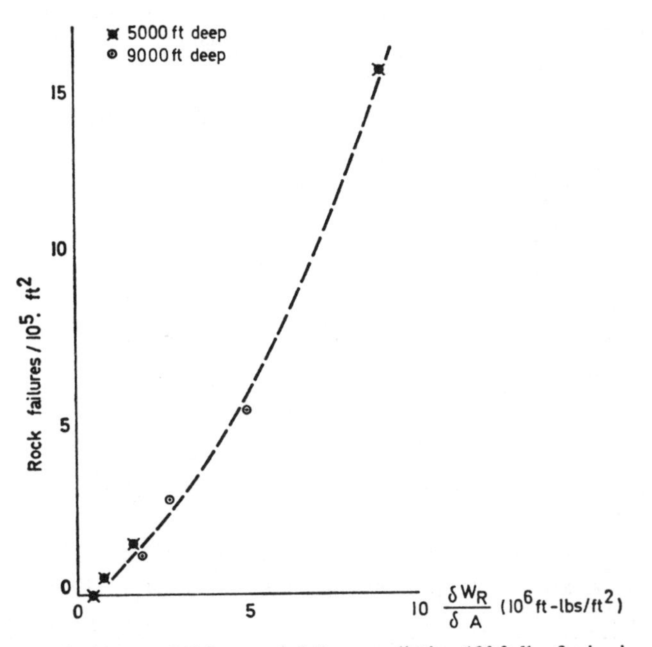

Fig. 18.4.2 The incidence of violent rock failures radiating 10^6 ft-lb of seismic energy or more as a function of the spatial rate of energy release at two deep mines.

hangingwall consisted of quartzite interbedded with conglomerates and weak shales. The incidence of violent rock failures was found to increase with the spatial rate of energy release in a very similar manner for both mines, Fig. 18.4.2. Furthermore, problems caused by rockfalls were found to follow a very similar pattern.

18.5 An analogue for determining displacements and stresses around tabular excavations

Frequently it is necessary to study the stresses and displacements in the immediate vicinity of tabular excavations in some detail. Where the shape of such excavations is complicated, simple geometrical approximations, such as those discussed in the preceding section, are inadequate, and a more general method of analysis is desirable.

In the case of deep, tabular deposits, the virgin rock stress across the plane of the deposit is often essentially normal, and only negligible shear stresses are induced on this plane by mining. This plane is approximately one of symmetry and is, therefore, a plane of principal stress across which only normal stress changes are induced by mining. The effect of these changes in stress may be calculated from the formulae of § 10.17 for normal load $p(x, y)$ applied to the surface of the semi-infinite region $z > 0$. In practice, only the vertical displacement w, closure, and the horizontal displacements u, v, ride, in the plane $z = 0$ are needed. The displacement w in the plane of the deposit $z = 0$ is by § 10.17 (44)

$$[w]_{z=0} = -\frac{\lambda + 2G}{4\pi G(\lambda + G)} \iint \frac{p(\xi, \eta)\, d\xi\, d\eta}{[(x-\xi)^2 + (y-\eta)^2]^{\frac{1}{2}}},$$

where the integral is taken over the plane of mining.
Now define

$$W = -\frac{\lambda + 2G}{4\pi G(\lambda + G)} \iint \rho^{-1} p(\xi, \eta)\, d\xi\, d\eta, \tag{1}$$

where $\rho = [(x-\xi)^2 + (y-\eta)^2 + z^2]^{\frac{1}{2}}$,

so that $[W]_{z=0} = [w]_{z=0}.$ (2)

W is the Newtonian potential of a distribution

$$-(\lambda + 2G)p(x, y)/4\pi G(\lambda + G)$$

on the surface $z = 0$, so that, as can be verified directly,

$$\nabla^2 W = 0, \tag{3}$$

and $$\left[\frac{\partial W}{\partial z}\right]_{z=0} = \frac{\lambda + 2G}{2G(\lambda + G)} p(x, y) = 2\sigma_z \,.\, (1 - \nu^2)/E. \tag{4}$$

The problem thus reduces to solving Laplace's equation in the region $z > 0$ with prescribed values of W or $\partial W/\partial z$ on the surface, and this can be done by the electrical analogue described below.

It should be made clear that W does not give the value of the displacement w at a general point x, y, z; it is a harmonic function which has the same value as w in the plane $z = 0$ only. It is proportional to the function Φ introduced in § 10.17. It should also be said that values for the plane $z = 0$ cannot be obtained from the general formulae § 10.17 (35)–(43) merely by putting $z = 0$ without further discussion, since, when $z = 0$, ρ vanishes at some point within the region of integration. It can be verified that the terms in § 10.17 (35)–(37) which contain ρ^{-3} do vanish when $z = 0$.

The same procedure may be used to calculate u and v in the plane $z = 0$. Considering u, let

$$U = \frac{1}{4\pi(\lambda + G)} \iint \frac{(x - \xi)}{\rho(z + \rho)} p(\xi, \eta)\, d\xi\, d\eta, \tag{5}$$

so that

$$[U]_{z=0} = [u]_{z=0}$$

and also

$$\nabla^2 U = 0.$$

It follows from (1) and (5) that

$$G\frac{\partial W}{\partial x} + (\lambda + 2G)\frac{\partial U}{\partial z} = 0,$$

so that

$$[U]_{z=0} = \frac{G}{\lambda + 2G} \int_0^\infty \frac{\partial W}{\partial x} dz, \tag{6}$$

so that U, and similarly V, can be obtained from a knowledge of W.

The induced stress normal to the plane of a tabular excavation can therefore be found from the derivative of the component of displacement.

Many physical phenomena are described by Laplace's equation. Consider an elementary brick of conductive material Δx, Δy, Δz with a specific resistance R. Let e_1 and e_2 be the electrical potentials in the middle of the opposite faces of this brick in the x direction, and let i_1 and i_2 be the currents flowing into the faces. The voltage gradient across each face is expressed in terms of the current across the face and the resistance, so that

$$\frac{\partial e_1}{\partial x} = \frac{i_1 R}{\Delta y \Delta z} \quad \text{and} \quad \frac{\partial e_2}{\partial x} = \frac{i_2 R}{\Delta y \Delta z} \tag{7}$$

and, as $\Delta x \to 0$,

$$\frac{1}{\Delta x}\left\{\frac{\partial e_1}{\partial x} - \frac{\partial e_2}{\partial x}\right\} = \frac{i_1 R}{\Delta x \Delta y \Delta z} - \frac{i_2 R}{\Delta x \Delta y \Delta z}. \tag{8}$$

Similar expressions can be derived for the y and x directions, and the sum of the currents flowing into the cube must be zero. Therefore,

$$\frac{\partial^2 e}{\partial x^2} + \frac{\partial^2 e}{\partial y^2} + \frac{\partial^2 e}{\partial z^2} = 0. \tag{9}$$

If the elementary brick has finite dimensions and its electrical properties are simulated by six resistors, R_1 through R_6 running from the middle of each face to a common node at the centre of the brick, the resistors must have values of

$$R_1 = R_2 = \frac{R}{2}\frac{\Delta x}{\Delta y \Delta z}, \tag{10}$$

with similar expressions for the y and z directions. Since the sum of the currents flowing into the central node at an electrical potential e_0 is zero,

$$(e_1 - e_0)/R_1 + (e_2 - e_0)/R_2 + (e_3 - e_0)/R_3 + (e_4 - e_0)/R_4 + (e_5 - e_0)/R_5 + (e_6 - e_0)/R_6 = 0. \tag{11}$$

Substituting the values from (10) in (11) and multiplying by

$$R/(2\Delta x \Delta y \Delta z)$$

yields

$$(e_1 - e_0)/\Delta x^2 + (e_2 - e_0)/\Delta x^2 + (e_3 - e_0)/\Delta y^2 + (e_4 - e_0)/\Delta y^2 + (e_5 - e_0)/\Delta z^2 + (e_6 - e_0)/\Delta z^2 = 0, \tag{12}$$

which is the finite-difference approximate to Laplace's equation.

A solid cube of electrically conductive material can be used, with appropriate boundary conditions, to provide an analogue solution to (3) and (4), Salamon *et al.* (1964). In practice, it is easier to approximate the conductive cube by a rectilinear network of resistors representing an agglomeration of many elementary cubes such as the one described above. If one face of the large composite cube is chosen to represent the plane containing the excavation it can be seen from (3) and (9) that voltages at points on this face are analogous to the displacements induced at corresponding points in the plane of the excavation and from (4) and (7) that current densities at points on this face are analogous to the normal stresses at corresponding points in the plane of the excavation. The factors of proportionality between voltage and displacement and current and stress depend upon the electrical properties of the resistance network, the mechanical properties of the rock, the virgin rock stresses, and the scale between the size of the excavation and its model on the face of the cube of resistors. To facilitate consideration of the boundary conditions, it is convenient to divide the plane of the excavation into four areas, namely: the area where the excavation is open, A; the area where its opposite surfaces have made contact, C; the area where it is subject to significant support loads, S; and the unmined remainder, B. In each of these areas the boundary conditions can be expressed in terms of the relative convergence between the opposite surfaces of the excavation, the stresses between these surfaces, or a functional relationship between these variables which is also amenable to simulation on the analogue. Thus, in the simplest case of an open excavation in hard rock, where the compression of the unmined ore is negligible compared with the closure in the excavation, the displacement, and hence the voltage, in B is zero, while the stress, and hence the current, in A is zero. When such an excavation closes over part of its area the boundary conditions are as above, save that the voltage in C is constant with a value proportional to the thickness of extraction. Where the compression of unmined seam, or of significant support, must be taken into account, this may be done by connecting areas B and C on the analogue to electrical components simulating the stress–strain properties of the seam or of the support.

The appropriate boundary conditions are conveniently obtained by applying some suitable voltage to the face of the cube of resistors opposite that representing the plane of the excavation and making provision so that the nodes of the resistors on the face representing the plane of the excavation can be short-circuited to ground, zero voltage, open-circuited, zero current, or connected to some intermediate impedance or voltage. In this case zero voltage corresponds to the displacement of the plane of the excavation before any mining, that is, zero induced displacement and the current when the whole face of the cube is short-circuited, that is, no mining, corresponds to the virgin rock stress. The factors relating voltage and current to displacement and stress are found by calculation or calibration in terms of a simple geometry having an analytical solution.

The amount of energy released by the extension of any excavation is given by the integral over the area of that extension of half the product of the stresses in that area before it is mined and the displacements of the same area after it is mined. These factors are obviously proportional to the nodal currents and voltages in the area of the analogue representing the extension, and the integrals can be approximated by the sum over the relevant nodes of half the product of current and voltage.

The stresses and displacements outside the plane of the excavation are found by application of the principal of superposition. In § 10.17 the stresses and displacements in a semi-infinite solid $z \geqslant 0$ due to a point load, P, and a distributed pressure, $p(x, y)$, normal to its surface are given. The stresses or displacements anywhere in the solid can be found by applying these formulae to the stresses and displacements in plane of the excavation obtained from the analogue. It is often convenient to work in terms of the induced stresses and displacements and to add these to the virgin conditions to obtain the total stresses and displacements. To facilitate computation it is also convenient to replace the integration by summation.

18.6 Digital methods for the analysis of three-dimensional excavations

By making the assumption that the relationship between stress and strain in the rock is linear, it becomes possible to solve for the stresses, strains and displacements induced by any excavation in the rock around it from the superposition of a number of elementary solutions. The use of superposition in elasticity is common but the advent of digital computers has made it practicable to extend this principle on an unprecedented scale; all that is required is that the functions in the set of elementary solutions be linearly independent and otherwise well-behaved, mathematically (Deist *et al.*, 1973; Salamon, 1974).

Specifically, consider a large, infinite or semi-infinite, continuous solid mass of rock fixed in space at infinity. Within this solid mass define only the boundary of the excavation by a number of plane surface elements. At the centroid of each surface element apply three components of traction each of unknown magnitude. The stresses and displacements caused by each component anywhere in the rock mass, and particularly on every surface element defining the boundary of the excavation, can be calculated in terms of its, unknown, magnitude. To determine the magnitudes of each component for every surface element, the sum of these components and of all the stresses, or displacements, generated by all the other components on this surface element must be equated to the boundary values on this surface element. The large number of equations which result can be solved with the aid of a digital computer. In the case of an open excavation, this sum must result in stresses on each surface element equal in magnitude but opposite in sign to the virgin rock stresses or the field stresses before the excavation was made. In general, the boundary conditions may be specified in terms of stresses or displacements or both. Having defined the boundary conditions fully in this way, it becomes a simple matter to find the stresses, strains or displacements induced by the excavation at any point in the rock mass around it. All that is necessary is to sum the contributions to the stresses, strains or displacements at that point from each surface element, using an appropriate elementary solution for the effect of a force in an extended elastic body and the magnitudes for the components of the traction vectors which satisfy the boundary conditions.

Having described the method of solution in principle, it can now be stated formally. Let the boundary of the excavation only be defined by a number, N, of plane surface elements in an extended, continuous elastic body fixed at infinity. Assume that a component of force $P_l{}^j$ is applied at point j, the centroid of a surface element. Define the components of displacement, $u_k{}^i$, produced by this force at point, i, the centroid of another surface element, as

$$u_k{}^i = h_k{}^i{}_l{}^j P_l{}^j, \tag{1}$$

where $i, j = 1, 2, \ldots N$ and $k, l = 1, 2$ and 3, the coordinate axes.

Denote the $3N \times 1$ vectors of the components of the displacements by matrix $[\mathbf{U}]$, and those of the components of force by matrix $[\mathbf{P}]$. Let $[\mathbf{H}] = [h_k{}^i{}_l{}^j]$ be the $3N \times 3N$ matrix defining the relationship between

[U] and [P]. The particular relationship for points i and j given in (1) can be written for every surface element as

$$[\mathbf{U}] = [\mathbf{H}][\mathbf{P}]. \tag{2}$$

The reciprocal theorem and Clapeyron's theorem, § 5.8, can be used to show that [H] is symmetrical and positive definite (Deist *et al.*, 1973). This matrix is determined fully if all three components of displacement at point, i, caused by all three components of the force at point, j, are known. Elements of this matrix for homogeneous and isotropic elastic bodies of infinite and semi-infinite extent have been derived by Lord Kelvin (Love, 1927) and Mindlin (Sokolnikoff, 1956).

If P_l^j is a unit force at the centroid, j, of one surface element, the components of stress induced by it at the centroid, i, of another surface element are σ_{kr}, where k, $r = 1$, 2 and 3. Let the components of a unit normal to the surface element around i and pointing towards the interior of the excavation be n_r^i, so that the components of traction on this element are $n_r^i\sigma_{kr}$. The summation of all the tractions over the element around i gives force components $b_k^i{}_l^j$. Let the 3N × 3N matrix $[\mathbf{B}] = [b_k^i{}_l^j]$.

As the body is fixed rigidly at infinity, any distribution of forces acting on the surface of the excavation must induce positive definite strain energy in it. The matrix [B] is, therefore, non-singular.

The problem has been defined by

$$[\mathbf{U}] = [\mathbf{H}][\mathbf{P}] \tag{3}$$

and

$$[\mathbf{F}] = [\mathbf{B}][\mathbf{P}]. \tag{4}$$

The solution involves the computation of the magnitudes of the components of each vector of [P] so that the boundary conditions are satisfied. The boundary conditions comprise the specification of [U] or of [F] or of a combination of [U] and [F]. In practice the number of surface elements is large and it is not practicable to invert [B] nor [H] direct, so that some iterative solution must be used. Once values for each component of each vector of [P] are known, the stresses, strains or displacements in the rock surrounding the excavation can be found by superposing 3N times the elementary solutions such as those derived by Kelvin and Mindlin.

18.7 Pillars and ribs

In many systems of mining, such as bord and pillar mining of coal, parts of the mineral body are left unmined to form pillars for the purposes of support. In other systems waste rock is brought in to the mined areas to fill them in parts or completely for the same purpose. Unmined pillars usually form a regular pattern of square or rectangular blocks, and waste is often packed to form long ribs.

It is relatively simple to set an upper limit to the load which a pillar or fill may be required to carry. Consider an extensive area which has been mined according to some regular pattern. Let A_p be the area of the pillars or fill and let A_m be the area which has been mined. The maximum load which can come on to the pillars or fill is equal to the entire load carried by the rock across the plane of the excavation before mining. If the virgin rock

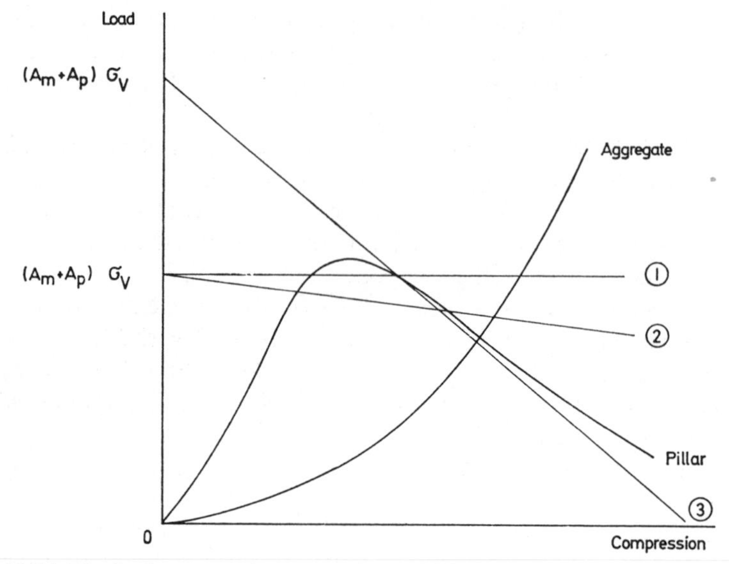

Fig. 18.7.1 The load–compression curves for an initially unbroken pillar and an aggregate fill, with three different load–convergence curves.

stress across the plane of the excavation is σ_v, then the average stress on the pillars or fill, σ_{av}, is

$$\sigma_{av} = (A_m + A_p)\sigma_v/A_p. \tag{1}$$

The resilience of the rock around an excavation and the compressibility of the pillars or fill combine to decrease the average stress to a value less than that given by (1). Consider the compression of a pillar or waste fill as mining proceeds away from it. Initially the load on a pillar corresponds to the virgin rock stress σ_v and that on a fill is zero. As mining proceeds, the pillar or fill is compressed by convergence between the hangingwall and footwall, and the load or stress on it increases. However, the hangingwall and footwall must possess finite strength and resilience if it is to bridge the spaces between pillars as is supposed in the derivation of (1). Some of the pillar load must be shed to neighbouring abutments in the process of convergence. The complete load–compression curves for a pillar and an aggregate fill are shown in Fig. 18.7.1, together with load–convergence curves for three different conditions for the convergence between hangingwall and footwall. The first condition (1) corresponds to the case where mining extends over an infinite area and no contact between hangingwall and footwall occurs, so that the load remains constant with convergence. The second condition (2) corresponds to the case where the area of mining is very extensive but some load transfer occurs with convergence. The third condition (3) corresponds to the case where abutments or barrier pillars exist in the vicinity and considerable load transfer occurs with convergence. Equilibrium holds where the load–compression curve of the pillar or fill intersects the load–convergence curve. Under the third condition unmined pillars provide stable support, even though they become crushed and their strength is less than the load calculated by (1). The slope of the load–compression curve of an aggregate fill is always positive, and such fills provide stable support under all conditions. The complete load–compression curves for initially solid pillars have not yet been investigated, but the strength of such pillars has been studied.

The axial compressive stress between a pillar and the hangingwall and footwall generates a compressive stress parallel to these surfaces which is directed radially inward towards the centre of the pillar. The constraint provided by this stress at the ends of a solid pillar has a considerable strengthening effect. Several investigators have attempted to assess the magnitude of this effect. Obert and Duvall (1967) propose that § 6.2 (1) can be used for this purpose. Field measurements, Greenwald *et al.* (1939), and laboratory measurements, Steart (1954), Holland and Gaddy (1957), and Holland (1962), of the strength of pillars of different sizes have led to a simple power relationship between height, h, and width, w, for the strength of rectangular coal pillars of the form

$$C_p = C_1 w^{\alpha}/h^{\beta}. \tag{2}$$

From a statistical analysis of case histories of pillars in South African collieries Salamon and Munro (1967) concluded that the most likely values for the constants in (2) are

$$\alpha = 0{\cdot}46, \quad \beta = 0{\cdot}66, \quad C_1 = 1{,}320 \text{ psi},$$

where C_1 is the nominal strength of a cubical pillar of 1 ft side. Note that, for (2) to be dimensionally consistent, w and h must be expressed as dimensionless multiples of the width and height of this cubical pillar, by normalizing them with respect to its dimensions of 1 ft.

The method of testing pillars underground described in § 15.8 has been applied to coal pillars in South Africa by Wagner (1974) using a 100 Meganewton jacking system designed and built by Cook *et al.* (1971). Several important results have emerged from these large-scale underground tests. First, complete load-compression curves for coal pillars of different

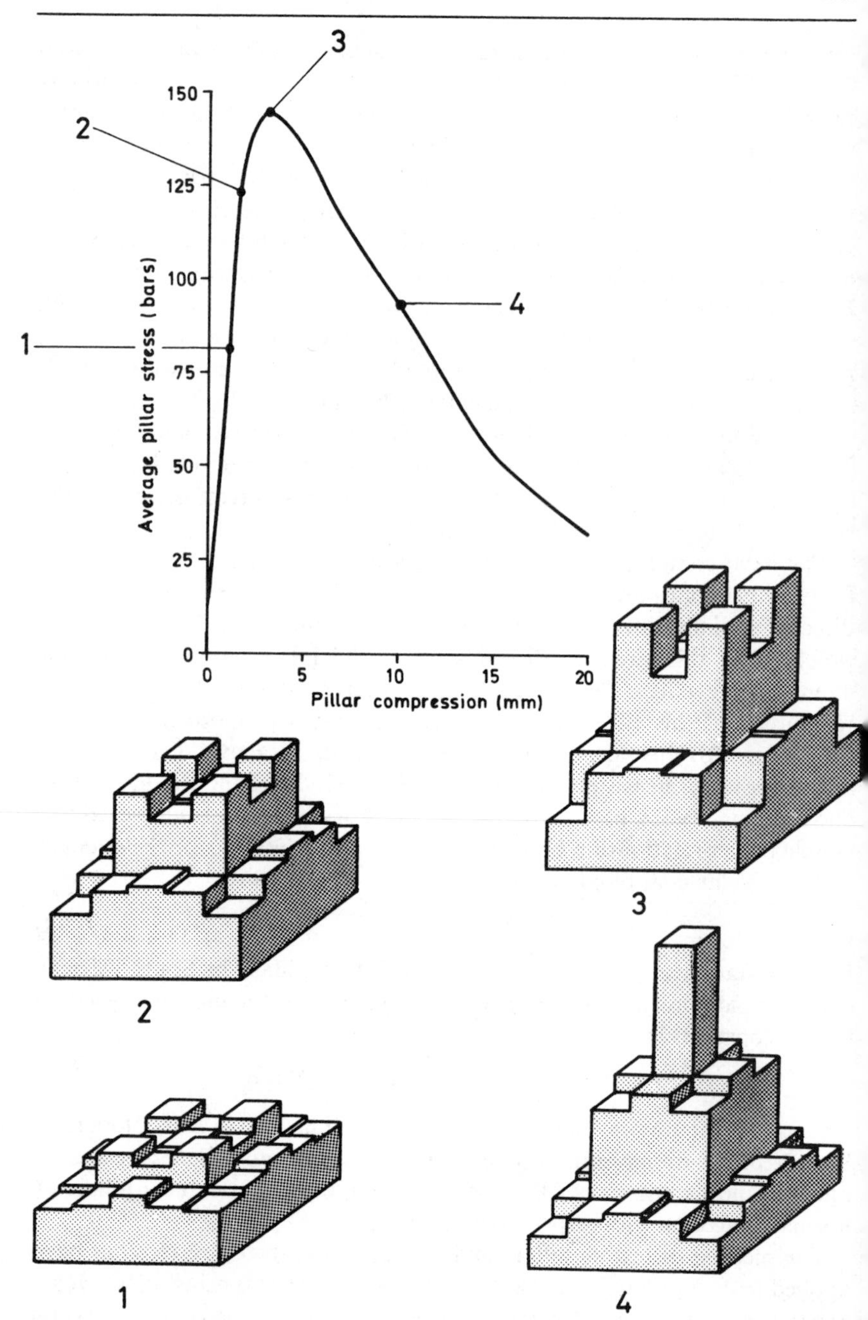

Fig. 18.7.2 A complete average stress–compression curve for a coal pillar with a width-to-height ratio of 2 and isometric histograms showing the vertical stress distribution at various stages of compression.

dimensions were measured; values of the brittle modulus, § 4.2, were found to be of the order of one-fifth of the elastic modulus. Second, the measured strengths of the larger pillars tested were found to increase as the square root of their width-to-height ratios, but their absolute strengths were found to be about fifty per cent more than those given by equation (2). Third, the distribution of stress across the pillars at different stages of compression was established. A typical complete average stress–compression curve for a pillar with a width-to-height ratio of 2 is illustrated in Fig. 18.7.2, together with isometric histograms showing the stress distribution at different stages of compression. These show that the circumference of a pillar is capable of carrying relatively little stress throughout its compression. However, this portion of a pillar provides lateral confinement which is important in increasing the strength of the centre of the pillar. The results from tests on pillars of rectangular rather than square plan section indicated that the strength of a long, strip pillar is virtually forty per cent greater than that of a square pillar of the same width. From this, it follows that long, strip pillars make more efficient use of coal for providing regional support of the overburden than do square pillars.

The load–compression characteristics of elongated rib pillars of waste aggregate with different properties have been investigated by Schumann (personal communication). The results for unconstrained pillars with sides initially at their angle of repose and different initial base-width to height ratios are shown in Fig. 18.7.3, together with the load–compression characteristic for the same aggregate with complete constraint. These results show that the compression of waste ribs approaches that of the constrained aggregate when the base-width-to-height ratio is about 10.

In addition to providing regional support of the hanging wall, pillars can be used to reduce the amount of energy released by making an excavation and the amount of energy stored around an excavation. The difference between the maximum amount of energy released by an extensive tabular excavation, § 18.4, and that which would be released if elongated pillars were left unmined on centres of $2S$ of § 18.4 (4) is

$$\Delta W_{R_p} = 2\sigma_v S\left[v_0 - \frac{4(\nu^2-1)}{\pi E}\ln\left(\cos\frac{\pi c}{2S}\right)\sigma_v S\right]. \tag{3}$$

The effectiveness of partial extraction in reducing the amount of energy released can be gauged by taking the ratio between the reduction in energy release (3) due to partial extraction and the maximum possible reduction that could be obtained by reducing the thickness of the excavation to zero, that is

$$\Delta W_{R_p}/W_R = 1 - 2(\nu^2-1)\ln(\cos \pi C/2S)/\pi E v_0. \tag{4}$$

A similar equation using the compressed thickness of the fill in place of the excavation thickness can be used to gauge the effectiveness of partial filling in relation to complete filling. Evaluation of (4), Cook (1967), shows that pillars occupying 10–20 per cent of the area of an excavation are 80–90 per cent effective in reducing the amount of energy released.

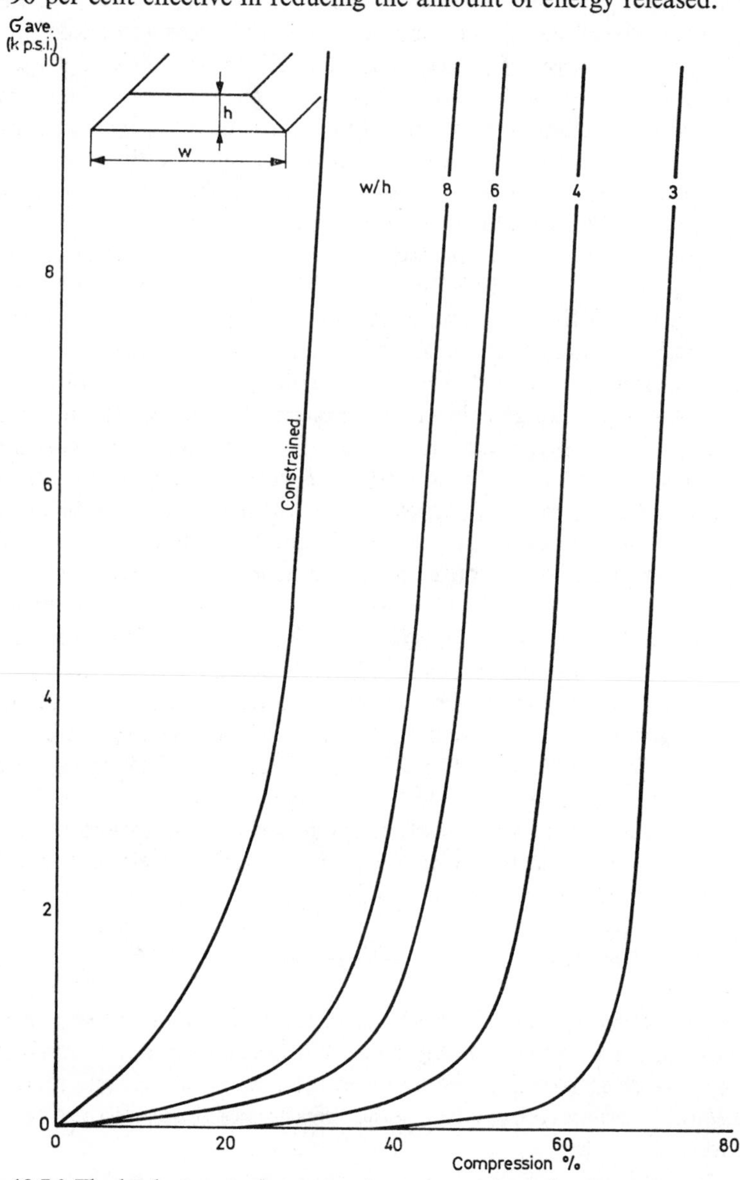

Fig. 18.7.3 The load–compression curves for unconstrained rib pillars of waste aggregate with different base-width-to-height ratios compared with the curve for completely constrained aggregate.

18.8 Support

The direct cause of damage to excavations is usually the separation of pieces of rock, bounded by geological discontinuities and fractures induced by excavation, from the rest of the rock around the excavation. Separation may be the result of sliding caused by stress, gravitational forces, or accelerations due to violent rock failures and blasting. The integrity of the rock around an excavation is readily destroyed by any movements in excess of those resulting from its behaviour as a continuous elastic material. If a piece of rock becomes loosened it is no longer capable of contributing to its own support. Opening, separation, and rotation destroy the strength due to interlocking inherent in even the most extensively jointed rock. The stability and safety of an excavation in rock is determined in the first place by the extent to which these disruptive displacements are prevented and in the second place by the extent to which they can be controlled.

Rock can be regarded as being strong, massive, and competent if its uniaxial strength is three or more times the field stress around an excavation and the spacing of any joints or planes of weakness in it is greater than the intermediate dimension of an excavation. Most rocks are laminated as a result of bedding or foliation and have little tensile strength normal to the laminae. Many rocks are bedded and jointed, or have joints running in several directions, with the result that the rock is divided into polyhedrons. If the pitch of bedding or jointing is smaller than the intermediate dimension of an excavation the rock must be regarded as incompetent. Competent rock can become incompetent as a result of fractures induced by making an excavation. Every effort must be made in the design of an excavation to retain the competence of the surrounding rock, and support must be used to provide such additional competence as is required.

It has been common practice to study bedded strata as if they were elastic beams, Obert and Duvall (1967). The difficulty with this approach is that the strength of a beam is determined by the maximum tensile stress it can withstand, which, in the case of a rock stratum, depends critically upon the compressive stress parallel to that stratum. A more realistic approach to the support of bedded and fractured strata is to accept that they must accommodate the displacements forced on them by the movement of the rock as a continuous elastic whole but that any displacements in excess of these result in separation and the formation of voids which are undesirable. Separation can be prevented if the support is designed so as to prevent tension occurring across any weakness in the rock. Regions of tensile stress in the rock can be found by calculating the stresses in the rock on the basis of the theory of elasticity. Where such regions may cause tensile separation support must be provided to prevent its occurrence. A useful criterion for the design of such support is that it generates loads sufficient to eliminate the tension. This should be applied literally to those situations where there is tension normal to the surfaces of an excavation, but it can be relaxed in connection with tension tangential to the surface.

Tension normal to the surface of an excavation arises in a number of ways. For example, the radial stress around a circular opening of radius R_1 in a material subject to a uniaxial compressive stress field is tensile along a radius parallel to the direction of the uniaxial stress in the interval $R_1 < r < 3^{\frac{1}{2}}R_1$, § 10.4 (10). Alternatively, tension may be due to gravitational effects. If the induced tension along the y-axis of a flat elliptical opening, § 10.11, is superposed on the vertical component of the virgin rock stress, which increases linearly with depth due to gravity, it is seen that the resultant normal stress is tensile within a region determined by the ratio between the width of the elliptical opening and its depth below surface, Fig. 18.8.1.

Most types of support can be divided into *active support* which imposes deliberate loads on the rock surface and *passive support* which generates loads as a result of its compression by convergence across an excavation. Examples of active support are hydraulic props used in mining and long rockbolts anchored into the solid elastic rock beyond the zones of failed rock and tension around an excavation. Arches, linings, and packs are examples of passive support which generate load only as a result of their compression, as is also rockbolting, where the rockbolts are anchored in the completely failed rock and form a lining of bolted aggregate, § 18.2. Rock surfaces should be supported as soon after they have been exposed

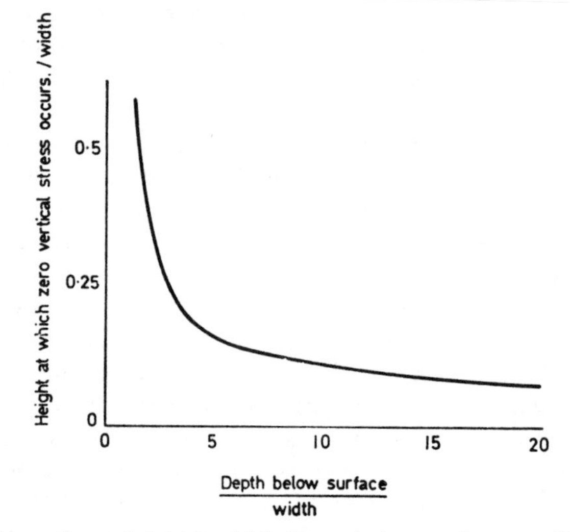

Fig. 18.8.1 The maximum height to which the vertical stress above an elliptical opening is tensile.

as is possible. For this reason active support is to be preferred and passive support must have the property of great initial stiffness so that it generates support loads with a minimum amount of displacement of the surrounding rock. However, this initial stiffness which gives support has the highly desirable property of being rapid-bearing, but must often be tempered with capacity to yield, once adequate support loads have been generated. Otherwise, further compression of the support by convergence across the excavation generates such high loads in the support that it either fails or damages the rock which it is supporting as a result of excessive loads.

The effect of the average support stress provided by rockbolts to an annulus of completely failed rock is discussed in § 18.2. Lang (1961) has studied the effect of rockbolting using photoelastic models and experiments in which the capacity of bolted aggregate in an open-ended square frame to carry load was measured. He concluded that the centre spacing of the bolts should not be more than seven times the size of the pieces of aggregate and that the bolt length should be at least twice the centre spacing of the bolts. Bolting in this pattern generates a more or less uniform compression normal to the axes of the bolts in the body of the aggregate between the anchors and washers of the bolts, the magnitude of which is about half the average compressive stress parallel to the axis of the bolts due to their tension. Shallow zones of tension are generated near the surface between the bolts. It is important that the rock in these zones of tension be kept in position, usually with relatively light wire netting, otherwise the bolted aggregate fails by an unravelling chain reaction following the loss of the first few pieces of aggregate. Properly bolted aggregate possesses a remarkable degree of ductile strength, Fig. 18.8.2. An extensive bolted surface can probably be regarded as a beam or plate with a tensile strength equal to the value of the uniform compressive stress generated by the rockbolts in a direction normal to their axes.

In hard rock partial failure of the rock around an excavation produces slabs which are prone to fail by buckling, Fairhurst and Cook (1966). Rockbolts provide an effective means of restraining this mode of failure. Their action is that of an elastic foundation or a resilient restraint, § 5.10 (21), (22), and depends upon their stiffness rather than their strength. When used to stabilize buckling, rockbolts should be long enough to be anchored in the solid elastic rock outside the zone of failed rock around an excavation. It should be noted that linings, arches, and annuli of bolted aggregate have considerable stiffness against uniform radial compression but much reduced strength and stiffness against concentrated loads. For example, the total radial load which an annulus of bolted aggregate can resist is considerably greater than the sum of the bolt tensions, whereas the strength of the bolted aggregate in Fig. 18.8.2 against a concentrated load appears to be about 15,000 lb compared with a total bolt tension of 45,000

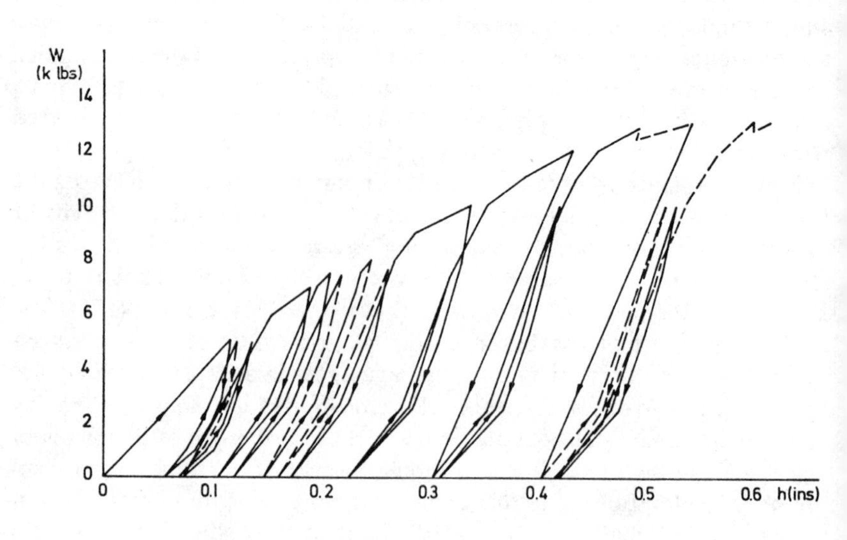

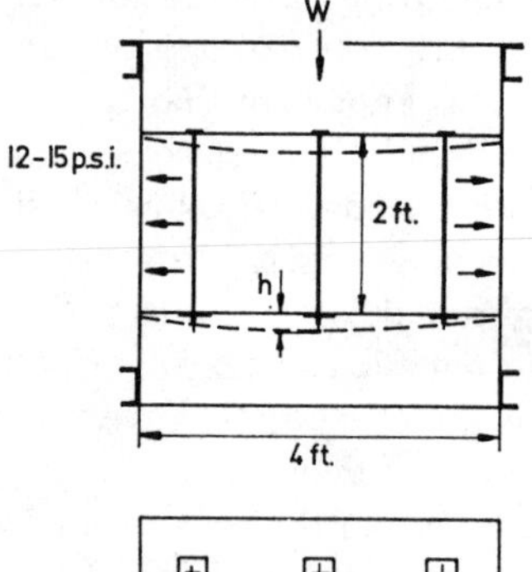

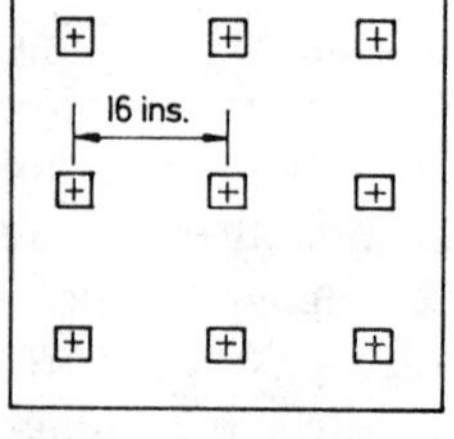

mean rock size 1.88 ins
bolt tension 5000 lbs

Fig. 18.8.2 The results of a load–compression test on a base of bolted aggregate (from Lang, 1961).

lb. A single bolt correctly placed and anchored into solid elastic rock is more effective in resisting concentrated loads than are arches or linings of pattern-bolted aggregate.

18.9 Multiple excavations

The problems of strata control depend primarily upon the extent to which the rock around a structure becomes fractured. The relationship between the amount of energy released by an excavation and the extent of these fractures has been discussed in connection with an isolated tunnel, § 18.2, and tabular excavations, § 18.4. In other cases the extent to which the rock fails around a number of contiguous excavations of complicated geometry, such as service excavations near a shaft, must be examined. These problems are usually not amenable to mathematical analysis, and Hoek (1967) has developed an analogue technique for treating them in two dimensions.

According to the Coulomb, § 4.6, or modified Griffith failure criteria, § 10.14, the strength of rock in compression is determined by a linear relationship between the shear and normal stresses. Equation § 4.6 (23) shows that this relationship can be written in terms of the mean normal stress, $\sigma_m = (\sigma_1 + \sigma_3)/2$, and the maximum shear stress $\tau_m(\sigma_1 - \sigma_3)/2$ as

$$\tau_m = S_0/(\mu^2 + 1)^{\frac{1}{2}} + \mu\sigma_m/(\mu^2 + 1)^{\frac{1}{2}}. \tag{1}$$

The isochromatic fringes around a photoelastic model of a section through any set of two-dimensional excavations provides a ready means for obtaining the distribution of τ_m around such excavations, § 10.19. It remains to assign values to μ and S_0 for the rock and to find the distribution of σ_m around the same excavations before the regions in which failure will occur can be delineated, in accordance with (1). Values for μ and S_0 can be determined from laboratory tests, large-scale tests, and experience. Equation § 5.5 (16) shows that the sum of the stresses in an elastic material satisfies Laplace's equation. It follows that the distribution of σ_m around a section through the excavations can be found with the aid of a conducting-sheet electrical analogue, Karplus (1958), using boundary conditions from the photoelastic model.

This technique is very useful for dealing with complicated geometries, but suffers from three limitations. First, it is applicable only to situations which can be studied in two dimensions or situations where a two-dimensional approximation provides adequate insight to the problem. Second, it ignores any effect that the value of the intermediate principal stress has on the strength of the rock, though (1) could be modified to take this into account, using values for the intermediate principal stress calculated from an assumption of plane strain. Third, the analysis is applicable only if the major and minor principal stresses lie in the plane of the section through the excavations used for the photoelastic and conducting model.

Let σ_x and σ_y be the principal stresses in the plane of this section and let σ_z be the principal stress normal to this plane. If σ_x is the major principal stress the analysis is valid, provided that $\sigma_z > \sigma_y$. For plane strain,

$$\sigma_z = \nu(\sigma_x + \sigma_y). \tag{2}$$

Assuming that σ_x and σ_y are the major and minor principal stresses, respectively, their values on the boundary of the failed region are given by (1) and can be expressed as

$$\sigma_x = q\sigma_y + 2q^{\frac{1}{2}}S_0 \tag{3}$$

using § 4.6 (9), (20), (21). Substituting from (3) in (2) yields

$$\sigma_z = \nu[(q+1)\sigma_y + 2q^{\frac{1}{2}}S_0]. \tag{4}$$

Since $S_0 \geqslant 0$, $\sigma_z > \sigma_y$ on and beyond the boundary of the failed region, provided that $\nu(q+1) > 1$. For most rocks $\nu > 0{\cdot}15$ and $q > 5$, so that this condition is often satisfied.

Theoretically, this technique could be extended to three-dimensional excavations using 'frozen-stress' three-dimensional photoelastic models and a conducting tank. However, the practical difficulties of doing this would be considerable.

General guidance for the disposition of adjacent excavations can be drawn from the results of theoretical analyses of cases with simple geometries. Howland (1935) has solved the problem of a row of circular holes, and Salamon (1964) has solved the analogous problem of a series of coplanar slits in an elastic body subject to stress. The latter solution is discussed in § 18.4, where it is noted that significant interaction between the slits occurs only when they cover more than three-quarters of their plane. Likewise, the maximum stress concentration around the circumference of each of a series of holes increases significantly only when the diameters of the holes cover more than three-quarters of the plane through their centres. Nevertheless, in both examples the average stress in the material between the openings must increase as $S/(S-c)$, where $2c$ is the width of the slits or the diameter of the holes and $2S$ is their centre distance. From this it can be concluded that the average stress between adjacent openings is increased to a greater extent than are the maximum stress concentrations around each of the openings. The extent of the failed region around an opening is determined largely by the magnitude of the tangential stress concentration around that opening and the tangential extent of the adjacent free surface of that excavation. Adjacent excavations should be separated by a distance sufficient to leave an adequate width of unfailed rock between them. This width can be estimated from the strength of the rock and the average stress concentration. The effective size of an excavation, that is, the size of the opening plus the size of the surrounding failed region, can be regarded as virtually the same for one of a number of adjacent excavations as it is for the same excavation in an isolated position. In general, the width of unfailed rock between adjacent excavations should not be less than a third of their centre spacing, and the failed region can be considered as extending ahead of the surface of an excavation a distance equal to its average radius of curvature. Thus, a series of circular openings should be spaced on centres of $2S = 6c$, that is, a centre spacing of no less than 3 diameters.

18.10 Delineation of regions of failure

Many problems in rock mechanics involve the prediction of those regions in which failure is likely to occur and an estimate of the degree to which the rock in such regions is likely to be fractured.

In Chapter 10 solutions to several problems in the theory of elasticity are given from which the elastic stress distributions can be calculated and in § 18.5 a method of determining the stresses around tabular excavations is described. Such calculations are valid only while the rock is elastic. The state of stress at which rock ceases to be elastic is given by the various failure criteria detailed in §§ 4.6, 4.7, 4.8, 4.9, 6.6, 6.15, 10.13, 10.15 and 12.4. Failure in brittle materials such as rock tends to be confined to regions of limited extent; the bulk of the surrounding rock continuing to deform in an elastic manner. This phenomenon makes it possible to use the theory of elasticity to delineate those regions of rock in which failure is likely to occur and to estimate the degree to which the rock will become fractured.

The procedure is to calculate the state of stress throughout the region of interest, assuming that all the rock behaves elastically and using any of the several methods which this assumption makes possible. The values of the stresses calculated in this way are then compared with those values of these stresses defined by the appropriate failure criterion as the limit for elastic behaviour of the rock. In particular, the difference between the value of the calculated major principal stress and the value of this stress given by the failure criterion for the state of stress corresponding to the values of the calculated minor principal stresses is defined as the *excess stress*. Where the excess stress is greater than zero, that is, where the value of the calculated major principal stress exceeds that allowed by the failure criterion, failure will probably occur. Thus, the contours of zero excess stress define the limits of the region within which failure will occur. Likewise, the contour of maximum excess stress defines the locus of the most severe failure or fracture. Where the excess stress is a small fraction, say a tenth, of the uniaxial compressive strength little alteration in the strength of the rock is likely to have taken place in consequence of failure but when it exceeds half the uniaxial compressive strength extensive fracturing and loss of strength will have occurred.

This procedure has been applied with success to several problems ranging from laboratory experiments to failure in the rock around highly-stressed, barrier pillars between mines. As an example, its application to a stamp-loading test, § 7.3, on quartzite is given. Using the methods of § 10.17, the principal stresses beneath a circular stamp applying a constant normal stress to a flat, rock surface can be calculated. Contours of maximum and minimum principal stress for a surface stress of 4000 bars are shown in Fig. 18.10.1 (*a*). Using the Coulomb failure criterion, § 4.6 (20), truncated at a constant tensile strength, Fig. 18.10.1 (*b*) with values of C_0, q and T_0 applicable to quartzite, contours of excess stress were calculated, Fig. 18.10.1 (*c*). The hatched area indicates that region where the calculated stresses outside of the elastic range are tensile, and its interpretation is problematical because the first tensile crack obviously destroys most of the tensile stress. The region where failure occurs due to excess compressive stress lies between the zero excess stress contours. Subsequent examination of a section through a specimen of quartzite loaded in this way shows close agreement between the predicted and observed regions of failure. However, failure is usually observed to be confined to a more narrow zone than that defined by the contours of zero excess stress. This has been attributed (Wagner and Schümann, 1971) to the additional confining stress resulting from the dilatation of rock in excess of Poisson expansion which occurs near and beyond failure. Fig. 4.2.5.

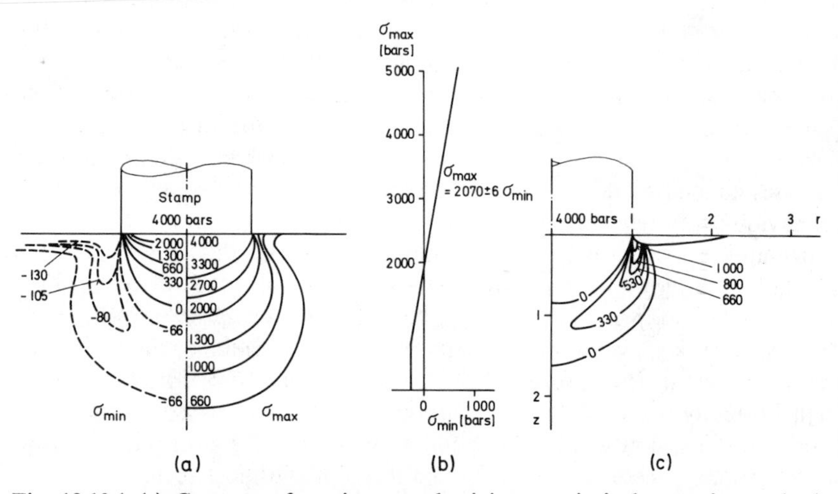

Fig. 18.10.1 (*a*) Contours of maximum and minimum principal stress in an elastic material beneath a circular stamp with a uniform normal pressure of 4000 bars. (*b*) Truncated Coulomb failure criterion for quartzite. (*c*) Contours of excess stress delineating the region within which failure occurs under one-half of the stamp.

18.11 Rockbursts

In general, rock failures induced by mining are referred to as rockfalls, rockbursts, bumps and outbursts. Rockfalls are relatively non-violent, though potentially dangerous, falls of loose rock under the influence of gravity whereas rockbursts are violent failures of rock caused by high stress which result in damage to excavations. Rockbursts have occurred in a wide variety of excavations including metalliferrous, coal and non-metallic mines and even in granite quarries at depths of less than 30 m below surface, presumably as a result of exceptionally high values of the horizontal component of the virgin rock stresses, § 14.3. However, it is mainly in the deep metalliferrous mines in which vein deposits in hard, strong brittle siliceous rocks of igneous or metamorphic origin are worked that they are most common. Outbursts usually occur in coal or salt and are violent ejections of rock caused by the sudden release, resulting from rock failure, of adsorbed or entrapped gas. In rockbursts and outbursts, the volume of rock moved into the excavation can range from a fraction of a cubic metre to thousands of cubic metres. The term 'bumps' is used usually to describe a violent failure or movement of rock which does not necessarily give rise to damage within an excavation. The phenomenology and terminology are discussed by Obert and Duvall (1967).

Rock failure and seismic activity are unavoidable concomitants of extensive mining deep below the surface of the earth. An early example of the relationship between the incidence of rockbursts and the rate of mining comes from the mines of the Brezove Hoty district of Czechoslovakia where rockbursts became a problem at the beginning of this century and where mining is now taking place at depths of 1 km to 2 km below surface (Sibec, 1963). The annual frequency of rockburst incidents and the rate of mining, in cubic metres per year, are plotted for the period 1913 through 1938 in Fig. 18.11.1; only 20 per cent of the rockbursts caused major damage underground. Similar phenomena occur in the Sudbury nickel mines of Canada, the Kolar gold fields in India, the gold mines of the Witwatersrand System in South Africa and in the zinc mines of the Cœur d'Alène district of the United States of America, among others.

Seismic records of tremors in Johannesburg have been kept since 1910. In analyzing these data Cazalet (1919–20) noted two major features, namely, a peak in the diurnal distribution on weekday afternoons at about the time that blasting takes place in the gold mines of the Witwatersrand and a maximum incidence on Thursdays. Later, Stokes (1936) found that the diurnal peak was absent on Sundays when no mining takes place. Gane (1939) made a statistical analysis of 14 830 tremors in the period mid-1910 through 1937, confirming these observations and indicating that mine blasting is an important factor in the initiation of tremors. Finsen (1950) made a similar analysis of 29 669 tremors for the period mid-1938

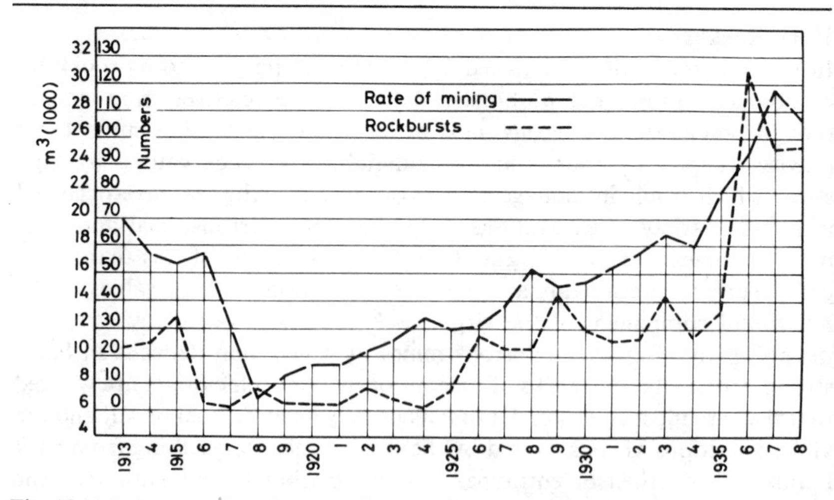

Fig. 18.11.1 A comparison between the rate of mining and the incidence of rockbursts in the Brezove Hoty district of Czechoslovakia (after Sibec, 1963).

through 1947. He confirmed the maximum daily incidence of all tremors on Thursdays, but noted that the maximum was a function of the size of the tremor, the maximum for small tremors occurring on Fridays and that for large tremors on Wednesdays. The incidence of tremors on Sundays was about half the weekday average for large tremors, but the decrease was less for smaller tremors.

The direct relationship between tremors felt in Johannesburg and gold mining was established first by Gane *et al.* (1946) using six seismic stations spread several kilometres apart on the surface. Subsequently, using surface stations on one mine Gane *et al.* (1953) found that the foci of tremors were in close proximity to places of active mining. In 1963 a three-dimensional array of seismometers was used underground to study the seismicity of about a kilometre of mine working face at a depth of about 2,5 kilometres (Cook, 1963).

It has been established that seismic events induced by mining radiate seismic energy from as little as 10^{-5} J for microseismic events to as much as 10^9 J for large rockbursts and bumps (Cook, 1963; Blake *et al.*, 1974) corresponding to seismic magnitudes of −6 to 5 using the Richter scale (Gutenberg and Richter, 1956). The frequencies of the radiated energy range from less than 1 Hz to over 10 kHz. That part of this frequency spectrum in which most of the energy is concentrated depends upon the size of the event; the frequency decreases with increasing size, and high frequencies tend to be attenuated rapidly with increase in distance away from the focus of an event.

Many thousands of rock failures with seismic magnitudes of up to 5 have occurred as a result of mining the gold-bearing reefs of the Witwatersrand System. Fortunately, only a small percentage of these events cause identifiable damage in the workings and few are therefore classed as rockbursts. However, seismic investigations of rock failure associated with mining have provided a wealth of data which has been used to improve the stability and safety of deep-level mining.

Locations of the foci of these events by underground seismic networks have shown that most of them occur close to faces which are being advanced (Cook, 1963; Joughin, 1965). The majority of the plan positions of the foci detected seismically on two different mines by Cook and Joughin were found to lie within less than 20 m of the positions of advancing stope faces, Fig. 18.11.2 (*a*), although the elevations of these foci relative to the reef were quite different. In the mine investigated by Cook, the quartzitic strata above the reef are a little less argillaceous than those below it, and the foci tended to fall a little above the elevation of the reef, Fig. 18.11.2 (*b*). In the other mine studied by Joughin, the strata adjacent to the reef are both highly quartzitic, but a weak layer of shale exists at an elevation of 80 m above the reef; the elevations of most of the foci fell within this thickness, Fig. 18.11.2 (*c*).

These locations can be compared with the zones in the rock where the stress concentrations induced by mining are sufficient to generate new fractures through solid rock or to cause movement along pre-existing fault planes. In Fig. 18.11.3 are shown contours of the maximum and minimum principal stresses in a section through an idealized longwall stope in the vicinity of the face. The spatial dimensions are expressed in terms of the stope half span = 1 and the magnitudes of the stresses as factors by which the value of the vertical component of the virgin rock stress is concentrated. The upper half of the figure shows the total stresses around the excavation and the lower half the induced stresses, §§ 18.3 and 18.4. Assuming that the strength of the Witwatersrand quartzites is typically given by $\sigma_1 = 200$ MPa $+ 6\sigma_3$ (Cook, 1962) and using the method of delineating fracture zones described in § 18.10, the total principal stress contours in Fig. 18.11.3 have been used to delineate the zone within which shear fracture of solid rock could occur in the proximity of a face at a depth of 3 km. In this example, the vertical component of the virgin rock stresses is 78 MPa and the horizontal component 39 MPa. It can be seen that the zone of potential shear fracture extends above and, of course, virtually symmetrically below, the reef and behind the face. If the half-span of the stope is 300 m, the vertical extent of this zone is almost 30 m. Fractures starting within this zone may extend beyond it, as they propagate due to the stress concentration at their tips. Also, this analysis does not account for the effects of the zone of fractured rock which is known to extend a few metres ahead of the face; this should have the effect of

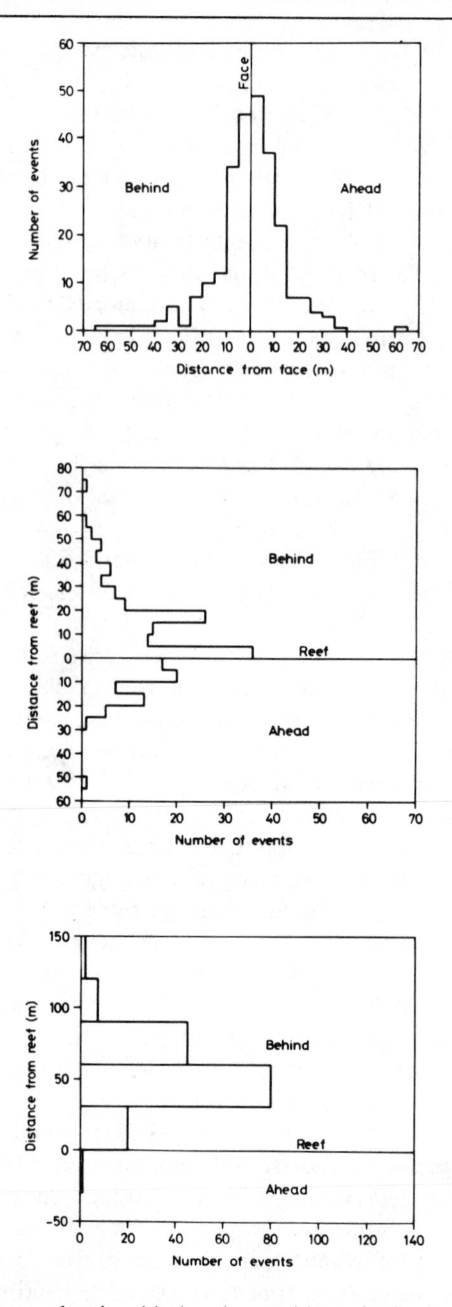

Fig. 18.11.2 Histograms showing (*a*) the plan position of seismic foci relative to the position of the face at the time of occurrence, (*b*) their elevation relative to the plane of the reef and (*c*) the elevation of foci relative to the reef where a weak shale layer exists 80 m above the reef.

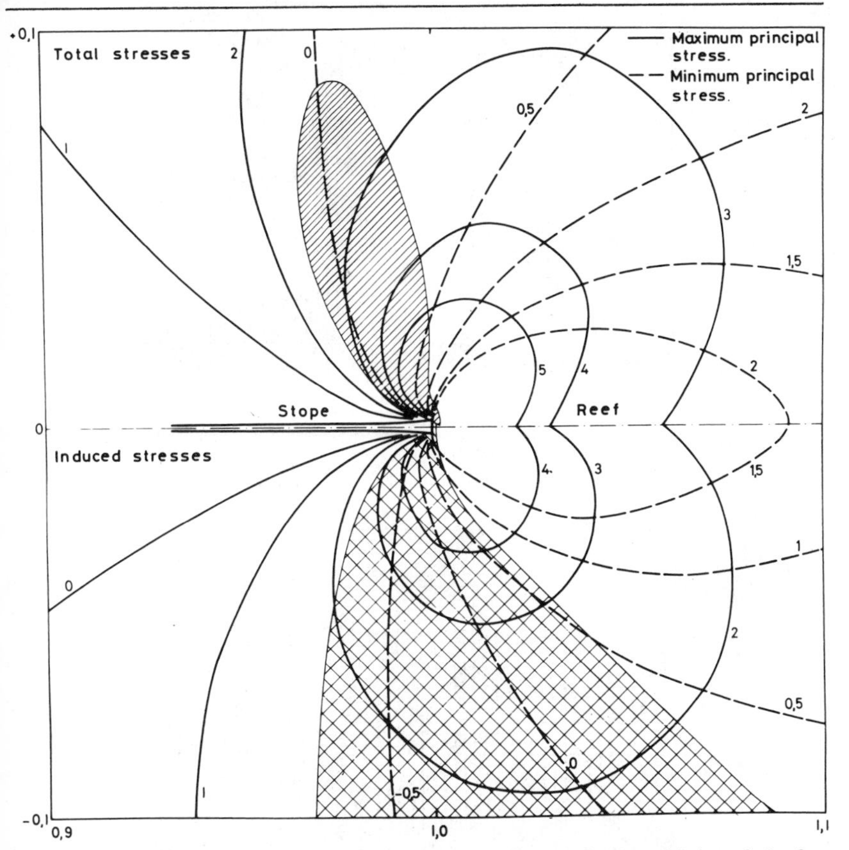

Fig. 18.11.3 A section through an idealized longwall stope in the vicinity of the face showing contours of the maximum and minimum principal total stress (upper half) and induced stresses (lower half). The dimensions are given in terms of the half-span of the stope = 1. The hatched area in the upper half shows the zone within which shear fracture of solid rock is possible at a depth of 3 km and the cross-hatched area in the lower half shows the zone within which sliding on pre-existing faults of suitable orientation may be initiated.

moving the potential fracture zone forward relative to the actual face position.

In the lower half of Fig. 18.11.3, the induced stresses for the same depth have been used to delineate the zone within which they would be sufficient to cause sliding along pre-existing faults of suitable orientation using the criterion $\sigma_1 \geqslant 6\sigma_3$. It can be seen that this region, which is also virtually symmetrical above the plane of the reef, straddles the contour of zero induced principal stress. It extends for an infinite distance away from the stope but the magnitudes of the induced stresses decrease rapidly with increasing distance from the stope, and so must also the probability of these stresses causing sliding.

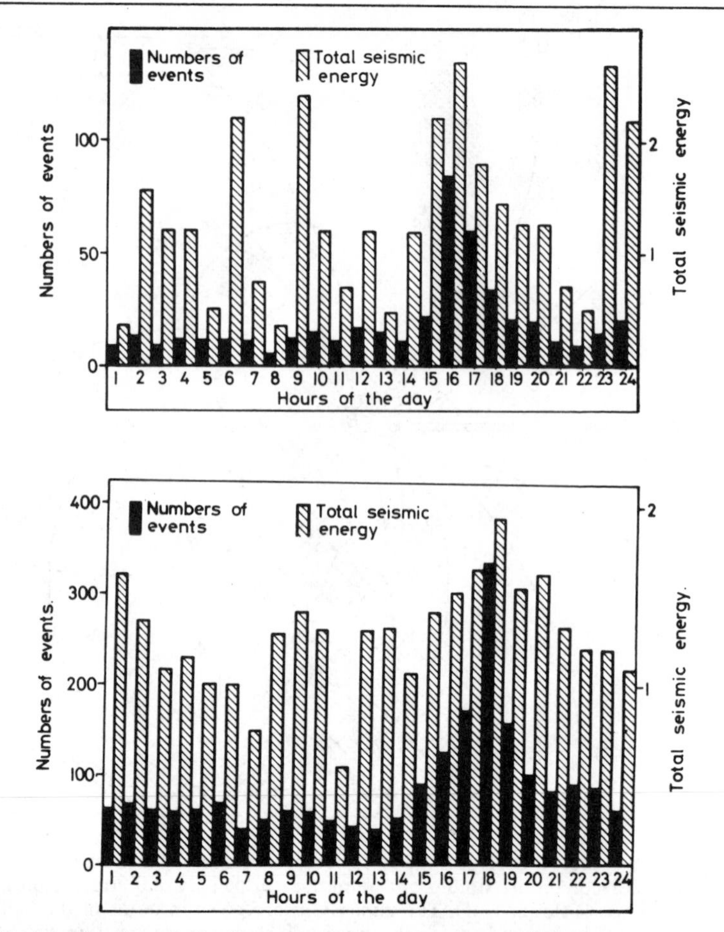

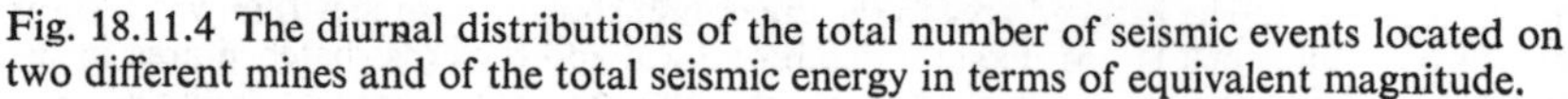

Fig. 18.11.4 The diurnal distributions of the total number of seismic events located on two different mines and of the total seismic energy in terms of equivalent magnitude.

To the left of both these zones, that is, above the open stope, in both vertical and horizontal directions, and below it, in the horizontal direction only, tensile stresses predominate. Interestingly, it is only in these regions of tensile stress and in the zone where shear fracture through solid rock can occur, that the maximum and minimum principal stresses lie in the plane of this section. Elsewhere, the minimum principal stress is normal to this plane.

An important point to note is that the magnitudes of the induced principal stresses in this plane are equal at the elevation of the reef, except immediately ahead of the stress-free stope face. Their effect is, therefore, to inhibit fracture or sliding in the rock ahead of the stope, except along planes near normal to the face and the reef.

This analysis suggests that the positions of rock failures induced by mining must be close to the face position in plan and above or below it in elevation, except for those fractures which occur in the rock immediately ahead of the free face. The results are in reasonable agreement with seismic observations of the positions of most fracture foci.

Traces of fracture planes caused by mining are revealed on the hanging- and footwalls of stopes and can be seen in tunnels above or below stopes. Such observations (Pretorius, 1966) have provided a clear picture of these fractures. They are near-vertical and tend to run parallel to the stope face or the general line of the longwall. The spacing between them varies from many millimetres to a metre and they may appear as hairline cracks or as slicken-sided fractures several millimetres thick; the latter are often assumed to be rockburst fractures. The mean length of these fractures is of the order of 10 m to 100 m although they do merge and branch. Their vertical extent varies from 5 m to 100 m; often they terminate on a bedding plane. These observations, too, are in fair agreement with the calculated position of the fracture zone and the locations of seismic foci.

The diurnal distribution of the numbers of seismic events and of their cumulative energy for two major seismic investigations on different mines are illustrated in Fig. 18.11.4. Although the incidence of events is greatest during and immediately after blasting, a large proportion of them occurs throughout the remainder of the day as does much of the total amount of energy released. This evidence, which is very similar to that from tremors observed on surface, indicates that these rock failures are time-dependent in terms of hours. Evidence of longer-term time-dependency is provided by the incidence of tremors on different days of the week, as illustrated by Fig. 18.11.5.

It must, therefore, be concluded that the seismic system comprises a rock mass which responds elastically to the overall changes brought about by mining. These changes involve the enlargement of excavations by blasting and the time-dependent formation of a zone of fractured rock immediately surrounding them.

Predictions of rockbursts or earthquakes may be based on recognizing a precursive pattern. The frequency with which rock failures occur in deep, extensive mining provides an excellent opportunity for studying such precursive phenomena.

Miners have long known that rock 'talks' and 'bumps' in response to the stress changes induced by mining and have used these phenomena in a varied and qualitative way to assess the safety of their workings. It is well-known that when rock is subjected to stresses approaching its strength, minor mechanical instabilities arise within the rock which give out microseismic radiation, § 6.5 (Obert and Duvall, 1957). Since 1938 many attempts have been made to measure the microseismic activity in mines for

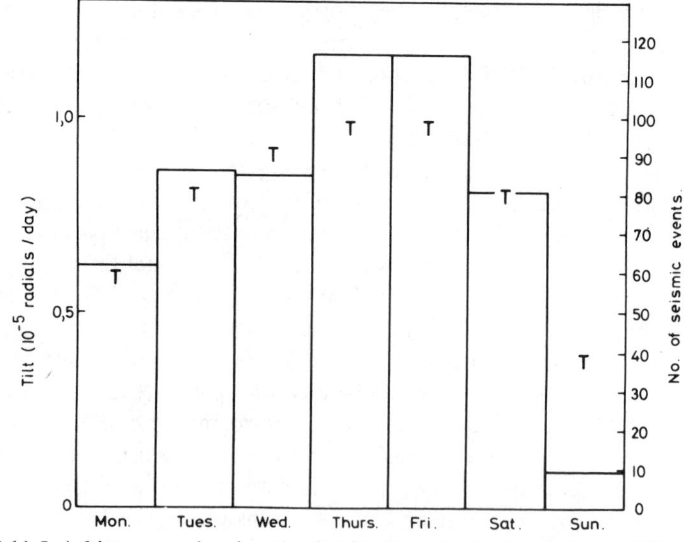

Fig. 18.11.5 A histogram showing the distribution of seismic events on different days of the week (there is no mining on Sundays) and of average daily rates of tilting of the solid rock mass in the vicinity of active mining.

the purpose of delineating and predicting rockfalls, rockbursts and outbursts (Obert and Duvall, 1957; Antsyferov, 1966). The rate of microseismic activity has been found to vary between wide extremes indicating that the stress in the rock is continually changing as a result of deformation and failure. Both Antsyferov (1966, p. 3), referring mainly to outbursts, and Obert and Duvall (1967, p. 606), referring mainly to rockbursts, concluded that a large increase in the rate of microseismic activity precedes most 'bursts' but that this rate is not in itself a satisfactory basis for prediction. The practical value of prediction hinges on being able to maintain a very high ratio of correct predictions to total predictions; Obert and Duvall were able to achieve a value of only 9 out of 14. The variability in the strength of the rock and relief of stress by failure and re-adjustment make it unlikely that this ratio can be improved upon under conditions in which rockbursts are common. Where failure is the exception rather than the rule, microseismic activity is an indicator of exceptionally high stress or weak rock. In such situations, it can be used to follow the progression of failure and may be of value in prediction.

Hodgson and Joughin (1967) showed that the seismicity of a region is a function of the spatial rate of energy release as the size of an excavation is increased, § 18.4, and that this term provides a good index of other problems associated with rock fracture in a mine; the rate of energy release is now in universal use among gold mines for planning the layout and sequence of stoping. Also, it has been found that the relationship between the

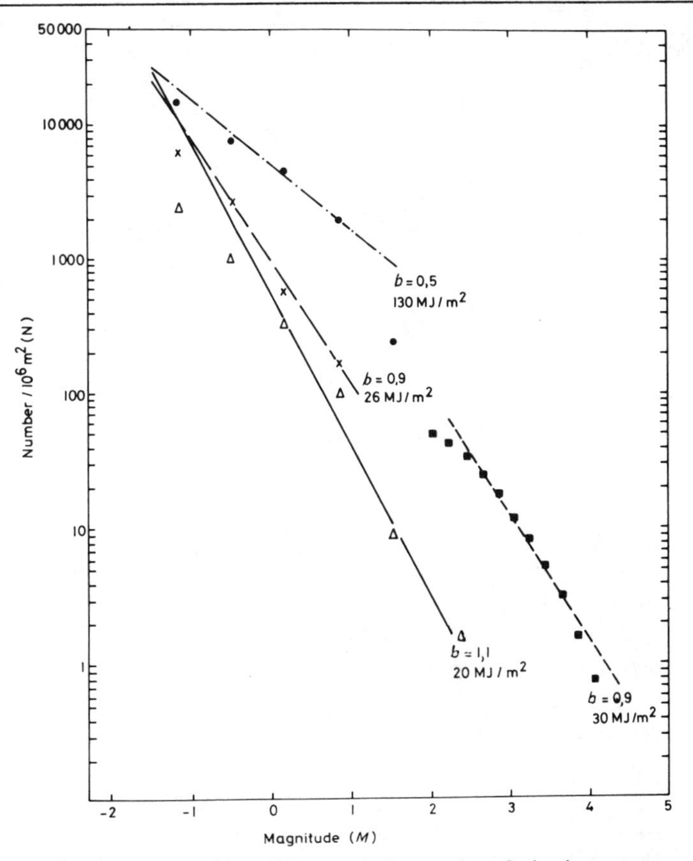

Fig. 18.11.6 Plots of the logarithm of the cumulative number of seismic events greater than a given magnitude as a function of magnitude. The three lines in the top left are derived from underground seismic observations and that in the bottom right from a surface seismic station (after Fernandez, 1973). Note that the slope of these lines, b, is a function of the spatial rate of energy release.

incidence of seismic events and their magnitude as shown by the plots of the logarithm of the total numbers of seismic events of greater than a given magnitude as a function of magnitude, Fig. 18.11.6, is determined by the rate of energy release. The greater the rate of energy release the greater is the proportion of large events. The relationship between the rate of energy release and the incidence of seismic events of different magnitudes allows statistical predictions to be made on the basis of a calculated rate of energy release, about the numbers of seismic events likely to occur in mining particular areas of reef, Fig. 18.11.7. However, it is interesting to note that the seismicity is contingent upon continued mining activity. If mining stops, most of the seismic activity stops soon afterwards, no matter how high is the calculated rate of energy release. The only exception to this is

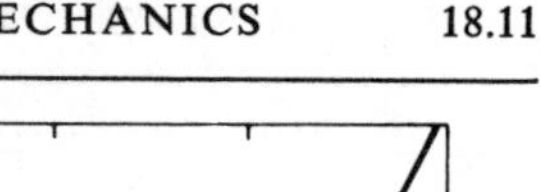

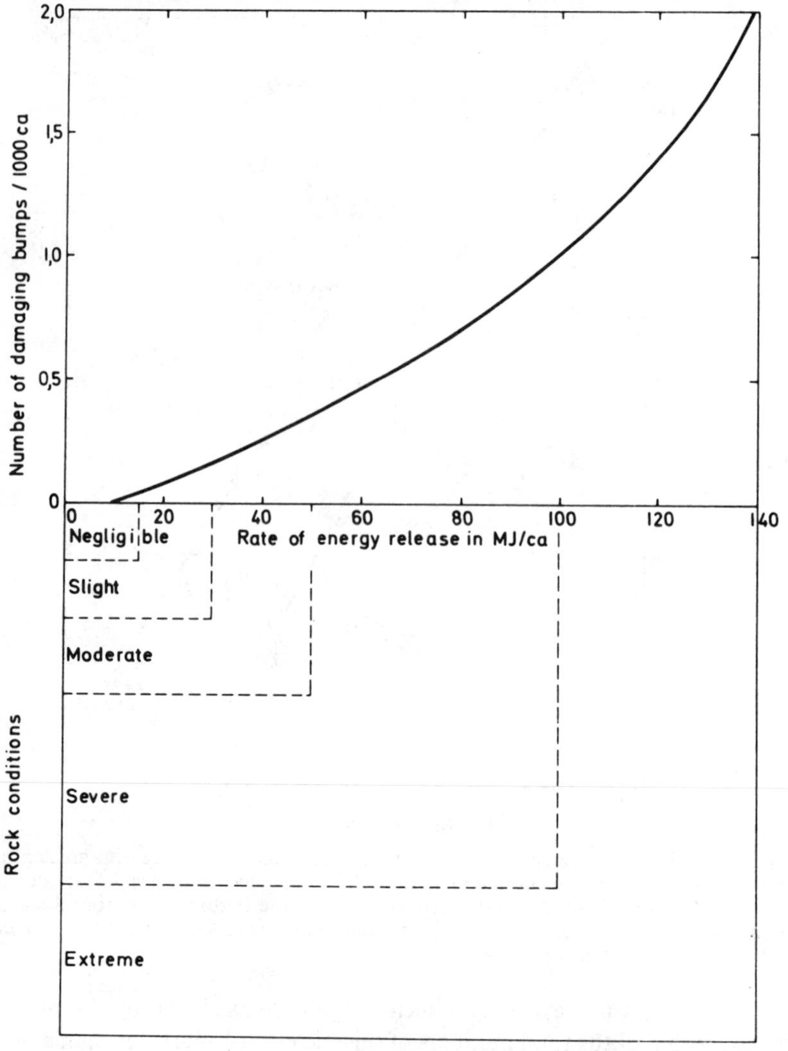

Fig. 18.11.7 A graph showing the increase in the incidence of damaging bumps with the rate of energy release caused by mining and their relationship to rock damage around stopes.

where continued mining elsewhere results in the stress on a remnant of unmined ground increasing to the point where it punches into the hanging- or footwall.

Two approaches must be used to obviate the problems of rockbursts. First, the layout of the mine excavations and the sequences of extraction must be so planned as to minimize the rates of energy release in stoping and the magnitudes of the field stresses to which service excavations are subjected. The methods described in §§ 18.4, 18.5, 18.6 and 18.7 provide a means for doing such planning well in advance of any mining. Second, most of the damage caused by rockbursts can be prevented by suitable support of the excavations (§ 18.8). In deep, hard rock stopes it has been found that the use of rapid-yielding hydraulic props virtually eliminates damage in stopes caused by most rockbursts because the action of the props keeps the fractured rock comprising the hanging in place even during the violent displacements of the hanging- and footwall (Hodgson *et al.*, 1971).

18.12 Determination of rock quality for engineering purposes

A measure of the properties of rock can be obtained from laboratory tests on small specimens, Chapter 6. Specimens for laboratory testing are generally prepared from drill core. Only the 'stronger' and more competent parts of the rock are recovered as core, and only the stronger pieces of core withstand specimen preparation. Laboratory measurements of strength, modulus, and seismic velocities are therefore biased towards the stronger portions of the rock traversed by any drill hole. Nevertheless, they provide an extremely useful upper limit to the strength and other properties of that rock.

A good core log, especially of fracturing and core recovery, provides an excellent measure of the variability of the rock traversed by the hole. If more than 95 per cent of the core is recovered and the pieces are 1 ft or more in length the rock can be considered very sound and its bulk properties will be close to those of small specimens. Deere (1964) has proposed a rock quality designation, RQD, based on the percentage of sound core recovered in pieces greater than 4 in. in length, shorter pieces being discarded. Where the core log indicates that the bulk properties of this rock are significantly different from those of small specimens the bulk properties must be found by large-scale tests or calculation.

The quality of a large volume of rock can also be estimated by comparing the velocities of propagation of seismic pulses in the field, § 13.5, with the velocities of propagation in laboratory specimens of the same rock, § 6.14, Onodera (1963). The ratio between these velocities provides a direct measure of the difference between the average elastic properties of the rock in the field and the properties of intact pieces of the stronger portions of that rock. In general, dynamic moduli such as can be calculated from seismic velocities, § 6.14, are greater than the corresponding static moduli. The static moduli of rock in the field can be measured in a number of ways.

Flat-jacks, § 15.2, can be used to find the static value for Young's modulus of the rock near any free surface in the field. Such measurements, however, embrace only a limited volume of rock and are primarily sensitive to the value of the modulus tangential to the free surface. Jointing and fracturing often tend to run parallel to a free surface, and it is likely that the value of Young's modulus normal to a surface is less than that parallel to it. If a suitable chamber is available, the static value of Young's modulus in a direction normal to the walls of the chamber can be found from plate-bearing tests, § 15.7, or pressure chamber tests, § 15.3. The former test again usually samples only a small volume of rock, but it can provide some indication of the strength as well as the modulus of the rock. If a chamber or a portion of a tunnel can be plugged and lined with an impermeable membrane its surface can be loaded with water under pressure and the resulting expansion of the chamber can be measured. The pressure-chamber test provides information concerning the modulus of a larger volume of rock, but is generally limited to modest stresses, well below the compressive strength of the rock. If the load in plate-bearing or pressure-chamber tests is cycled and the resulting hysteresis is observed the specific energy loss and the ratio between the loading and initial unloading moduli provide a measure of the extent of cracks and joints in the rock, § 12.3. Plate-bearing tests can be performed on surface by pulling a plate against the surface with a jack attached to cables anchored in relatively deep boreholes beneath the plate. With two sets of plates and cables this system can be used to apply tangential as well as normal loads to a surface, thereby making it possible to study anisotropy, Zienkiewicz and Stagg (1967).

18.13 Rock-slope stability

The stability of a rock slope can be investigated by considering the conditions necessary to prevent a slope, such as shown in Fig. 18.13.1, from sliding along some surface, *ab*, inclined at an angle, α, to the horizontal, which is less than the inclination of the slope. Let the weight of the portion of the slope above *ab* exert a vertical force P on the surface and let the coefficient of friction of the surface be μ. The shear force tending to cause sliding is

$$Q = P \sin \alpha, \tag{1}$$

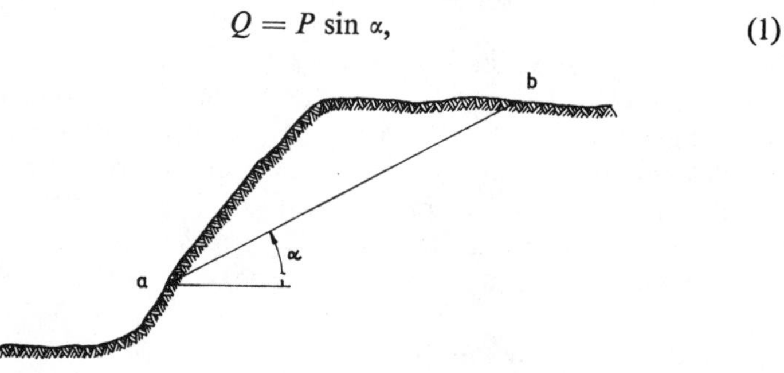

Fig. 18.13.1 A section through a rock slope with a potential slide surface, *ab*, inclined at α to the horizontal.

which is resisted by the cohesion of the surface AS_0 and the frictional resistance to sliding

$$F = \mu P \cos \alpha. \tag{2}$$

Sliding will not occur, provided that

$$(F + AS_0)/Q > 1. \tag{3}$$

Terzaghi (1962) has proposed that the cohesion can be taken as zero, and laboratory measurements by the U.S. Army Corps of Engineers (Lane, 1967) on natural and artificial joints and seams indicate that this is so. If the cohesion is taken to be zero the condition for slope stability becomes

$$\mu/\tan \alpha > 1, \tag{4}$$

which can be written as

$$\tan \phi/\tan \alpha > 1, \tag{5}$$

where $\phi = \tan^{-1} \mu$ is the angle of friction, § 3.6. Friction can be studied in triaxial tests, § 3.7, and the results of such tests, Table 3.4, show that the values of ϕ for rock range between 22° and 43°; most being concentrated around 30°. In general, sliding on more than one surface will be involved, and the case of plane surfaces is most simply treated by using vectors, Goodman (1967).

The simplest case is that of the sliding of a block on two intersecting planes. Suppose that ACB and ADB, Fig. 18.13.2 (*a*), are the regions of contact between the block and the two planes.

Let $\mathbf{R}$ be the resultant force on the block and let $\mathbf{w}_1$ and $\mathbf{w}_2$ be unit vectors normal to the two surfaces ACB and ADB and directed away from the block $ABCD$. The line of intersection AB of the planes is represented by the unit vector

$$\mathbf{u} = (\mathbf{w}_1 \times \mathbf{w}_2)/|\mathbf{w}_1 \times \mathbf{w}_2|. \tag{6}$$

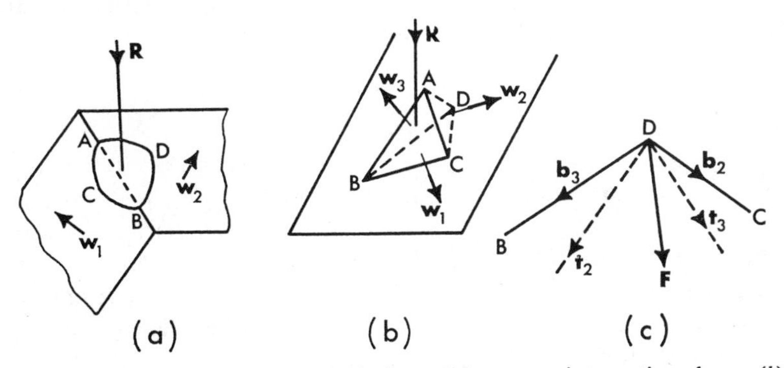

Fig. 18.13.2 (*a*) The conditions for a block to slide on two intersecting planes. (*b*) A tetrahedral block with a free surface. (*c*) The condition for sliding of the tetrahedral block on the surface BCD.

Let the normal reactions on the body from the planes ACB and ADB be $-N_1\mathbf{w}_1$ and $-N_2\mathbf{w}_2$. In limiting equilibrium, the frictional force will act in the direction $\mathbf{u}$ and will be $-(\mu_1 N_1 + S_{01}a_1 + \mu_2 N_2 + S_{02}a_2)\mathbf{u}$ where S_{01}, μ_1, a_1 are the shear strength, coefficient of friction and area of the contact ACB, and S_{02}, μ_2, a_2 are those for the contact ADB.

Thus the condition for the block to be in equilibrium is

$$\mathbf{R} - N_1\mathbf{w}_1 - N_2\mathbf{w}_2 - (\mu_1 N_1 + S_{01}\, a_1 + \mu_2 N_2 S_{02} a_2)\mathbf{u} = 0. \tag{7}$$

Multiplying this scalarly by $\mathbf{w}_1$, $\mathbf{w}_2$, and $\mathbf{u}$ gives

$$\mathbf{R} \,.\, \mathbf{w}_1 = N_1 + N_2\, \mathbf{w}_1 \,.\, \mathbf{w}_2,\ \mathbf{R} \,.\, \mathbf{w}_2 = N_2 + N_1\, \mathbf{w}_1 \,.\, \mathbf{w}_2, \tag{8}$$

$$\mathbf{R} \,.\, \mathbf{u} = \mu_1 N_1 + S_{01}\mathrm{a}_1 + \mu_2 N_2 + S_{02}\mathrm{a}_2. \tag{9}$$

If either of N_1 or N_2 given by (8) is negative or zero, it implies that contact with this plane is lost and the problem is that of simple sliding on the other.

If there is sliding on one plane only, say w_1, the direction of sliding will be in the plane whose normal is $\mathbf{R} \times \mathbf{w}_1$, and the condition for sliding is

$$|\, \mathbf{R} - (\mathbf{R} \,.\, \mathbf{w}_1)\mathbf{w}_1 \,| = S_{01}a_1 + \mu_1 \mathbf{R} \,.\, \mathbf{w}_1. \tag{10}$$

A slightly more complicated case is that of a tetrahedral block $ABCD$ with only one free surface ABC, Fig. 18.13.2 (b). Let the other faces BCD, CAD, ABD be described by unit vectors $\mathbf{w}_1$, $\mathbf{w}_2$, $\mathbf{w}_3$, respectively, normal to them and directed away from the tetrahedron. In this case motion towards any of these three faces is impossible, and a condition for this has to be written down. For example, for sliding on the plane BDC only, the resultant force $\mathbf{F}$ in this plane, which is $\mathbf{F} = \mathbf{R} - (\mathbf{R} \,.\, \mathbf{w}_1)\mathbf{w}_1$, must lie between DB and DC. The condition for this may be stated as follows: suppose that $\mathbf{b}_3$ is a unit vector along BD and $\mathbf{b}_2$ is one along CD and define $\mathbf{t}_2 = \mathbf{w}_1 \times \mathbf{b}_2$ and $\mathbf{t}_3 = \mathbf{b}_3 \times \mathbf{w}_1$ so that the situation in the plane BDC is as shown in Fig. 18.10.2 (c). The condition for $\mathbf{F}$ to lie between DB and DC is

$$\mathbf{F} \,.\, \mathbf{t}_2 \geqslant 0, \quad \mathbf{F} \,.\, \mathbf{t}_3 \geqslant 0, \tag{11}$$

and the condition for sliding is (10).

Numerous slopes exist which are inclined to the horizontal at angles greater than ϕ. This suggests that the rock in which these slopes exist must have finite cohesion. Slopes in competent rock where there are joints or fractures inclined to the horizontal at angles equal to, or greater than, ϕ obviously possess cohesion along potential slide surfaces. Irregular surfaces of rock also exhibit non-linear frictional characteristics which give them apparent cohesion. In mining it is often of the greatest economic importance to maintain a slope at the steepest possible angle. While the assumption of zero cohesion and intense jointing certainly provides a criterion sufficient for the design of stable slopes, it is once again neither economically acceptable nor necessary for the stability of slopes in many situations.

The effects of joints and other planes of weakness of different orientations on the strength of rock are discussed in §§ 3.6, 3.8. These concepts can be applied to the analysis of the stability of rock slopes, but require that the stresses in the slope be determined. Where detailed analysis of this kind is warranted, the geometrical complexity of individual slopes becomes significant, and it is necessary to use techniques such as finite-element analysis, § 10.20, to determine the principal stresses.

Most rock surfaces are neither smooth nor plane. Surface irregularities can have an important effect on the frictional properties of surfaces, which has been investigated by Patton (1966) and discussed by Deere *et al.* (1967). A cross-section through an irregular joint inclined at α to the horizontal is shown in Fig. 18.13.3 (*a*). When it slides, the block above the joint can move parallel to the surface irregularities or it can shear through these irregularities and move parallel to the average inclination of the joint surface. When the normal stress across the joint is small the block is likely to slide parallel to the surface irregularities; at high normal stresses the irregularities are likely to shear. Let the inclination of the surface irregularities on average be less than the inclination of the joint, α, by an amount, i. Investigation of the angle of friction of such a joint would show it to be $(\phi + i)$ for low normal stresses and small shear displacements, decreasing to ϕ as the normal stress and displacement increased and the irregularities become sheared. While such a joint has zero cohesion at zero normal stress, its behaviour for small displacements at greater normal stress is similar to that of a joint with an angle of friction ϕ and a cohesion S_0, Fig. 18.13.3 (*b*).

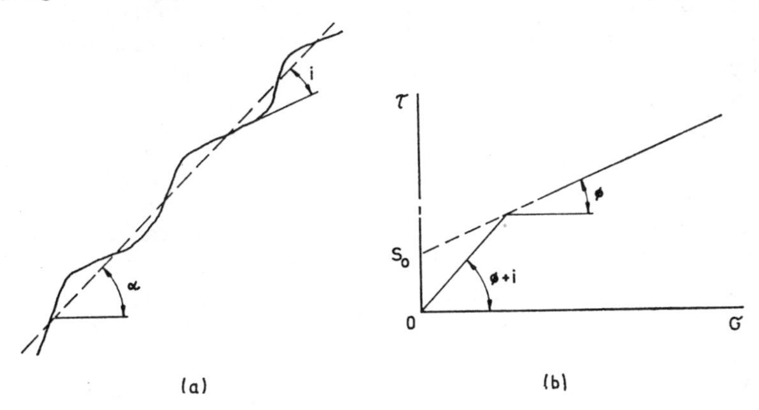

Fig. 18.13.3 (*a*) A section through a joint with an irregular surface inclined at an average angle α to the horizontal. (*b*) A Mohr diagram for sliding of an irregular surface.

As the stability of a slope is intimately connected with the frictional resistance to sliding generated by normal stresses, any factors which tend to reduce the friction or normal stress are potentially hazardous. The most important of these factors is pore or joint water pressure, §§ 8.8, 8.9. This reduces the normal stress across any joint by an amount equal to the water pressure, thereby diminishing the frictional resistance to sliding of the joint. Thin seams of clay frequently constitute potential slide surfaces. Many clays have very low angles of friction, particularly when wet, and deform in a completely ductile manner. Frequently, this deformation is time-dependent, being almost viscoplastic. Such seams are a potential hazard to the stability of any slope. The usual concepts of stability based on friction do not apply, and failure may be inevitable in the long term. Slides of this type frequently start very slowly, gathering significant velocity only just before catastrophic failure, Terzaghi (1950) and Müller (1964). Some calculations showing the effect of fluid pressure on sliding under gravity have been given in § 17.4.

If rock has multiple fractures in all directions the situation becomes similar to that considered in soil mechanics, and sliding is likely to occur on any surface inclined at $\alpha \geqslant \phi$. Slopes of intensely shattered or jointed rock with zero cohesion are stable only if they are inclined to the horizontal at an angle equal to or less than ϕ. For such slopes ϕ is also known as the *angle of repose*.

The previous discussion has referred to the case in which there is a set or sets of well-defined joint planes on which sliding may take place. For soils and rock with multiple fractures in all directions the situation is quite different since there is no *a priori* reason why failure should take place on a plane surface, and in fact many failures are known to take place on approximately cylindrical surfaces. In general, three different possibilities may be considered. These are, in increasing order or complexity: (i) a plane surface of failure may be assumed, in which case a generalization of (3) will be satisfied, and differentiating this with respect to α gives the most favourable inclination for failure; (ii) the other geometrically and mechanically simple surface of failure, a circle, may be assumed; in this case there is no general analytical solution, so that, in principle, numerical calculations are made for all possible circles, and the most favourable circle for slip is found. The procedure is discussed in detail in works on soil mechanics, Terzaghi (1943), Bishop and Morgenstern (1960), Morgenstern (1965). (iii) General failure of the material along a system of slip lines as in § 16.5 may be assumed.

18.14 Rock breaking

Rock breaking is an important part of most operations involving rock mechanics. It is convenient to distinguish between *excavation* in which different methods of rock breaking are used to remove pieces of rock from the ground mass and *comminution* in which rock that has been excavated is reduced in particle size.

Whereas Coulomb's empirical proposal concerning the strength of rock, § 4.6, was followed by Griffith's mechanistic formulation, §§ 4.8 and 10.13, hypotheses concerning the energy absorbed in comminution began with the mechanistic approaches of Rittinger (1867) and Kick (1885) which have been superseded, largely, by Bond's (1952) empirical approach.

According to Rittinger, the energy absorbed in comminution is proportional to the increase in the surface area of the fragments; the surface area of a given mass of rock increases inversely as the size of the fragments, assuming the shape of the fragments to be similar for different sizes. This can be expressed formally as

$$W_R = W_\alpha(1/P - 1/F), \tag{1}$$

where W_R = the energy required to comminute a unit volume of solid rock, or a unit mass of rock, from a nominal feed size F to a nominal product size P;

W_α = a constant, relating the surface energy of the fragments to their nominal size, times the comminution efficiency.

This law has been criticized on theoretical grounds, because the energy absorbed in comminution is many orders of magnitude greater than the increase in true surface energy, and on practical grounds, because the actual energy requirements for fine grinding are less than those indicated by it.

In essence, Kick (1885) assumed that particles of any size break at a constant value of the specific strain energy, that is, the particles have a constant strength, independent of their size. From this it follows that

$$W_K = W_S \log (F/P), \tag{2}$$

where W_K = the energy required to comminute a unit volume of solid rock, or a unit mass of rock, from a nominal feed size F to a product size $P = F/10$;

W_S = a constant, representing the strain energy required to effect a size reduction of one tenth, times the comminution efficiency.

These hypotheses can be explained further in terms of the Griffith locus, § 12.4. If a particle of material is such that the stress required to initiate fracture is above the point of minimum strain, Fig. 18.14.1, then the strain energy in the particle when fracture begins is sufficient to break it, as supposed by Kick. In detail, this implies that the particles contain many Griffith-type weaknesses of a similar but relatively small size; small be-

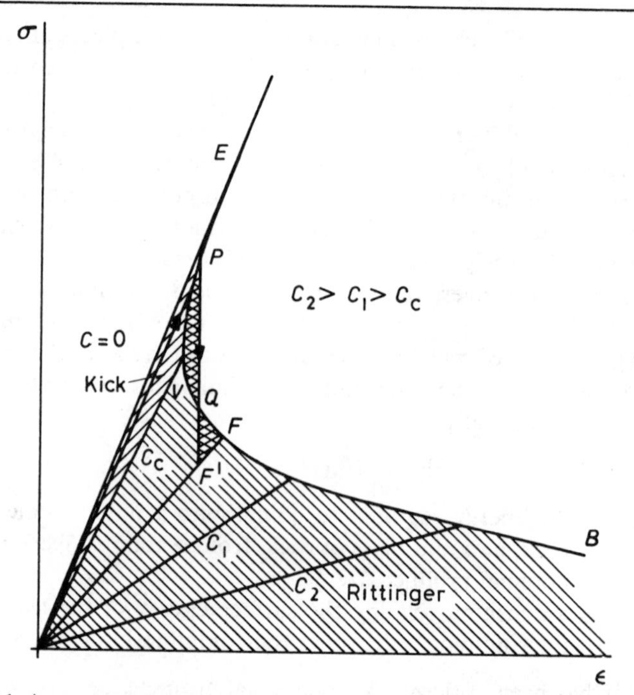

Fig. 18.14.1 A stress–strain diagram for material containing Griffith cracks, showing the Young's moduli for different lengths of cracks and the Griffith's locus. Materials with weaknesses smaller than the crack C_c, which corresponds to the point of minimum strain on the Griffith's locus, V, should fail in accordance with Kick's hypothesis and those with weaknesses larger than the crack C_c should fail in accordance with Rittinger's hypothesis.

cause the strength of the particles must be high and because the weaknesses must be smaller than the particles themselves. On the other hand, if a particle of material begins to break at a stress less than that corresponding to the point of minimum strain on the Griffith locus, external energy additional to the strain energy in the particle is required to propagate the fracture. This approaches the condition postulated by Rittinger. In detail, it is implied that the particles contain relatively large Griffith-type weaknesses. Clearly, a stage must be reached in the process of comminution when the particle size becomes less than that of these weaknesses and Rittinger's hypothesis is no longer valid, as is observed in practice.

Both the above hypotheses have been superseded in practice by Bond's (1952) proposal that the energy required for crushing and grinding can be found from

$$W_B = W_i(10/P^{\frac{1}{2}} - 10/F^{\frac{1}{2}}), \tag{3}$$

where P and F are the sizes of aperture in microns through which 80 per cent of the product and feed will pass, respectively, and W_i is the *work*

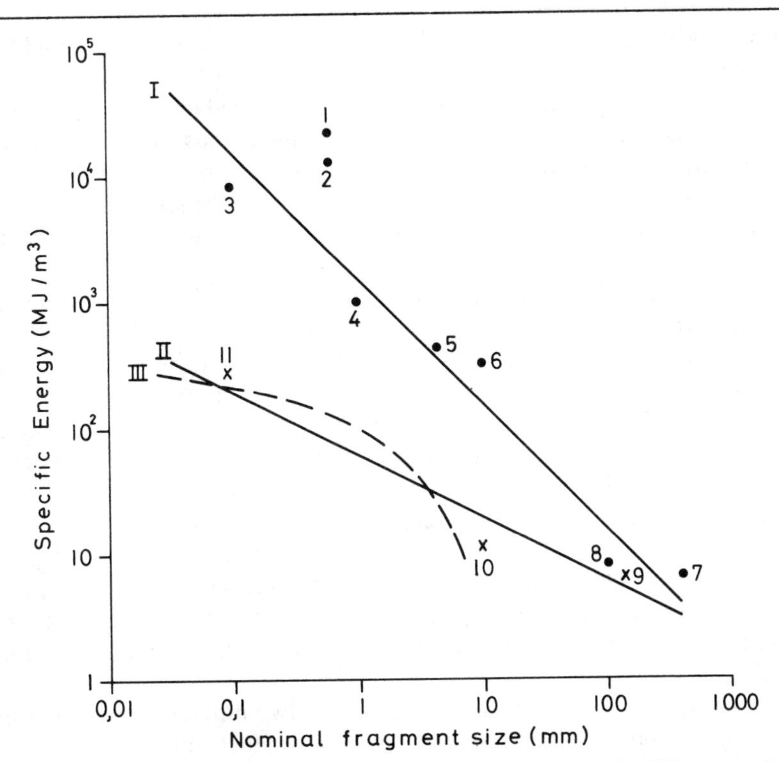

Fig. 18.14.2 Specific energies as a function of nominal particle size for different methods of breaking hard rock with a uniaxial compressive strength of about 200 MPa (after Cook and Joughin, 1970). (1) Flame jet piercing; (2) Water erosion jet; (3) Diamond cutting or drilling; (4) Percussive drilling; (5) Drag bit cutting; (6) Roller bit boring; (7) Impact-driven wedge; (8) Explosive blasting; (9) Jaw-crusher; (10) Gyratory crusher; (11) Milling. Line I, for methods of excavation, corresponds to Rittinger's hypothesis using an effective surface energy of 0,27 MJ/m²; line II for methods of comminution corresponds to Bond's relationship using a work index of 200 MJ/m³ or about 25 kWh/ton, and line III to Kick's hypothesis using a strength of 100 MJ/m³ or 100 MPa.

index. It has been found that W_i has a value of 10 kWh/ton to 20 kWh/ton for grinding most rocks to 80 per cent passing through a square mesh aperture of 0,1 mm (Johnson, 1968).

In Fig. 18.14.2 are shown the specific energies as a function of the nominal, that is, median by mass, particle size of the products for a number of methods of breaking hard rock with a uniaxial compressive strength of about 200 MPa (Cook and Joughin, 1970), and curves corresponding to the relationships defined by Rittinger, Bond and Kick. These show that direct mechanical techniques of rock breaking used for excavation tend to follow the Rittinger relationships and those for comminution that proposed by Bond. Indirect techniques such as flame and water jets appear to require about an order of magnitude more energy, presumably because of a low value for the coefficient of transfer of energy in the jet to that used for breaking the rock.

The information contained in Fig. 18.14.2 is of fundamental importance in evaluating practical methods of rock breaking. It shows that methods of comminution are relatively more efficient than methods used for excavation, especially for small sizes of fragments. Since the size of fragment is a nominal measure, this implies that methods of excavation probably produce a greater proportion of very fine particles and dissipate more energy as heat than do methods for comminution. A study of drag-bit cutting through quartzite (Chamber of Mines Research Review 1973) showed that only 0,2 per cent of the total energy went into creating new surfaces and the remainder appeared as heat; about 80 per cent in the cutting tool and 20 per cent in the rock fragments. For each method of excavation there is a maximum acceptable size of fragment; fragments produced by drilling must be sufficiently small to be removed easily from the hole, for example. However, the production of fragments smaller than this, especially of very fine particles, is wasteful and inefficient, and gives rise to other problems such as dust. It follows that the most direct way of improving the efficiency of any method of excavation is to minimize the production of unnecessarily small fragments and, especially, of fine particles.

In practice this approach can be realized in two different ways. First by devising methods of excavation which yield a product consisting mainly of large fragments. This is done intuitively in manual excavation, as has been shown by Baily and Dean (1967), and in winning coal by undercutting and in boring by kerf cutting. Second, the size of the cutting bits and the depths of cut can be increased to produce larger fragments direct. This necessitates the use of higher forces on the bits but, because of the reduction in specific energy resulting from the increase in the size of fragments, the amount of power needed to achieve a given rate of excavation actually may be reduced.

For hard rocks the forces do not increase in direct proportion to the area of the cutting bit in contact with the rock. Hodgson and Cook (1970) and Wagner and Schumann (1971) showed that the stamp bearing strength, §§ 7.3 and 15.7, decreases inversely as the area of the stamp for hard, brittle rocks. This effect of size on strength diminishes with decreasing hardness and increasing plasticity of the rock. The results observed for stamp bearing strengths on hard rock are in accord with those of the specific energy for breaking hard rock shown in Fig. 18.14.2. The strain energy per unit volume at failure in a stamp bearing test on hard rock is proportional to the square of the peak stress or strength. The strength varies inversely as the area of the stamp or as the square root of its diameter, so that the strain energy per unit volume varies inversely as the diameter of the stamp. If the specific energy for rock breaking is equated to the strain energy per unit volume at failure and the diameter of the stamp is proportional to the nominal fragment size, the relationship is the same as that given in Fig. 18.14.2 for methods of excavation.

Most practical methods of rock breaking, with the notable exception of explosives, § 18.15, use the stress generated by direct mechanical loading to fragment the rock locally. The actions of percussive drill bits, drag bits and roller bits are in the nature of that of a mechanical indenter. Breaking by indentation is made possible by the brittle, 'work-softening' behaviour of

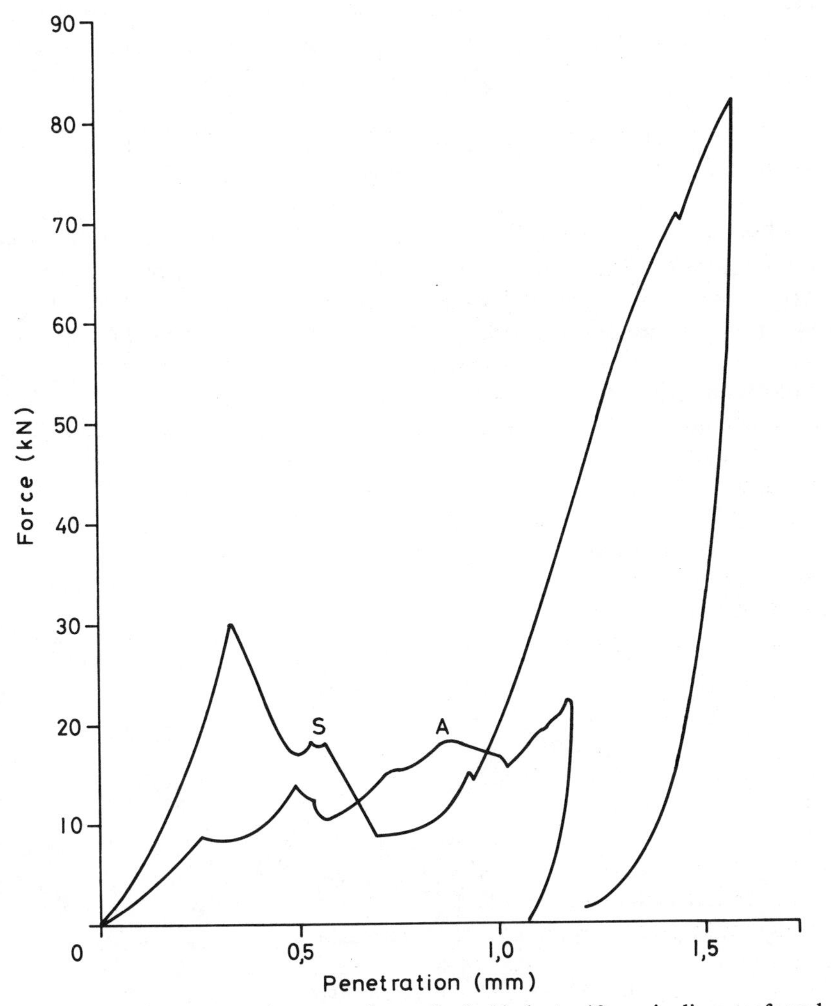

Fig. 18.14.3 Force–penetration curves for a spherical indenter 10 mm in diameter forced into a flat surface of an argillaceous quartzite, *A*, and a siliceous quartzite, *S*, of similar uniaxial compressive strengths of about 200 MPa.

rock § 4.2. Indentation does not remove material which is elastic-plastic or 'work-hardening'. In 'work-hardening' materials the resistance of the material to deformation increases with increasing deformation, so that any local concentration of deformation 'work-hardens' the material in its vicinity making it more resistant to further deformation which, therefore, spreads progressively throughout the weaker surround. In 'work-softening' materials the resistance to deformation decreases with increasing deformation so that any local concentration of deformation 'work-softens' the material adjacent to it and focuses further deformation into this region forming, ultimately, a cleavage or fracture surface.

Force–penetration curves for the indentation of two hard rocks, a siliceous quartzite and an argillaceous quartzite of similar uniaxial compressive strengths, by a hard, steel spherical indenter are shown in Fig. 18.14.3. Initially, indentation proceeds by Hertzian elastic deformation but, at some critical force, the rock around the indenter fails, and fragments spall from the rock surface in its vicinity. With further indentation, the force again builds up elastically to a second critical force, greater than the first, before the rock fails in a similar way for a second time. It is important to note that very little damage is done to the rock until the first critical force is reached and that this force can be very different for rocks of similar uniaxial compressive strength. For sediments of the Witwatersrand System a good relationship between the value of the first critical force and the mineralogical composition has been established, Fig. 18.14.4.

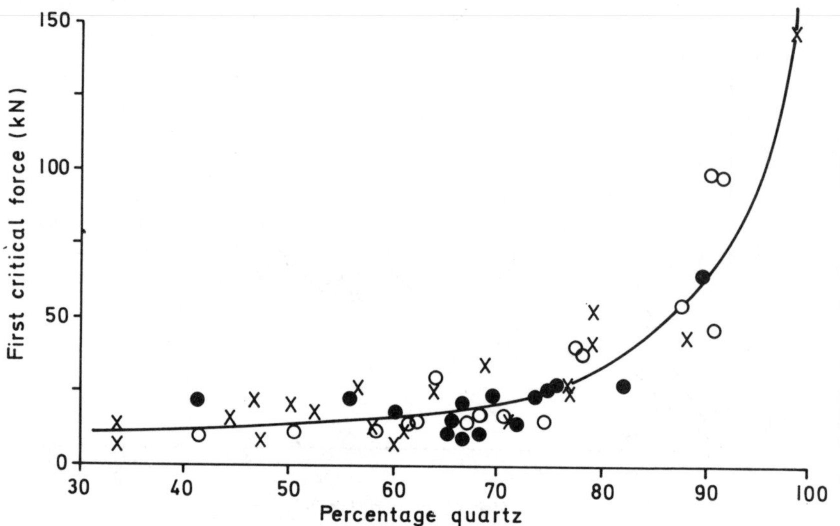

Fig. 18.14.4 The relationship between the value of the first critical force at which the rock around a spherical indenter fails and the total percentage of quartz in specimens of rock from three different gold fields of the Witwatersrand System.

There is no simple, reliable and comprehensive method of assessing the ease with which rock can be broken by mechanical devices. The specific energy of rock breaking is measured in MJ/m^3, which has the same dimensions, MN/m^2, as strength. The uniaxial compressive strength of rock provides a first estimate of specific energy and breakability. However, rocks of similar uniaxial strength but differing composition and structure show significant variations in breakability. To a large extent these are revealed by differences in the results of indentation tests. Differences in composition and structure can be obtained from petrographic analysis; rocks with a closely intergrown fabric are much more difficult to break than those in which the mineral grains are separated by a weak matrix. In addition, abrasiveness has an important bearing on bit life; the proportion of quartz present has a major effect on the abrasiveness.

The rate of rock excavation is given by

$$R = P/S, \tag{4}$$

where R = rate of excavation in the direction of advance of the bit or tool, mm/s;

P = the specific power delivered to the working face per unit area projected on to a plane normal to the direction of advance, kW/m^2;

S = the specific rock breaking energy, or strength of the rock, appropriate to the fragment size produced, MJ/m^3.

A convenient graphical representation of this relationship is given in Fig. 18.14.5 on which are plotted also typical operating points of different

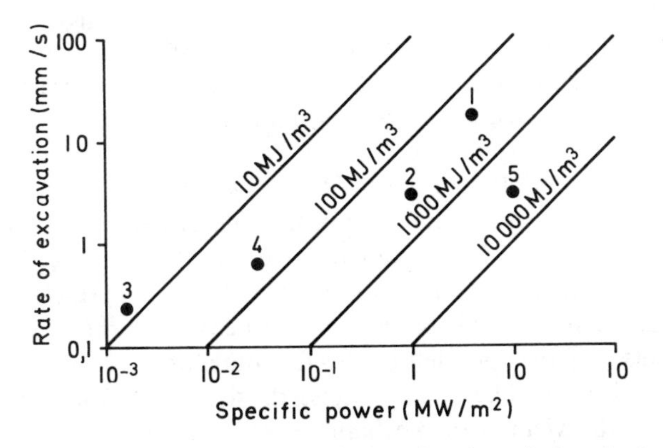

Fig. 18.14.5 A comparison of the characteristics of various devices for breaking rock showing the relationship between the rate of excavation, the specific power and the specific energy. (1) Percussive drills (small); (2) Rotary drills; (3) Drill-and-blast tunnelling; (4) Raise-and-tunnel-boring machines; (5) Flame jets.

rock breaking devices. This figure can be used to make quick estimates of the performance of any rock breaking device. It can be used to infer the specific energy, or strength, of rock in place from the measured performance of, say, a drill for which the specific power is known.

Also, it shows the principal characteristics of the different rock breaking devices. Small-hole percussive and rotary drills are highly developed and offer the greatest rates of excavation with moderate values of specific energy. Drilling and blasting is very efficient, having the lowest specific energy, but because of the cyclic nature of the operation, the average specific power is limited as is the rate of rock breaking. The novel methods are characterized by very high specific powers of the order of 10 MW/m^2 for flame jets, 10^4 to 10^5 MW/m^2 for water jets and electron beams and 10^9 MW/m^2 for lasers. However, the specific energies of these devices are also high, being of the order of 10^4 MJ/m^3. Nevertheless, the exceptional specific powers would yield high rates of excavation if the total power requirements were not so prohibitive. To limit the total power requirements to practicable values, these devices can be used only to cut very narrow kerfs comprising a very small proportion of all the rock broken. A comprehensive review of rock excavation by mechanical, hydraulic, thermal and electromagnetic means has been given by Cook and Harvey (1974).

Perhaps the high efficiency of drilling and blasting is the most striking feature on this figure. Drilling and blasting is a hybrid combination of two different rock breaking techniques. It seems likely that the key to further improvements in the performance of rock breaking devices lies in hybridization. This accomplishes two things; a reduction in specific energy and an increase in specific power. Potentially promising combinations are those of mechanical and thermal devices and water jets. Water jets remove loose and powdered, and fissured, porous and permeable material easily. The debris and damaged rock left by the passage of a mechanical tool or a thermal device are frequently of this nature, and insulate the unbroken rock beneath them from direct attack by the next pass of the tool or device. If this material were removed by a water jet in combination with the mechanical tool or thermal device, the rate of rock breaking would be enhanced not only by the amount removed by the water jet but also because a fresh surface of solid rock is exposed for direct attack by the other device. This fresh surface is often irregular because of the erosive effect of the water jets, and the voids left by this facilitate breakage of the rock under a mechanical tool. The highly stressed, deformed and cracked rock generated in the vicinity of mechanical tools also is removed readily by water jets. Thermal devices readily induce high stresses in hard, brittle rock such as is left exposed by a water jet or a mechanical tool, but not in weak fractured rock which is readily removed by them.

18.15 Blasting

Although drilling and blasting has been the principal method of excavating hard rock for over a century, the fundamentals of the processes of explosive rock breaking are but poorly understood.

Investigations of explosive rock breaking have been concerned with mathematical analyses of different physical models (Clark, 1967, 1968), theories concerning, and experiments on, spalling by the reflection of strain waves at a free surface, § 13.4 (Rinehart, 1960; Starfield, 1967), and field and model experiments using concentrated charges (Livingston, 1956). In practice, blasting usually is done in holes with a length of ten to a hundred times their diameter and separated from one another and from free boundaries by distances of about half the length of the hole. To a first approximation, this allows an analysis of the effects of the explosion in a blasthole in rock to be made by considering phenomena in a plane normal to the axis of the hole.

Using such a plane, consider first the phenomena which occur around a single, long hole in an extended volume of homogeneous rock which has been charged with a column of explosive. Immediately following detonation of the explosive, this hole is filled with gaseous detonation products at a pressure of the order of 1000 MPa and a temperature of the order of 3000 °K. If the explosive fills the hole completely, as is usually the case especially with prilled or slurried explosives, this pressure is applied immediately to the inside surface of the hole in the rock. The radial stress generated by this pressure is so great compared with the strength of the rock that an annulus of intensely crushed rock is formed around the hole by inter- and intra-granular cracking, the collapse of voids, differential compression of the particles and matrix of the rock, and other modes of microscopic deformation. This process takes place with a speed which is probably greater than the velocity of propagation of compressive elastic waves in rock, C_P, § 13.2, and is one of the few situations where true shock waves, § 13.7, occur in rock mechanics. The outer boundary of this crushed annulus then exerts a radial stress on the surrounding rock, which begins to move radially outwards. In consequence of the work done in forming the crushed annulus and of its expansion against the surrounding rock, the value of the stress at the outer boundary of the crushed annulus drops to a level where the short-period response of the rock beyond this boundary is initially elastic.

It is convenient to divide the processes which follow the formation of the crushed annulus into three different periods of time, Fig. 18.15.1.

During the first period of time, when elastic stress waves radiate from the outer boundary of the crushed annulus and a field of stresses is set up elastically in the rock beyond it, no fracturing other than the microfracturing within the annulus takes place because the speed of crack

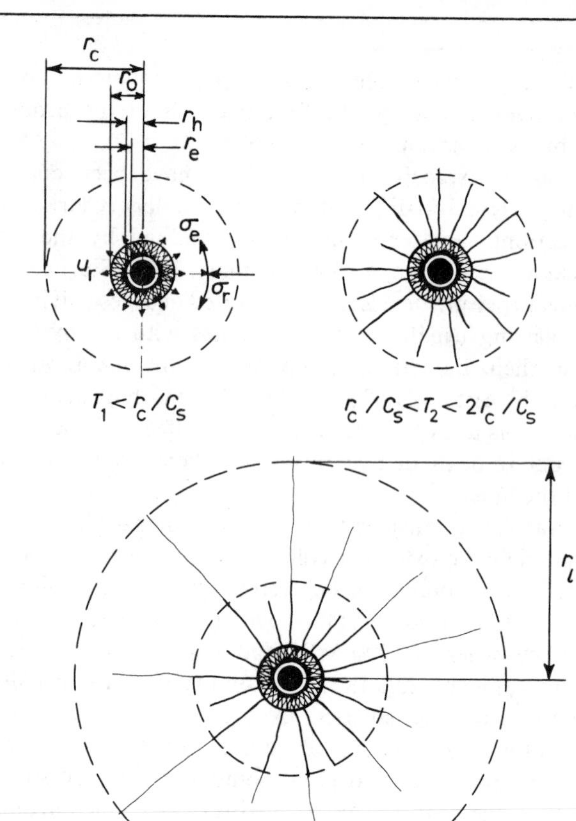

Fig. 18.15.1 Diagrams showing the processes which occur in the rock around a blasthole during each of three time periods, T_1, T_2 and T_3. C_S is the velocity of propagation of shear waves in the rock and r_e, r_h, r_o, r_c and r_l are the radii of the explosive, the hole, the crushed annulus, the cracked zone and the average length of radial cracks extended by gas pressure, respectively.

propagation, § 12.6, is less than the velocities of propagation of elastic compressive and shear waves. The stresses set up in this period can be thought to consist of a set of radial and tangential stresses in the rock outside the crushed annulus caused by the quasi-static pressure exerted by it and the contained gases. The radial stress, σ_r, is compressive and the tangential stress, σ_θ, is tensile but they are of the same value and this decreases in magnitude inversely as the square of the radial distance from the outside of the annulus, § 5.11.

During the second time period a radial compressive stress wave travels through an extended seismic zone, the value of its stress decreasing inversely as the square root of the radial distance, neglecting inelastic attenuation. In the same period both the tangential and radial stresses cause the rock outside the crushed annulus to be broken into sectors by radial cracks of different lengths and angular spacing.

Finally, during the third period of time, which is of much longer duration than the first two, gas under pressure enters the radial cracks, formed originally by these quasi-static stresses, and extends them. The extension of these cracks is controlled by a number of simple factors. Kutter (1967) has shown that the gas pressure required to extend the radial cracks increases with the number of these cracks, being least for two diametrically opposed cracks, and that longer cracks extend at lower values of gas pressure than are required to extend shorter cracks. Longer cracks extend preferentially at the expense of shorter cracks. Cracks which run into pre-existing joints or fractures are blunted by these features and cease to extend; the gas pressure spreads into the joint or fracture tending to open it. The effect of a pre-existing stress field around a blasthole is of great importance. Both the initiation and growth of the radial cracks, caused by the quasi-static stresses and their subsequent extension by gas under pressure, are influenced strongly by the magnitude and direction of any pre-existing stresses. Crack initiation, growth and extension occur preferentially in the direction of the maximum principal stress which promotes cracking parallel to the direction of this stress, but high values of the minimum principal stress normal to these cracks inhibit their growth and extension. If a state of great hydrostatic stress exists around a blasthole radial cracking may be eliminated virtually and damage may be confined to micro-cracking in the annulus of crushed rock immediately around the hole.

In practice, the crushed annulus around a blasthole is of little use and can be a real liability. It results in overbreak and unnecessary damage to the surfaces of an excavation which can impair its stability badly. The size of the crushed annulus depends upon the magnitude of the pressure exerted by the explosive in the drillhole. This pressure can be diminished by *decoupling* the explosive, that is, by using explosive charges smaller than the hole, (Atchison *et al.*, 1964; Atchison, 1968). The amount of decoupling for a circular hole is defined by the ratio of the diameter of the explosive to that of the hole. From field measurements Atchison *et al.* (1964) concluded that the seismic stress from a blast decreased as (explosive diameter/hole diameter)$^{1.5}$. Though they did not attempt to assess the effects decoupling had on the size of the crushed annulus, its effect on the size of the radially cracked zone was estimated from the change in period of the seismic signal.

The effect of decoupling on the pressure exerted by the gaseous explosion products on the walls of the drillhole may be estimated, roughly, from the gas laws. Let p_x be the gas pressure in the explosive column behind the detonation front. The pressure exerted by these gases when they have expanded to fill the hole is, approximately, $p_h = p_x$ (explosive diameter/hole diameter)^2.

Decoupling and splitting are used often to minimize overbreak and damage back of a line of holes forming the boundary of an excavation. In splitting, the holes in the row drilled along the line or curve defining the boundary are spaced between ten and twenty diameters apart. These holes are charged with decoupled sticks of explosive which often are spaced along the length of the hole also, so as to decrease further the total charge and increase the decoupling. To achieve a split along this boundary, all the holes and the charges spaced along them are exploded virtually simultaneously with detonating fuse. Pairs of radial cracks spread preferentially from each hole along the line or curve running through the centres of the holes, thereby splitting the rock along the boundary of the excavation. These pairs of cracks grow at the expense of other radial cracks as mentioned above and, together with the decoupling and low charge, minimize overbreak and back damage. In relatively stress-free rock, splitting can be done before the main blast used to remove the rock. This has the advantage that the split protects the rock behind it from damage by the main blast. It is virtually impossible to achieve an effective split between holes when a significant compressive stress exists normal to the direction of the proposed split. This is often the situation when making underground excavations at any significant depth. In such cases it is best to do the bulk of the excavation first, leaving only a small additional amount of rock around the circumference to be removed by a subsequent splitting blast. An additional advantage of this procedure is that the maximum principal stress around such excavations is parallel to their circumference and this stress promotes the formation and extension of radial cracks in directions parallel to it and hence parallel to the circumference of the excavation.

No discussion of blasting would be complete without reference to spalling caused by the tensile wave reflected off a free surface as the compressive seismic wave from the blast strikes it, § 13.4 (Rinehart, 1960; Atchison, 1968). The process is illustrated in Fig. 18.15.2 for the case of triangular seismic pulses. Whenever the sum of the incident compressive pulse and the reflected tensile pulse has a tensile value exceeding the strength of the rock at that position, a slab of rock spalls off. The details of this process depend upon the shape of the incident pulse. In general, the following rules apply. For a pulse with a very sharp front or rise-time, the front of the pulse is trapped in the slab and a number of successive slabs can be formed. If the rise-time is finite but less than the decay-time, a single slab will form at the point of maximum net tensile stress; immedi-

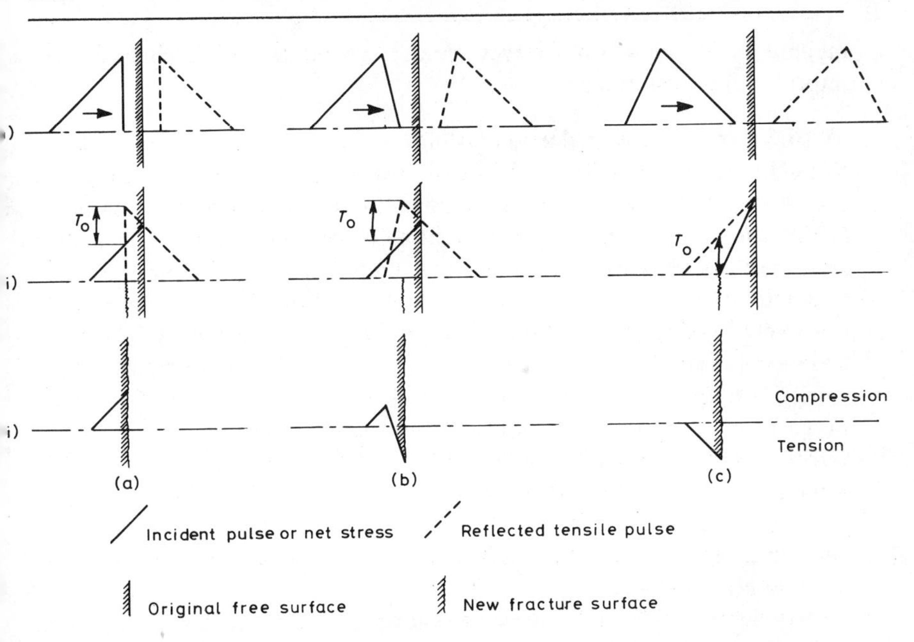

Fig. 18.15.2 Diagrams showing spalling by reflection from a free surface of an incident seismic pulse with (*a*) zero rise-time, (*b*) a rise-time less than the decay-time and (*c*) a rise-time greater than the decay-time; (i) before the incident pulse is reflected, (ii) at fracture and (iii) the net stress after fracture.

ately afterwards the tension adjacent to the fracture surface may be sufficient to cause a series of very thin slabs, or dust, to spall off. When the rise-time is greater than the decay-time only one relatively thick slab spalls off, after the toe of the reflected pulse has passed the heel of the incident pulse. The velocities with which the slabs fly off depend upon the magnitudes of the incident stress. Their values lie in the range given by $2\sigma_1 C_P/E$ to $2T_0 C_P/E$ where σ_1 is the value of the incident stress and T_0 is the tensile strength of the rock, E being the value of Young's modulus for the rock.

Finally, some general remarks concerning two fundamental factors which govern fragmentation by blasting must be made. First, the greater is the amount of explosive energy per unit volume of rock, or the *specific charge*, the finer is the degree of fragmentation. Second, for a given specific charge in a given rock, the smaller the ratio of hole spacing to hole burden the coarser will be the fragmentation. The *burden* of a hole is defined as the distance of the hole from the free surfaces to which it is expected to break. In practice, ratios of hole spacing to hole burden in the range 1 to 8 are used and the specific charge is usually less than 1 kg of explosive per cubic metre of rock. Time delays of several milliseconds are introduced between the detonation of each hole so as to reduce the magnitude of the seismic waves from large blasts; this technique also improves fragmentation.

18.16 Vibration damage due to blasting

When blasting is done in the vicinity of a structure it is necessary to design the blast so that the resulting vibrations do not cause structural damage.

Vibration damage might be thought to depend upon the magnitude of the accelerations. However, for a given amplitude, acceleration is proportional to frequency, and small high-frequency vibrations are obviously less likely to damage a structure than are longer low-frequency vibrations producing the same acceleration. Duvall *et al.* (1967) have summarized the work of a number of investigators and have concluded that over a frequency range from 3 to 300 c/s major structural damage occurs if the particle velocity exceeds 7·6 in. per second and no damage is likely if the particle velocity is less than 2 in. per second. The proportionality between particle velocity and stress or strain is established by § 13.3 (9), (10), (11), and it is not unreasonable to accept that it is the stress or strain which is the cause of structural damage.

Particle velocities can be measured with comparative ease. One approach to the design of blasts to avoid structural damage is to make measurements of particle velocities on a series of trial blasts, progressively increasing the size of the charge until the results can be extrapolated to determine the maximum charge which can be used without causing damage. It is useful to have a theoretical basis for making such an extrapolation and for predicting the particle velocities in circumstances where measurements may not have been made or it is inconvenient to make them. The charge size per shot and the distance between the shot and the structure, r, are obviously the two most important parameters in determining the particle velocity. The following relations between peak particle velocity, $\dot{A}_0$, charge size, W, and distance, r, can be expected on grounds of similitude for three-dimensional spherical propagation and two-dimensional cylindrical propagation, respectively

$$\dot{A}_0 = HW^{\frac{1}{3}}/r, \tag{1}$$

$$\dot{A}_0 = H(W/r)^{\frac{1}{2}}, \tag{2}$$

where H is a constant depending upon the type of explosive and the properties of the rock. In practice, particle velocities attenuate more rapidly than (1) and (2) suggest because of dissipation, § 13.6. On the basis of numerous tests, Duvall *et al.* (1967) propose that the peak particle velocity is well represented by

$$\dot{A}_0 = H(W^{\frac{1}{2}}/r)^n, \tag{3}$$

where $n = 1{\cdot}71$, $1{\cdot}67$, and $1{\cdot}14$ and $H = 84{\cdot}8$, $121{\cdot}1$, and $10{\cdot}5$ for the vertical, radial, and transverse components of vibration respectively. The difference between (3) and (1) or (2) suggests that greater charges generate vibrations more efficiently than smaller ones, and that dissipative attenuation is significant. The reciprocal of the factor in parentheses in (3) is known as the *scaled distance*. A scaled distance of 50 ft/lb$^{\frac{1}{2}}$ or more is most unlikely to produce particle velocities in excess of 2 in. per second. Where many shots are fired with millisecond delays the possibility of waves from different shots reinforcing one another in a particular direction needs to be examined.

Appendix

Units

Measurement involves the comparison of a quantity with some standard of the same kind. The specification of a physical quantity consists of a pure number, showing how the quantity compares with the standard, and a unit showing what standard is used. Various standards have been established by custom, agreement and law. Most of the units used in this text have followed those in the original sources from which the data have been drawn. If the data are of a scientific nature, usually they are given in terms of the *centimetre–gram–second* (cgs) system. On the other hand data for engineering purposes generally are given in the *foot–pound* (force)–*second* (ft lbf/s) system which is used extensively in America and Britain.

There is growing now a world-wide tendency to use the metric units of the Système International (SI) for both engineering and scientific purposes. This provides a coherent system of units based on the metre (m), kilogram (kg) and second (s): derived units being obtained by direct multiplication and division of these units without the need of factors, such as gravitational acceleration.

The basic SI units are defined in Table I.

Table I
Definitions of the basic SI units

Quantity	Unit	Symbol	Definition
Length	metre	m	The metre is that length equal to 1 650 763'73 wavelengths *in vacuo* of the radiation corresponding to the transition between the energy levels $2p_{10}$ and $5d_5$ of the krypton 86 atom
Mass	kilogram	kg	The kilogram is the mass of the international prototype kilogramme in the custody of the Bureau International des Poids et Mesures, Sevres, France
Time	second	s	The second is the duration of 9 192 631 770 oscillations of a transition between the two hyperfine levels $f = 4$, $m = 0$ and $f = 3$, $m = 0$ of the fundamental state $2S_{\frac{1}{2}}$ of an atom of caesium 133
Thermodynamic temperature*	kelvin	K	The kelvin, a unit of the thermodynamic temperature equal to the fraction 1/273'16 of the temperature of the triple point of water

* Degrees Celsius or centigrade (°C) can be used.
°C = K − 273'16

Some of the derived SI units have been given special names. Those of concern in mechanics are:

Table II
Derived SI units

Quantity	Unit	Symbol	Derivation
Force	newton	N	kgm/s^2
Pressure, stress	pascal	Pa	N/m^2
Work, energy	joule	J	Nm
Power	watt	W	Nm/s
Frequency	hertz	Hz	s^{-1}

The unit for the measurement of an angle is the radian (rad), but degrees, minutes and seconds are commonly used; 360° = 2πrad.

In practice it is sometimes convenient to use decimal multiples and fractions of the units, but these should be limited to factors of 10^3. The following prefixes and their symbols are accepted as standard.

Table III
Prefixes designating factors of 10^3

Prefix	Symbol	Factor
tera	T	10^{12}
giga	G	10^9
mega	M	10^6
kilo	k	10^3
milli	m	10^{-3}
micro	μ	10^{-6}
nano	n	10^{-9}

The relations between the more common mechanical SI units and those of the *length–mass–time* systems based on the *centimetre–gram–second* (cgs) and the *foot–pound–second* (ft lbm s)* the *length–force–time* system based on *foot–pound–second* (ft lbf s),* and the *length–mass–force–time* system (ft lbm lbf s),† which requires a constant of proportionality in Newton's second law of 32'174 ft/s^2, are given in Table IV.

To convert SI units to other units, multiply the quantity in SI units by the factor listed in the appropriate column under Table IV. Conversely, to convert other units to SI units, divide the quantity in the other units by the factor listed under the appropriate column in this Table.

* Because the pound is used both as a unit of force and a unit of mass, the abbreviations of lbf and lbm are used, respectively, to avoid confusion.

† A pressure of one pascal is very small by engineering standards whereas that of MPa = 145,038 psi is unusually large; 1 bar = 0,1 MPa.

Table IV

Relations between the mechanical units of the SI and other systems

Quantity	SI Unit	Equivalents in other systems			
		cgs	ft lbm s	ft lbf s	ft lbm lbf s
Length	1 m	10^2 cm	3,2808 ft	3,2808 ft	3,2808 ft
Mass	1 kg	10^3 g	2,2046 lbm	$68{,}521 \times 10^{-3}$ slug	2,2046 lbm
Time	1 s	1 sec	1 sec	1 sec	1 sec
Speed	1 m/s	100 cm/sec	3,2808 ft/sec	3,2808 ft/sec	3,2808 ft/sec
Acceleration	1 m/s^2	100 cm/sec^2	3,2808 ft/sec^2	3,2808 ft/sec^2	3,2808 ft/sec^2
Force	1 N	10^5 dyne	7,2330 poundal	0,22481 lbf	0,22481 lbf
Energy, heat	1 J	10^7 erg	23,730 ft poundal	0,73756 ft lbf	0,73756 ft lbf
work	1 kJ	0,23884 Kcal	0,94781 Btu		0,94781 Btu
Power	1 W	10^7 erg/sec	23,730 ft poundal/sec	$1{,}34410 \times 10^{-3}$ horsepower	0,73756 ft lbf/sec
Area	1 m^2	10^4 cm^2	10,764 ft^2	10,764 ft^2	10,764 ft^2
Volume	1 m^3	10^6 cm^3	35,315 ft^3	35,315 ft^3	35,315 ft^3
Pressure, Stress	1 Pa	10 dyne/cm^2	0,67196 poundal/ft^2	$20{,}886 \times 10^{-3}$ lbf/ft^2	$20{,}886 \times 10^{-3}$ lbf/ft^2 $0{,}14504 \times 10^{-3}$ psi
Density	1 kg/m^3	10^{-3} g/cm^3	$62{,}428 \times 10^{-3}$ lbm/ft^3	$1{,}9403 \times 10^{-3}$ slug/ft^3	$62{,}428 \times 10^{-3}$ lbm/ft^3
Specific heat	1 kJ/kg °C	0,23884 cal/g °C	0,23885 Btu/hft °F		0,23885 Btu/lbm °F
Thermal conductivity	1 W/m °C	10^5 ergs/cm sec °C	0,57778 Btu/hft °F		0,57778 Btu/hft °F
Viscosity:					
Dynamic	Pa . s	10 Poise	0,67197 lbm/ft sec	$20{,}886 \times 10^{-3}$ lbf sec/ft^2	$0{,}14504 \times 10^{-3}$ psi . sec
Kinematic	m^2/s	10×10^3 Stoke	10,764 ft^2/sec	10,764 ft^2/sec	$38{,}750 \times 10^3$ ft^2/h
Permeability	m^2/Pa . s	10^9 Darcy	16,019 ft^4/poundal sec	$0{,}51537 \times 10^3$ ft^4/lbf sec	$74{,}214 \times 10^3$ ft^2/s psi

References

ABRAMIAN, B. L., ARUNTIUNIAN, N. KH., and BABLOIAN, A. F. (1964). On the two-contact problem for an elastic sphere, *Fizika Metall.*, **28**, 622–9.

ADAMS, F. D. (1912). An experimental contribution to the question of the depth of the zone of flow in the Earth's crust, *J. Geol.*, **20**, 97–118.

ADAMS, L. H., and WILLIAMSON, E. D. (1923). On the compressibility of rocks and minerals at high pressures, *J. Franklin Inst.*, **195**, 475–529.

ALEXANDER, L. G. (1960). Field and laboratory tests in rock mechanics, *Third Aust.–N.Z. Conf. on Soil Mech*, 161–8.

ALLEN, N. P. (1959). The mechanism of the brittle fracture of metals, in *Fracture*, B. L. Averbach, D. K. Felbeck, G. T. Hahn and D. A. Thomas, Eds., New York, Wiley, pp. 123–146.

ANDERSON, E. M. (1935). The dynamics of the formation of cone-sheets, ring-dykes and caldron-subsidencies, *Proc. Roy. Soc. Edinb.*, **56**, 128–57.

ANDERSON, E. M. (1937). The dynamics of sheet intrusion, *Proc. Roy. Soc. Edinb.*, **58**, 242–51.

ANDERSON, E. M. (1951). *The Dynamics of Faulting and Dyke Formation with Applications to Britain*, 2nd edn., Edinburgh, Oliver and Boyd.

ANDERSON, O. L. (1959). The Griffith criterion for glass fracture, in *Fracture*, B. L. Averbach, D. K. Felbeck, G. T. Hahn and D. A. Thomas, Eds., New York, Wiley, p. 331.

ANDERSON, W. H. and DODD, J. S. (1967). Finite element method applied to rock mechanics, *Proc. First Congress International Society of Rock Mechanics*, Lisbon, 1966, 317–21.

ANDRADE, E. N. D A C. (1910), (1914). On the viscous flow in metals and allied phenomena; The flow in metals under large constant stresses, *Proc. Roy. Soc., London* **A 84**, 1–12; **90**, 329–42.

ARCHARD, J. F. (1958). Elastic deformation and the laws of friction, *Proc. Roy. Soc., London*, **A 243**, 190–205.

A.S.T.M. STAND. (1942). Compressive strength of natural building stones; tentative method. *ASTM. designation C170–41T, Part II*, 1102–4.

ATTEWELL, P. B. (1962a). Composite model to simulate porous rock, *Engineering*, **193**, 574–5.

ATTEWELL, P. B. (1962b). Response of rocks to high velocity impact, *Trans. Instn. Min. Metall.*, **71**, 705–24.

ATTEWELL, P. B. (1963). Dynamic fracturing of rocks, *Colliery Engng*, **40**, 203–10; 248–52; 289–94.

BARNARD, P. R. (1964). Researches into the complete stress–strain curve for concrete, *Mag. Concr. Res.*, **16** 203–10.

BECKER, G. F. (1893). Finite homogeneous strain, flow and rupture of rocks, *Bull Geol. Soc. Am*, **4**, 13–90.

BELL, R. J. T. (1920). *An Elementary Treatise on Coordinate Geometry of Three Dimensions*, 2nd edn., London, Macmillan.

BELLAMY, C. J. (1960). *Dept. Civil Engng. Res. Rept. R23, Univ. Sydney.*

BENIOFF, H. (1932). A new vertical seismograph, *Bull. Seism. Soc. Am.*, **22**, 155–69.

BERENBAUM, R. and BRODIE, I. (1959a). The tensile strength of coal, *J. Inst. Fuel*, **32**, 320–7.

BERENBAUM, R., and BRODIE, I. (1959b). Measurement of the tensile strength of brittle materials, *Br. J. Appl. Phys.*, **10**, 281–6.

BERRY, D. S. (1960). An elastic treatment of ground movement due to mining. I. Isotropic ground, *J. Mech. Phys. Solids*, **8**, 280–92.

BERRY, D. S., and FAIRHURST, C. (1965). Influence of rock anisotropy and time-dependent deformation on the stress-relief and high-modulus inclusion techniques of *in situ* stress determination, *Testing Techniques for Rock Mechanics, ASTM*, STP 402, *Am. Soc. Testing Mats.*, 190–206.

BERRY, D. S., and SALES. T. W. (1961). An elastic treatment of ground movement due to mining. II. Transversely isotropic ground, *J. Mech. Phys. Solids*, **9**, 52–62.

BERRY, D. S., and SALES, T. W. (1962). An elastic treatment of ground movement due to mining. III. Three-dimensional problem, transversely isotropic ground, *J. Mech. Phys. Solids*, **10**, 73–83.

BERRY, J. P. (1960a). Some kinetic considerations of the Griffith criterion for fracture. I. Equations of motion at constant force, *J. Mech. Phys. Solids*, **8**, 194–206.

BERRY, J. P. (1960b). Some kinetic considerations of the Griffith criterion for fracture. II. Equations of motion at constant deformation, *J. Mech. Phys. Solids*, **8**, 207–16.

BERRY, J. P. (1963). Determination of fracture surface energies by the cleavage technique, *J. Appl. Phys.*, **34**, 62–8.

BERTRAND, P. L. (1964). Note sur l'équilibre élastique d'un milieu indéfini percé d'une cavité cylindrique sous pression. *Annls. Ponts Chauss.*, **134**, 473–522.

BHARGAVA, R. D. and KAPOOR, O. P. (1966). Circular inclusion in an infinite elastic medium with a circular inhomogeneity, *Proc. Camb. Phil. Soc.*, **62**, 113–27.

BIENIAWSKI, Z. T. (1966). Contribution to discussion on a rigid 50-ton compression and testing machine, *S. Afr. Mech. Engr.*, **16**, 213.

BIOT, M. A. (1941a). General theory of three-dimensional consolidation, *J. Appl. Phys.*, **12**, 155–64.

BIOT, M. A. (1941b). Consolidation settlement under a rectangular load distribution, *J. Appl. Phys.*, **12**, 426–30.

BIOT, M. A. (1955). Theory of elasticity and consolidation for a porous anisotropic solid, *J. Appl. Phys.*, **26**, 182–5.

BIOT, M. A. (1957). Folding instability of a layered visco-elastic medium under compression, *Proc. Roy. Soc. London*, **A 242**, 444–54.

BIOT, M. A. (1961). Theory of folding of stratified visco-elastic media and its implication in tectonics and orogenesis, *Bull. Geol. Soc. Am.*, **72**, 1595–620.

BIOT, M. A. (1965). Theory of viscous buckling and gravity instability of multilayers with large deformation, *Bull. Geol. Soc. Am.*, **76**, 371–8.

BIOT, M. A. and CLINGAN, F. M. (1941). Consolidation settlement of a soil with an impervious top surface, *J. Appl. Phys.*, **12**, 578–81.

BIOT, M. A., ODÉ, H., and ROEVER, W. L. (1961). Experimental verification of the folding of stratified visco-elastic media, *Bull. Geol. Soc. Am.*, **72**, 1621–32.

BIRCH, F. (1960). The velocity of compressional waves in rocks to 10 kilobars, I. *J. Geophys. Res.*, **65**, 1083–102. II. *J. Geophys. Res.*, **66**, 2199–224.

BIRCH, F. (1961). Role of fluid pressure in mechanics of overthrust faulting: Discussion. *Bull. Geol. Soc. Am.*, **72**, 1441–3.

BIRCH, F., and BANCROFT, D. (1938a). The effect of pressure on the rigidity of rocks, I, II. *J. Geol.*, **46**, 59–87; **46**, 113–41.

BIRCH, F., and BANCROFT, D. (1938b). Elasticity and internal friction in a long column of granite, *Bull. Seism. Soc. Am.*, **28**, 243–54.

BIRCH, F., and BANCROFT, D. (1940). New measurements of the rigidity of rocks at high pressure, *J. Geol.*, **48**, 752–66.

BIRCH, F., SCHAIRER, F. J., and SPICER, H. C. (1942). Handbook of physical constants, *Spec. Pap. Geol. Soc. Am.*, No. 36, 1–319.

BISHOP, A. W. (1966). Strength of soils as engineering materials, *Géotechnique*, **16**, 91–130.

BISHOP, A. W., and BLIGHT, G. E. (1963). Some aspects of effective stress in saturated and partly saturated soils, *Géotechnique*, **13**, 177–97.

BISHOP, A. W., and HENKEL, D. J., (1962). *The Measurement of Soil Properties in the Triaxial Test*, 2nd edn., London, Edward Arnold.

BISHOP, A. W., and MORGENSTERN, N. (1960). Stability coefficients for earth slopes, *Géotechnique*, **10**, 129–50.

BLAIR, B. E., (1955, 1956). Physical properties of mine rock, Pts. III, IV. *U.S. Bureau Mines Rep. Inv.*, 5130, 5244.

BLOK, H. (1940). Fundamental aspects of boundary lubrication, *J. Soc. Automotive Engrs.*, **46**, 54–68.

BOAS, W., and HONEYCOMBE, R. W. K. (1947). The anisotropy of thermal expansion as a cause of deformation in metals and alloys, *Proc. Roy. Soc., London*, **A 188**, 427–39.

BÖKER, R. (1915). Die Mechanik der bleibenden Formänderung in Kristallinisch aufgebauten Körpern, *Ver. dt. Ing. Mitt. Forsch.*, **175**, 1–51.

BONNECHERE, F. (1967). A comparative study of *in situ* rock stress measurement techniques, M.Sc. Thesis, University of Minnesota.

BOTT, M. H. P. (1959). The mechanics of oblique slip faulting, *Geol. Mag.*, **96**, 109–17.

BOWDEN, F. P. (1954). The friction of non-metallic solids, *J. Inst. Petrol.*, **40**, 89–103.

BOWDEN, F. P., and LEBEN, L. (1939). The nature of sliding and the analysis of friction, *Proc. Roy. Soc., London*, **A 169**, 371–91.

BOWDEN, F. P., and TABOR, D. (1950). *The Friction and Lubrication of Solids*, vol. 1, Oxford, Clarendon Press.

BOWDEN, F. P., and TABOR, D. (1964). *The Friction and Lubrication of Solids*, vol. II, Oxford, Clarendon Press.

BRACE, W. F. (1960a). Behaviour of rock salt, limestone and anhydrite during indentation, *J. Geophys. Res.*, **65**, 1773–88.

BRACE, W. F. (1960b). An extension of Griffith theory of fracture to rocks, *J. Geophys. Res.*, **65**, 3477–80.

BRACE, W. F. (1961). Dependence of fracture strength of rocks on grain size, *Bull. Miner. Inds. Expr. Stn. Penn. St. Univ.*, No. 76, 99–103.

BRACE, W. F. (1961). Mohr construction in the analysis of large geologic strain, *Bull. Geol. Soc. Am.*, **72**, 1059–80.

BRACE, W. F. (1964a). Indentation hardness of minerals and rocks, *N. Jahrbuch F. Mineralogie Monatshefte*, 257–69.

BRACE, W. F. (1964b). Brittle fracture of rocks, in *State of Stress in the Earth's Crust*, W. R. Judd (Ed.), New York, Elsevier, 111–74.

BRACE, W. F. (1965). Some new measurements of linear compressibility of rocks, *J. Geophys. Res.*, **70**, 391–8.

BRACE, W. F., and BOMBOLAKIS, E. G. (1963). A note on brittle crack growth in compression, *J. Geophys. Res.*, **68**, 3709–13.

BRACE, W. F., and BYERLEE, J. D. (1966). Stick–slip as a mechanism for earthquakes, *Science*, **153**, 990–2.

BRACE, W. F., and BYERLEE, J. D. (1967). Recent experimental studies of brittle fracture in rocks, *Proc. Eighth Symposium on Rock Mechanics, University of Minnesota*, 1966, in *Failure and Breakage of Rock*, C. Fairhurst (Ed.), 1967, 58–81.

BRACE, W. F., ORANGE, A. S., and MADDEN, T. R. (1965). The effect of pressure on the electrical resistivity of water-saturated crystalline rocks, *J. Geophys Res.*, **70**, 5669–78.

BRACE, W. F., and WALSH, J. B. (1962). Some direct measurements of the surface energy of quartz and orthoclase, *Am. Miner.*, **47**, 1111–22.

BRADLEY, J. J. and A. NEWMAN FORT, Jr. (1966). Internal friction in rocks, in Handbook of Physical Constants, S. P. Clark, Jr., (Ed.), *The Geol Soc. Amer. Inc.*, New York Memoir 97.

BREDTHAUER, R. O. (1957). Strength characteristics of rock samples under hydrostatic pressure, *Trans. Amer. Soc. Mech. Engrs.*, **79**, 695–708.

BREKHOVSKIKH, L. M. (1960). *Waves in Layered Media* (translated D. Lieberman), London, Academic Press.

BRESLER, B., and PISTER, K. S. (1957). Failure of plain concrete under combined stresses, *Trans. Am. Soc. Civ. Engrs.*, **122**, 1049–68.

BRIDGMAN, P. W. (1912). Breaking tests under hydrostatic pressure and conditions of rupture, *Phil. Mag.*, **24**, 63–80.

BRIDGMAN, P. W. (1931). *The Physics of High Pressures*, 1st edn., London, Bell.

BROMWICH, T. J. I'A. (1926). *An Introduction to the Theory of Infinite Series*, London, Macmillan.

BROUTMAN, L. J., and CORNISH, R. H. (1965). Effect of polyaxial stress on failure strength of alumina ceramics, *J. Am. Ceram. Soc.*, **48**, 519–24.

BULLEN, K. E. (1963). *Introduction to the Theory of Seismology*, 3rd edn., Cambridge, University Press.

BURRIDGE, R., and KNOPOFF, L. (1967). Model and theoretical seismicity, *Bull. Seism. Soc. Am.*, **57**, 341–71.

BYERLEE, J. D. (1967a). Theory of friction based on brittle fracture, *J. Appl. Phys.*, **38**, 2928–34.

BYERLEE, J. D. (1967b). Frictional characteristics of granite under high confining pressure, *J. Geophys. Res.*, **72**, 3639–48.

CAREY, S. W. (1953). The rheid concept in geotectonics, *J. Geol. Soc. Aust.*, **1**, 67–117.

CARSLAW, H. S., and JAEGER, J. C. (1959). *Conduction of Heat in Solids*, 2nd edn., Oxford, Clarendon Press.

CARTER, N. L., CHRISTIE, J. M., and GRIGGS, D. T. (1964). Experimental deformation and recrystallization of quartz, *J. Geol.*, **72**, 687–733.

CASAGRANDE, A., and CARRILLO, N. (1953). Shear failure of anisotropic materials, in *Contributions to Soil Mechanics, Boston Soc. Civil Engrs.*, 122–35.

CAUCHY, A. L. (1828). Sur les équations qui expriment les conditions d'équilibre ou les lois de mouvement intérieur d'un corps solide, *Bulletin des Sciences à la Société philomathique.*

CAUCHY, A. L. (1829). Mémoire sur l'équilibre et le mouvement des corps élastiques, *Paris, Mém. de l'Acad., t.8.*

CHEATHAM, J. B., JR. (1964). Indentation analysis for rock having a parabolic yield envelope, *Int. J. Rock Mech. Min. Sci.*, **1**, 431–40.

CHEATHAM, J. B., and GNIRK, P. F. (1967). The mechanics of rock failure associated with drilling at depth, *Proc. Eighth Symposium on Rock Mechanics, University of Minnesota, 1966*, in *Failure and Breakage of Rock.* C. Fairhurst (Ed.), 1967, 410–39.

CHINNERY, M. A. (1961). The deformation of the ground around surface faults, *Bull. Seism. Soc. Am.*, **51**, 355–72.

CHINNERY, M. A. (1966a). Secondary faulting. I. Theoretical aspects, *Canad. J. Earth Sci.*, **3**, 163–74.

CHINNERY, M. A. (1966b). Secondary faulting. II. Geological aspects, *Canad. J. Earth Sci.*, **3**, 175–90.

CHUGH, Y. P., HARDY, H. R., JR., and STEFANKO, R. (1967). Microseismic activity in rocks under uniaxial tension, *Trans. Am. Geophys. Un.*, **48**, 204–5.

CHURCHILL, R. V. (1948). *Introduction to Complex Variables and Applications*, New York, McGraw-Hill.

CHURCHILL, R. V. (1958). *Operational Mathematics*, 2nd edn., New York, McGraw-Hill.

CLARK, S. P. JR. (1966). (Ed.) Handbook of physical constants, *The Geol. Soc. Amer. Inc.*, New York, Memoir 97.

CLOOS, E. (1947). Oolite deformation in South Mountain Fold, Maryland, *Bull. Geol. Soc. Am.*, **58**, 843–918.

CLOUGH, R. W. (1960). The finite element method in plane stress analyses, *Proc. 2nd A.S.C.E. Conf. on Electronic Computation*, Pittsburgh, 345–78.

CLOUGH, R. W. (1965). *Stress Analysis*, (Ed.) O. C. Zienkiewicz, G. S. Holister, Ch. 7, New York, Wiley.

CLOUGH, R. W., and WILSON, E. L. (1963). Stress analysis of a gravity dam by the finite element method, *Bull. R.I.L.E.M.*, 45–54.

COATES, D. F., and GYENGE, M. (1966). Plate-load testing on rock for deformation and strength properties, *Testing Techniques for Rock Mechanics, ASTM*, STP 402, *Am. Soc. Test. Mats.*, 19–35.

COKER, E. G., and FILON, L. N. G. (1957). *A Treatise on Photoelasticity*, revised by H. T. Jessop, Cambridge, University Press.

COLBACK, P. S. B. (1967). An analysis of brittle fracture initiation and propagation in the Brazilian test, *Proc. First Congress International Society of Rock Mechanics, Lisbon*, 1966, **1**, 385–91.

COLBACK, P. S. B., and WIID, B. L. (1965). The influence of moisture content on the compressive strength of rock, *Rock Mechanics Symposium (3rd), University of Toronto, 1965*, 65–83.

COLLINS, F. (1960). Plane compressional Voigt waves, *Geophys.*, **25**, 483–504.

COOK, N. G. W. (1962). A study of failure in the rock surrounding underground excavations. Thesis, University of the Witwatersrand.

COOK, N. G. W. (1963). The seismic location of rockbursts, *Proc. 5th Rock Mechanics Symposium*, Oxford, Pergamon Press, 493–516.

COOK, N. G. W. (1965). The failure of rock, *Int. J. Rock Mech. Min. Sci.*, **2**, 389–403.

COOK, N. G. W. (1967). The design of underground excavations. *Proc. Eighth Symposium on Rock Mechanics, University of Minnesota*, 1966, in *Failure and Breakage of Rock*, C. Fairhurst (Ed.), 1967, 167–93.

COOK, N. G. W., and HODGSON, K. (1965). Some detailed stress–strain curves for rock, *J. Geophys. Res.*, **70**, 2883–88.

COOK, N. G. W., HOEK, E., PRETORIUS, J. P. G., ORTLEPP, W. D., and SALAMON, M. D. G. (1966). Rock mechanics applied to the study of rockbursts, *Jl. S. Afr. Inst. Min. Metall.*, **66**, 435–528.

COOK, N. G. W., and HOJEM, J. P. M. (1966). A rigid 50-ton compression and tension testing machine, *S. Afr. Mech. Engr.*, **16**, 89—92.

CORNET, I., and GRASSI, R. G. (1961). A study of theories of fracture under combined stress, *J. Bas. Engng.*, **83**, 39–44.

COTTRELL, A. H. (1952). The time laws of creep, *J. Mech. Phys. Solids*, **1**, 53–63.

COTTRELL, A. H. (1953). *Dislocations and Plastic Flow in Crystals*, Oxford, Clarendon Press.

COTTRELL, A. H. (1965). Mechanics of fracture, *Tewksbury Symposium on Fracture*, Eng. Faculty, University of Melbourne, 1–27.

COULOMB, C. A. (1773). Sur une application des règles de Maximis et Minimis a quelques problèmes de statique relatifs à l'Architecture, *Acad. Roy. des Sciences Memoires de math. et de physique par divers savans*, **7**, 343–82.

COUTINHO, A. (1949). Theory of an experimental method for determining stresses not requiring an accurate knowledge of the elasticity modulus, *Third Congr. int. Ass. Bridge struct. Engng. Liege*, 1948.

CURRIE, J. B., PATNODE, H. W., and TRUMP, R. P. (1962). Development of folds in sedimentary strata, *Bull. Geol. Soc. Am.*, 73, 655–74.

D'ANDREA, D. V., FISHER, R. L., and FOGELSON, D. E. (1964). Prediction of compressive strength from other rock properties, *Colo. Sch. Mines, Q.*, Golden, Colorado, **59**, 623–40.

DARCY, H. (1856). *Les fontaines publiques de la ville de Dijon.*

DAVIES, R. M. (1948). A critical study of the Hopkinson pressure bar, *Phil. Trans. Roy. Soc., London*, **A 240**, 375–457.

DEERE, D. U. (1964). Technical description of rock cores for engineering purposes, *Rock. Mech. and Engr. Geol.*, **1**, 16–22.

DEERE, D. U., HENDRON, A. J., PATTON, F. D., and CORDING, E. J. (1967). Design of surface and near-surface construction in rock. *Proc. Eighth Symposium on Rock Mechanics, University of Minnesota*, 1966, in *Failure and Breakage of Rock*, C. Fairhurst (Ed.), 1967, 237–302.

DENKHAUS, H. G. (1958). The application of the mathematical theory of elasticity to problems of stress in hard rock at great depth, *Pap. Discuss. Ass. Mine Mgrs. S. Afr.*, 1958/59, 271–310.

DENKHAUS, H. G. (1966). Residual stresses in rock masses, *Proc. First Congress International Society of Rock Mechanics, Lisbon*, 1966.

DE SITTER, L. U. (1956). *Structural Geology*, New York, McGraw-Hill.

DIX, C. H. (1952). *Seismic Prospecting for Oil*, New York, Harper.

DONATH, F. A. (1961). Experimental study of shear failure in anisotropic rocks, *Bull. Geol. Soc. Am.*, **72**, 985–90.

DONATH, F. A. (1964). Strength variation and deformational behaviour in anisotropic rock, in *State of Stress in the Earth's Crust*, W. R. Judd (Ed.), New York, Elsevier, 281–97.

DONATH, F. A. (1966). A triaxial pressure apparatus for testing of consolidated or unconsolidated materials subject to pore pressure, *Testing Techniques for Rock Mechanics*, *ASTM*, STP 402, *Am. Soc. Test. Mats.*, **41**.

DONNELL, L. H. (1941). Stress concentrations due to elliptical discontinuities in plates under edge forces, *Theodore von Karman Anniversary Volume*, Pasadena, 293–309

DRUCKER, D. C. (1962). On the role of experiment in the development of theory, *Proc. Fourth U.S. National Cong. of Appl. Mech.*, 15–33.

DURELLI, A. J., and PARKS, V. (1962). Relationship of size and stress gradient to brittle failure stress, *Proc. Fourth U.S. National Cong. of Appl. Mech.*, 931–8.

DURELLI, A. J., PHILLIPS, E. A., and TSAO, C. H. (1958). *Introduction to the Theoretical and Experimental Analysis of Stress and Strain*, New York, McGraw-Hill.

DUVALL, W. I. (1953). Strain wave shapes in rocks near explosions, *Geophys.*, **18**, 310–23.

DUVALL, W. I., and ATCHISON, T. C. (1957). Rock breakage by explosives, *U.S. Bur. Mines Rept. Invest.*, 5356.

DUVALL, W. I., ATCHISON, T. C., and FOGELSON, D. E. (1967). Empirical approach to problems in blasting research. *Proc. Eighth Symposium on Rock Mechanics*, University of Minnesota, 1966, in *Failure and Breakage of Rock*, C. Fairhurst (Ed.), 1967, 500–23.

EDWARDS, R. H. (1951). Stress concentration around spheroidal inclusions and cavities, *J. Appl. Mech.*, **18**, 19–30.

EIRICH, F. (Ed.) (1956). *Rheology, Theory and Applications*, New York, Academic Press.

ELLIOTT, H. A. (1947). An analysis of the conditions for rupture due to Griffith cracks, *Proc. Phys. Soc. London*, **59**, 208–23.

ELY, R. E. (1965). Strength of graphite tube specimens under combined stresses. *J. Am. Ceram. Soc.*, **48**, 505–8.

EMERY, C. L. (1964). Strain energy in rocks, in *State of Stress in the Earth's Crust*, W. R. Judd (Ed.), New York, Elsevier, 235–60.

ENDERSBEE, L. A., and HOFTO, E. O. (1963). Civil engineering design and studies in rock mechanics for Poatina underground power station. *J. Inst. Eng. Aust.*, **35**, 187–209.

ERDOGAN, F., and SIH, G. C. (1963). On the crack extension in plates under plane loading and transverse shear, *J. Bas. Engng.*, **85**, 519–27.

ERNSBERGER, F. M. (1965). Strength-controlling structures in glass, *Tewksbury Symposium on Fracture*, University of Melbourne, 1965, 120–40.

ESHELBY, J. D. (1957). The determination of the elastic field of an ellipsoidal inclusion, and related problems, *Proc. Roy. Soc. London*, **A 241**, 376–96.

EVANS, I. (1961). The tensile strength of coal, *Colliery Engng.*, **38**, 428–34.

EVANS, I., and MURRELL, S. A. F. (1962). Wedge penetration into coal, *Colliery Engng*, **39**, 11–16.

EVANS, I., and POMEROY, C. D. (1958). The strength of cubes of coal in uniaxial compression, in *Mechanical Properties of Non-metallic and Brittle Materials*, W. H. Walton (Ed.), London, Butterworths, 5–28.

EVANS, I., POMEROY, C. D., and BERENBAUM, R. (1961). The compressive strength of coal, *Colliery Engng.*, **38**, 75–80; 123–7; 172–8.

EVANS, R. H., and WOOD, R. H. (1937). Transverse elasticity of natural stones, *Proc. Leeds phil. Lit. Soc.*, **3**, pt. 5, 340–52.

EVERLING, G. (1964). Comments upon the definition of shear strength, *Int. J. Rock Mech. Min. Sci.*, **1**, 145–57.

EVERLING, G. (1964). A contribution to '*In situ* determination of stress by relief techniques' by R. H. Merrill in *State of Stress in the Earth's Crust*, W. R. Judd (Ed.), New York, Elsevier, 377–8.

EVISON, F. F. (1960). On the growth of continents by plastic flow under gravity, *Geophys. J.*, **3**, 155–190.

EWING, W. M., JARDETZKY, W. S., and PRESS, F. (1957). *Elastic Waves in Layered Media*, New York, McGraw-Hill.

FAIRHURST, C. (1961). Laboratory measurement of some physical properties of rock. *Proc. Fourth Symposium Rock Mechanics, Penn. State. University*, 105–18.

FAIRHURST, C. (1964a). Measurement of *in situ* rock stresses, with particular reference to hydraulic fracturing, *Rock Mech. and Engng. Geol.*, **2**, 3–4, 129–47.

FAIRHURST, C. (1964b). On the validity of the 'Brazilian' test for brittle materials, *Int. J. Rock Mech. Min. Sci.*, **1**, 535–46.

FAIRHURST, C. (1965). On the determination of the state of stress in rock masses, *Pap. No. SPE 1062, Second Conference on Drilling and Rock Mechanics, University of Texas.*

FAIRHURST, C., and COOK, N. G. W. (1966). The phenomenon of rock splitting parallel to a free surface under compressive stress, *Proc. First Congress International Society of Rock Mechanics, Lisbon*, 1966, **1**, 687–92.

FILON, L. N. G. (1902). On the elastic equilibrium of circular cylinders under certain practical systems of load, *Phil. Trans. Roy. Soc.*, **A 198**, 147–233.

FILON, L. N. G. (1924). The stresses in a circular ring, *Sel. Engng. Pap. Instn. civ. Engr.*, No. 12.

FITZPATRICK, J. (1962). Biaxial device for determining the modulus of elasticity on stress-relief cores, *U.S. Bureau Mines*, RI 6128.

FLINN, D. (1956). On the deformation of the Funzie conglomerate, Fetlar, Shetland, *J. Geol.*, **64**, 480–505.

FLINN, D. (1962). On folding during three-dimensional progressive deformation, *Q. J. Geol. Soc., Lond.*, **118**, 385–433.

FÖPPL, A. (1900). *Mitt. mech-tech. Lab. Münch.*, T. Ackermann, Vol. 7.

FREUDENTHAL, A. (1951). The inelastic behaviour and failure of concrete, *Proc. First U.S. Nat. Congress of Appl. Mech., Illinois Institute of Technology*, 641–6.

FRIEDMAN, M. (1963). Petrofabric analysis of experimentally deformed calcite-cemented sandstones, *J. Geol.*, **71**, 12–37.

FRIEDMAN, M. (1964). Petrofabric techniques for the determination of principal stress directions in rocks, in *State of Stress in the Earth's Crust*, W. R. Judd (Ed.), New York, Elsevier, 451–550.

GADDY, F. L. (1956). A study of the ultimate strength of coal as related to the absolute size of the cubical specimens tested. Ser. 112, *Virginia Engng. Exp. Stn., Virginia Polytech. Inst.*

GALITZIN, B. (1914). *Vorlesungen über Seismometrie*, Leipzig, Teubner.

GALLE, E. M., and WILHOIT, J. C., JR. (1962). Stresses around a wellbore due to internal pressure and unequal principal geostatic stresses, *J. Soc. Petrol. Engrs.*, **225**, 145–60.

GANE, P. G., HALES, A. L., and OLIVER, H. O. (1946). A seismic investigation of the Witwatersrand earth tremors, *Bull. Seism. Soc. Am.*, **36**, 49–80.

GARDNER, M. F., and BARNES, J. L. (1942). *Transients in Linear Systems*, New York, Wiley.

GILMAN, J. (1959). Cleavage, ductility and tenacity in crystals, in *Fracture*, Technology Press, 193–224.

GILMAN, J. (1961). Direct measurements of the surface energies of crystals, *J. Appl. Phys.*, **31**, 2208–18.

GILVARRY, J. J. (1961). Fracture of brittle solids. I. Distribution function for fragment size in single fracture (theoretical). *J. Appl. Phys.*, **32**, 391–9.

GNIRK, P. F., and JOHNSON, R. E. (1964). The deformational behaviour of a circular mine shaft situated in a viscoelastic medium under hydrostatic stress. *Proc. Sixth Symposium Rock Mechanics*, Rolla, Missouri, 231–59.

GOLDSTEIN, S. (1926). The stability of a strut under thrust when buckling is resisted by a force proportional to the displacement, *Proc. Camb. Phil. Soc.*, **23**, 120–9.

GOODIER, J. N. (1932). Compression of rectangular blocks and the bending of beams by non-linear distributions of bending forces, *Trans. Am. Soc. Mech. Engrs.*, **54**, 173–83.

GOODIER, J. N. (1933). Concentration of stress around spherical inclusions and flaws, *Trans. Am. Soc. Mech. Engrs.*, **55**, 39–44.

GOODMAN, R. E. (1964). The resolution of stresses in rock using stereographic projection, *Int. J. Rock Mech. Min. Sci.*, **1**, 93–103.

GOODMAN, R. E. (1966). On the distribution of stresses around tunnels in non-homogeneous rocks, *Proc. First Congress International Society of Rock Mechanics*, Lisbon, **2**, 249–55.

GOODMAN, R. E., and TAYLOR, R. L. (1966). Methods of analysis for rock slopes and abutments: a review of recent developments, *Proc. Eighth Symposium on Rock Mechanics*, University of Minnesota, 1966, in *Failure and Breakage of Rock*, C. Fairhurst (Ed.), 1967, 303–20.

GORDON, J. E., MARSH, D. M., and PARRATT, MARGARET E. M. L. (1959). On the strength and structure of glass, *Proc. Roy. Soc. London*, **A 249**, 65–72.

GRANT, F. S., and WEST, G. F. (1965). *Interpretation Theory in Applied Geophysics*. International Series in the Earth Sciences, R. R. Shrock Ed., New York, McGraw–Hill.

GRASSI, R. C., and CORNET, I. (1949). Fracture of gray cast-iron tubes under biaxial stresses, *Trans. Am. Soc. Mech. Engrs.*, **71**, 178–82.

GREEN, A. E. (1939). Stress systems in aeolotropic plates, II. *Proc. Roy. Soc. London*, **A 173**, 173–92.

GREEN, A. E. (1942). Stress systems in aeolotropic plates, IV. *Proc. Roy. Soc. London*, **A 180**, 173–208.

GREEN, A. E. (1945). Stress systems in aeolotropic plates, V, VI, VII. *Proc. Roy. Soc., London*, **A 184**, 231–52; 289–300; 301–45.

GREEN, A. E. (1947). A concentrated force problem of plane strain or plane stress, *J. Appl. Mech.*, **14**, A 246.

GREEN, A. E., and TAYLOR, G. I. (1939). Stress systems in aeolotropic plates, I. *Proc. Roy. Soc.*, **A 173**, 162–72.

GREEN, A. E., and TAYLOR, G. I. (1945). Stress systems in aeolotropic plates, III. *Proc. Roy. Soc., London*, **A 184**, 181–95.

GREEN, A. E., and ZERNA, W. (1954). *Theoretical Elasticity*, Oxford, Clarendon Press.

GREENWALD, H. P., HOWARTH, H. C., and HARTMANN, I. (1939). Experiments on the strength of small pillars of coal in the Pittsburgh bed, *U.S. Bur. Min. Tech. Pap.* 605.

GREENWALD, H. P., HOWARTH, H. C., and HARTMAN, I. (1941). Progress Report; experiments on strength of small pillars of coal in the Pittsburgh bed, *U.S. Bur. Min., Rept. Inv.* 3575.

GREENWOOD, J. A., and TRIPP, J. H. (1967). The elastic contact of rough spheres, *J. Appl. Mech. Trans., A.S.M.E.*, **89**, 153–9.

GRESSETH, E. W. (1964). Determination of principal stress directions through an analysis of rock joint and fracture orientations, Star Mine, Burke, Idaho, *U.S. Bur. Min. Rept. Inv.* 6413.

GRETENER, P. E. (1965). Can the state of stress be determined from hydraulic fracturing data. *J. Geophys. Res.*, **70**, 6205–15.

GRIFFITH, A. A. (1921). The phenomena of rupture and flow in solids, *Phil. Trans. Roy. Soc., London*, **A 221**, 163–98.

GRIFFITH, A. A. (1924). Theory of rupture, *Proc. First International Congress Applied Mechanics*, Delft, 55–63.

GRIGGS, D. T. (1936). Deformation of rocks under high confining pressures, *J. Geol.*, **44**, 541–77.

GRIGGS, D. T. (1935). The strain ellipsoid as a theory of rupture, *Am. J. Sci.*, **230**, 121–37.

GRIGGS, D. T. (1939). Creep of rocks, *J. Geol.*, **47**, 225–51.

GRIGGS, D. T. (1940). Experimental flow of rocks under conditions favouring recrystallization, *Bull. Geol. Soc. Am.*, **51**, 1001–22.

GRIGGS, D. T. and HANDIN, J. (1960). Observations on fracture and an hypothesis of earthquakes, in *Rock Deformation*, *Geol. Soc. Am.* Mem. 79, 347–64.

GRIGGS, D. T., TURNER, F. J., and HEARD, H. C. (1960). Deformation of rocks at 500° to 800° C, in *Rock Deformation*, *Geol. Soc. Am.* Mem. 79, 39–104.

GUERNSEY, R., and GILMAN, J. (1961). Photoelastic study of the stress near a cleavage crack, *Proc. Soc. Exp. Stress Analysis*, **18**, 50–4.

GUEST, J. J. (1940). Yield surface in combined stress, *Phil. Mag.* Ser. 7, **30**, 349–69.

GURNEY, C., and ROWE, P. W. (1945a). The effect of radial pressure on the flow and fracture of reinforced plastic rods, *R. Aircraft Estab. Rep. Memo.* 2283.

GURNEY, C., and ROWE, P. W. (1945b). Fracture of glass rods in bending and under radial pressure, *R. Aircraft Estab. Rep. Memo.* 2284.

GUTENBERG, B. (1959). *Physics of the Earth's Interior*, New York, Academic Press.

HABIB, P., and MARCHAND, R. (1952). Mesures des pressions de terrains par l'essai de verin plat, *Annls Inst. tech. Bâtim. Série sols et fondations x*, 5th ann. No. 58, 967–71.

HAFNER, W. (1951). Stress distributions and faulting, *Bull. Geol., Soc. Am.*, **62**, 373–98.

HAKALEHTO, K. O. (1967). A study of the dynamic behaviour of rock using the Hopkinson split bar method, Thesis, University of Minnesota.

HANDIN, J. (1966). Strength and ductility, in *Handbook of Physical Constants*, S. P. Clark, Jr. (Ed.), *Geol. Soc. Amer. Memoir*, **97**, 238–89.

HANDIN, J., and FAIRBAIRN, H. W. (1955). Experimental deformation of Hasmark dolomite, *Bull. Geol. Soc. Am.*, **66**, 1257–73.

HANDIN, J., and HAGER, R. V., JR (1957). Experimental deformation of sedimentary rocks under confining pressure: tests at room temperature on dry samples, *Bull. Am. Ass. Petrol. Geol.*, **41**, 1–50.

HANDIN, J., and HAGER, R. V., JR (1958). Experimental deformation of sedimentary rocks under confining pressure: tests at high temperature, *Bull. Am. Ass. Petrol. Geol.*, **42**, 2892–934.

HANDIN, J., HAGER, R. V., JR., FRIEDMAN, M., and FEATHER, J. N. (1963). Experimental deformation of sedimentary rocks under confining pressure: pore pressure tests. *Bull. Am. Ass. Petrol. Geol.*, **47**, 717–55.

HANDIN, J., HEARD, H. C., and MAGOUIRK, J. N. (1967). Effects of the intermediate principal stress on the failure of limestone, dolomite and glass at different temperatures and strain rates, *J. Geophys. Res.*, **72**, 611–40.

HANDIN, J., HIGGS, D. V., and O'BRIEN, J. K. (1960). Torsion of Yule marble under confining pressure, in *Rock Deformation*, *Geol. Soc. Am.* Mem. 79, 245–74.

HANDIN, J., and STEARNS, D. W. (1964). Sliding friction of rock, *Trans. Am. Geophys. Un.*, **45**, 103.

HARDY, H. R., JR. (1959). Time-dependent deformation and failure of geologic materials, *Colo. Sch. Mines Q.*, **54**, 135–75.

HARDY, H. R., JR. (1966a). A loading system for the investigation of the inelastic properties of geologic materials, *Testing Techniques for Rock Mechanics, ASTM*, STP 402, *Am. Soc. Test. Mats.*, 232–65.

HARDY, H. R., JR. (1966b). Inelastic behaviour of geologic materials, *Canada Dept. Mines Tech. Surveys*, Rept. FMP 65, 66.

HARGREAVES, A. J. (1958). Instantaneous outbursts of coal and gas, *Proc. Austral. Inst. Min. Met.*, No. 186, 21–72.

HARGREAVES, A. J. (1967). Instantaneous outbursts of coal and gas in Australia, *Proc. Eighth Commonwealth Min. Congr.*, Australasian inst. min. and met., 1965.

HARZA, L. F. (1949). The significance of pore pressure in hydraulic structures, *Trans. Am. Soc. Civ. Engrs.*, **114**, 193–214.

HAST, N. (1958). The measurement of rock pressure in mines. *Årsb. Sver. geol. Unders.*, **52**, 3.

HAWKES, I. (1959). A study of stress waves in rock and the blasting action of an explosive charge, *Colliery Engng.*, **36**, 186–96.

HAWKES, I., and MOXON, S. (1966). The measurement of *in situ* rock stress, using the photoelastic biaxial gauge with the over-coring technique, *Int. J. Rock Mech. Min. Sci.*, **2**, 405–14.

HEARD, H. C. (1960). Transition from brittle to ductile flow in Solenhofen limestone as a function of temperature, confining pressure and interstitial fluid pressure, in *Rock Deformation, Geol. Soc. Amer.* Mem. 79, 193–226.

HEARD, H. C. (1963). Effect of large changes in strain rate in the experimental deformation of Yule marble, *J. Geol.*, **71**, 162–95.

HEARD, H. C., and RUBEY, W. W. (1963). Possible tectonic significance of transformation of gypsum to anhydrite plus water, *Spec. Pap. Geol. Soc. Am.*, No. 76, 77–8.

HEARMON, R. F. S. (1961). *An Introduction to Applied Anistropic Elasticity*, Oxford, University Press.

HENCKY, H., (1924). Zur Theorie plastischer Deformationen und der hierdurch im Material hervorgerufenen Nebenspannungen, *Int. Congr. Appl. Mech, 1st, Delft, 1924, Proc.*, 312–17.

HETÉNYI, M. (Ed.) (1950). *Handbook of Experimental Stress Analysis*, New York, Wiley.

HILL, R. (1950). *The Mathematical Theory of Plasticity*, Oxford (Engng. science series), Clarendon Press.

HINDE, P. B. (1964). Testing machine stiffness problem, *Engineer*, **217**, 1124–7.

HIRAMATSU, Y., NISHIHARA, M., and OKA, Y. (1954). A discussion on the methods of tension test of rock, *J. Min. Metall. Inst. Japan*, **70**, 285–9.

HIRAMATSU, Y., NIWA, Y., and OKA, Y. (1957). Measurements of stress in the field by application of photo-elasticity, *Tech. Rep. Engng. Res. Inst. Kyoto*, **7**, 49–63.

HIRAMATSU, Y., and OKA, Y. (1962). Stress around a shaft or level excavated in ground with a three-dimensional stress state. *Mem. Fac. Engng. Kyoto Univ.*, **24**, 56–76.

HIRAMATSU, Y., and OKA, Y. (1966). Determination of the tensile strength of rock by a compression test of an irregular test piece, *Int. J. Rock Mech. Min. Sci.*, **3**, 89–99.

HIRASAWA, T., and STAUDER, W. (1965). On the seismic body waves from a finite moving source, *Bull. Seism. Soc. Am.*, **55**, 237–62.

HOBBS, D. W. (1962). The strength of coal under biaxial compression, *Colliery Engng*, **39**, 285–90.

HOBBS, D. W. (1964a). The strength and stress–strain characteristics of Oakdale coal in triaxial compression, *J. Geol.*, **72**, 214–31.

HOBBS, D. W. (1964b). A simple method for assessing the uniaxial compressive strength of rock, *Int. J. Rock Mech. Min. Sci.*, **1**, 5–15.

HOBBS, D. W. (1964c). The tensile strength of rocks, *Int. J. Rock Mech. Min. Sci.*, **1**, 385–96.

HOBBS, D. W. (1965). An assessment of a technique for determining the tensile strength of rock, *Brit. J. Appl. Phys.*, **16**, 259–68.

HOBBS, D. W. (1966). A study of the behaviour of a broken rock under triaxial compression, and its application to mine roadways, *Int. J. Rock Mech. Min. Sci.*, **3**, 11–43.

HOBBS, D. W. (1967). Rock tensile strength and its relationship to a number of alternative measures of rock strength, *Int. J. Rock Mech. Min. Sci.*, **4**, 115–27.

HODGSON, K., and JOUGHIN, N. C. (1967). The relationship between energy release rate, damage and seismicity in deep mines, *Proc. Eighth Symposium on Rock Mechanics, University of Minnesota*, 1966, in *Failure and Breakage of Rock*, C. Fairhurst (Ed.), 1967, 194–203.

HODGSON, R. A. (1961). Classification of structures on joint surfaces, *Am. J. Sci.*, **259**, 493–502.

HOEK, E. (1963). Experimental study of rock-stress problems in deep level mining, *Expl. Mech.*, **3**, 177–94.

HOEK, E. (1964a). Rock fracture around mining excavations. *Proc. Fourth Internl. Conf. on Strata Control and Rock Mechanics, Columbia University*, New York, 334–48; *C.S.I.R. Pretoria, Rock Mech. Spec. Rept. No. 58, 1963.*

HOEK, E. (1964b). Fracture of anisotropic rock, *J. S. Afr. Inst. Min. Metall.*, **64**, 501–18.

HOEK, E. (1965). Rock fracture under static stress conditions, *Natl. Mech. Eng. Res. Inst., C.S.I.R., Pretoria, C.S.I.R. Rept. MEG 383.*

HOEK, E. (1967). A photoelastic technique for the determination of potential fracture zones in rock structures, *Proc. Eighth Symposium on Rock Mechanics, University of Minnesota*, 1966, in *Failure and Breakage of Rock*, C. Fairhurst (Ed.), 1967, 94–112.

HOEK, E., and BIENIAWSKI, Z. T. (1965). Brittle fracture propagation in rock under compression, *Int. J. Fracture Mech*, **1**, 137–55.

HOEK, E., and BIENIAWSKI, Z. T. (1966). Fracture propagation mechanism in hard rock. *Proc. First Congress of International Society of Rock Mechanics*, Lisbon, **1**, 243–9.

HOJEM, J. P. M., and COOK, N. G. W. (1968). The design and construction of a triaxial and polyaxial cell for testing rock specimens, *S. Afr. Mech. Engr.*, **18**, 57-61.

HOLLAND, C. T. (1962). Design of pillars for overburden support, *Min. Congr. J.*, **48**, 24–8; 66–71.

HOLLAND, C. T., and GADDY, F. L. (1957). Some aspects of permanent support of overburden on coal beds, *Proc. W. Va. Coal Min. Inst.*, 43–66.

HONDROS, G. (1959). The evaluation of Poisson's ratio and the modulus of materials of a low tensile resistance by the Brazilian (indirect tensile) test with particular reference to concrete, *Aust. J. Appl. Sci.*, **10**, 243–64.

HOOKER, V. E., and DUVALL, W. I. (1966). Stresses in rock outcrops near Atlanta, Ga., *U.S. Bur. Mines Rept. Inv.*, 6860.

HOPKINSON, B. (1912). *Collected Scientific Papers*, p. 423, Cambridge, University Press, 1921.

HOPKINSON, B. (1914). A method of measuring the pressure produced in the detonation of high explosives or by the impact of bullets, *Phil. Trans. Roy. Soc.*, **A 213**, 437–56.

HORIBE, T., and KOBAYASHI, R. (1960). Physical and mechanical properties of coal measure rocks under triaxial pressure. *Proc. Internl. Conf. Strata Control, Paris*, 175–86.

HORN, H. M., and DEERE, D. U. (1962). Frictional characteristics of minerals, *Géotechnique*, **12**, 319–35.

HOSKINS, E. R. (1966). An investigation of the flatjack method of measuring rock stress, *Int. J. Rock Mech. Min. Sci.*, **3**, 249–64.

HOSKINS, E. R. (1967). Field and laboratory experiments in rock mechanics, Thesis, The Australian National University.

HOSKINS, E. R., JAEGER, J. C., and ROSENGREN, K. J. (1968). A medium scale direct friction experiment, *Int. J. Rock Mech. Min. Sci.*, **5**, 143–54.

HOWLAND, R. C. J. (1935). Stresses in a plate containing an infinite row of holes, *Proc. Roy. Soc.*, **A 148**, 471–91.

HSU, T. T. C., SLATE, F. O., STURMAN, G. M., and WINTER, G. (1963). Microcracking of plain concrete and the shape of the stress–strain curve, *J. Am. Concrete Inst.*, **60**, 209–23.

HUBBERT, M. KING (1937). Theory of scale models as applied to the study of geologic structures, *Bull. Geol. Soc. Am.*, **48**, 1459–520.

HUBBERT, M. KING (1945). Strength of the earth, *Bull. Am. Ass. Petrol. Geol.*, **29**, 1630–53.

HUBBERT, M. KING (1951). Mechanical basis for certain familiar geological structures, *Bull. Geol. Soc. Am.*, **62**, 355.

HUBBERT, M. KING, and RUBEY, W. W. (1959), (1960), (1961). Role of fluid pressure in mechanics of overthrust faulting, *Bull. Geol. Soc. Am.*, **70**, 115–205; **71**, 617–28; **72**, 1581–94.

HUBBERT, M. KING, and WILLIS, D. G. (1957). Mechanics of hydraulic fracturing, *Trans. Am. Inst. Min. Engrs.*, **210**, 153–68.

HUGHES, D. S., and JONES, H. J. (1951). Elastic wave velocities in sedimentary rocks, *Trans. Am. geophys. Un.*, **32**, 173–8.

IDE, J. M. (1936). Comparison of statistically and dynamically determined Young's modulus of rocks, *Proc. Natn. Acad. Sci., U.S.A.*, **22**, 81–92.

INGLIS, C. E. (1913). Stresses in a plate due to the presence of cracks and sharp corners, *Trans. Instn. Nav. Archit., London*, **55**, 219–41.

IRWIN, G. R. (1957). Analyses of stresses and strains near the end of a crack traversing a plate, *J. Appl. Mech.*, **24**, 361–4.

IRWIN, G. R. (1958). Fracture, *Handbuch der Physik*, Vol. 6, 551–90, Berlin, Springer.

JACOBI, O., and BRÄNDLE, E. (1956). *Electric remote measuring instruments*, Glückauf, **92**, No. 13/14, 397–411.

JAEGER, J. C. (1959). The frictional properties of joints in rock, *Geofis. pura appl.*, **43**, 148–58.

JAEGER, J. C. (1960a). Rock failure at low confining pressures, *Engineering*, **189**, 283–4.

JAEGER, J. C. (1960b). Shear fracture of anisotropic rocks, *Geol. Mag.*, **97**, 65–72.

JAEGER, J. C. (1961a). *An Introduction to the Laplace Transformation*, 2nd edn., London, Methuen.

JAEGER, J. C. (1961b). The cooling of irregularly shaped igneous bodies, *Am. J. Sci.*, **259**, 721–34.

JAEGER, J. C. (1962a). *Elasticity, Fracture and Flow*, 2nd edn., London, Methuen.

JAEGER, J. C. (1962b). Punching tests on disks of rock under hydrostatic pressure, *J. Geophys. Res.*, **67**, 369–73.

JAEGER, J. C. (1963). Extension failures in rocks subject to fluid pressure, *J. Geophys., Res.*, **68**, 6066–7.

JAEGER, J. C. (1965). Fracture of rocks, *Tewksbury Symposium on Fracture*, 1963, Eng. Fac. University of Melbourne, 268–83.

JAEGER, J. C. (1967a). Failure of rocks under tensile conditions, *Int. J. Rock Mech. Min. Sci.*, **4**, 219–27.

JAEGER, J. C. (1967b). Brittle fracture of rocks, *Proc. Eighth Symposium on Rock Mechanics, University of Minnesota*, 1966, in *Failure and Breakage of Rock*, C. Fairhurst (Ed.), 3–58.

JAEGER, J. C., and COOK, N. G. W. (1963). Pinching-off and disking of rocks, *J. Geophys. Res.*, **68**, 1759–65.

JAEGER, J. C., and COOK, N. G. W. (1964). Theory and application of curved jacks for measurement of stresses, in *State of Stress in the Earth's Crust*, W. R. Judd (Ed.), New York, Elsevier, 381–95.

JAEGER, J. C., and HOSKINS, E. R. (1966a). Stresses and failure in rings of rock loaded in diametral tension or compression, *Br. J. Appl. Phys.*, **17**, 685–92.

JAEGER, J. C., and HOSKINS, E. R. (1966b). Rock failure under the confined Brazilian test, *J. Geophys. Res.*, **71**, 2651–9.

JEFFERY, G. B. (1921). Plane stress and plane strain in bipolar coordinates, *Phil. Trans. Roy. Soc. London*, **A 221**, 265–93.

JEFFREYS, H. (1952). *The Earth*, 3rd edn., Cambridge, University Press.

JEFFREYS, H. (1958). A modification of Lomnitz's law of creep in rocks, *Geophys. J.*, **1**, 92–95.

JENNINGS, J. E., and KIRCHMAN, P. F. (1968). Critical strain energy of distortion in soil at limiting yield, *submitted* to *Boston Soc. Civ. Engrs.*

JONES, R. (1952). A method of studying the formation of cracks in a material subjected to stress, *Br. J. Appl. Phys.*, **3**, 229–32.

JOUGHIN, N. C. (1966). An analysis of the growth and extent of the fracture zone around a deep-level tabular mine excavation, *Proc. First Congress International Society of Rock Mechanics, Lisbon*, 1966, 261–5.

JUDD, W. R. (1964). Rock stress, rock mechanics, and research, *State of Stress in the Earth's Crust*. W. R. Judd (Ed.), New York, Elsevier, 5–51.

JUDD, W. R., and HUBER, C. (1961). Correlation of rock properties by statistical methods, *Int. Symposium Min. Res., Rolla, MO.*

JURGENSON, L. (1934). The application of theories of elasticity and plasticity to foundation problems, *J. Boston Soc. Civ. Engrs.*, **21**.

KÁRMAN, TH. VON (1911). Festigkeitsversuche unter allseitigem Druck, *Z. Ver. dt. Ing.*, **55**, 1749–57.

KÁRMÁN, TH. VON, and DUWEZ, P. (1950). The propagation of plastic deformation in solids, *J. Appl. Phys.*, **21**, 987–94.

KARPLUS, W. J. (1958). *Analog Simulation: Solution of Field Problems*, New York, McGraw-Hill.

KEHLE, R. O. (1964). The determination of tectonic stresses through analysis of hydraulic well fracturing, *J. Geophys. Res.*, **69**, 259–73.

KING, L. V. (1912). On the limiting strength of rocks under conditions of stress existing in the earth's interior, *J. Geol.*, **20**, 119–38.

KNOPF, E. B. (1957). Petrofabrics in structural geology, *Colo. Sch. Mines Q.*, **52**, 99–111.

KNOPOFF, L. (1957). The seismic pulse in materials possessing solid friction: I. Plane waves, *Bull. Geol. Soc. Am.*, **68**, 1853–4.

KNOPOFF, L. (1958). Energy release in earthquakes, *Geophys. J.*, **1**, 44–52.

KNOPOFF, L., and MACDONALD, G. J. F. (1958). Attenuation of small amplitude stress waves in solids, *Rev. mod. Phys.*, **30**, 1178–92.

KOLSKY, H. (1949). An investigation of the mechanical properties of materials at very high rates of loading, *Proc. Phys. Soc., London*, **62B**, 676–700.

KOLSKY, H. (1960). Viscoelastic waves, *Int. Symp. on Stress Waves in Materials*, 59–91, New York, Interscience.

KOLSKY, H. (1963). *Stress Waves in Solids*, New York, Dover.

KOLSKY, H. (1965). The propagation of mechanical pulses in anelastic solids, in *Behaviour of Materials under Dynamic Loading*, N. J. Huffington, Jr. (Ed.), *Amer. Soc. Mech. Engrs.*, New York, 1–18.

LACHENBRUCH, A. H. (1961). Depth and spacing of tension cracks, *J. Geophys. Res.*, **66**, 4273–92.

LACHENBRUCH, A. H. (1962). Mechanics of thermal contraction cracks and ice-wedge polygons in permafrost, *Spec. Pap. Geol. Soc. Am.*, No. 70, 1–69.

LANE, K. S. (1967). Stability of reservoir slopes, *Proc. Eighth Symposium on Rock Mechanics, University of Minnesota*, 1966, in *Failure and Breakage of Rock*, C. Fairhurst (Ed.), 1967, 321–36.

LANE, K. S., and HECK, W. J. (1964). Triaxial testing for strength of rock joints, *Proc. Sixth Symp. Rock Mech., Rolla, MO*, 98–108.

LANG, T. A. (1961). Theory and practice of rock bolting, *Am. Inst. Min. Met. and Petrol. Engrs.*, **220**, 333–48.

LE COMTE, P. (1965), Creep in rock salt, *J. Geol.*, **73**, 469–84.

LEE, E. H. (1955). Stress analysis in visco-elastic bodies, *Q. Appl. Math.*, **13**, 183–90.

LEE, T. M. (1964). Spherical waves in visco-elastic media, *J. Acoust. Soc. Am.*, **36**, 2402–7.

LEEMAN, E. R. (1964). The measurement of stress in rock: I. The principles of rock stress measurement; II. Borehole rock stress measuring instruments; III. The results of some rock stress investigations, *J. S. Afr. Inst. Min. Metall.*, **65**, 45–114; **65**, 254–84.

LEEMAN, E. R., and GROBBELAAR, C. (1957). A compressometer for obtaining stress–strain curves of rock specimens up to fracture, *J. Scient. Instrum.*, **34**, 279–80; 503–5.

LEKHNITSKII, S. G. (1963). *Theory of Elasticity of an Anisotropic Elastic Body*, Translated by P. Fern, San Francisco, Holden-Day.

LELIAVSKY, S. (1958). *Uplift in Gravity Dams*, London, Constable.

LENSEN, G. J. (1958). Measurement of compression and tension: some applications, *N.Z. J. Geol. Geophys*, **1**, 565–70.

LEON, A. (1934). Über die Rolle der Trennungsbrüche ins Rahmen der Mohr'-schen anstrengungshypothese, *Bauingenieur*, **15**.

LEVENGOOD, W. D. (1959). Experimental method for developing minute flaw patterns in glass, *J. Appl. Phys.*, **30**, 378–86.

LING, C. B. (1947). On the stresses in a notched plate under tension, *J. Math. Phys.*, **26**, 284–9.

LING, C. B. (1948). On the stresses in a plate containing two circular holes, *J. Appl. Phys.*, **19**, 77–82.

LODE, W. (1926). *Proc. Int. Congress appl. Mech. 2nd, Zurich.*

LOMNITZ, C. (1956). Creep measurements in igneous rocks, *J. Geol.*, **64**, 473–9.

LOMNITZ, C. (1957). Linear dissipation in solids, *J. Appl. Phys.*, **28**, 201–5.

LOVE, A. E. H. (1911). *Some Problems of Geodynamics*, Cambridge University Press.

LOVE, A. E. H. (1927). *A Treatise on the Mathematical Theory of Elasticity*, 4th edn., New York, Dover.

LUDWIK, P. (1909). *Elemente der technologischen Mechanik*, Berlin, Springer.

LUNDBORG, N. (1967). The strength–size relation of granite, *Int. J. Rock Mech. Min. Sci.*, **4**, 269–72.

LURÉ, A. I. (1964). *Three-dimensional Problems of the Theory of Elasticity*, translated by J. R. Radok, New York, Interscience.

LUTSCH, A., and SZENDREI, M. E. (1958). The experimental determination of the extent and degree of fracture of a rock face by means of sonic and ultrasonic methods, *Pap. Discuss. Ass. Mine Mgrs. S. Afr.*, 1958/59, 465–89.

MACELWANE, J. B., and SOHON, F. W. (1936). *Theoretical Seismology*, Pts. I and II, New York, Wiley.

MACROBERT, T. M. (1917). *Functions of a Complex Variable*, London, Macmillan.

MARCUS, H. (1962). The permeability of a sample of an anisotropic porous medium, *J. Geophys., Res.*, **67**, 5215–25.

MARIN, J. (1935). Failure theories of materials subjected to combined stresses, *Proc. Am. Soc. Civ. Engrs.*, **61**, 851–67.

MASON, W. P. (1958). *Physical Acoustics and the Properties of Solids*, New York, D. van Nostrand.

MATSUSHIMA, S. (1960). On the flow and fracture of igneous rocks, *Bull. Disast. Prev. Res. Inst., Kyoto Univ.*, **36**, 1–9.

MATTICE, H. C., and LIEBER, P. (1954). On the attenuation of waves produced in viscoelastic materials, *Trans. Am. Geophys. Un.*, **35**, 613–25.

MAUNSELL, F. G. (1936). Stresses in a notched plate under tension, *Phil. Mag., 7th Ser.*, **21**, 765–73.

MAURER, W. C. (1965). Shear failure of rock under compression, *J. Soc. Petrol. Engrs.*, 167–76.

MAY, A. N. (1959). Instruments to measure the stress conditions existing in the rocks surrounding underground openings, *Canada, Dept. Mines and Tech. Surveys, Fuels Div. Tech. Mem. 102/59–MIN.*

MAYER, A., HABIB, P., and MARCHAND, R. (1951). Conférence internationale sur les pressions de terrains et le soutènement dans les chantiers d'exploration, Liège, du 24 au 28 avril, 1951, 217–21.

MAZENOT, P. (1965). Interprétation de nombreuses mesures de déformations exécutées sur massifs rocheux par E.D.F., *Annls. Inst. tech. Bâtim.*, Série Sds et Foundations, 46, 234–45.

MCCLINTOCK, F. A., and WALSH, J. B. (1962). Friction on Griffith cracks under pressure, *Fourth U.S. Nat. Congress of Appl. Mech., Proc.*, 1015–21.

MCDONAL, F. J., ANGONA, F. A., MILLS, R. L., SENGBUSH, R. L., VAN NOSTRAND, R. G., and WHITE, J. E. (1958). Attenuation of shear and compressional waves in Pierre Shale, *Geophys*, **23**, 421–39.

MCHENRY, D. (1948). The effect of uplift pressure on the shearing strength of concrete, *Trans. Int. Congr. large Dams, 3rd, Stockholm, 1948*, Vol. 1.

MCKINSTRY, H. E. (1953). Shears of the second order, *Am. J. Sci.*, **251**, 401–14.

MCNAMEE, J., and GIBSON, R. E. (1960a). Displacement functions and linear transforms applied to diffusion through porous elastic media, *Q. J. Mech. Appl. Math.*, **13**, 98–111.

MCNAMEE, J., and GIBSON, R. E. (1960b). Plane strain and axially symmetric problems of the consolidation of a semi-infinite clay stratum, *Q. J. Mech. Appl. Math.*, **13**, 210–27.

MCWILLIAMS, J. R. (1966). The role of microstructure in the physical properties of rock, *Testing Techniques for Rock Mechanics, ASTM*, STP 402, *Am. Soc. Test. Mats.* 175–89.

MCWILLIAMS, J. R. (1967). The brittle fracture of rocks: contribution to discussion, *Proc. Eighth Symposium on Rock Mechanics, University of Minnesota*, 1966, in *Failure and Breakage of Rock*, C. Fairhurst (Ed.), 142–5.

MEANS, W. D., and PATERSON, M. S. (1966). Experiments on preferred orientation of platy minerals, *Contr. Min. and Petrol.*, **13**, 108–33.

MERRILL, R. H. (1964). *In situ* determination of stress by relief techniques, in *State of Stress in the Earth's Crust*, W. R. Judd (Ed.), New York, Elsevier, 343–69.

MERRILL, R. H. (1967). Three-component borehole deformation gage for determining the stress in rock, *U.S. Bur. Mines Rept. of Invest.* RI 7015.

MERRILL, R. H., and MORGAN, T. A. (1950). Method of determining the strength of a mine roof, *U.S. Bur. Mines Rept. Invest.*, 5406.

MERRILL, R. H., and PETERSON, J. R. (1961). Deformation of a bore hole in rock, *U.S. Bur. Mines Rept. Invest.*, 5881.

MICHELL, J. H. (1900). Elementary distribution of plane stress, *Proc. London math. Soc.*, **32**, 35–61.

MICHELL, J. H. (1902). The inversion of plane stress, *Proc. London Math. Soc.*, **34**, 134–42.

MICHELSON, A. A. (1917), (1920). The laws of elastico-viscous flow. I, *J. Geol.*, **25**, 405–10; II, *J. Geol.*, **28**, 18–24.

MIELENZ, R. C. (1948). Petrography and engineering properties of igneous rocks. *Engng. Monogr. U.S. Reclam. Bur.*, No. 1.

MILLARD, D. J., NEWMAN, P. C., and PHILLIPS, J. W. (1955). The apparent strength of extensively cracked materials, *Proc. Phys. Soc. London*, **B 68**, 723–8.

MINDLIN, R. D. (1936). Force at a point in the interior of a semi-infinite solid, *Physics*, **7**, 195–202.

MINDLIN, R. D. (1940). Stress distribution around a tunnel, *Trans. Am. Soc. Civ. Engrs.*, **105**, 1117–53.

MISES, R. VON (1913). Mechanik der festen Körper im plastich deformablen Zustand, *Nachr. Ges. Wiss. Göttingen, Mathematisch-physikalische Klasse*, 582–92.

MOGI, K. (1966). Some precise measurements of fracture strength of rocks under uniform compressive stress. *Felsmechanik und Ingenieurgeologie*, **4**, 41–55.

MOHR, H. F. (1956). Measurement of rock pressure, *Mine Quarry Engng.*, 178–89.

MOHR, O. (1900). Welche Umstände bedingen die Elastizitätsgrenze und den Bruch eines Materials? *Z. Ver. dt. Ing.*, **44**, 1524–30; 1572–77.

MOHR, O. (1914). *Abhandlungen aus dem Gebiete der technische Mechanik*, 2nd edn., Ernst und Sohn, Berlin.

MOODY, J. D., and HILL, M. J. (1956). Wrench fault tectonics, *Bull. Geol. Soc. Am.*, **67**, 1207–46.

MOORE, W. L. (1961). Role of fluid pressure in mechanics of overthrust faulting: a discussion, *Bull. Geol. Soc. Am.*, **72**, 1581–94.

MORGAN, F., MUSKAT, M., and REED, D. W. (1941). Friction phenomena and the stick–slip process, *J. Appl. Phys.*, **12**, 743–52.

MORGENSTERN, N. R. (1965). The analysis of the stability of general slip surfaces, *Géotechnique*, **15**, 79–93.

MORLEY, A. (1923). *Strength of Materials*, London, Longmans.

MORRISON, R. G. K., and COATES, D. F. (1955). Soil mechanics applied to rock failure in mines, *Bull. Canad. Inst. Min. Metall.*, **48**, 701–11.

MOSSAKOVSKII, V. I., and RYBKA, M. T. (1965). An attempt to construct a theory of fracture for brittle materials based on Griffith's criterion, *Prikl. Mat. Mekh.*, **29**, 291–6.

MOTT, B. W. (1956). *Micro-indentation Hardness Testing*, London, Butterworth.

MOTT, N. F. (1948). Brittle fracture in mild-steel plates, *Engineering*, **165**, 16–18.

MUEHLBERGER, W. R. (1961). Conjugate joint sets of small dihedral angle, *J. Geol.*, **69**, 211–19.

MÜLLER, L. (1964). Application of rock mechanics in the design of rock slopes, in *State of Stress in the Earth's Crust*, W. R. Judd (Ed.), New York, Elsevier, 575–98.

MURRELL, S. A. F. (1958). The strength of coal under triaxial compression, in *Mechanical Properties of Non-metallic Brittle Materials*, W. H. Walton (Ed.), London, Butterworths, 123–53.

MURRELL, S. A. F. (1963). A criterion for brittle fracture of rocks and concrete under triaxial stress and the effect of pore pressure on the criterion, *Proc. Fifth Rock Mechanics Symposium, University of Minnesota*, in *Rock Mechanics*, C. Fairhurst (Ed.), Oxford, Pergamon, 563–77.

MURRELL, S. A. F. (1964a). The theory of the propagation of elliptical Griffith cracks under various conditions of plane strain or plane stress, Pt. 1, *Br. J. Appl. Phys.*, **15**, 1195–210.

MURRELL, S. A. F. (1964b). The theory of the propagation of elliptical Griffith cracks under various conditions of plane strain or plane stress, Pts. II, III, *Br. J. Appl. Phys.*, **15**, 1211–23.

MURRELL, S. A. F. (1965). The effect of triaxial stress systems on the strength of rocks at atmospheric temperatures, *Geophys. J.*, **10**, 231–81.

MURRELL, S. A. F., and MISRA, A. K. (1961–2). Time-dependent strain or 'creep' in rocks and similar non-metallic materials, *Trans. Instn. Min. Metall.*, **71**, 353–78.

MUSKAT, M. (1937). *The Flow of Homogeneous Fluids through Porous Media*, New York, McGraw-Hill.

MUSKHELISHVILI, N. I. (1953). *Some Basic Problems of the Mathematical Theory of Elasticity*, 4th edn., translated by J. R. M. Radok, Groningen, Noordhoff.

NADAI, A. (1938). The influence of time upon creep, the hyperbolic sine creep law, *Timoshenko Anniversary Volume*, New York, Macmillan, 155–70.

NADAI, A. (1950). *Theory of Flow and Fracture of Solids*, Vol. 1, 2nd edn., New York, McGraw-Hill.

NADAI, A. (1963). *Theory of Flow and Fracture of Solids*, Vol. 2, New York, McGraw-Hill.

NEFF, T. L. (1966). Equipment for measuring pore pressure in rock specimens under triaxial load, *Testing Techniques for Rock Mechanics, ASTM*, STP 402, *Am. Soc. Test. Mats.*, 3–17.

NEUBER, H. (1933). Elastich-streng Lösungen zur Kerbwirkung bei Scheiben und Umdrehungskörpern, *Z. angew, Math. Mech.*, **13**, 439–42.

NEUBER, H. (1958). *Theory of Notch Stresses*, Ann Arbor, Edwards; Berlin, Springer, 1958.

NICHOLLS, H. R. (1961). *In situ* determination of the dynamic elastic constants of rock, *Rep. Invest. U.S. Bur. Mines*, RI 5888.

NISHIHARA, M. (1958). Stress–strain–time relations of rocks, *Doshisha Engng. Rev.*, **8**, 32–55; 85–115.

NOSE, M. (1964). Rock test *in situ*, conventional tests on rock properties and design of Kurobegawa No. 4 dam based thereon, *Eighth International Congress on Large Dams, Trans.*, Edinburgh, **1**, 219–52.

NYE, J. F, (1951). The flow of glaciers and ice sheets as a problem in plasticity, *Proc. Roy. Soc. London*, **A 207**, 554–72.

NYE, J. F. (1957). *Physical Properties of Crystals*, Oxford, Clarendon Press.

OBERT, L., and DUVALL, W. I. (1957). Microseismic method of determining the stability of underground openings, *U.S. Bur. Mines Bull.* 573.

OBERT, L., and DUVALL, W. I. (1961). Seismic methods of detecting and delineating sub-surface subsidence, *U.S. Bur. Mines. Rept. Invest.*, 5882.

OBERT, L., and DUVALL, W. I. (1967). *Rock Mechanics and the Design of Structures in Rock*, New York, Wiley.

OBERT, L., MERRILL, R. H., and MORGAN, T. A. (1962). Borehole deformation gauge for determining the stress in mine rock, *U.S. Bur. Mines Rept. Invest.* 5978.

OBERT, L., and STEPHENSON, D. E. (1965). Stress conditions under which core discing occurs, *Soc. Min. Engrs., Trans.*, **232**, 227–34.

OBERT, L., WINDES, S. L., and DUVALL, W. I. (1946). Standardised tests for determining the physical properties of mine rock. *U.S. Bur. Mines Rept. Invest.*, 3891.

OBERTI, G. (1960). Experimentelle Untersuchungen über die Charakteristika der Verformbarkeit der Felsen. *Geologie Bauwes.*, **25**, 95–113.

ODÉ, H. (1956). A note concerning the mechanism of artificial and natural hydraulic fracturing systems, *Colo. Sch. Mines Q.*, Golden, Colorado, **51**, 19–29.

ODÉ, H. (1957). Mechanical analysis of the dyke pattern of the Spanish Peaks area, Colorado, *Bull. Geol. Soc. Am.*, **68**, 567–76.

ODÉ, H. (1960). Faulting as a velocity discontinuity in plastic deformation, *Rock Deformation, Geol. Soc. Am.*, Mem 79, 293–321.

O'DRISCOLL, E. S. (1962). Experimental patterns in superimposed similar folding, *J. Alberta Soc. Petrol. Geol.*, **10**, 145–67.

O'DRISCOLL, E. S. (1964a). Interference patterns from inclined shear fold systems, *Bull. Canad. Petrol Geol.*, **12**, 279–310.

O'DRISCOLL, E. S. (1964b). Rheid and rigid rotations, *Nature*, **203**, 832–5.

OFFICER, C. B. (1958). *Introduction to the Theory of Sound Transmission*, New York, McGraw-Hill.

OMORI, F. (1901). Results of the horizontal pendulum observations of earthquakes, July 1898 to December 1899, Tokyo. *Publ. Earthq. Invest. Comm., Tokyo*, **5**, 1–82.

ONODERA, T. F. (1963). Dynamic investigation of foundation rocks *in situ*, *Proc. Fifth Symposium on Rock Mechanics, University of Minnesota*, C. Fairhurst (Ed.), New York, Pergamon Press, 517–33.

OROWAN, E. (1949). Fracture and strength of solids, *Repts. on Progress in Physics*, **12**, 185–232.

OROWAN, E. (1952). Fundamentals of brittle behaviour in metals, in *Fatigue and Fracture of Metals*, New York, Wiley, 139–67.

OROWAN, E. (1959). Classical and dislocation theories of brittle fracture, in *Fracture*, R. L. Averbach (Ed.), New York, Wiley, 147–60.

ORTLEPP, W. D., and COOK, N. G. W. (1964). The measurement and analysis of the deformation around deep, hardrock excavations. *Proc. Fourth International Conference on Strata Control and Rock Mechanics, Henry Crumb School of Mines, Col. Univ.*, New York, 140–50.

ORTLEPP, W. D.; and NICOLL, A. (1964). The elastic analysis of observed strata movement by means of an electrical analogue, *J. S. Afr. Inst. Min. Metall.*, **65**, 214–35.

PALLISTER, G. F. (1968). The measurement of virgin rock stresses. M.Sc. Thesis, University of the Witwatersrand.

PANEK, L. A. (1961). Measurement of rock pressure with a hydraulic cell, *Am. Inst. Min. Met. and Petrol. Engng, Trans. AIME*, **220**, 287–90.

PANEK, L. A., and STOCK, J. A. (1964). Development of a rock stress monitoring station based on the flat slot method of measuring existing rock stress, *U.S. Bur. Mines Rept. of Invest.*, 6537.

PAPOULIS, A. (1962). *The Fourier Integral and its Applications*, New York, McGraw-Hill.

PARASNIS, D. S. (1960). The compaction of sediments and its bearing on some geophysical problems, *Geophys. J.*, **3**, 1–28.

PATERSON, M. S. (1958). Experimental deformation and faulting in Wombeyan marble, *Bull. Geol. Soc. Am.*, **69**, 465–76.

PATERSON, M. S. (1964). Triaxial testing of materials at pressures up to 10,000 kg./sq.cm., *Journ. Inst. Engrs. Aust.*, Jan.–Feb., 23–9.

PATERSON, M. S., and WIESS, L. E. (1966). Experimental deformation and folding in phyllite, *Bull. Geol. Soc. Am.*, **77**, 343–74.

PATTON, F. D. (1966). Multiple modes of shear failure in rock, *Proc. First Congress International Society of Rock Mechanics, Lisbon*, **1**, 509–13.

PAUL, B. (1961). Modification of the Coulomb-Mohr theory of fracture, *J. Appl. Mech.*, **28**, 259–68.

PAUL, B., and GANGAL, M. (1967). Initial and subsequent fracture curves for biaxial compression of brittle materials, *Proc. Eighth Symposium on Rock Mechanics, University of Minnesota*, 1966, in *Failure and Breakage of Rock*, C. Fairhurst (Ed.), 1967, 113–41.

PAULDING, B. W., JR. (1966). Techniques used in studying the fracture mechanics of rock, *Testing Techniques for Rock Mechanics, ASTM, Am. Soc. Test. Mats.*, 73.

PENMAN, A. D. M. (1953). Shear characteristics of a saturated silt measured in triaxial compression, *Géotechnique*, **3**, 312–28.

PERKINS, T. K., and BARTLETT, L. F. (1963), Surface energies of rocks measured during cleavage, *J. Soc. Petrol. Engrs.*, **3**, 4.

PHILLIPS, D. W. (1930–1). The nature and physical properties of some coal-measure strata, *Trans. Instn. Min. Engrs.*, **80**, 212–42.

PHILLIPS, D. W. (1931–2). Further investigation of the physical properties of coal-measure rocks and experimental work on the development of fractures. *Trans. Instn. Min. Engrs.*, **82**, 432–50.

PHILLIPS, D. W. (1948). Tectonics of mining, *Colliery Engng.*, **25**, 199–203; 278–82.

PHILLIPS, F. C. (1954). *The Use of the Stereographic Projection in Structural Geology*, London, Arnold.

POCHHAMMER, L. (1876). *Z. reine angew Math.*, **81**, 324.

POLAK, E. J. (1963). The measurements of, relation between, and factors affecting the properties of rocks, *Proc. Fourth Aust.–N.Z. Conference on Soil Mechanics and Found, Engng.*, 220–4.

POMEROY, C. D. (1956). Creep in coal at room temperature, *Nature*, **178**, 279–80.

POMEROY, C. D., and MORGANS, W. T. A. (1956). The tensile strength of coal, *Br. J. Appl. Phys.*, **7**, 243–6.

POOLLEN, H. K. VAN (1957). Theories of hydraulic fracturing, *Colo. Sch. Mines Q.*, **52**, 113–25.

POŠCHL, T. (1921). Über eine partikuläre Losung des biharmonischen Probleme für den Aussenraum der Ellipse, *Math. Z.*, **11**, 89–96.

POTTS, E. L. J. (1954). Stress distribution, rock pressures and support loads, *Colliery Engng.*, 333–9.

POTTS, E. L. J. (1957). Underground instrumentation, *Colo. Sch. Mines, Q.*, **52**, 135–82.

POTTS, E. L. J. (1964). The *in situ* measurement of rock stress based on deformation measurements, in *State of Stress in the Earth's Crust*, W. R. Judd (Ed.), New York, Elsevier, 397–407.

PRAGER, W., and HODGE, P. G. (1951). *Theory of Perfectly Plastic Solids*, New York, Wiley.

PRANDTL, L. (1924). Spannungsverteilung in plastischen körpern, *Int. Congr. appl. Mech., 1st, Delft, 1924, Proc.*, 43–54.

PRICE, N. J. (1958). A study of rock properties in conditions of triaxial stress, in *Mechanical Properties of Non-metallic Brittle Materials*, W. H. Walton (Ed.), London, Butterworths, 106–22.

PRICE, N. J. (1959). Mechanics of jointing in rocks, *Geol. Mag.*, **96**, 149–67.

PRICE, N. J. (1960). The strength of coal-measure rocks in tri-axial compression, *National Coal Board, MRE. Rept. No. 2159.*

PRICE, N. J. (1962). The tectonics of the Aberystwyth grits, *Geol. Mag.*, **99**, 542–57.

PRICE, N. J. (1964). A study of the time–strain behaviour of coal-measure rocks, *Int. J. Rock Mech. Min. Sci.*, **1**, 277–303.

PRICE, N. J. (1966). *Fault and Joint Development in Brittle and Semi-brittle Rock*, Oxford, Pergamon.

PROTODIAKONOV, M. N. (1964). Methods for evaluation of cracks and strength of rock systems in depth, *Fourth Int. Conf. Strata Control and Rock Mechs., Henry Crumb School of Mines, Columbia University, New York, Addendum.*

PULPAN, H., and SCHEIDEGGER, A. E. (1965). Calculations of tectonic stresses from hydraulic well-fracturing data, *J. Inst. Petrol.*, **51**, 169–76.

RAE, D. (1963). The measurement of the coefficient of friction of some rocks during continuous rubbing, *J. Scient. Instrum.*, **40**, 438–40.

RALEIGH, C. B., and GRIGGS, D. T. (1963). Effect of toe in the mechanics of overthrust faulting, *Bull. Geol. Soc. Am.*, **74**, 819–30.

RALEIGH, C. B., and PATERSON, M. S. (1965). Experimental deformation of serpentinite and its tectonic implications, *J. Geophys., Res.*, **70**, 3965–85.

RAMBERG, H. (1964). Selective buckling of composite layers with contrasted rheological properties: a theory for simultaneous formation of several orders of folds, *Tectonophys.*, **1**, 307–41.

RAMBERG, H. (1967). *Gravity, Deformation and the Earth's Crust*, London, Academic Press.

RAMSAY, J. G. (1967). *Folding and Fracturing of Rocks*, New York, McGraw-Hill.

RAYLEIGH, LORD (1885). On waves propagated along the plane surface of an elastic solid. *Proc. London Math. Soc.*, **17**, 4–11.

RAYLEIGH, LORD (1934). The bending of marble, *Proc. Roy. Soc., London*, **A 144**, 266–79.

REE, F. H., REE, T., and EYRING, H., (1960). Relaxation theory of creep of metals, *Proc. Am. Soc. Civ. Engrs., J. Eng. Mech. Div.*, **86**, 41–59.

REICHMUTH, D. R. (1963). Correlation of force–displacement data with physical properties of rock for percussive drilling systems, *Proc. of Fifth Rock Mech. Symp., University of Minnesota*, In *Rock Mechanics*, C. Fairhurst (Ed.), Oxford, Pergamon, 33–59.

REINER, M. (1947). *Deformation and Flow*, London, Lewis.

REINER, M. (1949). *Twelve Lectures on Theoretical Rheology*, Amsterdam, North-Holland.

REUSS, E. (1930). Berücksichtigung der elastischen Förmanderungen in der Plastizitätstheorie, *Z. für Angewandte Mat. und Mech.*, **10**, 266–74.

REYES, S. F., and DEERE, D. U. (1966). Elastic plastic analysis of underground openings by the finite element method. *Proc. First Congress International Society of Rock Mechanics, Lisbon*, **2**, 477–83.

RINEHART, J. S. (1960). On fractures caused by explosions and impact, *Colo. School Mines Quart.*, **55**, 4.

RINEHART, J. S., FORTIN, JEAN-PIERRE, and BURGIN, LORRAINE (1961). Propagation velocity of longitudinal waves in rocks. Effect of state of stress, stress level of the wave, water content, porosity, temperature stratification and texture, *Proc. Fourth Symp. on Rock Mech., Penn. State Univ.*, 1961, 119–35.

RIPPERGER, E. A., and DAVIDS, N. (1947). Critical stresses in a circular ring, *Trans. Am. Soc. Civ. Engrs.*, **112**, 619–35.

ROBERTS, A., EMERY, C. L., CHAKRAVARTY, P. K., WILLIAMS, F. T., and HAWKES, I. (1962). Photoelastic coating technique applied to research in rock mechanics, *Trans. Instn. Min. Metall.*, **71**, 581–617.

ROBERTS, D. K., and WELLS, A. A. (1954). Velocity of brittle fracture, *Engineering*, **178**, 820–1.

ROBERTS, J. C. (1961). Feather-fracture, and the mechanics of rock-jointing, *Am. J. Sci.*, **259**, 481–92.

ROBERTSON, E. C. (1955). Experimental study of the strength of rocks, *Bull. Geol. Soc. Am.*, **66**, 1275–314.

ROBERTSON, E. C. (1960). Creep of Solenhofen limestone under moderate hydrostatic pressure, *Geol. Soc. Am., Mem.*, **79**, 227–44.

ROBERTSON, E. C. (1964). Viscoelasticity of rocks, in *State of Stress in the Earth's Crust*, W. R. Judd (Ed.), New York, Elsevier, 181–233.

ROBINSON, L. H., JR. (1959). Effect of pore and confining pressure on the failure process in sedimentary rocks, *Colo. Sch. Mines Q.*, **54**, 177–99.

ROBSON, G. R., and BARR, K. G. (1964). The effect of stress on faulting and minor intrusions in the vicinity of a magma body, *Bull. volcan.*, **27**, 315–30.

ROCHA, M. (1964). Mechanical behaviour of rock foundations in concrete dams, *Trans. Eighth Congress on Large Dams, Edinburgh*, 785–831.

ROCHA, M., SERAFIM, J. L., and DA SILVEIRA, A. F. (1955). Deformability of foundation rocks, *Proc. Fifth Internat. Congress on Large Dams, Paris*, 1955, **3**, 531–61.

ROSENGREN, K. J., and JAEGER, J. C. (1968). The mechanical properties of an interlocked low-porosity aggregate, *Géotechnique*, **18**, 317-26.

ROTHERHAM, L. A. (1951). *Creep of Metals, Inst. of Phys.*, London.

RUIZ, M. D., and PIRES DE CAMARGO, F. (1966). A large-scale shear test on rock, *Proc. First Congress International Society of Rock Mechanics, Lisbon*, **1**, 257–61.

RYDER, J. A., and OFFICER, N. C. (1964). An elastic analysis of strata movement in the vicinity of inclined excavations, *J. S. Afr. Inst. Min. Metall.*, **64**, 219–44.

SACK, R. A. (1946). Extension of Griffith's theory of rupture to three dimensions, *Proc. Phys. Soc., London*, **58**, 729–36.

SADOWSKY, M. A., and STERNBERG, E. (1947). Stress concentration around an ellipsoidal cavity in an infinite body under arbitrary plane stress perpendicular to the axis of revolution of the cavity, *J. Appl. Mech.*, **14**, 191–201.

SADOWSKY, M. A., and STERNBERG, E. (1949). Stress concentration around a triaxial ellipsoidal cavity, *J. Appl. Mech.*, **16**, 149–57.

SAINT-VENANT, B. DE (1870). Mémoire sue l'establissement des équations différentielles des mouvements intérieurs opérés dans les corps solides ductiles au delá des limites où l'élasticité pourrait les ramener à leur premier état. *C.r. bedb. Séanc. Acad. Sci., Paris*, **70**, 473–80.

SALAMON, M. D. G. (1961). Some theoretical aspects of stress-meter design, *King's College, Univ. of Durham, Newcastle Research Reports, vol.* **9**, *Bull. No.* **2** *Series: Strata Control Res. No. 13.*

SALAMON, M. D. G. (1964). Elastic analysis of displacements and stresses induced by the mining of seam or reef deposits, Pt. I. *J. S. Afr. Inst. Min. Metall.*, **64**, 128–49.

SALAMON, M. D. G. (1964). Elastic analysis of displacements and stresses induced by the mining of reef deposits, Pt. II. *J. S. Afr. Inst. Min. Metall.*, **64**, 197–218.

SALAMON, M. D. G. (1964). Elastic analysis of displacements and stresses induced by the mining of seam or reef deposits, Pt. III. *J. S. Afr. Inst. Min. Metall.*, **64**, 468–500.

SALAMON, M. D. G. (1965). Elastic analysis of displacements and stresses induced by the mining of seam or reef deposits. Pt. IV. *J. S. Afr. Inst. Min. Metall.*, **65**, 319–38.

SALAMON, M. D. G., and ORAVECZ, K. I. (1966). Displacement and strains induced by bord and pillar mining in South African collieries, *Proc. First Congress International Society of Rock Mechanics, Lisbon*, **2**, 227–31.

SALAMON, M. D. G. (1967). A method of designing bord and pillar workings., *J. S. Afr. Inst. Min. Metall.*, **68**, 68–78.

SALAMON, M. D. G., and MUNRO, A. H. (1967). A study of the strength of coal pillars, *J. S. Afr. Inst. Min. Metall.*, **68**, 55–67.

SALAMON, M. D. G., RYDER, J. A., and ORTLEPP, W. D. (1964). An analogue solution for determining the elastic response of strata surrounding tabular mining excavations, *J. S. Afr. Inst. Min. Metall.*, **65**, 115–37.

SANFORD, A. R. (1959). Analytical and experimental study of simple geological structures, *Bull. Geol. Soc. Am.*, **70**, 19–51.

SAVIN, G. N. (1961). *Stress Concentration around Holes*, translated by E. Gros, Oxford, Pergamon.

SAX, H. G. J. (1946). De tectoniek van het Carboon in het Zuid-Limburgsche mijngebied, *Meded. Geol. Stricht.*, *Ser. C-I-I* No. 3.

SCHEIDEGGER, A. E. (1960). On the connection between tectonic stresses and well fracturing data, *Geofis. pura appl.*, **46**, 66–76.

SCHEIDEGGER, A. E. (1962). Stresses in the earth's crust as determined from hydraulic fracturing data, *Geol. u. Bauwesen.*, **27**, 45–53.

SCHEIDEGGER, A. E. (1963). Geometrical significance of isallostress, *N.Z. Jl. Geol. Geophys.*, **6**, 221–7.

SCHEIDEGGER, A. E. (1966). Isallostress prospecting, *Z. Geophys.*, **32**, 183–99.

SCHOLZ, C. H. (1967). Frequency–magnitude relation of micro-fracturing events during the triaxial compresison of rock, *Trans. Am. geophys. Un.*, **48**, 1, 205.

SCOTT, R. F. (1963). *Principles of Soil Mechanics*, London, Addison–Wesley.

SECOR, D. T. (1965). Role of fluid pressure in jointing, *Am. J. Sci.*, **263**, 633–46.

SEDLACEK, R., and HOLDEN, R. A. (1962). Method for tensile testing of brittle materials, *Rev. Scient. Instrum.*, **33**, 298–300.

SELDENRATH, R., and GRAMBERG, J. (1958). Stress–strain relations and the breakage of rocks, in *Mechanical Properties of Non-metallic Brittle Materials*, W. H. Walton (Ed.), London, Butterworths, 79–105.

SERAFIM, J. L. (1961). Internal stresses in galleries, *Proc. Seventh Congr. on Large Dams, 26th June–1st July, 1961, Rome*, Sec. R. 1, vol. 2, 141–65.

SERAFIM, J. L. (1964). Rock mechanics considerations in the design of concrete dams, in *State of Stress in the Earth's Crust*, W. R. Judd (Ed.), New York, Elsevier, 611–45.

SERATA, S. (1964). Theory and model of underground opening and support system, *Proc. Sixth Symposium on Rock Mechanics, Rolla, Missouri*, 260–92.

SERDENGECTI, S., and BOOZER, G. D. (1961). The effects of strain rate and temperature on the behaviour of rocks subjected to triaxial compression, *Proc. Fourth Symposium on Rock Mechanics, Penn. State Univ.*, 83–97.

SERDENGECTI, S., BOOZER, G., and HILLER, K. H. (1962). Effects of pore fluids on the deformation behaviour of rocks subjected to triaxial compression, *Proc. Fifth Symposium on Rock Mechanics, Univ. of Minnesota*, Pergamon, 579–625.

SEZAWA, K. (1933). The effect of local heterogeneity on the stress distribution in solids, *Engineering*, **135**, 695–6.

SEZAWA, K., and NISHIMURA, G. (1931). Stresses under tension in a plate with a heterogeneous insertion, *Rep. Aero. Res. Inst., Tokyo*, **6**, 25–43.

SHARPE, J. A. (1942). The production of elastic waves by explosion pressures. I. Theory and empirical field observations, *Geophys.*, **7**, 144–54.

SHELLEY, J. F., and YI-YUAN YU (1966). The effect of two rigid spherical inclusions on the stresses in an infinite elastic solid, *J. Appl. Mech.*, **33**, 68–74.

SIEBEL, E., and POMP, A. (1927). Die Ermittlung der Formänsweunfadwarigkeit von Metallen durch den Stauchversuch, *Mitt. K.-Wilhelm-Inst., Eisenforsch.* Düsseld, **9**.

SIMMONS, G. (1965). Ultrasonics in geology, *Proc. Instn. Elect. Engrs.*, **53**, 1337–45.

SIMMONS, G., and BRACE, W. F. (1965). Comparison of static and dynamic measurements of compressibility of rocks, *J. Geophys. Res.*, **70**, 5649–56.

SKEMPTON, A. W. (1944). Notes on the compressibility of clays, *Q. J. Geol. Soc. London*, **100**, 119–35.

SKEMPTON, A. W. (1960). Effective stress in soils, concrete and rocks, in *Pore Pressure and Suction in Soils*, London, Butterworths.

SMEKAL, A. (1936). Die Festigkeitseigenschaften spröder Körper, *Ergebn. exakt. Naturw.*, **15**, 106–88.

SNEDDON, I. N. (1946). The distribution of stress in the neighbourhood of a crack in an elastic solid, *Proc. Roy. Soc., London*, **A 187**, 229–60.

SNEDDON, I. N. (1951). *Fourier Transforms*, New York, McGraw-Hill (Int. Ser. in Pure and Appl. Maths.)

SNEDDON, I. N., and ELLIOTT, H. A. (1946). The opening of a Griffith crack under internal pressure, *Q. Appl. Math.*, **4**, 262–7.

SOKOLNIKOFF, I. S. (1956). *Mathematical Theory of Elasticity*, 2nd edn., New York, McGraw-Hill.

SOKOLOVSKII, V. V. (1965). *Statics of Granular Media*, translated by J. K. Lusher, Oxford, Pergamon.

SOUTHWELL, R. V. (1941). *An Introduction to the Theory of Elasticity*, 2nd edn., Oxford University Press.

SPRY, A. (1961). The origin of columnar jointing, particularly in basalt flows, *J. Geol. Soc. Aust.*, **8**, 191–216.

STACEY, F. D. (1963). The theory of creep in rocks and the problem of convection in the Earth's mantle, *Icarus*, **1**, 304–13.

STARFIELD, A. M., and PUGLIESE, J. M. (1968). Compression waves generated in rock by cylindrical explosive charges: A comparison between a computer model and field measurements, *Int. J. Rock Mech. Min. Sci.*, **5**, 65–77.

STARR, A. T. (1928). Slip in a crystal and rupture in a solid due to shear, *Proc. Camb. Phil. Soc.*, **24**, 489–500.

STASSI-D'ALIA, F. (1959). A limiting condition of yielding and its experimental confirmation, *Industria Grafica Nazionale, Palermo.*

STEART, F. A. (1955). Strength and stability of pillars in coal mines, *J. Chem. Metall. Min. Soc. S. Afr.*, **56**, 22–3.

STEINHART, J. S., and MEYER, R. P. (1961). *Explosion Studies of Continental Structure*, Carnegie Inst. Publ. 622, Washington.

STEKETEE, J. A. (1958). Some geophysical applications of the elasticity theory of dislocations, *Canad. J. Phys.*, **36**, 1168–98.

STERNBERG, E., and ROSENTHAL, F. (1952). The elastic sphere under concentrated loads, *J. Appl. Mech.*, **19**, 413–21.

STEVENSON, A. C. (1945). Complex potentials in two-dimensional elasticity, *Proc. Roy. Soc., London*, **A 184**, 129–79.

STROH, A. N. (1954). The formation of cracks as a result of plastic flow, *Proc. Roy. Soc., London*, **A 223**, 404–14.

STROH, A. N. (1955). The formation of cracks in plastic flow. II. *Proc. Roy. Soc., London*, **A 232**, 548–60.

SULLY, A. H. (1956). Recent advances in knowledge concerning the process of creep in metals, *Prog. Metal Phys.*, **6**, 135–80.

SWAIN, R. J. (1962). Recent techniques for determination of *in situ* elastic properties and measurement of motion amplification in layered media, *Geophys*, **27**, 237–41.

SYMONDS, P. S. (1946). Concentrated-force problems in plane strain, plane stress and transverse bending of plates, *J. Appl. Mech.*, **13**, 183–97.

TABOR, D. (1951). *The Hardness of Metals*, Oxford, Clarendon Press.

TABOR, D. (1959). Junction growth in metallic friction, *Proc. Roy. Soc., London*, **A 251**, 378–93.

TALOBRE, J. (1957). *La mécanique des roches*, Paris, Dunod.

TALOBRE, J. (1958). Dix ans de mesures de compression interne des roches: progrès et résultats pratiques, *Geologie Bauwes.*, **25**, 148–65.

TAYLOR, G. I. (1934). Faults in a material which yields to shear stress while retaining its volume elasticity, *Proc. Roy. Soc., London*, **A 145**, 1–18.

TAYLOR, G. I., and QUINNEY, H. (1931). The plastic distortion of metals, *Phil. Trans. Roy. Soc., London*, **A 230**, 323–62.

TERRY, N. B. (1959). Dependence of the elastic behaviour of coal on the microcrack structure, *Fuel*, **38**, 125–46.

TERZAGHI, K. VAN (1923). Die Berechnung der Durchlassigkeitsziffer des Tones aus dem Verlauf der hydrodynamischen Spannungserscheinungen, *Sber. Akad. Wiss. Wien*, **132**, 105.

TERZAGHI, K. VAN (1943). *Theoretical Soil Mechanics*, New York, Wiley.

TERZAGHI, K. VAN (1945). Stress conditions for the failure of saturated concrete and rock, *Proc. Am. Soc. Test. Mater.*, **45**, 777–801.

TERZAGHI, K. VAN (1950). Mechanism of landslides, in *Application of Geology to Engineering Practice*, Berkey vol., Geol. Soc., Amer., 83–123.

TERZAGHI, K. VAN (1962). Stability of steep slopes on hard unweathered rock, *Géotechnique*, **12**, 251–70.

TERZAGHI, K. VAN, and RICHART, F. E. (1952). Stresses in rocks about cavities, *Géotechnique*, **3**, 57–90.

THOMSON, SIR WILLIAM (LORD KELVIN) and TAIT, P. G. (1962). *Principles of Mechanics and Dynamics*, New York, Dover.

TIMOSHENKO, S. (1958). *Strength of Materials. I. Elementary Theory and Problems*, 3rd edn., Princeton, N.J., D. Van Nostrand.

TIMOSHENKO, S., and GERE, J. M. (1961). *Theory of Elastic Stability*, 2nd edn., New York, McGraw-Hill.

TIMOSHENKO, S. P., and GOODIER, J. N. (1951, *Theory of Elasticity*, 2nd edn., New York, McGraw-Hill.

TINCELIN, M. E., (1951). Conférence internationale sur les pressions de terrains et le soutènement dans les chantiers d'exploritation, Liège, du 24 au 28 avril, 1951, 158–75.

TINCELIN, M. E. (1952). Mesures des pressions de terrains dans les mines de fer de l'est, *Ann. Inst. Tech. Bat. et Trav. Publ., Sols et Fond,* **x**, *Ann. No. 58,* 972–90.

TOCHER, D. (1957). Anisotropy in rocks under simple compression, *Trans. Am. Geophys. Un.*, **38**, 89–94.

TOMKEIEFF, S. I. (1940). The basalt lavas of the Giant's Causeway district of Northern Ireland, *Bull. Volcan.* **6**, 89–143.

TRANTER, C. J., and CRAGGS, J. W. (1945). The stress distribution in a long circular cylinder when a discontinuous pressure is applied to the curved surface, *Phil. Mag.*, **36**, 241–50.

TRESCA, H. (1868), Mémoire sur l'ecoulement des corps solides, *Mém. prés. div. Sav. Acad. Sci., Inst. Fr.*, **18**, 733–99.

TURNER, F. J. (1949). Preferred orientation of calcite in Yule marble, *Am. J. Sci.*, **247**, 593–621.

TURNER, F. J., GRIGGS, D. T., CLARK, R. H., and DIXON, ROBERTA, H. (1956). Deformation of Yule marble. Pt. VII. Development of oriented fabrics at 300° C–500° C, *Bull. Geol. Soc. Am.*, **67**, 1259–94.

TURNER, F. J., GRIGGS, D. T., and HEARD, H. (1954). Experimental deformation of calcite crystals, *Bull. Geol. Soc. Am.*, **65**, 883–934.

TURNER, F. J., and WEISS, L. E. (1963). *Structural Analysis of Metamorphic Tectonites*, New York, McGraw-Hill.

TURNER, P. W., and BARNARD, P. R. (1962). Stiff constant strain rate testing machine, *Engineer*, **214**, 146–8.

U.S. BUREAU OF RECLAMATION (1953). Physical properties of some typical foundation rocks, *Concrete Laboratory Report No. SP-39*, Engineering Laboratories Branch.

USHER, M. J. (1962). Elastic behaviour of rocks at low frequencies, *Geophys. Prosp.*, **10**, 119–27.

VERSIAC REPORT (1962). Proceedings of colloquium on detection of underground nuclear explosions, *Spec. Pap., Univ. of Michigan*, 477 pp.

VOIGHT, B. (1966a). Beziehung zwischen grossen horizontalen spannungen im Gebirge und der Tektonik und der Abtragung, *Proc. First Congress International Society of Rock Mechanics*, Lisbon, **2**, 51–6

VOIGHT, B. (1966b). Restspannungen im Gestein, *Proc. First Congress International Society of Rock Mechanics*, Lisbon, **2**, 45–50.

VOIGHT, B. (1967). On photoelastic techniques, *in situ* stress and strain measurement and the field geologist, *J. Geol.*, **75**, 46–58.

VOIGHT, W. (1910). *Lehrbuch der Krystallphysik*, Leipsig, Teubner.

WAGNER, H., and SCHÜMANN, E. H. R. (1971). The stamp-load bearing strength of rock: An experimental and theoretical investigation, submitted to *Rock Mechanics*, Springer-Verlag.

WALSH, J. B. (1965a). The effect of cracks on the compressibility of rock, *J. Geophys. Res.*, **70**, 381–9.

WALSH, J. B. (1965b). The effect of cracks on the uniaxial elastic compression of rocks, *J. Geophys. Res.*, **70**, 399–411.

WALSH, J. B. (1965c). The effect of cracks in rocks on Poisson's ratio, *J. Geophys. Res.*, **70**, 5249–57.

WALSH, J. B. (1966). Seismic wave attenuation in rock due to friction, *J. Geophys. Res.*, **71**, 2591–9.

WALSH, J. B., and BRACE, W. F. (1964). A fracture criterion for brittle anisotropic rock, *J. Geophys. Res.*, **69**, 3449–56.

WALSH, J. B., and BRACE, W. F. (1966). Cracks and pores in rocks, *Proc. First Congress International Society of Rock Mechanics, Lisbon*, **1**, 643–6.

WALSH, J. B., and DECKER, E. R. (1966). Effect of pressure and saturating fluid on the thermal conductivity of compact rock, *J. Geophys, Res.*, **71**, 3053–61.

WAWERSIK, W. R. (1967). The brittle fracture of rocks: contribution to discussion, *Proc. Eighth Symposium on Rock Mechanics, University of Minnesota*, 1966, in *Failure and Breakage of Rock*, C. Fairhurst (Ed.), 158–60.

WAWERSIK, W. R. (1968). Experimental study of the fundamental mechanisms of rock failure in static uniaxial and triaxial compression, and uniaxial tension. Thesis, University of Minnesota.

WEIBULL, W. (1938). Investigations into strength properties of brittle materials, *Ingvetensk. Akad. Handl.*, No. 149.

WEIBULL, W. (1939a). A statistical theory of the strength of materials, *Ingvetensk. Akad. Handl.*, No. 151.

WEIBULL, W. (1939b). The phenomenon of rupture in solids, *Ingvetensk. Akad. Handl.*, No. 153.

WEIBULL, W. (1951). A statistical distribution function of wide applicability, *J. Appl. Mech.*, **18**, 293–7.

WEIBULL, W. (1952). A survey of statistical effects in the field of material failure, *Appl. Mech. Rev.*, **5**, 449–51.

WERFEL, A. (1965). Mohr's circle as an aid to the transformation of symmetrical second order tensors and to the solution of some other problems, in *Topics in Appl. Mech.*, Elsevier, 299–311.

WESTERGAARD, H. M. (1939). Bearing pressures and cracks, *J. Appl. Mech.*, **6**, A49–53.

WESTWOOD, A. R. C., and HITCH, T. T. (1963). Surface energy of potassium chloride, *J. Appl. Phys.*, **34**, 3085–9.

WHITE, J. E. (1965). *Seismic Waves: radiation, transmission and attenuation*, New York, McGraw-Hill.

WIEBOLS, G. A., and COOK, N. G. W. (1968). An energy criterion for the strength of rock in polyaxial compression, submitted to *Int. J. Rock Mech. Min. Sci.*

WIEBOLS, G. A., JAEGER, J. C., and COOK, N. G. W. (1968). Rock property tests in a stiff testing machine, to be submitted to *Tenth Rock Mechanics Symposium*, Rice University, Houston, 1968.

WIECHART, E. (1903). Theorie der automatischen Seismographen, *Abh. Ges. Wiss., Göttingen, Math.-Phys. Klasse*, **2**, 1–128.

WILLIAMS, A. (1959). A structural history of the Girvan district, S.W. Ayrshire, *Trans. Roy. Soc., Edinb.*, **63**, 629–67.

WILLMORE, P. L., HALES, A. L., and GANE, P. G. (1952). A seismic investigation of crustal structure in the Western Transvaal, *Bull. Seism. Soc. Am.*, **42**, 53–80.

WILSON, A. H. (1961). A laboratory investigation of a high modulus borehole plug gage for the measurement of rock stress, *Proc. Fourth Symp. Rock Mech., Penn. State Univ.*, 185–95.

WILSON, J. T. (1965). A new class of faults and their bearing on continental drift, *Nature*, **207**, 343–7.

WINDES, S. L. (1949, 1950). Physical properties of mine rock. *U.S. Bur. Mines Rept. Invest.* 4459, 4727.

WUERKER, R. G. (1938–48). Annotated tables of strength and elastic properties of rocks, *Trans. Am. Inst. Min Engrs.*, **1–11** (6).

WYLLIE, M. R. J., GREGORY, A. R., and GARDNER, L. W. (1956). Elastic wave velocities in heterogeneous and porous media, *Geophys.*, **21**, 41–70.

WYLLIE, M. R. J., GREGORY, A. R., and GARDNER, G. H. F. (1958). An experimental investigation of factors affecting elastic wave velocities in porous media, *Geophys.*, **23**, 459–93.

YU, YI-YUAN (1952). Gravitation stresses on deep tunnels, *J. appl. Mech.*, **19**, 537–42.

ZENER, C. (1948). *Elasticity and Anelasticity of Metals*, Chicago, University of Chicago.

ZIENKIEWICZ, D., and CHEUNG, Y. K. (1966). Application of the finite element method to problems of rock mechanics, *Proc. First Congress International Society of Rock Mechanics, Lisbon*, **1**, 661–6.

ZIENKIEWICZ, O. C., and STAGG, K. G. (1967). Cable method of *in situ* rock testing, *Int. J. Rock Mech. Min. Sci.*, **4**, 273–300.

ZISMAN, W. A. (1933). Comparison of the statistically and seismologically determined elastic constants of rocks, *Proc. Natn. Acad. Sci., U.S.A.*, **19**, 680–6.

ZIZICAS, G. A. (1955). Representation of three-dimensional stress distributions by Mohr circles, *J. Appl. Mech.*, **22**, 273–4.

Additional References

ANTSYFEROV, M. S. (Ed.) (1966). *Seismo-Acoustic Methods in Mining*, New York, Consultants Bureau.

ATCHISON, T. C., DUVALL, W. I., and PUGLIESE, J. M. (1964). Effect of decoupling on explosion-generated strain pulses in rock, *U.S. Dept. Interior. Bureau of Mines*, **RI** 6333.

ATCHISON, T. C. (1968). Fragmentation principles, in *Surface Mining*, E. P. Pfleider (Ed.), Mudd Series, A.I.M.E., 355–72.

BAILEY, J. J., and DEAN, R. C. (1967). Rock mechanics and the evolution of improved rockcutting methods, *Proc. Eighth Symposium on Rock Mechanics, University of Minnesota, 1966*, in *Failure and Breakage of Rock*, C. Fairhurst (Ed.).

BENIOFF, H. (1964). Earthquake source mechanisms, *Science*, **143**, 1399.

BERG, C. A. (1967). A note on the mechanics of seismic faulting, *Geophys. J.*, **14**, 89.

BLAKE, W., LEIGHTON, F., and DUVALL, W. I. (1974). Microseismic techniques for monitoring the behaviour of rock structures, *U.S. Bur. Mines Bull.*, 665.

BOND, F. C. (1952). The third theory of comminution, *A.I.M.E. Trans.*, **193**, 484.

BRACE, W. F. (1971). Micromechanics in rock systems, *Proc. Civil Engineering Materials Conference, Southampton*, in *Structure, Solid Mechanics and Engineering Design*, Part 1, M Te'Eni (Ed.), New York, Wiley–Interscience, 187–204.

BRACE, W. F. (1972). Laboratory studies of stick–slip and their application to earthquakes, *Tectonophysics*, **14** (3/4), 189.

BRACE, W. F. (1975). The physical basis for earthquake prediction, *Technology Review*, March/April, M.I.T., 26–9.

BRACE, W. F., PAULDING, B. W., and SCHOLTZ, C. (1966). Dilatancy in the fracture of crystalline rocks, *J. Geophys. Res.*, **77**, 3939.

BRACE, W. F., and ORANGE, A. S. (1966a). Electrical resistivity changes in saturated rocks during fracture and frictional sliding, *J. Geophys. Res.*, **73**, 1433–45.

BRACE, W. F., and ORANGE, A. S. (1966b). Further studies of the effects of pressure on electrical resistivity of rocks, *J. Geophys. Res.*, **73**, 5407–20.

BRIDGMANN, P. W. (1945). Polymorphic transitions and geological phenomena, *Am. J. Sci.*, **243 a**, 90.

BROCK, G. (1962). Concrete: Complete stress–strain curves, *Engineering* (*London*), **193** (5011), 606–8.

BRUNE, J. N. (1968). Seismic moment, seismicity and rate of slip along major fault zones, *J. Geophys. Res.*, **73**, 777–84.

BRUNE, J. N. (1970). Tectonic stress and the spectra of seismic shear waves from earthquakes, *J. Geophys. Res.*, **75**, 4997.

BYERLEE, J. D. (1970). The mechanics of stick-slip, *Tectonophysics*, **9**, 475

BYERLEE, J. D., and BRACE, W. F. (1968). Stick–slip, stable sliding and earthquakes, Part 1. Effect on rock type, pressure, strain-rate and stiffness, *J. Geophys. Res.*, **73**, 6031.

CAZALET, P. (1919–20). Notes on the study of records of earth tremors on the Central Rand, *J. S. Afr. Instn Engrs*, **18**, 211.

CHAMBER OF MINES OF SOUTH AFRICA (1973). Research Review.

CHINNERY, M. A. (1969). Theoretical fault models, a *Symposium on Processes in the Focal Region*, publ. of the Dominion Observatory, 211.

CLARK, G. B. (1967). Blasting and dynamic rock mechanics, *Proc. Eighth Symposium on Rock Mechanics, University of Minnesota, 1966*, in *Failure and Breakage of Rock*, C. Fairhurst (Ed.).

CLARK, G. B. (1968). Explosives, in *Surface Mining*, E. P. Pfleider (Ed.), Mudd Series, A.I.M.E., 341–54.

COOK, N. G. W. (1970). An experiment proving that dilatancy is a pervasive volumetric property of brittle rock loaded to failure, *Rock Mechanics*, **2**, 181–8.

COOK, N. G. W., and HARVEY, V. R. (1974). An appraisal of rock excavation by mechanical, hydraulic, thermal and electromagnetic means, *Proc. Third Congress of the International Society for Rock Mechanics*, Denver, in *Advances in Rock Mechanics*, Vol. 1, Part B, 1599–1615, National Academy of Sciences, Washington D.C.

COOK, N. G. W., HODGSON, K., and HOJEM, J. P. M. (1971). 100-MN jacking system for testing coal pillars underground, *J. S. Afr. Inst. Min. Metall.*, **71** (11), 215–24.

COOK, N. G. W., and JOUGHIN, N. C. (1970). Rock fragmentation by mechanical, chemical and thermal methods, *Proc. Sixth International Mining Congress*, Madrid.

DEIST, F. H., SALAMON, M. D. G., and GEORGIADIS, E. (1973). A new digital method for three-dimensional stress analysis in elastic media, *Rock Mechanics*, **5**, 189–202.

DIETERICH, J. H. (1969). Mathematical modelling of fault-zone tectonics and seismicity (Abs.), *Geol. Soc. Am.*, **7**, 47.

DIETERICH, J. H. (1972). Time-dependent friction in rocks, *J. Geophys. Res.*, **77**, 3690–94.

EVANS, D. M. (1966). The Denver area earthquakes and the Rocky Mountain arsenal disposal well, *Mountain Geologist*, **3**, 23–36.

FERNANDEZ, L. M. (1973). Seismic energy released by the deep mining operations in the Transvaal and Orange Free State during 1971, *S.A. Dept. Mines Geol. Survey*.

FINSEN, W. S. (1950). Union Observatory Circular No. 110.

GANE, P. G. (1939). A statistical study of the Witwatersrand earth tremors. *J. Chem. Metall. Min. Soc. S. Afr.*, **40**, 155.

GANE, P. G., SELIGMAN, P., and STEPHEN, J. H. (1953). Focal depths of Witwatersrand tremors, *Bull. Seism. Soc. Am.*, **42**, 239–50.

GRIGGS, D. T., and BAKER, D. W. (1969). The origin of deep focus earthquakes, in *Properties of Matter Under Unusual Conditions*, H. Mark and S. Fernbach (Eds.), Wiley Publications Co., N.Y.

GUTENBERG, B., and RICHTER, C. F. (1956). Magnitude and energy of earthquakes, *Ann. Geof.*, **9**, 1–15.

HADLEY, KATE (1975). V_p/V_s anomalies in dilatant rock samples. (To be published.)

HALLAM, A. (1973). *A Revolution in the Earth Sciences from Continental Drift to Plate Tectonics*, London, Oxford University Press.

HALLBAUER, D. K., WAGNER, H., and COOK, N. G. W. (1973). Some observations concerning the microscopic and mechanical behaviour of quartzite specimens in stiff, triaxial compression tests, *Int. J. Rock Mech. Min. Sci. & Geomech. Abstr.*, **10**, 713–26.

HANCOCK, P. (1969). Fracture patterns in the Cotswold Hills, *Proc. Geol. Ass.*, 80.

HANDIN, J., and RALEIGH, C. B. (1972). Man-made earthquakes and earthquake control. Percolation through fissured rocks, *Symposium International Society for Rock Mechanics*, Stuttgart, published by Deutsche Gesellschaft fur Erd und Grundbau, W. Wittke (Ed.).

HODGSON, K., and COOK, N. G. W. (1970). The effects of size and stress gradient on the strength of rock, *Second Congress International Society of Rock Mechanics*, Belgrade.

HODGSON, K., and COOK, N. G. W. (1971). The mechanism, energy content and radiation efficiency of seismic waves generated by rockbursts in deep-level mining, *Dynamic Waves in Civil Engineering*, D. A. Howells *et al.* (Eds.), New York, Wiley-Interscience, 121–35.

HOJEM, J. P. M., COOK, N. G. W., and HEINS, C. (1975). A stiff, two meganewton testing machine for measuring the 'work-softening' behaviour of brittle materials, *S.A. Mech. Engr.* **25,** 250-70.

HUDSON, J. A., BROWN, E. T., and FAIRHURST, C. (1971). Optimising the control of rock failure in controlled laboratory tests, *Rock Mechanics*, **3**, 217–24.

IIDA, K. (1959). Earthquake energy and earthquake fault. *J. Earth Sci.*, Nagoya Univ., **7**, 98–107.

IIDA, K. (1965). 'Earthquake magnitude, earthquake fault and source dimensions,' *J. Earth Sci.*, Nagoya Univ., **13**, 115–132, cited in Nur, A. (1974).

JAEGER, J. C., and COOK, N. G. W. (1971). Friction in granular materials, *Proc. of the Civil Engineering Materials Conference*, Southampton, in *Solid Mechanics and Engineering Design*, Part 1, A. M. Te'eni (Ed.), 257–66, London, John Wiley.

JENSEN, V. P. (1943). The plasticity ratio of concrete and its effect on the ultimate strength of beams, *J. Am. Concr. Inst.*, **39**, 565–82.

JOHNSON, E. R. (1968). Crushing and ore loading, in *Surface Mining*, E. P. Pfleider (Ed.), Mudd Series, A.I.M.E., 727–30.

JOUGHIN, N. C. (1965). The measurement and analysis of earth motion resulting from underground rock failure. Ph.D. Thesis submitted to the Department of Electrical Engineering of the University of the Witwatersrand.

KEILIS-BOROK, V. I. (1960). Investigation of the mechanism of earthquakes, *Sov. Res. Geophys.*, **4**, 201 (Trans. Tr. Geofiz. Inst., 40, 1957), American Geophysical Union Consultants Bureau, New York.

KICK, F. (1885), *Das Gesetz der proportionalen Widerstande und Seine Anwendung*, Leipzig.

KUTTER, H. K. (1967). The interaction between stress wave and gas pressure in the fracture process of an underground explosion in rock, with particular application to pre-splitting. Ph.D. Thesis submitted to the University of Minnesota.

LIVINGSTON, C. W. (1956). Fundamental concepts of rock failure, *Proc. First Symposium Rock Mechanics*, Colorado School of Mines, 1–11.

LIVINGSTON, C. W. (1956). Theory of fragmentation in blasting, *Sixth Annual Drilling and Blasting Symposium*, University of Minnesota.

NUR, A. (1974). Tectonophysics: The study of relations between deformation and forces in the earth. *Proc. Third Congress of the International Society for Rock Mechanics*, Denver, in *Advances in Rock Mechanics*, Vol. 1, Part A, 243–317. National Academy of Sciences, Washington D.C.

OROWAN, E. (1960). The mechanism of seismic faulting, *Geol. Soc. Am. Mem.*, **79**, 232.

PRESS, F. (1965). Displacements, strains and tilts at teleseismic distances, *J. Geophys. Res.*, **70**, 2395.

PRETORIUS, P. G. D. (1966). Contribution to 'Rock mechanics applied to the study of rockbursts', *J. S. Afr. Inst. Min. Metall.*, **66**, 705–13.

PRICE, N. J. (1974). The development of stress systems and fracture patterns in undeformed sediments, *Proc. Third Congress of the International Society for Rock Mechanics*, Denver, in *Advances in Rock Mechanics*, Vol. 1, Part A, 487–96, National Academy of Sciences, Washington D.C.

RAISTRICK, A., and MARSHALL, C. (1939). The nature and origin of coal and coal seams, Edinburgh University Press, London.

REID, H. F. (1910). The mechanics of the earthquake, in The California earthquake of April 18, 1906, *Rept. State Earthquake Invest. Com.*, Carnegie Inst., Washington D.C.

RITTINGER, P. VON (1867). *Lehrbuch der Aufbereitungskunde*, Berlin, Ernst and Korn.

SALAMON, M. D. G. (1974). Rock mechanics of underground excavations, *Proc. Third Congress of the International Society for Rock Mechanics*, Denver, in *Advances in Rock Mechanics*, Vol. 1, Part B, 951–1099, National Academy of Sciences, Washington D.C.

SCHOLTZ, C. H. (1972). Static fatigue of quartz, *J. Geophys. Res.*, **77**, 2104.

SIBEK, V. (1963). On securing the safety of operation ore mines exposed to rockburst hazards, *Proc. Third International Mining Congress*, Salzburg, in *Safety in Mining*, Oxford, Pergamon.

SPAETH, W. (1935). Einflus der Federung der Serreifmaschine auf das Spannungs-Dehnungs-Schaubild, *Arch. Eisenhuettenwesen*, **6**, 277–83.

STARFIELD, A. M. (1967). Strain wave theory in rock blasting. *Proc. Eighth Symposium on Rock Mechanics, University of Minnesota, 1966*, in *Failure and Breakage of Rock*, C. Fairhurst (Ed.).

STOKES, R. S. G. (1936). Recent developments in mining practice on the Witwatersrand, *Trans. Instn Min. Metall.*, **45**, 77.

SYKES, L. R. (1967). Mechanism of earthquakes and nature of faulting on the mid-oceanic ridges. *J. Geophys. Res.*, **72**, 2131.

WAGNER, H. (1974). Determination of the complete load-deformation characteristics of coal pillars, *Proc. Third Congress of the International Society for Rock Mechanics*, Denver, in *Advances in Rock Mechanics*, Vol. 2, Part B, 1076–81, National Academy of Sciences, Washington D.C.

WAWERSICK, W. R., and FAIRHURST, C. (1970). A study of brittle rock fracture in laboratory compression experiments, *Int. J. Rock Mech. Min. Sci.*, **7**, 561–75.

WEERTMAN, J. (1969). Continuum distribution of dislocations on faults with finite friction, *Bull. Seism. Soc. Am.*, **54**, 1035.

WHITNEY, C. S. (1943). Discussion on paper by V. P. Jensen, *J. Am. Concr. Inst.*, **39**, 584–6.

WIDEMAN, C. J., and MAJOR, M. W. (1967). Strain steps associated with earthquakes, *Bull. Seism. Soc. Am.* **57**, 1429.

WILSON, J. T. (1965). A new class of faults and their bearing on continental drift, *Nature*, **207**, 343.

WYSS, M. (1970). Stress estimates for South American shallow and deep earthquakes, *J. Geophys. Res.*, **75**, 1529–44.

Author Index

Subject Index